Marine Geology and Oceanography of the Arctic Seas

Marine Geology and Oceanography of the Arctic Seas

Edited by

YVONNE HERMAN

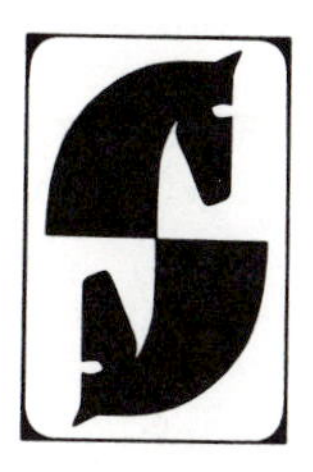

Springer-Verlag New York · Heidelberg · Berlin
1974

Library of Congress Cataloging in Publication Data
Rosenberg-Herman, Yvonne.
Marine geology and oceanography of the arctic seas.
1. Geology—Arctic Ocean. 2. Oceanography—Arctic Ocean. I. Title.
QE70.R67 551.4′68 73-22236

Printed in the United States of America.

ISBN 0-387-06628-4 Springer-Verlag New York · Heidelberg · Berlin
ISBN 3-540-06628-4 Springer-Verlag Berlin · Heidelberg · New York

Preface

Lorsqu'il n'est pas en notre pouvoir de discerner les plus vraies opinions, nous devons suivre les plus probables.

—René Descartes

When, in the early 1960's I undertook to study Arctic Ocean deep-sea cores, I did not anticipate that 10 years later the climatic history of the north polar basin would be still a matter of debate. Although much new data have accumulated in various fields of Arctic geology and oceanography during the past decade, many questions remain to be answered. The paleo-oceanography and the past atmospheric circulatory patterns are still open to controversy, as are the structure and evolution of the crust beneath this ocean; furthermore, the origin and mode of dispersal of sediments is still not fully understood. The current status of many of these problems is discussed in the present volume.

Since data on Arctic research is scattered through North American, Soviet, and European literature, it seemed timely to bring together, under one cover, important current studies from several areas of research. Because of the wide topical range and the multi-authored nature of this book, it was not possible to insure precise balance; moreover, certain subjects could not be covered, due to limitations imposed by a single volume.

Although not comprehensive, it is hoped that this book will provide an insight into the current status of Arctic research and will also serve as a reference for investigators studying the Arctic and subarctic seas.

It is with pleasure that I acknowledge the following people for helping in various ways during the preparation of this book: Philip E. Rosenberg, David M. Hopkins, C. Hans Nelson, Horace G. Richards, Richard C. Allison, Peter Barnes, Edwin C. Buffington, Joe S. Creager, D. A. McManus, Joseph H. Kravitz, G. Vilks, Thomas D. Hamilton, Ronald J. Echols, G. D. Sharma, J. Valentine, and Henry Grosshans.

I also wish to acknowledge the cooperation I received from all contributing authors and from the publisher, Springer-Verlag New York Inc. and Dr. Konrad F. Springer.

YVONNE HERMAN

Contributors

KNUT AAGAARD, Department of Oceanography, University of Washington, Seattle, Washington 98105, U.S.A.

JOHN A. ANDREW, Shell Oil Company, P.O. Box 127, Metairie, Louisiana 70004, U.S.A.

OTIS E. AVERY, U.S. Naval Oceanographic Office, Washington, D.C. 20373, U.S.A.

PETER BUURMAN, Department of Soil Science and Geology, Agricultural University, Wageningen, The Netherlands.

LAWRENCE K. COACHMAN, Department of Oceanography, University of Washington, Seattle, Washington 98105, U.S.A.

JOE S. CREAGER, Department of Oceanography, University of Washington, Seattle, Washington 98105, U.S.A.

RONALD J. ECHOLS, Department of Oceanography, University of Washington, Seattle, Washington 98105, U.S.A.

YURI B. GLADENKOV, Geological Institute of the Academy of Sciences of the U.S.S.R., Pyzhevsky per 7, Moscow, U.S.S.R.

YVONNE HERMAN, Department of Geology, Washington State University, Pullman, Washington 99163, U.S.A.

EDWARD PETER JACOBUS VAN DEN HEUVEL, Sterrewacht "Sonnenborgh," Rijksuniversiteit, Utrecht, The Netherlands and Astrophysical Institute, Vrije Universiteit, Brussels, Belgium.

MARK L. HOLMES, Department of Oceanography, University of Washington, Seattle, Washington 98105, U.S.A.

DAVID M. HOPKINS, U.S. Geological Survey, 345 Middlefield Road, Menlo Park, California 94025, U.S.A.

HARLEY J. KNEBEL, Marine Biological Laboratory, Woods Hole, Massachusetts 02543, U.S.A.

JOSEPH H. KRAVITZ, U.S. Naval Oceanographic Office, Washington, D.C. 20390, U.S.A.

HUBERT HORACE LAMB, Climatic Research Unit, School of Environmental Sciences, University of East Anglia, Norwich, England.

ANGI SATYANARAYAN NAIDU, Institute of Marine Science, University of Alaska, Fairbanks, Alaska 99701, U.S.A.

FREDERIC P. NAUGLER (deceased), Pacific Oceanographic Laboratories, NOAA, University of Washington, Seattle, Washington 98105, U.S.A.

C. HANS NELSON, U.S. Geological Survey, 345 Middlefield Road, Menlo Park, California 94025, U.S.A.

DAVID W. SCHOLL, U.S. Geological Survey, 345 Middlefield Road, Menlo Park, California 94025, U.S.A.

GHANSHYAM D. SHARMA, Institute of Marine Science, University of Alaska, Fairbanks, Alaska 99701, U.S.A.

NORMAN SILVERBERG, Centre d'Etudes Universitaires de Rimouski, Université du Quebec, Rimouski, Quebec, Canada.

SERGEI L. TROITSKIY, Institute of Geology and Geophysics, Siberian Division, Academy of Sciences of the U.S.S.R., Novosibirsk, U.S.S.R.

PETER R. VOGT, U.S. Naval Oceanographic Office, Washington, D.C. 20373, U.S.A.

Contents

Chapter 1
Physical Oceanography of Arctic and Subarctic Seas[1]

L. K. COACHMAN[2] AND K. AAGAARD[2]

"From all these considerations it appears unquestionable that the sea around the Pole is fed with considerable quantities of water, partly fresh, . . . partly salt, . . . proceeding from the different ocean currents."

F. Nansen, Naturen, *March 1891*

I. Introduction

The Arctic Ocean (Fig. 1) surrounds the North Pole and is bordered by Europe, Siberia, Alaska, Canada, and Greenland. It is a large basin (9.5×10^6 km^2), in area about four times larger than the Mediterranean Sea, connected primarily with the Atlantic Ocean via the major water bodies of the Greenland and Norwegian Seas and Baffin Bay, which also lie in the Arctic. Technically, the Arctic Ocean is probably not an ocean but rather a mediterranean sea of the Atlantic, as recognized by the oceanographic pioneers from Norway (in Norwegian: Nordpolarhavet = North Polar Sea). The various Arctic basins do, however, attain depths similar to those of the oceans ($\sim$4000 m), and time and convention have dictated our present nomenclature.

A feature contributing to the oceanographic uniqueness of the Arctic is the ice of the sea. Sea ice covers an average of 26×10^6 km^2 of the earth's surface, about 7% of the total ocean area, and of this about 10×10^6 km^2 (40%) lies in the Arctic. However, large seasonal fluctuations create about 3×10^{19} g of new sea ice each year, of which about 40% is in the northern hemisphere (Shumskiy et al., 1964). The average yearly thickness of new ice added to the perennial ice cover is about 50 cm (Untersteiner, 1964), or 0.4×10^{19} g, and thus the remaining two-thirds, 0.8×10^{19} g, is formed from open sea water. This growth takes place largely in the areas peripheral to the central Arctic basin, as these areas become open in summer, but some ice is formed on leads that occur throughout the pack even during winter (Zubov, 1945).

The general oceanographic consequences of a perennial or seasonal ice cover are:

(1) The water temperature of the near-surface layer in the presence of ice is always maintained close to the freezing point for its salinity by the change-of-phase process.

(2) Salt is excluded from the ice to a varying extent, but the water under the ice is always enriched in salt by any ice growth. The dependence of water density on temperature and salinity (Fig. 2) is such that close to the freezing point density is almost solely a function of salinity. Therefore, ice formation can increase the density locally and some vertical convection may result.

(3) In the transfer of momentum from the atmosphere to the ocean, the wind must act on the sea through the intermediary of the ice.

[1] Contribution No. 626 from the Department of Oceanography, University of Washington.

[2] Department of Oceanography, University of Washington, Seattle, Washington 98105, U.S.A.

Methodology

The presence of ice requires that the methodology of arctic oceanographic investigations be somewhat different from those in the open sea. This comes about not only in a prohibitive sense, with the ice impeding both ship and surface travel; but also in an enabling sense, with the ice providing an endless supply of steady economical observational platforms.

The use of such platforms can historically be categorized in three ways. The first is the establishment of single drifting stations intended for long-term occupancy. This has been done either with a ship embedded in the ice, such as the *Fram* (Nansen, 1902), or by using an actual ice flow, such as with *North Pole I* (Shirshov, 1944). Lack of mobility and of extended spatial coverage on a short time scale are major drawbacks to such a scheme, although permanent drifting stations have certainly not outlived their usefulness.

The second category of ice platform use contains the attempts to circumvent the difficulties mentioned above by using multiple mobile scientific parties, ferried about by aircraft. This technique was pioneered by the Soviet high-latitude aerial expeditions (A.N.-I.I., 1946) and has recently flowered also in the western world (Coachman and Smith, 1971).

The third category involves the placement of automated data-gathering equipment on the ice. The technique is still in its infancy, the only system presently in general use being the Soviet

Fig. 1. Bathymetry of the Arctic Ocean (depths in fathoms) and location of the six stations of Fig. 4 (from Sater, 1969).

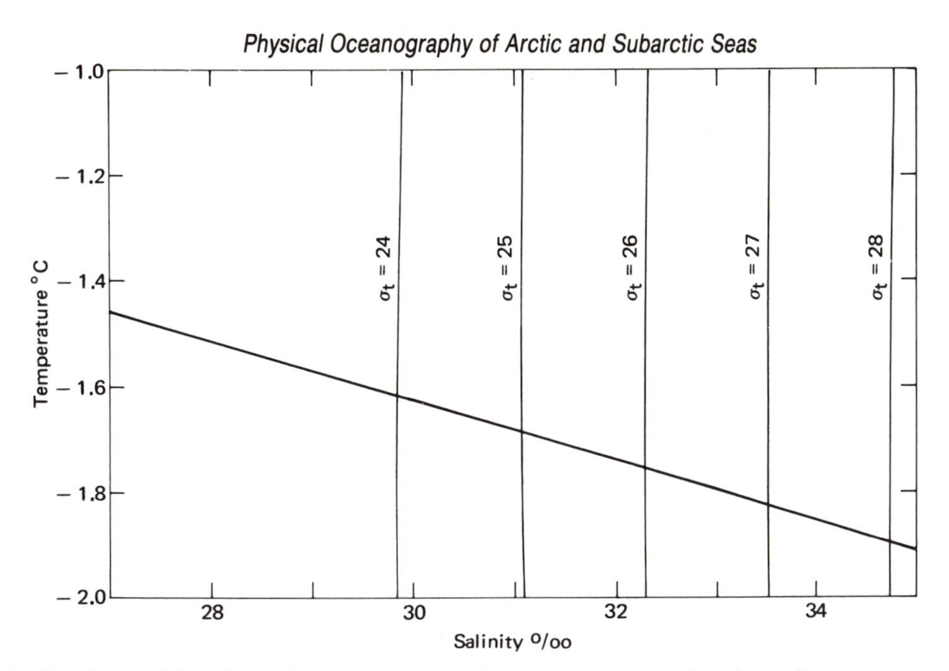

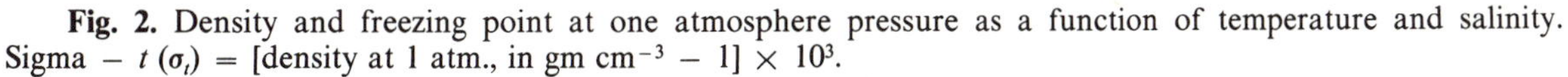
Fig. 2. Density and freezing point at one atmosphere pressure as a function of temperature and salinity. Sigma $- t$ (σ_t) = [density at 1 atm., in gm cm^{-3} $-$ 1] $\times$ 10^3.

DARMS (Drifting Automatic Radiometeorological Stations) (Olenicoff, 1971).

A liability of any drifting observational site is of course that the observations lie neither in a Lagrangian nor an Eulerian frame of reference. Nevertheless, the optimal use of the ice as an observational platform for oceanographic purposes has thus far not been approached.

No surface vessel can operate freely in the central Arctic, whereas submarines for many purposes would be eminently suitable for oceanographic research because of their size, speed, and navigational capability (Lyon, 1961). The practicability of making oceanographic observations from a pressurized submarine compartment was in fact demonstrated already in 1931, on the *Nautilus* expedition (Sverdrup, 1933).

As in the rest of the world ocean, we may expect that a variety of methodologies will be required to effect solutions to the various oceanographic problems. An assessment of the applicability of possible platforms to physical oceanographic research in the Arctic has been made by Coachman (1968).

II. Physical Features

Central Arctic

The bathymetry of the Arctic basins (Fig. 1) is now generally known, though in certain areas, notably the north edge of the continental shelf within the Arctic Ocean, detailed information is lacking. The major physiographic features, with suggested standard nomenclature, are reported by Beal et al. (1966).

The continental shelf on the North American side of the Arctic Ocean is narrow (50 to 90 km), whereas the half bounded by Europe and Siberia is very broad (>800 km) and shallow, with peninsulas and islands splitting it into five marginal seas—the Barents, Kara, Laptev, East Siberian, and Chukchi. Even though these marginal seas occupy 36% of the area of the Arctic Ocean, they contain only 2% of its volume of water. All the major continental rivers reaching the Arctic Ocean, with the exception of the Mackenzie in northern Canada and the Yukon, which enters via the Bering Sea and the Bering Strait, flow into these seas. Thus these shallow seas, with a high ratio of exposed surface to total volume and with a substantial input of fresh water in summer, greatly influence surface water conditions in the Arctic Ocean.

The continental shelf is indented by numerous submarine canyons, including the very large Svataya Anna and Voronin Canyons in the northern Kara Sea and the much smaller Herald and Barrow canyons in the Chukchi Sea (Fig. 1). The northern edge of the continental shelf, particularly in the Laptev and East Siberian Seas and along the Canadian Archipelago, is poorly surveyed, and undoubtedly there are canyons yet to be defined. The oceanographic importance of these canyons is that they act as preferential pathways for egress of water from the relatively warm Atlantic layer onto areas of the continental shelf, where it comes within the influence of strong mixing processes, and hence can locally modify the surface waters and ice cover (see below).

The Lomonosov Ridge, with a sill depth of about 1400 m, divides the Polar Basin into two deep basins, the Eurasian toward Europe (4200 m) and the Canadian* toward North America (3800 m). The latter is further subdivided by the Alpha Rise, which runs parallel to the Lomonosov Ridge at 500 km distance.

Greenland and Norwegian Seas *(Fig. 3)*

The continental shelf along eastern Greenland south to 77°N is broad (300 km) and contains a system of banks of less than 200 m depth. South of 77°N the shelf narrows until at 75°N it is less than 100 km wide, and except in the Denmark Strait it remains thus to the southern tip of Greenland. The shelf is marked by several deep indentations.

At 71°N the Jan Mayen Ridge extends toward Greenland, with a maximum depth between Greenland and Jan Mayen of about 1500 m. East of Jan Mayen the ridge (in this area called the Mohn Rise) continues northeastward toward Svalbard, with a sill depth of about 2600 m. This ridge system effects the separation of the Greenland Sea to the north from the Norwegian Sea to the south. The central Greenland Sea has two deep basins, the northernmost about 3200 m deep and the southernmost about 3800 m. (Eggvin's [1963] chart shows one very deep sounding of 4845 m in the northernmost basin.) Physiography and nomenclature of the Greenland Sea is given in Johnson and Eckhoff (1966).

A ridge also extends south from Jan Mayen, effectively separating a basin to the west with depths over 2200 m from the larger main Norwegian Sea basin to the east. The area between Iceland and Jan Mayen is sometimes referred to as the Iceland Sea (Stefansson, 1962). The deep basin of the Norwegian Sea is compound, with the northern part extending to about 3500 m and the southern to over 3900 m.

* Here we take exception to the nomenclature of Beal et al. (1966) in preferring "Canadian" (or "Canada") to "Amerasian" for this basin.

The Advection Boundaries

The Bering Strait is the only connection between the Arctic system and the Pacific. It is very shallow, with a sill depth of about 45 m, and narrow, about 85 km. The Strait is open in summer, but from October until June it is typically ice-clogged so that navigation is possible only by icebreaker—and even these vessels have not always been able to navigate in the area in winter. Conducting oceanography from the ice surface does not appear to be a realistic consideration, because the ice is broken and in rapid motion for considerable distances north and south of the Strait. Attempts have been made to assess the transport through the eastern channel by means of electromagnetic induction in a cable (Bloom, 1964), and attempts are also continuing to maintain bottom-mounted current, temperature, and salinity sensors (Bloom, personal communication). During summer, current measurements have been made from anchored vessels and moored buoys (Coachman and Aagaard, 1966).

The Canadian Arctic Archipelago consists of some 16 passages connecting the Arctic and Atlantic Oceans, ranging in width from 10 to 120 km, and in depth to over 700 m (Collin, 1963). In northern Baffin Bay the number of passages is reduced to three (see, e.g. Muench, 1971). Two of these—Nares Strait (sill depth 250 m) and Lancaster Sound (sill depth 130 m)—seem to be important for the exchange of water, while the third, Jones Sound, is of lesser importance. These areas are navigable only in summer. They are probably not navigable even by icebreaker in winter, except within the semipermanent polynya (northwater) at the south end of Nares Strait. Moorings with current meters attached have been made here (Day, 1968). It seems probable that oceanography could be done from the ice surface in winter in most of these areas.

On the Eurasian side of the Arctic system the passage between Spitsbergen and Greenland is both wide (about 600 km) and deep (sill depth about 2600 m). The eastern side, west to about the prime meridian, is navigable throughout much of the year, though probably considerable difficulty would be encountered in winter. The western side is perennially ice-covered, though vessels have penetrated north of 80°N to near Greenland on at least two occasions from August to September (the *Ob'* in 1956 and the *Edisto* in 1964). Winter oceanographic observations were obtained during

Fig. 3. Bathymetry of the Greenland, Norwegian, and Iceland Seas (depths in fathoms) (from an unpublished chart by G. L. Johnson).

1938 (Shirshov, 1944) and 1965 (Tripp and Kusunoki, 1967) from drifting ice stations.

Between Spitsbergen and Franz Josef Land the sill depth is over 200 m, and between Franz Josef Land and Novaya Zemlya it is also about 200 m. No winter work has been attempted in these areas, though it might be possible to effect penetration by icebreaker. Some oceanographic stations have been made in summer (cruises of the *Quest* in 1931, the *Southwind* in 1970, and the *G. O. Sars* in 1971).

The Barents Sea itself is a deep shelf sea which is indented in the southern and western parts by depths of 300 to 400 m. It is shallower in the northern and eastern portions, but the bottom is quite irregular. There are several prominent canyons in the western part. Numerous oceanographic data are available, particularly from summer, due to intensive efforts by European fisheries research groups under the aegis of I.C.E.S. (see, e.g., Dickson et al., 1970). Current measurements have also been made. Again, though, direct current measurements for the Barents–Norwegian Sea boundary (Nordkap–Bear Island–Spitsbergen) are nonexistent, except for some time-series measurements near Bear Island made by the University in Bergen.

Denmark Strait, between Greenland and Iceland, is about 275 km wide and has a sill depth of 650 m. The eastern portion of the Strait is generally navigable throughout the year, while the western portion presents genuine navigational difficulties to icebreakers only during winter and spring. Many oceanographic data are available from the Strait, particularly due to the efforts of the Icelandic oceanographers (see, e.g., Stefansson, 1968), and some summer current measurements have been made (Malmberg et al., 1967). Moored current measurements have also been successfully accomplished in winter (Worthington, 1969).

The passage between Iceland and the Faeroe Islands is about 420 km wide. The submarine ridge therein has a broad sill with no distinct channel, but limiting depths seem to be about 480 m. The ridge connecting the Faeroe and Shetland Islands (320 km apart) is indented to the southeast and south of the Faeroes by a narrow deep channel with a sill depth of 820 to 830 m (the Faeroe-Bank Channel). The oceanography of these major connections between the Norwegian Sea and the Atlantic Ocean has in the last decade been vigorously pursued by a consortium of European groups under the I.C.E.S. (see, e.g., Tait, 1967). Oceanographic research in the Iceland–Faeroe–Shetland channels is not restricted by ice, only by occasionally severe weather. Recent work has included direct current measurements both from shipboard and moored buoys.

Fresh Water

Precipitation over the Arctic basin seems to average about 100 mm/yr (Mosby, 1962b), though it is extremely difficult to measure accurately. However, this is a small value, leading to a fresh-water influx of only about 0.03 Sv (1 Sv = 1 Sverdrup = 1×10^6 m^3 sec^{-1}), which is negligible. Estimates of evaporation do not agree well, ranging from 4 cm yr^{-1} (Mosby, 1962b) to 30 cm yr^{-1} (Fletcher, 1966). In any event these values also lead to negligible effect on the water budget, even though the obviously rapid evaporation from open leads may have significant influence on the convection and micrometeorology, particularly in winter.

Table 1 River Discharge to the Arctic Basin (in 10^3 m^3 sec^{-1})*

Rivers	Maximum	Minimum	Average
Ob'	20.8	13.5	16.3
Yenisei	20.5	16.8	19.2
Others			6.2
Total: Kara Sea			41.7 ± 6
Lena	19.9	13.6	16.2
Others			8.4
Total: Laptev Sea			24.6 ± 5
Kolyma	5.6	2.4	4.2
Others			3.7
Total: East Siberian			7.9 ± 2
Total: Chukchi			2.5 ± 1
Mackenzie			7.9
Total: all rivers			84.6

* Sources: Siberian rivers (Antonov, 1964). Mackenzie River (Cram, 1968).

Larger inflows of fresh water to the basin come from land runoff. The discharge of the large rivers is given in Table 1. The average total fresh-water flux is probably about one order of magnitude larger than that due to the combined effects of precipitation and evaporation over the sea. Most of the runoff discharges into the continental shelf seas which border the Siberian side of the Arctic basin. This physical arrangement, together with large seasonal variation in inflow, leads to a significant influence on the surface and subsurface water conditions obtaining in the basin.

There is no significant quantity of fresh-water discharge from land into the Greenland–Norwegian sea system.

III. Arctic Ocean: Water Masses

Figure 4 shows typical vertical distributions of temperature and salinity from the various Arctic basins (positions shown in Fig. 1). The remarkable similarity in vertical structure shown by stations taken at widely diverse times and locations leads to the conclusions that over the long term the Arctic basins are dynamically in a steady state and that the observed distributions of properties are a result of continuing processes within the basin (Coachman and Barnes, 1961). Thus, the water deeper than 200 m reflects its common origin (the Atlantic) and the surface

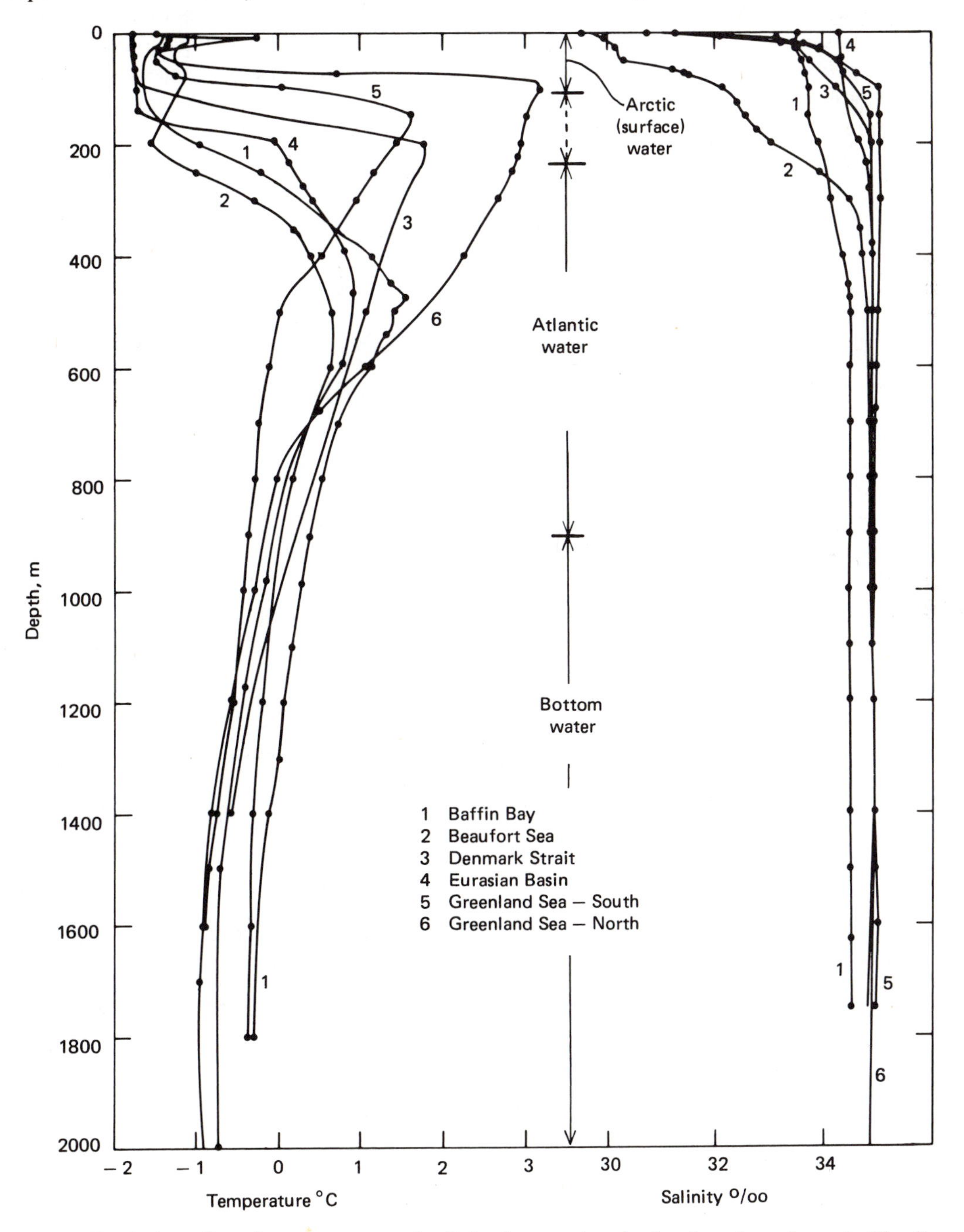

Fig. 4. Vertical profiles of temperature and salinity from various basins (locations shown in Fig. 1).

water reflects the similarity of local modifying processes (primarily Arctic climatic factors). Based on temperature, three water masses are defined:

(1) Surface water (Arctic water), from the surface down to about 200 m, has varying characteristics. In ice-covered areas the temperature is close to that of freezing for the salt content. In the usually ice-free areas (eastern Greenland Sea; along west Greenland north through Davis Strait) temperatures may be a few degrees above freezing. Areas ice-free in summer (Chukchi Sea, nearshore areas of other peripheral seas, most of Baffin Bay) may seasonally exhibit surface temperatures of 1 to 2°C or more. Temperatures below the surface are typically always cold, except in the Canadian Basin of the Arctic Ocean, where there may appear a small temperature maximum (−1.0°C) in the 75-to-100-m layer. This is attributable to influx of Bering Sea water (see below). The salinity of the surface layer may be uniform down to 50 m or so, and then increase until at 200 m it is ~34.5‰, or it may begin to increase nearer the surface. The surface salinity values do exhibit spatial variation (see below).

(2) The layer below the Arctic water, from about 200 to 900 m, called the Atlantic layer, has temperatures above 0°C, with a maximum at 300 to 500 m, again depending on location. Salinities continue to increase over the surface values until, by 400 m, but in many instances shallower, they attain in the Arctic Ocean and most of the Greenland Sea, nearly uniform values in the range 34.9 to 35.0‰, and in Baffin Bay, 34.5 to 34.6‰.

Fig. 5. Surface (5 m) salinity of the Arctic basins, summer (from A.I.N.A., 1967).

(3) Below the Atlantic layer lies bottom water, with temperatures below 0°C and the same uniform salinities attained in the Atlantic layer. Deep temperatures vary slightly from basin to basin, being in the Canadian Basin about −0.5°C, the Eurasian Basin −0.9°C, the Greenland Sea −1.2°C, and Baffin Bay −0.45°C.

It should be noted that on the basis of density (which in vertical distribution closely resembles that of salinity) the Arctic waters exhibit essentially a two-layer system, with a thin less dense surface layer separated from the main body of water of quite uniform density by a strong pycnocline. The pycnocline restricts vertical motion and the vertical transfer of heat and salt, and hence the surface layer acts as a "lid" over the large masses of warmer water below.

Surface (Arctic) Water

The most important processes conditioning and modifying the surface layer are:

(1) addition of mass (fresh water) from land, primarily from the large Siberian rivers;

(2) additions of fresh water locally through melting of ice;

(3) heat gain through absorption of solar radiation in non-ice-covered areas during summer;

(4) concentration of salt, and hence increase of density of surface water, through freezing of ice;

(5) heat loss to the atmosphere through any open water surface, including leads in the central Arctic pack ice;

(6) inflow and subsequent mixing of Atlantic and Pacific waters.

The advective processes [(1), (6) primarily] give rise to geographic variations in surface salinity values (Fig. 5). The total range is about 7‰; the lowest values (27 to 30‰) are encountered closest to the sources of fresh-water influx (Mackenzie River, Siberian Seas, Bering Strait), and the highest values (33 to 34.5‰) are associated directly with advection into the Arctic of North Atlantic surface water through the eastern Norwegian–Greenland Seas and eastern Baffin Bay.

Figure 6 shows longitudinal sections of salinity and temperature, from the Greenland Sea over the North Pole to the Chukchi Sea. The distributions are typical for a basin in higher latitudes, in which there is a surface outflow of relatively fresh water with some saline water entrained with it and an inflow of more saline water at depth to maintain the salt balance. In such basins there is an increase in surface salinities from the head of the basin, close to the source of fresh water, toward the mouth, owing to progressive mixing between the surface layer and the underlying more saline water. These conditions are fulfilled by the Arctic Ocean and also Baffin Bay. The salinity is lowest along the periphery of the Arctic Ocean from the Mackenzie River delta west to the Kara Sea, because major rivers discharge throughout this area (water flowing north through the Bering Strait is also of low salinity because of the Yukon River discharge). However, the salinities are never much below about 27‰, except very close to a river mouth, because this fresh water is efficiently mixed with more saline deeper water on the shallow continental shelf by the action of winds, tides, and currents.

Salinities in general progressively increase in the directions of flow of the surface waters (see below) and attain maximum values in those areas where currents are directly introducing North Atlantic surface water into the Arctic basins. The North Atlantic surface water is relatively dense and sinks within the basins to subsurface levels.

All the processes give rise to small seasonal variations. Processes (1), (2), and (3) lead to decreases in the density of the water affected, and they operate only during the short summer (June to September). Thus in summer the surface waters exhibit somewhat lower salt content than in winter, and in ice-free areas (e.g., parts of Baffin Bay, parts of the Canadian Archipelago, along the Siberian and Alaskan coasts) surface temperatures may remain near the freezing point because the added heat goes to melting the ice rather than warming the water. Figure 2 shows the relationship of freezing point to salinity (note that maximum density of sea water occurs at a temperature colder than the freezing point).

Processes (4) and (5), of greatest import in winter, lead to increases in the density of surface water. In some areas and under some conditions these affected waters sink to subsurface levels, but in general in the Arctic Ocean these concentration mechanisms are not adequate to create water of sufficient density that it can replace waters lying deeper than 200 m. Thus all intermediate and deep water in the Arctic Ocean is advected in from adjacent areas. As the connection with the Pacific Ocean is so limited (Bering Strait: width about 85 km, depth about 45 m), the deeper

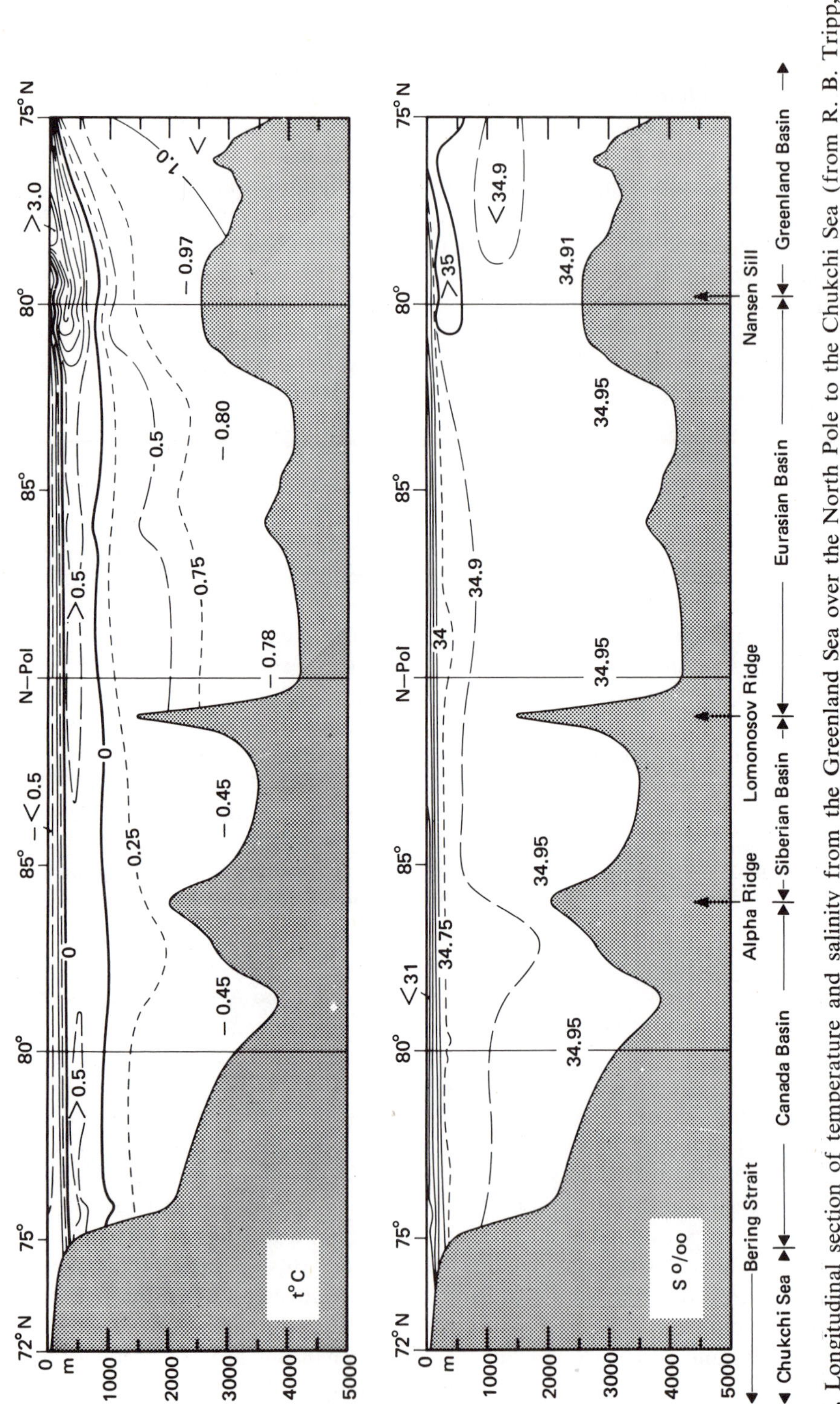

Fig. 6. Longitudinal section of temperature and salinity from the Greenland Sea over the North Pole to the Chukchi Sea (from R. B. Tripp, unpublished).

waters of the Arctic Ocean, as well as Baffin Bay, essentially originate in the North Atlantic.

Figure 5 represents summer conditions. Surface salinities in winter can be higher by 0.5 to 1‰ in the deep basins of the Arctic Ocean and by 1 to 2‰ in the peripheral seas, because during this season melt and runoff are not contributing fresh water, but mixing with more saline waters is continuous. However, the general pattern of distribution remains similar.

The general distribution of surface temperature is much more uniform than that of salinity, and there is little spatial or temporal variation throughout most of the Arctic, because of the thermal buffering by the ice. Only those areas that are ice-free for extended periods exhibit temperatures significantly above freezing. They are either permanently ice-free areas influenced by currents carrying warmer water into the Arctic (eastern Greenland Sea–West Spitsbergen Current; eastern Baffin Bay–West Greenland Current); or areas that are typically ice-free seasonally (July to September), such as the nearshore portions of the peripheral seas, the Chukchi Sea, and northern Baffin Bay. These areas all show seasonal temperature fluctuations. Areas in which major currents carry Arctic water toward the North Atlantic (western Baffin Bay–Canadian Current; western Greenland Sea—East Greenland Current) remain perennially ice-covered and hence have temperatures close to the freezing point all year round.

The Subsurface Layers*

Below the upper 30 to 50 m, which at most locations exhibit relatively uniform values of temperature and salinity throughout the layer, there are in the two major basins of the Arctic Ocean two distinctly different subsurface layers:

(1) In the Eurasian basin the salinity increases below 30 to 50 m so that by about 200 m, the bottom of the Arctic water mass, uniform deep values of 34.9 to 35.0‰ are reached. In contrast the temperature remains uniformly cold to deeper than 100 m before beginning its rise toward values above 0°C (Fig. 4, curve 4).

(2) In the Canadian Basin the salinity also increases with depth below 30 to 50 m, but not so rapidly as in the Eurasian Basin, so that the isohaline water is not reached until about 300 m. The temperature curves over much of the basin show a slight maximum between 50 to 100 m depth, then decrease to close to −1.5°C at 150 m depth, before the major temperature increase to values above 0°C (Fig. 4, curve 2). These subsurface layers, even though part of the surface (Arctic) water mass, each have particular advective origins.

The Eurasian Basin subsurface layer has been researched by Coachman and Barnes (1962). The temperature-salinity relationships clearly show that this subsurface layer is not a simple mixture of surface waters with underlying Atlantic water (Fig. 7). Isopycnal analyses (an example is given in Fig. 8) demonstrate the intimate connection between this water mass and the basin's peripheries, particularly the large submarine canyons Sv. Anna and Voronin in the northern Kara Sea.

The cold (close to freezing point) and saline water is advected over the Eurasian Basin from the areas of formation, which are over the continental slope and primarily over the canyons indenting the slope. Dynamically speaking, the submarine canyons are sunken estuarine-like features providing egress locally into the peripheral seas. Highly saline (>34.5‰) Atlantic water moves along the Eurasian cintinental slope and enters the canyons. Surface water in general flows north above the canyons and off the continental slope, and through dynamics similar to those in estuaries the Atlantic layer in these localities rises to shallower depths. At the shallower depths the Atlantic water is mixed with surface water and comes under the influence of climatic conditions. The temperature of this mixture, which is still relatively saline, is reduced to its freezing point, and then the mixture feeds back into the Eurasian Basin as the lower part of the subsurface layer. The submarine canyons thus serve as primary sources of the subsurface water.

One important consequence of this formation process is that in regions where a significant amount of heat from Atlantic water is lost upward to the surface the normal winter growth of sea ice is inhibited. Not only is the ice on the average thinner in these regions (Eskin, 1960), but even semipermanent open-water areas (polynyas) exist through the winter. One important polynya is indicated in Fig. 8. Another extensive polynya occurs in Smith Sound, northern Baffin Bay (the North Water; Dunbar, 1969), though the wind probably plays a more important role in its formation than does heat loss from the water column (Muench, 1971). Yet another large po-

* See Note 1 at the end of this chapter.

lynya occurs from north of the New Siberian Islands to north of the Laptev Sea (the Great Siberian Lead; Laktionov and Shamont'yev, 1957). This lead occurs along the boundary between fast ice and the polar pack, but to our knowledge has not been subject to definitive study. Polynyas are of considerable importance locally to operations and climate.

The Canadian Basin subsurface water has been researched by Coachman and Barnes (1961) and Gushenkov (1964). Water from the Bering Sea flows north almost continually into the Arctic Ocean (Coachman and Aagaard, 1966) and has a wide range of characteristics. In summer, salinity values range from <31‰ up to about 33‰ maximum, and temperatures range from about 1°C to as high as 10°C, though the range 2 to 6°C is more usual. In winter, salinity values are higher (32 to 34‰), while temperatures are uniform and close to the freezing point (−1.5 to −1.7°C). This water intruding into the Arctic Ocean contributes to the subsurface layer over much of the Canadian Basin.

The relationship is best shown in a temperature-salinity diagram (Fig. 9). The water transiting Bering Strait loses heat as it crosses the Chukchi Sea, probably in part through the surface and in part by mixing with local shelf water of nearly the same density (cf. data from *Maud*, eastern Chukchi Sea). Hence, within the Canadian Basin the subsurface layer is not purely Bering Sea water but rather consists of 10 to 20% Bering Sea water and 80 to 90% shelf water. The water formed in summer, less dense than the winter water primarily because of somewhat lower salinity values, remains above the winter water, and its warmer temperatures create and maintain the temperature maximum (always <0°C) encountered in the 50- to 100-m layer. The winter intrusion lies above the Atlantic layer, and as it is cold it creates and maintains a temperature minimum centered at 150 m depth over much of the basin. Because of the

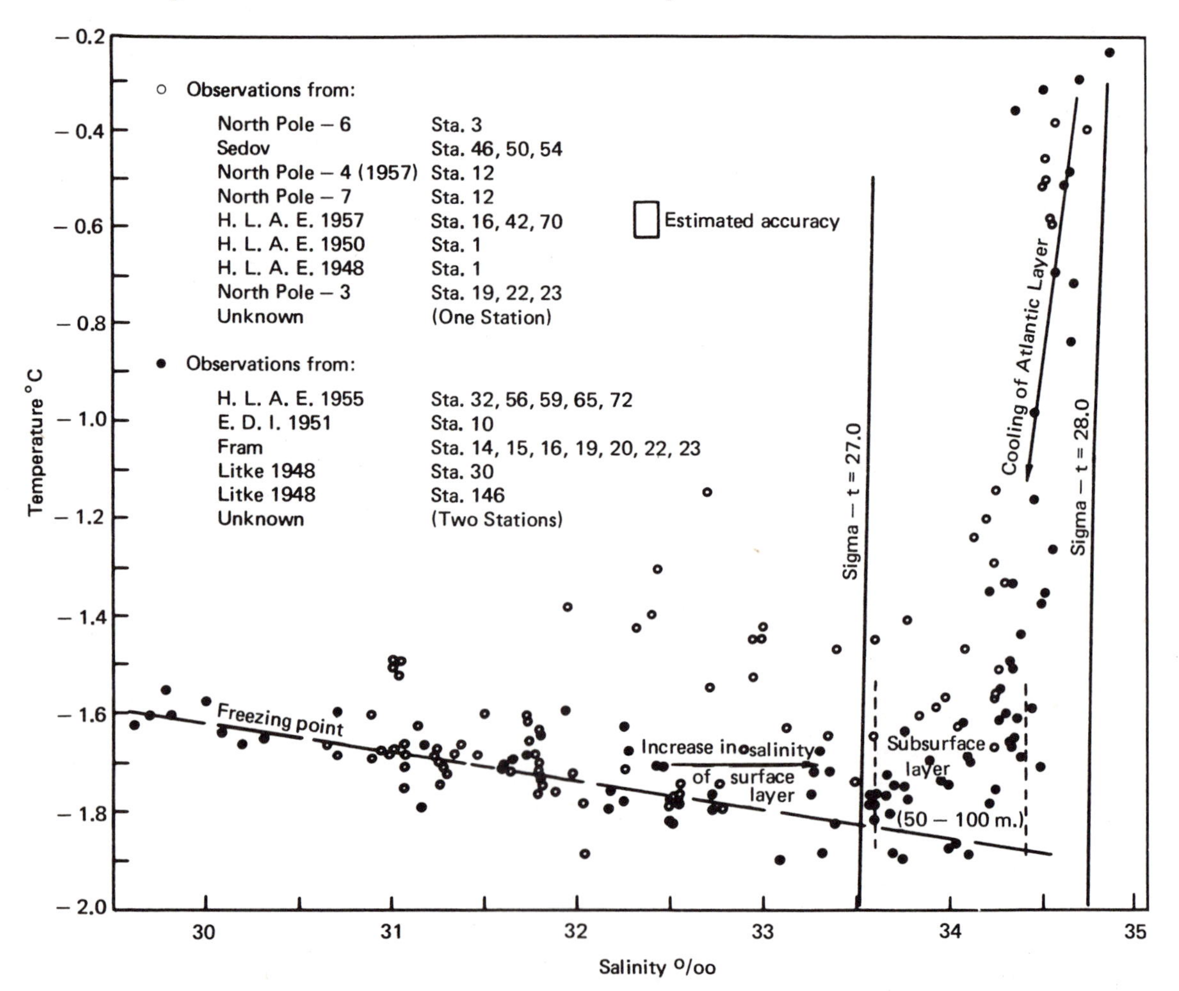

Fig. 7. T-S diagram of Eurasian Basin subsurface water (from Coachman and Barnes, 1962).

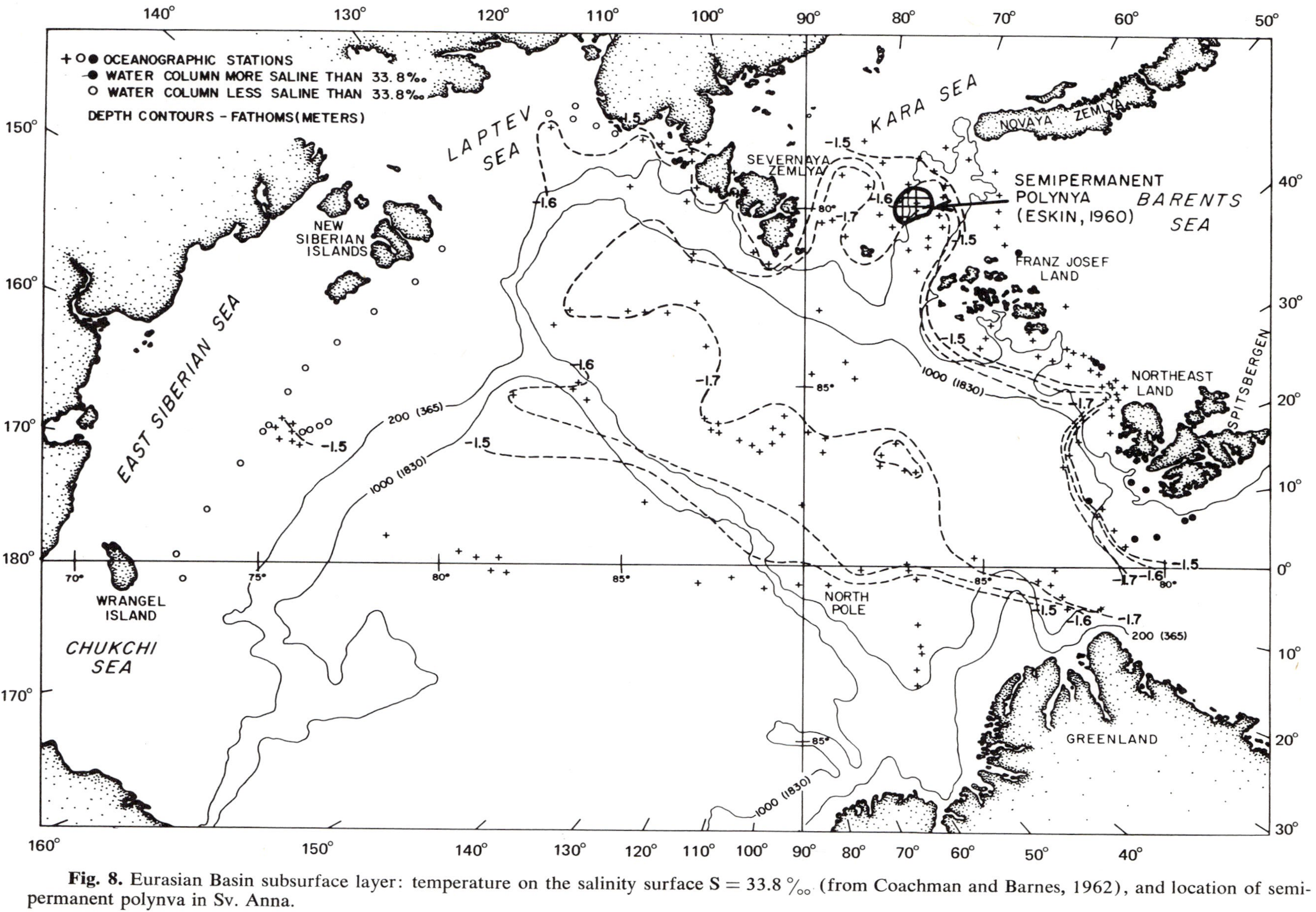

Fig. 8. Eurasian Basin subsurface layer: temperature on the salinity surface S = 33.8 ‰ (from Coachman and Barnes, 1962), and location of semipermanent polynva in Sv. Anna.

uniformity of temperature of the incoming water and, apparently, a very slow vertical diffusion of heat, the temperature minimum is very uniformly close to −1.5°C.

Within the Arctic Ocean the distribution of the Bering Sea water has been followed by tracing the temperature maximum in the 50- to 100-m layer. The layer has been thought to circulate in consonance with the general surface circulation in the basin deduced from the movement of ice (see below): from north of the Chukchi Sea north toward the Pole and clockwise around the Beaufort Sea gyre. We are at present engaged in a more detailed analysis, aided by many new data ob-

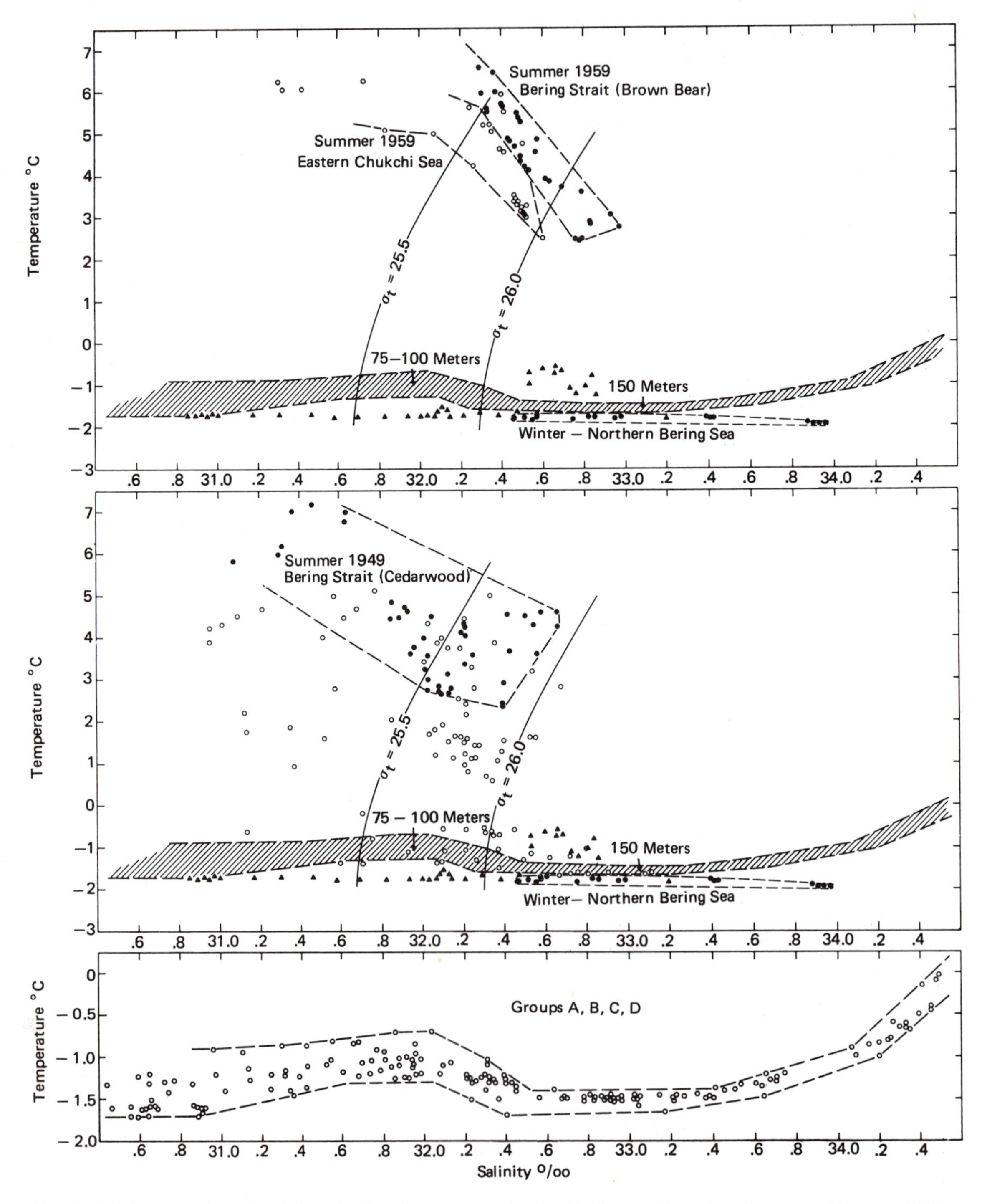

Fig. 9. T-S diagrams for elucidating Bering Sea water in the Arctic Ocean (from Coachman and Barnes, 1961).

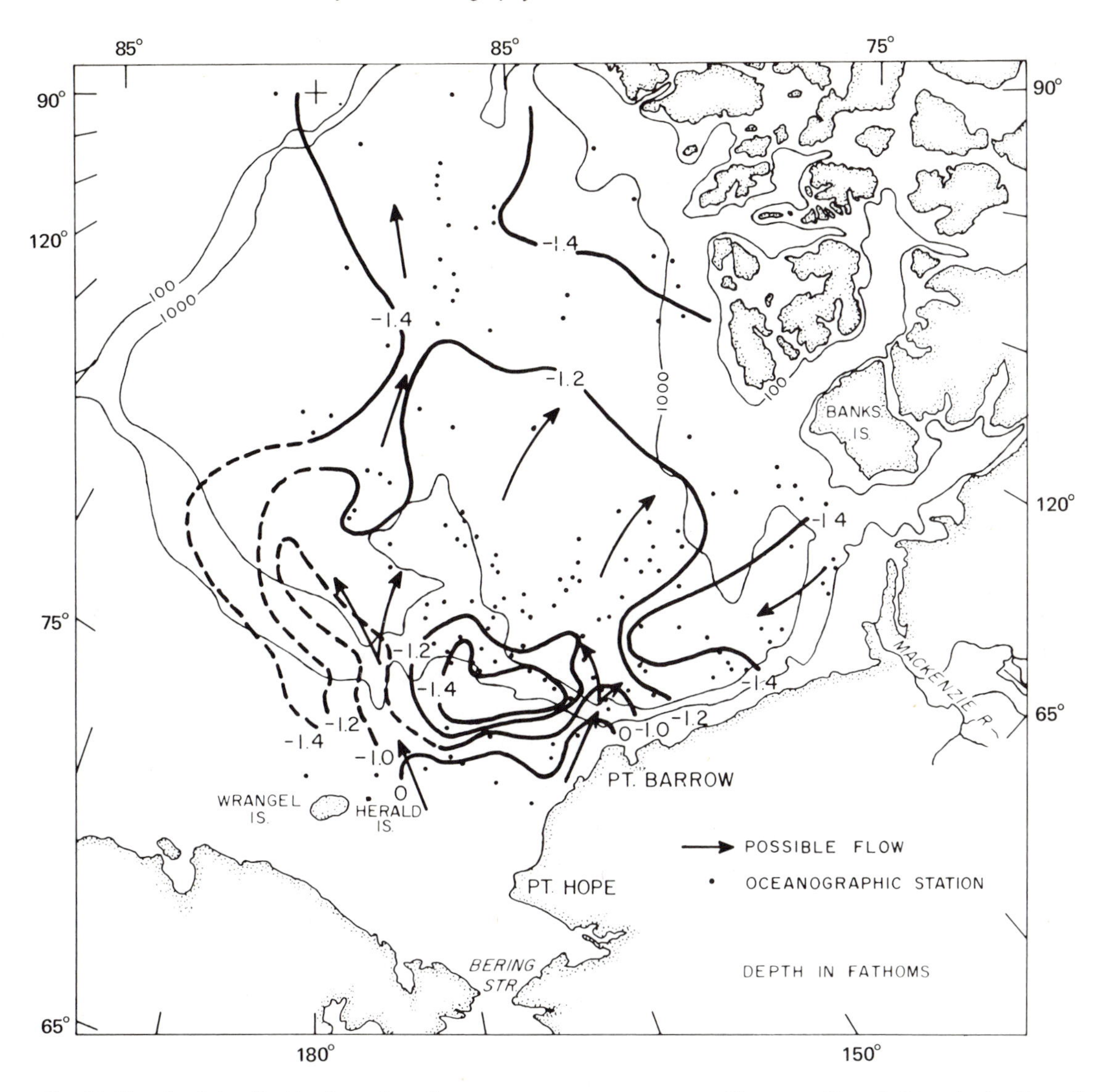

Fig. 10. Sketch of the distribution of maximum temperature encountered between 50 and 100 m in the Canadian Basin. At shallow stations (west of Point Barrow) it is the temperature at S $\approx 32‰$. Dashed isotherms are from Gushenkov (1964). Arrows suggest a possible flow pattern.

tained in recent years, and the distribution appears to be more complicated than previously reported.

Figure 10 shows isotherms of the maximum temperature encountered between 50 and 100 m, or where the stations are shallower than 100 m, the temperature at S $\sim 32‰$. As the stations were obtained at all seasons and over many years, some smoothing was required to obtain a coherent picture. The isotherms west of 180° are after Gushenkov (1964). The figure should be considered as only a preliminary analysis. The Bering Sea water follows two routes into the Arctic Ocean (cf. Aagaard, 1964; Gushenkov, 1964). After leaving Point Hope, the main branch of the north flow proceeds north and west toward Herald Island, and enters the general arctic circulation west of Herald Shoal. Another branch flows north and east along the Alaskan coast and enters the general circulation north of Point Barrow. The dynamics of these flows may be related to the presence of Herald and Barrow Canyons (Fig. 1), in a manner similar to that discussed for the Eurasian Basin subsurface layer (see above); or it may be influenced by the presence of Herald Shoal.

Within the Canadian Basin the layer is advected in part north toward the Pole, in keeping with the surface circulation. Herald Shoal and the Chukchi Cap north of it deflect the general westerly flow northward. However, the movement of the layer within the basin is not everywhere directly related to the surface circulation (cf. Figs. 10 and 16), because there appears to be a general spreading of the layer eastward over the whole central Beaufort Sea. East of Point Barrow the layer penetrates eastward along the Alaskan continental slope on occasion (cf. Johnson, 1956), but our tentative interpretation is that another subsurface water mass, formed on the shelf among the Canadian islands, is in general flowing west with the surface circulation. Hence, positively identified Bering Sea water is only occasionally encountered east of 145°W, and then usually as an isolated patch (cf. Coachman and Barnes, 1961, Fig. 4, the dashed curve in group K). We visualize eddies of Bering Sea water being occasionally advected into the region.

The temperature and thickness of the layer varies temporally as well as spatially, due both to a diffusive loss of heat away from the temperature maximum and to large seasonal variations in the inflowing water. Temperature and layer thickness variations were investigated in numerous and detailed data from *North Pole-12* when it was drifting north of Wrangel Island from May 1963 to October 1964 (Shpaikher et al., 1966). Though the analysis must be discounted as accurately defining temporal variations, because of the impossibility of separating time- and space-dependence from the drifting platform (NP-12 drifted about 840 km in the period), temperature values of the maximum ranged between −0.6 and −1.2°C and the layer thickness between 25 and 55 m. Maximum temperature and layer thickness was observed during July and August, and minima during September and November. These data were later combined (Shpaikher and Yankina, 1969) with data from the Bering Strait and the Chukchi Sea during the decade 1956 to 1966 (sources not cited) to derive a picture of the propagation of Bering Sea water into the Canadian Basin (Fig. 11). We feel that, as a first approximation, the net advection of the Bering Sea water is reasonably depicted, though the speed of advance may be low (Table 2).

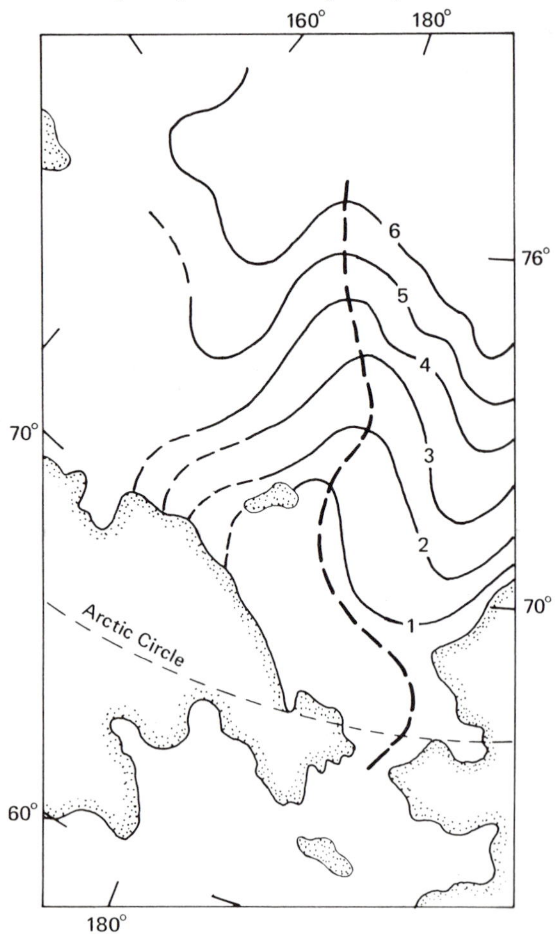

Fig. 11. Propagation of Bering Sea water into the Canadian Basin (after Shpaikher and Yankina, 1969). Dashed line is interpreted flow axis, and isochrones are years.

Table 2 Net Speed of Advance of Bering Sea Water into the Canadian Basin (after Shpaikher and Yankina, 1969); see Fig. 12.

Year	Speed (cm/sec^{-1})
1	2
2	0.5
3–6	0.2

One consequence of the intrusion of Bering Sea water as a subsurface layer over much of the Canadian Basin is that it carries with it certain properties and organisms that are not indigenous to the Arctic, nor to the Atlantic, which provides most of the water to the Arctic Basin. As examples, certain copepods netted at 85°N are definitely native to Pacific waters (Johnson, 1963), and the high phosphate concentrations found in bottom water on the shelf in the East Siberian Sea can be accounted for only by considering a contribution from the Bering Sea (Codispoti and Richards, 1968).

Another consequence of this advected subsurface layer is similar to that of the subsurface layer in the Eurasian Basin: as the advected water

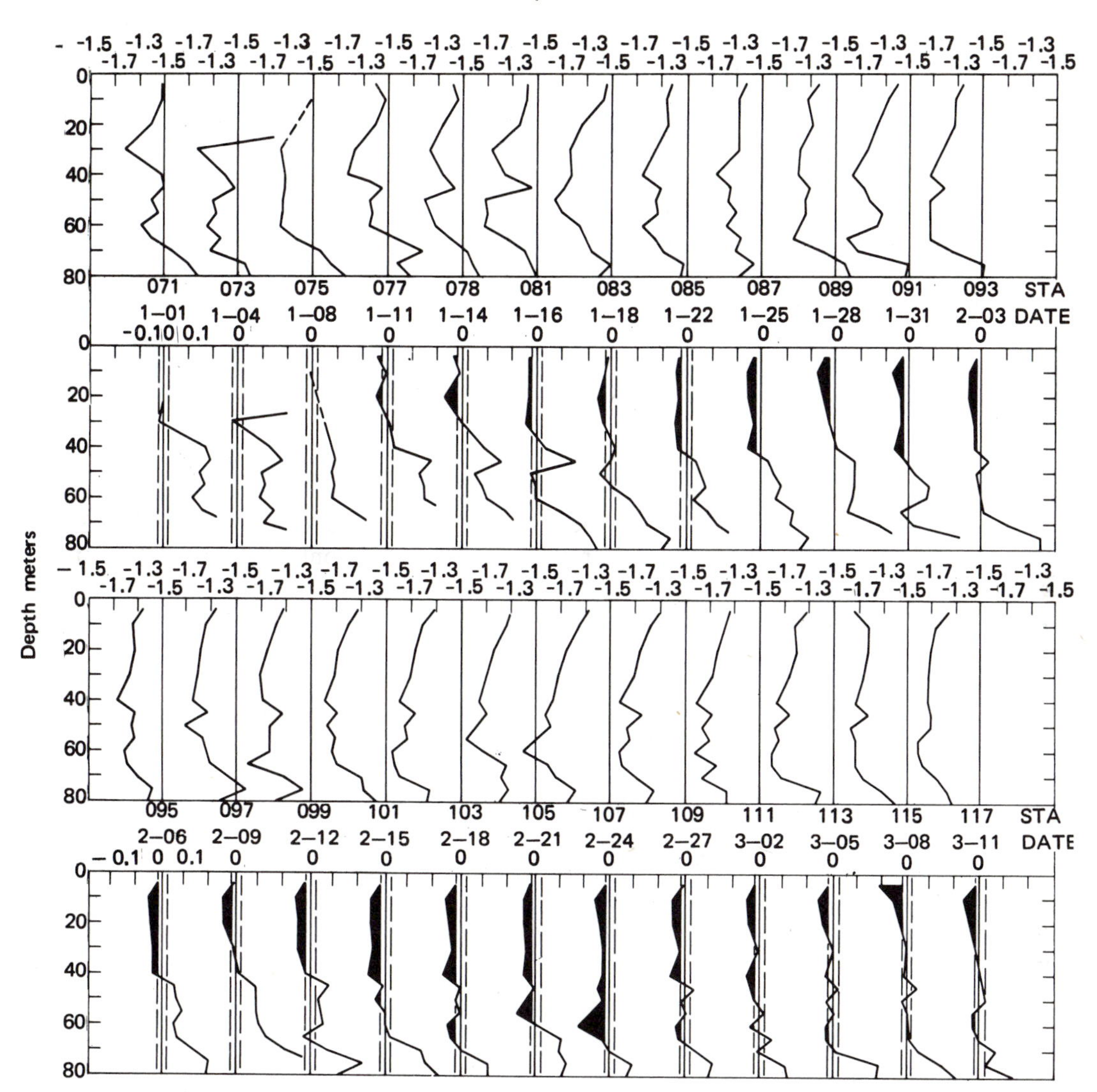

Fig. 12. Vertical temperature profiles and deviations from the freezing point (at 1 atm) from Arlis-I (from Coachman, 1966). Supercooling >0.2°C is shaded.

is of higher salinity, and hence density, than the surface water, it creates and maintains a relatively strong pycnocline over the Atlantic water. The pycnocline inhibits the depth of penetration of vertical convection generated primarily by the freezing process. *Vertical convection* is a dynamic process of profound importance to the formation of deep water in the world oceans. It is particularly significant in polar oceans because of the extreme cooling and the salt separation through freezing, although deep convection may be locally important in certain other areas, for example, the western Mediterranean (Voorhis and Webb, 1970).

Ice growing from sea water excludes salt, which added to the ambient water increases its density, and gravitational convection results. This phenomenon has been researched by Foster both theoretically (1968) and experimentally (1969). Field experiments under continuous growing ice sheets have been reported by Lewis and co-workers (Lake and Lewis, 1970; Lewis and Walker, 1970; Lewis and Lake, 1971). The freezing process generates convective streamers which have relatively small horizontal dimensions (the order of 1 cm), and these streamers carry the excess salt to depth. They also are cold compared to the ambient water and act as heat sinks. There must be compensatory upward movement of water elsewhere, thus providing a mechanism by which heat can be fluxed upward through the convective layer. For example, in Cambridge Bay (Canadian Archipelago) residual summer heat is, over the course of winter, slowly removed from the water column by this process. The depth of the layer affected by the process seems to be rather strictly limited by the density structure, as the streamers do not form in the presence of a pycnocline. Hence it appears that this process does not provide significantly deep convection.

Nevertheless during winter, when large temperature differentials between air and water are present, there is evidence that convection occurs throughout the surface layer and to some degree penetrates into the underlying subsurface layer. The evidence is the widespread occurrence of "supercooled" water, the source of which can only be the surface, with its very rapid heat exchange. The water is in fact not supercooled at depth (pressure depresses the freezing point), but only relative to surface pressure. As pointed out by Lewis and Lake (1971), *in situ* supercooling under natural conditions is only transient, and observations that show *in situ* supercooling are probably suspect. The temperature serves as a tracer of the "supercooled" water at depth because the density is essentially salinity-controlled. The most comprehensive evidence for the distribution of the convected water at depth so far obtained is that from drifting ice station ARLIS-I (Brayton, 1962; Fig. 12). The first instances of the "supercooled" water at depth (30 m) occurred in October and lasted into March, during which time the depth of penetration doubled. The temperature of the layer decreased slightly over the winter, and a halocline was present within the layer.

A model has been proposed (Coachman, 1966) to explain these observations, which in our opinion cannot be accounted for by the convection associated with sea ice growth. The hypothesis calls for a larger-scale convection, the size of the cells being tied to the sizes of the leads. Heat loss in winter in the Arctic through an open water surface is at least 2 orders of magnitude greater than through ice (Badgley, 1966), so that freezing and sinking would create local vertical circulations of the same horizontal scale as the leads (from 1 to 100 m). The model is shown schematically in Fig. 13. The sinking, which near the surface is close to the leads, probably entrains some near-surface water which has been involved in the convection induced by sea ice growth. The sinking water would tend to spread and produce lateral homogeneity. Excess salt carried down by the process would develop and maintain the halocline within the convective layer.

The depth to which such convection can reach is determined by the difference between the salinity at the sea surface and that at depth. Almost everywhere in the Arctic, surface salinity values are so low compared with those at depth (32 to 34‰ versus 34.9 to 35.0‰) that convection generated under present conditions of temperature and ice growth cannot reach into layers

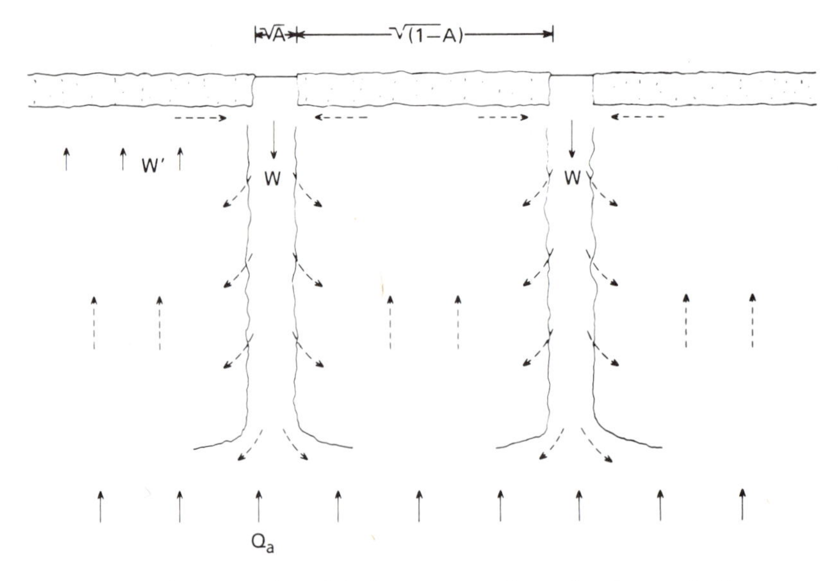

Fig. 13. Proposed model of convection (from Coachman, 1966).

with salinities much greater than 1‰ above those prevailing at the surface. Hence, nowhere in the Arctic Basin does convection reach as deep as the Atlantic layer, but only into the subsurface layers.

A major consequence of the model is an upward compensating flow providing a mechanism by which heat from the underlying Atlantic layer is transported to the surface. Thus, seasonal variation in vertical fluxes of heat and salt is anticipated. Supporting evidence is provided by calculations of heat loss from the Atlantic layer (Panov and Shpaikher, 1964; Table 3) in which 50% more heat seems to be lost in winter than in summer. It should be noted, though, that this heat flux is so small that given the present state of knowledge, variations can neither be accurately identified, nor can their significance, for example on ice growth, be assessed (cf. Maykut and Untersteiner, 1971). Clearly, convection in the surface layers is an extremely important topic for contemporary Arctic research.

Atlantic Water

Below the Arctic water there is a thermocline in which the temperature increases to a maximum of about 0.5°C. The maximum lies between 250 and 500 m depth, depending on location, below which the temperature decreases to values <0°C. In contrast, the salinity, which began increasing in the subsurface layer, achieves in the Atlantic layer values (34.92 to 34.99‰) which continue nearly uniform to the bottom (Fig. 4). Since the density is largely salinity-controlled, this nearly isohaline water, which comprises over 90% of the water in the Arctic Ocean, is of nearly uniform potential density. The ultimate source of this water is the North Atlantic, via the Norwegian–Greenland seas. However it is convenient to define two distinct water masses: the Atlantic water, warmer than 0°C, and the bottom water, colder than 0°C.

The temperature-salinity relationships, which clearly define the characteristics of the Atlantic water mass and its geographic variations, are shown in Fig. 14. Close to Spitsbergen, where the West Spitsbergen current carrying this water enters the Arctic Ocean, temperatures may be +3°C or higher and salinities may be slightly above 35‰. These values are maximal, and only short distances away from the Spitsbergan–Greenland passage the maximum defining the core of the Atlantic water is <1.5°C, while salinities lie between 34.85 and 34.95‰.

The warmest core water is found along the continental slope of the Eurasian Basin, where values between 1.0 and 1.5°C are typical. Over the rest of the ocean, core temperatures are about 0.5°C: the coldest maximum encountered is about 0.38°C, north of Ellesmere Island. The loss of heat from the layer of maximum temperature, as evidenced by decreasing temperature, is the basic method that has been used to determine the general circulation of the Atlantic layer (see below).

The depth of the upper boundary of the Atlantic layer (0°C) is definitely shallower in the Eurasian Basin (100 to 200 m) than in the Canadian Basin (200 to 300 m). Likewise, the core of the layer is relatively shallow (200 to 250 m) in the Eurasian Basin and deeper (400 to 500 m) in

Table 3 Vertical Heat Transfer from the Atlantic Layer (Data of Panov and Shpaikher, 1964)

(a) Transfer of heat by Atlantic waters (Panov and Shpaikher, Tables 2, 3, 4)

		Heat loss upward (cal cm^{-2} sec^{-1})		
Location	Area* (cm^2)	November–May	June–October	Annual
Eurasian Basin	2.1×10^{16}	27.2×10^{-5}	18.9×10^{-5}	23.8×10^{-5}
Canadian Basin	3.2×10^{16}	9.9×10^{-5}	6.9×10^{-5}	8.6×10^{-5}

(b) Heat of Atlantic waters received upward, 1933–1961 (Panov and Shpaikher, Table 5)

		Heat received upward (cal cm^{-2} sec^{-1})		
Location	Area* (cm^2)	Maximum	Minimum	Average
Eurasian Basin	2.1×10^{16}	17.3×10^{-5}	9.9×10^{-5}	13.6×10^{-5}
Canadian Basin	3.2×10^{16}	6.3×10^{-5}	3.6×10^{-5}	5.0×10^{-5}

* Calculated from data on ice decrease of Panov and Shpaikher, 1964, Table 5.

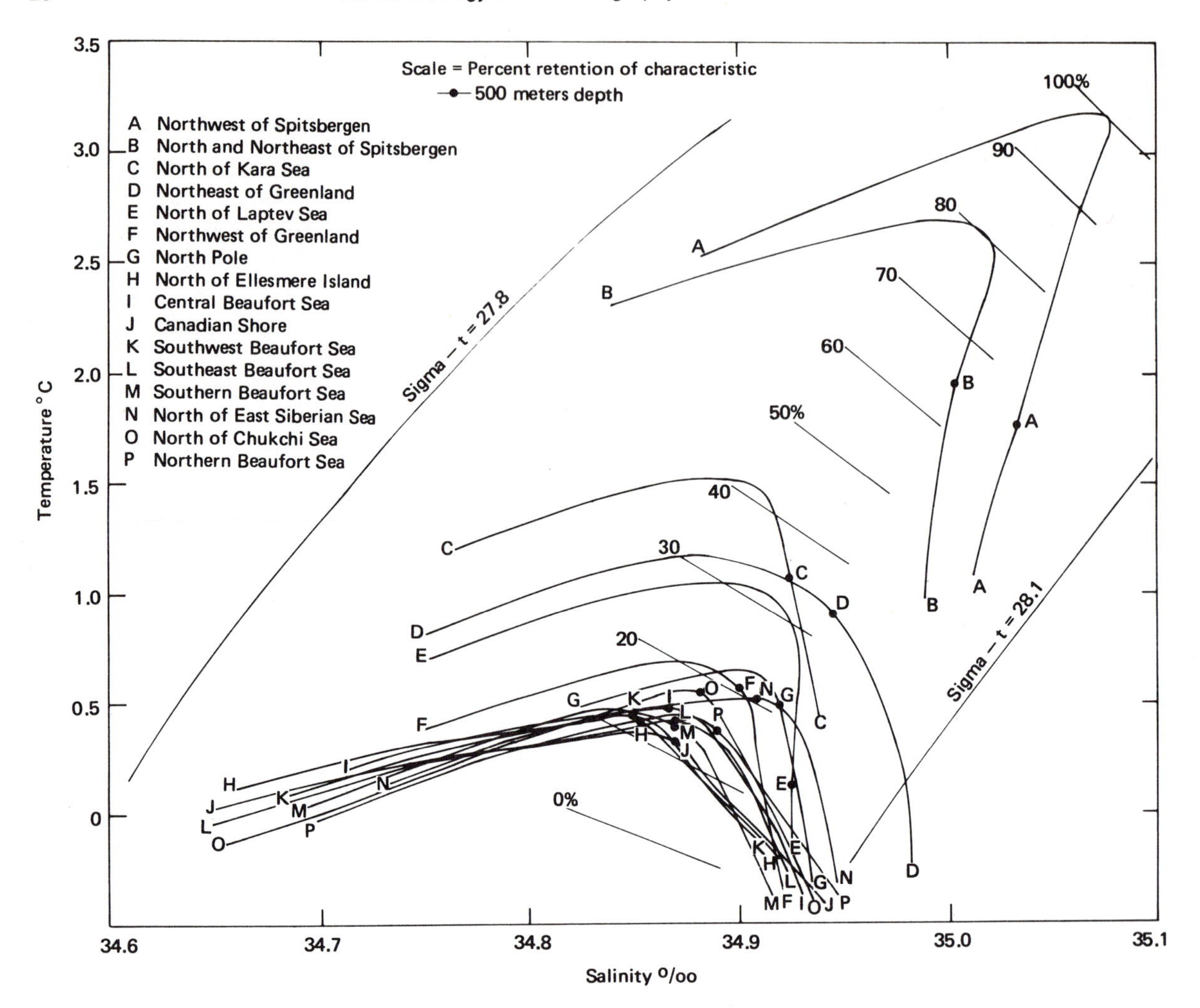

Fig. 14. T-S diagram of Atlantic water (from Coachman and Barnes, 1963).

the Canadian. These geographic variations are probably a result of two causes: (1) as the temperature gradients are always greater above the maximum, vertical diffusion of heat upward is probably greater than downward, and the temperature maximum is eroded downward as a result; and (2) the water mass actually sinks somewhat in the course of its travel. Though data as yet are too sparse to make a definitive analysis, it is likely that the core water sinks relatively abruptly on crossing the Lomonosov Ridge into the Canadian Basin.

Temporal variations occur in the Atlantic layer, though again insufficient data preclude a rigorous analysis. The best information available giving an idea of temporal variations to be encountered were obtained in a quasi-synoptic survey along the core of Atlantic water (Timofeyev, 1958). Stations were made during spring 1955 by the High Latitude Aerial Expedition of that year, from north of Spitsbergen to the Laptev Sea, along a line paralleling the continental slope. The mean temperature of the layer, the vertical distance between 0°C isotherms, and the heat content along the section are shown in Fig. 15. The variations seem to be about 2°C in temperature and 50% in layer thickness. Variations would be much less in the Canadian Basin because of the

The significance of the Atlantic water mass in the Arctic is that it provides a major heat source within the ocean. Quite probably ample heat is available in the Atlantic water to prevent formation of an ice cover. However, the upward diffusion of heat is extremely slow in the deep ocean. Hence, the Atlantic layer never loses all its heat, its temperature remaining everywhere at least 0.4°C higher than that of the other layers, even though its residence time in the Arctic Ocean may be of the order of 10 years (see below).

Attempts have been made to estimate the upward heat flux, both by calculation from the observed vertical temperature distributions and from heat budget conditions applied at the ice-water interface. An example of the former are the fluxes reported by Panov and Shpaikher (1964; Table 3). The vertical heat flux is given by $Q = \rho c_p K_\theta(\delta\theta/\delta z)$, and with ρ, c_p, K_θ all of order unity, with $\delta\theta/\delta z$ in the Canadia Basin observed to be (1sC/100*m*), $Q = 10 \times 10^{-5}$ cal cm^{-2} sec^{-1}, so that Panov and Shpaikher's results are reasonable. While they are the best presently available estimates, they can only be considered reliable to about one-half order of magnitude, mainly because of uncertainties in values of K_θ.

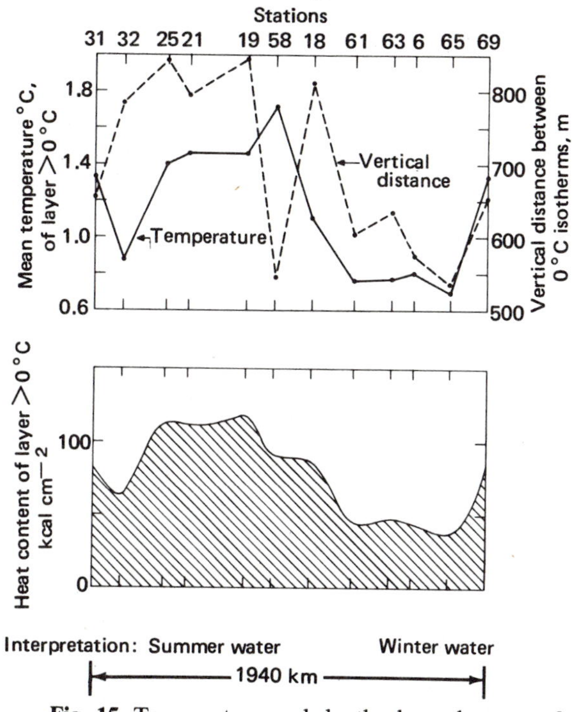

Fig. 15. Temperature and depth along the core of Atlantic water from north of Spitsbergen (left) to the east (after Timofeyev, 1958). Upper: mean temperature of the layer >0°C and the vertical distance between 0°C isotherms. Lower: heat content of the layer.

Budget calculations at the sea surface, on the other hand, require less heat flux from below to maintain the equilibrium ice conditions observed today. For example, the ice model of Maykut and Untersteiner (1971) calls for an equilibrium flux of 3×10^{-5} to 6×10^{-5} cal cm^{-2} sec^{-1} to maintain present conditions. If the flux directly affecting the ice were as great as 15 to 20 $\times$ 10^{-5} cal cm^{-2} sec^{-1}, values within the range of those calculated by Panov and Shpaikher, the ice would vanish in their model. We conclude that there is no discrepancy; rather, as noted by Maykut and Untersteiner and suggested by our model of convection, probably more heat is carried into the basin than is effective in melting or preventing ice formation, but the excess is lost directly through leads.

Bottom Water

The bottom water mass, constituting some 60% of the water in the Polar Basin, has temperatures everywhere below 0°C. In the Eurasian Basin they decrease with depth to about 2500 m, where very uniform values between −0.8 and −0.9°C occur, and then increase slightly. In the Canadian Basin the minimum is encountered at about 2000 m, with values between −0.4 and −0.5°C. As noted previously, salinities are everywhere very uniform between 34.92 and 34.99‰.

The source of the bottom water is the same as that of Atlantic water, that is, the Norwegian–Greenland seas (see below). There is also probably some degree of mixing between the bottom water and the lower Atlantic water. As with the Atlantic water, data are insufficient to be definitive about the quantities involved.

The temperature difference of 0.5°C between the two basins (see Fig. 6) is due to the presence of the Lomonosov Ridge. Temperature observations from stations extending deeper than 3000 m (Fig. 16) were analyzed by Coachman (1963). Two conclusions can be drawn: (1) bottom water of the Canadian Basin comes from the 1400- to 1500-m level of the Eurasian Basin, which probably represents the depth of saddles in the Lomonosov Ridge, and (2) the deep temperature increase of about 0.1°C/1000 m in each basin corresponds closely to the adiabatic increase to be expected (Fig. 16: curve labeled "Ekman") and is due to the compression of a single water type. These conclusions, together with other arguments presented by Coachman (1963), show bottom water to come from a single source (the Greenland–Norwegian seas). Other sources, e.g., the freezing of ice in certain localities along the edge of the Eurasian Basin, can produce only negligible amounts.

IV. Arctic Ocean: Field of Motion

Our knowledge of the field of motion in Arctic waters is much less well developed than that of temperature and salinity distributions. This is due in part to the fact that there is much more "noise" in the velocity field than in the fields of temperature and salinity, and in part to the scarcity of current measurements, as noted previously.

Surface Layers*

The surface circulation has been largely deduced from the observed drift of various manned ice islands, floe stations, and ships. Current measurements have shown that on the average the drifting stations and the surface water tend to move in similar directions at similar speeds, though there may be considerable differences over short periods. In addition, a dynamic topography of the Arctic Ocean surface has been constructed from over 300 stations. The resulting circulation would represent at best a long-term mean flow, as the stations were taken over many years and at different seasons.

* See Note 2 at the end of this chapter.

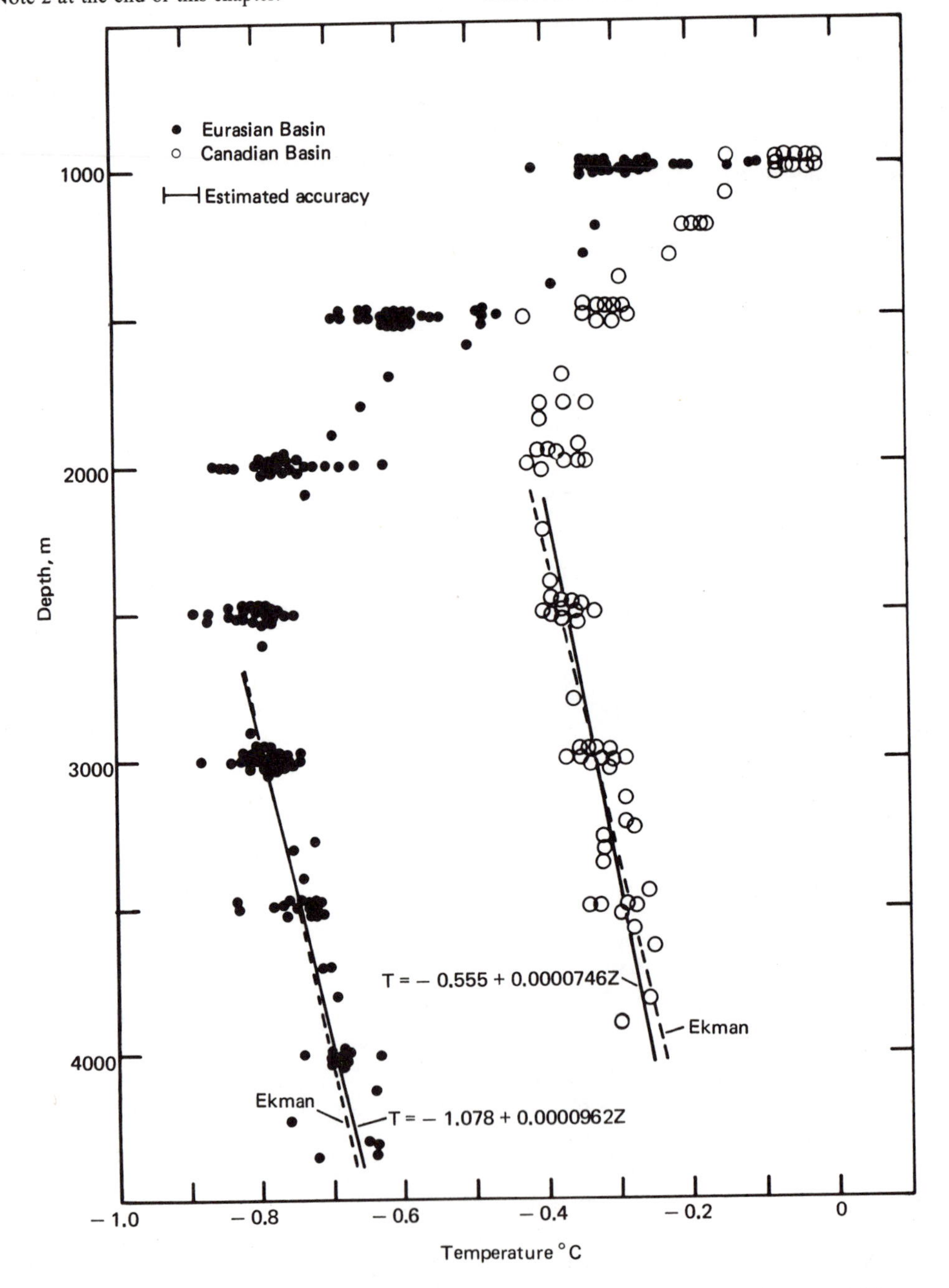

Fig. 16. Temperature observations from stations >3000 m (from Coachman, 1963).

A composite picture of the surface circulation based on these results is presented in Fig. 17. The drift station vectors are long-term averages, which can best be compared to the dynamic topography. The schematic circulations shown in Baffin Bay and the Greenland Sea are mainly from distributions of temperature and salinity; circulation in the Greenland Sea is discussed in detail later.

The surface waters from the whole Eurasian side of the Arctic Ocean tend to move toward the North Pole. Speeds are slow, perhaps on the average 2 to 3 cm sec^{-1}, but after passing the region of the Pole the flow becomes more concentrated and then exits from the basin as part of the East Greenland Current. In the Beaufort Sea, surface waters have a clockwise movement, and the center of the gyre, about 80°N and 140°W, coincides very nearly with the center of the mean atmospheric pressure anticyclone (Felzenbaum, from Campbell, 1965). The close correspondence between mean drifts and the dynamic topography leads to the conclusion that, on a long-term average basis, surface motion in the Arctic Ocean is in geostrophic balance.

Spatial variations in the long-term flow pattern were investigated by Coachman (1969). Mean monthly drift speeds from north of Siberia toward the North Pole (the transpolar drift stream) are faster by about 1 to $1\frac{1}{2}$ km day^{-1} (1 km day^{-1} = 1.2 cm sec^{-1}) than the southwesterly drifts along the Canadian Archipelago. Drifts near north and east Greenland are still faster, as mentioned. North of Ellesmere Island is a region where the mean speed is very slow, as evidenced by the behavior of the ice island WH-5 (Nutt, 1966). Nonetheless the ultimate fate of both this ice island and of NP-7 was to drift south through Nares Strait (Dunbar, 1962).

Temporal variations, both seasonal and longer, were also identified by Coachman (1969). For example, the mean monthly drift speeds are faster in late summer and fall (3 to 4 km day^{-1}) than in the middle of winter (1 to 2 km day^{-1}). Furthermore, in some years the drift speeds are definitely faster (1959 to 1962) or slower (1967 to 1968) than normal for the location and season. This difference is as much as 1 km day^{-1}. All these variations seemed to be closely correlated with similar variations in intensity of the appropriate surface atmospheric pressure gradients, and therefore with the prevailing winds.

We feel that the surface circulation averaged over periods of a month or longer gives every indication of being primarily driven by the prevailing wind field (cf. Campbell, 1965). That this momentum exchange is being done through the intermediary of the ice appears to make very little difference, so that dynamically the ice is simply part of the water at large spatial and temporal scales. However, in considering the oceanic velocity field on smaller scales, or near land, the unique mechanical properties of the ice are undoubtedly of major importance. The thermodynamic properties of the ice are of course important in all heat exchange considerations.

On time scales shorter than one month ice motion is much more complex. The ice responds in part directly to the local wind. Reed and Campbell (1960) examined 107 periods of unaccelerated drift of ice station Alpha. The data for wind speeds up to ~7 knots showed the ice to move at $2\frac{1}{2}$ to $3\frac{1}{2}$% of the wind speed and 40 to 45° to the right of the wind. For stronger winds (up to ~20 knots) the ice moved at about 2% of the wind speed but only 25 to 40° to the right. However, there was large scatter in the data, including a few cases of movement to the left of the wind. Hunkins (1966) reported on 13 cases of equilibrium drift with winds up to 7 knots and arrived at a speed factor of 2% and a deviation of 45° to the right of the wind. His data were also from the Alpha drift and so are probably included in those of Reed and Campbell, though perhaps selected on a different basis.

A major difficulty in studies of this kind has been the observational inability to separate the direct effect of the wind on the ice from the tendency of the ice to move with the ocean currents. This is undoubtedly why at most only one-half the drift of Camp-200 during March 1970 could be accounted for by direct wind effects, the remainder being associated with the movement of the underlying water (Coachman and Newton, in press). These ocean currents are largely wind-driven, both in the Ekman and geostrophic modes, and their response to wind changes appears to be rather rapid. Thus the observational requirements to determine the direct local wind effect on the ice are quite stringent; for example, extensive current measurements are required. If one additionally wishes to examine the effects of the wind on the ocean currents, then the field of wind stress needs much better spatial definition than has thus far been accomplished.

The advent of satellite navigation, providing accurate (~0.5 km or better) positions every 1 to 2 hr, independent of weather, allows definition of ice motion on time scales as short as a few hours. Satellite navigation equipment was installed on T-3 in 1966, and Fig. 18 shows the drift track from summer 1967. Analysis of this and other drift tracks shows that the ice motion in the Arctic Ocean is characterized by certain time scales of a month or less:

(1) There is a great deal of energy in motions with characteristic times of $2\frac{1}{2}$ weeks or more. These motions are not persistently periodic, although the numerically filtered drift records may show harmonic-like components existing over several cycles.

For example, during summer 1967 the drift of T-3 showed a very pronounced spectral peak at a period somewhat in excess of 3 weeks. Numerical filtering of the drift showed this spectral band to be rather complex, with varying periodicities and phasing. The average period was close to 3 weeks, and the east component of drift velocity led the north by about $3\frac{1}{2}$ days, or 60° in phase. The latter

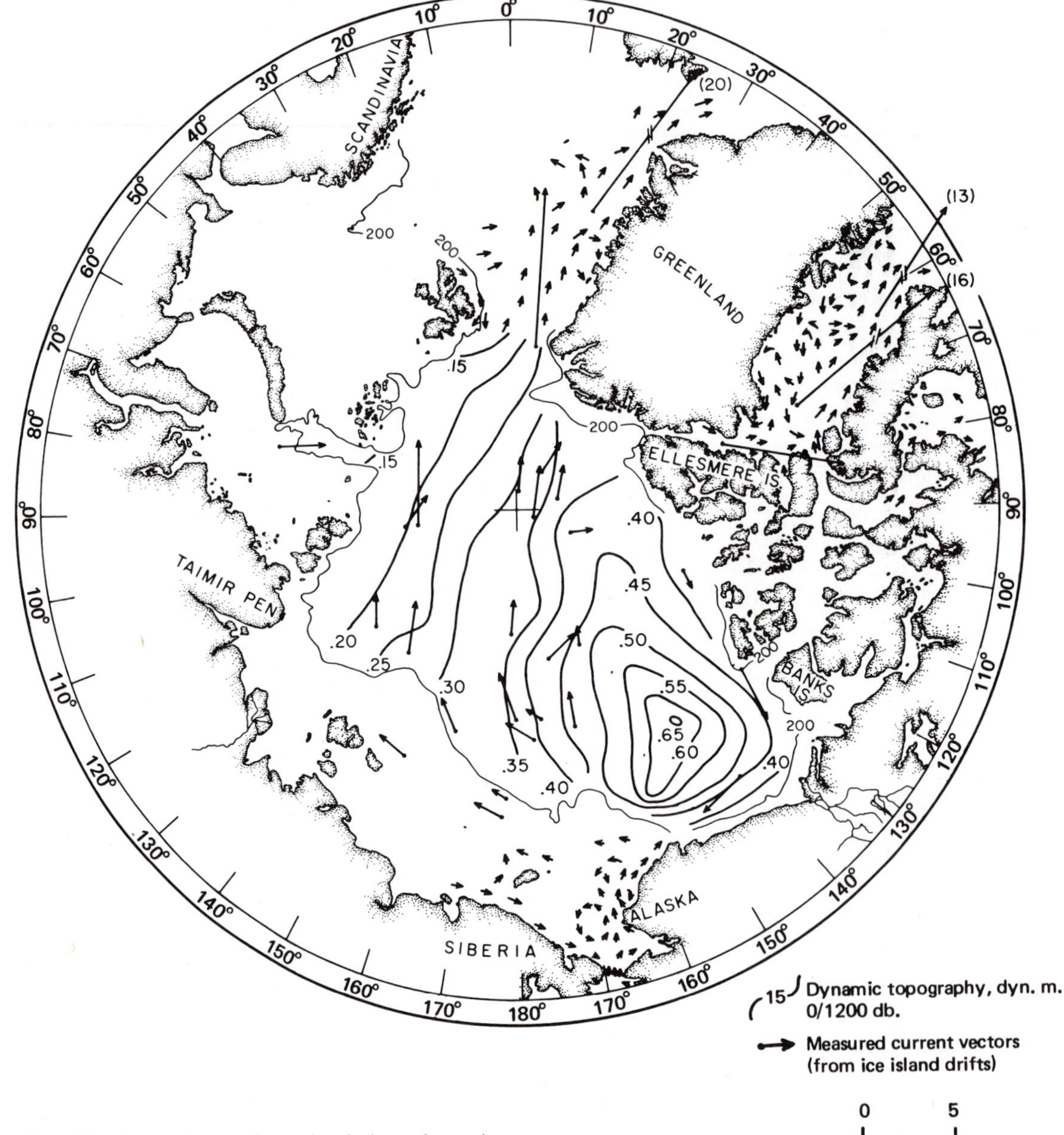

Fig. 17. Composite surface circulation, dynamic topography, and long-term mean (~ one year) station drifts (from A.I.N.A., 1967). The data for the drift vectors are from Sater (1969).

compares favorably with the value 70° derived from phase spectral analysis. We should therefore expect to see in the drift track large counterclockwise looping tendencies, with a completion time of about 3 weeks. Since the filtered drift velocity components had amplitudes of 5 to 10 cm sec^{-1} (contrasted to a mean drift over the summer of only 1.2 cm sec^{-1}), these motions should be a dominant feature of the drift and have a characteristic radius of order 7.5 cm $sec^{-1} \times (1/2\pi) \times 3$ weeks = 22 km. From Fig. 18 we see that such motions were indeed quite prominent, particularly during the latter part of the summer.

We cannot presently make a definitive causal statement regarding these motions. Although it is reasonable to try to relate them to atmospheric events, particularly to changes in the wind field, one should not *a priori* expect a universally applicable mechanism.

(2) The power spectra of the ice drift all show multiple peaks in the frequency band corresponding to periods of 4 to 10 days. While there generally seems to be less energy in these motions than in the long-period ones, this is not always the case. For example, during summer 1968 the drift of T-3 was dominated by the 10-day time scale, with the long-period motions relegated to a secondary role. This mid-frequency band of time scales from $\frac{1}{2}$ to $1\frac{1}{2}$ weeks appears very complex, with greatly varying amplitudes, periods, and phasing. We are unable to suggest the causal mechanisms for these motions, though we intuitively feel that they must be related to changes in the wind field. There is evidence that the ocean responds nearly geostrophically to such changes, with adjustment times on the order of a few days or less (cf. Coachman and Newton, in press).

(3) All the drifts exhibit a semi-diurnal periodicity, typically varying in amplitude from near-zero to about 10 to 15 cm sec^{-1}. There is frequently also a smaller diurnal variation. Both Galt (1967) and Hunkins (1967) have interpreted the semi-diurnal variations as inertial oscillations rather than tidal, but in our opinion the relative importance of inertial and tidal oscillations has not yet been resolved.

The upper ~50 m of the surface water mass probably has a mean motion similar to that of the surface, although at any instant the currents may deviate markedly from those at the surface. This upper layer provides the frictional coupling between the air and ice and the deeper water; it may

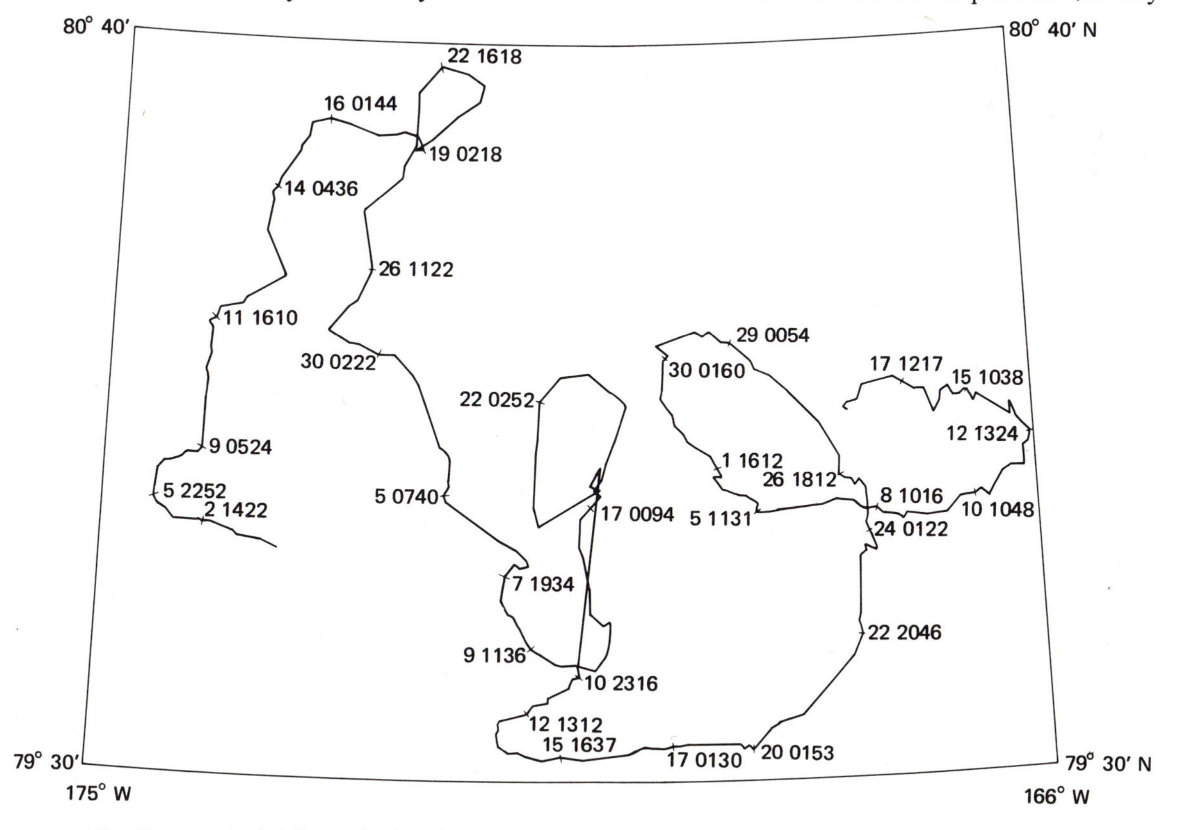

Fig. 18. A typical drift track (T-3: beginning June 1 (on the left) and ending September 20, 1967).

dynamically be considered as two layers: a frictional boundary layer immediately under the ice, a few meters thick, and an Ekman layer extending below that to several tens of meters (Smith, 1971). The flow in these layers is in general both nonuniform and unsteady (Smith, 1971), and therefore the currents will exhibit complicated behavior. A close correspondence to the Ekman flow field has been measured under certain ideal conditions by Hunkins (1966), who found that the frictional boundary layer was about 1 m thick and the Ekman layer, 18 m.

In contrast, measurements made during the March 1971 Aidjex Pilot Study showed the Ekman layer to be about 40 m deep; in addition, the current did not change direction to the right uniformly with increasing depth, but on occasion reversed this trend (Smith, personal communication).

Thus, we conclude that currents in the surface layer are slow, with speeds similar to the ice motion, i.e., 10 cm sec^{-1} or less. The average direction of motion is also similar, at least in the upper approximately 50 m. Below about 50 m it is possible that the mean flow direction is somewhat different than in the upper layer, but this remains to be investigated. In any case it is clear that the surface waters experience speed fluctuations as much as an order of magnitude greater than the mean flow and that there are large variations in direction associated with these fluctuations.

Pycnocline Region*

The significant increase of density with depth occurs between about 50 and 250 m. Within this layer occur the swiftest currents yet observed in the Arctic Ocean. These currents appear as pulses with durations of a few days to perhaps two weeks. For example, Galt (1967) measured the decreasing portion of such a pulse from T-3 in August 1965: the current relative to T-3 (which had a maximum drift rate during the measurement period of about 9 cm sec^{-1}) was greatest at 150 m, where it decreased from 57 cm sec^{-1} to 7 cm sec^{-1} during one week; there were several intermediate speed maxima during this time, rather than a simple monotonic decrease. Two somewhat similar pulses were observed from North Pole-2 (Somov, 1954 to 1955), one of 2 days duration (60 cm sec^{-1} maximum relative speed at 150 m) and the other containing several intermediate speed maxima and lasting 6 days (21 cm sec^{-1} maximum at 150 m). The NP-2 measurements provided little information on the vertical distribution of current below 150 m, however. All these currents rotated with time. During the 1970 Aidjex Pilot Study two pulses of each 3 days' duration and separated by 4 days were observed (about 30 cm sec^{-1} maximum relative speed), but the current measurements did not extend below 150 m (Coachman and Newton, in press).

The high-speed currents in the pycnocline appear to be occasional, transient responses to a resonant forcing mechanism. Probably they are associated with changes in the wind field which excite a barotropic response, and this is in turn amplified baroclinically by the pycnocline. The pulses observed during March 1970 were strongly correlated with ice motion which in turn was related to the wind field, and the pulses of current (at 150 m depth) actually preceded corresponding pulses in ice motion by about one-half day. This apparently occurred because the baroclinic rate of adjustment in the ocean was slower than the barotropic by a factor of about 3, so that immediately after surface readjustment the barotropic and baroclinic modes reinforced each other. The currents appeared to be nearly in geostrophic equilibrium at all times.

The mean circulation in the pycnocline layer is similar to that of the subsurface water masses discussed earlier. All the above observations of swift pulsating currents have been in the Canadian Basin, and therefore are short-term fluctuations in the flow of the Bering Sea water. In the absence of numerous long-term measurements, the circulation must be interpreted from the distribution of properties (see, e.g., Fig. 10). It appears that the circulation of the Bering Sea water is in part similar to the surface circulation: after entering the ocean north and west of Point Barrow the general motion is northerly toward the Pole, and in the southeastern Beaufort Sea it is to the west. However, over the central Beaufort Sea the temperature distribution suggests a more general spreading, that is, greater easterly component, than seems to be the case with the surface water (cf. Fig. 17).

In the Eurasian Basin the long-term circulation of the subsurface water must likewise be inferred from the distribution of properties (see, e.g., Fig. 8). Comparison of the temperature distribution with the surface circulation (Fig. 10) suggests that the long-term motion of the two

* See Note 3 at the end of this chapter.

layers is quite similar. We should also anticipate short-term rapid motion similar to that observed in the Canadian Basin because the pycnocline is equally well developed.

Atlantic Layer

Circulation of the Atlantic layer was inferred by Coachman and Barnes (1963) from the temperature and salinity observations using the "core-layer" method of Wüst (1935). The arbitrary scale of percent retention of characteristics used is shown in Fig. 14. The inferred circulation is shown in Fig. 19.

Atlantic water enters the Arctic Ocean west of Spitsbergen as an extension of the West Spitsbergen Current. It then descends to subsurface levels, and its main flow parallels the continental slope of Eurasia to the Laptev Sea. Some of the water in this flow is continually being lost to the north toward the Lomonosov Ridge, and some to the south, particularly into submarine canyons in the slope. Through this movement the Atlantic water flows through all of the Eurasian Basin relatively rapidly, the T-S characteristic of the layer being everywhere greater than 20% on the arbitrary scale (cf. Fig. 14), which is larger than in most of the Canadian Basin. The water in the

Fig. 19. Composite of mid-depth current measurements and circulation of the Atlantic water inferred from the temperature field (from Galt, 1967). Isochrones (years) after Timofeyev (1961).

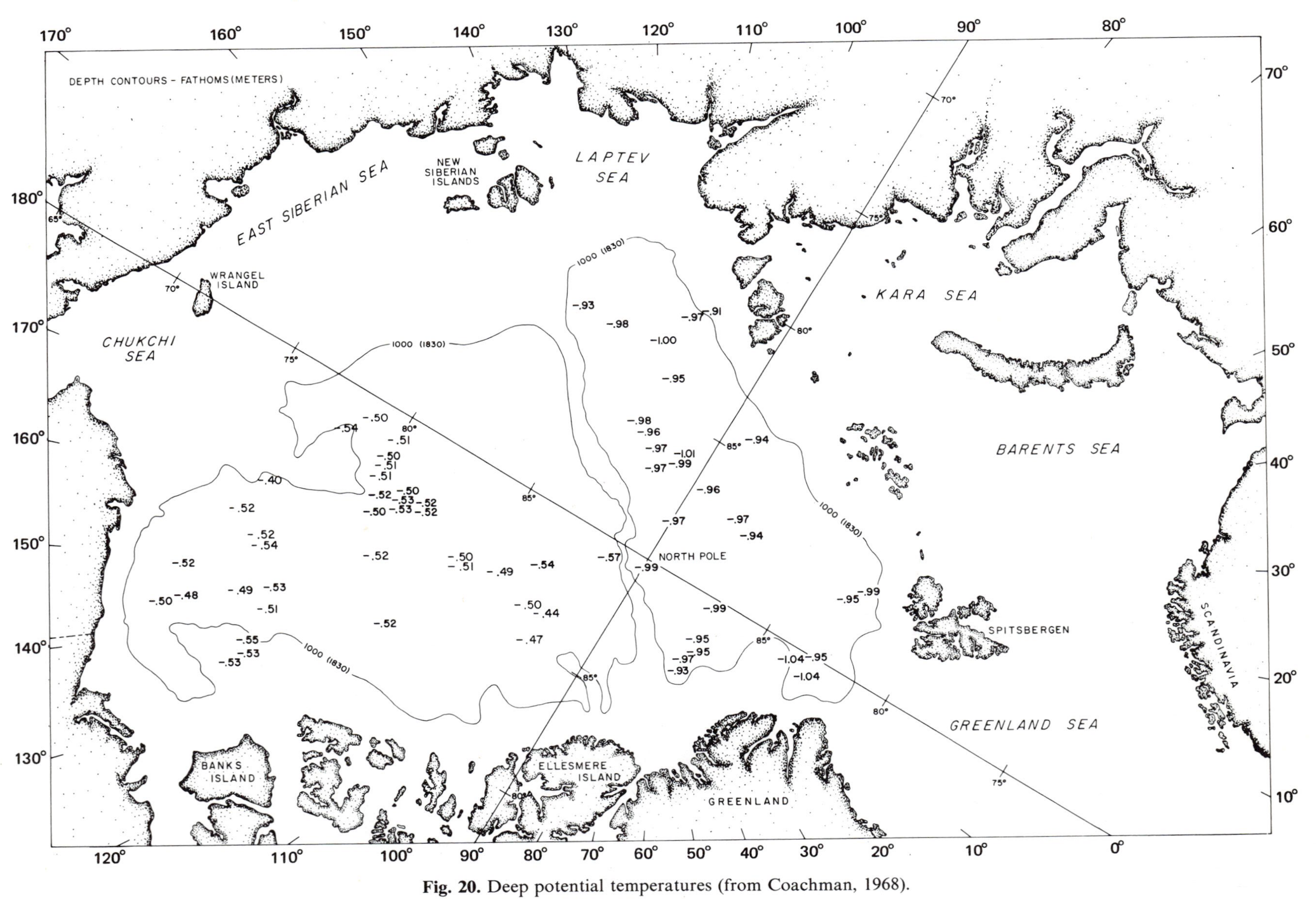

Fig. 20. Deep potential temperatures (from Coachman, 1968).

Eurasian Basin farthest removed in time from the entrance lies in the area northwest of Greenland (the Lincoln Sea). Outflow of the Atlantic water is along East Greenland as an under-layer of the East Greenland Current (see below).

The Canadian Basin is supplied with water that crosses the Lomonosov Ridge on a broad front from the areas north of the Kara and Laptev seas. The flow parallels the continental slope north of the East Siberian and Chukchi Seas and then moves into the Beaufort Sea, penetrating more or less directly to the area northeast of Alaska and west of Banks Island. The flow turns gradually northward and, last of all, moves eastward along the continental slope of the Canadian Archipelago and Ellesmere Island, ultimately recrossing the Lomonosov Ridge and joining the outflow from the Eurasian Basin.

North of the Chukchi Sea a prominent rise (the Chukchi Cap) apparently prevents direct penetration of the layer into the region immediately north and east of Point Barrow; a more detailed analysis suggests that this region is supplied by water that has crossed the central Beaufort Sea and recurved south and west along the continental slope off Alaska (Coachman and Barnes, 1963). The circulation described in general agrees with a similar but independent interpretation by Timofeyev (1961); a major difference is the presence in our interpretation of the clockwise eddy immediately north of Alaska.

Timofeyev (1961) also estimated the time required for the Atlantic water to transit the ocean (isochrones of Fig. 19). The isochrones suggest long-term average flow speeds of 1 to 3 cm sec^{-1}.

Twenty-nine current measurements representing averages over 58 to 291 hr at 750 or 1000 m depth were reported by Nikitin and Dem'yanov (1965). These vectors together with some more recent measurements (Galt, 1967, and some unpublished) are superimposed on the inferred Atlantic circulation in Fig. 19. There is general agreement between the relatively few observations and the inferred water movement in both speed and direction. Measured speeds in the layer are low, 0 to 5 cm sec^{-1}, and the directions conform in the main to those deduced from the mass field. The only area of major disagreement, centered around 83°N, 150°E, is probably one in which bathymetric effects are important: currents at 750 and 1000 m would be influenced by the presence of the Lomonosov Ridge, whereas the T-S analysis is applicable to the level of the temperature maximum (300 to 500 m), where the ridge influence would be less.

Bottom Water

Circulation of the bottom water, except for the primary separation effected by the Lomonosov Ridge, is unknown. Deep (>2000 m) potential temperatures (Fig. 20) show the extreme uniformity of conditions obtaining in each basin. Values differ by only 0.1°C, part of which is undoubtedly uncertainties in measurement, and therefore inferring the circulation from the mass field seems impossible at present.

Galt (1967) reported two direct deep (2 m above bottom) measurements from the Canadian Basin. Speeds were 1.5 and 2.6 cm sec^{-1}, and both the speeds and directions conformed to other current measurements from intermediate depths (500 to 1500 m).

Hunkins et al. (1969) reported five bottom current measurements in the Canadian Basin. Four were on the Mendeleyev Ridge and showed speeds of 4 to 6 cm sec^{-1}, with a southerly component in each case. This is a region in which current measurements in the Atlantic layer (Nikitin and Dem'yanov, 1965) have all shown southerly components. The fifth measurement reported by Hunkins et al. was on the abyssal plain and indicated motion less than 1 cm sec^{-1}.

We conclude, from the meager evidence available, that averaged over time the whole Arctic Ocean water column below about 400 m, within which there is essentially no density gradient, moves as a unit without significant shear. Significant but slow speeds occur to the bottom.

Waves*

The types of waves that are found in all oceans, e.g. surface, internal, tidal, storm surges, etc., occur in the Arctic Ocean. However, the spectra for *surface waves* differ from those for other oceans, due largely to the presence of the ice cover (Hunkins, 1962), which effectively damps the higher-frequency surface oscillations. Only in open areas, which may occur in the marginal seas in summer, can there be visible wind waves much larger than capillary waves, but even there the fetch is normally so limited that large wind waves and swell do not develop. These waves are quickly

* See Note 4 at the end of this chapter.

damped when they propagate into ice fields. Propagation of waves into pack ice was studied by Robin (1963), and was modeled by Evans and Davies (1968) and Henry (1968).

Hunkins (1962), using a recording gravimeter and a seismometer, was able to identify oscillations with periods between 5 and 60 sec and others with periods between 10 and 100 min. A general relationship of increasing amplitude with increasing period was shown, though actual amplitudes in all cases were minuscule. The higher-frequency waves appeared to be correlated with the local wind, with a threshold at 10 to 12 m sec^{-1}; that is, winds stronger than the threshold seemed to excite the oscillations. The longer-period waves remain unexplained, though there was evidence that the amplitudes of those with periods of several minutes were enhanced by the wind. Possible causes include seiches or edge waves generated by atmospheric pressure variations, and flexural-gravity waves in the ice.

Internal waves in the Arctic Ocean were investigated by Yearsley (1966) from T-3. He recorded temperature as a function of time, using thermistors at 60 and 125 m, corresponding to the upper and lower thermoclines associated with the Bering Sea water intrusion into the Canadian Basin. Power spectra of the records showed a peak with amplitudes of 0.25 to 0.5 m at $\approx$8-min period, a period not much longer than the limit imposed by the density distribution (1/N = 1/Brunt-Väisälä frequency = 5.3 min). As with surface oscillations, at greater frequencies a general rapid decrease of amplitude with increasing frequency was observed. On occasion very-high-frequency oscillations were also observed, with amplitudes of $\approx$0.5 m at periods $\approx$20 sec.

We remark the generally small amplitudes of internal waves in the Arctic Ocean compared to those measured in other oceans, where values of several meters are not uncommon (see, e.g., LaFond, 1962). This probably reflects a lack of energy available to generate oscillations of this sort, presumably due to ineffective coupling by the ice cover of the atmosphere and ocean at short time scales. Though observations are scarce, even internal waves of tidal period, common in other oceans, do not seem to be of any significance. In this case, however, tides in the Arctic are small (see below), so that the external generating mechanism is weak.

A curious internal wave phenomenon which was most commonly encountered in Arctic waters was studied by Ekman (1904). In the peripheral seas in summer, ice melt and runoff can combine to produce a thin (2 to 3 m) fresh surface layer separated from the underlying sea water by a very sharp pycnocline. Internal waves can propagate on such a pycnocline at speeds of 1 to 3 knots (depending on the particular conditions). In the time of sailing vessels many captains reported that in a light breeze their vessels seemed to "stick" in the water, behaved sluggishly, and made little headway. Ekman explained that a vessel moving slowly in water with this particular density structure, in which the draft of the vessel was approximately the same as the pycnocline depth, expended energy in generating internal waves on the pycnocline, which detracted from energy available for propulsion. The phenomenon is termed "dead water," but is rarely encountered today because power vessels normally travel faster than 3 knots.

Tidal wave behavior in the Arctic Ocean was first elucidated by Harris (1911). Many data have accumulated since from tidal stations around the peripheries, and a recent compilation showing the type of tide, cotidal lines, and range of spring tides is given in the Oceanographic Atlas of the Polar Seas, Part 2 (USNHO, 1958).

An interesting feature of Harris's analysis was the necessity for postulating a mid-ocean tract of land or relatively shallow water in order to reconcile the tidal observations. Over the next 40 years this theory was generally discredited, as none of the expeditions discovered land in the central basin. However, further arguments for the presence of a submarine ridge dividing the Arctic Ocean into two basins were presented by Fjeldstad (1936) from tidal observations and Worthington (1953) from oceanographic observations. Finally, in the late 1950's sufficient reliable echo soundings accumulated to demonstrate conclusively the Lomonosov Ridge.

The tidal waves come from the Atlantic Ocean, transit the Norwegian–Greenland seas, and then sweep counterclockwise around the Arctic Basin. Since the waves come from the Atlantic, the tide type is semi-diurnal; only close to the Bering Strait does some diurnal inequality (typical of Bering Sea tides) appear in the records.

Tide ranges along Norway are typically 2 to 3 m, but at Spitsbergen they are only 1 to 1.5 m. Within the basin the range is reduced even further, so that at Point Barrow, for example, the range is $\sim \frac{1}{4}$ to $\frac{1}{3}$ m. The waves are refracted by the shallow water of the continental shelf, and hence in the

Kara, Laptev, East Siberian, and Chukchi Seas the tidal waves move from north to south. The wave heights are amplified somewhat in shoaling water, and observed ranges can be somewhat greater along the shores in these seas than they are along the Arctic Ocean shores of Alaska and the Canadian Archipelago.

Confirmation of the small tidal range within the Arctic Ocean has come from measurements made from the ice. Gudkovich and Sitinskii (1965) recorded for one month in the Canadian Basin the tilt relative to horizontal of ice station North Pole-10. They reported slopes of 10^{-6} associated with the semi-diurnal tide wave, from which the tide range was estimated to be ≈ 60 cm. These results must be viewed with caution because of the extreme difficulty of accurately measuring such small slopes. However, Hunkins (1962) obtained reliable measurements from T-3 when it was grounded 130 km northwest of Point Barrow. The measured spring tide range was 18 cm.

The largest variations in sea level, at least along the coasts, are aperiodic and occur on time scales of a few days to weeks. For example, along the shores of the Chukchi Sea we have seen sea level in summer change by ≈ 1 m in one week, whereas the tidal range was about 0.2 m. The cause of these *surges* lies in the local wind and/or atmospheric pressure fields and is illustrated by the data presented in Fig. 21 (Matthews, 1970): the passage of an atmospheric pressure system resulted in a sea-level rise of over 3 m in 6 days.

On rare occasions the proper combination of wind and open water can lead to catastrophic results. A recent example was the storm of October 3, 1963, during which the sea level at Point Barrow rose ≈ 3 to 4 m during one day, upon which was superimposed waves of 2 to 3 m height (Hume and Schalk, 1967). The winds were northerly, with a maximum 1-min average of 55 mph, and there was an effective fetch of about 120 km of open water to the north. Such waves of course cannot occur in winter, when the water is ice-covered, but large sea-level variations do show in the tidal records (Matthews, 1970), and can probably account for the heavy ice ridging typically encountered along the coast.

There is also an annual variation in sea level along the coast of 25 to 30 cm, with the highest levels in summer (August through September) and the lowest in winter (February through March) (Beal, 1968). Beal argued that most of the annual

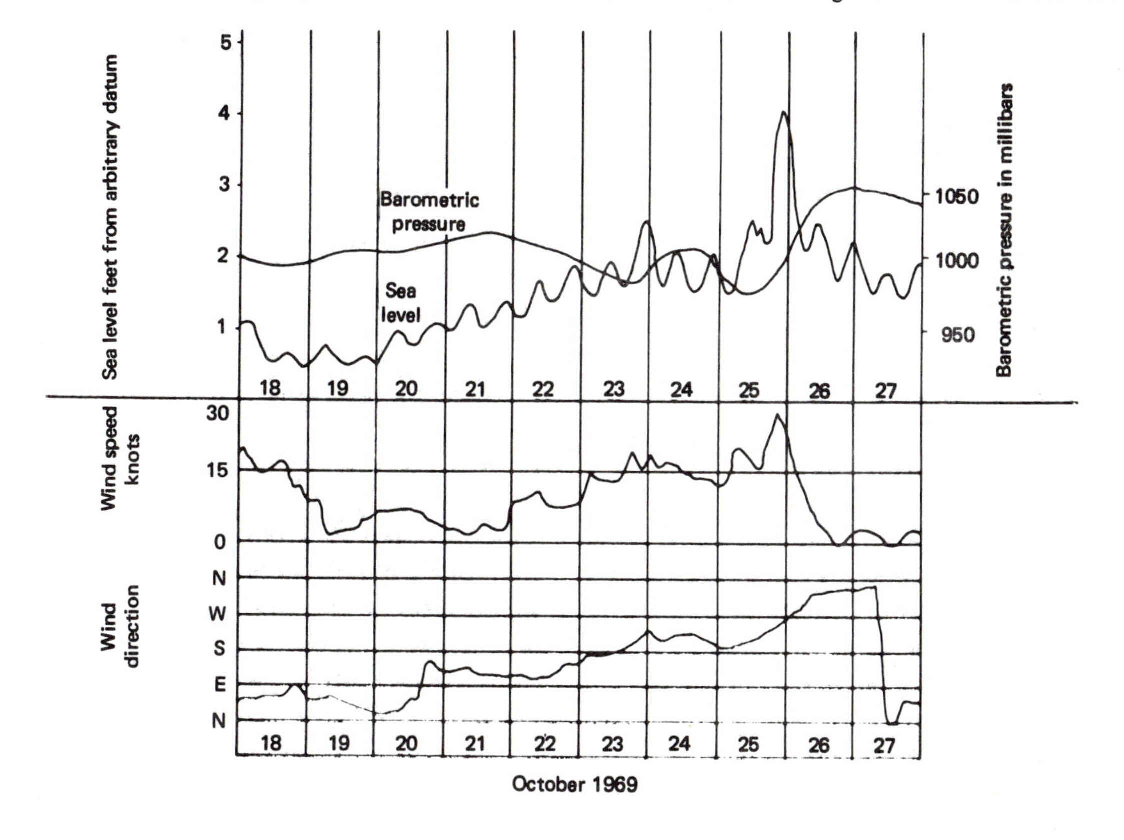

Fig. 21. Sea level, wind, and pressure records from Point Barrow (from Matthews, 1970).

variation could be a result of atmospheric pressure changes (on the average 10 mb higher in winter than summer) and steric variations in the water column (a warm low-salinity mass of water occupies a greater volume than a cold saline one of the same mass). Calculations showed that the mechanisms would contribute about equally to the variation.

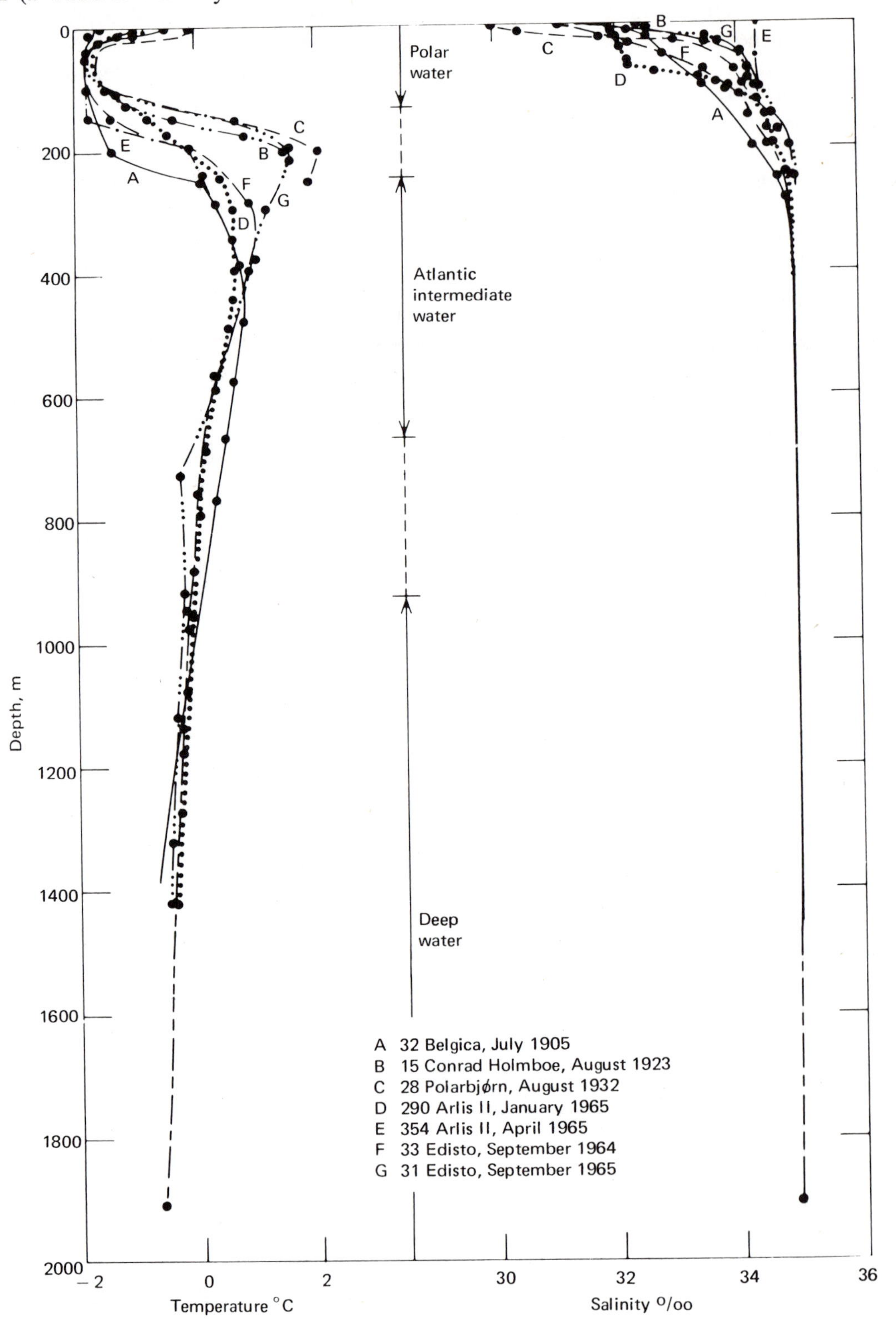

Fig. 22. Water masses of the East Greenland Current (from Aagaard and Coachman, 1968a).

V. Greenland Sea: Water Masses

The complex of basins between Greenland, Iceland, Norway, and Spitsbergen has had many names attached. In its entirety it was referred to by the early Norwegian oceanographers as the Northern Sea (Nordhavet), but after the turn of the century it became the Norwegian Sea. In recent years American oceanographers in particular have tended to distinguish the portion between Greenland, Jan Mayen, and Spitsbergen as the Greenland Sea. In addition, Icelandic oceanographers denote the region between Greenland, Iceland, and Jan Mayen as the Iceland Sea. The remaining portion is always referred to as the Norwegian Sea. Both in terms of ocean circulation and basin configuration, it seems useful to retain this Caesarean division into three parts.

The principal connection between the waters of the Polar Basin and the rest of the world ocean is through the Greenland Sea. Not only does this sea constitute the single most important avenue for exchange of water and heat, but it also serves as the principal northern hemisphere source of deep and bottom water for the world ocean.

Water Masses of the East Greenland Current

On the western side of the Greenland Sea is found a layering of water masses (Fig. 22) very similar to that of the Arctic Ocean (Aagaard and Coachman, 1968a, b):

(1) Polar water extends from the surface down to a mean depth somewhat greater than 150 m. The temperature varies between the freezing point (in the upper layers during winter) and 0°C (at the bottom of the Polar water). Normally a strong halocline is present: the salinity at the surface may be less than 30‰, but at the bottom of the Polar water it is nearly 34‰ or more.

(2) Atlantic intermediate water underlies the Polar water, extending down to approximately 800 m. The temperature is greater than 0°C, and a temperature maximum between 200 and 400 m is present throughout the year. The salinity increases from the Polar water downward until it attains a value between 34.88 and 35‰; this usually occurs above 400 m. Below 400 m the salinity is practically uniform.

(3) Deep water is found below about 800 m. The temperature of this water is less than 0°C, and the salinity is between 34.87 and 34.95‰.

The Polar water has its origin in the Arctic Ocean, from whence it has been carried by the transpolar drift stream merging into the East Greenland Current. Contrary to conditions prevailing in the surface water of the Arctic Ocean, however, the temperatures of the upper layers of the Polar water seasonally deviate significantly from the freezing point.

Figure 23 shows the temperatures and salinities at eight stations in the upper 300 m of the East Greenland Current, each set of stations separated by about 5 degrees of latitude. Summer and winter conditions are markedly different. During summer, when the ice is melting, the near-surface salinity is lower due to the addition of melt water. This decrease is often 5‰ or more and establishes a very strong halocline in the upper layers. Simultaneously the temperature in these layers increases, and it is not unusual to find temperatures in the upper 20 m which are 0.5°C or more above the freezing point, in ice-covered waters carrying multi-year polar ice. This heating from above usually leaves a temperature minimum near 50 m, which may be anywhere from a few hundredths to one-half degree above the freezing point, because the temperature minimum is at the same time being eroded from below by the upward heat flux from the Atlantic intermediate water. The net effect is such that noticeable seasonal temperature changes may occur over a depth range exceeding 100 m.

Thermal equilibrium between the ice and the upper layers of Polar water is achieved only during winter, when the water tends toward homogeneity, particularly with respect to temperature and to a lesser extent salinity. The vertical homogeneity is conditioned by the freezing process, and the temperature in at least the upper 50 to 75 m is uniformly near the freezing point during winter. This isothermal condition persists even where there is a moderately strong increase in salinity (Fig. 23), for the freezing point varies only slowly with salinity (Fig. 2). During winter the isothermal layer extends down to at least 50 m, and as is seen in Fig. 24 it may on occasion extend to 150 m. On the other hand the salinity in Fig. 24 is uniform only through the upper 10 m. The salinity, unlike the temperature, is therefore not a useful index of the depth to which there is active convection.

The achievement of deep convection, such as is indicated by the temperature structure in Fig. 24, and the simultaneous maintenance of a halocline (and hence pycnocline) depends upon either bringing into contact with the cold atmosphere water of higher salinities, or increasing the salinity of the near-surface water through addition of brine formed by the freezing process. The most saline (and hence densest) water would eventually assume a position near the bottom of the upper layer (cf. Coachman, 1966). The actual surface cooling need not occur at the same location for each parcel of water in the water column; the

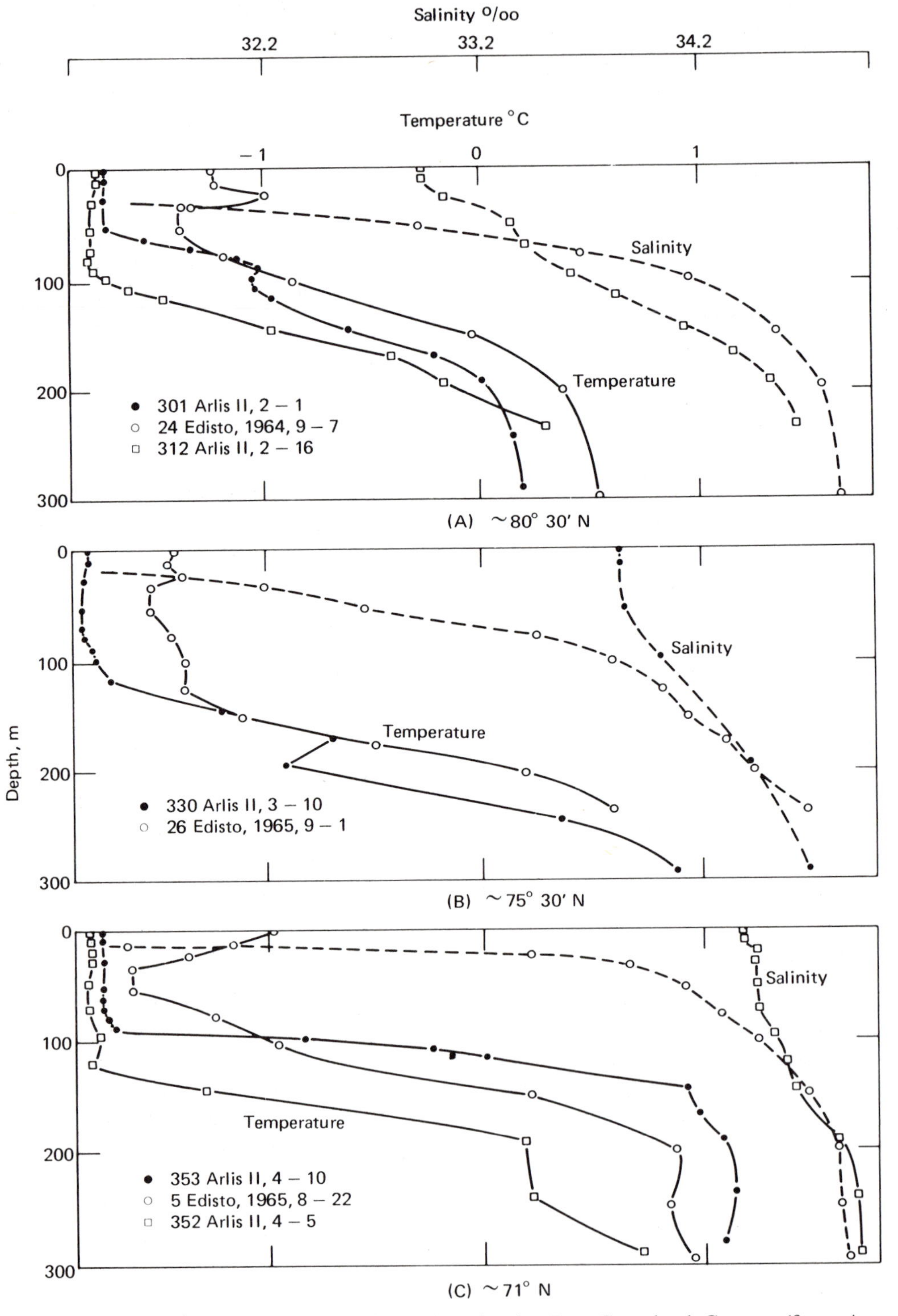

Fig. 23. Temperatures and salinities at eight stations in the East Greenland Current (from Aagaard and Coachman, 1968b).

observed structure could result from advection at subsurface levels of water of differing salinities which had been cooled at other surface locations. Whatever the details of the convective mechanism, whether local or influenced by lateral advection, it appears that the Polar water of the East Greenland Current is, under present climatological conditions, much more affected by the sea-

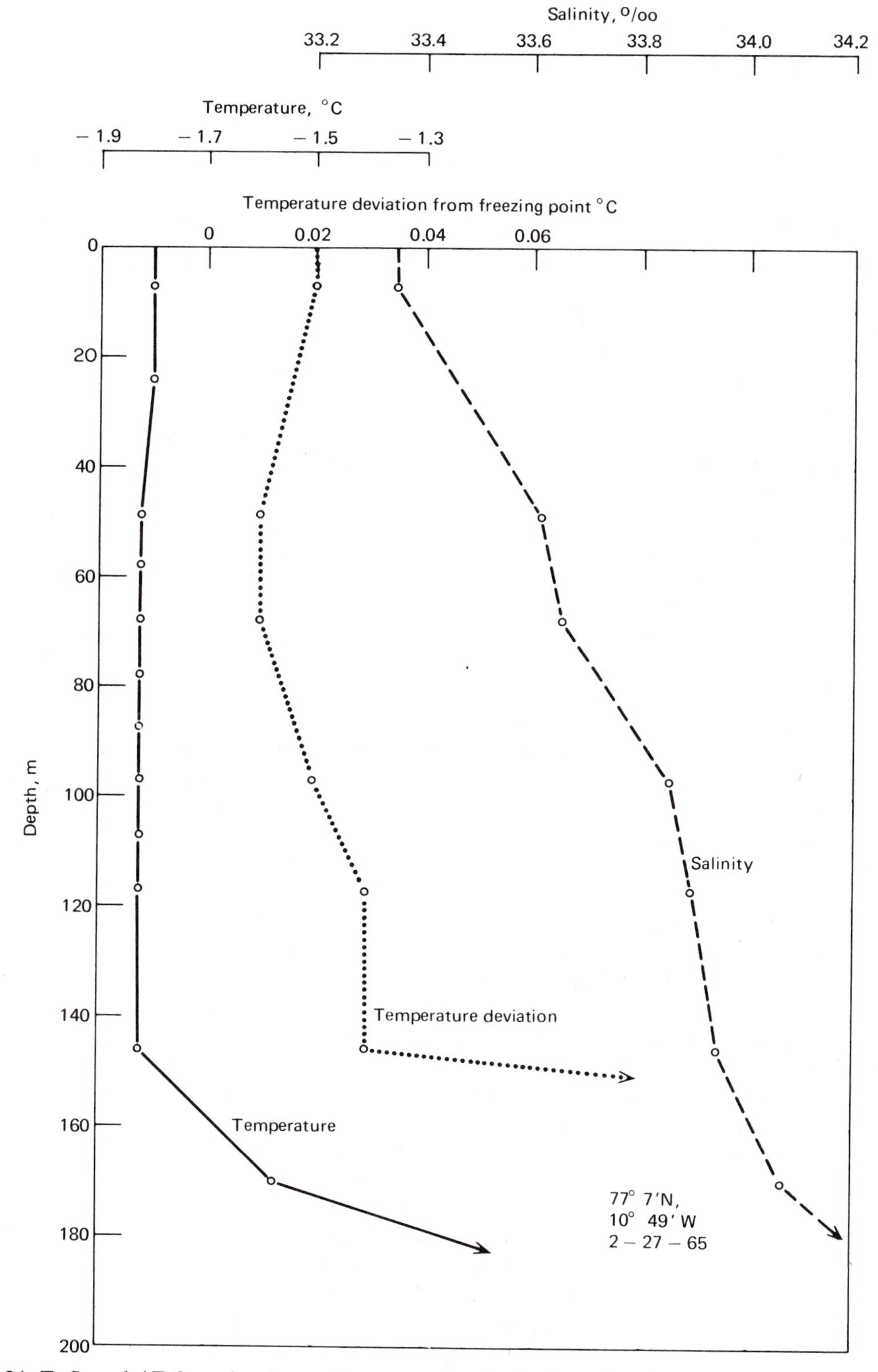

Fig. 24. T, S, and ΔT from freezing point at a station in the East Greenland Current (from Aagaard and Coachman, 1968b).

sonal cycles of cooling and heating, and freezing and thawing than is the Arctic Ocean surface water.

The Atlantic intermediate water was recognized in 1891 as having its origin in the West Spitsbergen Current, as attested to by the rather high maximum temperatures (frequently >2°C). In contrast, the Atlantic layer in the Arctic Ocean immediately northeast of Greenland is not warmer than about 1.2°C. Undoubtedly part of the Atlantic intermediate water with temperatures somewhat lower than the maximum represents outflow from the Polar basin, for there is a continual refluxing of the Atlantic layer in the basin back toward the Greenland–Spitsbergen passage (cf. Coachman and Barnes, 1963). However, because of the very large transport of Atlantic intermediate water in the East Greenland Current, on the order of 20 Sv (see next section), most of this water must represent a contribution from the West Spitsbergen Current that has not entered the Arctic Ocean.

This is indicated also by a quasi-isentropic analysis of the temperature field. Figure 25 shows the temperature on the density surface $\sigma_t = 28$, which lies close to the temperature maximum of the intermediate water. The inferred flow parallels the isotherms, except where there is strong mixing. The implication of Figure 25 is therefore that north of 75°N there is a broad sweep of warm water from the West Spitsbergen Current across the Greenland Sea.

We further note that at about 73°N the isotherms are directed eastward with clearly identifiable Atlantic intermediate water being found east of 5°W. A similar configuration of the isotherms is found in the Polar water, and a portion of both these water masses does move eastward in this region. In the surface layer the movement is manifested in the presence of a tongue of ice extending eastward from the East Greenland Current, and long familiar to sealers and whalers, the so-called "isodden." The eastward flow of Polar water is sometimes referred to as the Jan Mayen Polar Current.

The deep water is the overwhelming hydrographic feature of the Greenland Sea. It is an enormous dome of water which, when defined by the 0°C isotherm, in the center of the sea extends to within 50 m of the sea surface and at the peripheries within about 800 m (Fig. 26). It is remarkably nearly homogeneous. For example, a vertical hydrographic profile near the center during summer is typically isothermal to within 0.1°C and isohaline to within 0.04‰ from 400 to 500 m down to the bottom at about 3700 m. During winter, the same near-homogeneity extends upward to within 100 m of the surface. There are, however, small but systematic spatial and temporal variations in temperature and salinity, and we discuss these in subsequent sections.

Water Masses of the West Spitsbergen Current

On the eastern side of the Greenland Sea the predominant surface feature is not permanent ice cover, but rather permanent open water. As one crosses the Greenland Sea from west to east, the oceanographic features change from polar to subarctic, and the continually open and relatively warm water west of Spitsbergen is unique in the high-latitude regions of the world. Two water masses are usually distinguished in the area of the West Spitsbergen Current:

(1) The Atlantic water extends from the surface down to about 800 m, its lower boundary being defined by the 0°C isotherm. The temperature in the upper 100 m is frequently >5°C, and even during winter, surface temperatures immediately off the continental shelf are typically 2 to 4°C. The salinity increases rapidly downward in the upper 100 m, and at a depth that is usually somewhat below the sharpest portion of the halocline, achieves a maximum value of about 35.0 to 35.2‰. Below this the salinity decreases on the order of 0.1‰.

(2) The deep water which underlies the Atlantic layer is part of the vast, nearly isohaline water mass which occupies most of the Greenland Sea. Its temperature is between 0° and about −1°C, and its salinity west of Spitsbergen is somewhat >34.9‰.

The Atlantic water has been studied by a number of investigators, beginning with the Norwegian work in the first decade of the century (Helland-Hansen and Nansen, 1909, 1912; Nansen, 1915); the continuing northward expansion of commercial fisheries has provided the impetus for more recent research. Figure 27 shows the temperature-salinity correlations at four stations west of Spitsbergen. Most of the halocline is situated in the upper 100 m or less. It is particularly sharp near southern Spitsbergen (curves A, B), where the addition of low-salinity water from the East Spitsbergen Current occurs. During winter this low-salinity water is greatly reduced in temperature and carries sea ice, so that while the southern Spitsbergen coast may be blocked by ice the northern coast is open. It is for this reason that the first whaling stations on Spitsbergen were established on the extreme northwestern coast.

The main body of Atlantic water is not so strongly affected by low winter air temperatures, so that away from the influence of the East Spitsbergen Current, surface temperatures of 3 to 4°C are common during winter (curve C). There are, however, considerable variations in the temperature and salinity maxima of the Atlantic water, on the order of 1 to 2°C and 0.1 to 0.2‰, which are not seasonal in nature. Some of these variations appear to be changes in the average

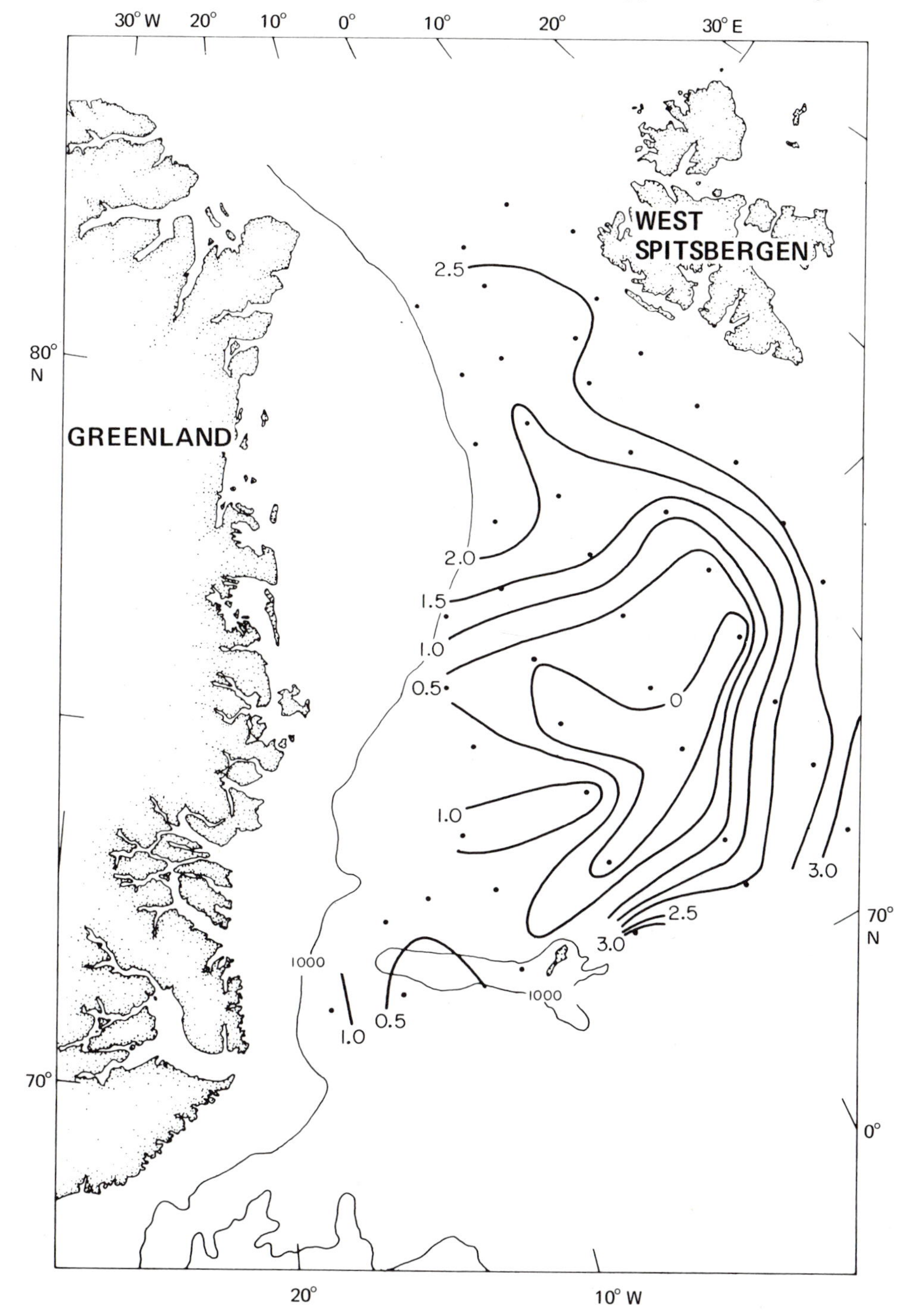

Fig. 25. Temperatures on σ_t = 28 (from Aagaard and Coachman, 1968b).

conditions of the water mass from year to year (e.g., cf. Mosby, 1938). Other variations seem related to a patchiness in the geographical distributions of temperature and salinity. Such conditions are common in the ocean, and they have been documented for the western Greenland Sea by Aagaard and Coachman (1968b). The important conclusion is that the Atlantic water does not represent a steady source of water with constant high temperature and salinity; rather, the Atlantic water flows both into the Arctic Ocean and southward along East Greenland with considerable variations in its initial characteristics. These variations are subsequently reduced in magnitude by diffusive processes, but their presence persists in a marked fashion over long distances (e.g., cf. Jakhelln, 1936).

VI. Greenland Sea: Field of Motion

General Features

The major features of the surface water circulation in the Greenland Sea have for centuries been known qualitatively: northward flow of warm saline water on the eastern side (West Spitsbergen Current) and southward flow of cold low-salinity water on the western side (East Greenland Current). However, it was in a number of classic papers by Mohn (1887), Nansen (1906), Helland-Hansen and Koefoed (1907), Helland-Hansen and Nansen (1909), and Kiilerich (1945) that the general circulation was systematically and quantitatively examined. Among the important features noted was the eastward movement north of Jan Mayen of a portion of the polar water of the East Greenland Current (the so-called Jan Mayen Polar Current), and the westward movement in the northern Greenland Sea of a portion of the Atlantic water of the West Spitsbergen Current. The net result is a large cyclonic surface gyre occupying most of the Greenland Sea.

From dynamic computations it was inferred that the speed decreased with depth at a very rapid rate within the East Greenland Current and somewhat less rapidly in the West Spitsbergen Current. For example, while the surface speed of the East Greenland Current might typically be 20 cm sec^{-1}, the calculations indicated speeds below 200 m of 2 cm sec^{-1} or less. The strongest flows were thought to occur above the continental slope rather than on the shelf. Transport estimates based on dynamic computations varied from less than 1 Sv (Lee, 1961) to more than 4 Sv (Timofeyev, 1963) in the West Spitsbergen Current, and a similar range of estimates has been applied to the East Greenland Current (cf. Aagaard and Coachman, 1968a).

A cyclonic flow pattern was also thought to prevail at intermediate depths, with Atlantic water moving northward along Spitsbergen, a portion then turning westward across the northern Greenland Sea and finally flowing southward as an underlayer of the East Greenland Current. Speeds and transports were deemed quite small, typically no more than 2 cm sec^{-1} and a fraction of one Sv. Whether or not this cyclonic movement was closed in the southern Greenland Sea was not known. The motion of the deep water was supposed negligibly small.

Direct Measurements*

During winter 1965 the ice island Arlis II drifted out of the Arctic Ocean and south along East Greenland, providing the platform for the first available set of direct current measurements in the Greenland Sea. The motion of the water relative to the ice island was determined by current meters, and the motion of the island relative to the sea floor by a drift meter using a transducer resting on the sea floor and three hydrophones on the ice. Vectorial addition of the relative current and the drift then gave the water motion relative to the sea floor. These measurements showed large short-period fluctuations in velocity that required supression by averaging, although such averaging from a drifting platform is of course neither strictly Eulerian nor Lagrangian. Figure 28 shows the velocity vectors for the three water masses of the East Greenland Current averaged over every 30 minutes of latitude. We note three features, each of which has been statistically confirmed:

(1) The mean current speed increases toward the southern portion of the drift, from about 4 cm sec^{-1} near 78°N to about 14 cm sec^{-1} near 70°N. This may not have been purely an effect of decreasing latitude, however, for the drift of the ice island in part constituted a gradual movement laterally across the continental shelf and out over the slope, where the current speeds are known to be higher than farther inshore.

* See Note 5 at the end of this chapter.

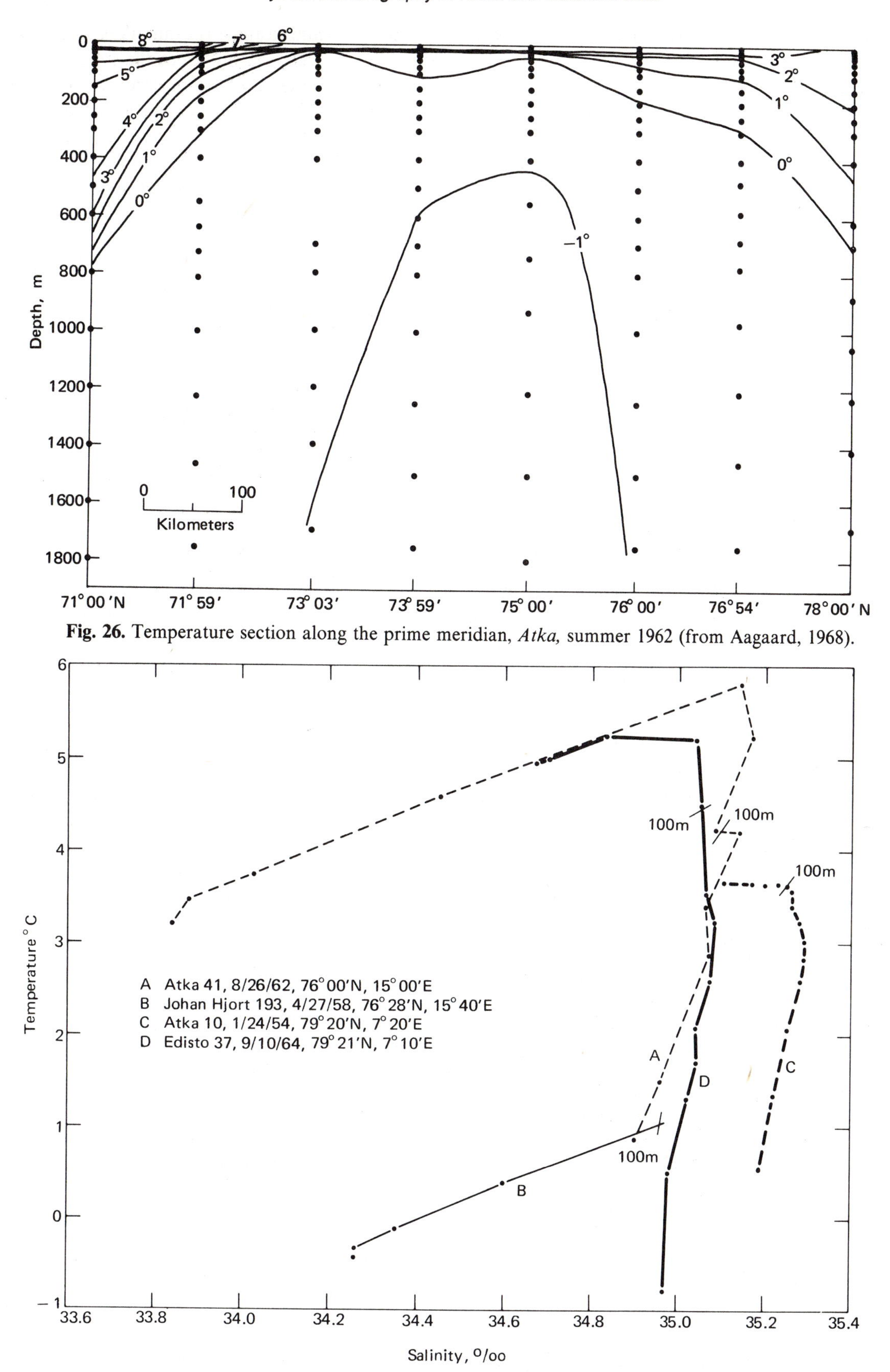

Fig. 26. Temperature section along the prime meridian, *Atka,* summer 1962 (from Aagaard, 1968).

Fig. 27. T-S correlations for four stations west of Spitsbergen.

(2) The current speed does not in general decrease significantly with depth, in marked contrast to earlier views.

(3) With increasing depth there is an eastward rotation of the mean current vectors. A detailed analysis strongly suggests this to have been the result of prevailing northerly winds creating a westward Ekman transport in the Polar water. For example, the vectorial difference in mean velocity between the Polar and Atlantic layers was 6 cm sec^{-1} toward 282° true, which is 90° to the right of the mean wind vector over the same period (6 m sec^{-1} from 012° true).

Since the ice island appeared to experience a

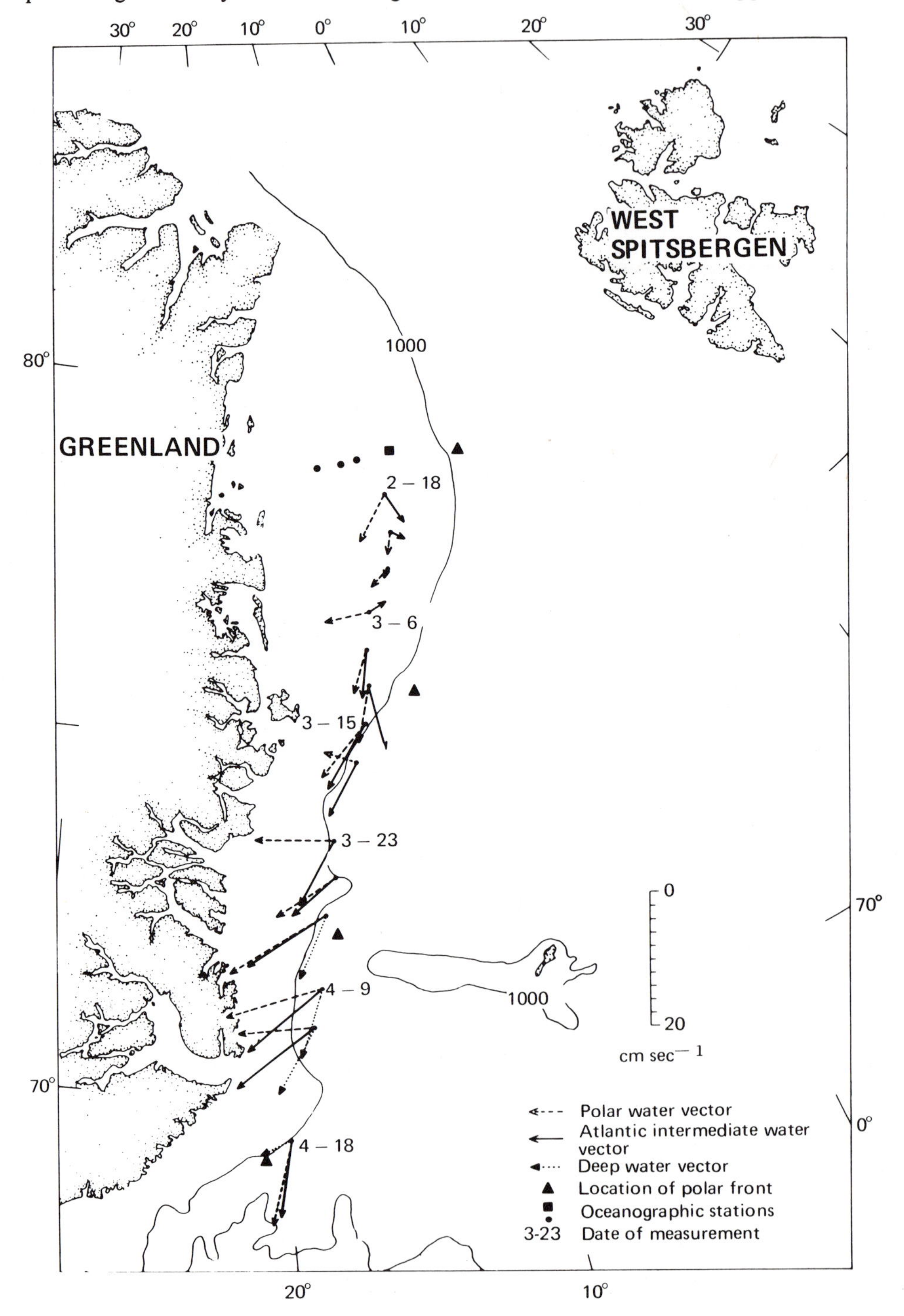

Fig. 28. Velocities of the water masses of the East Greenland Current, spring 1965 (from Aagaard and Coachman, 1968a).

gradual cross-current drift, the total volume transport across the drift line would give a minimum estimate of the transport of the East Greenland Current during the time of the drift. The computed transports, based on over 175 reliable current measurements extending from near-surface to 1220 m, are given in Table 4. We estimate that about 75% of the flow of the East Greenland Current had been measured during the drift, and we therefore conclude that during late winter of 1965 the total volume transport of the East Greenland Current was 35 to 40 Sv. This is a full order of magnitude greater than earlier estimates. We note that two-thirds of the total transport is in the flow of Atlantic water, previously thought negligible.

Table 4 East Greenland Current Transport Estimates February–April 1965

Water mass	Volume transport across drift track between 78–69°N, Sv
Polar water	7.7
Atlantic intermediate water	21.3
Deep water	2.5
Total	31.5

It appears to us unrealistic to assume that such a large flow either issues from the Arctic Ocean through the Greenland–Spitsbergen passage or exits through Denmark Strait. While it is true that most, if not all, of what we have called Polar water must come from the Arctic Ocean, an export of approximately 10 Sv of this water mass from the Polar Basin, while not an impossibly high value, is about three times larger than previous estimates. However, these estimates were based on summer hydrographic sections with some assumed subsurface reference level of no motion, and we have seen that at least during winter there is no such level.

It is quite another matter to account for 25 to 30 Sv of Atlantic water exiting from the Arctic Ocean. Not only is this 1½ orders of magnitude greater than earlier estimates, but the temperature distributions in the Greenland Sea point toward the recent origin of the major portion of this water in the West Spitsbergen Current rather than the Arctic Ocean. Our conclusion is therefore that the major portion of the East Greenland Current, particularly the flow of Atlantic water, represents a circulation internal to the Greenland Sea, or more probably to the combined Greenland–Iceland–Norwegian Sea system.

The measured constancy with depth of the current speed is difficult to explain. Summer hydrographic sections show that the isopycnals within the East Greenland Current frequently have an inclination exceeding 1 m/km. The mode of motion associated with this baroclinic contribution to the pressure gradient would tend toward geostrophic equilibrium. We have examined numerous pairs of synoptic hydrographic stations, and the calculated geostrophic velocity differences between the surface and 200 m typically exceed 10 cm sec^{-1}. Furthermore, in contrast to earlier opinions, the geostrophic shear is not negligible, even in the deeper layers, being on the order of 1 cm sec^{-1} per 100 m in both the Atlantic and deep water. It is possible that the current is nowhere near equilibrium, so that dynamic calculations are meaningless, but both the persistence of the internal mass distribution and a detailed examination of the dynamics, indicating small frictional and inertial effects, make this seem highly unlikely (Aagaard and Coachman, 1968b). The more probable explanation for the discrepancy between a pronounced baroclinicity, shown by summer hydrographic data and a lack of vertical shear in the measured winter currents, is that the baroclinic contribution to the pressure gradient is relatively minor in winter, but not in summer. A comparison of the very few winter hydrographic stations with those from summer indicate that this may indeed be the case (Aagaard and Coachman, 1968b), but observations are far too scarce to permit a meaningful estimate of such possible seasonal changes.

The next line of inquiry is the cause of the large transports. Krauss (1955, 1958) has argued that the circulation is haline, being driven by runoff from the bordering lands. However, calculations show that the precipitation required to drive even a very modest circulation would be on the order of 100 m per year; this is high, even for western Norway. This is not to say that the thermohaline circulation is unimportant, for it is essential to the maintenance of the deep water mass, and its proper role will be discussed in a subsequent section. Rather, the implication is that we must look to the wind field as the primary source of energy for maintaining the horizontal circulation.

The Wind-driven Circulation

The dominant features of climatological surface pressure charts of the Greenland and Norwegian Seas are the Greenland high on the western side and the trough of low pressure directed southwest-northeast in the central or eastern part. The latter results from the passage of cyclones from the Irminger Sea toward the Barents Sea. Associated with this pressure distribution there is in the mean a predominance of northerly or northeasterly winds on the western side of the ocean; these decrease toward the east, and along the Norwegian coast give way to southerly or southwesterly winds. The net effect is to provide a field of cyclonic wind stress curl over most of the sea between Greenland and Europe.

Under these conditions one would *a priori* expect certain circulation features to obtain. When averaged over some appropriately long time, ocean circulation dynamics (except for the western side of the ocean) appear to first order to be governed by a vorticity balance between the curl of the wind stress and the net meridional transport of planetary vorticity (Sverdrup dynamics: $\beta M_y = \text{curl } \tau$, where β is the variation with latitude of the Coriolis parameter, M_y is the vertically integrated meridional transport, and τ is the wind stress applied to the sea surface). In a region of cyclonic wind stress curl one therefore expects northward transport, except along the western boundary, where a southward flow closes the circulation. If the current is not negligibly small close to the bottom, then one also expects modification of the flow by bathymetric features. Such modification would include some degree of blocking by submarine ridges oriented transversely to the flow, and an opposite sense of circulation around islands. In general, there would be a tendency for the flow to follow the isobaths. In view of the bathymetry between Greenland and Europe, and the prevailing field of wind stress curl, we should see a tendency toward:

(1) A broad northward flow in the Norwegian Sea, turning more toward the northeast as it encounters the Mohn Rise, and then continuing northward along Spitsbergen in the eastern Greenland Sea;

(2) A southerly flow along Greenland;

(3) Some degree of blocking of meridional flow by the Greenland–Jan Mayen Ridge, and hence a tendency toward two separate cyclonic gyres in the Greenland and Iceland seas;

(4) Additional blocking by the Denmark Strait sill, and hence a tendency toward a separate cyclonic gyre in the Irminger Sea southwest of Iceland;

(5) Anticyclonic flow around Iceland.

All these features have indeed been observed, thus giving some factual support to our purely heuristic arguments.

To provide a more rigorously based understanding, we have computed the wind stress field for the Greenland and Norwegian seas in 1965 (cf. Aagaard, 1970). The computations were based on six-hourly surface charts and of necessity computerized. Figure 29 shows the results in terms of streamlines of the vertically integrated transport, averaged over the entire year. The streamlines were presumably closed by the return transport along East Greenland. There are two notable features:

(1) The circulation is primarily internal to the Greenland–Iceland–Norwegian Sea system, rather than simply representing transit through the region, which is in accord with the conclusions based on flow measurements in the East Greenland Current. It is particularly interesting that north of about 75°N the transport is largely westward, in agreement with our earlier conclusions from the temperature distributions in the Atlantic layer; flow into the Polar Basin appears to come from the northeastern Norwegian Sea, as suggested by Metcalf (1960).

We now also see the solution to an old problem, namely that of explaining the isolation of the deep water of the Greenland Sea from the Polar Basin. Beginning with Nansen (1902), numerous investigators have shown that the deep water of the Polar Basin is never colder than about −0.9°C, while the Greenland Sea deep water is several tenths degree colder than this. Nansen therefore proposed that a submarine ridge, which had been observed to extend westward from Spitsbergen at about 80°N, actually continued across to Greenland, thus restricting the northward movement of deep water beneath sill depth, which was estimated as 1200 to 1500 m. However, in the late 1950's it became clear from bathymetric investigations that there is no such restrictive sill (Laktionov, 1959). We can now postulate the following explanation of the deep water isolation: northward from the southern Greenland Sea the mean wind stress curl decreases greatly, and this results in a pressure gradient force directed southward. This force is

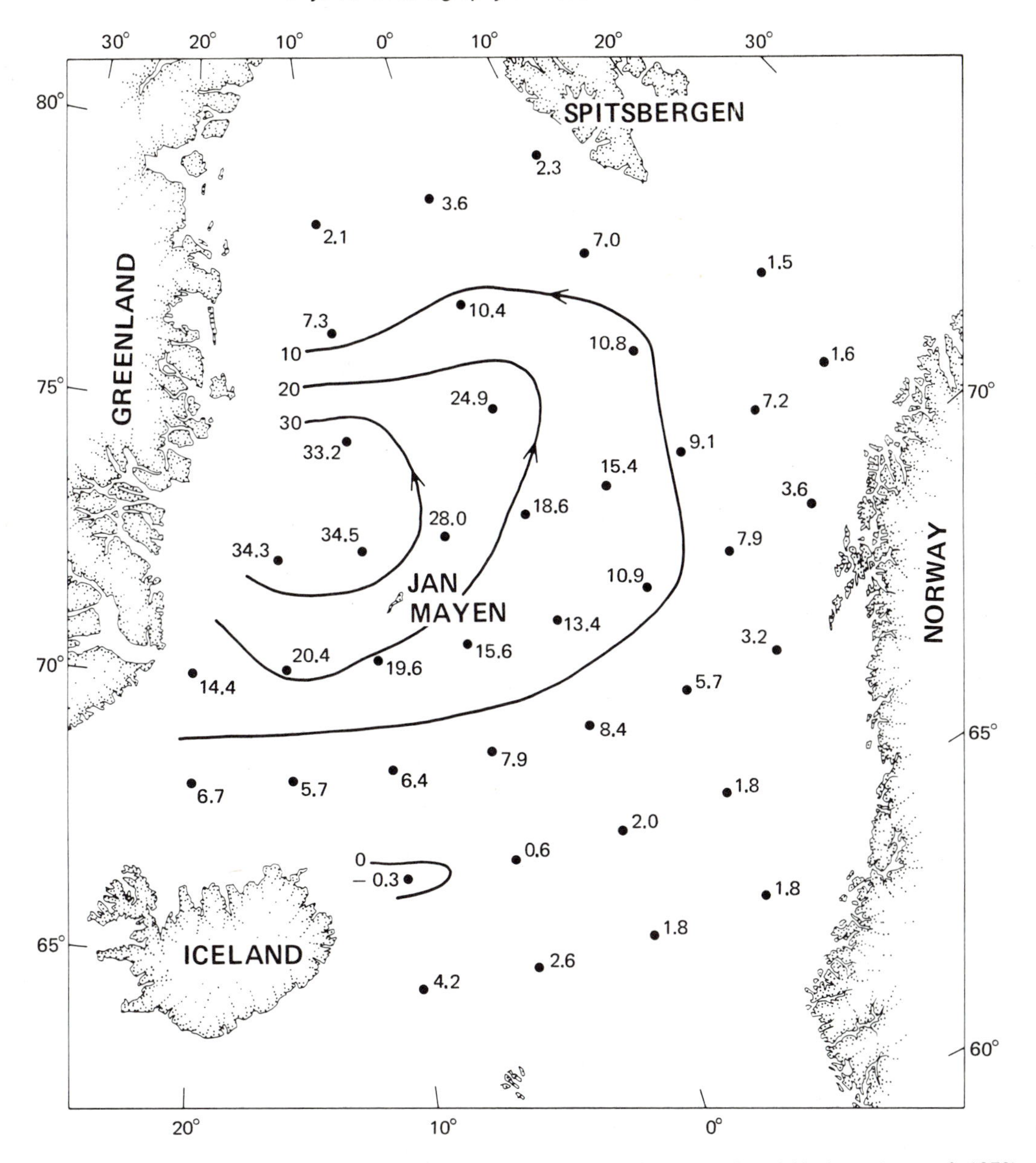

Fig. 29. Streamlines of vertically integrated transport (Sv), annual average for 1965 (from Aagaard, 1970).

the dynamic equivalent of the submarine ridge between Spitsbergen and Greenland first proposed by Nansen.

(2) The calculated transports are in remarkably good quantitative agreement with those actually measured. For example, the East Greenland Current transport required to close the streamlines between 72 and 74°N is nearly 35 Sv. While this near-agreement with the Arlis II current measurement is more due to circumstance than to understanding, it does show that transports of this magnitude can be maintained by the prevailing winds. Furthermore, the calculated wind-driven exchange of water with the Arctic Ocean and the North Atlantic is very close to that estimated from long-term averages of dynamic sections. For example, the mean transport of Atlantic water into the Polar Basin through the Greenland–Spitsbergen passage was estimated by Timofeyev (1962) as 3.3 Sv, whereas Fig. 29 shows a northward transport at the same latitude and over the same meridional expanse as Timofeyev's sections of 3.6 Sv. Similarly the mean transport of Atlantic water into the Norwegian Sea through the Faeroe-Shetland Channel was estimated by Tait (1955) as 2.3 Sv, whereas Fig. 28 shows a total northward transport at 64°N between the Norwegian coast and 5°W of 2.6 Sv.

We are left with the impression that the calculated annual mean wind-driven transports are of realistic magnitude. On the other hand, our calculations of mean monthly wind fields and the theoretically expected associated ocean circulation have shown enormous variations from month to month (Aagaard, in press, a), but it is unknown how the ocean circulation actually responds to these variations. Certain anomalies in the wind field during recent years seem to have resulted in a heavy influx of Polar water to northern Iceland, with disastrous results for shipping because of heavy ice. The probable dynamics of this situation have recently been discussed by Aagaard (in press, b).

The Thermohaline Circulation*

Heat budget calculations show that in the seas between Greenland and Europe the transfer of heat from the ocean to the atmosphere is on the order of 60 kcal cm^{-2} yr^{-1} (e.g., cf. Mosby, 1962b; Worthington, 1970). This is 1 order of magnitude greater than the transfer through the surface of the Arctic Ocean. On the other hand the static stability in both the Greenland and Norwegian seas in summer (when the stability is maximal) is only about 250×10^{-8} m^{-1} in the upper 200 m; whereas in the upper 200 m of the Arctic Ocean, even in winter (when it is minimal), it exceeds 1000×10^{-8} m^{-1}, that is, one-half order of magnitude greater. We should therefore expect a vastly more vigorous convective regime to obtain in the Greenland and Norwegian seas than in the Arctic Ocean. Vertical convection in the Norwegian current appears to be effective to 200 to 300 m (Mosby, 1970), but it is in the Greenland Sea that the convection achieves its most impressive proportions.

We have already indicated that the maintenance of the huge dome of cold deep water in the Greenland Sea depends upon the winter flux of buoyancy across the sea surface. The classic view of this mechanism is due to Nansen (1906), who proposed that the characteristically low temperatures of the deep water in the Greenland, Iceland, and Norwegian seas are maintained by the cooling and subsequent sinking of surface water during late winter, primarily in the Greenland Sea.

Certain features of the process have been elaborated by Metcalf (1955, 1960) and Mosby (1959, 1962a, 1967). Based on the fact that winter hydrographic stations in the proposed formation area did not show vertical homogeneity in the upper layers, Metcalf argued that the cooled water did not sink nearly vertically, but rather along only very slightly inclined isopycnal surfaces. He felt that a homogeneous water column probably played no part in bottom water formation, but agreed with Nansen as to place and time of formation. Mosby, on the other hand, has provided a qualitative model of the process which is intermediate in nature between Nansen's and Metcalf's in that it assigns a primary role to vertical motion and homogeneity, but also stresses the lateral motion of the sinking water. According to Mosby the cooling of the surface water first causes deep-reaching convection near the center of the Greenland Sea, because below about 30 to 50 m the static stability is least near the center. As cooling continues, this convection occurs over a larger and larger area and reaches deeper and deeper, forming an inverted cone of cold water with a downward-moving apex and a horizontally expanding base. At the same time the pressure gradient force due to the average density near the center of the area being greater than nearer the periphery, tends to move the denser water radially. The latter argument does not take into account the effect of a sloping sea surface. In the Greenland Sea such a slope is normally directed radially inward toward the center, the lowest sea level being at the center. There would be a net outward pressure gradient force at depth only if the component due to the internal mass distribution were strong enough to overcome the inward component due to the sloping sea surface.

A few years ago we presented a simple mathematical model showing how such an inverted cone-shaped isotherm structure might develop from the initial dome of cold water, along with a hydrographic section across the Greenland Sea during April 1958, which showed the inverted-cone structure (Aagaard, 1969). However, the problem of the deep-water formation remains essentially unsolved. Not only is this the case quantitatively, that is, there has been no determination of how much deep water is produced under a certain set of circumstances or its rate of formation, but it is also true in a qualitative way; for example, one does not understand even approximately the details of the buoyancy flux and convective mechanism. That the site of the convection is determined by the horizontal circulation, or more specifically by the accompanying

* See Note 6 at the end of this chapter.

density structure, seems reasonably certain to us. We do not believe, however, that the deep water is actually formed at the surface. For example, with the possible exception of a few of Metcalf's stations from 1951 and 1952, no winter hydrographic data show surface salinities high enough to form deep water, even if cooled to the freezing point. We feel instead that the deep water acquires its characteristics at a subsurface level, and we are presently engaged in a careful analysis of all available hydrographic stations in the Greenland Sea, from which we expect to gain better understanding of the deep water formation process.

It seems reasonably certain that the severity of winter cooling in the atmosphere is reflected in small but systematic variations in the deep-water temperatures (cf. Aagaard, 1968). The temperatures observed at 2000 m during 1901 to 1910 and in June 1954 are shown in Figs. 30 and 31, respectively. When one considers the nearly isothermal conditions that prevail in the main body of the deep water during any given year, the difference between the two distributions is rather striking, and a tentative hypothesis would be that during the intervening years the deep water had been warmed approximately 0.2°C. From 1952 onward there have been numerous temperature observations of the deep water, and these temperatures show slight variations from year to year. These variations have been analyzed statistically, and the results are given in Table 5. Several temperature trends are apparent:

(1) Some time between the first and sixth decades of the century there was a net warming of the deep water, and temperatures as low as those prevailing during the first decade have not been observed since.

(2) Beginning about 1956 the deep water was warmed further, so that the highest temperatures observed in the Greenland Sea deep water were during April 1958 to September 1959.

(3) Some time after 1959 a cooling trend began, the first hints coming in 1962, and the trend being firmly established by 1965; by 1966 the temperatures were reduced to those of the early 1950's.

Table 5 Temperature Variations in the Greenland Sea at 1500 and 2000 m
During period in column A, the deep water was (colder than) (warmer than) (of the same temperature as) during period in row B.

A \ B	June, 1953	Jan.–Feb., 1954	June, 1954	July, 1955	June, 1956	April, 1958	June, 1958	Sept., 1958	June, 1959	Sept., 1959	Aug., 1962	Aug.–Sept., 1965	April, 1966
1901–1910			Cx-10			Cx-10	Cx-10	Cx-5					
			Cy-10			Cy-5	Cy-10	Cy-2.5					
June, 1952	Sx		Cx-0.5	Sx	Cx-10	Cx-5	Cx-0.5	Cx-0.5					Sx
	Sy		Sy	Sy	Sy	Cy-5	Cy-2.5	Cy-0.5			Sy		Sy
June, 1953		Sx	Sx	Sx		Cx-1	Cx-0.5	Cx-0.5			Sx		Sx
		Sy	Sy	Sy		Cy-10	Cy-0.5	Cy-0.5			Sy		Sy
Jan.–Feb., 1954							Cx-5	Cx-1			Sx		
			Sy				Cy-10	Cy-5			Sy		
June, 1954				Sx		Cx-10	Cx-0.5	Cx-0.5			Sx		
				Sy	Cy-5	Cy-10	Cy-0.5	Cy-1			Cy-5		Sy
July, 1955							Cx-1						
							Cy-10						Wy-10
June, 1956							Cx-5	Cx-5					
							Cy-5						
April, 1958							Sx	Sx			Wx-10		Wx-2.5
							Sy	Sy		Sy	Sy		Wy-2.5
June, 1958								Sx			Wx-5	Wx-10	Wx-1
								Sy			Sy	Sy	Wy-1
Sept., 1958											Sx	Wx-5	Wx-1
										Sy	Sy	Wy-5	Wy-1
May, 1959									Sy	Sy			
June, 1959										Sy			

Key: C = colder than. W = warmer than. S = of the same temperature as. x = 1500 m. y = 2000 m. The number following x or y indicates the percent significance level of the difference in temperature, as determined from the sign test.

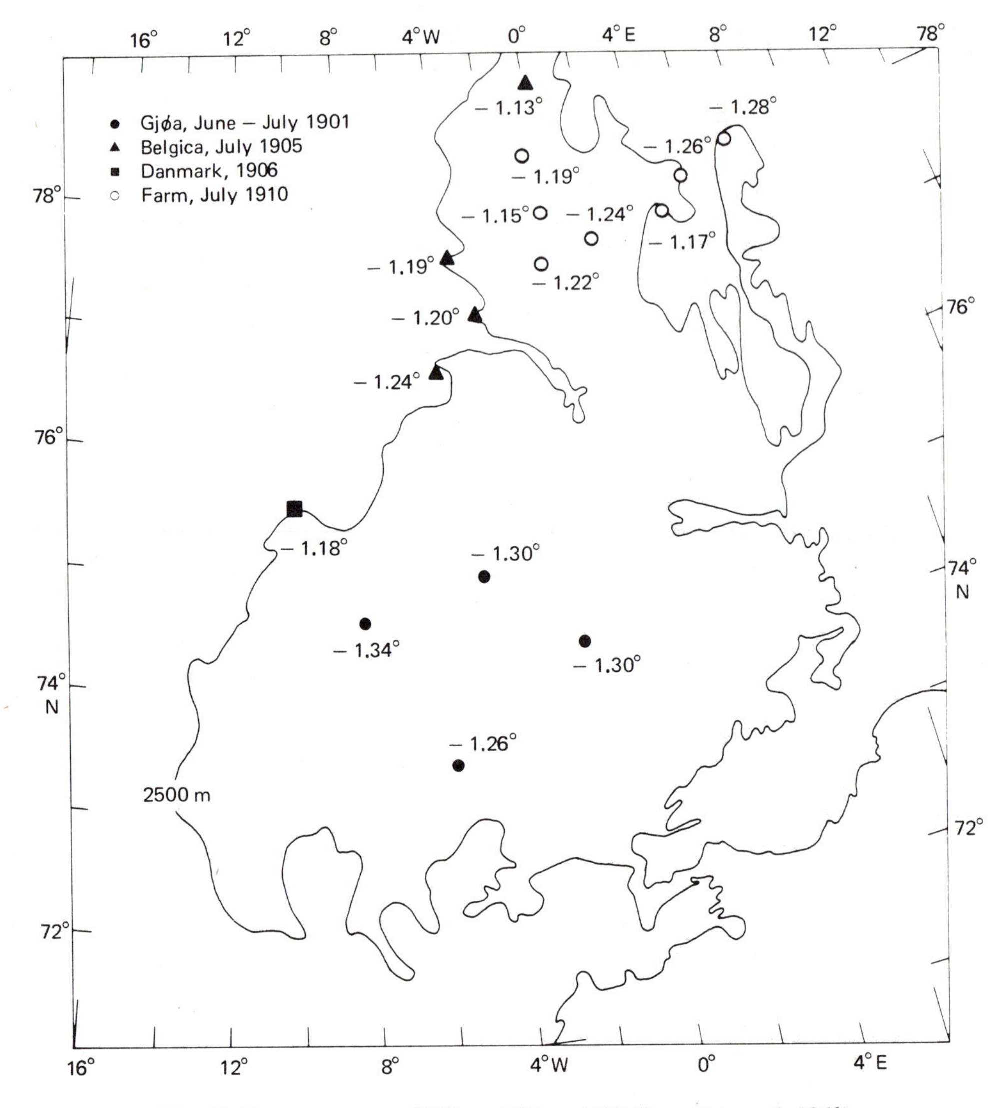

Fig. 30. Temperatures at 2000 m, 1901 to 1910 (from Aagaard, 1968).

Whether the deep water is formed at the surface or at some subsurface level, the formation must ultimately be dependent upon atmospheric cooling. As a working hypothesis we might assume that air temperatures lower than about $-1\,°C$ would tend to form deep water. We have therefore calculated the sum $\sum_n |T_n + 1\,°C|$ as a measure of the potential for deep-water formation during a given winter, where T_n is the mean monthly air temperature at Jan Mayen during months when T_n was less than $-1\,°C$. Figure 32 shows this cooling index from 1950 to 1966. The deep-water temperature trends are clearly associated with similar trends in the Jan Mayen air temperatures. For example, the relatively constant deep-water temperatures from 1952 to 1955 coincided with a period of nearly normal winter air temperatures, and the maximum deep-water temperatures in 1958 followed two unusually mild winters. The two winters from 1960 to 1962 had a cooling index near or only slightly below the mean for the period 1950 to 1966, and this coincided with the cooling trend in the deep water observed in 1962. The abnormally cold winters during the following three out of four years were accompanied by further reduction in the deep-water temperatures in 1965 and 1966. We cannot correlate the warming of the deep water during the first half of the century with air temperatures, for these are not available from Jan Mayen prior to 1921. However, between 1921 and 1935 the calculated cooling index was always less than the mean value for 1950 to 1966. It has also been shown that beginning about 1915, both Spitsbergen and northern

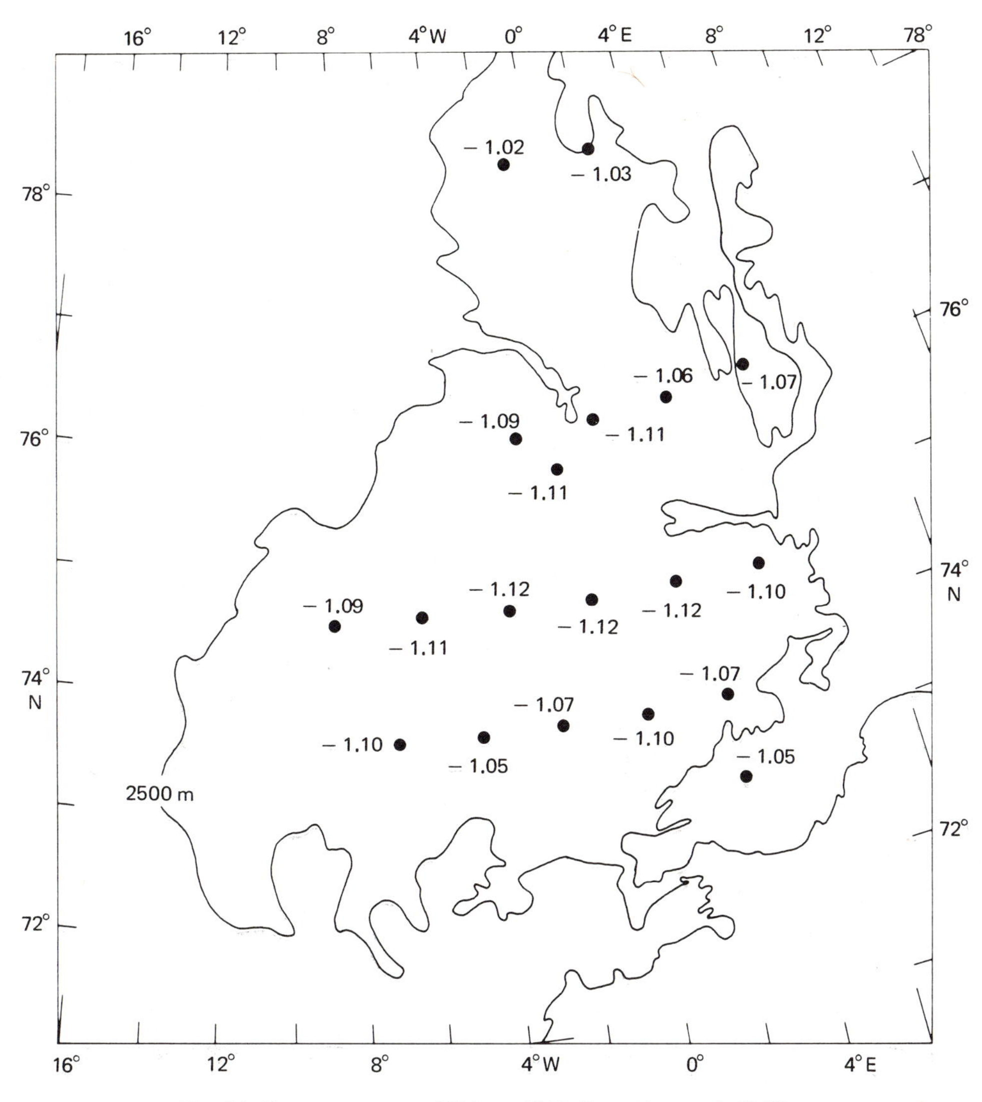

Fig. 31. Temperatures at 2000 m, 1954 (from Aagaard, 1968).

Norway experienced a very large increase in winter air temperatures (Hesselberg and Birkeland, 1940), so that a warming trend in the deep water some time after 1910 appears reasonable. We are left with the strong impression, then, that the major trends in the deep-water temperatures reflect, and perhaps rather rapidly respond to, the intensity of the winter cooling over the Greenland Sea.

In recent years great interest has been attached to the outflow of deep water ("overflow") into the North Atlantic. We suggest that a matter of equal interest, but hitherto not investigated, is the deep outflow to the Polar Basin; we shall in the next section discuss both of these matters. While we intuitively feel that such outflows somehow reflect the thermohaline circulation in the Greenland Sea, the nature of the connection is not known. For example, it is unknown how the overflow in a given year is related to the amount of deep water formed that year or its mean density.

VII. Budgets for the Arctic

There have been numerous attempts recently to construct heat budgets for the Arctic Ocean and adjacent seas (e.g., Vowinkel and Orvig, 1961; Mosby, 1962b, 1963; Timofeyev, 1963; Worthington, 1970). The preparation of definitive budgets takes on added importance now that the Arctic is recognized to play a key role in controlling many specific aspects of the climate of the Northern Hemisphere (Fletcher, 1965; Hare, 1970).

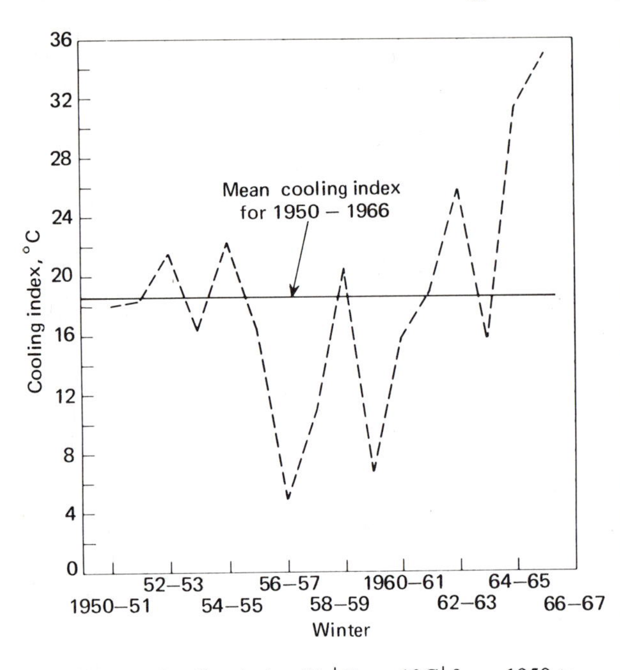

Fig. 32. Cooling index $\sum_n |T_n + 1°C|$ from 1950 to 1966. (form Aagaard, 1968).

Among the heat budgets, there seems to be general agreement as to the order of magnitude of the various flux terms. In budgets for both the Arctic Ocean and the Greenland–Norwegian Seas, of the seven first-order terms, four are due to oceanic advection. Hence, accurate assessment of the heat advected through the particular boundaries must be a *sine qua non* for a definitive Arctic heat budget. Evidence of our abysmal lack of knowledge of heat advected by ocean currents in the Arctic appears in a recent thermodynamic model of sea ice (Maykut and Untersteiner, 1971): they found it impossible to reconcile recent observations of the amount of open water, where the major heat transfer to the atmosphere occurs, with the oceanographic information available on advected heat (the amount of heat advected was too little and/or the area of open water was grossly overestimated).

Oceanic heat advection is normally defined by $Q = \rho C_p TV$, where Q is the heat flux, ρ the density, C_p the specific heat, T the temperature (usually referenced to 0°C), and V the water transport (volume/time). For heat budget purposes ρ and C_p can be considered constants, so that to define Q, measurements of temperature and transport are required. Ocean temperatures are much easier to measure than transports, and in the Arctic good temperature measurements have been accumulating for over 70 years. Transport values, however, depend on current measurements, and these are still very sparse and much more difficult to obtain and interpret. Thus, knowledge of the heat fluxes through the Arctic Ocean system can be much improved by concentrating future research on the measurement of selected oceanic transports.

Tables 6 and 7 summarize recent estimates of advection and temperatures used to calculate heat budgets. There is good agreement among investigators on some transports (e.g., the Bering Strait) and poor agreement on others (e.g., the Atlantic inflow to the Norwegian Sea), reflecting the wide variation in the quality of present knowledge regarding each individual transport. Also, each budget was balanced in itself by adjusting the least well-known flows to those considered better known.

The annual water budgets should balance closely because over recent years sea level in the Arctic has shown no significant variation other than the hydrostatic and steric ones (Beal, 1968). The area of the Arctic Ocean is 9.5×10^6 km^2. If there occurred an imbalance in the advection of 0.1 Sv for six months, sea level would be altered by $\approx$16 cm, while an imbalance of 1 Sv would cause a 1.6 m change in level. There is no evidence that such changes occur, and we conclude that water budgets for periods of a few months and longer must balance within $\approx$0.1 Sv.

As accurate values of transport through the advection boundaries of the Arctic basins are critical to accurate heat budgets, present knowledge of transport through the various advection boundaries is assessed in the following sections. We point out that no budgets have considered the effects of horizontal turbulent diffusion on the exchange of salt and heat. Such omission will in general lead to overestimates of volume transports when trying to construct budgets. However, we feel that at the present state of the art, the introduction into the problem of horizontal diffusion is an unwarranted sophistication which serves no useful purpose.

Bering Strait*

A comprehensive set of direct current measurements was made across the Bering Strait from the *Northwind* in 1964, the vessel being anchored at each station (Coachman and Aagaard, 1966). The transport was northward into the Arctic Ocean at 1.4 Sv. In the period 1967 to 1969 the

* See Note 7 at the end of this chapter.

University of Washington conducted further extensive measurements in the Strait (as yet unpublished). Figure 33 presents the cross-sections of flow based on the direct current measurements from this work.

The four sections occupied in one week (Fig. 33, right) illustrate the variability to be expected in flow through even a small strait such as this. The first section shows a northward transport of 2.5 Sv, and no southward flow at all. One day later there appeared a little southward flow in the immediate surface layers (a northerly wind had begun to blow), and the net transport was reduced to 1.6 Sv. At the time of the next measurements the wind effect was less and so was the southward flow; the northward transport had gone back up to about 2 Sv. During the time of the last section the southward flow was greater than during the previous measurements, and now the net northerly transport was reduced to 1.4 Sv.

Thus, a factor of 2 in variability in the transport occurred during one week. That the variability can be even greater is illustrated by the detailed section made from the *Northwind* during

Table 6 Water Balance of the Arctic Basins
The main components, compiled from recent sources. Units are Sverdrup (1 Sv = 10^6 m^3 sec^{-1}).
Variations by 0.1 Sv may be due to rounding and conversion.

ARCTIC OCEAN

Inflow	Fletcher (1965)	Mosby (1962b)	Timofeyev (1963)	Vowinkel and Orvig (1961)
West Spitsbergen Current	3.2	[d]2.0	4.2 [a] { 2.8 5.1	[b] { 3.0 3.8
Bering Strait	1.0	1.2	1.0 [a] { 0.5 1.7	1.1 [b] { 1.0 1.2
River runoff	0.1	0.1	0.1	0.1
Barents Sea	1.0			
Outflow				
East Greenland Current	4.0	2.0	[c]4.1	[c] { 3.0 3.8
Canadian Archipelago	1.3	1.1	1.3	1.2 [b] { 1.0 1.4
Barents Sea		0.05		
Meltwater and ice		0.15		

GREENLAND–NORWEGIAN SEAS

Inflow	Vowinkel and Orvig (1961)	Worthington (1970)	Fletcher (1965)	Mosby (1963)	Timofeyev (1963)
Atlantic (Faeroe Bank)	4.5 [b] { 3.0 4.9	8	9.5	3.6	9.3 [a] { 6.6 12.9
Arctic (E.G.C.)	[c] { 3.0 3.8	3	4.0	2.0	
Atlantic (Denmark Strait)		1			
Barents Sea			1.0		0.95
Meltwater and ice				0.1	
Outflow					
Denmark Strait overflow		4			
Denmark Strait (E.G.C.)	4.5 [b] { 3.6 5.2	3	4.0	3.6	4.1
Iceland–Scotland overflow		2	5.3		
Arctic Ocean	[b] { 3.0 3.8	3	3.2	2.0	4.2 [a] { 2.8 5.1
Barents Sea			1.9		1.9 [a] { 1.5 2.2
Meltwater and ice				0.1	

[a] Extremes of mean monthly values.
[b] Range deduced from various literature estimates.
[c] By difference.
[d] Considered in two parts: Atlantic water (1.4 Sv) and bottom water (0.6 Sv).

Table 7 Temperatures [°C] of Advected Flows (Table 6) Used in Calculating Heat Budgets

	Vowinkel and Orvig (1961)	Mosby (1962b, 1963)	Worthington (1970)
ARCTIC OCEAN			
Inflow			
West Spitsbergen Current (bottom water)	1.62	3.25 (−0.9)	
Bering Strait	0.93		
Outflow			
E.G.C.	−0.5	−1.8	
Canadian Archipelago	−0.7	−1.8	
GREENLAND–NORWEGIAN SEAS			
Inflow			
Atlantic	7.8	8.9	9
Arctic (E.G.C.)	−0.5	−1.8	[a]
Denmark Strait			6
Outflow			
Denmark Strait (E.G.C.)	0, −1	−1.8	[a]
West Spitsbergen Current	1.62	3.25 / −0.9	3
Meltwater and ice		−0.3	
Iceland–Scotland Ridge overflow			1
Denmark Strait overflow			1

[a] Heat transport by the East Greenland Current in from the north and out through the Denmark Strait was considered to balance.

July 21–22, 1967 (Fig. 33, left), when the transport was 0.2 Sv southward. This, incidentally, is the only documented incidence of net southward transport through the Bering Strait.

We feel that the variability in transport is related to the atmospheric regime, but the details of the mechanism are not clear. On the one hand, at times a reversal of the normal northward surface flow is associated with local northerly winds and the net northward transport is reduced by over 50%; but attributing variations in transport in excess of 1 Sv to local winds is unrealistic, for oceanic Ekman transports through such small cross-sections simply are not that large. On the other hand, the dominant cause of the generally northward flow is a surface slope of order 10^{-6} downward to the north (Coachman and Aagaard, 1966), and major modifications of this slope must be caused by changes in the regional wind field. Variations in transport which appear to be purely reflections of changes in the surface slope do not alter the general features of the internal mass field and relative velocity distributions. Such variations can easily be as large as those observed during times of local northerly winds.

All the measurements on which good transport calculations can be based were made in summer. Soviet oceanographers for many years have reported that the transport in winter is only $\frac{1}{3}$ to $\frac{1}{4}$ that in summer (cf. Fedorova and Yankina, 1963), but without presentation of supporting data. For example, Antonov (1968) shows a relatively smooth annual cycle with a minimum in March (0.4 Sv) and maximum in August (1.6 Sv). The idea of a winter minimum seems to have originated with Maksimov (1945). Measurements in the Bering Strait in April 1969 (Fig. 34; cruise of U.S.C.G.C. *Staten Island*, data as yet unpublished) show currents and current shears similar to those prevailing during the times of highest transports in summer (Fig. 33). We conclude that there is no good evidence for an annual cycle in Bering Strait transport, but that variations of 50% or more can be expected on a time scale of less than a week. We estimate the mean annual transport at 1.5 Sv, which is somewhat higher than earlier estimates (1 Sv) based on much more limited data.

Canadian Archipelago

The only reasonable area for measurement of water exchange between the Arctic and Atlantic through the Canadian Island passages is in northern Baffin Bay. There, either the flows through Smith, Jones, and Lancaster Sounds can be separately measured and summed, or a net section made between Cape York on Greenland and Bylot Island.

Oceanographic sections providing sufficient detail for transport calculations are available from numerous summer cruises during the 1950's (Bailey, 1957) and from the *Godthaab* expedition of 1928 (Kiilerich, 1939). However, there is always considerable doubt as to the validity of geostrophic transport calculations without independent measurements of the flow to aid in locating a suitable reference level. The only hydrographic section concurrent with current measurements has been the 1963 *Evergreen* section across the narrow part of Smith Sound (Franceschetti et al., 1964). The current measurements, from moored buoy systems, were reported by Day (1968). Only two meters provided data over the period of station occupation: one at 350 m at the center mooring, with a mean flow of 4.3 cm sec^{-1} south, and one at 150 m at the east mooring (0.8 cm sec^{-1} north). Figure 35 shows the flow calculated from the hydrographic data adjusted to the current measurements. The agreement between dynamic calculations and current measurements was very good when taking the flow at the 150-db reference level to be 1 cm sec^{-1} northward. The net transport was 0.14 Sv south.

It appears that geostrophic calculations may give a reasonable estimate of the transport through Smith Sound, at least at certain times. A

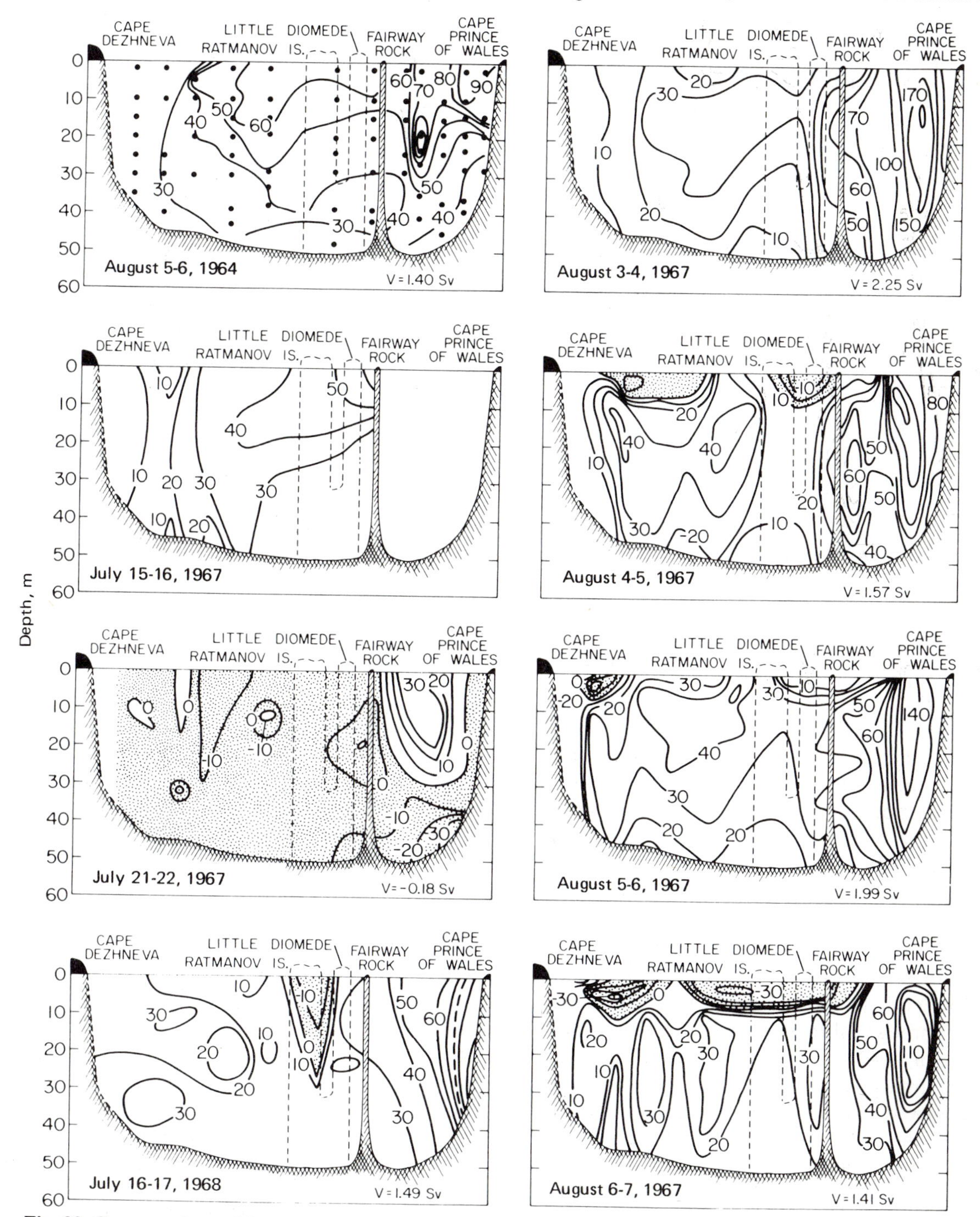

Fig. 33. Cross-sections of flow through Bering Strait. Isotachs in cm sec^{-1} with + indicating northerly flow and indicating southerly flow; current meter locations indicated by *X*.

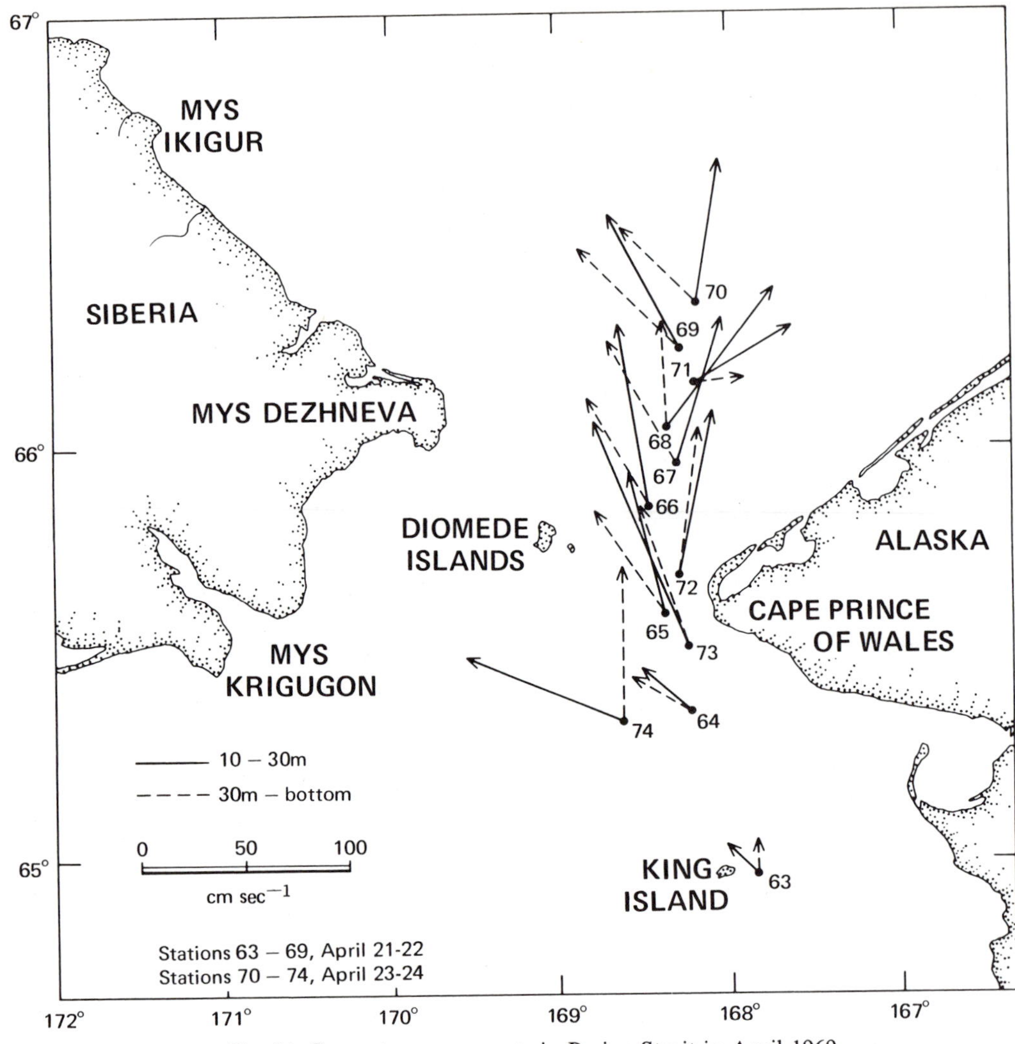

Fig. 34. Current measurements in Bering Strait in April 1969.

similar analysis cannot be made for sections across the eastern end of Lancaster Sound or any other section in Baffin Bay, as no current measurements are available. Future studies should address this question. However, with the evidence that geostrophic calculations may have some validity, variability in transports can be examined.

Muench (1971) summarized all available values for sections across northern Baffin Bay from Cape York to Bylot Island (which includes the flow through all three sounds) for the years 1961 through 1963 and 1966, and concluded that a water transport of 2.0 ± 0.7 Sv to the south is representative of summer conditions.

If the transports through the three sounds are assessed separately and then added to achieve an estimate of net flow, the results are more variable. For example, Collin (1963) reported net south flows of 1.40 Sv for the 1928 sections and 0.67 for the 1954 data. Even greater variability is apparent from the 1966 *Edisto* cruise, when the following sections were made: four across different parts of Nares Strait, two each across Jones and Lancaster Sounds, and the inclusive section from Cape York to Bylot Island, all within three weeks (Palfrey, 1968). The calculated transports are presented in Fig. 36, in which transports through Jones and Lancaster Sounds have been deducted from sections made south of them. It appears that the transport southward through Nares Strait fluctuated between 0 and 2 Sv in the three-week period. The two sections across eastern Lancaster Sound gave 0.3 and 0.6 Sv, and the Jones Sound section 0.43 and 0.20 Sv. Westward flow in Jones Sound (−0.39 Sv) and northward flow in Smith Sound (−0.42 Sv) has been reported for 1954 (Bailey, 1957). It appears that the transport values through the individual straits can exhibit variations as great as or greater than the total southward flow, over time scales of the order of a week. It is quite clear that good estimates of Arctic-Atlantic exchange through the Archipelago cannot be obtained by summing the Smith, Jones, and Lancaster Sound transports unless they are based on synoptic observations.

Another question is the persistence of the flow patterns in the straits. If the relative velocity distributions remained essentially unchanged with time, it would be possible to monitor the total flow by using observations at a few key locations; this may be feasible for the Bering Strait. We have recalculated the 1966 *Edisto* dynamic section K – K′ across the narrow part of Smith Sound (Palfrey, 1968) using the same reference level as in the 1963 *Evergreen* section. The sections from both years show the same relatively complicated flow pattern (cf. Fig. 35): southerly flow in the upper 100-m layer, with the swiftest current near the eastern shore; northerly flow in the 100-to-200-m layer throughout the strait, but extending deeper along both shores; and southerly flow in the central part from 200 m to the bottom. The southward transport of the 1966 section was, however, much greater (~1.2 Sv) than that of the 1963 section (~0.1 Sv). This suggests that the major variations in flow are in the barotropic mode, as in the much shallower Bering Strait.

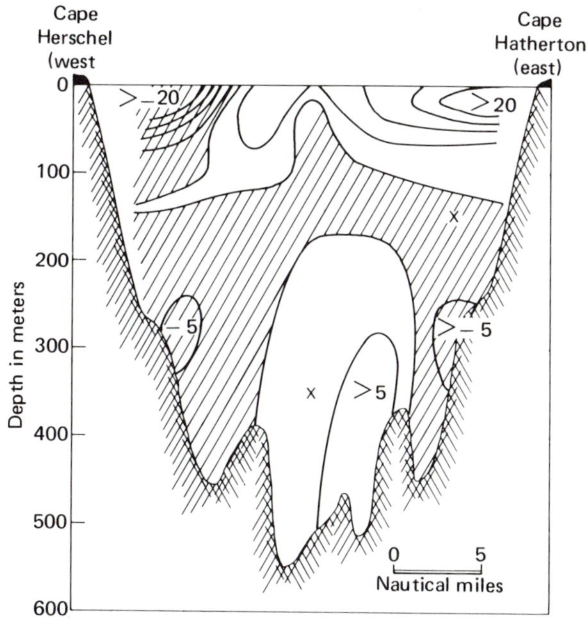

Fig. 35. Calculated flow through Smith Sound (July 30, 1963) adjusted to direct current measurements. Isotachs in cm sec^{-1}, with + indicating northerly flow and − indicating southerly flow; current meter locations indicated by *X*.

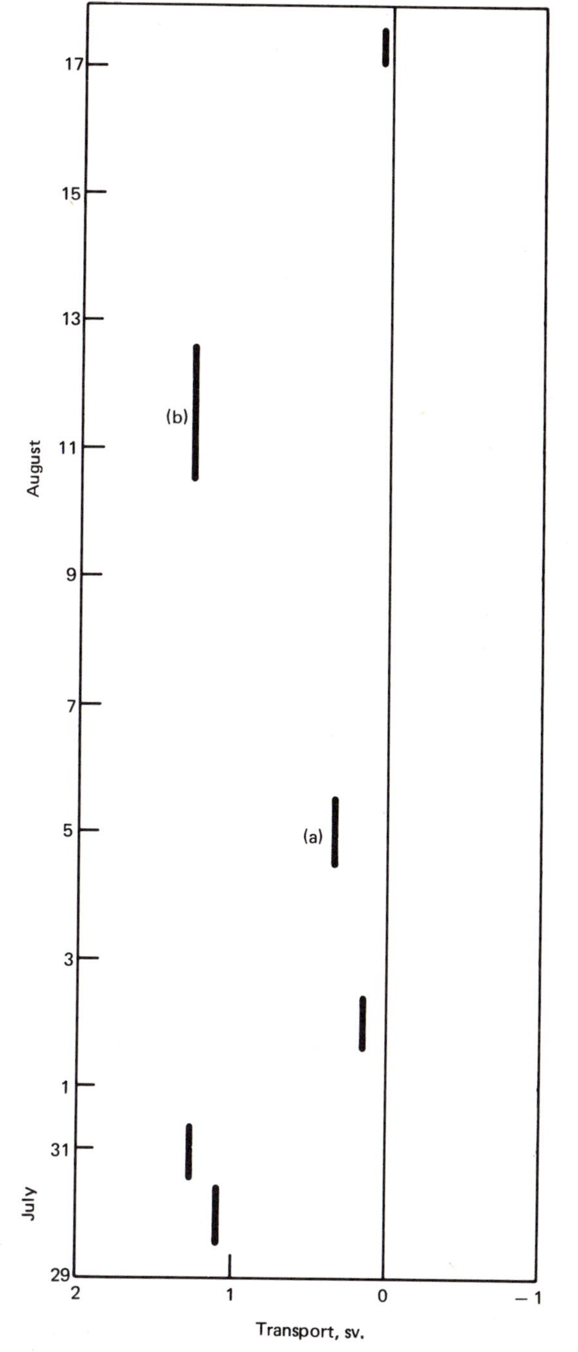

Fig. 36. Calculated south transport through Smith Sound, summer 1966. Flow through Jones Sound subtracted in (a), and flow through Jones and Lancaster Sounds subtracted in (b).

We feel that the best estimates of trans-Archipelago exchange available are those reported by Muench (1971) based on sections extending across northern Baffin Bay. These number only four and are from summer (July to October); the transports at other times are unknown. We are therefore forced to conclude purely on the basis of summer observations that there is a mean transport from the Arctic to Atlantic of about 2 Sv, variable by a factor of two or more on a time scale of 1 to 2 weeks.

In searching for causes of the predominant southward flow from the Arctic Ocean to Baffin Bay, Muench (1971) presented data on the mean steric levels in the upper 250 m of the Arctic Ocean basins and Baffin Bay (Table 8). These indicate the likelihood of the sea surface in the Canadian Basin standing higher than in the Eurasian Basin or in Baffin Bay, subject, of course, to the verity of the 250-db surface being approximately level.

Table 8

Dynamic Depth Anomalies Relative to 250-db Level

Location	D (dyn. m.)	Year(s)	Platform
Arctic Ocean (Canadian Basin)	0.57	1965	T-3
		1966	T-3
		1967	T-3
Arctic Ocean (Eurasian Basin)	0.27	1964	Arlis-II
Northern central Baffin Bay	0.25	1961	Labrador
		1962	Labrador
		1963	Labrador
		1966	Edisto

Barents Sea Boundaries

There have been practically no direct measurements of the currents between Spitsbergen and Franz Josef Land and between Franz Josef Land and Novaya Zemlya—respectively, the northern and eastern boundaries of the Barents Sea. In fact, there were very few available hydrographic data from these areas, apart from the basic work done by Mosby on the *Quest* in 1931 (Mosby, 1938), until the very recent cruises by the *Southwind* in 1970 and the *G. O. Sars* in 1971 (as yet unpublished). By far the greatest amount of work accomplished in the Barents Sea has been by Soviet oceanographers, but by and large their data are not available, nor are their methods reported in detail sufficient to check the validity of stated results.

Three passages separate the Barents Sea from the Arctic Ocean between Spitsbergen and Franz Josef Land. The westernmost, between Northeast Land and White Island (Kvitøya) is about 56 km wide and 300 m deep; between White and Victoria islands the passage is 46 km and >200 m deep; and the easternmost passage between Victoria Island and Alexandra Land (Franz Josef Land) is 100 km and 400 m deep.

Mosby (1938) made an anchored current station extending over 32 hr in the eastern channel. The measured currents (see Table 9), after removing the tidal components (typically ± 15 cm sec^{-1}), could not be adjusted very satisfactorily to geostrophic currents calculated from a hydrographic section made at the same time. Combining all results, including careful water mass analysis of sections, both across the three passages and to the north of them, Mosby concluded that about 0.05 Sv of Atlantic water was moving south from the Arctic Ocean into the Barents Sea, about 0.01 Sv of which passed between White and Victoria Islands and the remainder between Victoria Island and Franz Josef Land. This net southward transport may have been countered to some extent by a northward flow of more surficial Arctic water; there was evidence that this was occurring between Spitsbergen and White Island and on the east side of the Victoria Island—Franz Josef Land passage. Indeed, the 50-m residual current at the anchor station (Table 9) was north.

By chance a 28-hr anchored current station was made from the *Southwind* in 1970, located 15 km south of the *Quest* station in the same water depth (that is, both were on the west side of the deep part of the channel). The residual currents (treated in the same way as Mosby's) are compared in Table 9.

Table 9 Residual Currents in the Victoria Island–Franz Josef Land Passage, in cm sec^{-1} Positive toward 155°T

Depth (m)	1931, July 23–24	1970, August 27–28
10	0.2	5.3
50	−2.1	4.6
100		1.3
150	1.7	

A recent Soviet depiction of the currents in the Barents Sea is reproduced in Fig. 37 (Tantsura, 1959), to which has been added the locations of the *Southwind* anchor stations. For the Spitsbergen–Franz Josef Land passage, one can infer either no net exchange, or only a very small net southward flow. Tentatively, then, Mosby's conclusion that there is only a very small net transport into the Barents Sea from the Arctic Ocean seems to be confirmed by more recent evidence. These results, of course, apply only to summer; what the exchange might be at other times remains unknown.

The transport through the Franz Josef Land–Novaya Zemlya passage is essentially unknown. Timofeyev (1963) reported that the mean annual flux, based on eight traverses between 1932 and 1960 (presumably geostrophic calculations), was 0.55 Sv from the Barents to the Kara Sea. However, Timofeyev's report is inconsistent in that his total water balance shows a mean annual transport through the passage of 1.0 Sv into the Kara Sea. He did not reconcile the discrepancy.

The inference about exchange to be drawn from Tantsura's current chart (Fig. 37) is that there would be little if any net exchange through this passage. In 1970 the *Southwind* occupied a 27-hr anchored current station (marked in Fig. 37). The net flow of the 10-to-30-m layer was southeast at 2 cm sec^{-1}. This is in conflict with Tantsura's presentation, as the current measurements were located in a flow denoted the "Makarov Current" from the Kara into the Barents. Rather, the current measurements are more consistent with a net flow from the Barents to the Kara. We conclude that the transport through the Franz Josef Land–Novaya Zemlya passage remains unknown, but that the probability is of a small annual net flow (<1 Sv) from the Barents to the Kara Sea.

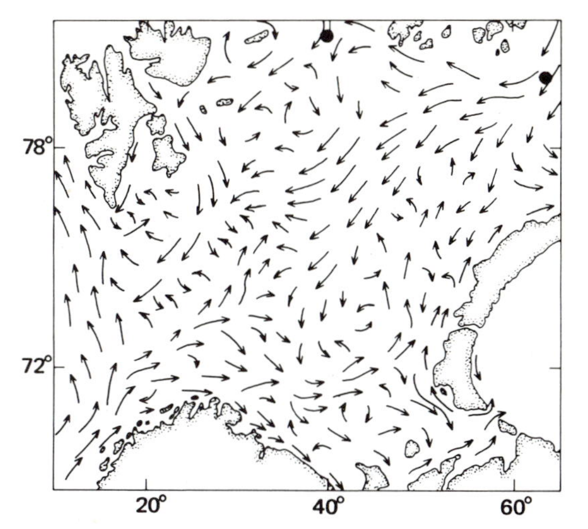

Fig. 37. Surface currents of the Barents Sea (after Tantsura, 1959). Location of two current measurement stations shown by solid circles.

Denmark Strait

A counter-current flow regime, first elucidated by the *Ingolf* expedition results (Knudsen, 1899), obtains in this strait between Iceland and Greenland. The East Greenland Current flows southwest along the Greenland side, with highest velocities over the continental slope, and carries with it masses of ice. A northward flow of relatively warm Atlantic water (Irminger Sea water) occurs along the eastern side, again with highest velocities over the continental slope.

Oceanographic sections across the East Greenland Current are difficult to obtain due to pack-ice being transported south by the current, while sections northwest from Iceland across the Atlantic water inflow are more readily obtainable. Stefansson (1962) reported transports of Atlantic inflow from nine sections made in June, July, or August between 1949 and 1955. The transports calculated from these hydrographic data ranged between 0.2 and 1.0 Sv, with a mean of 0.6 Sv.

The joint Icelandic–Norwegian expeditions of 1963 (Gade et al., 1965) and 1965 (Malmberg et al., 1967) were able to obtain hydrographic sections across the entire Denmark Strait from August to September, together with direct current measurements. The results showed significant flows close to the bottom along both the Greenland and Icelandic continental slopes, in the same directions as the surface flows. Thus, there would seem to be considerable uncertainty involved in choosing a suitable reference level for dynamic calculations, which casts doubt on the validity of transport values calculated from hydrographic sections without supporting direct current measurements.

From the results of the joint expeditions, transports of the East Greenland current southwest over the Greenland shelf based only on geostrophic profiles were ~1.5 Sv, but when these were adjusted to the direct current measurements, the transports were ~3 Sv. The current was composed of two filaments, one inshore over the shelf and the other over the upper part of the slope. Each filament transported about 1.5 Sv, but the transport by the filament on the shelf was much more accurately predicted by the geostrophic calculations.

On the east side of the strait, on the other hand, the authors could not obtain even qualitative agreement between the flow regime calculated from the station data and the current measurements, because the current measurements showed strong vertical shear in the water column over the slope, while the geostrophic profiles gave a vertical shear of one-tenth the magnitude and opposite sign. Northward transport of Atlantic water, based on current measurements, was estimated between 3 and 4 Sv. Thus, the authors concluded that while assessment of the flow regime can be obtained from hydrographic data in the East Greenland Current, this seems impossible in the region of the Icelandic slope.

In view of the above results, it does not seem valid to interpret the earlier sections reported by Stefansson (1962) as good estimates of the actual transport into the Iceland Sea of Atlantic water; rather, it is probable that the calculated transports considerably underestimate the actual transport. We tentatively conclude that the summer transport of the East Greenland Current in Denmark Strait is about 3 Sv, but may vary by, say, a factor of 2. Inflow of Atlantic water along the Icelandic

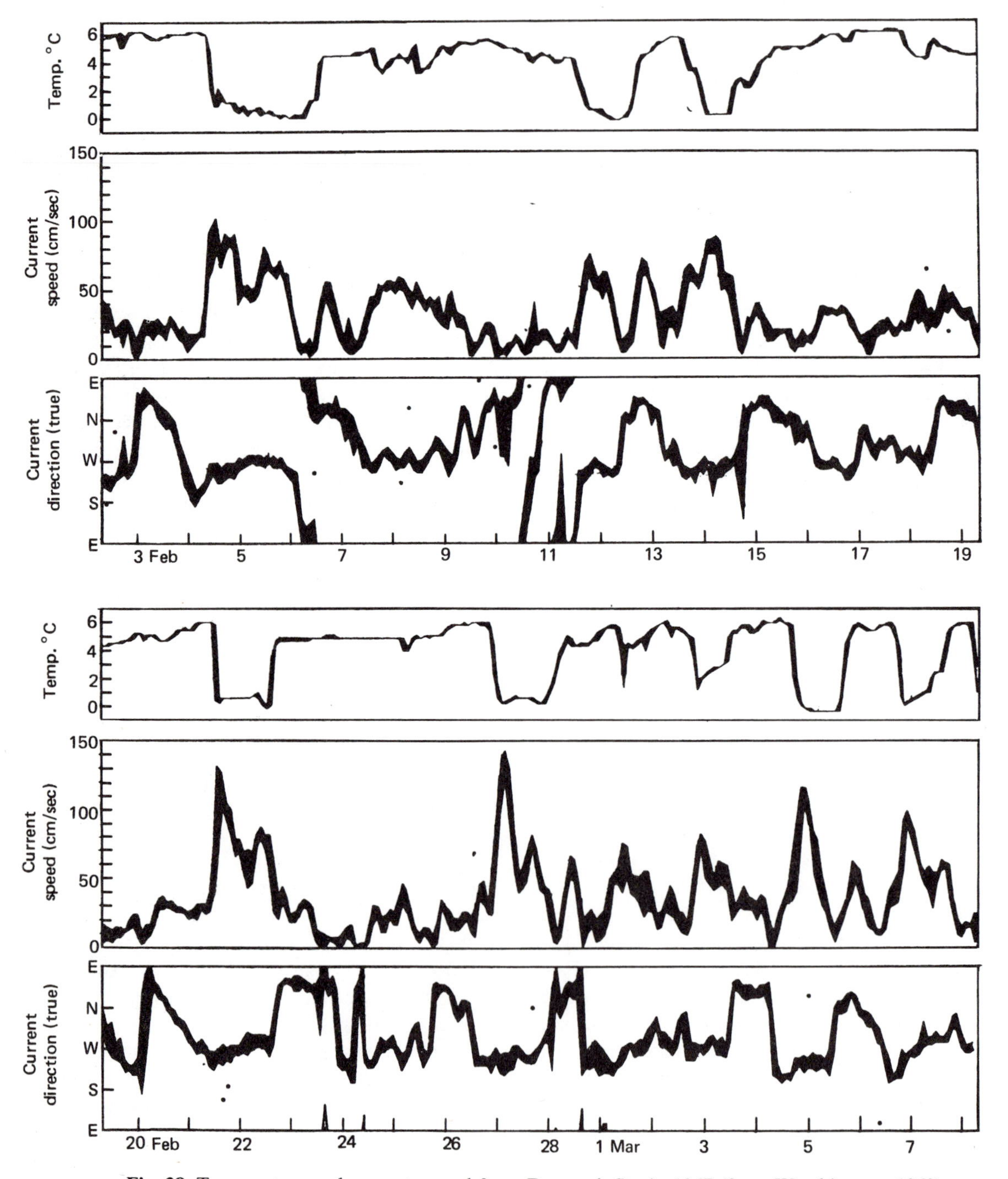

Fig. 38. Temperature and current record from Denmark Strait, 1967 (from Worthington, 1969).

continental slope is even less well known, but is probably of the order 2 to 4 Sv, with considerable variability.

In addition to the counter-current flow regime there is a third flow, namely a deep overflow out of the Iceland Sea, which is intermittent but of significant volume transport. Knudsen (1899) first reported that what was called Norwegian Sea deep water overflowed the sills into the North Atlantic and thereby contributed to some extent to North Atlantic bottom water. This view was supported by Nansen (1912). Cooper (1955) revived interest in the problem through his studies in the Bay of Biscay, which suggested that the overflow was quantitatively significant. It is now well established that overflow water, from both the Denmark Strait and the Iceland–Faeroe–Shetland passage (see below), is the primary source of deep water in the North Atlantic (shown graphically in Worthington and Wright, 1970).

A first attempt to measure the volume transport of the overflow through the Denmark Strait was made from February to March 1967 (Worthington, 1969). Six subsurface moorings with 30 current meters were anchored for 34 days across the deep portion of the Strait just south of the sill. Only two of the moorings were recovered, and of the 10 current meters on the two moorings only one produced reliable results. This current meter, which also had a temperature recorder attached, was located at 760 m depth in the center of the deepest part of the passage. The temperature and current records are shown in Fig. 38.

There is an obvious correlation between low temperature (0°C) and west or southwest flow, and high temperature (4 to 6°C) and north or northeast flow. The cold water is undoubtedly Norwegian Sea bottom water; whereas the warm water is the Irminger Sea water flowing northward around the Icelandic continental slope, of which part continues north through the Denmark Strait as an Atlantic inflow to the Iceland Sea and part recurves westward and southwestward in the Denmark Strait. The overflow is seen to occur intermittently as distinctive pulses, or bursts, which build up strength quite suddenly—in a matter of a few hours—and cease suddenly. Thus, Cooper's (1955) hypothesis, later reinforced by Stefansson (1962), that the overflow occurs intermittently as discrete "boluses" is confirmed by direct measurements.

There were eight distinct pulses during the 34 days, one of 2 days' duration and the rest of about 1 day each. Each pulse was accompanied by a rapid increase to high current speeds (>100 cm sec^{-1}). At the conclusion of each burst the warm water usually flowed rapidly but briefly toward the northeast before resuming its more usual pace, between 20 and 50 cm sec^{-1}. Further analysis showed that even though cold water overflow was occurring only about 25% of the time, about twice as much cold water flowed past the meter out of the Iceland Sea as did inflowing warm water. Also, some warm water flowed southwesterly, which is evidence of the recurving flow mentioned above.

Other observations were available to aid in arriving at an estimate of the transport of the overflow. These included records from two temperature sensors on the same mooring as the current meter, at 655 and 815 m, from which the thickness of the overflow layer could be estimated. The records showed large changes in thickness of the layer—over 50% of the time a layer at least 25 m thick was present, whereas only about 25% of the time did the layer reach to 80 m above bottom where the current meter was located. On occurrence of a pulse, the layer steadily increased in thickness, achieving at least 190 m thickness about one-half day after onset. Pulses terminated more abruptly, as warm water returned to all three sensors at nearly the same time.

The other useful datum was a hydrographic and current section made just prior to the launching of the moorings but extending landward on both ends. The section showed that the core of cold water lay to the west of the channel center (where the retrieved mooring was located), extending up the Greenland slope. The highest water velocities were associated with the cold core. The interpretation by Worthington was that the section was obtained during a relatively slack period. Even so, the measured transport out of the Iceland Sea was at least 2.7 Sv and more probably about 4 Sv. The transport during the pulses was undoubtedly very much larger.

The overflow water is close to 0°C and 34.90‰, and hence is closely related to the Greenland Sea bottom water, which is formed primarily during winter (see above). It is tempting to relate the transport of the overflow to the quantities of bottom water formed, but we submit that this would be too simplistic a view of the processes involved. We conclude that overflow transport through the Denmark Strait is significant and that it is of the same order of magnitude as transport by currents in the upper layers, but that reliable estimates of the average quantities and variations are not possible at this time. Water and heat budgets for the Norwegian–Greenland–Iceland Sea system must therefore include overflow water; the only budget presently available which does so is that of Worthington (1970). We also note that the variability of the net flux through the Denmark Strait, in part due to the intermittent overflow, is likely to be much greater than for some of the other boundaries, e.g., the Bering Strait; that is, perhaps the variability is an order of magnitude rather than a factor of 2.

Iceland–Faeroe–Shetland Passage

This passage is the primary entrance for Atlantic water into the Arctic system. The eastern passage, the Faeroe–Shetland Channel, carries the bulk of the inflow, while the western passage, between the Faeroes and Iceland, has flow in both directions. Some overflow of Norwegian Sea deep water to the Atlantic occurs in selected places.

Tait (1955, 1957) reported on transports calculated from 69 standard hydrographic sections across the Faeroe–Shetland Channel made between 1927 and 1952. Net transports of Atlantic water into the Norwegian Sea ranged between 0.4 and 6.5 Sv, with a mean of 2.3 Sv and a median of 2.0 Sv. In a later paper Tait (1962) discussed in detail various experimental confirmations of his assumption of the 35‰ isohaline as the reference level. For example, agreement between the measured and calculated currents was such as to lend considerable confidence to the transport values. He also discussed other evidence that lends support to the validity of the calculated flows: in 1927 the Fishery Board for Scotland had established a standard hydrographic section along the threshold of the Wyville-Thomson Ridge, from the Butt of Lewis northwest across the Faeroe Bank, so that it lay 150 to 200 km southwest of sections between the Shetland and Faeroe Islands. Frequently the standard section was occupied contemporaneously with one or more sections in the Faeroe–Shetland passage, and agreement in values between 22 paired sections was remarkably close over a wide range (0.4 to 5.6 Sv) of transports. Possible biasing of the calculations by internal waves of tidal period was not treated quantitatively, but because in most cases the time required for occupation of a section was very nearly a multiple of 12 hr and because of the close agreement between paired sections any possible significant bias seemed to be discounted.

Sufficient sections were available from various times of the year to make it possible to estimate seasonal variations in transport. Lee (1963) pointed out that summer transport values, with one exception, were less than the first quartile value of 1.2 Sv, while the autumn-winter values were higher and mostly exceeded the third quartile value of 3.3 Sv. A statistical analysis showed the values obtained during June, July, and August to have a mean of 2.2 Sv and a standard deviation of 1.3 Sv, while those obtained from September through March had a mean of 3.6 Sv and a standard deviation of 1.2 Sv. The means were significantly different at a >99% confidence level. Tait (1962) cautioned, however, that the phenomenon of greater autumn-winter than spring-summer transports was not invariably an annual occurrence, but rather seemed to occur in groups of years—for example, all values taken during the mid-1930's appear to be low, while the period 1949 to 1951 contained the highest values (but also the largest fluctuations). The magnitude of seasonal variations seemed to be typically 40% per month for two or three months during periods of marked change, but between August and September 1951 the transport increased 270%.

Timofeyev (1963) reported results from 79 hydrographic sections made between 1902 and 1959. These data presumably included most of those analyzed by Tait. He did not discuss any evidence to aid in assessing the validity of the calculated transports, nor did he state the reference level used for the calculations other than that the calculations were based on one method, presumably meaning that the same reference level was used for each section.

The transports ranged between 12.9 and 6.6 Sv, with an annual average value of 9.5 Sv. We remark the considerably higher values reported by Timofeyev as contrasted to those of Tait. Moreover, the sense of possible seasonal trends suggested by the results were different. Even though the average for June, July, and August (9.9 Sv) was slightly higher than for September through March (8.8 Sv) (individual values were not reported, so that the difference could not be tested statistically), the minimum values in Timofeyev's transports occurred in February and March (7.3 and 6.6 Sv, respectively, two cases each month). These compare with Tait's values for February and March of 3.3 and 3.5 Sv, one and three cases, respectively, which are high among his values. Table 10 compares the results for spring-summer months, when the greatest number of sections have been occupied.

Table 10 Comparison of the Atlantic Water Flow through the Faeroe–Shetland Channel from Results of Tait (1962) and Timofeyev (1963)

Month	Number of sections	Tait (1962) Transport, Sv	Number of sections	Timofeyev (1963) Transport, Sv
May	14	1.8	16	10.0
June	18	2.5	16	9.6
July	8	2.4	9	7.6
August	6	1.0	13	9.2

The flow regime in the Faeroe Islands–Iceland part of the passage has been less intensively studied. Those of Tait's sections that extended this far west showed the ratio of the volume transport into the Norwegian Sea to the west of the Faroes to that through the Faeroe–Shetland Channel to vary from 1:4 to 2.5:1 (Lee, 1963). Even though these results suggest that the net transport between the Faeroes and Iceland seems to be highly variable, a basic flow pattern appears to exist in general, with inflow of Atlantic water to the Norwegian Sea on the eastern side (west of the Faeroe Islands) and an outflow of mixed water to the Atlantic east of Iceland (see, e.g., Alekseev and Istoshin, 1960).

In 1960 a consortium of European oceanographic institutions, under the aegis of I.C.E.S., began a systematic study of the exchange through this passage. The first expedition of nine vessels from five countries included numerous direct current measurements as well as 18 quasi-synoptic hydrographic sections (Tait, 1967). Most of the sections were parallel to the ridge, both north and south of it, and were repeated three times at about one-week intervals. The motion field calculated from the hydrographic sections showed the flow on both sides of the ridge to be predominantly parallel to the ridge (Bogdanov et al., 1967). This flow pattern for the water above the bottom layer was confirmed by the direct current measurements (Joseph, 1967). Transports on the north side were about 5 Sv to the east and south, part of a circulation in the Norwegian Sea, while south of the ridge transports were west at about 14 Sv. Actual net transport across the ridge could be estimated as only about 1 Sv, with strong mixing between the different water masses passing over the ridge.

The data achieved by this comprehensive program allowed an assessment of the validity of geostrophic transport calculations. The calculations based on sections across the Faeroe Bank Channel can give realistic values, while the method seems to be quite inadequate along the Faeroe–Iceland Ridge. However, in any case care must be exercised in selection of a suitable reference level: concurrent current measurements would appear to be a necessity. Thus, the weight of evidence leads us to conclude that the inflow of Atlantic water takes place largely through the Faeroe–Shetland Channel, but that the annual mean transport is not better known than within about one-half order of magnitude. We are unable to resolve the discrepancies between the different transport calculations. On the one hand there is the masterfully documented work of Tait, with its relatively low transports, and these agree very well with our wind-driven transport calculations for 1965. On the other hand, we intuitively feel that Tait's transports may be low. While the actual transports may not be as high as Timofeyev's values, which are inadequately controlled and could easily be in error by, say, a factor of 2 (due, for example, to use of a fixed-depth reference level), we should at present feel most comfortable with a transport of at least twice that of Tait. We definitely feel that some of the conclusions that have been drawn from Timofeyev's data, for example, an annual cycle of inflow with a maximum ($\sim$12 Sv) in January and February and a minimum ($\sim$7 Sv) in July (Antonov, 1968) are not justified. The variability of the inflow is probably about 50% on a time scale of a month, but many more observations are required before the cycles can be clearly identified.

We also conclude that the inflow across the Faeroe–Iceland Ridge is much less, in general, than through the Faeroe–Shetland Channel, being perhaps only one-fifth to one-tenth as great. Thus, Lee's conclusion (1963) that the ratio of the flows may on occasion be such that 2.5 times more Atlantic water flows in west of the Faeroes than east of them is highly suspect. A similar conclusion about the smallness of the Faeroes–Iceland inflow is implicit in the various water budgets that have been calculated (e.g., Mosby, 1962b; Timofeyev, 1963; Worthington, 1970), none of which include Atlantic water inflow west of the Faeroe Islands.

As in the Denmark Strait, significant amounts of Norwegian Sea water flow out over the sills of the Iceland–Faeroe–Shetland passage. This phenomenon has been known for some time, but recognition of its significance, and hence appropriate research programs to ascertain its magnitude and timing, began only in the last two decades (cf. Lee and Ellett, 1965). Because these overflows take place in layers following the bottom, detailed knowledge of the topography is very important. This has been provided by Harvey (1965) for the southwestern end of the Faeroe Bank Channel, where its cross-section is most restrictive, and so this channel seems now to be adequately described in detail. Its axis trends northeast-southwest between the Shetland and Faeroe islands, but south of the passage it bends sharply to the northwest and runs into the North Atlantic basin to the north of Faeroe Bank (between Faeroe Bank and the Faeroe Islands). In this region occurs the minimum cross-sectional area: it is about 28 km wide at 400 m depth, with a sill depth of 830 m.

Recently Crease (1965) summarized all the available observations providing quantitative estimates of the deep transport. He concluded that overflow through this channel was a major contribution to Norwegian Sea water in the Atlantic, but that unlike the overflows in the other areas (the Denmark Strait, the Faeroe–Iceland Ridge) which were very likely intermittent in nature, there was little doubt that through the Faeroe Bank

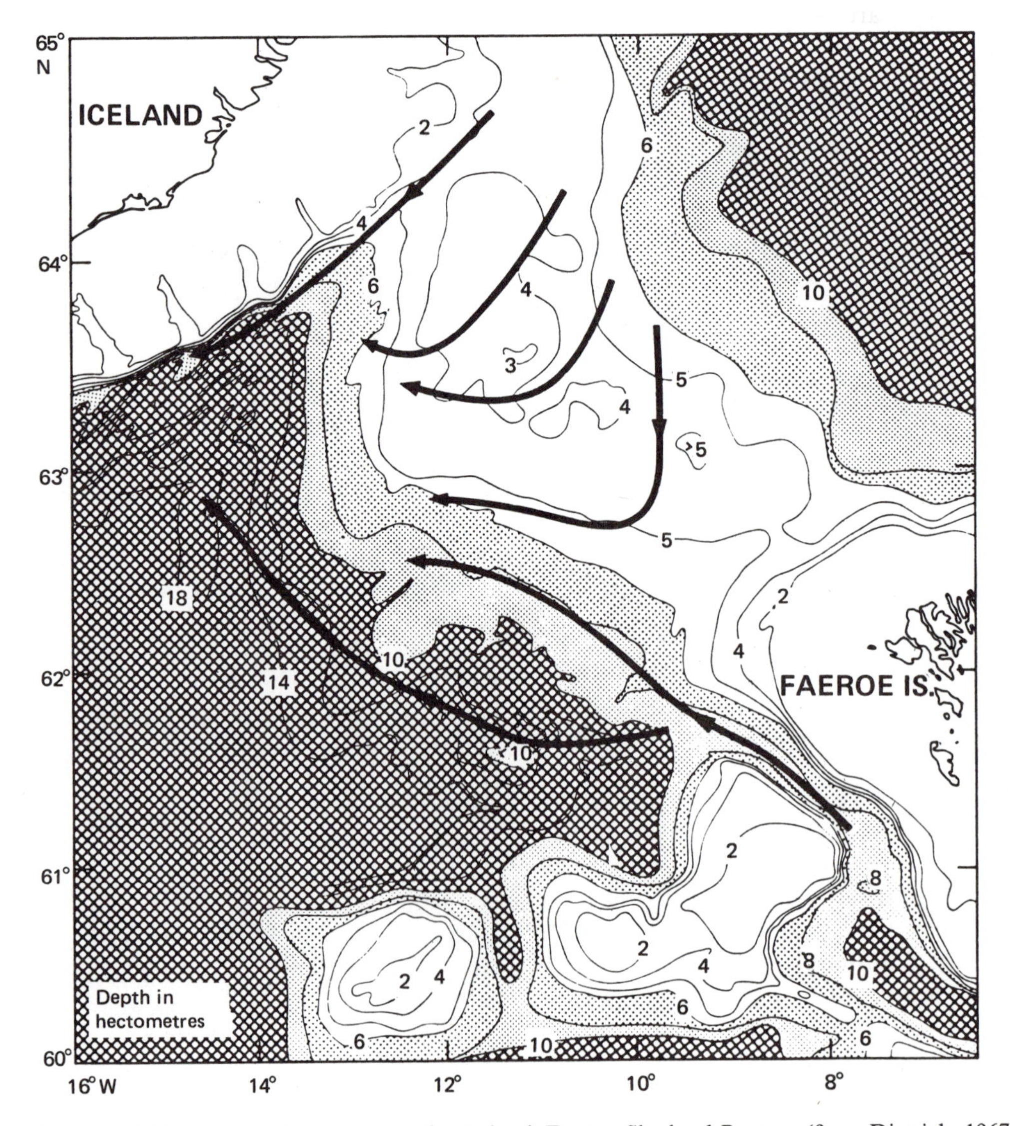

Fig. 39. Cores of the main overflow through the Iceland–Faeroe–Shetland Passage (from Dietrich, 1967).

Channel it was continuous. The magnitude of the transport was 1.3 to 1.8 Sv, composed of two-thirds undiluted Norwegian Sea water (34.92‰, −0.5°C) and one-third overlying North Atlantic water (35.33‰, 9°C). This estimate seems firmly based on hydrographic data and current measurements. The results from the 1960 Overflow Expedition gave high and low estimates of 1.9 and 0.9 Sv, with a mean of 1.4 Sv (Hermann, 1967). The results from Norwegian investigations in June and July, 1965, using hydrographic data and current measurements, gave transports of 1.8 Sv beneath the 35.2‰ isohaline and 1.1 Sv beneath the 35.0‰ isohaline (Saetre, 1967).

The 1960 Overflow Expedition is the only program so far undertaken to include sufficient measurements to define the flow over the Iceland–Faeroe Ridge quantitatively. The overflow was found to exist as four cores (Fig. 39), located above the deep places (400 to 500 m) in the relatively rough ridge topography. Each core was estimated to transport between $\frac{1}{4}$ and $\frac{1}{3}$ Sv, giving total high and low estimates of 1.4 and 0.7 Sv, and a mean of 1.1 Sv. The positions of the cores did not change materially between the three surveys of the expedition. The magnitudes did exhibit some fluctuation (for example, between surveys one and two the overflow in the northwestern part of the section increased, while that through the Faeroe Bank Channel decreased), but the nature of the pulsations could not be determined (Dietrich, 1967).

After transiting the ridge, the flows turn to the right following the bottom contours. Ultimately the overflows apparently combine and flow southwest parallel to the Icelandic continental slope. Steele et al. (1962) measured the transport of this flow in October 1960 to be 1.4 Sv. However, Hermann (1967) pointed out that their measurement applied only to pure Norwegian Sea deep water, whose source would have been the Faeroe Bank Channel; therefore the value was probably an underestimate of the total overflow, because the more mixed character of the Iceland–Faeroe Ridge overflow water was neglected.

We conclude that a more or less continuous overflow through the Iceland–Faeroe–Shetland passage seems well established. Its magnitude seems to be about 2.5 Sv, with somewhat more transiting the Faeroe Bank Channel (~1.5 Sv) than the Iceland–Faeroe Ridge (~1 Sv). These overflows appear to fluctuate much less than the highly intermittent pulsations characteristic of the Denmark Strait overflow. They do vary in time, though, perhaps by a factor of 2, but more definitive conclusions must await further studies of the type undertaken by the 1960 Overflow Expedition.

Greenland–Spitsbergen Passage*

About 600 km separate Greenland and Spitsbergen along the parallel 79°N. The depth exceeds 2000 m over 40% of this distance, so that in total cross-section the passage is by far the most important connection between the Arctic Ocean and the rest of the world ocean. It is also the only deep connection (~2600 m), for the next deepest is Nares Strait, with a sill depth of only 250 m. Finally, all heat and mass budgets agree on the Greenland–Spitsbergen passage being the primary avenue of exchange with the Arctic Ocean. In the Soviet literature the passage is usually referred to as Fram Strait, but this term is not in use elsewhere.

In estimating the exchange of water through the passage, there is a recurring difficulty, namely the continual recurving westward and southward of water from the West Spitsbergen Current. This makes it practically impossible to combine, into a single budget, dynamic sections of the East Greenland and West Spitsbergen currents taken at different latitudes. Rather, what is required are sections across the entire passage, and these are rare.

Let us first consider the East Greenland Current. To our knowledge no budgets thus far constructed have considered outflow of Atlantic water from the Arctic Ocean with this current. We have already seen that most of the Atlantic water transport by the East Greenland Current represents an internal Greenland–Norwegian Sea circulation. However, it is inconceivable to us that there is not some outflow of Atlantic water from the Arctic Ocean also (Fig. 19), and we find it difficult to believe that such a transport is completely negligible in the mass and heat budgets. For example, even if only 10 to 20% of the transport of Atlantic water estimated from the Arlis II data represented Arctic Ocean outflow, this would amount to 3 to 6 Sv; if this is the case, it is larger than the probable total flow through any other connection with the Arctic Ocean. The point is that while we may in reality have no valid

* See Note 8 at the end of this chapter.

estimate of this transport, it most certainly cannot *a priori* be considered negligible.

All the budgets, however, consider the transport of Polar water to be important. The estimates vary considerably, but are typically 2 Sv (Mosby, 1962b) to 4 Sv (Timofeyev, 1963). In a number of cases the East Greenland Current transport estimate has been obtained by difference, i.e., by treating it as the unknown quantity in the steady-state budget equations. There are also some summer hydrographic sections across the current, and while these indicate transports of the above magnitude, they also point toward complexities in the dynamic topography that greatly complicate transport interpretations.

As an example, Fig. 40 (Palfrey, 1967) shows transport calculations based on the 0/600-db dynamic topography, using the *Edisto* observations from August and September, 1964. There was a large anticyclonic eddy clearly developed in the northwesternmost part of the sea, which had a calculated transport considerably larger than the southward flow across 78°N or 79°N between the prime meridian and 14°W. Had the only section been the one along 80°N, we should have been faced with trying to account for an excessive calculated outflow from the Arctic Ocean of 5 Sv in the upper 600 m. That northward counterflow exists in the area was first pointed out by Riis-Carstensen (1938), and Laktionov et al. (1960) argued for the presence of the eddy as a semi-permanent feature.

We are also faced with synthesizing the budgets in a manner consonant with the Arlis II observations. These showed a Polar water transport of 7.7 Sv, but since the traverse of the East Greenland Current was incomplete, the actual transport could easily have been near 10 Sv. Most, if not all, of this water must represent outflow from the Arctic Ocean. It is quite possible that seasonal transport fluctuations are large, and Helland-Hansen and Koefoed (1907), Trolle (1913), Chaplygin (1960), Laktionov et al. (1960), and Timofeyev (1962) have all discoursed on the subject. Our opinion, however, is that at present one can say nothing substantial about seasonal variations. From consideration of month-to-month variations in the wind stress curl during 1965, which showed a very different wind regime prevailing during May through July than during the rest of the year (Aagaard, in press, a), one might conceivably argue that the transport during summer should be much less than during winter. However, without firm oceanographic evidence we wish only to point out that even if such a seasonal variation prevailed, an annual mean outflow of

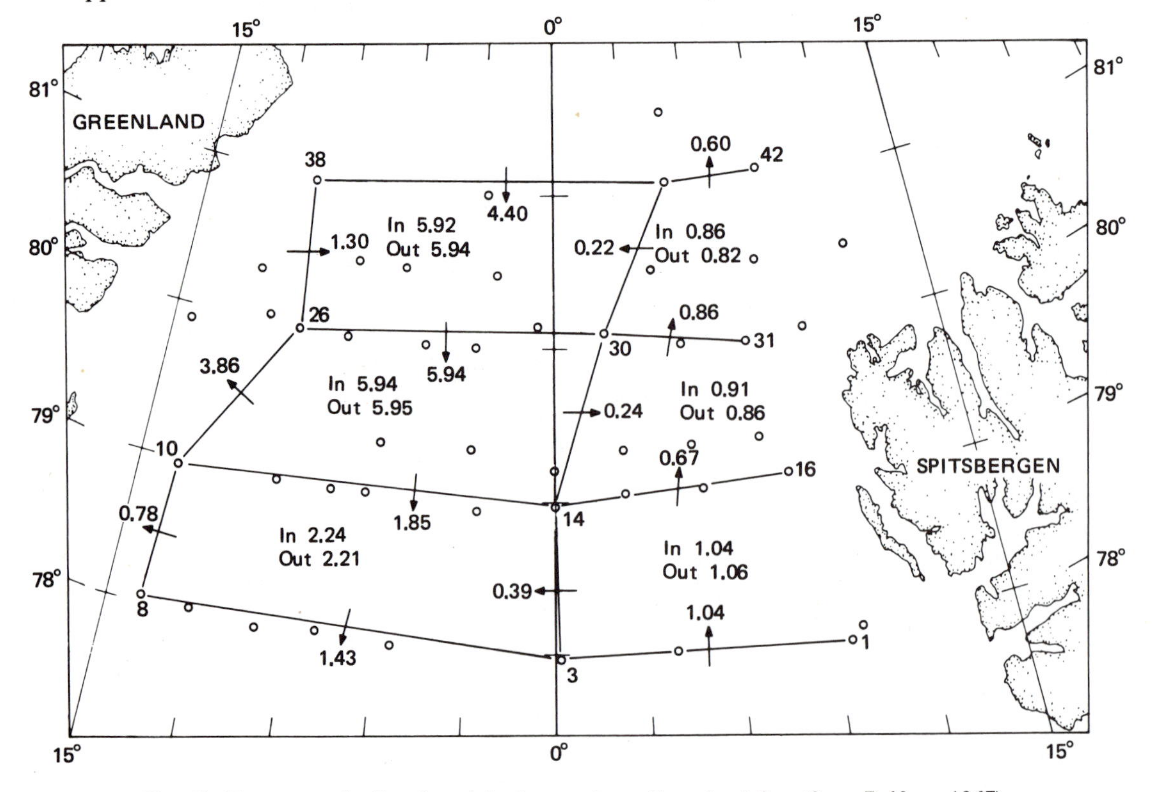

Fig. 40. Transports (in Sverdrups) in the northern Greenland Sea (from Palfrey, 1967).

Polar water from the Arctic Ocean of 6 Sv (8 months at 8 Sv and 4 months at 2 Sv) is conservatively low. If, in addition, we suppose an outflow of Atlantic water of at least 1 to 2 Sv, then one might tentatively expect a total annual outflow through the Greenland–Spitsbergen passage of at least 7 to 8 Sv, or about 2 to 3 times greater than present estimates.

The inflow to the Arctic Ocean of water via the West Spitsbergen Current is equally difficult to assess quantitatively. Budgets are unanimous in assigning it the primary role in the total inflow, and hence it must very closely match the outflow via the East Greenland Current, for the net exchange through the other passages to the Arctic Ocean appears to be quite small. There are a relatively large number of hydrographic sections across the current, and the transport estimates based on dynamic computations vary considerably. Even if we restrict ourselves to those sections made at latitude 78°N or higher, to minimize the problem of a recurving westward flow, the transport estimates still vary from 1.4 Sv (Mosby, 1938, whose dynamic calculations were adjusted to direct current measurements) to over 4 Sv (e.g., Timofeyev, 1963, whose calculations represent multi-year means of dynamic sections). If, on the other hand, the outflow on the western side of the passage is as great as 7 to 8 Sv, then either the inflow on the eastern side is also about the same, or there is a very large inflow through the Spitsbergen–Franz Josef Land–Novaya Zemlya passages. The former alternative seems to us more probable.

The most definitive conclusion possible at this time is that the inflow and outflow through the Greenland–Spitsbergen passage are about equal, but whether their annual mean values lie closer to 4 Sv than to 8 Sv cannot presently be determined.

Finally, there is the problem of deep water inflow to the Arctic Ocean. Already in 1902, Nansen had concluded that the bottom water of the Arctic Ocean entered through the Greenland–Spitsbergen passage. Arguments have since been advanced purporting to show that significant quantities of bottom water could be formed within the Polar Basin (e.g., Shirshov, 1944), but these are demonstrably fallacious (cf. Coachman, 1963), and there is no doubt but that the Arctic Ocean bottom water enters through the Greenland–Spitsbergen passage. The amount of inflow is again problematic, and the only published estimate is that due to Mosby (1962b) of 0.6 Sv; it was incorporated into his budgets. We have embarked on a program of current measurements in the passage and hope to be able in the future to elucidate these transports, but at the present time any estimates of the flow of deep water are purely conjectural.

We do, however, feel that the deep flow through the passage may be of an intermittent nature, perhaps much like the Denmark Strait overflow. The evidence is sparse and indirect, but tantalizing: during the August 1962 *Atka* cruise three observations were made of abnormally cold water near the bottom north of 78°N. Figure 41 shows an example, depicting the temperatures along the prime meridian from the Mohn Rise to the northern edge of the Greenland Sea Basin. The presence of an isolated (at least in the plane of the section) patch of abnormally cold water at 79°N is striking. Water of this temperature (corresponding to a potential temperature of −1.16°C) was characteristic of the core of the deep water farther south. All three observations of such cold patches during the *Atka* cruise were made within about 100 m of the bottom, and could be interpreted as showing cold pulses moving northward into the Arctic Ocean. The matter certainly merits further investigation.

Addendums

Note 1. During July 1972, the *Oshoro Maru* (data as yet unpublished) made three detailed sections of anchored stations from the Alaskan coast westward across the Chukchi Sea. One was across Bering Strait, one west from Cape Lisburne toward Cape Schmidt, and one between Point Barrow and Herald Island. The current measurements showed the northward flow of Bering Sea water to be branching near Point Hope; at that time the transport through the Cape Lisburne section was 1.3 Sv, with approximately one-third moving northwest toward Herald Canyon and two-thirds northeast toward Barrow Canyon.

Transport through the Barrow-Herald Island section was 1.9 Sv into the Arctic Ocean. Of this, 0.5 Sv represented water from the East Siberian Sea (cold, but relatively saline) moving through the western part of the section. In the eastern part of the Barrow section the deeper layers were not of Bering Sea characteristics as in the Lisburne section, but rather closely akin in T-S correlation to the Arctic Ocean subsurface water north of

Point Barrow. Our tentative explanation is that the general northerly flow along the Alaskan coast is intermittent; and that during some prior period, southerly flow had brought subsurface Arctic Ocean water into the region southwest of Point Barrow where it was somewhat warmed. At the time of the *Oshoro Maru* measurements this water was returning to the Arctic Ocean. We have now begun a program utilizing long-term moored current meters to investigate the time-dependent flow in Barrow Canyon.

The *Oshoro Maru* results thus provide confirmation of the pattern of entry of Bering Sea water into the Arctic Ocean as described earlier. The speed of advance deduced from the water property changes (Table 2) appears, however, to be much too low.

Note 2. The oceanographic experiments undertaken as part of the AIDJEX Pilot Studies of 1971 and 1972 have provided many new data on the hydrographic field and associated currents on scales of 10 to 100 km (cf. articles by Newton and Coachman in *AIDJEX Bulletin*, Nos. 18 and 19, 1973).

The results confirm the high degree of geostrophic balance achieved by both the surface water and the ice on time scales of two days and longer. Calculations of force balance on the ice showed the horizontal pressure gradient force and Coriolis force to be always nearly equal and opposite. On days when accelerations indicated them to differ by >10%, adjustments were such as to restore the balance within two days.

In the direction of ice motion the wind stress was presumably balanced by the total ice resistance; however, the relationship between the various forces differed markedly in the two years. In 1971, when the study area was 100 km west of Banks Island, winds were relatively strong, and ice resistance was also great. The ice moved at only about 1% of wind speed, the wind stress was about five times greater than the Coriolis and pressure gradient forces, and the ice motion and wind stress were nearly aligned. In 1972, 300 km north of Point Barrow, ice speeds fluctuated around 2% of wind speed, the stress vector was about the same magnitude as the Coriolis and pressure gradient vectors, and its direction averaged 30° to the left of the motion.

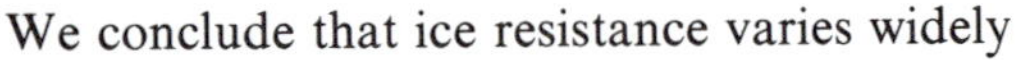

We conclude that ice resistance varies widely

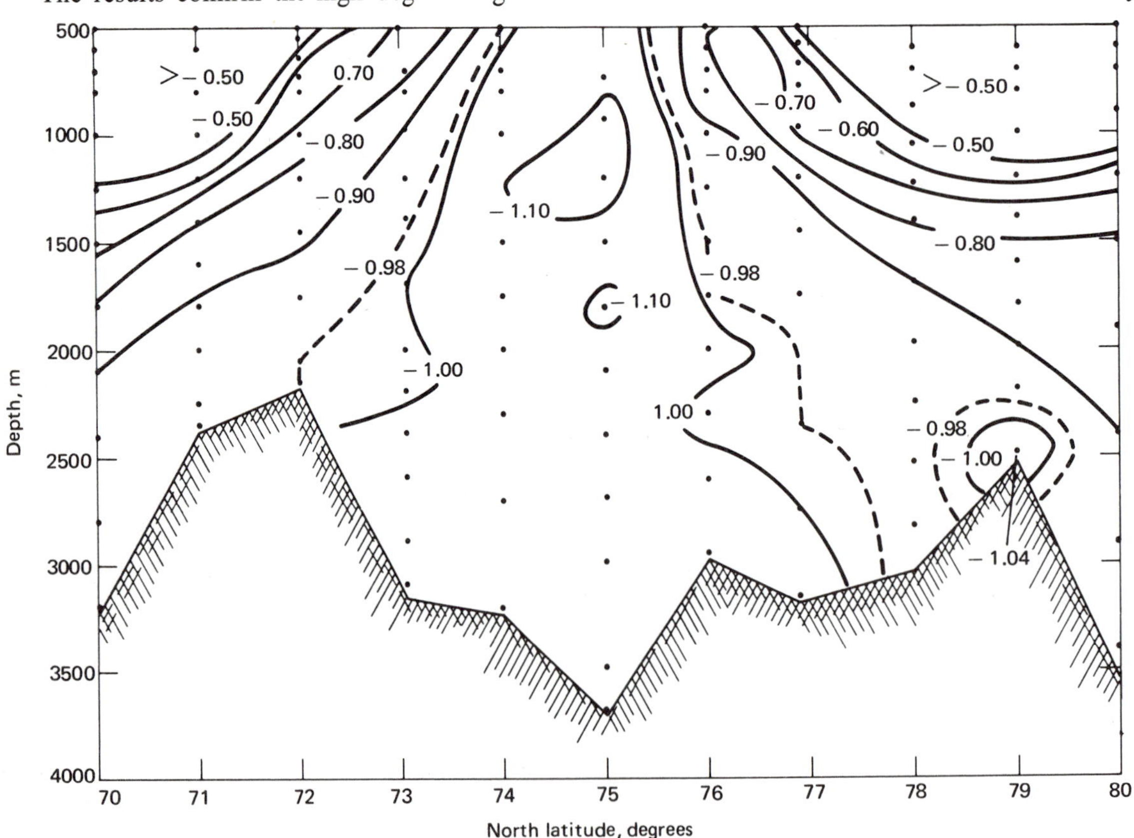

Fig. 41. Temperature section along the prime meridian, *Atka,* summer 1962.

over the ocean, and that this in turn has an influence not only on the ice motion, but probably also on the effectiveness of stress coupling between wind and water. The importance of ice resistance may in part account for the fact that Arctic Ocean currents appear to be slower in general than currents in other oceans.

Note 3. One of our graduate students, John Newton, has derived a new concept of the relatively high-speed currents occasionally observed in the pycnocline. Data from the 1972 Pilot Study revealed three occurrences of the high-speed flows, one as great as 60 cm sec^{-1}. The water moving at higher speeds differed in its T-S correlation from the ambient water, suggesting lenses imbedded in the pycnocline. These lenses are associated with baroclinic pressure gradients and appear as quasi-geostrophic eddies. The time-dependent appearance of the flow in the ice station records derives from the advection of the current meters across the eddies; i.e., the eddies change location and characteristics at a much slower rate than the movement of the measurement platform.

All recorded occurrences of high-speed pycnocline flow have been reviewed in this light, and the data are consistent with the model. The eddies have diameters of a few tens of kilometers. The isopycnals of the upper and lower pycnocline region are appropriately bowed up and down across the eddies, giving rise to baroclinic currents with speeds and directions corresponding to those observed. The eddies are apparently being advected around the Canadian Basin with the mean flow.

We have not as yet been able to elucidate the sources and/or causes of these eddies. There appear to be two possibilities: time-dependent injection of appropriate density water near the basin perimeter, and/or the growth of lenses within the basin through some dynamic instability.

The presence of pycnocline eddies could be associated with the occasionally observed presence of Bering Sea water in the southeastern part of the Beaufort Sea. In spring 1972 the three observed eddies were more or less aligned from southwest to northeast, in effect forming a trough in the dynamic topography of the pycnocline region. The trough, much like the spoke of a wheel radiating from the center of the Beaufort Sea gyre, could conceivably provide a preferential pathway for the ingress of Bering Sea water injections from the southwest.

Note 4. From recent current measurements in the Chukchi Sea (Coachman and Tripp: Currents north of the Bering Strait in winter, *Limnology and Oceanography*, 15(4):625–632, 1970) and the Greenland–Spitsbergen passage (cited in Note 5), all of which show rather strong diurnal cycles, we now believe that the diurnal tidal waves from the Atlantic are quite important to understanding the total Arctic tidal regime. For example, the deep tidal currents (at 1360 m) between Greenland and Spitsbergen, which is the major tidal wave entry into the Arctic Ocean, typically show a diurnal inequality of 50% or more. An even more extreme example was found at the other end of the Arctic, in the eastern Chukchi Sea, where the diurnal variation completely dominated the tidal currents during the four-day measurement period. Conceivably, the current meter in that instance was near an amphidromic point for the M_2 wave, thus distorting the relative amplitudes; but this does not negate the fact that we can in general expect important diurnal tidal effects. South of the Bering Strait the diurnal effects are frequently dominant; but in that region the waves come from the Pacific.

In proofreading the original manuscript, we note a perhaps unfortunate choice of phrase in discussing the tidal range in the vicinity of the shelf. Certainly the primary effect of the wide and shallow shelves must be to frictionally dampen the tidal wave; the diminishing tidal range in the direction of propagation of the wave on the Siberian shelf has been shown by Sverdrup (Dynamic of tides on the north Siberian shelf, *Geofysiske Publikasjoner*, 4(5), 1926). Our original reference was to another effect, also shown by Sverdrup; namely, that in certain areas, notably north of the New Siberian Islands, the tide travels southeastward with a component parallel to the shelf, and its amplitude resembles that of a Kelvin wave, diminishing exponentially away from the shelf. There also appear to be local convergence effects.

Note 5. In August 1972, we recovered four current meters deployed the previous summer in the Greenland–Spitsbergen passage at about 79°N. (Aagaard, Darnall and Greisman: Year-long current measurements in the Greenland–Spitsbergen passage, *Deep-Sea Research*, in press.) Two meters (at 150 and 500 m depth) were located on the continental slope within the West Spitsbergen Current. The maximum mean hourly speed of 86 cm sec^{-1} was recorded at the upper meter. The annual average current at that meter was 19.4 cm sec^{-1} to the north, and at 500 m typically 40% less. The mean monthly currents all set north, with an absolute maximum at 150 m in November (31.2 cm sec^{-1}), a relative minimum in February (8.4 cm sec^{-1}), a secondary maximum in April (22.7 cm sec^{-1}), and an absolute minimum in August (1.4 cm sec^{-1}, but averaged over only 10 days). These variations are in disagreement with all previous discussion of seasonal cycles of currents in the area, but there is, of course, no guarantee that the observed variations are representative of a "normal" yearly cycle. While the average monthly current was always northerly, periods of southerly flow lasting from a few hours to several days occurred during every month except September. The southerly velocity component was as high as 40 cm sec^{-1}, and frequently above 20 cm sec^{-1}; such flow reversals appear to be rather coherent vertically. Hydrographic stations taken during the early part of the current record indicate approximate geostrophic equilibrium in the West Spitsbergen Current.

Two other current meters (at 120 and 1360 m depth) were located in the middle of the passage. The maximum mean hourly speed at the upper meter was 60 cm sec^{-1} and at the deep meter 31 cm sec^{-1}. The annual average deep current was 5.6 cm sec^{-1} to the southeast, but the mean monthly vector varied considerably, with maxima in November (11.5 cm sec^{-1} toward the southeast) and June (11.1 cm sec^{-1} toward the south southeast) and an absolute minimum in January (1.9 cm sec^{-1} toward the east). The monthly mean speeds at the upper meter were on the whole comparable, although the directions were frequently quite different and sometimes even opposite. The mean set toward the southeast cannot be reconciled with the prevailing idea that this region constitutes the northern portion of a cyclonic circulation, unless a purely local phenomenon was being registered.

We are presently involved in an analysis not only of these current measurements, but also of 13 sets of drogue drifts distributed throughout the Greenland Sea during summer 1972. The drogues were set as deep as 1000 m, were deployed in pairs at different depths for direct vertical shear determination, and were typically of two days' duration. Triads of hydrographic stations around the drift tracks were made for geostrophic shear computation.

Note 6. We have recently completed an analysis of the Greenland Sea hydrography, with special emphasis on the deep water formation process (Carmack and Aagaard: On the deep water of the Greenland Sea, *Deep-Sea Research*, in press). A volumetric T-S analysis indicates a seasonal variation in the deep water volume of 30×10^3 km^3, corresponding to a summer outflow rate of 3 Sv. Presumably the outflow of deep water in late winter is even larger.

The indications mentioned earlier, that the deep water is formed at a subsurface level, are substantiated by a wide variety of evidence, including volumetric analysis, heat budgets, oxygen budgets, core analysis, and mixing histories. The formation process appears to proceed as follows. During winter increased amounts of Polar water are transported eastward in the surface layers of the Greenland Sea by the Jan Mayen Polar Current. Presumably this is a response to the increased cyclonic wind torque during winter. The winter data also indicate a subsurface movement of Atlantic water into the central Greenland Sea from the southeast. Whether the inward movement of Atlantic water represents a thermally forced advection, a decoupling of the normal peripheral flow from the bathymetry by a deep countercurrent, or some other mechanism is at present a matter for speculation. The fact is that in winter a layer of cold, low-salinity Polar water overlies relatively warm, saline Atlantic water in the central Greenland gyre. Static stability is maintained by a sharp halocline between the two layers. Core analysis then shows that the Atlantic water, identified by its salinity maximum, gradually loses much of its heat, but only a relatively small amount of salt. The upper freely convecting layer effectively transfers the heat to the atmosphere without the Atlantic water ever being at the surface. The new deep water is thus formed from the Atlantic water which, having lost heat and thus increased in density during its travel toward the center of the gyre (but largely retaining its oxygen and salinity), finally becomes dense enough to drive the deep convection. The mechanism for differential transfer of salt and heat is no doubt ultimately associated with the difference in molecular diffusivity of the two properties (double-diffusion), but the exact dynamics are unknown, for a strict application of present theoretical and laboratory results to the problem is not appropriate. Nevertheless, it is of interest that the ratio of effective turbulent diffusion coefficients for salt and heat calculated from a core analysis of the winter hydrographic data is about 0.3.

With these ideas in mind we might consider the Greenland and Norwegian Seas to be a thermodynamic couple containing a partially closed thermohaline circulation. In the Greenland Sea, Atlantic water is transformed into deep water; and in the Norwegian Sea, deep water is transformed into Atlantic water in accordance with Mosby's (1962a) model. The circulation is only partially closed since there is considerable exchange of each water mass with the adjacent seas, particularly the North Atlantic.

Note 7. The 1972 *Oshoro Maru* cruise completed our ninth detailed Bering Strait section. We have also calculated the transports through all closed cross-sections between Alaska and Siberia based on the measurements previously cited, and the values are assembled in Table 11. All values can be treated as comparable because runoff into the region between St. Lawrence and Cape Lisburne is at least three orders less than the oceanic transport. Our previous estimate that the average north transport is >1 Sv seems to be correct.

Table 11 Transport (Sv; + north) through All Closed Sections Between Alaska and Siberia

Date		Transport, Sv	Location
1964	August 5–6	+1.4	Bering Strait
1967	July 14	+2.0	Bering Strait
	July 15	+0.3	Shishmaref
	July 16–17	0	Point Hope—C. Serdtse-Kamen
	July 18	+0.7	King Island
	July 19	−0.2	Bering Strait
	July 20–21	0	Shishmaref
	July 22	−0.1	Point Hope—C. Serdtse-Kamen
	August 4	+2.2	Bering Strait
	August 5	+1.6	Bering Strait
	August 6	+2.0	Bering Strait
	August 7	+1.4	Bering Strait
1968	July 9–11	+1.9	Strait of Anadyr; St. Lawrence—C. Romanof
	July 13–14	+1.2	C. Rodney
	July 15	+1.4	King Island
	July 17–18	+1.5	Bering Strait
	July 18–19	+2.2	Shishmaref
1972	July 24–25	+1.7	Bering Strait
	July 27–28	+1.3	C. Lisburne—C. Schmidt

The conclusion that major variations in transport are associated with major modifications in surface slope through the strait are confirmed by calculations based on the 1967 data. For the section of July 19, 1967 (Fig. 34), when transport was 0.2 Sv south (using the same dynamical terms as for the 1964 data; Coachman and Aagaard, 1966), the surface slope was about 6×10^{-7} down to the south.

We are now investigating the regional atmospheric pressure conditions vis-a-vis the observed transport fluctuations. The conditions obtained during times of extra-large north transport (>2 Sv) were: (a) pressure in the north (Cape Schmidt) at least 10 mb lower than in the south (Nome); and (b) pressure decreasing westward across both the Chukchi and northern Bering Seas. The conditions for reduced northerly flow were the reverse of those for large flow, and in addition the negative transports occurred when the regional pressure had fallen precipitously to low (>1000 mb) values. It appears that an effective mechanism for reducing the pressure head on the system is a strong atmospheric low southeast of Nome such that large amounts of water are swept south out of the northern Bering Sea. For increasing the pressure head, a low over the East Siberian Sea such that water is swept north out of the Chukchi Sea appears to be very effective. We are attempting to relate these factors quantitatively and to elucidate the dynamics.

Note 8. Our recent current measurements (quoted in Note 5) indicate an annual mean northward transport through the eastern part of the Greenland–Spitsbergen passage above 1000 m of 8 Sv. This result argues strongly for our earlier conclusions that

(1) the inflow through the eastern side of this passage is about the same as the outflow through the western side (and therefore the net inflow through the Spitsbergen–Franz Josef Land–Novaya Zemlya passages must be about an order of magnitude smaller), and

(2) the annual mean inflows and outflows are each about 8 Sv, with the outflow being partitioned between Polar and Atlantic water about as envisioned earlier.

Acknowledgments

It is a genuine pleasure for us to acknowledge the support over many years of Codes 480 and 415 of the Office of Naval Research. Not only has ONR financially supported our Arctic oceanographic research, but there has also been a flow of advice, encouragement, and guidance which continues to be deeply appreciated. This paper was prepared under contract N-00014-67-A-0103-0014.

References

Aagaard, K. 1964. Features of the physical oceanography of the Chukchi Sea in the autumn. Masters Thesis, University of Washington, Seattle. 41 pp.

———. 1968. Temperature variations in the Greenland Sea deep water. *Deep-Sea Res.* 15:281–296.

———. 1969. Circulation of the Greenland Sea. *Proceedings, 5th U.S. Navy Symposium on Military Oceanography. Oceanographer of the Navy* 1:335–363.

———. 1970. Wind-driven transports in the Greenland and Norwegian Seas. *Deep-Sea Res.* 17:281–291.

———. in press a. The wind-driven circulation of the Greenland and Norwegian Seas and its variability. *Rapp. Proc.-Verb. Cons. Int. Explor. Mer.*

———. in press b. On the drift of the Greenland pack ice. In: *Proceedings of the International Sea Ice Conference, Reykjavik, Iceland.*

———, and L. K. Coachman. 1968a. The East Greenland Current north of Denmark Strait: Part I. *Arctic* 21(3):181–200.

———, and L. K. Coachman. 1968b. The East Greenland Current north of Denmark Strait: Part II. *Arctic* 21(4):267–290.

Alekseev, A. P., and B. V. Istoshin. 1960. Nekotoriye resultati okeanograficheskikh issledovanii v Norvezhskom i Grenlandskom moryakh. *Sovetskiye Ribokh. Issled. v Moryakh Evropeiskogo Severa.* Moscow: Vniro-Pinro. pp. 23–37.

Antonov, V. S. 1964. Noviye danniye o velichine zhidkogo stoka Sibirskikh rek, vpadayushikh v Arkticheskie morya. *Problemy Arktiki i Antarktiki* 17:73–6.

———. 1968. The nature of water and ice movement in the Arctic Ocean. *Ark. i Antark. Nauchno-Issled. Inst., Trudy* (transl.) 285:154–182.

Arctic Institute of North America. 1967. Oceanography. In: *Naval Arctic Manual ATP-17(A). Part I: Environment.* Montreal: Department of National Defense.

Arkticheskii Nauchno-Issledovatel'skii Institut. 1946. Expeditsiya na samolete "SSR-N-169" v raion "polyusa nedostupnosti." *Nauchniye resul'taty.* Moskva: Glavsevmorput. pp. 116–118.

Badgley, F. I. 1966. Heat balance at the surface of the Arctic Ocean. *Proceedings Symposium on the Arctic Heat Budget and Atmospheric Circulation.* Lake Arrowhead, California, January 31–February 4, 1966. Santa Monica: The Rand Corporation, Memoir RM-5233-NSF. pp. 267–278.

Bailey, W. B. 1957. Oceanographic features of the Canadian Archipelago. *J. Fish. Res. Bd. Canada* 14:731–769.

Beal, M. A. 1968. The seasonal variation in sea level at Barrow, Alaska. In: J. E. Sater, Coord. *Arctic Drifting Stations.* Washington, D.C.: Arctic Institute of North America. pp. 327–341.

———, F. Edvalson, K. Hunkins, A. Molloy, and N. Ostenso. 1966. The floor of the Arctic Ocean: geographic names. *Arctic* 19(3):215–219.

Bloom, G. L. 1964. Water transport and temperature measurements in the eastern Bering Strait, 1953–1958. *J. Geophys. Res.* 69:3335–3354.

Bogdanov, M. A., G. N. Zaitsev, and S. I. Potaichuk. 1967. Water mass dynamics in the Iceland–Faeroe Ridge area. *Rapp. Proc.-Verb. Cons. Int. Explor. Mer* 157:150–156.

Brayton, G. E. 1962. Station Arlis-I Oceanography. Scientific Report, University of Washington Department of Meteorology and Climatology, Parts 1, 2.

Campbell, W. J. 1965. The wind-driven circulation of ice and water in a polar ocean. *J. Geophys. Res.* 70:3279–3301.

Chaplygin, E. I. 1960. O dinamike vod vostochno-Grenlandskogo techeniya. *Problemy Arktiki i Antarktiki* 5:15–19.

Coachman, L. K. 1963. Water masses of the Arctic. *Proceedings of the Arctic Basin Symposium.* Washington, D.C.: Arctic Institute of North America. pp. 143–167.

———. 1966. Production of supercooled water during sea ice formation. *Proceedings of the Symposium on the Arctic Heat Budget and Atmospheric Circulation,* Lake Arrowhead, California, January 31–February 4, 1966. Santa Monica: The Rand Corporation Memoir RM-5233-NSF. pp. 497–529.

———. 1968. Physical oceanography of the Arctic Ocean: 1965. In: J. E. Sater, Coord., *Arctic Drifting Stations.* Washington, D.C.: Arctic Institute of North America. pp. 255–280.

———. 1969. Physical oceanography in the Arctic Ocean: 1968. *Arctic* 22(3):214–224.

———, and C. A. Barnes. 1961. The contribution of Bering Sea water to the Arctic Ocean. *Arctic* 14(3):146–161.

———, and C. A. Barnes. 1962. Surface water in the Eurasian Basin of the Arctic Ocean. *Arctic* 15(4):251–277.

———, and C. A. Barnes. 1963. The movement of Atlantic water in the Arctic Ocean. *Arctic* 16(1):8–16.

———, and K. Aagaard. 1966. On the water exchange through Bering Strait. *Limnol. Oceanog.* 11(1):44–59.

———, and J. D. Smith. 1971. University of Washington oceanographic studies. *Aidjex Bull.* 8:21–29.

———, and J. L. Newton. Water and ice motion in the

Beaufort Sea, Spring 1970. Ottawa: Marine Sciences Branch, Ms. Report Series. In press.

Codispoti, L. A., and F. A. Richards. 1968. Micronutrient distributions in the East Siberian and Laptev seas during summer 1963. *Arctic* 21(2):67–83.

Collin, A. E. 1963. The waters of the Canadian Arctic Archipelago. *Proceedings of the Arctic Basin Symposium, October, 1962.* Washington, D.C.: Arctic Institute of North America. pp. 128–136.

Cooper, L. H. N. 1955. Deep-water movements in the North Atlantic as a link between climatic changes around Iceland and biological productivity of the English Channel and Celtic Sea. *J. Mar. Res.* 14:347–362.

Cram, J. S. 1968. *Water: Canadian Needs and Resources.* Montreal: Harvest House. 184 pp.

Crease, J. 1965. The flow of Norwegian Sea water through the Faroe Bank Channel. *Deep-Sea Res.* 12(2):143–150.

Day, C. G. 1968. Current measurements in Smith Sound, summer 1963. *U.S.C.G. Oceanog. Rep.* 16:75–84.

Dickson, R. R., L. Midttun, and A. I. Mukhin. 1970. The hydrographic conditions in the Barents Sea in August–September 1965–1968. *I.C.E.S. Coop. Res. Rep. Ser. A.* (18) 3–24.

Dietrich, G. 1967. The International "Overflow" Expedition (I.C.E.S.) of the Iceland–Faroe Ridge, May–June 1960. A review. *Rapp. Proc.-Verb. Cons. Int. Explor. Mer* 157:268–274.

Dunbar, M. 1962. The drift of North Pole 7 after its abandonment. *Can. Geog.* 6(3–4):129–142.

———. 1969. The geographical position of the North Water. *Arctic* 22(4):438–441.

Eggvin, J. 1963. Bathymetric chart of the Norwegian Sea and adjacent areas. Bergen: Fiskeridirektoratets Havforskningsinstitutt.

Ekman, V. W. 1904. On dead water. *Norwegian North Polar Expedition 1893–1896 Sci. Res.* 5 (15). 152 pp.

Eskin, F. I. 1960. On the question of the influence of Atlantic water on the upper horizons of the Arctic sea (translated). Vestnik Leningrad University No. 6, *Ser. Geol. Geog.* 1:153–158.

Evans, D. V., and T. V. Davies. 1968. Wave-ice interaction. *Stevens Inst. Tech., Davidson Lab. Rep. 1313.* 102 pp.

Fedorova, Z. P., and Z. S. Yankina. 1963. The passage of Pacific Ocean water through the Bering Strait into the Chukchi Sea (translated). *Okeanologiya* 3(5):777–784.

Fjeldstad, J. E. 1936. Results of tidal observations. *Norwegian North Polar Expedition "Maud," 1918–1925. Sci. Res.* 4(4). 88 pp.

Fletcher, J. O. 1965. *The Heat Budget of the Arctic Basin and Its Relation to Climate.* Santa Monica, California: The Rand Corporation, R-444-PR. 179 pp.

———. 1966. The Arctic heat budget and atmospheric circulation. *Proceedings of the Symposium on the Arctic Heat Budget and Atmospheric Circulation,* Lake Arrowhead, California, January 31–February 4, 1966. The Rand Corporation Memoir RM-5233-NSF. pp. 25–43.

Foster, T. D. 1968. Haline convection induced by the freezing of sea water. *J. Geophys. Res.* 73(6):1933–1938.

———. 1969. Experiments on haline convection induced by the freezing of sea water. *J. Geophys. Res.* 74(28):6967–6974.

Franceschetti, A. P., D. A. McGill, N. Corwin, and E. Uchupi. 1964. Oceanographic observations in Kennedy Channel, Kane Basin, Smith Sound, and Baffin Bay. *U.S.C.G. Oceanographic Rep. 5.* 98 pp.

Gade, H., S.-A. Malmberg, and U. Stefansson. 1965. Report on the joint Icelandic–Norwegian expedition to the area between Iceland and Greenland, 1963(preliminary results. *NATO Subcommittee on Oceanographic Research Technical Report 22.*

Galt, J. A. 1967. Current measurements in the Canadian Basin of the Arctic Ocean, summer, 1965. *University of Washington Department of Oceanography Technical Report 184,* 17 pp.

Gudkovich, Z. M., and A. D. Sitinskii. 1965. Nekotoriye resul'taty nablyudenii nad prilivnimi yavleniyami v Arkticheskom Basseine pri pomoshchi naklonomerov. *Okeanologiya* 5(5):819–824.

Gushenkov, E. M. 1964. Rasprostraneniye i metamorfizatsiya Tikhookeanskikh vod v Arkticheskom Basseine. *Okeanologiya* 4(1):36–42.

Hare, F. K. 1970. The case for further heat budget investigations in the Arctic. In: *Toward an Operational Technology for the Arctic Basin.* ARPA Workshop, Arlington, Virginia, October 12–14, 1970. Arctic Institute of North America. pp. 29–36.

Harris, R. A. 1911. *Arctic Tides.* Washington, D.C.: U.S. Government Printing Office. 103 pp.

Harvey, J. 1965. The topography of the southwestern Faroe Channel. *Deep-Sea Res.* 12(2):121–127.

Helland-Hansen, B., and E. Koefoed. 1907. Hydrographie. In: *Croisière océanographique accomplie à board de la Belgica dans la Mer du Grönland 1905.* Bruxelles: Duc D'Orléans. pp. 275–343.

———, and F. Nansen. 1909. The Norwegian Sea. Its physical oceanography based upon the Norwegian researches 1900–1904. *Rept. Norw. Fish. Mar. Invest.* Mallingske, Christiania, 2, Pt. 1, (2). 390 pp.

———. 1912. The sea west of Spitsbergen. The oceanographic observations of the Isachsen Spitsbergen Expedition in 1910. *Vid.-Selskap. Skrifter, I, Mat.-Naturv. Kl.* Dybwad, Christiania 2(12). 89 pp.

Henry, C. J. 1968. Wave-ice interaction model experiments. *Stevens Inst. Tech., Davidson Lab. Rep. 1314.* 44 pp.

Hermann, F. 1967. The T-S diagram analysis of the water masses over the Iceland–Faroe Ridge and in the Faroe Bank Channel. *Rapp. Proc.-Verb. Cons. Int. Explor. Mer* 157:139–149.

Hesselberg, T., and B. J. Birkeland. 1940. Säkulare schwankungen des klimas von Norwegen. I Teil. Die luftemperatur. *Geofys. Publ.* 14(4). 106 pp.

Hume, J. D., and M. Schalk. 1967. Shoreline processes near Barrow, Alaska: a comparison of the normal and the catastrophic. *Arctic* 20(2):86–103.

Hunkins, K. 1962. Waves on the Arctic Ocean. *J. Geophys. Res.* 67(6):2477–2489.

———. 1966. Ekman drift currents in the Arctic Ocean. *Deep-Sea Res.* 13(4):607–620.

———. 1967. Inertial oscillations of Fletcher's Ice Island (T-3). *J. Geophys. Res.* 72(4):1165–1174.

———, E. M. Thorndike, and G. Mathieu. 1969. Nepheloid layers and bottom currents in the Arctic Ocean. *J. Geophys. Res.* 74(28):6995–7008.

Jakhelln, A. 1936. Oceanographic investigations in East Greenland waters in the summers of 1930–1932. *Skrifter om Svalbard og Ishavet*, 67. 79 pp.

Johnson, G. L., and O. B. Eckhoff. 1966. Bathymetry of the north Greenland Sea. *Deep-Sea Res.* 13(6, Part 1):1161–1173.

Johnson, M. W. 1956. The plankton of the Beaufort and Chukchi Sea areas of the Arctic and its relation to the hydrography. *Arctic Institute of North America Technical Paper 1.* 32 pp.

———. 1963. Zooplankton collections from the high Polar Basin with special reference to the copepoda. *Limnol. Oceanog.* 8(1):89–102.

Joseph, J. 1967. Current measurements during the international Iceland–Faroe Ridge Expedition, 30 May to 18 June, 1960. *Rapp. Proces-Verb. Reunions.* 157:157–172.

Kiilerich, A. B. 1939. A theoretical treatment of the hydrographical observational material. The *Godthaab* expedition. *Meddelelser om Grønland* 78(5). pp. 1–148.

———. 1945. On the hydrography of the Greenland Sea. *Meddelelser om Grønland* 144(2). 63 pp.

Knudsen, M. 1899. Hydrography. *The Danish Ingolf-Expedition* 1(2):23–161.

Krauss, W. 1955. Zum system der meeresströmungen in den höheren breiten. *Dt. Hydrogr. Z.* 8:102–111.

———. 1958. Untersuchungen über die mittleren hydrographischen verhältnisse an der meeresoberfläche des nördlichen Nordatlantischen Ozeans. *Wiss. Erg. dt. atlant. Exped. Meteor, 1925–1927* 5:251–410.

La Fond, E. C. 1962. Internal waves. In: M. N. Hill, ed. *The Sea. Vol. 1.* London: Interscience. pp. 731–751.

Lake, R. A., and E. L. Lewis. 1970. Salt rejection by sea ice during growth. *J. Geophys. Res.* 75(3):583–597.

Laktionov, A. F. 1959. Bottom topography of the Greenland Sea in the region of Nansen's sill (translation). *Priroda*, 10:95–97.

———, and V. A. Shamont'yev. 1957. Primeneniye aviatsii v okeanograficheskikh issledovaniyakh v Arktike. *Problemy Arktiki* 2:19–31.

———, V. A. Shamont'yev and A. V. Yanes. 1960. Okeanograficheskii ocherk severnoi chasti Grenlandskogo morya. *Sovietskie Ribokh. Issled. v Moryakh Evropeiskogo Severa.* Moskva: Vniro-Pinro. pp. 51–65.

Lee, A. J. 1961. Hydrographic conditions in the Barents Sea and Greenland Sea during the IGY compared with those in previous years. *Rapp. Proc.-Verb. Cons. Int. Explor. Mer* 149:40–43.

———. 1963. The hydrography of the European Arctic and sub-Arctic seas. *Oceanogr. Mar. Biol. Ann. Rev.* 1:47–76.

———, and D. Ellett. 1965. On the contribution of overflow water from the Norwegian Sea to the hydrographic structure of the North Atlantic Ocean. *Deep-Sea Res.* 12(2):129–142.

Lewis, E. L., and R. A. Lake. 1971. Sea ice and supercooled water. *J. Geophys. Res.* 76(24):5836–5841.

———, and E. R. Walker. 1970. The water structure under a growing ice sheet. *J. Geophys. Res.* 75(33):6836–6845.

Lyon, W. 1961. Ocean and sea-ice research in the Arctic Ocean via submarine. *Transactions of the New York Academy of Science. Ser. 2.* 23:662–674.

Maksimov, I. V. 1945. Determining the relative volume of the annual flow of Pacific water into the Arctic Ocean through Bering Strait (translation). *Problemy Arktiki* 2:51–58.

Malmberg, S.-A., H. G. Gade, and H. E. Sweers. 1967. Report on the second joint Icelandic-Norwegian expedition to the area between Iceland and Greenland in August–September 1965. *NATO Subcommittee on Oceanographic Research.* Technical Report 41. 44 pp.

Matthews, J. B. 1970. Tides at Point Barrow. *Northern Eng.* 2(2):12–13.

Maykut, G. A., and N. Untersteiner. 1971. Some results from a time-dependent thermodynamic model of sea ice. *J. Geophys. Res.* 76(6):1550–1575.

Metcalf, W. G. 1955. On the formation of bottom water in the Norwegian Basin. *Transactions of the American Geophysical Union* 36(4):595–600.

———. 1960. A note on water movement in the Greenland–Norwegian Sea. *Deep-Sea Res.* 7(3):190–200.

Mohn, H. 1887. Nordhavets dybder, temperatur og strømninger. *Den Norske Nordhavs-Expedition 1876–1878.* Grøndahl, Christiania, 18. 212 pp.

Mosby, H. 1938. Svalbard waters. *Geofys. Publ.* 12(4). 85 pp.

———. 1959. Deep water in the Norwegian Sea. *Geofys. Publ.* 21(3). 62 pp.

———. 1962a. Recording the formation of bottom water in the Norwegian Sea. In: *Proceedings of the Symposium on Mathematical-Hydrodynamical Methods of Physical Oceanography.* Inst. Meereskde. University of Hamburg. pp. 289–296.

———. 1962b. Water, salt and heat balance of the

North Polar Sea and of the Norwegian Sea. *Geofys. Publ.* 24(11):289–313.

———. 1963. Interaction between the Polar Basin and peripheral seas, particularly the Atlantic approach. In: *Proceedings of the Arctic Basin Symposium October 1962.* Arctic Institute of North America, June 1963. pp. 109–116.

———. 1967. Bunnvannsdannelse i havet. *Norw. Acad. Sci. Lett.* The Nansen Memorial Lecture, October 11, 1965.

———. 1970. Atlantic water in the Norwegian Sea. *Geofys. Publ.* 28(1). 59 pp.

Muench, R. D. 1971. The physical oceanography of the northern Baffin Bay region. *The Baffin Bay-North Water Project, Scientific Report 1.* Arctic Institute of North America, January 1971. 150 pp.

Nansen, F. 1902. The oceanography of the North Polar Basin. *The Norwegian North Polar Expedition 1893–1896, Scientific Results,* 3(9). 427 pp.

———. 1906. Northern waters: Captain Roald Amundsen's oceanographic observations in the Arctic seas in 1901. *Vid.-Selskap Skrifter I. Mat.-Naturv. Kl.* Dybwad, Christiania, 1(3). 145 pp.

———. 1912. Das Bodenwasser und die Abkühlung des Meeres. *Int. Rev. d. ges. Hydrobiol. Hydrograph.* 5(1):1–42.

———. 1915. Spitsbergen waters. *Vid.-Selskap. Skrifter, I. Mat.-Naturv. Kl.,* 2. 132 pp.

Nikitin, M. M., and N. I. Dem'yanov. 1965. O glubinnikh techeniyakh Arkticheskogo Basseina. *Okeanologiya* 5(2):261–263.

Nutt, D. C. 1966. The drift of ice island WH-5. *Arctic* 19(3):244–262.

Olenicoff, S. M. 1971. The Soviet DARMS. *Aidjex Bull.* 7:5–23.

Palfrey, K. M., Jr. 1967. Physical oceanography of the northern part of the Greenland Sea in the summer of 1964. M.S. Thesis, University of Washington, Seattle. 52 pp.

———. 1968. Oceanography of Baffin Bay and Nares Strait in the summer of 1966. *U.S.C.G. Oceanographic Rep.* 16:1–74.

Panov, V. V., and A. O. Shpaikher. 1964. Influence of Atlantic waters upon some features of the hydrology of the Arctic Basin and adjacent seas (translation). *Deep-Sea Res.* 11(2):275–285.

Reed, R. J., and W. J. Campbell. 1960. Theory and observations of the drift of ice station Alpha. University of Washington Department of Meteorology and Climatology. Final Report to Office of Naval Research, NR 307–250, December 1960. 255 pp.

Riis-Carstensen, E. 1938. Fremsaettelse af et dynamisk-topografisk kort over Østgrønlandsstrømmen mellem 74° og 79°N. Br. paa grundlag af hidtidig gjorte undersøgelser i disse egne. *Geografisk Tidsskrift* 41(1):25–51.

Robin, G. de Q. 1963. Ocean waves and pack ice. *Polar Record* 11(73):389–393.

Saetre, R. 1967. Report on the Norwegian investigations in the Faeroe Channel 1964–65. *NATO Subcommittee on Oceanographic Research Technical Report* 38. 27 pp.

Sater, J. E. 1969. Oceanography: submarine topography and water masses. In: J. E. Sater, coordinator. *The Arctic Basin.* Washington, D.C.: Arctic Institute North America. pp. 14–25.

Shirshov, P. P. 1944. Scientific results of the drift of station North Pole (translation). *Akad. Nauk S.S.S.R., Obshchee Sobranie,* February 1944. pp. 110–140.

Shpaikher, A. O., L. N. Belyakov, and V. V. Izmailov. 1966. K voprosy o vliyanii Tikhookeanskikh vod na gidrologicheskii rezhim pritikhookeanskoi chasti Arkticheskogo Basseina. *Problemy Arktiki i Antarktiki* 22:35–42.

———, and Z. S. Yankina. 1969. Formation of anomalies of winter hydrological characteristics of East Siberian and Chukchi seas (translation). *Problemy Arktiki i Antarktiki,* 31.

Shumskiy, P. A., A. N. Krenke, and I. A. Zoltikov. 1964. Ice and its changes. In: Research in Geophysics. Vol. 2: Solid Earth and Interface Phenomena. Cambridge: MIT Press.

Smith, J. D. 1971. Aidjex oceanographic investigations. *Aidjex Bull.* 4:1–7.

Somov, M. M., ed. 1954–1955. *Observational data of the scientific research drifting station of 1950–1* (translation). Leningrad: Morskoi Transport, Vol. 1, (2):48–170, and 1(3):180–403.

Steele, J. H., J. R. Barrett, and L. V. Worthington. 1962. Deep currents south of Iceland. *Deep-Sea Res.* 9:465–474.

Stefansson, U. 1962. North Icelandic waters. *Rit. Fiskideildar,* Vol. III. 269 pp.

———. 1968. Dissolved nutrients, oxygen and water masses in the Northern Irminger Sea. *Deep-Sea Res.* 15(5):541–575.

Sverdrup, H. U. 1933. Scientific results of the *Nautilus* expedition, 1931. Part 2: Oceanography. *Pap. Phys. Oceanog. Meteorol.* 2:16–63.

Tait, J. B. 1955. Long-term trends and changes in the hydrography of the Faroe–Shetland Channel region. *Deep-Sea Res.* 3(Suppl.):482–498.

———. 1957. Hydrography of the Faroe–Shetland Channel, 1927–1952. *Mar. Res.*, Scottish Home Department, 2. 309 pp.

———. 1962. Oceanic water-mass transport through the Faroe–Shetland Channel. In: *Proceedings of the Symposium on Mathematical-Hydrodynamical Methods of Physical Oceanography, September 1961.* Inst. Meereskunde University of Hamburg. pp. 339–351.

———, ed. 1967. The Iceland–Faroe Ridge International (ICES) "Overflow" expedition, May–June 1960. *Rapp. Proc.-Verb. Cons. Int. Explor. Mer.* 274 pp.

Tantsura, A. I. 1959. O techeniyakh Barentseva Morya. *Trudy PINRO* 11:35–53.

Timofeyev, V. T. 1958. O "vozraste" Atlanticheskikh vod v Arktichesom Basseine. *Problemy Arktiki* 5:27–31.

———. 1961. *Vodniye Massy Arkticheskogo Basseina.* Leningrad: Gidromet. Izdat. 190 pp.

———. 1962. The movement of Atlantic water and heat into the Arctic Basin (translation). *Okeanologiia* 1(3):407–411.

———. 1963. Interaction of waters from the Arctic Ocean with those from the Atlantic and Pacific (translation). *Okeanologiya* 3(4):569–578.

Tripp, R. B., and K. Kusunoki. 1967. Physical, chemical, and current data from Arlis II: eastern Arctic Ocean, Greenland Sea, and Denmark Strait areas, February 1964–May 1965. Vol. 2. *University of Washington Department Oceanography Technical Report.* 185. 140 pp.

Trolle, A. 1913. Hydrographical observations from the Danmark expedition. *Meddelelser om Grønland* 41:275–426.

United States Navy Hydrographic Office. 1958. *Oceanographic Atlas of the Polar Seas. Part II: Arctic.* Hydrographic Office Publication, 705. 149 pp.

Untersteiner, N. 1964. Calculations of temperature regime and heat budget of sea ice in the central Arctic. *J. Geophys. Res.* 69(22):4755–4766.

Voorhis, A. D., and D. C. Webb. 1970. Vertical currents in a winter sinking region. *Trans. Amer. Geophys. Un.* 51(4)(abstract). p. 315.

Vowinkel, E., and S. Orvig. 1961. Water balance and heat flux of the Arctic Ocean. *McGill University Publication in Meteorology*, 44. 35 pp.

Worthington, L. V. 1953. Oceanographic results of Project Skijump I and Skijump II. *Trans. Am. Geophys. Un.* 34(4). pp. 543–551.

———. 1969. An attempt to measure the volume transport of Norwegian Sea overflow water through the Denmark Strait. *Deep-Sea Res.* 16(Suppl.): 421–432.

———. 1970. The Norwegian Sea as a mediterranean basin. *Deep-Sea Res.* 17(1):77–84.

———, and W. R. Wright. 1970. *North Atlantic Ocean Atlas.* The Woods Hole Oceanographic Institution Atlas Series, Vol. 2.

Wüst, G. 1935. Die Stratosphäre des Atlantischen Ozeans. *Deutsche Atlantische Exped. Meteor,* 1925–1927. Wiss. Erg., Bd. VI, 1. Teil 2. Lief. 288 pp.

Yearsley, J. R. 1966. Internal waves in the Arctic Ocean. University of Washington M. S. Thesis. 62 pp.

Zubov, N. N. 1945. *Arctic Ice.* (Translated from Russian, issued in 1965, U.S.N. Oceanographic Office and American Meteorological Society.)

Chapter 2
Topography of the Arctic Ocean

YVONNE HERMAN[1]

Introduction

The world's largest land-locked ocean, the Arctic, occupies a broad elliptical basin, covering an area of about 12 million km^2 and containing 13 million km^3 of water. Although it constitutes over 3% of the world's ocean area, the Arctic holds only 1% of its volume; this is accounted for by the broad continental shelves which underlie almost 70% of its area (Beal, 1969).

The permanent cover of sea-ice has greatly hampered the scientific exploration of the North Polar Basin, and early studies were restricted to the more accessible shallow continental margins, peripheral to the permanent pack-ice. It was not until Nansen's historic voyage across the Arctic at the turn of the century that the presence of a deep basin was established, thus dispelling the previously held opinions that sea-ice covers scattered islands separated by shallow water (Nansen, 1904). Post World War II topographic maps published by United States and Russian scientists (e.g., Emery, 1949) differ little from Nansen's charts. However, during the last two decades through international cooperative studies a vast amount of significant data concerning major topographic features, structures as well as sediment composition and distribution have become available. The bathymetric chart compiled by Dietz and Shumway (1961), still reproduced with minor modifications and additions shows diagrammatically the major features (Fig. 1). Beal's physiographic diagram (1969) reproduced here (Fig. 2) and based on bathymetric and geophysical data reveals the unique and complex topography of this ocean.

Outline of Physiography

In an attempt to review the information available on the geomorphic provinces of the Arctic Ocean, many publications have been drawn upon. Only provinces that are not discussed elsewhere in this volume are treated here in some detail.

Within the major physiographic provinces, numerous smaller features have been recognized and described; for simplicity they are combined into three major geomorphic provinces, namely: (1) continental margins, (2) plains, and (3) rise and ridges.

The nomenclature of the major geomorphic features has become increasingly confused as new provinces were named by geologists from various countries who claimed priority for discovering

[1] Department of Geology, Washington State University, Pullman, Washington 99163, U.S.A.

Fig. 1. Major geomorphic provinces of the Arctic Ocean Basin (after Dietz and Shumway, 1961).

them. Table 1 has been compiled to show the equivalence of different names applied to the same topographic features.

Continental Margins

Along its perimeter the Arctic is bordered by "normal margins" which include the shelf-slope-rise sequences, by the intersection of rise and ridges and by detached continental fragments separated by deep saddles from the adjacent shelf.

Two-thirds of the Arctic is underlain by *continental shelves,* which are broad and shallow along Europe and Asia and become much narrower, deeper, and more irregular off Greenland and the American continent (Fig. 2).

Geological and geophysical information indicates that the shelves are floored with sedimentary strata with thicknesses ranging from several meters to >18 km (Demenitskaya and Karasik, 1971). During the Quaternary low sea levels most shelves appear to have been affected by glacial, river, and wave erosion.

The main topographic features of the *continental slopes* are the submarine canyons, some of which have been studied in detail (e.g., Carsola, 1954; Carsola et al., 1961; Beal, 1969 and bibliography therein). The canyons have gradients ranging from 1 to 15°, those off eastern Alaska being the gentlest (Beal, 1969).

The *continental rises* are poorly developed, except along the Canadian Basin (Fisher, et al., 1958; Dietz and Shumway, 1961).

Continental Shelves

The 900-km wide *Barents Sea* continental shelf covers an area of approximately 1,300,000 km^2, extending between the Spitsbergen Archipelago, the Norwegian coast and Sea, and the Kara Sea (Fig. 2). Depths vary from 100 to 350 m and drop to 600 m near the Norwegian Sea; the shelf break is at 400 to 500 m. The rough bottom relief, dominated by numerous basins, depressions, and rises, appears to be controlled by the same tectonic processes that shaped the surrounding continent and islands. Continental and insular rocks include folded Precambrian metamorphics, granites, dolomites, slates, tillites, and Paleozoic slates, shales, limestones and sandstones, as well as Mesozoic and Tertiary clastics, some slightly carbonaceous. The upper Mesozoic clastics are interbedded with several basalt flows (Klenova, 1966). Two distinct submarine terraces are recognized: one at 200 m, the other at 70 m; each is believed to correspond to a Pleistocene glacial stage (Klenova, 1966). Above 200 m the submarine slopes are strongly dissected, possibly by subaerial erosion; the younger 70-m terrace is rather flat and featureless (Klenova, 1966).

The amount of terrigenous clasts carried to sea by rivers is limited, inasmuch as only one major river, the Pechora, flows directly into this sea. Small amounts of fine, land-derived clasts are also carried by winds, and are distributed by currents. Sandy-clays predominate in the low energy, large, flat areas of the shallow, <100-m zones; however, coarse sediments ranging from boulder to sand-size are also common. Studies of sediment cores by Soviet geologists reveal that in some regions strong current scour prevents sedimentation and that erosion by bottom currents exposes pre-Holocene deposits on the shelf floor (Kulikov et al., 1970). Where sediments accumulate they do so at very slow rates. Holocene silty-clays, several centimeters thick, overlie 2 to 3 m of gray, bedded, silty-clays, which are believed to have been deposited during late glacial time. These deposits are underlain by several-decimeter-thick grayish-brown and bluish-gray, poorly sorted mixtures ranging from clay to "coarse-size" fragments, laid down during upper Pleistocene glaciations (Kulikov et al., 1970).

The floor of the *Kara Sea*, with an average depth of 100 m and a maximum of 640 m, is occupied by a series of broad terraces stepping down from east to north and west (Fairbridge, 1966; Andrew and Kravitz, this volume). A deep north-south trending trough between Franz Joseph Island and the northern tip of Novaya Zemlya borders the sea on one side, and the smaller Voronin Trough crosses the sea west of Severnaya Zemlya (Fairbridge, 1966; Andrew and Kravitz, this volume, Fig. 2).

The wide and shallow (10 to 40 m) *Laptev Sea* is cut by numerous transverse submarine valleys of erosional and tectonic origin which can be traced into landward extensions (Ionin, 1966a; Holmes and Creager, this volume). The deep Sadko Trough, recognized by Heezen and Ewing (1963) to be an extension of the seismic Nansen Ridge, crosses the Laptev continental slope (Fig. 2).

Table 1 Equivalence of Names of Major Topographic Features Used in Various Publications

<table>
<tr><td rowspan="2">Alpha Rise
Alpha Cordillera
Alpha Ridge
Central Arctic Rise
Mendeleev Rise
Alpha-Mendeleev Rise
Alpha Mendeleev Cordillera</td><td>Canada Basin*
Laurentian Basin
Hyperborean Basin</td></tr>
<tr><td>Makarov Basin
Fletcher Abyssal Plain</td></tr>
<tr><td>Chukchi Rise* †
Chukchi Cap
Chukchi Plateau
Northwind Escarpment*
Northwind Seahigh
Northwind Cordillera</td><td>Fram Basin* with Hakkel Plain
Eurasia Basin
Amundsen Basin
Pole Abyssal Plain</td></tr>
<tr><td>Nansen Ridge*
Gakkel Ridge
Mid-Arctic Ridge
Arctic Mid-Oceanic Ridge</td><td>Nautilus Basin* with Sverdrup or Barents Abyssal Plain
Nansen Basin</td></tr>
</table>

* Generally accepted and widely used name.

† Chukchi Rise may include both Northwind Escarpment and Chukchi Cap.

The *East Siberian Sea* is the widest (800 km) among the shallow epicontinental seas bordering the Siberian coast (Fig. 2). The major topographic features of the shelf floor were shaped by grounded ice and subaerial erosion during glacial low sea level stands, as well as by river sedimentary discharges and tectonic movements (Ionin, 1966b; Naugler et al., this volume).

Extending from Wrangel Island to Point Barrow, Alaska, the *Chukchi Sea* covers a shallow submerged platform which links the continents of Asia to North America (Fig. 2). Over most of its surface the monotonously flat floor is covered by several meters of poorly sorted clay to gravel-size recent sediments which have masked the subsurface topographic irregularities. In the eastern sector of the sea a thin veneer of unconsolidated sediments (up to 3 m) overlies bedrock, whereas off Kotzebue Sound and between Cape Lisburne and Ice Cape recent deposits attain 14 m (Moore, 1964). Calcitic, arenaceous, and siliceous skeletal remains of organisms generally constitute less than 1% of the recent sediments (Dietz et al., 1964). Depths range from 40 to 60 m, except in the two wide and shallow bays, Kotzebue Sound on the Alaskan side and Kolyuchin Bay on the Siberian side, where depths are less than 20 m. The outer margin of the shelf, which breaks at about 200 m, exhibits minor topographic irregularities. Its floor is dissected by two submarine canyons: Barrow Canyon parallels the Alaskan coast, then crosses the shelf north of Point Barrow and leads into the Canada Basin; Herald Canyon cuts the shelf along the 175°W meridian and heads into the Chukchi Plain (Carsola, 1954; Dietz et al., 1964). Off eastern Alaska the continental slope has a gradient of 1 to 2°, whereas all other Arctic slopes appear to range between 10 and 15° (Beal, 1969).

Considerable variation in the extent and depth of the continental shelf surrounding the *Canadian Archipelago* exists; two major physiographic provinces are recognized: the inner narrow part consisting of the submerged coastal zone and the broader and rather uniform outer portion. The shelf break is at variable depths 80 to 150 km offshore (Pelletier, 1962, 1966a, 1966b; de Leeuw, 1967).

The Archipelago contains numerous channels

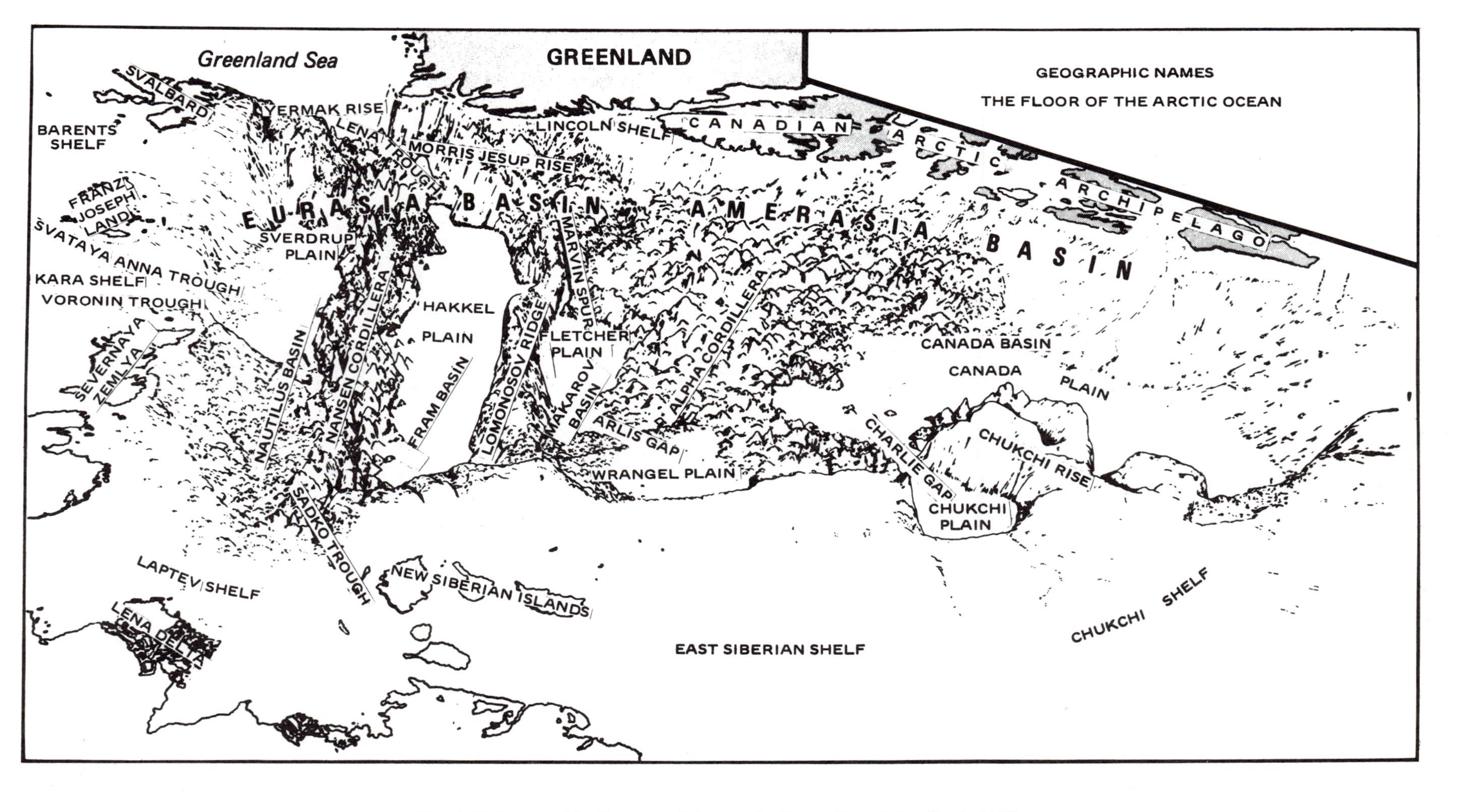

Fig. 2. Physiographic diagram of the Arctic Ocean floor (after Beal, 1969).

that range in width from 10 to 120 km and in depth from few meters to over 700 m; submerged longitudinal ridges divide some of these submarine valleys (Pelletier, 1966a). To the east and west of Baffin Bay the shelf is dissected by U-shaped submarine valleys which are contiguous with similar coastal and land features. These submarine troughs are believed to have been carved in pre-Pleistocene time by rivers and subsequently eroded by glaciers before the Postglacial submergence (Pelletier, 1966a).

Marginal Rises

Separated by a saddle from the continental shelf, the *Chukchi Rise* projects 700 km northward into the Canada Basin and is flanked by abyssal plains (Figs. 1, 2, and 3). The Rise is subdivided into the Chukchi Plateau or Cap, a circular structure, about 100 km in diameter and 300 m deep, and the Northwind Escarpment, which lies parallel to and east of the Cap and is separated from it by a prominent gap (Fisher et al., 1958; Dietz and Shumway, 1961; Beal, 1969). The Rise is thought to be a continental fragment, and Shaver and Hunkins (1964) suggested that it was detached from the shelf by horizontal movement resulting from the expansion of the earth, but the idea of an expanding earth is discredited by most geologists today.

North of Greenland and separated from the Lincoln shelf by a broad trough, the *Morris Jessup Rise* extends into the Fram Basin paralleling the Lomonosov Ridge (Fig. 2). Along the crest the average recorded depths are 500 to 700 m, and its surface is marked by a series of scarps (de Leeuw, 1967; Beal, 1969).

The Yermak Rise appears to be an extension of the continental shelf north of Vest Spitzbergen (Fig. 2). Echograms indicate that its surface is smooth on the western and irregular on the eastern flank, and that the relief of the latter resembles the fractured plateau of the Alpha Rise. A minimum depth of 500 m was measured near the northern edge of the Rise (Beal, 1969).

Central Basin

The Central Arctic Ocean is divided by three parallel mountain ranges and three marginal plateaus into four major basins, each nearly 4000 m deep, and two plains of intermediate depth. The major topographic features are illustrated in Figs. 1, 2, and 3.

Plains

Canada Basin

The most extensive abyssal plain in the Arctic is found in the Canada Basin, which is surrounded by steep continental slopes along most of its perimeter and by the Alpha Rise to the north (Figs. 1, 2, and 3). With the exception of a few knolls in the northern part, the basin has a rather smooth floor and uniform depth of about 3800 m sloping westward gradually (Beal, 1969). Sediment thickness is greatest near the continent and thins toward the Alpha Rise. This gradient suggests that clasts are derived mainly from the Canadian continental shelf via submarine canyons. The Charlie Gap connects the deep Canada Basin with the smaller and relatively shallow (2200 m) Chukchi Plain (de Leeuw, 1967); two submarine canyons lead into the Chukchi Plain (Beal, 1969).

Geophysical data (Panov, 1955a, 1955b) and aeromagnetic evidence suggest that the Canada Basin is underlain by great volumes of sediment with the crystalline basement sloping downward to the east (Ostenso, 1966a).

The narrow *Makarov Basin*, flanked by the Lomonosov, Marvin, and Alpha ranges (Fig. 2), is remarkably flat; in the central part of the basin the sea floor attains a depth of 4000 m and sediment thickness is >2 km. The deep abyssal plain is separated from the *Wrangel Plain* by a basement ridge dissected by several channels through which sediment is transported from the shallower Wrangel Plain (2700 m) into the Makarov Deep (Kutschale, 1966; de Leeuw, 1967). Seismic reflections suggest that the Wrangel Plain is underlain by at least 3.5 km of subhorizontal stratified sediments (Kutschale, 1966).

The *Fram Basin*, separated from the Makarov and Nautilus Basins by the Lomonosov and Nansen Ridges, respectively, is floored by the deepest Arctic plain (Fig. 2). Echo soundings indicate the presence of minor "bumps" and scarps and an increase in depth from the Lomonosov to the Nansen Ridge and from Eurasia toward Greenland; depths greater than 4000 m have been recorded along the Nansen Ridge (Ostenso, 1966a; de Leeuw, 1967; Beal, 1969). Sediment thickness is on the order of 2 to 3 km. In the central part of the basin, sediments directly overlie the Nansen Ridge basalts (Demenitskaya and Karasik, 1971).

The *Nautilus Basin*, the smallest of the four deep Arctic plains, is bound by the Barents Sea and the Nansen Ridge (Fig. 2). Its floor shows minor irregularities, sloping southward gradually; depths range between 3800 and 4000 m. The sediment fill is about 2 to 3 km and the underlying structure resembles that of the Fram Basin (Beal, 1969; Demenitskaya and Karasik, 1971).

Rises and Ridges

Three major essentially continuous ranges cross the ocean: the *Alpha Rise* named by Hope (1959) and subsequently referred to as the Central Arctic Rise (Dietz and Shumway, 1961) and the Mendeleev Ridge by the Russians, is an arcuate elevation, 250 to ~1000 km wide and about 1800 km long, joined by broad plateaus to the Siberian and Canadian shelves (Figs. 1, 2, and 3). The crest is 1200 to 1500 m deep, giving the Rise 2800 m of relief above the surrounding basins. Seismic reflections suggest that 300 to >500 m of unconsolidated sediments blanket the Rise (Crary and Goldstein, 1957). Numerous valleys and seamounts were recorded by geophysical techniques; among the seamounts, several appear to be topped by circular, concave peaks believed to be volcanic craters (Dietz and Shumway, 1961; Beal, 1969). On both sides of the Rise the floor drops gradually to the plains of the Canada and Makarov Basins.

Seismic reflection profiles allow identification of topographic features similar to those of the Mid-Atlantic Ridge (Beal, 1969). The Alpha Rise is aseismic, and is characterized by broad magnetic anomalies. Both topographic and magnetic signatures are roughly symmetrical to the assumed axis of the Rise (King et al., 1966; Rassokho et al., 1967).

Various hypotheses have been suggested for the origin of the Alpha Rise; on the basis of bathymetric and seismic information Hunkins (1961) proposed block faulting; others suggested that it is a "fossil" oceanic ridge which became dormant about 30 to 40 million years ago, when the locus of rifting and sea-floor spreading shifted to the Nansen Ridge (Johnson and Heezen, 1967; Beal, 1969; Demenitskaya and Karasik, 1969a, 1969b). According to Vogt (this volume), its genesis is still unresolved.

The *Lomonosov Ridge*, discovered by Soviet geologists, is a narrow aseismic and slightly asymmetric mountain range which crosses the Arctic Ocean from Ellesmere Island, where it merges with the Alpha Rise, to the New Siberian Islands (Figs. 1, 2, and 3). The 1800 km long and 40 to 200 km wide ridge is separated from the Siberian shelf by a structural depression filled with thick sequences of sedimentary rocks (Demenitskaya et al., 1968). The depth of the ridge varies between 850 and 1600 m, giving it a relief of about 3000 m above the surrounding basin floor; echograms reveal that its summit is extremely flat and that elongate steps or valleys parallel the flanks of the ridge. Geophysical data suggests that the aseismic Lomonosov Ridge, a fragment of the former Eurasian continental margin, composed of folded sedimentary rocks with few mafic crystalline elements, split from the continent, and was rafted to its present location by sea-floor spreading (Beal, 1969; Demenitskaya and Karasik, 1971; Vogt, this volume).

The seismically active *Nansen Ridge* is the northernmost continuation of the Mid-Atlantic Ridge; it enters the Arctic between Greenland and Svalbard, parallels the Barents Sea, and impinges on the Laptev continental slope north of the Lena Delta (Fig. 2). Its existence was postulated by Emery (1949) and Heezen and Ewing (1961, 1963) among others, on the basis of soundings and of a well-defined pattern of earthquake epicenters. Subsequently, the topographic form of the ridge was described by Dietz and Shumway (1961). The 2000 km long range is remarkably straight and narrow, and individual peaks rise to 1500 m above the adjacent plains. Its crest is marked by a deep median rift valley whose axis is offset in many places by fracture zones which intersect the ridge at approximately right angles. The most prominent change in direction is off the northeastern tip

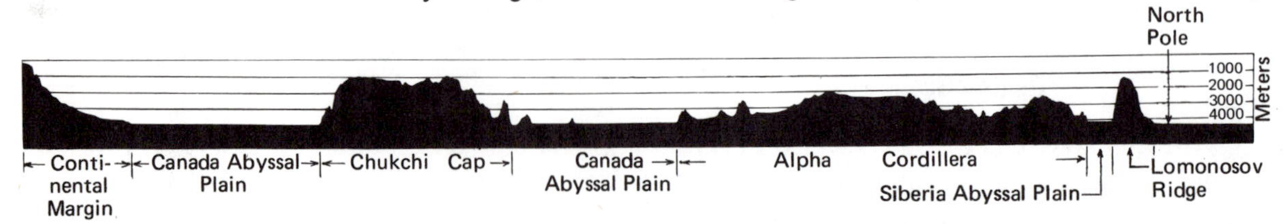

Fig. 3. Bathymetric profile across the Arctic Ocean floor along the 155°W meridian from the Alaskan Shelf to the North Pole. Vertical exaggeration is about × 26 (after Dietz and Shumway, 1961).

of Greenland (Johnson and Heezen, 1967). There is no detectable sediment fill in the central rift valley, and on the flanks the distribution is patchy, ranging from several meters to 400 m (Dementitskaya and Karasik, 1971). Sykes (1965) demonstrated that the Nansen Ridge is currently an active spreading center and that the earthquake foci within the Arctic region are confined to a narrow belt coinciding with the axis of the ridge. The belt of earthquake epicenters associated with the Mid-Atlantic Ridge continues along the Nansen Ridge, crosses the Laptev Shelf, and encroaches on the valley of the Lena River.

Rassokho et al. (1967) found linear magnetic anomaly patterns, typical of spreading ridges associated with the Nansen Ridge. The intersection of central rift by faults and basins offsets the linear pattern of magnetic anomalies; furthermore this pattern is asymmetrical about the ridge axis (Demenitskaya and Karasik, 1971).

Summary

Three major geomorphologic provinces are recognized: (1) the *continental margins,* which include (a) the shelf-slope-rise provinces, (b) the regions of intersection between submarine ranges and the shelf, and (c) the detached continental fragments separated by deep saddles from the shelf proper; (2) the *plains;* and (3) the *rise* and *ridges.*

(1) The continental shelves underlie almost two-thirds of the Arctic; they are broad and shallow along Europe and Asia and become much narrower and deeper off North America and Greenland.

The main topographic features of the continental slopes are canyons through which terrestrial and shallow water detritus is transported to the deep plains.

The continental rises are poorly developed features, except along the Canadian Basin.

Three marginal rises have been described: The Chukchi, the Morris Jessup, and the Yermak Rises.

(2) Four major deep plains and two plains of intermediate depths are known; they are characterized by smooth floors and nearly uniform depths. Sediment thickness is greatest near continents, suggesting that terrestrial and shallow-water minerals and biogenic elements transported principally via canyons are distributed by deep currents.

(3) Three major essentially continuous ranges cross the ocean: The Alpha Rise, whose genesis is still unresolved; the Lomonosov Ridge, regarded to be a fragment of the former Eurasian continental margin which split from the continent and was rafted to its present location by sea-floor spreading; and the Nansen Ridge, considered to be an active spreading center.

Acknowledgments

I thank Philip E. Rosenberg for critically reviewing an early version of the manuscript. This research was supported by W.S.U. Graduate School Development Fund 14N-2940-0020. Patricia M. Wilson typed the manuscript.

References

Beal, M. A. 1969. Bathymetry and structure of the Arctic Ocean. Doctoral thesis, Oregon State University, Corvallis. 204 pp.

———, F. Edvalson, K. Hunkins, A. Molloy, and N. Ostenso. 1966. The floor of the Arctic Ocean: geographic names. *Arctic* 19:215–219.

Carsola, A. J. 1954. Submarine canyons of the Arctic slope. *J. Geol.* 62:605–610.

———, R. L. Fisher, C. J. Shipek, and G. Shumway. 1961. Bathymetry of the Beaufort Sea. In: G. O. Raasch, ed. *Geology of the Arctic*, Vol. 1. Toronto: University of Toronto Press. pp. 678–689.

Crary, A. P. 1954. Bathymetric chart of the Arctic Ocean along the route of T-3. *Geol. Soc. Amer. Bull.* 65:709–712.

———, R. D. Cotell and J. Oliver. 1952. Geophysical studies in the Beaufort Sea, 1951. *Trans. Amer. Geophys. Un.* 33:211–216.

———, and N. Goldstein. 1957. Geophysical studies in the Arctic Ocean. *Deep-Sea Res.* 4:185–201.

Creager, J. S., and D. A. McManus. 1965. Pleistocene drainage patterns on the floor of the Chukchi Sea. *Marine Geol.* 3:279–290.

———, and D. A. McManus. 1967. Geology of the floor of Bering and Chukchi Seas—American studies. In: D. M. Hopkins, ed. *The Bering Land Bridge.* Stanford: Stanford University Press. pp. 7–31.

Cromie, W. J. 1961. Preliminary results of investigations on Arctic drift station Charlie. In: G. O. Raasch, ed. *Geology of the Arctic*, Vol. 1. Toronto: University of Toronto Press, pp. 690–708.

De Leeuw, M. M. 1967. New Canadian bathymetric chart of the western Arctic Ocean, north of 72° *Deep-Sea Res.* 14:489–504.

Demenitskaya, R. M. 1958. The structure of the Earth's crust in the Arctic. *Informational Bulletin of the Institute of the Geology of the Arctic*, Leningrad (7):42–49.

———, and A. M. Karasik. 1966. Magnetic data confirm that the Nansen-Amundsen Basin is of normal oceanic type. In: W. H. Poole, ed. *Continental Margins and Island Arcs.* Geological Survey of Canada Paper 66-15. pp. 191–196.

———, A. M. Karasik, Yu. G. Kiselev, I. N. Litvinenko, and S. A. Ushakov. 1968. Transition zone between the Eurasian continent and the Arctic Ocean. *Can. J. Earth Sci.* 5:1125–1129.

———, and Yu. G. Kiselev. 1968. The characteristic features of the structure, morphology and sedimentary cover of the central part of the Lomonosov Ridge from seismic investigations. In: *Geophysical Methods of Prospecting in the Arctic*, Leningrad 5:33–46.

———, and A. M. Karasik. 1969a. Mid-oceanic ridges. In: *Geotectonics of Continents and Oceanic Basins*, Nauka: Moscow (in Russian).

———, and A. M. Karasik. 1969b. The active rift system of the Arctic Ocean. *Tectonophysics* 8:345–351.

———, and K. L. Hunkins. 1970. Shape and structure of the Arctic Ocean. In: Arthur E. Maxwell, ed. *The Sea*, Vol. 4. New York, New York: Wiley-Interscience. pp. 223–249.

———, and A. M. Karasik. 1971. Problems of the genesis of the Arctic Ocean. In: *Istoriia Mirovogo okeana* (translation). Moscow: Akad. Nauk SSSR, 58–76.

Dietz, R. S., and G. Shumway. 1961. Arctic basin geomorphology. *Geol. Soc. Amer. Bull.* 72:1319–1330.

———, A. J. Carsola, E. C. Buffington, and C. J. Shipek. 1964. Sediments and topography of the Alaskan shelves. In: R. L. Miller, ed. *Papers in Marine Geology.* New York, New York: Macmillan. pp. 241–256.

Eardley, A. J. 1961. History of geologic thought on the origin of the Arctic basin. In: G. O. Raasch, ed. *Geology of the Arctic*, Vol. 1. Toronto: Toronto University Press. pp. 607–621.

Egiazarov, B. Kh., I. P. Atlasov, M. G. Ravich, G. E. Grikurov, R. M. Demenitskaya, G. A. Znachko-Yavorsky, A. P. Puminov, B. S. Romanovich, and D. S. Solovyev. 1971. Tectonic map of Arctic polar regions of earth. *Second International Symposium on Arctic Geology*, San Francisco, 1971 (abstract). p. 17.

Emery, K. O. 1949. Topography and sediments of the Arctic basin. *J. Geol.* 57(5):512–521.

Fairbridge, R. W. 1966. Kara Sea. In: R. W. Fairbridge, ed. *Encyclopedia of Oceanography*, Vol. 1. New York, New York: Reinhold Publishing Corporation. pp. 430–432.

Fisher, R. L., A. J. Carsola and G. Shumway. 1958. Deep-sea bathymetry north of Point Barrow. *Deep-Sea Res.* 5:1–6.

Hamilton, W. 1971. Continental drift in Arctic. *Second International Symposium on Arctic Geology.* San Francisco, 1971 (abstract). p. 24.

Heezen, B. C., and M. Ewing. 1961. The mid-oceanic ridge and its extension through the Arctic basin. In: G. O. Raasch, ed. *Geology of the Arctic*, Vol. 1. Toronto: University of Toronto Press, pp. 622–642.

———, and M. Ewing. 1963. The mid-oceanic ridge. In: M. N. Hill, ed. *The Sea.* New York, New York: Interscience Publishers, 3:388–410.

Herman, Y. 1970a. Late Cenozoic Arctic oceanography. *Transactions of the American Geophysical Union* (abstract). p. 333.

———. 1970b. Arctic paleo-oceanography in late Cenozoic time. *Science* 169:474–477.

Hope, E. R. 1959. Geotectonics of the Arctic Ocean and the great Arctic magnetic anomaly. *J. Geophys. Res.* 64:407–427.

Hopkins, D. M., and D. W. Scholl. 1971. Tectonic development of Beringia, late Mesozoic to Holocene. *Second International Symposium on Arctic Geology*, San Francisco, 1971 (abstract). pp. 26–27.

Hunkins, K. 1961. Seismic studies of the Arctic Ocean floor. In: G. O. Raasch, ed. *Geology of the Arctic*, Vol. 1. Toronto: University of Toronto Press. pp. 645–665.

Ionin, A. S. 1966a. Laptev Sea. In: R. W. Fairbridge, ed. *Encyclopedia of Oceanography*, Vol. 1. New York, New York: Reinhold Publishing Corporation. pp. 442–445.

———. 1966b. East Siberian Sea. In: R. W. Fairbridge, ed. *Encyclopedia of Oceanography*, Vol. 1. New York, New York: Reinhold Publishing Corporation. pp. 243–245.

Johnson, G. L., and B. C. Heezen. 1967. The Arctic mid-oceanic ridge. *Nature* 215:724–725.

———, and D. B. Milligan. 1967. Some geomorphological observations in the Kara Sea. *Deep-Sea Res.* 14(1):19–28.

King, E. R., I. Zietz, and L. R. Alldredge. 1966. Magnetic data on the structure of the Central Arctic region. *Geol. Soc. Amer. Bull.* 77(1):619–646.

Klenova, M. V. 1966. Barents Sea and White Sea. In: R. W. Fairbridge, ed. *Encyclopedia of Oceanography*, Vol. 1. New York, New York: Reinhold Publishing Corporation. pp. 95–101.

Kulikov, N. N., N. N. Lapina, IU P. Semenov, N. A. Belov, and M. A. Spiridonov. 1970. Stratification and rate of accumulation of bottom deposits in the Arctic Seas of the USSR. *The Arctic Ocean and its Shores during the Cenozoic.* Leningrad: Hydrometeorological Publ. pp. 34–41.

Kutschale, H. 1966. Arctic Ocean geophysical studies: the southern half of the Siberian Basin. *Geophysics* 31:683–710.

Linkova, T. I. 1965. Some results of paleomagnetic study of Arctic Ocean floor sediments. In: *The Present and Past of the Geomagnetic Field* (translated). Moscow: Nauka Press. pp. 279–291.

McManus, D. A., J. C. Kelley, and J. S. Creager. 1969. Continental shelf sedimentation in an Arctic environment. *Geol. Soc. Amer. Bull.* 80(10):1961–1984.

Moore, D. G. 1964. Acoustic-reflection reconnaissance of continental shelves: Eastern Bering and Chukchi Seas. In: Robert L. Miller, ed. *Papers in Marine Geology.* New York: Macmillan Company. pp. 319–362.

Nansen, F. 1904. The bathymetrical features of the North Polar Seas. In: Fridtjof Nansen, ed. *The Norwegian North Polar Expedition, 1893–1896.* London: Longmans, Green, and Co., 4(13). pp. 1–232.

Oliver, J., M. Ewing, and F. Press. 1955. Crustal structure of the Arctic regions from the log phase. *Geol. Soc. Amer. Bull.* 66:1063–1074.

Ostenso, N. A. 1966a. Arctic Ocean. In: R. W. Fairbridge, ed. *Encyclopedia of Oceanography*, Vol. 1. New York, New York: Reinhold Publishing Corporation. pp. 49–55.

———. 1966b. Beaufort Sea. In: R. W. Fairbridge, ed. *Encyclopedia of Oceanography*, Vol. 1. New York, New York: Reinhold Publishing Corporation. pp. 119–121.

Panov, D. G. 1955a. Neotectonic movements in the Arctic region (translation). *Dok. Akad. Nauk* 104(3):462–465.

———. 1955b. Tectonics of the Central Arctic (translation). *Dok. Akad. Nauk* 105(2):339–342.

Pelletier, B. R. 1962. Submarine geology program, Polar Continental Shelf Project, Isachsen, District of Franklin. *Geol. Surv. Canada Pap.*, 61–21:1–10.

———. 1966a. Canadian Arctic Archipelago (Bathymetry and Geology). In: R. W. Fairbridge, ed. *Encyclopedia of Oceanography*, Vol. 1. New York, New York: Reinhold Publishing Corporation. pp. 160–168.

———. 1966b. Development of submarine physiography in the Canadian Arctic and its relation to crustal movements. In: G. D. Garland, ed. *Continental Drift*, Vol. 9. Toronto: University of Toronto Press (Royal Society of Canada). pp. 77–101.

Rassokho, A. I., L. I. Senchura, R. M. Demenitskaya, A. M. Karasik, Yu. G. Kiselev, and N. K. Timoshenko. 1967. The Mid-Arctic Range as a unit of the Arctic Ocean mountain system. *Dok. Akad. Nauk SSSR* 172(3):659–662.

Saks, V. N., N. A. Belov, and N. N. Lapina. 1955. Our present concepts of the geology of the central Arctic. *Priroda* (translated in: Defense Research Board of Canada, Translation no. T196R, 1955, 11 pp.), 7:13–22.

Shaver, R., and K. Hunkins. 1964. Arctic Ocean geophysical studies: Chukchi Cap and Chukchi Abyssal Plain. *Deep-Sea Res.* 11:905–916.

Sykes, L. R. 1965. The seismicity of the Arctic. *Seismological Soc. Amer. Bull.* 55:501–518.

Treshnikov, A. L. 1960. The Arctic discloses its secrets: new data on the bottom topography and waters of the Arctic Basin. *Priroda* (translated in: Defense Research Board of Canada, Translations no. T357R, August, 1961), 2:25–32.

———, L. L. Balakshin, N. A. Belov, R. M. Demenitskaya, V. D. Dibner, A. M. Karasik, A. O. Shpeiker, and N. D. Shurgayeva. 1967. A geographic nomenclature for the chief topographic features of the Arctic Basin floor. *Problems of the Arctic and Antarctic* (in Russian), 27.

Vogt, P. R., and N. A. Ostenso. 1971. Geophysical studies in Barents and Kara Seas. *Second International Symposium on Arctic Geology*, San Francisco, 1971 (abstract). pp. 55–56.

Chapter 3

Tectonic History of the Arctic Basins: Partial Solutions and Unsolved Mysteries

PETER R. VOGT[1] and OTIS E. AVERY[1]

Abstract

As yet inaccessible to deep-sea drilling and shipborne surveying, the ice-locked Arctic Basin has been slow to give up details of crustal genesis and later modification. Extensive geophysical surveying both in the Arctic Basin and the North Atlantic confirms sea-floor spreading as the only probable mode of crustal genesis for the Eurasia Basin. Magnetic anomalies, although less clear than elsewhere, suggest spreading rates of approximately 0.5 to 1 cm/yr since 10 m.y.b.p.; the basin was born perhaps 60 m.y.b.p., with the separation of the Lomonosov Ridge from the Eurasian margin. The Eurasia Basin terminates rather abruptly, still 1000 to 2000 km from the plate rotation pole. It is unknown exactly how northern Siberia has accommodated up to 500 km crustal extension in post-Cretaceous times. Other mysteries include the abnormally low magnetic anomaly amplitudes and deep basement levels of the Eurasia Basin; the spreading axis is 0.5 to 1.0 km below the -2.8 km average for the Mid-Oceanic Ridge. This topographic anomaly seems part of a much larger, regional pattern centered on Iceland, where the crustal elevation is over 3 km above the norm. Crustal, and probably also older basement elevations decrease monotonically northward and southward. Elevations of the adjacent continental margins appear to mirror this pattern. Current speculations link the regional topographic high to an asthenospheric bulge pumped from the lower mantle by a convection plume under Iceland. The abrupt beginning, or rejuvenation, of the Iceland plume about 70 to 60 m.y.b.p. and its subsequent fluctuations may have played a primary role in cracking and then forcing the plates apart.

Much less can be said about the Amerasia Basin beyond the Lomonosov Ridge. Although the Alpha-Mendeleev Cordillera resembles the Mid-Atlantic Ridge in some ways, earlier speculations that the Alpha Ridge is a fossil-spreading axis, perhaps active until early Tertiary times, still remain untested for lack of detailed data. The Alpha Ridge exhibits magnetic and topographic peculiarities quite the opposite from those of the present Nansen (Gakkel) Ridge: With a -2 or -2.5 km basement depth the Alpha-Mendeleev Cordillera is 2 to 3 km too shallow for a spreading center inactive since 40 m.y.b.p.; the magnetization (or layer thickness) is twice the typical value to explain magnetic anomalies. Limited seismic refraction work supports neither of two alternate hypotheses: No thick oceanic crustal layer, suggestive of an Icelandic-type aseismic ridge, nor continental velocities and thicknesses, suggestive of foundered continental crust, have been reported.

Least of all is known about the crust below the Canada Basin. Ancient ocean floor spreading, possibly of the "inter arc" type that occurs behind island arcs, appears the most plausible mechanism.

Introduction

Tectonics of the oceanic Arctic (Figs. 1a, b, c, and 2) is a subject that desperately needs new

[1] U.S. Naval Oceanographic Office, Washington, D.C. 20373, U.S.A.

primary data, not more review papers. With this in mind, and at the risk of imparting a certain personal prejudice, we shall concentrate on the major results and some problems at the frontier between understanding and speculation. For excellent and comprehensive reviews of the existing data base and evolution of ideas about the Arctic Basin, the reader is referred to Demenitskaya and Hunkins (1971), Demenitskaya and Karasik (1971), Hunkins (1969), Ostenso (1972), and Johnson and Vogt (1972). Churkin (1972) analyzes the extension of the active plate boundary into Siberia, and discusses circum-Arctic geology from a plate tectonic viewpoint. The evolution of the Greenland, Norwegian, and Labrador seas has been treated by Pitman and Talwani (1972), Vogt and Avery (1973), Laughton et al. (1972), and Le Pichon et al. (1971).

Although the ice-covered Arctic Basin should receive primary emphasis here, the realities of plate tectonics (Morgan, 1968; Le Pichon, 1968) demand that the North Atlantic down to the latitude of the Azores be considered as well. There is only one plate boundary—a branch of the mid-oceanic ridge—currently passing through the high Arctic. The story of sea-floor spreading from the Nansen Cordillera (Gakkel Ridge) in the Eurasia Basin must echo the story of spreading from Mohns Ridge in the Greenland Basin or Reykjanes Ridge in the Iceland Basin. Magnetic anomalies and fracture zones have left a fairly clear record of the speeds and directions of sea-floor spreading that opened all these basins simultaneously, largely in the last 60 million years (Pitman and Talwani, 1972). Yet, there are some peculiar features of this history, and these will be considered in the first part of this paper (for examples, the magnetic anomalies in the Eurasia Basin are of exceptionally low amplitude, and the basement is exceptionally deep).

Recent thinking places a mantle convection plume under Iceland (Morgan, 1972). The location of this plume and past fluctuations in discharge (Vogt, 1971) seem to provide some fertile possibilities for understanding the evolution of the Arctic and the sub-Arctic.

Following the development and application of the mantle plume concept, the next problem to be discussed concerns the nature of plate motion beyond the rather blunt terminus of the Eurasia Basin. This is perhaps the only good example of an oceanic rift approaching its own rotation pole in a continent.

The northern edge of the Pacific plate does reach Arctic latitudes along the Aleutian arc and southern Alaska; however, this region cannot be fitted into the framework of a regional Arctic review unless the scope of this paper were extended to cover much of the Pacific as well. Thus, plate tectonics offers clear guidelines for regional discussions—a considerable improvement over the more or less arbitrary geographical boundaries of an earlier era.

If the Eurasia Basin and the Lomonosov Ridge—probably a detached fragment of the old Eurasian margin—are regarded as "understood" from the standpoint of genesis, then the Amerasia Basin must be labeled "unknown." Although surface wave dispersion (Hunkins, 1963), earthquake body waves (Gutenberg and Richter, 1954), and a seismic refraction station (Hunkins, 1961) long ago pointed to oceanic mantle and crust, the existing magnetic, bathymetric, and seismic reflection data have not unambiguously demonstrated an origin by sea-floor spreading. This may partly be explained by the relative paucity of data: Detailed aeromagnetic surveys carried out by the U.S.S.R. over the Eurasia Basin (Karasik, 1968) stand in contrast to the wide-spaced reconnaissance data available from the Amerasia Basin (Ostenso and Wold, 1971; Wold and Ostenso, 1971; Riddihough et al., 1973). However, if magnetic lineations were as well developed over the Amerasia Basin, as they are over many (although not all) oceanic regions created by spreading, they would have been identified, even with the scarce existing data. If this presumably older portion of the Arctic Basin were connected in some obvious geometric fashion with dated oceanic crust elsewhere, this absence of dated lineations would not be so critical. As mentioned previously, the origin and history of the Eurasia Basin could have been elucidated entirely from magnetic lineations charted in coeval basins to the south (Pitman and Talwani, 1972). However, it seems that no similar coeval oceanic crust exists for the Amerasia Basin. There is one possible exception to this: The Alpha-Mendeleev Ridge may possibly form a northward extension of the Mid-Labrador (or Ran) Ridge and Labrador Basin, where spreading probably occurred between about 80 and 40 m.y.b.p. (Le Pichon et al., 1971; Keen et al., 1972; Laughton and Berggren et al., 1972; Vogt and Avery, 1973) but is nearly or totally extinct today. If this connection were real, the Alpha-Mendeleev Cordillera would be a "fossil" mid-oceanic ridge

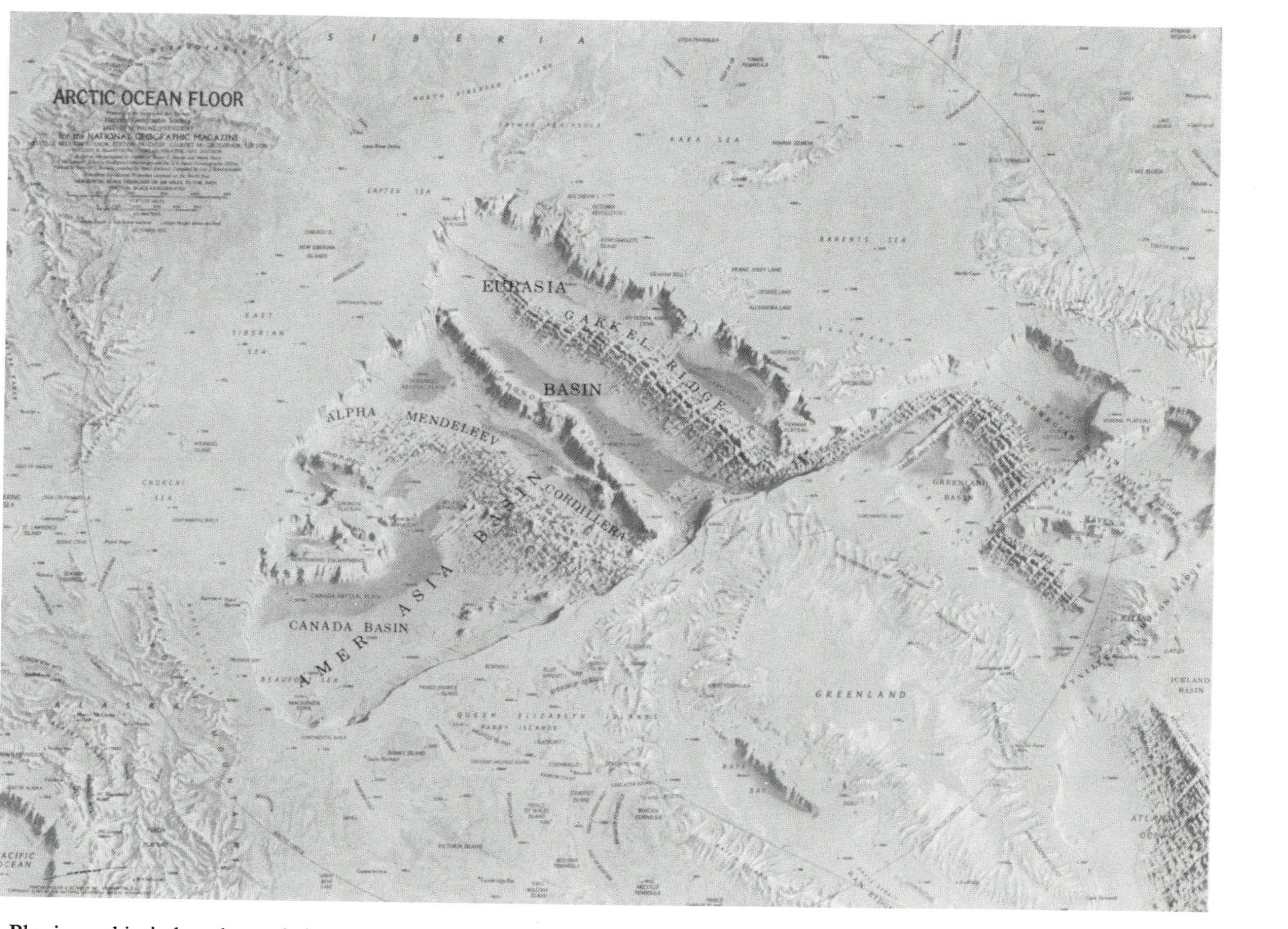

Fig. 1a. Physiographic index chart of the Arctic and subarctic ocean floors, modified from the National Geographic chart (Anonymous, 1971) by addition of the following place-names: Amerasia Basin, Alpha-Mendeleev Cordillera, Eurasia Basin, Gakkel Ridge, Spitsbergen fracture zone, Greenland Basin, Mohns Ridge, Norwegian Basin, Kolbeinsey Ridge, Jan Mayen Ridge, Aegir Ridge, Iceland Basin, and Ran Ridge. The chart is a reasonable interpretation of limited available data; all the larger features no doubt exist approximately as shown; details, such as the numerous transverse fractures shown cutting Gakkel, Mohns, and Kolbeinsey ridges and to a more subdued degree the Alpha-Mendeleev Cordillera, are for the most part highly speculative.

whose evolution could be analyzed with the help of data from the Labrador Sea (Vogt and Ostenso, 1970). However, there are several conflicting clues about the origin of the Alpha-Mendeleev Cordillera; as will become clear in the discussion below, this ridge is still as mysterious as any on the ocean floors.

Between the Alpha-Mendeleev Cordillera and the continental slopes of Alaska and eastern Siberia lies the Canada Basin; there is very little

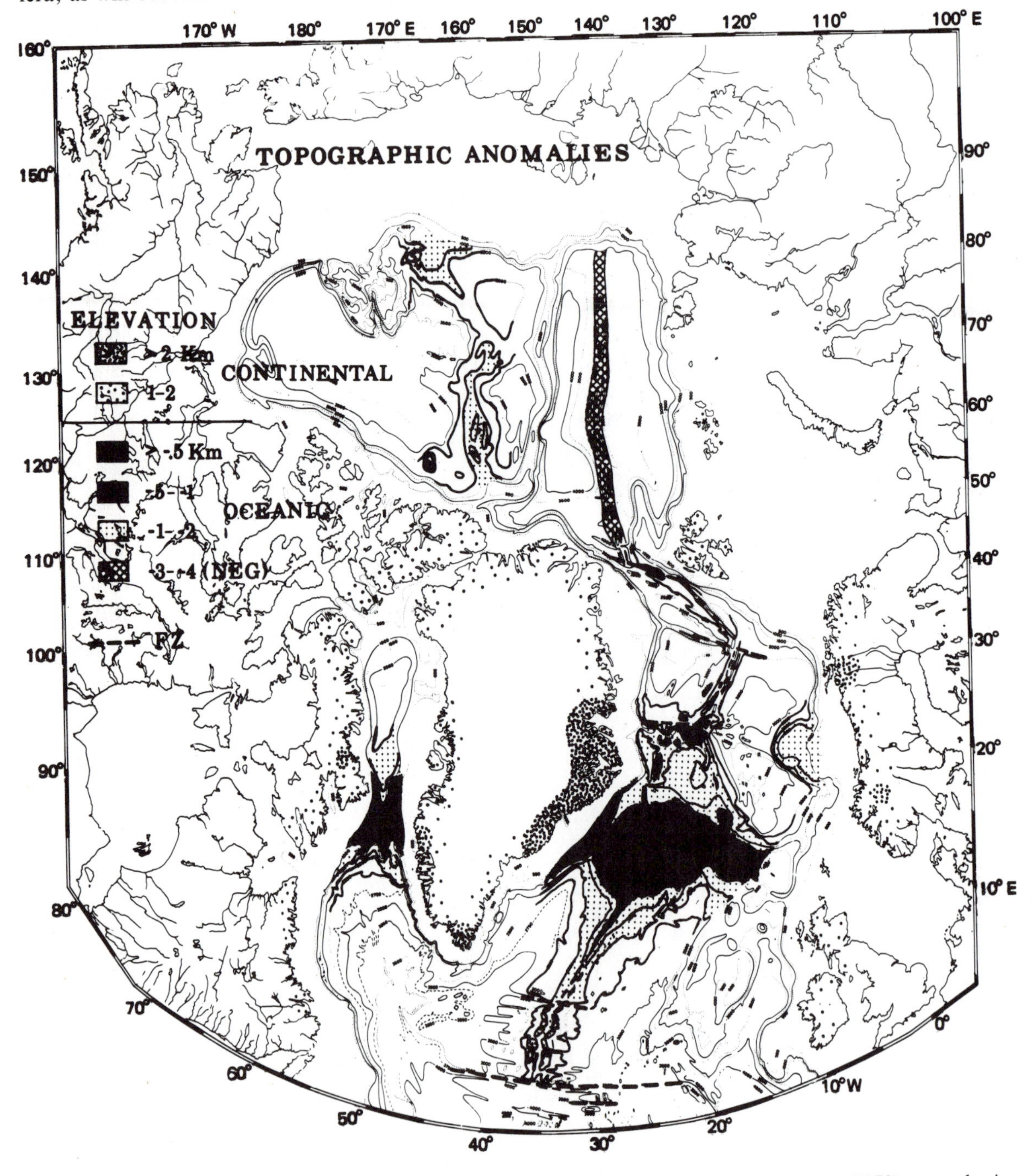

Fig. 1b. Bathymetry of the Arctic and sub-Arctic basins, modified from Johnson and Vogt (1972) to emphasize regional extent of Iceland regional topographic anomaly. Normal depth at crest of active mid-oceanic ridge is -2.5 to -3 km (Sclater et al., 1971). The crest of the Gakkel (Nansen) Ridge is 0.5 to 1 km below normal, whereas crest of the Alpha-Mendeleev Cordillera is at least 3 km too shallow for a mid-oceanic ridge inactive since 40 m.y.b.p. (Vogt and Ostenso, 1970) or earlier. Dotted patterns show maximum epeirogenic elevations, excluding glacier ice, on adjacent continents: wide dot spacing indicates maximum elevations of 1 to 2 km; maximum elevations exceed 2 km in regions of close dot spacing.

that can be said about the age and origin of this part of the Amerasia Basin that is not outright speculation. The Canada Basin will appropriately be discussed last.

No introduction to an Arctic review paper would be complete if it left the impression that Arctic geophysics is the usual brand of marine geophysics, but carried out in northern seas. It is readily apparent that research has been hampered and made expensive by the severe climate; the main handicap is sea ice, however, which precludes surface ship oceanography. This inaccessibility has discouraged nations other than the U.S.S.R., the United States, and Canada from undertaking major geophysical experiments in the Arctic Basin. American research of the Arctic Basin generally has been more modest than the Soviet effort, and, if anything, American efforts have declined. There are few American research results known today that were not already available five years ago. It is doubtful whether such an unhappy assessment could be made about another major region of the world ocean. Again because of its ice cover, the Arctic Basin has been bypassed by the Deep Sea Drilling Program. Results of deep drilling in most parts of the world ocean have

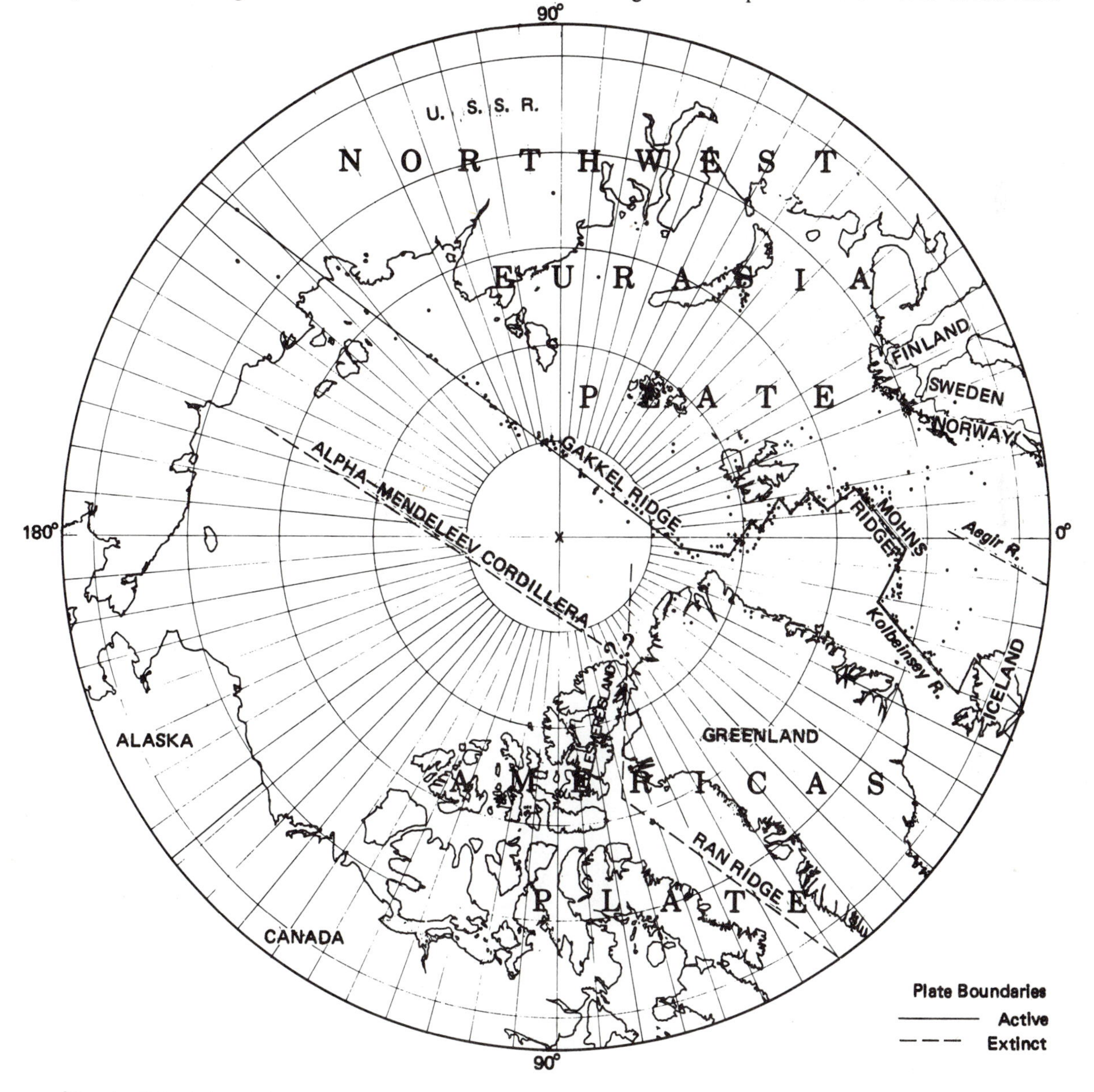

Fig. 1c. Seismicity of the Arctic (1961 to 1968) with approximate active and extinct plate boundaries. Modified from Vogt et al. (1970a).

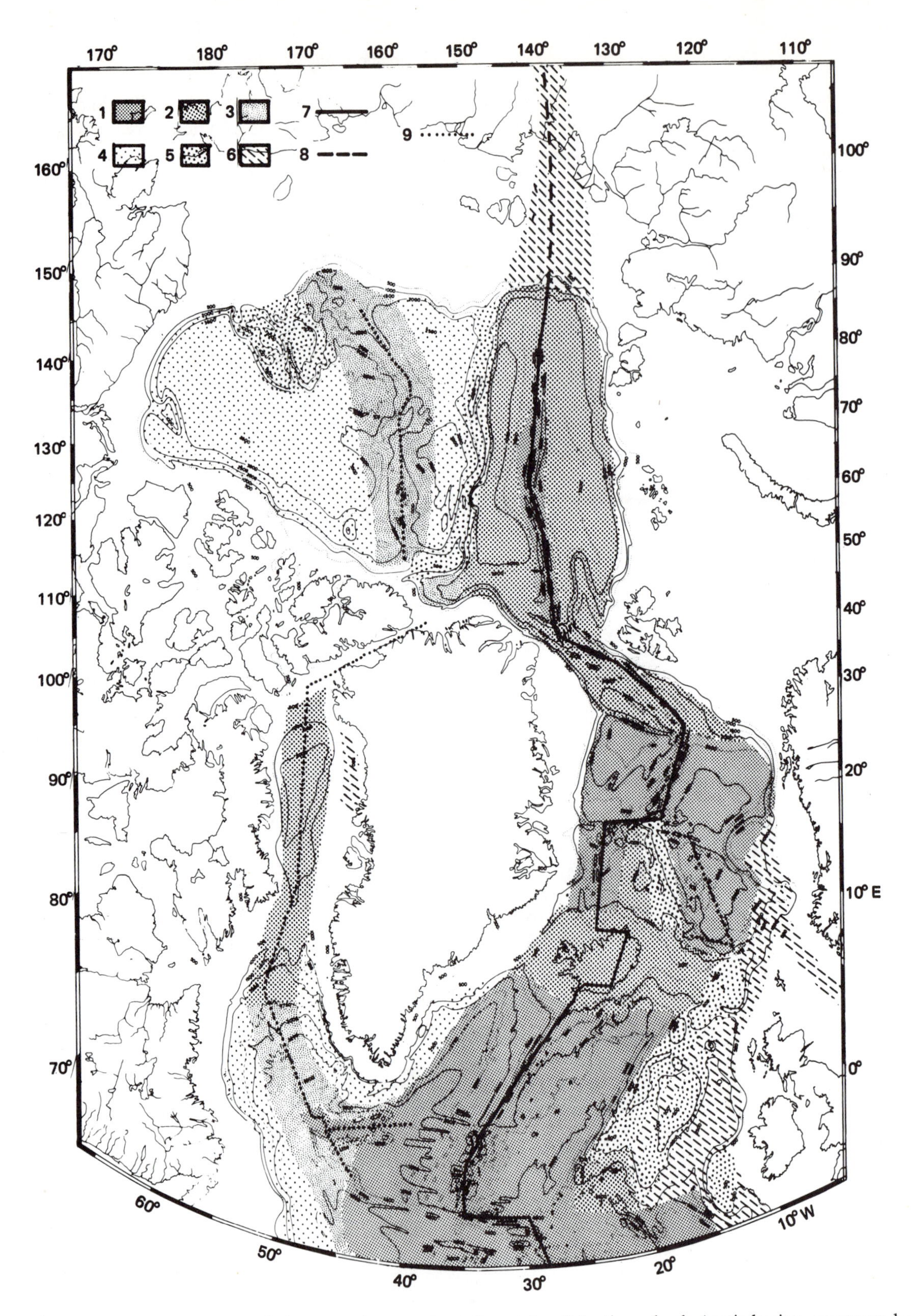

Fig. 2. Present state of knowledge or conjecture about the origin of Arctic and sub-Arctic basins, superposed on bathymetry (Fig. 1b): (1) ocean crust in which magnetic lineations due to sea-floor spreading have been charted or can be projected with high probability; (2) oceanic crust with magnetic lineations not well developed; age of crust predicted with plate tectonic methods; (3) probable oceanic crust, in which magnetic lineations appear to exist, but whose age is disputed; (4) probable oceanic crust of uncertain age or origin; magnetic anomalies subdued or not well developed; (5) continental or transitional crustal structure demonstrated or probable; (6) subsided or rifted

given immense stimulation to marine geology and geophysics. Research in the Arctic Basin needs a similar stimulus; it appears likely to us that surviving problems of age, crustal genesis, and evolution will remain unsolved until the technological barriers to deep drilling there are overcome.

Sea-Floor Spreading in the Northeast Atlantic and Eurasia Basin—Effects of the Iceland Mantle Plume?

1. History of Spreading

The essence of current thinking about plate movements in the Arctic and sub-Arctic (Pitman and Talwani, 1972) and the Iceland mantle hot spot (Morgan, 1972) was anticipated almost a decade ago by J. T. Wilson (1963). The idea of continental drift in this region is of course much older (Wegener, 1915).

The recent application of plate tectonic methods to the history of relative motion between the American and west-Eurasian lithospheric plates in a sense began with the aeromagnetic survey (Fig. 3) conducted by Project MAGNET over Reykjanes Ridge in 1963 (Baron et al., 1965). The significance of the magnetic anomaly pattern, which demonstrated symmetrical spreading at rates of 1 cm/yr, was not realized until later (Vine, 1966). About the time of the formulation of a global plate tectonic theory (Morgan, 1968; Le Pichon, 1968) came detailed magnetic surveys that described linear anomalies of sea-floor spreading origin in the Norwegian Sea (Fig. 3; Avery et al., 1968), Eurasia Basin (Fig. 4; Karasik, 1968), the Iceland Basin east of Reykjanes Ridge (Figs. 5 and 6), and the southern Labrador Sea (Fig. 7; Vogt and Avery, 1973). The results of these surveys have been extended by additional work on the Reykjanes Ridge (Talwani et al., 1971; Fleischer, 1971), a detailed survey of the Iceland-Faeroe aseismic ridge (Fleischer, 1971), and a semi-detailed survey of the Iceland–Jan Mayen (Kolbeinsey) Ridge (Johnson et al., 1972a), which has been supplemented by a detailed geophysical survey (Meyer et al., 1972). Considerable magnetic data, although inadequate for contouring and tracing of lineations, have also been recovered from the Davis Strait and Baffin Bay (Keen et al., 1972). Older data from the northeast Atlantic include a semi-detailed aeromagnetic survey of Iceland (Serson et al., 1968) and a semi-detailed shipborne survey of Mohns Ridge (Johnson and Heezen, 1967a).

The complexities of spreading in the Reykjanes Ridge area (Vogt and Avery, 1973) are shown in a series of schematic reconstructions (Fig. 8). By rotating successive well-defined magnetic isochrons in a trial-and-error fashion, using the method of Bullard et al. (1965), Pitman and Talwani (1972) have reconstructed the relative motion of the Greenland, American, and west Eurasian plates (Fig. 9). Le Pichon et al. (1971) and Laughton et al. (1972) also made plate reconstructions of the region of junction between the Labrador Sea and the northeastern Atlantic; Keen et al. (1972) identified oceanic crust under Baffin Bay and made appropriate reconstructions of Canada and Greenland. These various reconstructions agree with Fig. 8, except for details. Many other workers have contributed to unraveling this story; a more comprehensive bibliography is to be found in Vogt and Avery (1973). Figure 2 shows, in simplified fashion, the current state of knowledge about the various basins of the Arctic and sub-Arctic.

Although some details remain in dispute, this is the essence of the story: Prior to 80 m.y.b.p., both the Labrador Sea and the Eurasian, Greenland, and Iceland basins were certainly closed, although early continental rifting and shallow seaways, perhaps as old as Jurassic, may have already existed in what was to become the Labrador Sea (Johnson et al., 1972b; Watt, 1969) and in the Rockall-Hatton, Rockall, and Porcupine troughs (Bailey et al., 1971; Laughton and Berggren et al., 1972; Vogt and Avery, 1973), and possibly elsewhere (Fig. 2). Between 80 and 60 m.y.b.p. true ocean-floor spreading in the North Atlantic had advanced as far north as the southern edge of Rockall Bank, and a continuation of this rift had probably opened half the present width of the Labrador Sea (Le Pichon et al., 1971). However, magnetic anomalies of that age have not been certainly identified in the Labrador Basin. According to Pitman and Talwani (1972), the period 81 to 63 m.y.b.p. was characterized by compressional tectonics in the Arctic.

continental or transitional crust, or undated oceanic crust deeply buried along continental margins or intracontinental basins; (7) active mid-oceanic ridge axis; (8) continental extension of axis; (9) probable or conjectural extinct spreading axes.

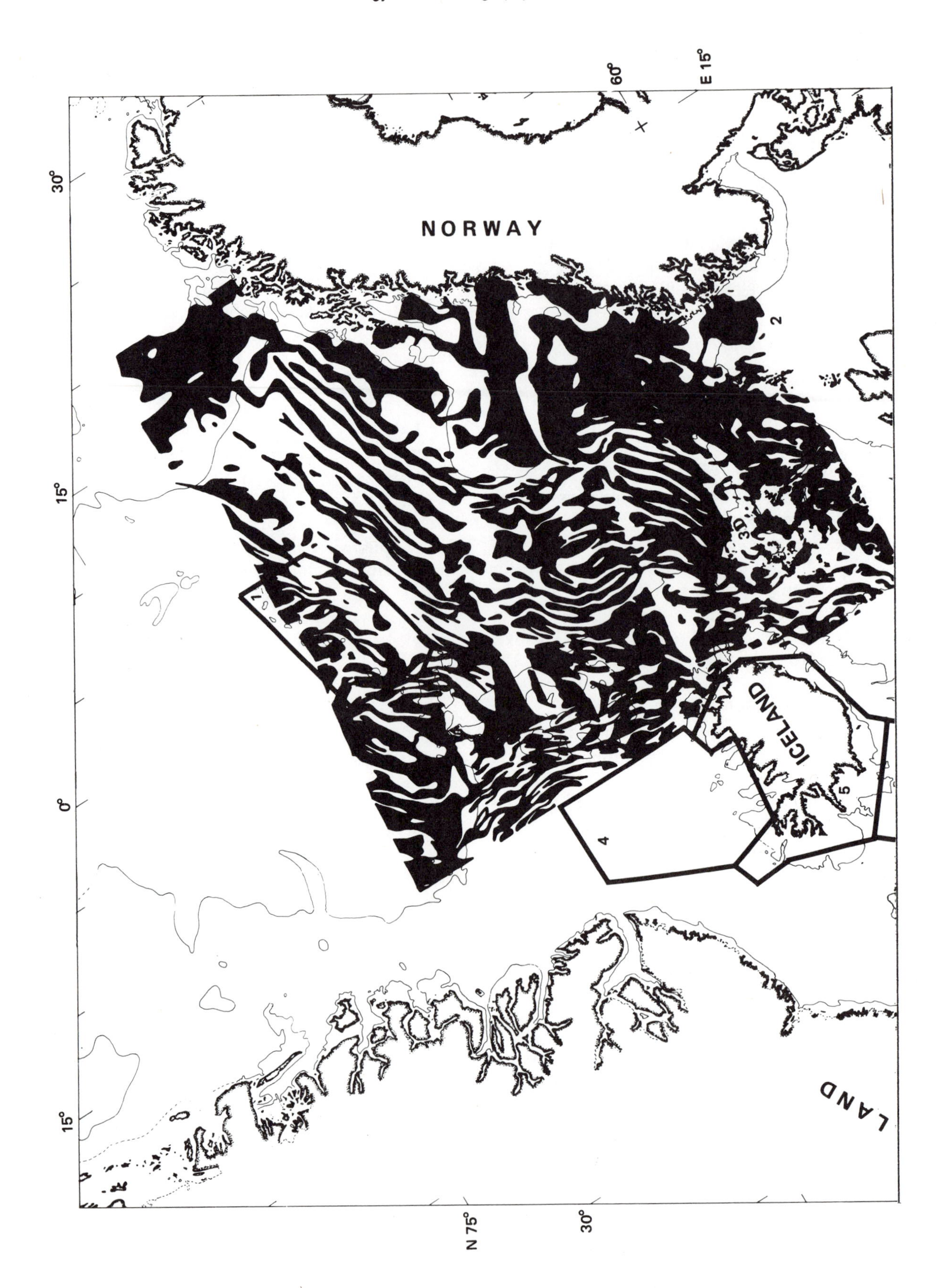
NORWAY
ICELAND
LAND
60°
E 15°
30°
15°
0°
15°
N 75°
30°
1
2
3D
4
5

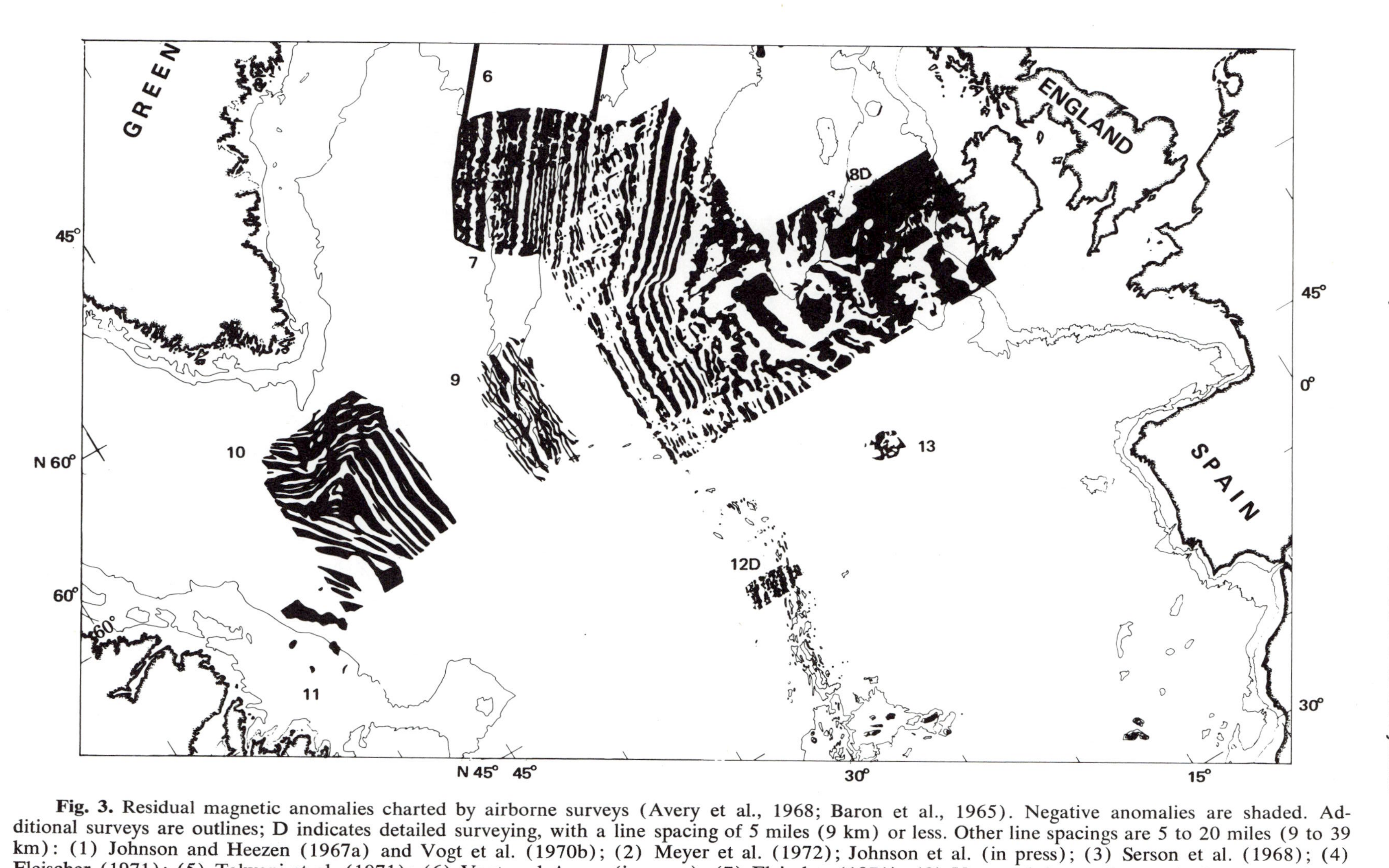

Fig. 3. Residual magnetic anomalies charted by airborne surveys (Avery et al., 1968; Baron et al., 1965). Negative anomalies are shaded. Additional surveys are outlines; D indicates detailed surveying, with a line spacing of 5 miles (9 km) or less. Other line spacings are 5 to 20 miles (9 to 39 km): (1) Johnson and Heezen (1967a) and Vogt et al. (1970b); (2) Meyer et al. (1972); Johnson et al. (in press); (3) Serson et al. (1968); (4) Fleischer (1971); (5) Talwani et al. (1971); (6) Vogt and Avery (in press); (7) Fleischer (1971); (8) Vogt and Avery (in press). Survey no. 3 was aeromagnetic; all other numbered areas were shipborne surveys. The earlier reconnaissance aeromagnetic survey of Godby et al. (1968) extends across Reykjanes Ridge from Greenland to Rockall-Hatton Bank.

About 60 m.y.b.p. Greenland began to separate from Europe, while Lomonosov Ridge separated from the Eurasian shelf (Fig. 9). Both this new ridge and the old Mid-Labrador Ridge, also called Ran Ridge (Johnson and Vogt, 1972) continued to spread until about 40 m.y.b.p., when the latter became essentially extinct. Almost simultaneously the spreading axis between Iceland and the Jan Mayen fracture zone (Aegir Ridge) was abandoned, and one or even two subsequent rifts, the later one called Kolbeinsey Ridge (Johnson et al., 1972), thereupon developed along the continental margin of Greenland, splitting off, it is believed, the Jan Mayen Ridge as a southern analogue to the Lomonosov Ridge. In the period 60 to 40 m.y.b.p. spreading rates in the basin west of Reykjanes Ridge fell from about 1.7 to 0.9 cm/yr, while the corresponding value in the Labrador Basin was on the order of 0.5 cm/yr. There is still an occasional earthquake epicenter in the Labrador Sea, and Le Pichon et al. (1971) believe Ran Ridge may yet be spreading at ultra-slow rates to this day. Both fossil ridges locally exhibit rift valleys along their axes, although no rift valley has been found under Baffin Bay and the Davis Straits. Anomalies of age 60 to 40 m.y.b.p. have not been identified with certainty in the Eurasia Basin, but the wavelength increases about midway between the axis of Gakkel Ridge and the edges of the basin (Karasik, 1968; Demenitskaya and Karasik, 1971; Fig. 4). Because such a change in wavelength occurs at about 40 m.y.b.p. in the surveyed areas to the south (Figs. 3 and 5), this is very likely the 40 m.y.b.p. isochron in the Eurasia Basin.

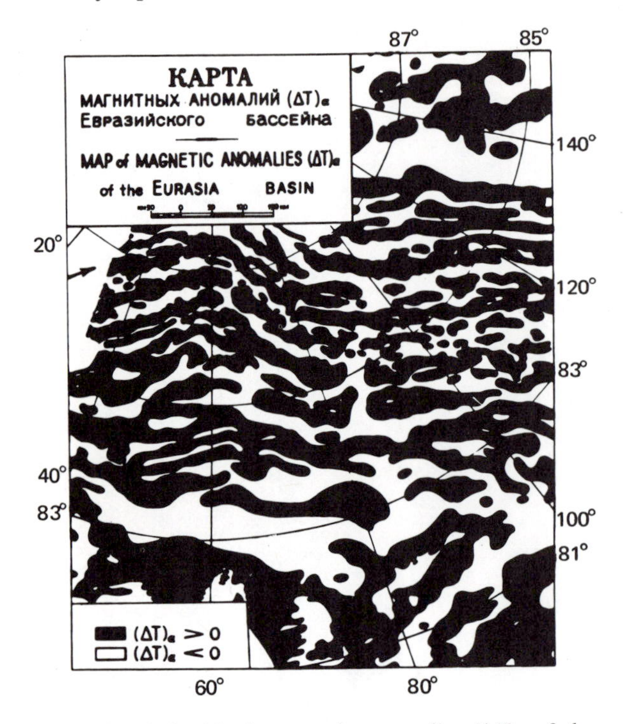

Fig. 4. Residual magnetic anomalies $(\Delta T)_a$ of the Eurasia Basin (after Karasik, 1968; and Demenitskaya and Karasik, 1971). Arrows show approximate location of spreading axis of the Gakkel (Nansen) Ridge.

Between about 40 and 20 m.y.b.p. spreading rates along the mid-oceanic ridge between Greenland and Europe had fallen to values of about 0.6 to 0.8 cm/yr (Vogt and Avery, 1973). Magnetic anomalies from this period are difficult to identify; close-spaced fracture zones had formed about 40 m.y.b.p. on the Reykjanes Ridge, probably as a response to a change in relative motion between Greenland and Europe (Fig. 8). About 20 m.y.b.p. these fracture zones began to be replaced by oblique spreading, while between 20 and 10 m.y.b.p. spreading rates began once more to increase. Spreading rates since 10 m.y.b.p., measured in the direction of relative plate motion, have been summarized by Johnson et al. (1972a). They range from about 1.4 cm/yr at 45°N (Pitman and Talwani, 1972) to possibly 0.55 cm/yr in the Eurasia Basin (Karasik, 1968; Demenitskaya and Karasik, 1971). The latter rate is the average value required to open the Eurasia Basin during the 60-m.y. period of opening of the Norwegian and Iceland basins.

2. The Iceland Mantle Plume

The development of a new rift at 60 m.y.b.p., followed by a slowing of plate motion toward the middle Tertiary and a renewed acceleration in the late Tertiary, has been attributed by Vogt (1971) to a similar fluctuation in discharge from a convection plume (Morgan, 1972a, 1972b) located under Iceland. It is now thought that the flow pattern set up in the asthenosphere by mantle plumes may help drive the lithospheric plates (Morgan, 1972a, 1972b). The fluctuations were measured by estimating the amount of excess basaltic crust produced per unit time as a function of age along the Greenland–Iceland–Faeroe aseismic ridge (Fig. 10). Time-transgressive V-shaped ridges on the Reykjanes Ridge appear to document a 10 to 20 cm/yr southwestward component of asthenosphere flow below the Reykjanes Ridge and directed away from Iceland (Vogt, 1971; Figs. 10 and 11). Although the measured speed may reflect kinematic waves or propagating fluid instabilities, it seems preferable to regard the speed as a

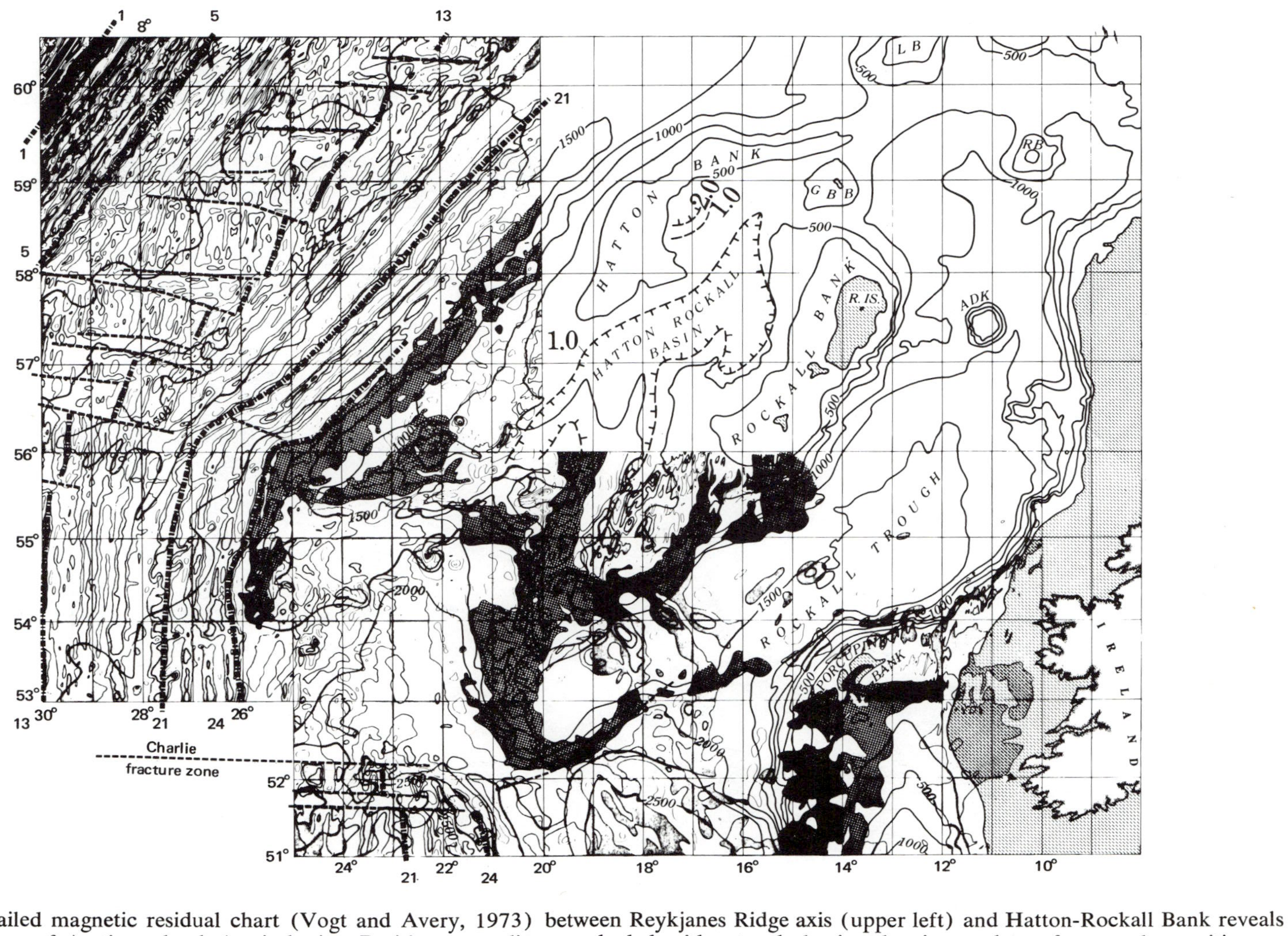

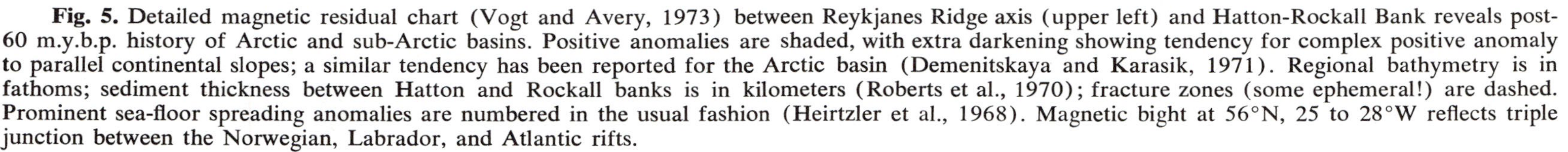
Fig. 5. Detailed magnetic residual chart (Vogt and Avery, 1973) between Reykjanes Ridge axis (upper left) and Hatton-Rockall Bank reveals post-60 m.y.b.p. history of Arctic and sub-Arctic basins. Positive anomalies are shaded, with extra darkening showing tendency for complex positive anomaly to parallel continental slopes; a similar tendency has been reported for the Arctic basin (Dementskaya and Karasik, 1971). Regional bathymetry is in fathoms; sediment thickness between Hatton and Rockall banks is in kilometers (Roberts et al., 1970); fracture zones (some ephemeral!) are dashed. Prominent sea-floor spreading anomalies are numbered in the usual fashion (Heirtzler et al., 1968). Magnetic bight at 56°N, 25 to 28°W reflects triple junction between the Norwegian, Labrador, and Atlantic rifts.

measure of mantle convection. As the bathymetric chart shows (Fig. 1b), there is a vast topographic anomaly centered on Iceland. A related gravity high shows that the excess mass of the topographic bulge overcomes the effect of a density deficit of the hot, rising plume (Morgan, 1972a, 1972b). It seems possible that the asthenosphere below this large region has been swelled by an enormous "laccolith" of mantle material disgorged from the mantle plume and presumably still in the process of spreading out. On Iceland the ridge crest is more than 3 km above the normal level of the Mid-Oceanic Ridge (Sclater et al., 1971). The ridge crest then descends northward and south-

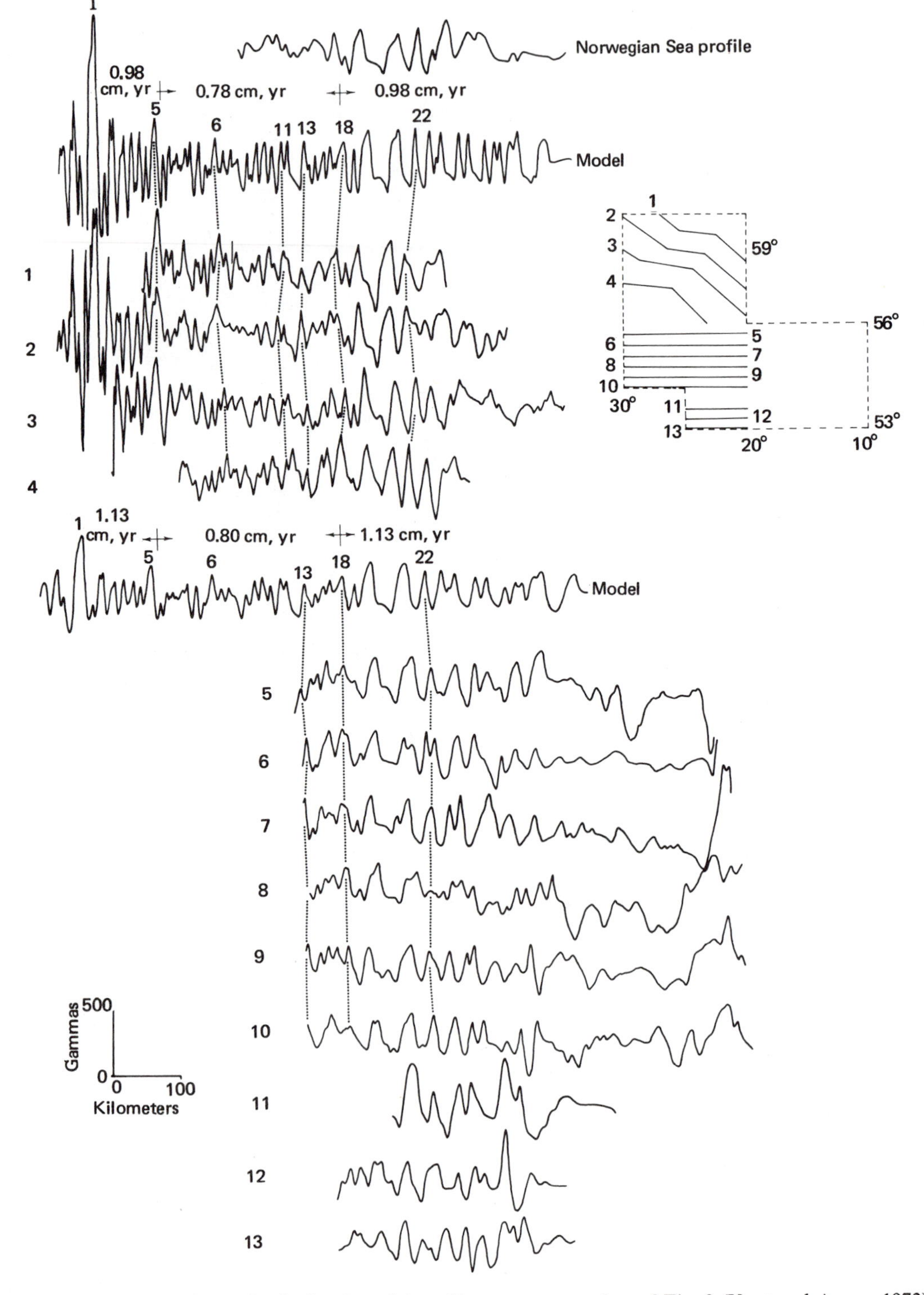

Fig. 6. Selected profiles and calculated model profiles across a portion of Fig. 3 (Vogt and Avery, 1973).

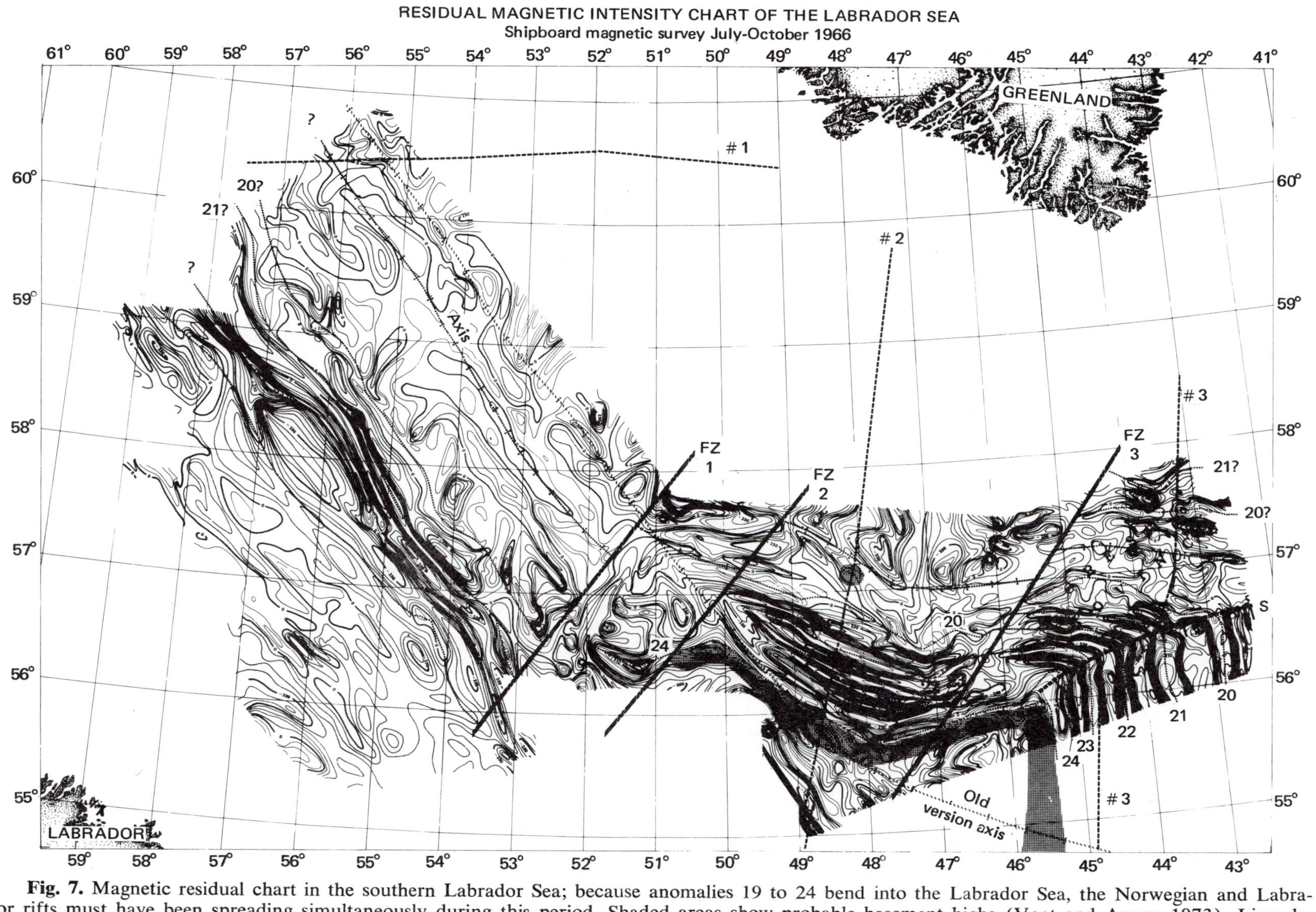

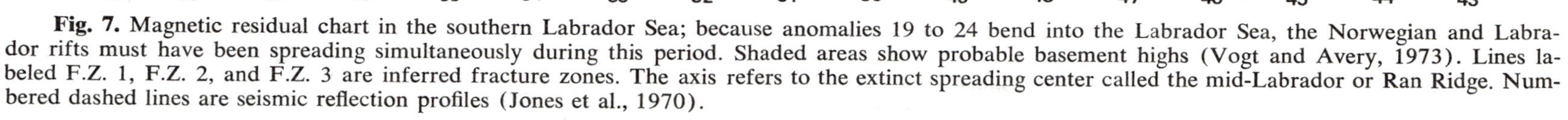

Fig. 7. Magnetic residual chart in the southern Labrador Sea; because anomalies 19 to 24 bend into the Labrador Sea, the Norwegian and Labrador rifts must have been spreading simultaneously during this period. Shaded areas show probable basement highs (Vogt and Avery, 1973). Lines labeled F.Z. 1, F.Z. 2, and F.Z. 3 are inferred fracture zones. The axis refers to the extinct spreading center called the mid-Labrador or Ran Ridge. Numbered dashed lines are seismic reflection profiles (Jones et al., 1970).

ward with a slope of about 10^{-3} to 10^{-4}. Discontinuous drops in elevation of the spreading axis occur at the Charlie Gibbs (53°N) and Jan Mayen (72°N) fracture zones (Figs. 1a and 1b), and Iceland is itself bounded by two major transform fractures (Fig. 10). These discontinuities could indicate that asthenosphere flow (Vogt, 1971) is fairly shallow (approximately 10 to 50 km below the ocean floor at the axis) and concentrated under the spreading axis rather than flowing

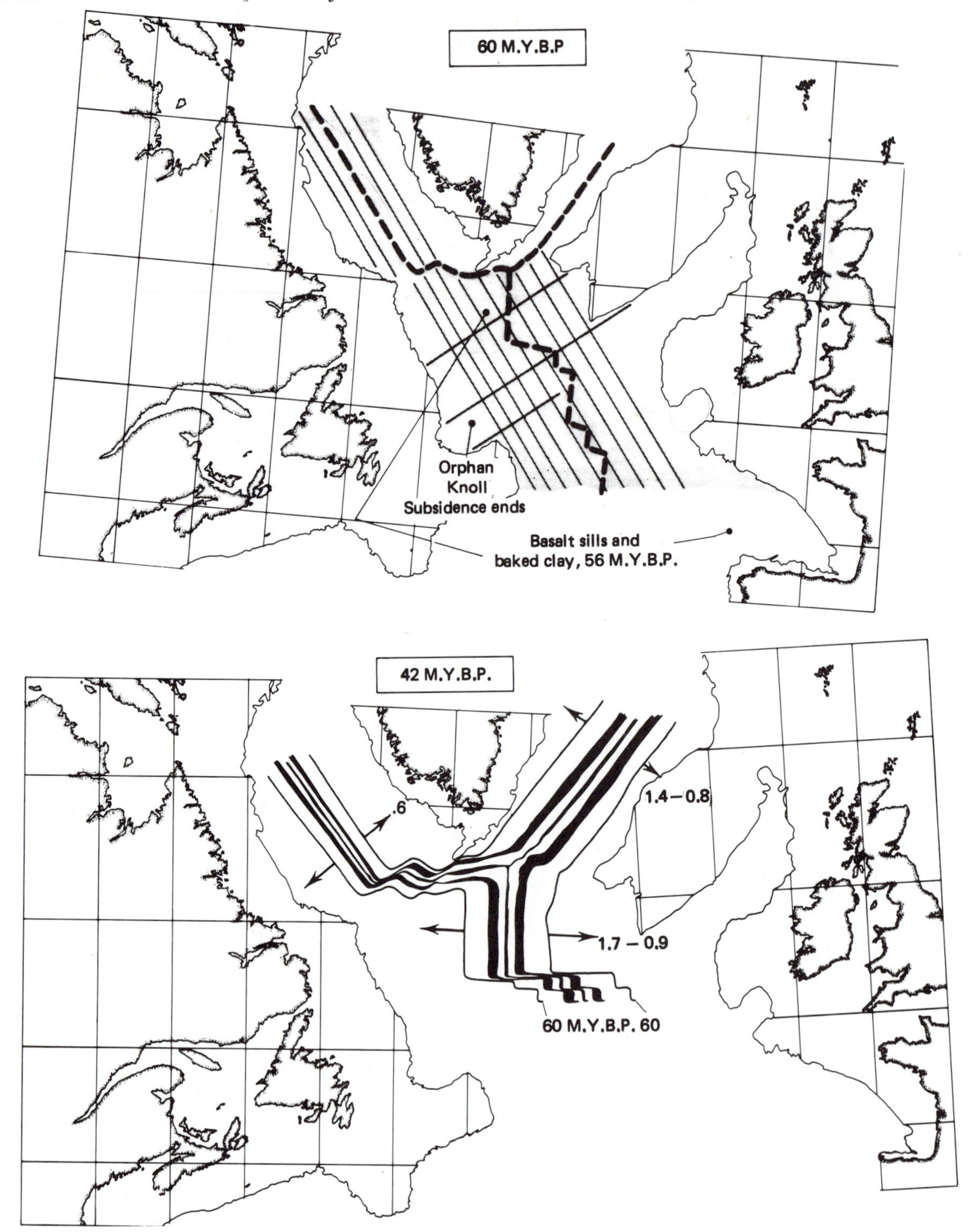

Fig. 8a. Four-stage semi-schematic evolution of the southern Labrador Sea and Reykjanes Ridge (Vogt and Avery, 1973).

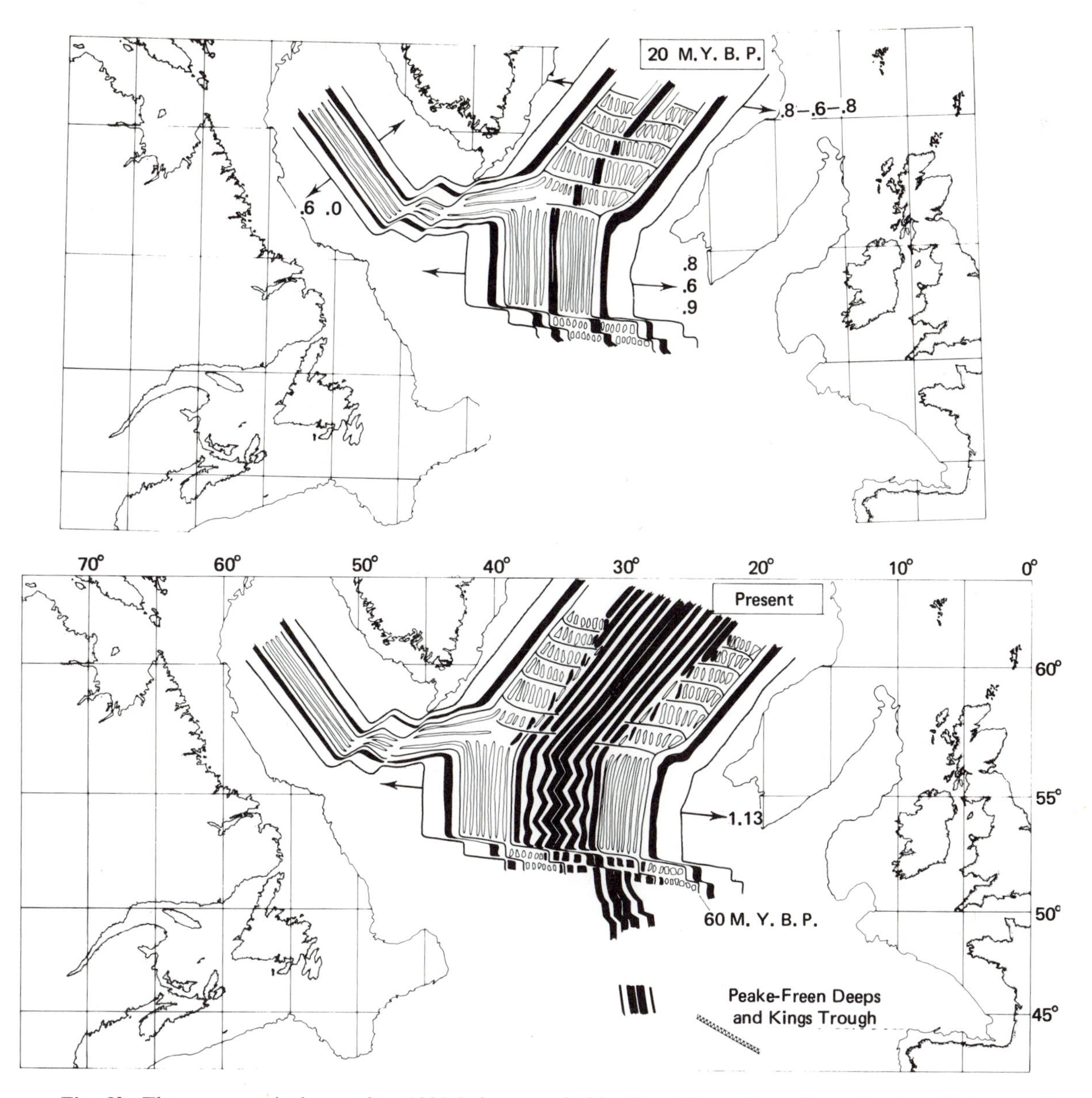

Fig. 8b. The gray area is deeper than 1000 fathoms and older than 60 m.y. Spreading rates at each stage apply to the period starting at the previous stage.

equally in all directions. The axially concentrated asthenosphere flow would be partially dammed wherever it encountered a barrier of abruptly thickened lithosphere at the transform fault. This might explain the discontinuous drops in regional depth.

If mantle density is fairly constant along the flow, the regional slope along the spreading axis in both directions away from the Iceland plume (about 10^{-3} to 10^{-4}) implies a horizontal pressure gradient directed southwest in the asthenosphere below the ridge crest. If this gradient is forcing the asthenosphere to flow at observed rates of the order 10 cm/yr (Vogt, 1971), the viscosity can be estimated by the Poiseuille equation. If the flow is approximated by a pipe of radius 25 km, and the 10 cm/yr is assumed to represent the mean speed within the pipe, a viscosity of about 10^{19} poises results. Although this is somewhat lower than the 4×10^{20} poises attributed to the typical asthenosphere (Cathles, 1971), a greater amount of partial melting below the spreading axis makes the difference understandable. Indeed, Einarsson (1966) has used postglacial rebound to show that subcrustal viscosity is an order of magnitude lower under Iceland than under Scandinavia.

As the bathymetry shows (Fig. 1b), Mohns Ridge is only slightly above the average −2.5 to −3.0 km elevation of the crest of the Mid-Oceanic Ridge (Sclater et al., 1971). Still further north, in the Eurasia Basin, the Gakkel (Nansen) Ridge continues the trend of northward deepening. This ridge barely rises above the −3.8 to −4.2 km level of the flanking abyssal plains. Although some peaks rise to −2.5 km, the bottom of the otherwise typically Mid-Atlantic rift valley in some areas drops below −5 km (Demenitskaya and Karasik, 1971). The average crestal elevation is apparently some 0.5 to 1.0 km *below* the norm for the Mid-Oceanic Ridge. The Paleogene flanks of the Gakkel Ridge are buried by 2 to 3 km of sediment (Demenitskaya and Kiselev, 1968). Thus the basement of lower Tertiary age appears to have subsided 3 or 4 km. This suggests that once the crust is generated by the Gakkel Ridge, it subsides in a normal manner, as described by Sclater et al. (1971). It is only the absolute elevation, not the profile, that is abnormal.

The regional variations of elevation along the crest of the Mid-Oceanic Ridge find a curious although somewhat less regular parallel on the coasts of adjacent continents (Fig. 1b). Maximum elevations exceeding 2 km are found in Norway and Greenland, in the general environs of the Iceland plume. Similarly, the highest elevations flanking the Baffin-Labrador rift are near the Davis Strait aseismic ridge, presumably a lesser cousin of the one produced by the Iceland plume. Toward the north and south, the edges of the continent descend; the abnormally deep Eurasia

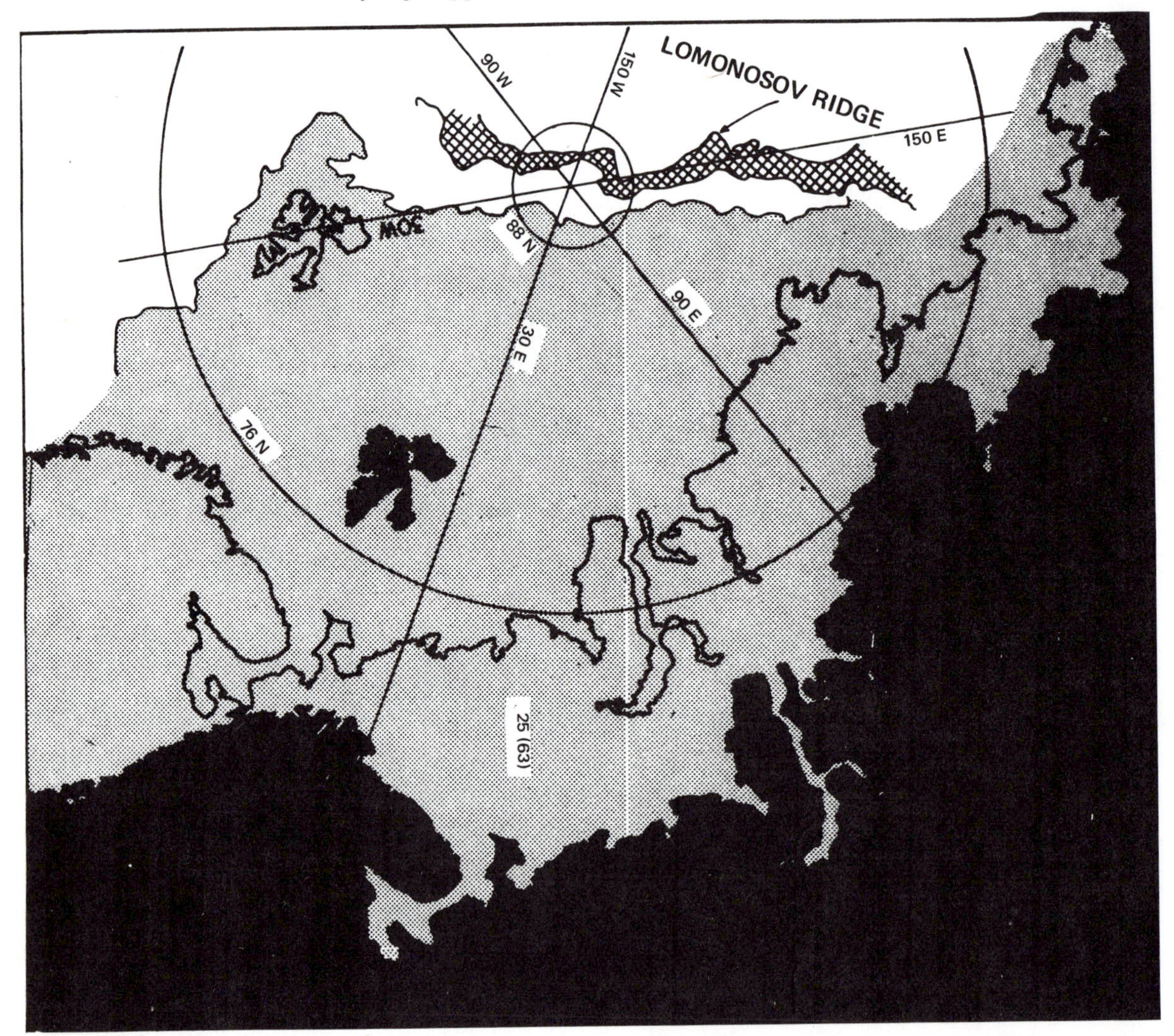

Fig. 9. Closure of post-63 m.y.b.p. ocean floor in the northeastern Atlantic also closes the Eurasia Basin, bringing the Lomonosov Ridge against the Eurasian margin. The present position of Eurasia is shown in black; the Lomonosov Ridge is arbitrarily held fixed. Reproduced from Pitman and Talwani (1972).

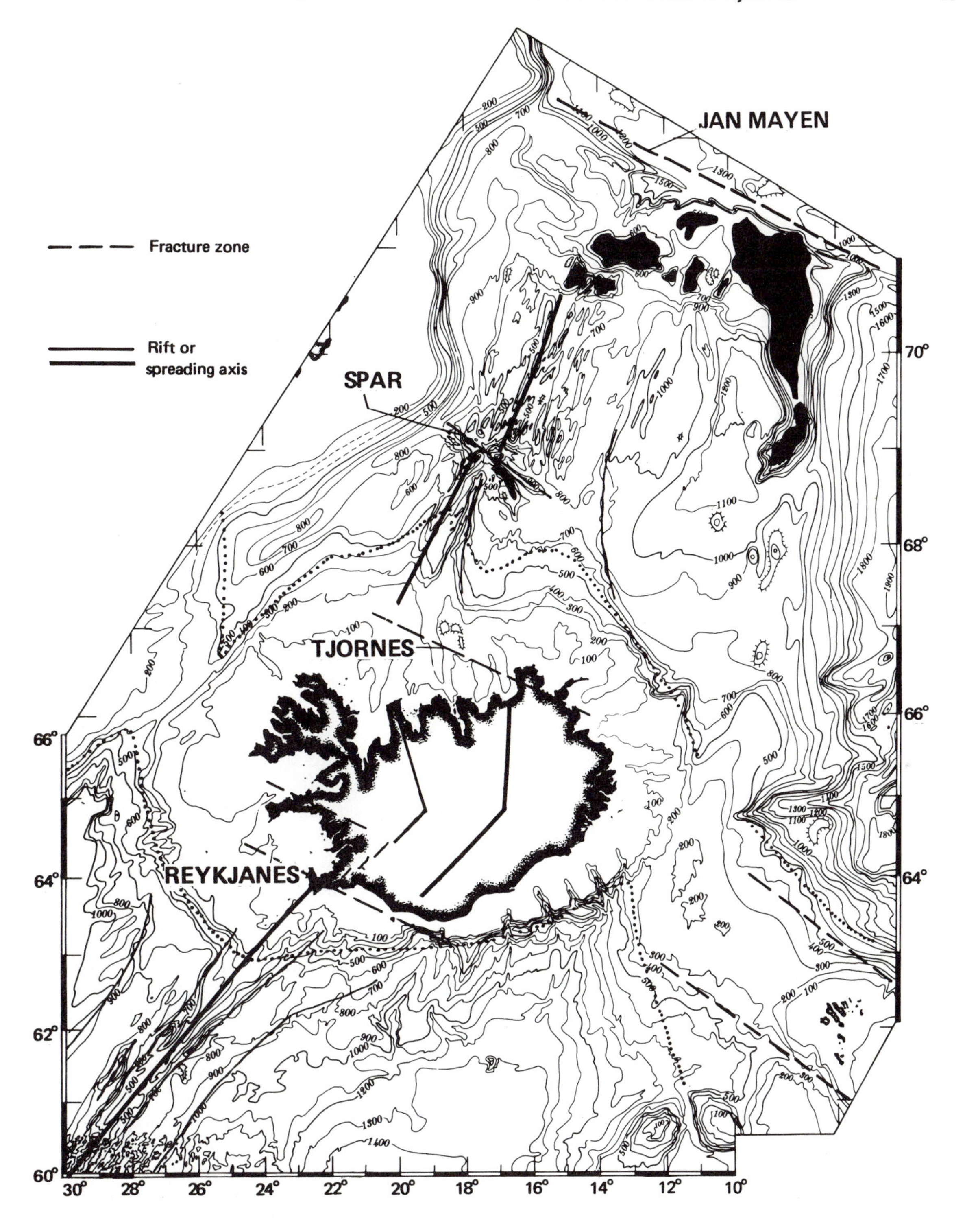

Fig. 10. Bathymetry and tectonic elements in the vicinity of Iceland (in fathoms), modified from Johnson and Vogt (1972) and Johnson et al. (1972). The dotted line shows variations in the width of the aseismic ridge, suggesting high plume discharge in early Tertiary, and again in late Tertiary times (Vogt, 1971). The light solid lines show time transgressive trends, suggesting asthenosphere motion away from the plume center; details of such trends on Reykjanes Ridge are shown in Fig. 11. Regions shallower than 500 fathoms north of the Iceland aseismic ridge are shaded black. The tectonic fabric on Iceland is simplified from Ward (1971); fractures near the Faeroes are deduced from a detailed magnetic survey of Fleischer (1971).

Basin is also flanked by abnormally deep margins. The Lomonosov Ridge is entirely below sea level, and the Barents–Kara–Laptev continental shelf is the world's widest, submerged except for isolated archipelagos. The connection between mid-oceanic and continental elevations could be explained in several ways; to the extent that the epeirogenic uplifts in Scandinavia, Greenland, and elsewhere are late Tertiary, the uplift could reflect asthenosphere pumped from the Iceland plume during its late Tertiary rejuvenation (Vogt, 1971). Some of the elevation could also have been derived from hot or otherwise relatively light materials (of a low enough density to reduce the average density of the plate) injected into the continental crust and lithosphere when the continents were still close together about 60 m.y.b.p. After all, volcanism at that time was spread out over an area at least 2000 km in diameter (Fig. 13). Whatever the explanation, this great variation in elevation of presumably coeval continental margins presents a difficulty for the simple thermal contraction model of Sleep (1971).

The anomalous depths of the Eurasia Basin and its margins are puzzling. Perhaps this regional depression is quite unrelated to processes in the crust and upper mantle, but instead represents a regional collapse of the earth's surface as a result of material being removed from the lower mantle and fed into the Iceland plume.

A second peculiar feature of the Eurasia Basin is the low amplitude of its magnetic anomalies, compared to what is typical for the Mid-Oceanic Ridge (Karasik, 1968; Demenitskaya and Karasik, 1971). This effect cannot be entirely accounted for by the slower spreading rate and greater water depth; either the magnetized layer is only about half as thick or the magnetization intensity is half as great as normal (Vogt et al., 1971). In a sense, the Gakkel Ridge can be said to be generating a magnetic smooth zone. A variety of hypotheses have been advanced for such phenomena (Vogt et al., 1970b). Some possibilities, such as spreading in low magnetic latitudes, are obviously not applicable here. Perhaps the magma is compositionally different from the "virgin" basalts disgorged from the Iceland plume, or some unknown parameter of the magnetization process is different. Some such mechanism must be postulated to explain regional amplitude variations of the same order produced by the Mid-Oceanic Ridge south of Australia (Weissel and Hayes, 1971). Vogt et al. (1970b) also suggested the following mechanism for producing smooth zones: When the spreading axis is inundated by sediments, highly magnetized (rapidly cooled) pillow lavas are replaced by much less intensely magnetized sills; concurrently the sediment blanket allows reheating and demagnetization in the high heat flow zone near the spreading axis. The Gakkel Ridge barely rises above the surrounding abyssal plains; near the southeastern end of the basin the spreading axis is actually buried. Demenitskaya and Karasik (1971) make an observation that seems very favorable to the "sediment burial" explanation for the low magnetic amplitudes of the Eurasia Basin: They state that the axial anomaly is well developed where the rift valley is a deep narrow canyon, but poorly developed where it is U-shaped. A U-shaped valley presumably indicates that sediments from the abyssal plain have breached the low rift mountains and flooded the rift valley. Thus, the low apparent magnetization of the crust below the Eurasia Basin may be indirectly related to its low elevation: If ridge height were normal, sediments might not reach the spreading axis.

Ocean crust produced by the Mid-Labrador Sea (Ran Ridge) may well have suffered from a similar process. Magnetic lineations end well south of the Davis Straits. North of these, seismic refraction data suggest an oceanic crust, but magnetic anomalies are subdued and lack pronounced lineation (Keen et al., 1972). The sediment burial hypothesis (Vogt et al., 1970b) may account for the subdued character of the lineations in this basin also.

Other than regional elevation and magnetic amplitudes, the Eurasia Basin appears to be much like the central Mid-Atlantic Ridge: a well-developed rift valley 20 to 30 km wide; high heat flow; and rift mountains with slopes of 3 to 25°, rising 2.5 km above the rift valley floor (Demenitskaya and Hunkins, 1971; Demenitskaya and Karasik, 1971; Johnson and Heezen, 1967b).

It is the ridge within 900 km of Iceland that is abnormal with respect to seismicity (Fig. 1c) and topographic roughness (Fig. 12). In the most elevated part of the Iceland topographic high, earthquakes are less common, fractures widely spaced, a rift valley generally absent, and flank topography is smooth. In all these ways this large strip of ridge crest (Province A in Fig. 12) resembles the fast-spreading East Pacific Rise more than it does the slow-spreading Mid-Atlantic Ridge. It is reasonable to suppose that it is the hot,

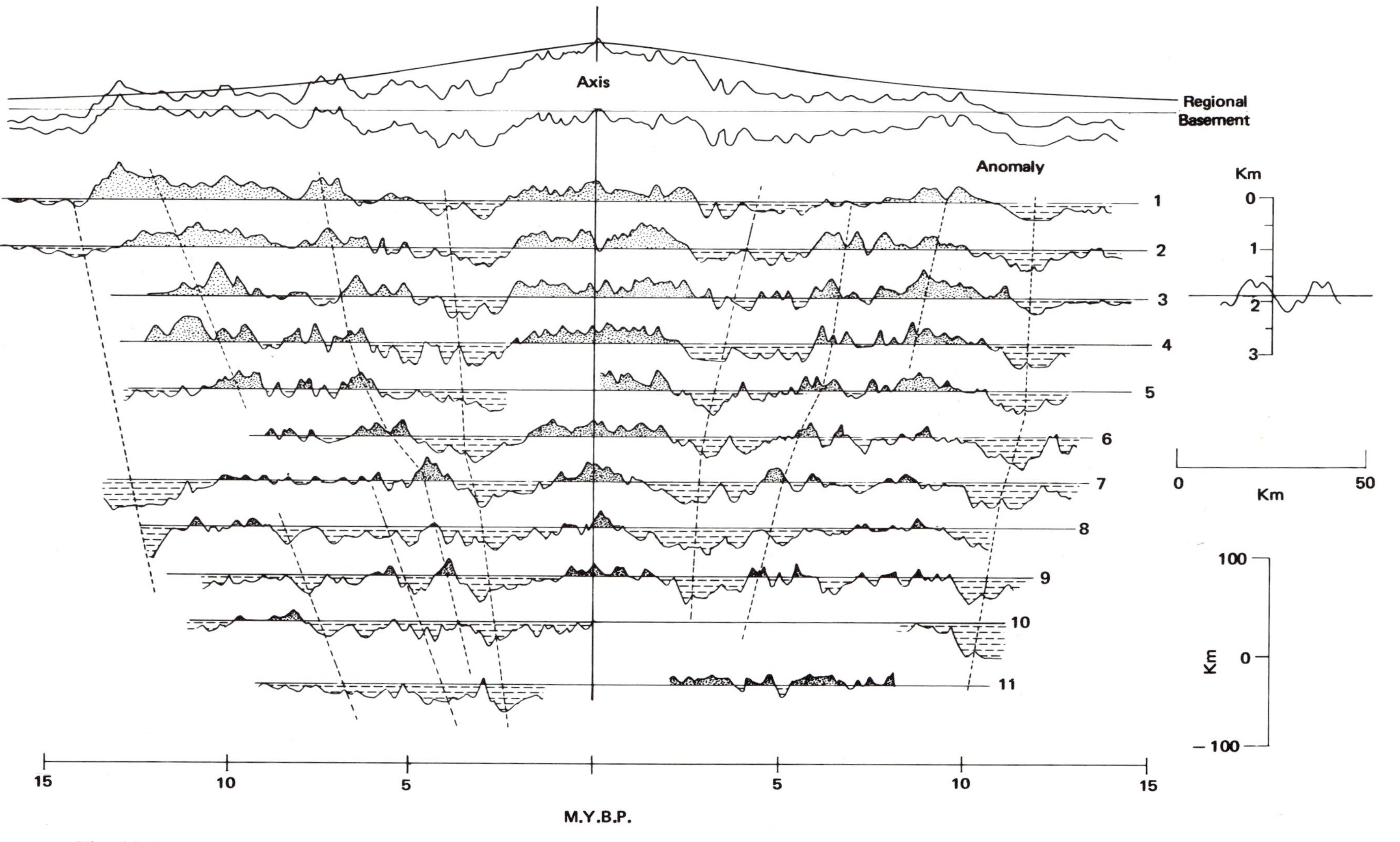

Fig. 11. Evidence for asthenosphere motion, adapted from Vogt (1971): When the normal regional subsidence curve (Sclater et al., 1971) is removed from basement profiles across the Reykjanes Ridge (Talwani et al., 1971), time-transgressive trends under the Reykjanes Ridge become apparent. Implied mantle flow rates are about 20 cm/yr (Vogt, 1971). North is at the top, and isochrons parallel the axis. Time-transgressive "events" are dashed; two prominent pairs of scarps also plotted as light black lines in Fig. 10. The absolute level of the basement is at about −1.8 km, 1 km above the norm of Sclater et al. (1971).

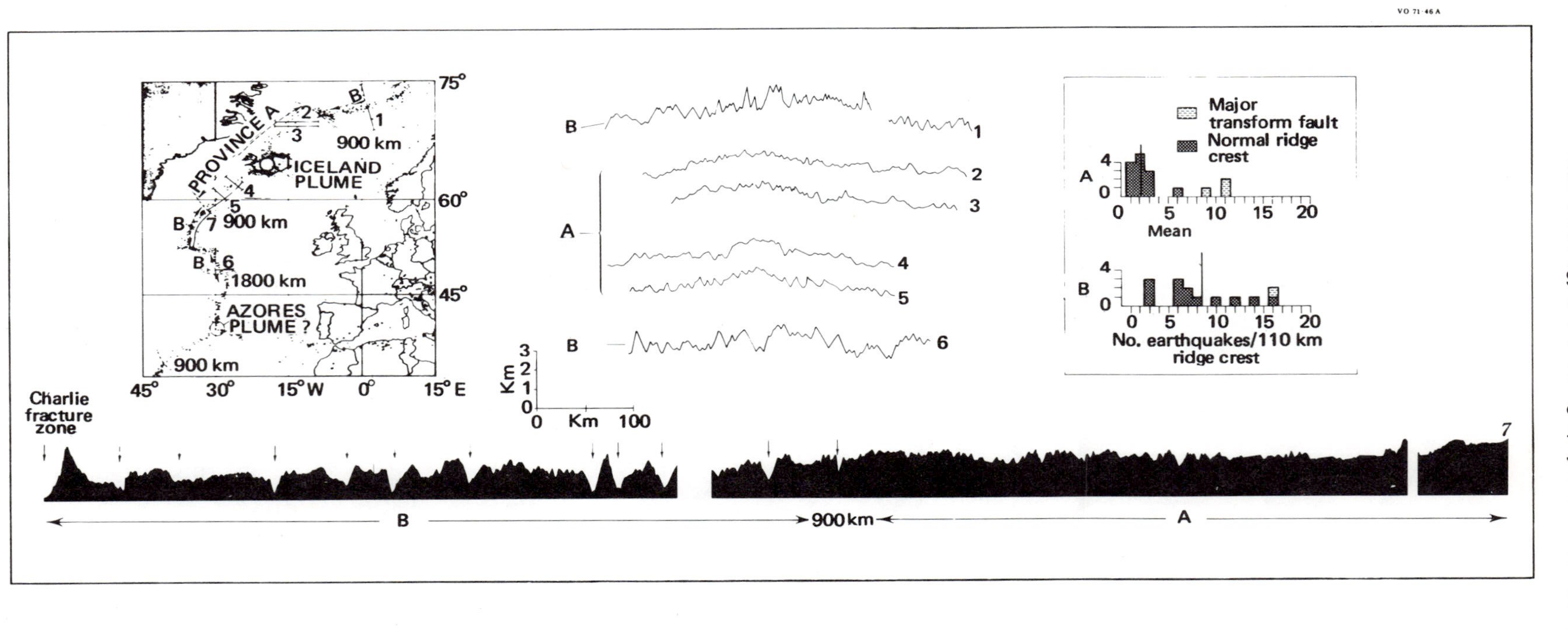

Fig. 12. Basement profiles and seismicity (1961–1969; anonymous, 1970) show provinces symmetrically disposed with respect to the Iceland mantle plume. As shown by histogram, the number of earthquakes per 110 km of ridge crest, excluding major transform fractures, was about two within 900 km of Ireland (province A) and eight between 900 and 1800 km. Basement roughness in province B is also significantly greater, both on transverse profiles (1 and 6) and longitudinally (profile 7) than it is in province A. Profiles 4 and 5 are from Talwani et al. (1970). The Gakkel (Nansen) Ridge is of type B, except for greater regional depth. Reduced seismicity also characterizes the Mid-Atlantic Ridge up to 900 km southwest of the Azores plume.

basalt-rich asthenosphere from the mantle plume that makes the ridge near Iceland behave like a fast-spreading one: Higher temperatures and plentiful basalt would allow ultrabasics to be injected in a slushy state, thus removing the conditions required for the formation of a rift valley and the accompanying higher seismicity (Vogt et al., 1969). The Gakkel Ridge seems to be of type "B," like Mohns Ridge and the Mid-Atlantic Ridge south of about 57°N (Figs. 1c and 12).

In summary, the complex history of sea-floor spreading since 60 m.y.b.p. is becoming reasonably well understood, and the mantle plume hypothesis (Morgan, 1972a, 1972b; Vogt, 1971) seems to be a promising way to account for many of the peculiar details. The plate tectonic events before 60 m.y.b.p., and especially before 80 m.y.b.p., are still largely guesswork. There may have been a very lengthy prelude of intracontinental rifting, shallow seaways, and igneous activity prior to the sea-floor spreading of Tertiary times. The record of this prelude may go back at least to Jurassic time (Watt, 1969; Johnson et al., 1972) and now mostly lies deeply buried under the continental margins.

A major unsolved question concerns the existence and location of an Iceland plume prior to the late Cretaceous. Another mystery about the Iceland plume involves the widespread occurrence of igneous activity about 60 m.y.b.p. Some of the known volcanic occurrences are shown in Fig. 13, but the area could be extended even further. For example, seamount groups occur preferentially on ocean crust 60 to 80 m.y.b.p. in age, both north and south of the Azores; on the margins of the Eurasia Basin, dike swarms of Cretaceous age occur on Franz Josef Land. Does this mean one large plume, many small plumes, or one plume plus secondary effects? Whatever the explanation for such widespread volcanism, it seems premature to use only the igneous activity in the British Isles and the Faeroes to recover absolute plate motion, as Duncan et al. (1972) have done. According to their compilation of isotope dates, the British Isles area moved rapidly southward in the 10 or 15 m.y. prior to the beginning of

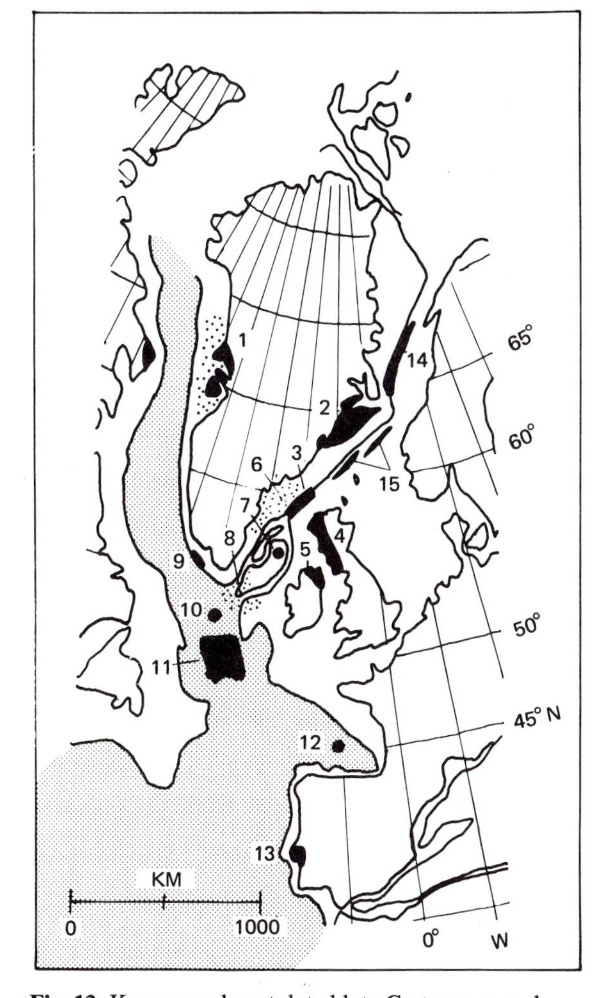

Fig. 13. Known and postulated late Cretaceous and early Tertiary igneous activity, plotted on 60 m.y.b.p. reconstruction of continents, modified from Fitch (1965). Black areas are known igneous occurrences or probable extra thicknesses of oceanic basaltic layer. Dotted areas show magnetic anomaly patterns, suggesting additional areas of early Tertiary igneous activity. (1) Davis Straits (Park et al., 1971; Manchester et al., 1972); (2) east Greenland (Beckinsale et al., 1970); (3) Faeroe Islands (Tarling and Gale, 1968); (4) northwestern Scotland (various sources summarized by Duncan et al., 1972); (5) northern Ireland (Purdy et al., 1972); (6) magnetic anomalies on Greenland shelf (Vogt, 1970); (7) Rockall Island and truncated volcanic center (Miller and Mohr, 1965; Roberts, 1969); (8) magnetic anomalies near the southern end of Rockall-Hatton Bank and on the ocean floor to the south (Vogt and Avery, 1973); (9) undated basalt, believed to be *in situ*, dredged on continental margin (G. L. Johnson, personal communication, 1972); (10) basalt sill encountered by deep-sea drilling (Laughton et al., 1972); (11) basement ridge, probably due to thickened crustal layer (Vogt and Avery, 1973); (12) basalt still encountered by deep-sea drilling (Laughton et al., 1972); (13) basaltic "Lisbon volcanics" (Watkins and Richardson, 1968); (14) and (15) are basement ridges and/or steep east-facing basement escarpments that formed along the Norwegian margin during the early Tertiary beginning of sea-floor spreading (Talwani and Eldholm, 1972). We interpret these as constructional volcanic piles similar to (11). (14) Voring escarpment; (15) Faeroe-Shetland escarpment.

spreading (55 or 70 m.y.b.p.). However, the age difference between the igneous activity of northern Ireland and that of the Faeroes is not great enough, considering the uncertainty in dating, to rule out coeval activity along this entire belt.

Rifting from the Gakkel Ridge to the Plate Rotation Pole

The continuation of the epicenter belt (Fig. 1c) from the Eurasia Basin across the Laptev Sea and into northeastern Siberia has been known for some time (Sykes, 1965). With the advent of plate tectonics (Morgan, 1968; Le Pichon, 1968) it should have become evident that this rather weakly developed epicenter pattern (Fig. 1c) must delineate the plate boundary between the west Eurasia plate and the Americas plate, including eastern Siberia (Chukotka).

It is therefore strange that Le Pichon (1968), and more recently Hamilton (1970), projected the plate boundary from the end of the Gakkel Ridge southeastward into the Bering Straits. There is no epicenter belt to support such a projection, and the implied discordance between the geology of Alaska and Chukotka seems nonexistent (Churkin, 1972).

A straight projection of the Gakkel Ridge passes near the Lena River delta and into the Verkhoyansk Mountains (Fig. 14), where Wilson (1963) and Morgan (1968) placed the plate boundary. Churkin (1972) argues that the Verkhoyansk foldbelt is only slightly seismic and lacks throughgoing faults. He proposes that the plate boundary lies along the east side of the Cherskiy Mountain system, in the Moma-Zyryansk basin and range province that lies just west of the Verkhoyansk (Fig. 14). Normal block faulting, seismicity, and one Quaternary volcano are the evidence for placing the plate boundary there. Although this is all very plausible, it is not necessary to look for a sharp plate boundary in continental regions. Tectonic and seismic activity in East Africa and the western United States, as well as the seismicity of the rift extension beyond the Gakkel Ridge (Fig. 1c) suggest broadly distributed crustal strains and, in the case of Africa and the United States, the existence of fairly stable microplates or blocks within the overall broad plate boundary.

Ever since a tensional regime has existed in the Moma-Zyryansk region (64 to 68°N, 138 to 150°E), the plate rotation pole must have been (and presumably is today) further south in Siberia. This stands in contradiction to the early computed late Tertiary rotation pole for the west Eurasian and American plates. Both Morgan (1968) and Le Pichon (1968) calculated poles at about 78°N, 102°E, near the Lena River delta. The pole of average motion (i.e., since 60 m.y.b.p.) to close the separation of Greenland and Europe was placed at 73°N, 96.5°E, by Bullard et al. (1965). Although this position contradicts the Moma-Zyryansk tensional regime, possibly rotation poles have wandered enough since 60 m.y.b.p. such that the Bullard pole is the average one, while the late Tertiary pole is further to the south. It is more likely that the Bullard pole is in error, however, because according to A. M. Karasik (Churkin, 1972), the pole that best closes the Eurasia Basin, and which should agree with Bullard's pole, is at 64°N, 138°E (Fig. 14).

Using the mantle plume concept to analyze absolute and relative plate motions, Morgan (1972b) placed the late Tertiary pole at 60°N, 135°E; close by is the late Tertiary pole computed by Pitman and Talwani (1972) using the method of fracture zones (Morgan, 1968; Le Pichon, 1968): 56.25°N, 141.44°E. This pole produces a standard deviation of 1.19° in the trend of sub-Arctic fracture zones. Needless to say, it is of little value to calculate these quantities to tenths and hundredths of a degree, when one considers that only four fracture zones were used, with trends hardly measurable to within one degree. This remark seems to be underscored by the authors themselves, who do not regard the difference between the fracture zone pole and the pole of best fit for anomaly 5 (68°N, 137°E) as being significant. The only true "instantaneous" plate rotation pole is one calculated from fault plane solutions on the Spitsbergen fracture zone (de Geer Line) by Horsfield and Maton (1970): 66° ± 5°N, 124° ± 5°E.

It is fair to say, based on this variety of more recent results, that both average and late Tertiary to Recent poles are in latitude 60 to 64°N, far enough south to accord with geological evidence for a tensional plate boundary in the Moma-Zyryansk region (Churkin, 1972).

Having the pole this far south requires that a vast area has been added to Siberia by stretching and rifting in the last 60 million years. Although surely complex in actual detail, the area added must be equivalent to that of a pie-shaped slice whose apex is at the rotation pole and whose base, about 500 km across, is the blunt southeastern terminus of the Eurasia Basin (Fig. 1a). The total area to be added to Siberia is thus of the order 0.4 million km². In this respect Churkin's analogy between the Eurasia Basin and the Gulf of California is not valid: The Pacific-American plate boundary north of the Gulf is primarily a transform fault, and no new crust need be created. How the stretching of Siberia was accommodated by the upper crust is an interesting and unsolved question, particularly because the Verkhoyansk foldbelt has apparently been *folded* in Pliocene and Holocene times (Churkin, 1972). It is possible that the anomalously smooth magnetic field of the Verkhoyansk region (Belyaev, 1970) represents

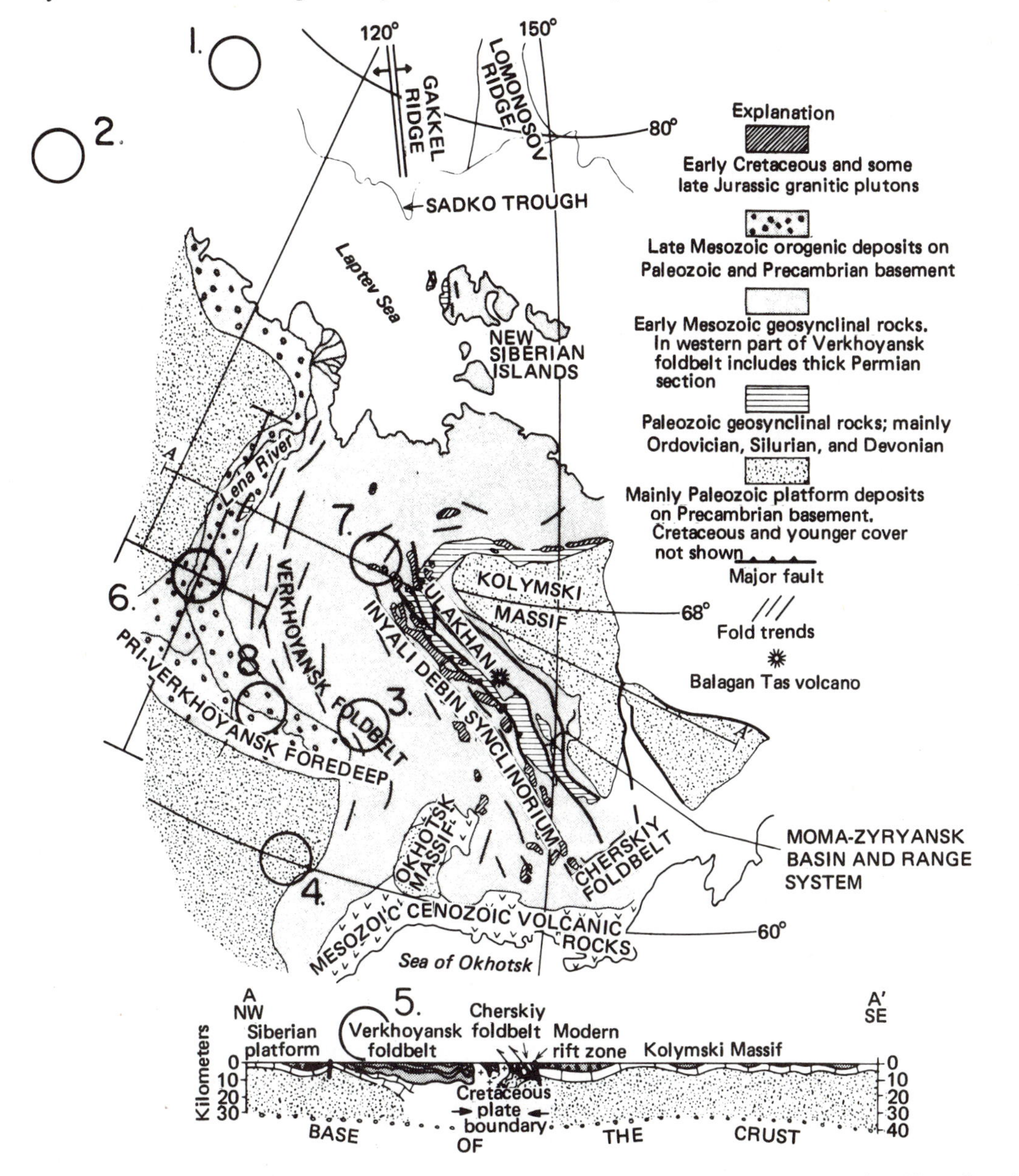

Fig. 14. Major geologic features of Yakutia (Churkin, 1972) with various plate rotation poles: (1) early estimate of late Tertiary pole position by Morgan (1968) and Le Pichon (1968); (2) pole of average motion to reassemble Greenland, North America, and Europe (Bullard et al., 1965); (3) pole of average motion to close Eurasia Basin, derived by A. M. Karasik (Churkin, 1972); (4) late Tertiary pole (Morgan, 1972b); (5) late Tertiary fracture zone pole (Pitman and Talwani, 1972); (6) pole of present instantaneous motion, based on fault plane solutions (Horsfield and Maton, 1970); pole to close post-9 m.y.b.p. (anomaly 5) sea floor north of Azores (Pitman and Talwani, 1972); (8) pole to close 9 to 38 m.y.b.p. (anomaly 13) sea floor (Pitman and Talwani, 1972). Older finite rotation poles for magnetic anomalies lie well outside the chart area.

intrusion of hot homogeneous mantle derivatives, and concomitant demagnetization of older rocks, associated with the required stretching process. This does not explain the apparent lack of surface stretching, however. Perhaps extension by block faulting in the Moma-Zyryansk region has been sufficient, but what about the region between the Moma-Zyryansk and the Laptev Sea?

A final problem about the extension of the rift beyond the Gakkel Ridge is why spreading stopped where it did, progressing only little beyond the original margin of Eurasia (Fig. 1a). This is probably not fortuitous, for two reasons: (1) It could well be easier to split off a section of continental margin than to penetrate a continent, whose structural complexity allows multiple fracturing rather than the development of a single straight fissure. (2) Further penetration of the spreading process into Eurasia may have been arrested by sediments supplied voluminously enough from both sides of the rift and from the Lena River to bury the developing basin—perhaps represented by the Sadko Trough—as quickly as it widened. By contrast comparatively little sediment could be supplied to the Eurasia Basin by the Lomonosov Ridge and the wide, largely submerged Barents shelf. Because the sedimentation hypothesis gives the Laptev Sea an essentially oceanic crust, while the first hypothesis (structural control) does not, geophysical tests are possible.

The geology of northeastern Siberia in the general vicinity of the Verkhoyansk belt (Fig. 14) should offer some sensitive tests for any proposed plate-polar wandering curves for the last 60 million years. Only relatively slight changes in the position of this pole would change the tectonic regime of the Cherskiy foldbelt, for example, from compression to tension, or vice versa. It is only in eastern Siberia that a boundary between two great crustal plates passes close to their pole of relative rotation.

The present continuation of rifting beyond the region of sea-floor spreading may help to explain the various troughs and shallow seaways of pre-60 m.y.b.p., but probably Mesozoic age, that occur in the Labrador Sea and the northeastern Atlantic (Fig. 2). Examples include the Porcupine Seabight, Rockall-Hatton Basin, and Rockall Trough (Bailey et al., 1971; Laughton et al., 1972; Vogt and Avery, 1973). In Cretaceous and possibly Jurassic times these regions may well have similarly lain in the tensional region between the plate rotation pole and the limit of actual sea-floor spreading. The various basins and troughs are presently and have long been topographic lows, whereas the extension of the Gakkel rift is near or at sea level. However, in the case of the presently active rift the lithosphere is undoubtedly thermally expanded; if this plate boundary were to become inactive, well-documented precepts of thermal subsidence (Sclater et al., 1971) would seem to dictate that deep sedimentary basins will develop north of about 60°N in the Cherskiy-Verkhoyansk region (Fig. 14). It thus seems potentially fruitful to compare the early history of the present basins with the late Tertiary and subsequent history of the Gakkel Ridge continuation into Siberia.

The Alpha-Mendeleev Cordillera—Conflicting Clues of Crustal Genesis

Somewhat asymmetrically located within the Amerasia Basin and roughly parallel to the Lomonosov and Gakkel ridges lies a complex rise, 250 to ~1000 km wide, with crestal depths locally shoaling to about −1.5 km (Demenitskaya and Karasik, 1971). A digression on the subject of nomenclature is definitely called for at this point. The rise or ridge has been endowed with a confusing diversity of names, a strong argument for standardized nomenclature. American writers have called the entire feature the Alpha Cordillera, Rise, or Ridge (Vogt and Ostenso, 1970), while some Russians called it the Mendeleev Ridge (Rassokho et al., 1967). The National Geographic Society calls it Fletcher Ridge in its 1966 atlas (anonymous, 1966). Other papers use a deep trough (Cooperation Gap) in the center of the ridge to separate it into the Mendeleev Ridge on the Siberian side and the Alpha Ridge on the Canadian side (Treshnikov et al., 1967). In a recent binationally authored publication, Dementskaya and Hunkins (1971) refer to the entire feature as Alpha Cordillera, reserving the term Mendeleev Ridge for the southwestern spur, as did Treshnikov et al. (1967). The term "cordillera" was introduced by a panel of Arctic experts (Beal et al., 1966) in a noble but partially futile attempt to give the Arctic Basin a scientifically reasonable nomenclature. Wishing to avoid the "mid-oceanic," hence generic, implication of the word "ridge," the authors introduced the term "cordillera," not only for the Alpha Ridge, but, surprisingly, also for the Nansen Ridge (the Gakkel

Ridge according to Demenitskaya and Hunkins, 1971), which at the time was already known to be a branch of the mid-oceanic ridge. The terms "ridge" and "cordillera" now appear to be used interchangeably as proper nouns for both Alpha and Nansen (Gakkel) Ridges (Ostenso, 1972). In this paper we use the name Gakkel Ridge, following Demenitskaya and Hunkins (1971), but the longer, least confusing, and even slightly euphonic term Alpha-Mendeleev Cordillera is preferred to indicate that the entire ridge is meant and that its mid-oceanic nature, suggested by Vogt and Ostenso (1970) and other authors (Johnson and Heezen, 1967b; Beal, 1969) is debatable. Alpha Rise, Alpha Ridge, and Alpha Cordillera are assumed to be substitute names in good standing.

Speculations concerning the origin of the Alpha-Mendeleev Cordillera have not been fenced in by detailed survey data; so it is possible that the approximately 500-km-wide band of high-amplitude magnetic anomalies associated with this ridge was used to suggest, first, that the ridge consists of subsided continental shield (King et al., 1966) and, contrarily, that it is a fossil mid-oceanic ridge which ceased spreading about 40 m.y.b.p. (Vogt and Ostenso, 1970; Ostenso and Wold, 1971).

The "foundered continent" view of the Alpha Ridge and other parts of the Arctic Basin was held by Eardley (1961) and various Soviet geologists (e.g., Atlasov et al., 1967). Although this concept has itself foundered somewhat with the advent of the plate tectonic revolution, a "mixed mode" origin is still held possible by some: The Eurasia Basin was formed by spreading, and the Amerasia by subsidence (King et al., 1966; Rassokho et al., 1967; Demenitskaya and Hunkins, 1971). Among the more unattractive aspects of the subsidence hypothesis is a mandatory corollary about "oceanization": Some mysterious process must be called upon to give the Amerasia Basin the ocean-like crustal and upper mantle structure demanded by existing geophysical data (Hunkins, 1961, 1963; Gutenberg and Richter, 1954). A further difficulty is that seismic reflection profiles (Hall, 1970) show no evidence of the flat truncated basement erosion surfaces that would be expected, had the Alpha Ridge subsided through sea level.

The fossil-spreading-center hypothesis owes much to seismic reflection profiles, which show a rough basement topography and a regional profile quite similar, save for a sediment blanket some 300 to <1000 m thick, to the basement topography and regional profile of parts of the Mid-Atlantic Ridge (Hall, 1970). There is even a suggestion of a central rift valley, and soundings taken along the erratic path of T-3 can be interpreted in terms of a family of nearly parallel close-spaced transform fractures (Hall, 1970). Although clearly complex in detail and not easily correlatable (Demenitskaya and Karasik, 1971; Vogt and Ostenso, 1970), the few available aeromagnetic profiles can be viewed as having resulted from sea-floor spreading at about 1 cm/yr, ending about 40 m.y.b.p. (Vogt and Ostenso, 1970; Ostenso and Wold, 1971). Hall's (1970) interpretation of magnetic measurements taken on T-3 suggest lineations parallel to his speculative fractures and other lineations, possibly due to geomagnetic reversals, paralleling the tectonic axis of the ridge.

Riddihough et al. (1973) report an extensive aeromagnetic survey of the Canadian Arctic, including that part of the Alpha Ridge east of 140° W. They confirm that "the diagnostic linear magnetic structure of sea-floor spreading does seem to be present, together with some possible transverse discontinuities which may be interpreted as transform faults.

There are a number of major and minor difficulties with the fossil mid-oceanic ridge hypothesis for the Alpha-Mendeleev Cordillera. The identification of magnetic anomalies as lying between 60 and 40 m.y.b.p. on the reversal time scale was put forth as a tenuous suggestion, considering the wide variability of the anomalies encountered on different profiles (Vogt and Ostenso, 1970). The recent recovery of late Cretaceous (70 m.y.b.p.?) microfossils from a site within 100 km of the crest of the Alpha Ridge (Ling et al., 1973) suggests an older age. Pitman and Talwani (1972) assert from their plate tectonic reconstructions that the Nansen (Gakkel) Ridge (also called mid-Arctic by those authors) has been the only locus of sea-floor spreading in the Arctic Basin during Cenozoic time. The argument is based on their conclusion that the tectonic regime was compressional between 81 and 63 m.y.b.p. and that a plate reconstruction to 63 m.y.b.p. (anomaly 25) places the Lomonosov Ridge adjacent to Eurasia (Fig. 9). While this is an elegant solution to an Arctic tectonic problem (no Arctic data were used), the quantitative conclusions of Pitman and Talwani suffer from the lack of a suitable quantitative index of probable error or uncertainty. One test for their method is how well transform fracture trends are predicted. This test is well met south of the Azores (Fig. 2 of Pitman and Talwani), but between the Azores and Iceland bathymetric and magnetic data collected by the U.S. Naval Oceanographic Office show substantial errors in their predicted transform trends, or flow lines, on crust formed between 38 and 63 m.y.b.p. It is thus debatable whether or not the results of Pitman and Talwani (1972) concerning Cenozoic spreading in the Arctic can be regarded as firm proof.

The suggestion that the Alpha Ridge is an extension of the Mid-Labrador or Ran Ridge (Vogt and Ostenso, 1970; Ostenso, 1972) also conflicts with some plate kinetic analyses (Le Pichon et al., 1971; Pitman and Talwani, 1972). The history of spreading between Greenland and North America is highly uncertain, however; probably it was confined largely to the period 80 to 40 m.y.b.p. (Le Pichon et al., 1971; Laughton et al., 1972; Vogt and Avery, 1973), although to date only anomalies 40 to 60 million years in age have been positively identified in the southernmost Labrador Sea (Vogt and Avery, 1973). The slow spreading rates, of the order 0.5 cm/yr between 60 and 40 m.y.b.p., may explain the complex character of the spreading anomalies, whose known northern termination is south of Davis Strait. Similarly there are only a few, poorly charted transform fractures in the southern Labrador Basin (Le Pichon et al., 1971; Vogt and Avery, 1973). It is true that the Greenland-America motion can be recovered by vector subtraction of Greenland-Europe and Europe-America motions. It may still be asked how certain a conclusion can be reached from the subtraction of two relatively long, imperfectly defined vectors. Supporting the existence of a continuous Labrador-Alpha rift is their approximate physical continuity, the approximately similar widths of the Labrador Basin and the Alpha-Mendeleev Cordillera, and the approximate parallelism between Hall's (1970) proposed transform fractures offsetting the Alpha Ridge, and the supposed Nares Strait transform fault between Greenland and Ellesmere Island.

Several other aspects of available data from the Alpha Ridge indicate that if this feature is (or was) a branch of the Mid-Oceanic Ridge, it was a peculiar one. First, the basement depths of the ridge crest (-2 km; Hall, 1970) are much too shallow. Sclater et al. (1971) show that most oceanic crust is formed at depths of about -2.5 to -3 km and then subsides, probably by thermal contraction of the lithospheric plate, with a time constant of the order 50 million years. An extinct ridge subsides in similar fashion. If the Alpha Ridge stopped spreading 40 m.y.b.p. (Vogt and Ostenso, 1970), basement depths should be about -4.7 km; if anything, the ridge is even older (Pitman and Talwani, 1972) and so should be still deeper. Actually the apparent lack of guyots and flat-topped basement ridges require a subsidence of the Alpha-Mendeleev Cordillera by no more than 2 km. True, some parts of the Mid-Oceanic Ridge, such as the Iceland region (Fig. 1), are many kilometers above the subsidence curve of Sclater et al. (1971). Even if it was such an abnormal mid-oceanic ridge, it must have subsided by more than 2 km, and again the apparent absence of sea-level erosion surfaces presents a problem. Moreover, abnormally elevated portions of the mid-oceanic ridge tend to produce high aseismic ridges more perpendicular than parallel to the spreading axis; instead, the Alpha Ridge is elevated all along its length. Nor does the evidence support an alternate hypothesis that the Alpha Cordillera is itself an aseismic ridge, running normal to crustal isochrons: Not only is the ridge too wide and too shallow, but the one available refraction station (Hunkins, 1961) showed a crustal layer of rather normal oceanic thickness. The Iceland–Faeroe aseismic ridge, by contrast, exhibits thicknesses from 8 to 13 km near Iceland (Palmason, 1970) and 15 to 20 km near the Faeroes (Bott et al., 1971).

Magnetic anomalies over the ridge (Vogt and Ostenso, 1970) are also not entirely typical. Their amplitude is so high that the magnetized layer would have to be about twice as thick or twice as intensely magnetized as normal, nonaxial oceanic crust (Vogt et al., 1971). True, there are areas of normal ocean crust, such as south of Australia (Weissel and Hayes, 1971) where amplitudes vary regionally by a similar magnitude, after differences of basement depths are considered. Exceptional amplitudes were also produced in a 50-km-wide band along the lower Cretaceous "Bermuda Discontinuity" of the central North Atlantic (Vogt et al., 1971). Furthermore, magnetic anomalies over Iceland are complex, high in amplitude, and difficult to correlate (Serson et al., 1968). These are rare exceptions, however, and do not make a spreading origin for the Alpha Ridge very attractive.

Finally, even heat flow data from the ridge are anomalous: Lachenbruch and Marshall (1966) derived an average of 0.8 HFU (heat flow units; 1 HFU = 10^{-6}cal/cm^2 sec) for the lower flank of the Alpha Ridge. Although this may still be within the range of variation for normal ocean crust, considering the number of measurements, there is usually a correlation between regional elevations of the ocean floor and high heat flow (Sclater et al., 1971). The low heat flow of the Alpha Ridge is an enigma. Lachenbruch and Marshall (1966) did derive a thermal conductivity distribution that resembles the low *density* distribution under the Mid-Atlantic Ridge. However, the meaning of this comparison is obscure because variable conductivity models for the Mid-Atlantic Ridge have not been attempted.

In summary, the Alpha-Mendeleev Cordillera resembles the Mid-Atlantic Ridge in only a few ways, and its origin is as baffling as any feature on the ocean floors. Possibly the ridge was related to subduction or at least compression of earlier ocean floor. A block-faulted regional uplift of oceanic crust would account for some of the features; no known ridge elsewhere is exactly comparable. Perhaps the Bermuda Rise is most similar, especially its axial band of high magnetic anomalies and fractured eastern flank (Vogt et al., 1971). However, the Bermuda Rise, while not itself a spreading axis, does contain magnetic lineations caused by sea-floor spreading. Detailed magnetic surveys and more seismic refraction lines might solve the problem of the Alpha-Mendeleev Cordillera; possibly a resolution awaits some type of Arctic deep-sea drilling program.

Origin of the Canada Basin and Remarks on Rifted Margins of the Arctic

Probably the oldest and certainly the least known part of the Arctic is the Canada Basin (Figs. 1a, 1b, 2, and 15). No crustal structure measurments have been made; only surface wave dispersion (Hunkins, 1963) gives direct evidence for an oceanic crust. Most of the basin lies between -3 and -3.8 km depths; seismic reflection has not penetrated to the basement, but at least locally the sediment cover is known to exceed 2 km (Hall, 1970). There are at least 3.5 km stratified sediments under the Wrangel Abyssal Plain (Kutschale, 1966). Estimates of depth to magnetic basement suggest that sediment thickness is not likely to be very much thicker (Ostenso and Wold, 1971). Thus the basement appears to be about 6 km deep, except under the continental slope and rise. This is a typical value for old ocean basins; virtually all basins with such basement depth exhibit oceanic or quasi-oceanic crustal structure, and it seems safe to predict that the Canada Basin does also. Free air gravity averages near zero, if a slight northward increase reflects an inadequacy of the international gravity formula (Wold and Ostenso, 1971). Thus, low-density subsided continental crust below the basin would demand exceptionally high mantle densities—a rather unlikely situation. A few heat flow measurements yield an average of 1.4 HFU (Lachenbruch and Marshall, 1966). This observation does not help to identify the crustal type; 1.4 HFU is well within the range of normal variation for both continental and oceanic crust, considering the sample sizes.

Although many kinds of processes have been conjectured to produce an oceanic crustal structure, only two are understood with reasonable certainty. The first is axial accretion (a term perhaps preferable to sea-floor spreading when crustal genesis is meant) along the Mid-Oceanic Ridge. It is now safe to state that the crust flooring virtually all the Norwegian, Greenland, and Eurasia basins—and in fact most other ocean basins of the world—have been produced by axial accretion at the Mid-Oceanic Ridge. A second process, less well understood but probably quite similar, produces a rather similar crust behind island arcs by a kind of episodic inter-arc or behind-the-arc spreading (Karig, 1971).

Magmas liberated from the melting downthrust slab of oceanic lithosphere, aided by a back-eddy set up in the asthenosphere by the motion of the slab (Sleep and Toksöz, 1971) and by the sinking of the slab, seem to provide the tension which drives the outer arcs toward the oceanic plate being subducted. Inter-arc spreading probably also involves a narrow central rift axis, but spreading rates probably vary rapidly with time and with distance parallel to the island arc. Such complex spreading may not produce a good "recording" of geomagnetic reversals. This may explain the fact that correlatable, identifiable magnetic lineations have not commonly been found in small ocean basins behind island arcs.

It seems unlikely that the Canada Basin crust was not formed by one or the other of the above processes, and the question is which one. The older view that continental subsidence produced the basin (Eardley, 1961) cannot be finally disproved, but neither does it in our opinion deserve serious consideration. Similarly, the proposal that the Canada Basin is a sphenochasm (Hamilton,

1970) no longer seems promising. Although Carey's (1958) type of continental drift has been replaced by sea-floor spreading and plate tectonics, his concept of oroclines and sphenochasms still finds favor among some. Thus, Hamilton (1970) believes that the Alaskan tectonic arcs are related to the opening of a Canada Basin about an axis perpendicular to the arcs and to all three ridges of the Arctic Basin.

It must never be forgotten that the *present shape* of the basin is not necessarily the same as it was originally (although it might be), nor need there be any connection between the processes that created the crust and the processes that subsequently modified the basin to its present outlines. Subsequent rifting, subduction, and transcurrent faulting along old plate boundaries are the main processes that could have made major geographical changes in the Arctic; marginal subsidence and sediment prograding generally do not shift the physiographic margin of the continent by more than 10 to 100 km, even after periods of the order 100 to 200 million years, to judge from the eastern margin of North America (Vogt, 1972). Studies of other continental margins suggest that this is quite typically true. Marginal magnetic anomalies, similar to those described by Taylor et al. (1968) along the eastern continental margin off North America, also rim the margins of the Arctic Basin. Individual magnetic maxima follow the outer edge of the continental slope, forming a chain of anomalies paralleling the slope (Demenitskaya and Karasik, 1971). Magnetic and gravity slope anomalies occur locally north of Alaska and in northwestern Canada (Wold and Ostenso, 1971; Ostenso and Wold, 1971). More detailed data come from the sub-Arctic basins. A relatively sharp magnetic boundary anomaly exists over the continental margin of Newfoundland (Keen et al., 1971; Fig. 3); magnetic and gravity anomaly patterns also suggest a relatively sharp boundary between continental and oceanic crust near the 1000-m isobath in Baffin Bay (Keen et al., 1972). In the northeastern Atlantic (Vogt and Avery, 1973) a conspicuous positive magnetic anomaly marks the physiographic boundary between oceanic crust associated with identifiable magnetic lineations and the continental crust underlying Rockall-Hatton Bank (Fig. 5).

The continental margin off Norway (Talwani and Eldholm, 1972) is more atypical: The landward border of identifiable magnetic lineations lies some 50 to 100 km east of the edge of the continental shelf. The border lies near the 2000-m isobath or deeper, except where it crosses the shallower Voring Plateau. This border is marked by an asymmetrical basement ridge with a steep landward-facing escarpment. Talwani and Eldholm (1972) believe that this escarpment marks the border between subsided continental crust on the east and oceanic crust on the west. They attribute the escarpments to isostatic adjustment of the newly formed oceanic crust with respect to the pre-existing continental crust. In our opinion the escarpments and the boundary ridges represent exceptionally high volcanic discharge associated with the late Cretaceous/early Tertiary discharge peak of the Iceland mantle plume (Fig. 13). The ridges and landward-facing escarpments are similar in form and dimension to those formed intra-oceanically where the early Cretaceous Bermuda discontinuity intersects the outer edge of the Grand Banks, and south of Hatton Bank (item 11 of Fig. 13; Vogt and Avery, 1973; Johnson and Vogt, 1973). In the latter area there is a 50- to 100-km wide basement ridge, with minimum depths of 2.5 to 3 km and a steep eastward facing escarpment. The basement east of the escarpment lies at depths of about 5 km, if the isostatic loading by sediment is removed. This older basement is associated with relatively subdued magnetic lineations which have been charted in detail and are supposed to be anomalies 25 to 31 (Laughton and Berggren et al., 1972; Vogt and Avery, 1973). By comparison, the ridge bordering the Voring escarpment is 100 km wide and rises to depths of 1.5 to 2 km. The ridge seems to be identical in age to the one south of Hatton Bank, viz., 55 to 60 m.y.b.p., the time of great volcanic activity in a vast region (Fig. 13). The basement east of the Voring escarpment is known to be pre-Tertiary in age and at least 4 km deep. Talwani and Eldholm (1972) believe it lies at depths of 5 to 8 km. If the basement level is adjusted to remove the effect of isostatic loading, it would lie at about the same depth as the one deduced for the old ocean crust south of Hatton Bank. Magnetic anomalies of reduced amplitude occur landward of both escarpments (Fig. 6, right 40% of profiles 5 through 10), although the field is smoother in the Norwegian case. Magnetic smooth zones may have various explanations (Vogt et al., 1970b); continental crust is not required. When all these observations are put together, it does not seem necessary to postulate subsided continental crust between the continental margin of Norway and the Voring escarpments. The basement may well be oceanic crust of Mesozoic age, perhaps coeval with that below Rockall Trough. Basement depths east of the Voring escarpment and east of the escarpment south of Hatton Bank are compatible with a late Cretaceous crustal age (Sclater et al., 1971). It is the younger basement, immediately east of these escarpments, that is anomalously elevated, rising 2 to 4 km above the norm for crust of comparable age (Sclater et al., 1971). The abnormal elevation in both cases seems most readily explained by a large basaltic fraction in a lithosphere formed in the late Cretaceous/early Tertiary by exceptional discharge from the Iceland plume.

These observations suggest that in general the present physiographic margins of the Arctic and sub-Arctic basins are close to the tectonic boundary. Even off Norway, where Talwani and Eldholm (1972) have advanced the debatable view

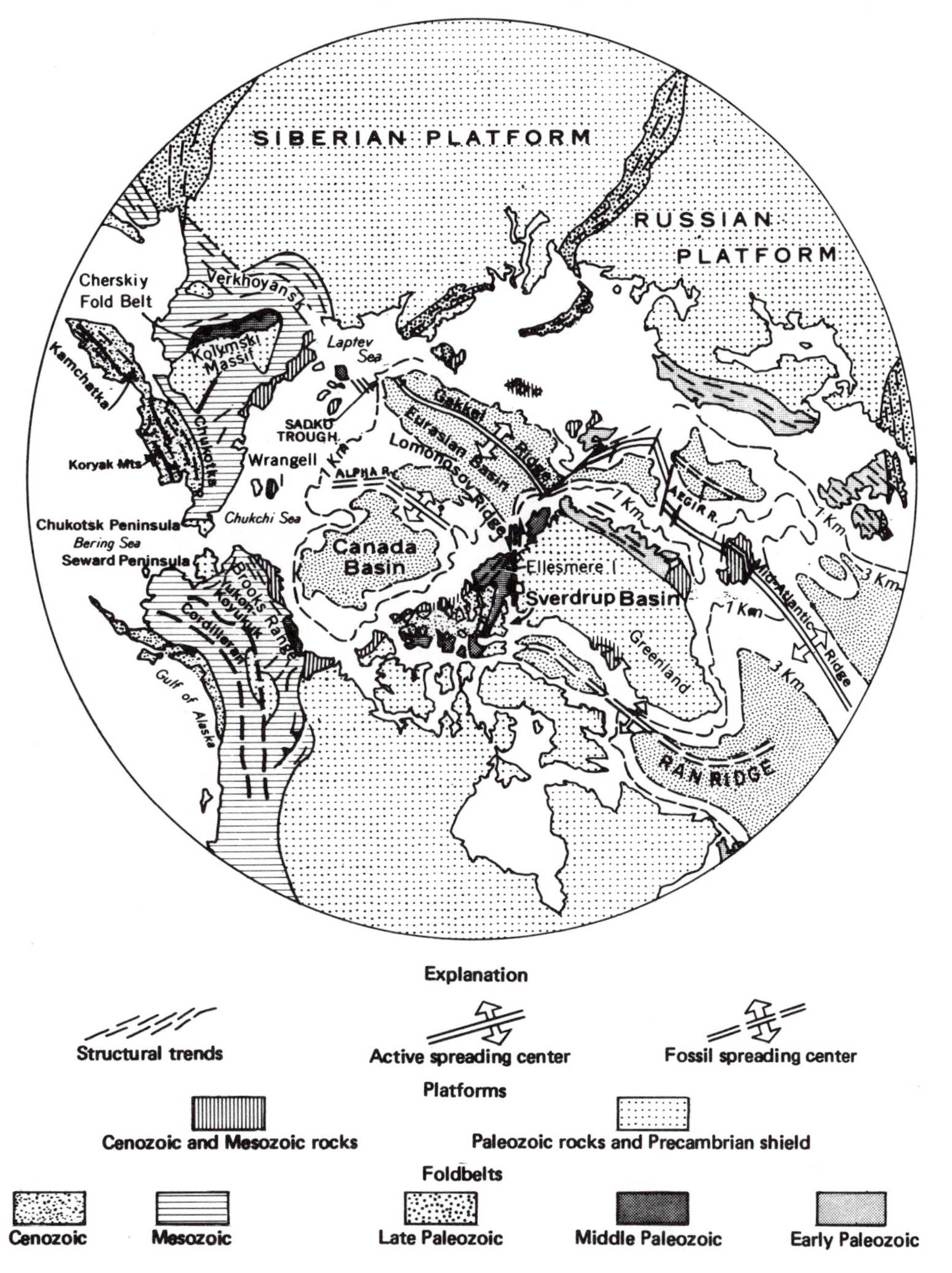

Fig. 15. Major tectonic features of the Arctic. From Churkin (1972), with minor corrections in the positions of the spreading centers.

that continental crust lies between the continental shelf break and the Voring escarpment, the distance between the supposed tectonic margin and the physiographic margin is only about 50 to 100 km. In the case of the Eurasia, Norwegian, Greenland, and Iceland basins and along much of the Baffin-Labrador rift, the process that created the crust (axial accretion, i.e., sea-floor spreading) also created the margins of the basin. It is thus possible that the margins of the Canada Basin are also rifted margins formed in the initial stages of sea-floor spreading at some ancient time. The apparent lack of conspicuous magnetic lineations in the basin (Ostenso and Wold, 1971) does not help this hypothesis, but not all oceanic crust known to have been formed at the Mid-Oceanic Ridge exhibits clear lineations.

The geology of the circum-Arctic (Churkin, 1972; Hamilton, 1970) places some rather severe constraints on the hypothesis that the crust below the Canada Basin was formed by normal (rather than inter-arc) type spreading. In the first place the Canada Basin, and possibly the entire Amerasia Basin, is not presently the northern extension of one or more oceanic crustal plates, in the sense that the Eurasia Basin is obviously an extension of the Cenozoic basins of the North Atlantic. It follows that if there was contiguous ocean crust coeval with the Canada Basin crust, it must have been subducted sometime after it was formed. Any possible extension could not have entered Canada, Alaska, or easternmost Siberia: The apparent continuity of crustal geologic provinces seems to forbid Phanerozoic rift oceans or other plate boundaries transecting these regions (Churkin, 1972). There are several possible now-closed oceanic rifts through the remainder of Eurasia, however: (1) An ocean may have existed between the Russian and Siberian platforms; its closure by late Paleozoic time produced the Uralides (Hamilton, 1970). (2) A proto-Atlantic ocean probably existed between the Canadian shield and the Russian platform; the closure of this ocean, probably completed by Devonian time, formed the Caledonides (Hamilton, 1970). A third fossil plate boundary has been located by Churkin (1972) in eastern Siberia, where the Paleozoic Cherskiy foldbelt on the east borders the Mesozoic Verkhoyansk foldbelt on the west. Subduction was going on there as late as Early Cretaceous, judging from extensive faulting, high-pressure metamorphism, and granite batholith intrusion.

The Canada Basin could thus be a surviving remnant of a pre-130 m.y.b.p., a pre-220 m.y.b.p., or even a pre-350 m.y.b.p. ocean. A pre-Cretaceous age for the crust under the Canada Basin therefore seems rather certain if it was created by sea-floor spreading. This accords with the observation that Mesozoic and younger sediments on the Alaska coastal plain are undisturbed (Churkin, 1972). However, even if the Canada Basin was a portion of any of the three former oceans, plate tectonics demands at least three continuing plate boundaries, one for each paleo-ocean. Evidently the roughly 100 million years age differences between the closing of the three oceans are too large to permit one ocean having been a continuation of the next. Conceivably later rifts and the subduction zones that closed the oceans followed and therefore in part obliterated the record of earlier oceans. This would mean, for example, that the Cherskiy-Verkhoyansk plate boundary could have masked or obliterated the continuations of the Uralide or Caledonide plate boundaries. Whatever the answer, the search for the missing plate boundaries, possibly hidden below the extensive Arctic continental shelves, should be a stimulating problem for Arctic geology, one that may at the same time clarify the origin of the Canada Basin.

An alternative hypothesis for the basin, inter-arc spreading, demands no such connecting ocean basins. The size, shape, and location of the Canada Basin inside a nested series of tectonic belts in Alaska and Chukotka (Churkin, 1972) is rather suggestive of numerous small ocean basins in the Indonesian Archipelago, the western, southwestern, and northern Pacific, the Scotia Sea in the south Atlantic, and the Caribbean. The oldest of the tectonic belts, possibly marking an ancient northern plate boundary for a proto-Pacific plate, is revealed by a discontinuous foldbelt of middle Paleozoic age that rims the north coasts of Alaska, Chukotka, and Canada (Churkin, 1972). The crust below the Canada Basin would then be of Carboniferous age. At that time it would have resembled the present Aleutian Basin, although some of the original Canada Basin crust may have subsequently been subducted under the Beaufort Sea continental margin (Ostenso, 1972). Continuing northward subduction of the Pacific plate then caused the southward growth, by accretion of originally oceanic and island arc materials, of Alaska and Chukotka.

Paleozoic inter-arc spreading means that the Canada Basin has always been an isolated feature;

this accords with the evolutionary model proposed by Ostenso (1972), who uses the euphonic name "Hyperborean Basin" for the ancestor of the Canada Basin.

Other examples of isolated small ocean basins in similar tectonic settings (the Gulf of Mexico and the Black Sea, for example) probably are in the same genetic category, and detailed geologic comparisons could prove fruitful in the quest for the correct genetic model, whether it be normal or inter-arc spreading or some still unanticipated process. It is also possible that the entire Amerasian Basin, including the Alpha Ridge, was formed by the same mechanism, with a much later process uplifting the ridge to its present configuration. Although more geophysical data from the Canada Basin are desperately needed, their acquisition does not guarantee a solution to the basic problems of origin. The Gulf of Mexico has been much more extensively probed; the reasonably close-spaced magnetic profiles show a smooth, low-amplitude, weakly lineated anomaly pattern, while refraction sections suggest a deeply buried oceanic crust. Despite the relatively plentiful data and the better-known geology of its coasts, neither mid-oceanic nor inter-arc type spreading nor crustal age have been conclusively determined for the Gulf of Mexico.

Conclusion

Tectonic problems the world over are now being studied within the framework of the plate tectonic concept (Morgan, 1968; Le Pichon, 1968). The Arctic and sub-Arctic basins are no exceptions in this regard. The future of Arctic research will undoubtedly see further refinements in the reconstruction of relative plate motion. This has been, and will probably continue to be, accomplished by rotation of magnetic isochrons (Pitman and Talwani, 1972), rotation of continental edges (Bullard et al., 1965), the use of plume and paleomagnetic "absolute" poles (Morgan, 1972b), pole positions computed from fracture zones (Morgan, 1968; Le Pichon, 1968; Pitman and Talwani, 1972), and instantaneous pole positions computed from fault plane solutions (Morgan, 1968; Horsfield and Maton, 1970). As the detailed magnetic survey coverage (Avery et al., 1968; Vogt and Avery, 1973; Karasik, 1968; Fleischer, 1971; Meyer et al., 1972) is gradually extended to cover the remainder of the Arctic and sub-Arctic basins, there will be little room for debating the exact evolution of regions in which magnetic lineations due to geomagnetic reversals and ocean-floor spreading (Heirtzler et al., 1968) are identified. North of about 53°N, this will probably include all ocean crust formed since 60 m.y.b.p. It is also possible, but by no means certain, that detailed surveying in the Amerasia Basin and northern and marginal parts of the Labrador Sea will disclose identifiable lineation patterns. If not, the genesis and evolution of this part of the Arctic will very likely remain conjectural, awaiting the development of a suitable deep-drilling technology. The same may be said of several depressions, for example, the basin east of the Voring escarpment (Talwani and Eldholm, 1972) and Rockall Trough. Even if deeply buried oceanic crust exists in such areas, and some of the original remanent magnetization survives, it is doubtful whether a narrow band of magnetic lineations could ever be identified with certainty. As suggested in this paper, some of these marginal phenomena may be ancient rifts—analogues to the present extension of the Gakkel Ridge into Siberia. If so, plate tectonic methods, using coeval ocean-floor isochrons (Pitman and Talwani, 1972) could provide some quantitative insight, even if the marginal rifts contain no oceanic crust in their depths. In all probability the origin of marginal troughs and plateaus will remain speculative, even with unlimited surface ship data, unless deep drilling is carried out.

A proper reconstruction of plate motion will need more than magnetic-bathymetric surveys and fault-plane solutions, for these only define motion between plates. Paleomagnetic methods, classically applied only to subaerial rocks but now being extended to cores of the oceanic sediments and basement, will continue to improve the resolution of absolute plate motion, i.e., motion with respect to the geomagnetic dipole. Provided the confidence ovals of magnetic poles can be significantly reduced, the method may help identify those interplate motions not legibly recorded by the oceanic crust. If hot spots, or mantle plumes (Morgan, 1972), are indeed fixed with respect to the earth's rotation axis, the igneous traces they leave on the plates will provide a powerful supplement to paleomagnetism. In a first attempt to combine the paleomagnetic and plume methods, Duncan et al. (1972) analyzed absolute motion of the two Arctic plates—western Eurasia and North America—over the Iceland (Thulean) and Eifel plumes. Their data indicate that either the magnetic pole or the Eifel and Iceland plumes have wandered with respect to the earth's spin axis. Intrigued with the plume hypothesis, the authors would rather believe in true polar wandering than admit to moving mantle plumes. Unfortunately the absolute (and relative) plate motions have been slow, and the late Cretaceous/early Tertiary volcanism has been widespread (Fig. 13). In view of the present accuracy of radiometric dating, the Iceland area would seem to be a difficult one in which to test the stationarity of mantle plumes against that of the geomagnetic dipole. It seems more reasonable that the British and Faeroe late Cretaceous/early Tertiary igneous activity was part of a large, more or less coeval igneous outburst (Fig. 13), a phenomenon that raises some difficulties for the thin plume hypothesis (Morgan, 1972a, b). In that case only the late Tertiary portion of the polar wander curve deduced from plumes (Duncan et al., 1972) need be considered, and during this period of time the uncertainty in paleomagnetic and Iceland-Eifel plume data prohibit a definite conclusion regarding stationariness of plumes or magnetic poles with respect to the spin axis.

A more speculative aspect of the mantle plume hypothesis is the idea that viscous traction by the plume-generated flow from the plumes helps drive the lithospheric plates (Morgan, 1972a, b). The discovery of time transgressive basement ridges (Fig. 11) raises the possibility that such flow has actually been recorded by ocean crust—at least a component of flow under and along the Reykjanes Ridge (Vogt, 1971). Detailed bathymetric, magnetic, seismic reflection and gravity surveys in the Arctic and sub-Arctic basins will be required to search for such time-transgressive features elsewhere. Recovery of the flow history of the Eifel, Iceland, and other plumes might one day lead from the present kinetic models of plate tectonics to a truly dynamic one.

The mantle plume hypothesis also offers some fertile possibilities to explain the peculiar seismicity (Figs. 1c and 12), fracture spacing, basement roughness (Fig. 12), and regional oceanic and continental topographic anomalies (Figs. 1b and 10) of the post-late Cretaceous Arctic and sub-Arctic. It may also explain the jumps of the spreading axis north of Iceland (Vogt et al., 1970a). Despite all this promise, the plume hypothesis is still based in a large measure on conjecture; it is definitely premature to regard it as a panacea for plate tectonics.

Acknowledgments

We thank G. L. Johnson and N. A. Ostenso for access to unpublished data and manuscripts. N. Hunt, B. Wells, B. Grosvenor, and S. Edwards assisted in manuscript preparation. This research was partially funded by the Office of Naval Research.

References

Anonymous. 1966. Atlas of the World. M. B. Grosvenor, ed. Washington, D.C.: National Geographic Society. 343 pp.

Anonymous. 1970. World seismicity 1961–1969. Washington, D.C. U.S. Department of Commerce, ESSA, NEIC 3005.

Anonymous. 1971. Arctic Ocean Floor. M. B. Grosvenor, ed. Washington, D.C. National Geographic Society.

Atlasov, I. P., B. Kh. Egiazarov, V. D. Dibner, B. S. Romanovich, A. V. Zimbin, V. A. Vakar, R. M. Demenitskaya, D. V. Levin, A. M. Karasik, Ya. Ya. Gakkel, and V. M. Litvin. 1967. "Tectonic Map of the Arctic and Subarctic," Tectonic Maps of the Continents at the XXII International Geological Congress. Moscow: Nauka. pp. 154–165.

Avery, O. E., G. D. Burton, and J. R. Heirtzler. 1968. An aeromagnetic survey of the Norwegian Sea. *J. Geophys. Res.* 73:4583–4600.

Bailey, R. J., J. S. Buckley, and R. H. Clarke. 1971. A model for the early evolution of the Irish continental margin. *Earth Planet. Sci. Lett.* 13:79–84.

Baron, J. G., J. R. Heirtzler, and G. R. Lorentzen. 1965. An airborne geomagnetic survey of the Reykjanes Ridge 1963. U.S. Naval Oceanographic Office Hydrographic Surveys Department, *Informal Report NO H-3-65.* 23 pp.

Beal, M. A. 1969. *Bathymetry and structure of the Arctic Ocean.* Corvallis: Oregon State University Doctoral thesis. 204 pp.

———, F. Edvalson, K. Hunkins, A. Molloy, and N. Ostenso. 1966. The floor of the Arctic Ocean: Geographic names. *Arctic* 19:215–219.

Beckinsale, R. D., C. K. Brooks, and D. C. Rex. 1970. K-Ar ages for the Tertiary of East Greenland. *Bull. Geol. Soc. Denmark* 20:27–37.

Belyaev, I. V. 1970. Osnovnye geologicheskie rezul'taty regional'nykh geofizicheskikh rabot (Major geological results of regional geophysics). In: A. V.

Sidorenko, chief ed. Geologiya SSSR, Tom 30, Severo-Vostok SSSR, Geologicheskoe opisanie, Kniga 2, Moscow, Ministry of Geology, Nedra. pp. 236–246.

Bhattacharyya, P. J., and D. I. Ross. 1972. Mid-Atlantic Ridge near 45°N. Computer interpolation and contouring of bathymetry and magnetics. *Marine Science Paper 11*, Marine Sciences Directorate, Department of Environment, Ottawa. 9 pp.

Bott, M. P. H., C. W. A. Browitt, and A. P. Stacey. 1971. The deep structure of the Iceland-Faeroe Ridge, Marine Geophysical Researches, 1:328–351.

Bullard, E., J. E. Everett, and A. G. Smith. 1965. The fit of the continents around the Atlantic. In: *A Symposium on Continental Drift: Royal Soc. London Philos. Trans., ser. A*, 258, 1088:41–51.

Carey, S. W. 1958. A tectonic approach to continental drift. In: S. W. Carey, ed. *Continental Drift—A Symposium.* Hobart, Australia Tasmania University. pp. 177–355.

Cathles, L. 1971. The viscosity of the earth's mantle. Doctoral thesis, Princeton, New Jersey. 552 pp.

Churkin, M. 1972. Western boundary of the North American continental plate in Asia. *Geol. Soc. Amer. Bull.* 83:1027–1036.

Demenitskaya, R. M., and Yu. G. Kiselev. 1968. The characteristic features of the structure, morphology, and sedimentary cover of the central part of the Lomonosov Ridge from seismic investigations. In: *Geophysical Methods of Prospecting in the Arctic*, Leningrad, 5:33–46.

———, and K. L. Hunkins. 1971. Shape and structure of the Arctic Ocean. *The Sea,* Vol. 4, Part 2. A. Maxwell, ed. New York: Wiley and Sons. pp. 223–249.

———, and A. M. Karasik. 1971. Problemy genezisa Severnogo Ledovitogo okeana. In: *Istoriia Mirovogo okeana.* Moscow: Akad. Nauk SSSR, Moskovskoe Obshchestvo Ispytatelei Prirody, Izdatel'stvo "Nauka": 58–76. Translated by Dorothy B. Vitaliano.

Duncan, R. A., N. Petersen, and R. B. Hargraves. 1972. Mantle plumes, movement of the European plate, and polar wandering. *Nature* 239:82–86.

Eardley, A. J. 1961. History of geologic thought on the origin of the Arctic Basin. In: *Geology of the Arctic*, G. O. Raasch, ed. University of Toronto Press. pp. 607–621.

Einarsson, T. 1966. Late and post-glacial rise in Iceland and subcrustal viscosity. *Jökull*, 16:157–166.

Ewing, J. I., and C. H. Hollister. 1972. Regional aspects of deep-sea drilling in the western North Atlantic. In: C. D. Hollister, et al., 1972. *Initial Reports of the Deep-Sea Drilling Project, Vol. 11,* Washington, D.C.: U.S. Government Printing Office. pp. 951–973.

Fitch, F. J. 1965. The structural unity of the reconstructed North Atlantic continent: A symposium on continental drift. *Trans. Royal Soc. Philadelphia* 258:191–193.

Fleischer, U. 1971. Gravity surveys over the Reykjanes Ridge and between Iceland and the Faeroe Islands. *Marine Geophys. Res.* 1:314–327.

Godby, E. A., P. J. Hood, and M. E. Bower. 1968. Aeromagnetic profiles across the Reykjanes Ridge southwest of Iceland. *J. Geophys. Res.* 73:7637–7650.

Gutenberg, B., and C. F. Richter. 1954. *Seismicity of the earth.* Princeton, New Jersey: Princeton University Press. 310 pp.

Hall, J. K. 1970. Arctic Ocean geophysical studies: The Alpha Cordillera and Mendeleyev Ridge. Lamont-Doherty Geological Observatory of Columbia University, I.R. No. 2. 125 pp.

Hamilton, W. 1970. The Uralides and the motions of the Russian and Siberian platforms. *Geol. Soc. Amer. Bull.* 81:2553–2576.

Heirtzler, J. R., G. O. Dickson, E. M. Herron, W. C. Pitman, III, and X. Le Pichon. 1968. Marine magnetic anomalies, geomagnetic field reversals and motions of the ocean floor and continents. *J. Geophys. Res.* 73:2119–2136.

Horsfield, W. T., and P. I. Maton. 1970. Transform faulting along the De Geer Line. *Nature* 226:256–257.

Hunkins, K. 1961. Seismic studies of the Arctic Ocean floor. In: G. O. Raasch, ed. *Geology of the Arctic*, Vol. 1. Toronto: University of Toronto Press. pp. 645–665.

———. 1963. Submarine structure of the Arctic Ocean from earthquake surface waves. In: *Proceedings of the Arctic Basin Symposium, October 1962.* Washington, D.C.: The Arctic Institute of North America. pp. 3–8.

———. 1969. Arctic geophysics. *Arctic* 22:225–232.

Johnson, G. L., J. R. Southall, P. W. Young and P. R. Vogt. 1972a. Origin and structure of the Iceland Plateau and Kolbeinsey Ridge. *J. Geophys. Res.* 77:5688–5696.

———, J. Campsie, M. Rasmussen, and F. Dittmer. 1972b. Mesozoic rocks from the Labrador Sea. *Nature Phys. Sci.* 236:86–87.

———, and B. Heezen. 1967a. The morphology and evolution of the Norwegian-Greenland Sea. *Deep-Sea Res.* 14:755–771.

———, and B. Heezen. 1967b. The Arctic Mid-Oceanic Ridge. *Nature* 215:724–725.

———, and P. R. Vogt. 1972. Marine geology of the Atlantic north of the Arctic Circle. *American Association of Petroleum Geologists Memoir, Second International Arctic Symposium*, in press.

———, and P. R. Vogt. 1973. The Mid-Atlantic Ridge from 47° to 51°N. *Bull. Geol. Soc. Am.,* in press.

Jones, E. J. W., M. Ewing, J. Ewing, and S. L. Eittreim. 1970. Influences of Norwegian overflow water on sedimentation in the northern North Atlantic and Labrador sea. *J. Geophys. Res.* 75:1655–1680.

Karasik, A. M. 1968. Magnetic anomalies of the Gakkel Ridge and origin of the Eurasia Subbasin of the Arctic Ocean. In: *Geophysical Methods of Prospecting in the Arctic*, 5, Leningrad, 8–19.

Karig, D. E. 1971. Structural history of the Mariana Island arc system. *Geol. Soc. Am. Bull.* 82:232–344.

Keen, M. J., B. C. Loncarevic, and G. N. Ewing. 1971. Continental margin of eastern Canada: Georges Bank to Kane Basin. In: *The Sea*, Vol. 4, Part 2. A. Maxwell, ed. New York: Wiley Interscience. pp. 251–292.

Keen, C. E., D. L. Barrett, K. S. Manchester, and D. I. Ross. 1972. Geophysical studies in Baffin Bay and some tectonic implications. *Can. J. Earth Sci.* 9:239–256.

King, E. R., I. Zietz, and L. R. Alldredge. 1966. Magnetic data on the structure of the central Arctic region. *Bull. Geol. Soc. Am.* 77:10–12.

Kutschale, H. 1966. Arctic Ocean geophysical studies: The southern half of the Siberian Basin. *Geophysics* 31:683–710.

Lachenbruch, A. L., and B. V. Marshall. 1966. Heat flow through the Arctic Ocean floor: The Canada Basin—Alpha Rise boundary. *J. Geophys. Res.* 71:1223–1248.

Laughton, A. S., W. A. Berggren, et al. 1972. Initial reports of the deep-sea drilling project, Vol. 12. Washington, D.C.: U.S. Government Printing Office. pp. 1181–1189.

Le Pichon, X. 1968. Sea-floor spreading and continental drift. *J. Geophys. Res.* 73:3661–3698.

———, R. D. Hyndman, and G. Pautot. 1971. Geophysical study of the opening of the Labrador Sea. *J. Geophys. Res.* 76:4724–4743.

Ling, H. Y., L. M. McPherson, and D. L. Clark. 1973. Late Cretaceous (Maestrichtian?) Silicoflagellates from the Alpha Cordillera of the Arctic Ocean. *Science* 180(4093):1360–1361.

Loncarevic, B. D., and R. L. Parker. 1971. The Mid-Atlantic Ridge near 45°N, Part VII, Magnetic anomalies and ocean-floor spreading *Can. J. Earth Sci.* 8:883–898.

Manchester, K. S., M. J. Keen, D. E. Clarke, and D. I. Ross. 1972. Geologic structure of Baffin Bay and Davis Strait determined by various geophysical techniques. *Second International Symposium of Arctic Geology, American Association of Petroleum Geologists Memoir*, in press.

Matthews, D. H., A. S. Laughton, D. T. Pugh, E. J. W. Jones, J. Sunderland, M. Takin, and M. Bacon. 1969. Crustal structure and origin of Peake and Freen deeps, Northeast Atlantic. *Geophys. J. Royal Astr. Soc.* 18:517–542.

McMillan, N. J. 1973. Surficial geology of Labrador and Baffin Island shelves, *Earth Science Symposium Offshore Eastern Canada*, Geol. Surv. Can. Paper 71-23, 451–468.

Meyer, O., D. Vappel, U. Fleischer, H. Closs, and K. Gerke. 1972. Results of bathymetric, magnetic, and gravimetric measurements between Iceland and 70°N. *Dt. Hydrogr. Z.* 25:193–203.

Miller, J. A., and P. A. Mohr. 1965. Potassium-argon determinations on rocks from St. Kilda and Rockall. *Scotland J. Geol.* 1:93.

Morgan, W. J. 1968. Rises, trenches, great faults, and crustal block. *J. Geophys. Res.* 73:1959–1982.

———. 1972a. Deep mantle convection plumes and plate motions. *Bull. Am. Assoc. Petroleum Geol.* 56:203–213.

———. 1972b. Plate motions and deep mantle convections. *Geol. Soc. Am. Mem.*, 132, Hess Volume, R. Shagam, ed. in press.

Ostenso, N. A. 1972. Sea-floor spreading and the origin of the Arctic Basin. In: *Continental drift, Sea-floor spreading and plate tectonics.* D. H. Tarling and S. K. Runcorn, eds. London: Academic Press, in press.

———, and R. J. Wold. 1971. Aeromagnetic survey of the Arctic Ocean: techniques and interpretations. *Marine Geophys. Res.* 1:178–219.

Palmason, G. 1970. Crustal structure of Iceland from explosion seismology. Science Institute, Union of Iceland and National Energy Authority, Reykjavik, Iceland. 239 pp.

Park, I., D. B. Clarke, J. Johnson, and M. J. Keen. 1971. Seaward extension of the west Greenland Tertiary volcanic province. *Earth Planet. Sci. Lett.* 10:235–238.

Pitman, W. C., and M. Talwani. 1972. Sea-floor spreading in the North Atlantic. *Bull. Geol. Soc. Am.* 83:619–646.

Purdy, J. W., A. E. Mussett, S. R. Charlton, M. J. Eckford, and H. N. English. 1972. The British Tertiary igneous province: Potassium-argon ages of the Antrim basalts. *Geophys. J. Royal Astron. Soc.* 27:327–335.

Rassokho, A. I., L. I. Senchura, R. M. Demenitskaya, R. M. Karasik, A. M. Karasik, Y. G. Kiselev, and N. K. Tomoshonko. 1967. The Mid-Arctic Range as a unit of the Arctic Ocean mountain system. *Dokl. Acad. Nauk SSSR* 172:659–662.

Riddihough, R. P., G. V. Haines, and W. Hannaford, Regional magnetic anomalies of the Canadian Arctic. 1973. *Can. J. Earth Sci.* 10:157–163.

Roberts, D. G. 1969. New Tertiary volcanic centre on Rockall Bank, eastern North Atlantic Ocean. *Nature* 223:819–820.

———, D. G. Bishop, A. S. Laughton, A. M. Ziolkowski, R. A. Scrutton, and D. H. Matthews. 1970. New sedimentary basin on Rockall Plateau. *Nature* 225:170–172.

Sclater, J. G., R. N. Anderson, and M. L. Ball. 1971.

Elevation of ridges and evolution of the central Eastern Pacific. *J. Geophys. Res.* 76:7888–7915.

Serson, P. H., W. Hannaford, and G. V. Haines. 1968. Magnetic anomalies over Iceland. *Science* 162:355–357.

Sleep, N. H. 1971. Thermal effects of the formation of Atlantic continental margins by continental breakup. *Geophys. J. Royal Astron. Soc.* 24:325–350.

———, and M. N. Toksöz. 1971. Evolution of marginal basins. *Nature* 233:548–550.

Sykes, L. R. 1965. The seismicity of the Arctic. *Seismological Soc. Am. Bull.* 55:501–518.

Talwani, M., C. Windisch, M. Langseth, and J. R. Heirtzler. 1971. Reykjanes Ridge crest: A detailed geophysical study. *J. Geophys. Res.* 76:473–517.

———, and O. Eldholm. 1972. Continental margin off Norway: A geophysical study. *Geol. Soc. Am. Bull.* 83:3575–3606.

Tarling, D. H., and N. H. Gale. 1968. Isotope dating and paleomagnetic polarity in the Faeroe Islands. *Nature* 218:1043.

Taylor, P. T., I. Zietz, and L. S. Dennis. 1968. Geologic implications of aeromagnetic data for the eastern continental margin of the United States. *Geophysics* 33:755.

Treshnikov, A. F., L. L. Balakshin, N. A. Belov, R. M. Demenitskaya, V. D. Dibner, A. M. Karasik, A. O. Shpeiker, and N. D. Shurgayeva. 1967. A geographic nomenclature for the chief topographic features of the Arctic Basin floor. *Prob. Arctic Antarctic* 27:5–15.

Vine, F. J. 1966. Spreading of the ocean floor: New evidence. *Science* 154:1405–1415.

Vogt, P. R. 1970. Magnetized basement outcrops on the southeast Greenland continental shelf. *Nature* 226:743–744.

———. 1971. Asthenosphere motion recorded by the ocean floor south of Iceland. *Earth Planet. Sci. Lett.* 13:153–160.

———. 1972. Early events in the evolution of the North Atlantic. *Continental drift, Sea-floor spreading and plate tectonics.* D. H. Tarling and S. K. Runcorn, eds. London: Academic Press, in press.

———, and N. A. Ostenso. 1970. Magnetic and gravity profiles across the Alpha Cordillera and their relation to the Arctic sea-floor spreading. *J. Geophys. Res.* 75:4925–4937.

———, and O. E. Avery. 1973. Detailed magnetic surveys in the northeast Atlantic and Labrador Sea. *J. Geophys. Res.*, in press.

———, and G. L. Johnson. 1973. A longitudinal seismic reflection profile of the Reykjanes Ridge. Part 1. Implications for the mantle hot spot hypothesis. *Earth Planet. Sci. Lett.*, in press.

———, E. D. Schneider, and G. L. Johnson. 1969. The crust and upper mantle beneath the sea. In: *The earth's crust and upper mantle*, Washington, D.C.: American Geophysical Union. pp. 556–617.

———, N. A. Ostenso, and G. L. Johnson. 1970a. Magnetic and bathymetric data bearing on sea-floor spreading North of Iceland. *J. Geophys. Res.* 75:903–920.

———, C. N. Anderson, D. R. Bracey, and E. D. Schneider. 1970b. North Atlantic magnetic smooth zones. *J. Geophys. Res.* 75:3955–3967.

———, C. N. Anderson, and D. R. Bracey. 1971. Mesozoic magnetic anomalies, sea-floor spreading and geomagnetic reversals in the southwestern North Atlantic. *J. Geophys. Res.* 76:4796–4823.

Ward, P. L. 1971. New interpretation of the geology of Iceland. *Geol. Soc. Am. Bull.* 82:2991–3012.

Watkins, N. D., and A. Richardson. 1968. Paleomagnetism of the Lisbon volcanics. *Geophys. J. Royal Astron. Soc.* 15:287–304.

Watt, W. S. 1969. The coast-parallel dike swarm of southwest Greenland in relation to the opening of the Labrador Sea. *Can. J. Earth Sci.* 6:1320–1321.

Wegener, A. 1915. *Die Enstehung der Kontinente und Ozeane*. Braunschweig: F. Vieweg. 94 pp.

Weissel, J. K., and D. E. Hayes. 1971. Asymmetric sea-floor spreading south of Australia. *Nature* 231:518–522.

Wilson, J. T. 1963. Hypothesis of earth's behavior. *Nature* 198:925–929.

Wold, R. J., and N. A. Ostenso. 1971. Gravity and bathymetry survey of the Arctic and its geodetic implications. *J. Geophys. Res.* 76:6253–6264.

Chapter 4
Tectonic Setting and Cenozoic Sedimentary History of the Bering Sea[1]

C. HANS NELSON,[2] DAVID M. HOPKINS[2]
AND DAVID W. SCHOLL[2]

Abstract

The Bering Sea consists of (a) an abyssal basin that apparently was isolated from the Pacific Ocean by the development of the Aleutian Ridge near the end of Cretaceous Period and (b) a large epicontinental shelf area that first became submerged near the middle of the Tertiary Period. We postulate that the sediment eroded from Alaska and Siberia during Cenozoic time. It (1) was trapped in subsiding basins on the Bering shelf and in abyssal basins during the Tertiary, (2) collected mostly in continental rise and abyssal plain deposits of the Bering Sea during periods of low sea level of the Pleistocene, and (3) has been transported generally northward from the Bering shelf through the Bering Strait into the Arctic Ocean during periods of high sea level in Pleistocene and Holocene time.

Filling of subsiding basins on the shelf apparently was dominated by continental sedimentation in the early Tertiary and by marine deposition in the later Tertiary. River diversions caused by Miocene uplift of the Alaskan Range appear to have more than doubled the drainage area of the Yukon River. This change established the Yukon as the dominant source of river sediments (~90%) reaching the Bering Sea and must have greatly accelerated sedimentation in the basins.

The Quaternary Period presumably has been a time of alternation between two modes of sedimentation. When sea level was glacio-eustatically lowered, the Yukon and other rivers extended their courses across the continental shelf and delivered most of their sediment to the abyssal basin; an incidental result was the cutting of some of the world's largest submarine canyons. Occurrence of the thin Holocene marine deposits near the present Yukon delta, of mainly relict sediments in the northern Bering Sea north of St. Lawrence Island and of several meters of Holocene sediment in the Chukchi Sea, suggests that when sea level was high, currents evidently swept much of the sediment from the Yukon River northward through the Bering Strait to be deposited on the continental shelf of the Chukchi Sea.

[1] Publication authorized by the director, U.S. Geological Survey

[2] U.S. Geological Survey, 345 Middlefield Road, Menlo Park, California 94025, U.S.A.

Introduction

The historical section of this paper attests to the wealth of geologic information on the Bering Sea that has been published recently by Russian and American investigators. This paper attempts to synthesize research on the Cenozoic sedimentary and tectonic history and, more specifically, it summarizes the three reports (Nelson, 1970a; Hopkins and Scholl, 1970; Scholl and Buffington, 1970) presented orally by the authors at the Second International Symposium on Arctic Geology.

Major changes in the pattern of sedimentation can be outlined for the Bering Sea during the Cenozoic Era. Radical shifts can be inferred in the position of major deposition sites that changed between the shallow half and the deep half of the sea floor. These shifts can be attributed to tectonic events that affected both the shelf and the Alaskan and Siberian hinterlands, and to eustatic sea-level changes. We suggest a new sedimentation model for a marginal sea based on the major displacement of marine depositional sites away from the main shore-line sources and on the long-term history of sediment dispersal delimited in this paper. This model may help geologists in reconstructing ancient sediment dispersal systems and defining paleogeography in this and similar areas. The most critical application of the model is in the search for petroleum and mineral resources in the Bering Sea, where it may assist in establishing the location and sequence of history of the Cenozoic sedimentary bodies.

History of Investigations

The Bering Sea has been studied by Russian and American oceanographers, geologists, and sedimentologists since the middle of the 19th century. An excellent review of the earlier studies was prepared by Lisitsyn (1966). The state of knowledge resulting from Soviet and American studies of the marine geology, shore-line geology, and paleontology of the Bering Sea region between 1948 and 1965 is summarized in *The Bering Land Bridge* (Hopkins, 1967a).

The modern era of marine geologic study began in the late 1940's with a series of cruises by scientists of the Navy Electronics Laboratory, during which data were gathered for the first published maps of the bottom sediments (Dietz et al., 1964) and the first seismic reflection profiles were obtained (Moore, 1964). Cruises during the 1950's by scientists of the Institute of Oceanology, Academy of Science, U.S.S.R., resulted in a general discussion of the bottom morphology of Bering Sea (Udintsev et al., 1959) and in a monographic treatment of the sediments and sedimentary processes of the western Bering Sea (Lisitsyn, 1966). Cruises during the late 1950's and early 1960's by scientists of the Institute of Fisheries and Oceanography (VNIRO—Ministry of Fisheries, U.S.S.R.) resulted in much new data on the sediments of the eastern part of the Bering Sea (Gershanovich, 1967, 1968, and papers cited therein). In the 1950's and 1960's the sediments and topography of the Chirikov Basin and Norton Sound areas (Fig. 1) of the northern Bering Sea were investigated by scientists of the Department of Oceanography of the University of Washington, working in cooperation with the U.S. Coast Guard and the U.S. Geological Survey (Creager and McManus, 1967; Creager et al., 1970; McManus and Smyth, 1970; Grim and McManus, 1970). Sediment and seismic profiling studies by the University of Washington have recently been extended to the south-central and southwestern part of the continental shelf of the Bering Sea (Knebel and Creager, 1970; Kummer and Creager, 1971). Recent Soviet geophysical studies oriented toward a study of the oil and gas potential of the continental shelf of the Gulf of Anadyr are reviewed by Verba et al. (1971). Bottom sediments of the Bristol Bay area were studied by G. D. Sharma (1970) of the University of Alaska. Studies of the deep structure of the abyssal basin of the Bering Sea were pioneered by Shor (1964). The Bowers Bank–Shirshov Ridge system has been studied by scientists of Lamont-Doherty Geological Observatory (Ewing et al., 1965; Ludwig et al., 1971a; Kienle, 1970) working in cooperation with a group of Japanese investigators (Ludwig et al., 1971b).

Knowledge of the Bering Sea has been further advanced by several types of offshore research by scientists of the U.S. Geological Survey working in cooperation with the U.S. Navy Undersea Research and Development Center, the U.S. National Oceanic and Atmospheric Administration, and the University of Washington. These cooperative studies have resulted in the acquisition of some 35,000 km of deep-penetration seismic-reflection profile (SRP) records (Fig. 2); shipborne magnetometer and bathymetric records for most of this distance; 4200 km of high-resolution SRP records (Fig. 2); 35 dredge samples from the area of the continental slope; and about 1000 large grab samples and 200 box cores from the continental shelf (Figs. 3 and 4).

Research by U.S. Geological Survey scientists in the Bering Sea has focused on and has helped to delineate (1) the structure and origin of the abyssal basin and major topographic features in the deep Bering Sea, including the Umnak Plateau, the Bowers Bank, and the Shirshov Ridge (Scholl et al., in press) (Fig. 1); (2) the nature and history of the continental margin (Scholl et al., 1968; Hopkins et al., 1969; Scholl et al., 1970); (3) the Tertiary tectonic pattern and the distribution, configuration, thickness, and probable age of Tertiary sedimentary basins on the shelf (Scholl

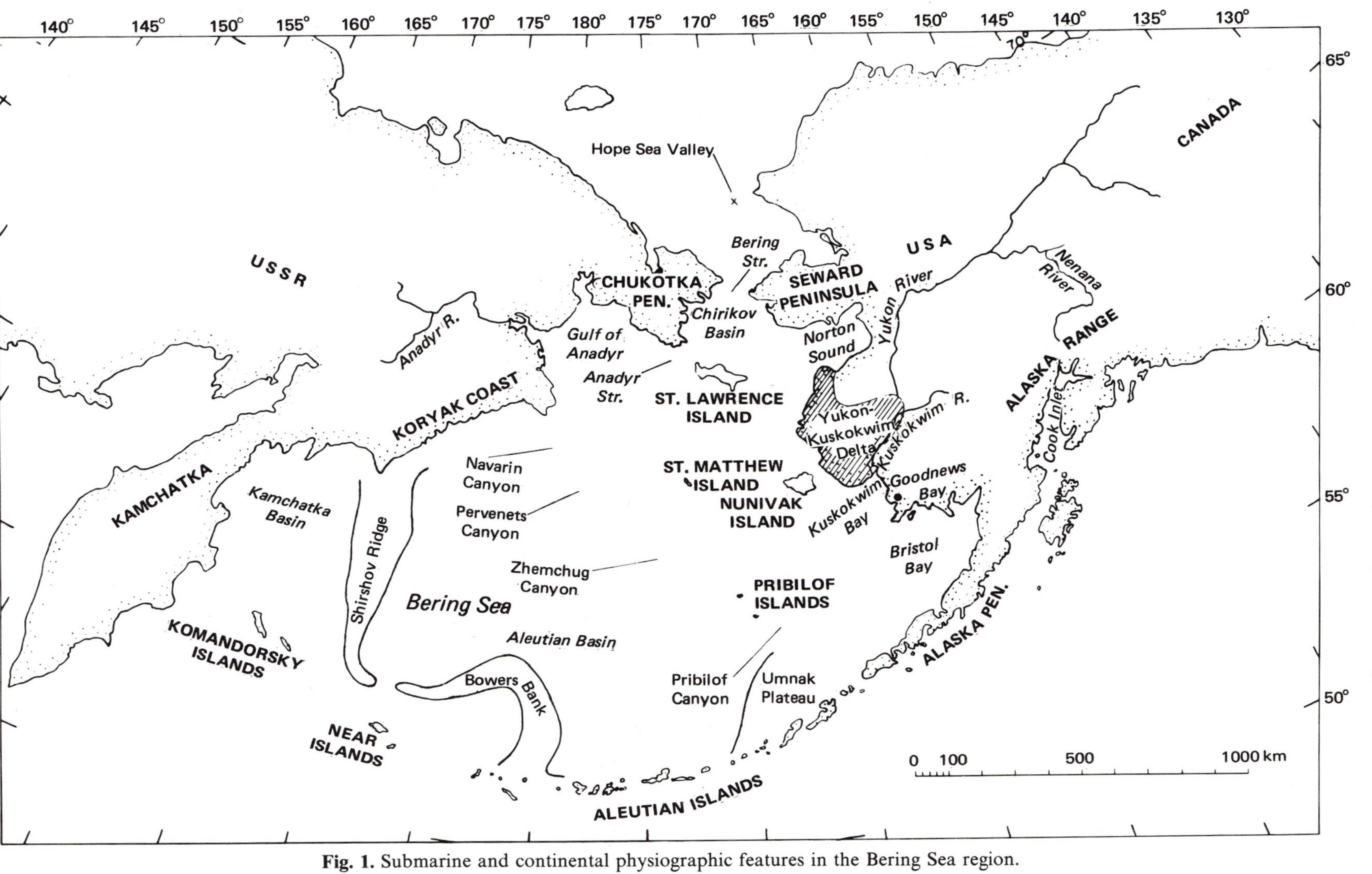

Fig. 1. Submarine and continental physiographic features in the Bering Sea region.

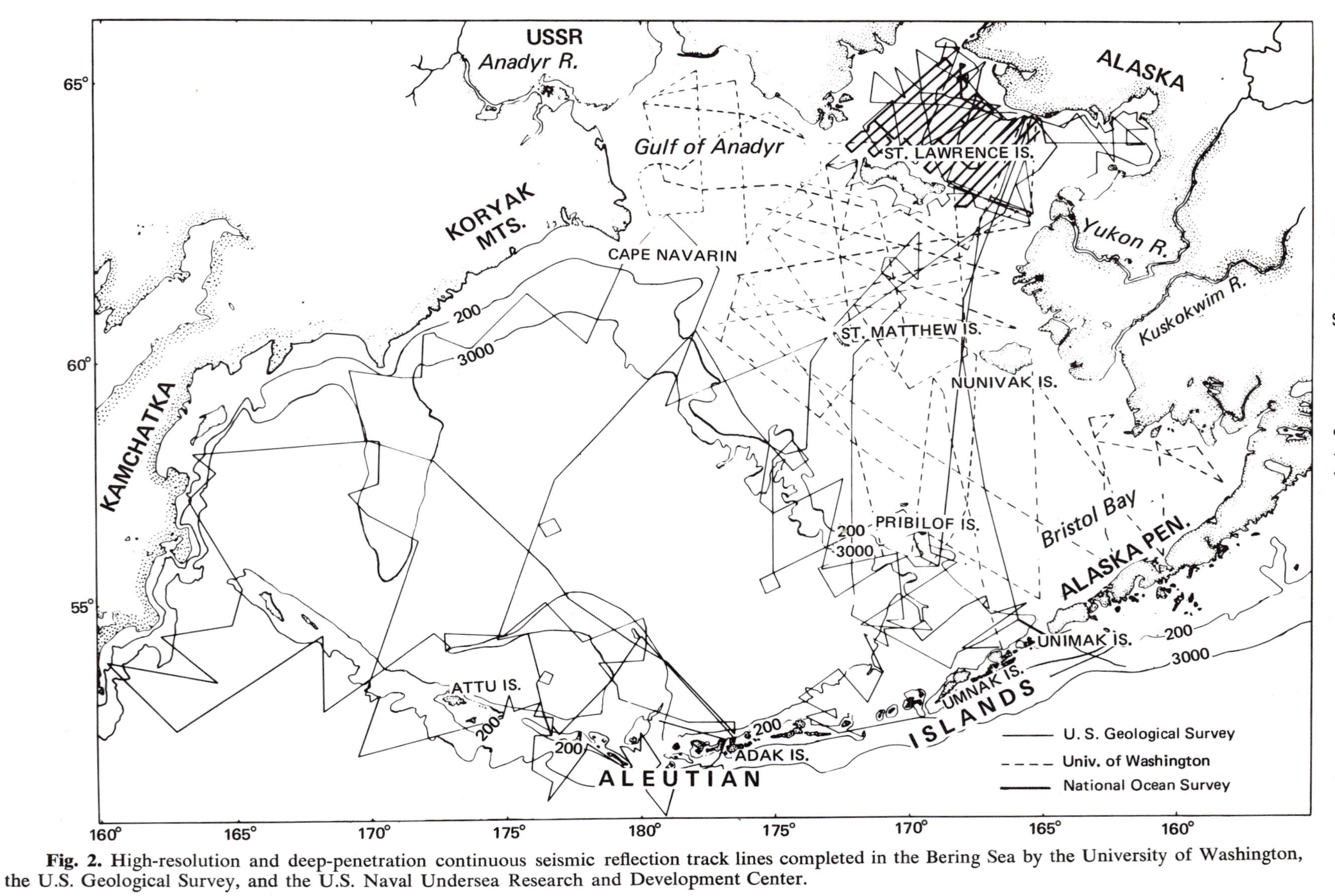

Fig. 2. High-resolution and deep-penetration continuous seismic reflection track lines completed in the Bering Sea by the University of Washington, the U.S. Geological Survey, and the U.S. Naval Undersea Research and Development Center.

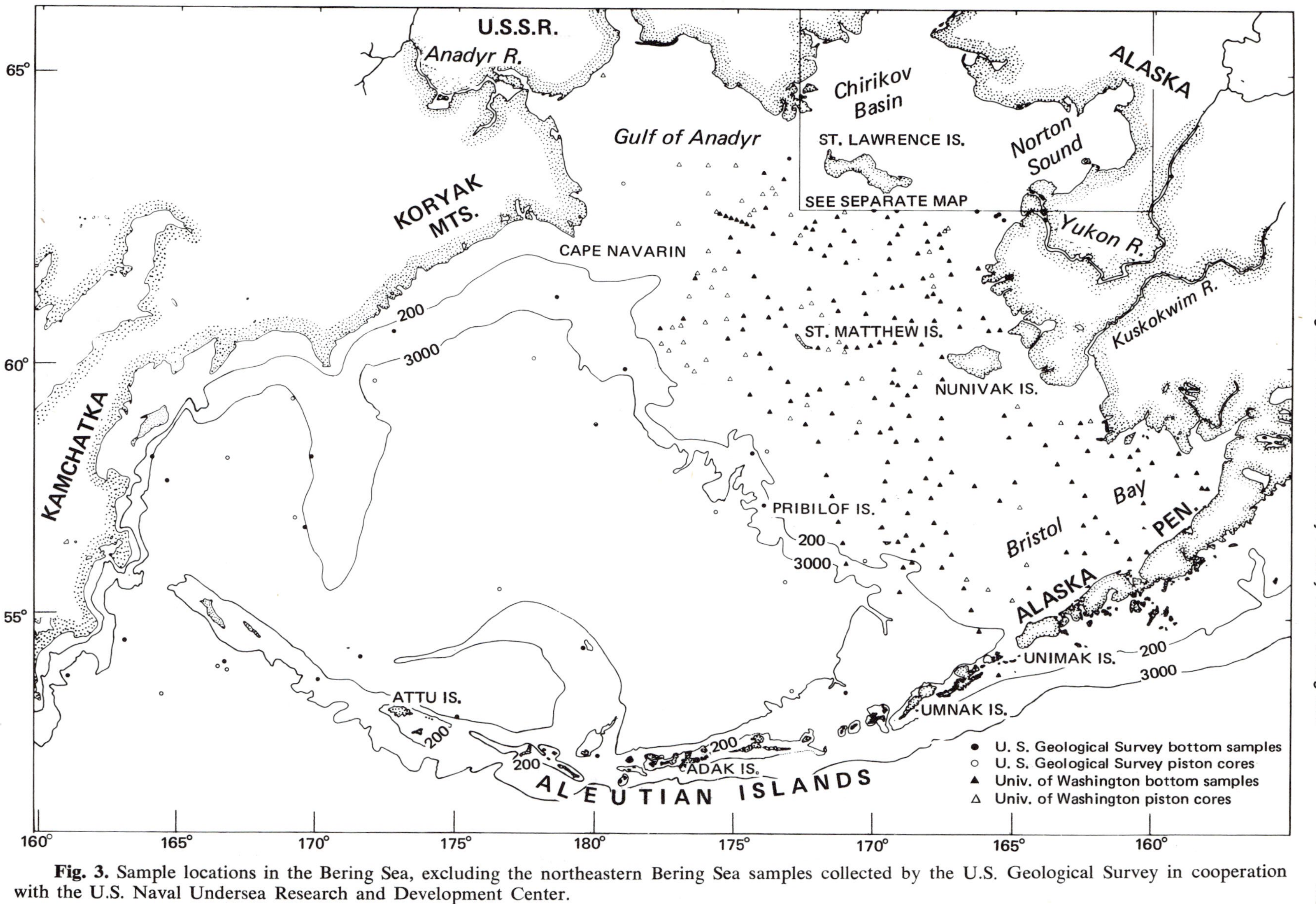

Fig. 3. Sample locations in the Bering Sea, excluding the northeastern Bering Sea samples collected by the U.S. Geological Survey in cooperation with the U.S. Naval Undersea Research and Development Center.

and Hopkins, 1969; Walton et al., 1969; Scholl and Marlow, 1970; Hopkins and Scholl, 1970); (4) the general distribution and nature of the unconsolidated sediments and the resource potential of detrital minerals in the northern Bering Sea and the Goodnews Bay area (Silberman, 1969; Greene, 1970; Moll, 1970; Tagg and Greene, 1971; Nelson, 1970b; Venkatarathnam, 1971); Sheth, 1971; Nelson and Hopkins, 1972; Tagg and Barnes, unpublished data, 1972); and (5) the broad outlines of the Quaternary paleogeography in the northern Bering Sea (Hopkins, 1967a,b, 1972; Hopkins et al., 1972). Studies have also been conducted in the adjacent Chukchi Sea to the north (Grantz et al., 1970a, 1970b) and in the Aleutian Trench–Gulf of Alaska region to the south (von Huene and Shor, 1969; Marlow et al., 1970; Holmes et al., 1972).

A new phase of research began with the first deep-sea drilling in the Bering Sea during the cruise of the *R/V Glomar Challenger* in 1971, available in a preliminary account (Scholl, et al., 1971).

Tertiary Geologic History

Tectonic Framework and Plate Motions

The continental shelf of the Bering Sea has long been known to be an area of continental crust that connects Asia and North America (Hopkins, 1959); the deep part of the sea is a small abyssal ocean basin, floored by oceanic crust, and overlain by a thick sequence of undeformed sediments (Shor, 1964; Menard, 1967; Ludwig et al., 1971a). Recent studies have suggested that the continental shelf of the Bering Sea is a submerged part of a large continental plate that includes all of North America and the northeastern part of Siberia (Churkin, 1972), and the abyssal basin of the Bering Sea is thought to be a northern embayment of the Pacific Ocean that became isolated by the development of the Aleutian Ridge prior to late Eocene time but no earlier than late Cretaceous time (Scholl, et al., 1970; Scholl and Buffington, 1970; Scholl, et al., in press). The interaction of oceanic and continental crust evidently was localized at the base of the continental slope of the Bering Sea during most of the Mesozoic Era. A shift in the zone of subduction southward away from Beringia[1] to the present site of the Aleutian trench–arc system evidently took place near the beginning of the Tertiary Period.

The tectonic history is complicated by the fact that prior to Miocene time the Pacific plate did not impinge directly upon the North American–Siberian plate, but was separated from it by the Kula spreading center and the Kula plate (Grow and Atwater, 1970). The floor of the abyssal basin of the Bering Sea is probably a remnant of the oceanic Kula plate that became incorporated into the North American–Siberian plate when the subduction zone shifted southward to the Aleutian Trench in the early Tertiary. The Kula plate continued to interact with the North American–Siberian plate until it was completely consumed by subduction along the Aleutian arc during the Miocene Epoch.

Much of the deeper part of the Bering Sea is occupied by two abyssal plains—the Aleutian Basin and the Kamchatka Basin (Fig. 1). These grade at their northern edges into a continental rise consisting of a series of coalescing deep-sea fans (Scholl et al., 1968). Seismic reflection profiles across the continental margin indicate that the continental basement is downflexed and extends some distance seaward beneath the Aleutian Basin; this may in part reflect foundering of the continental margin as an isostatic response to loading by terrigenous debris deposited on the continental rise. The Umnak Plateau, a large area of intermediate depth occupying the southeastern corner of the abyssal basin (Fig. 1) may also be a mass of foundered continental crust. An enigmatic feature of the abyssal portion of the Bering Sea is the Shirshov Ridge–Bowers Bank complex (Fig. 1), which consists of two large, nearly connecting aseismic ridges extending in an S-shape southward from the Siberian coast to the Aleutian Ridge. The ridge complex isolates the Kamchatka Basin from the Aleutian Basin.

The tectonic outline of continental Beringia is dominated by a gigantic structure composed of two sharp flexures concave toward the Pacific Ocean—the Alaskan orocline on the east and the Chukotkan orocline on the west. The broad intervening faulted flexure that is submerged on the continental shelf and concave toward the Arctic Ocean is shown on Fig. 5. Hopkins and Scholl (1970) believe that large transcurrent faults, developed mostly during late Cretaceous and Paleocene time, are the strongest expression of the flexure system which probably reflects the bending of Siberia relative to North America as a conse-

[1] Beringia includes the present area of western Alaska, northeastern Siberia, and the shallow parts of the Bering and Chukchi seas.

quence of rifting in the Atlantic and Arctic basins. The southward bulge of the Alaskan part of the North American continental margin may have resulted from this oroclinal folding. The shift in the site of the subduction zone from the continental margin in the Bering Sea to the Aleutian Trench may have been a response to bending of the continental margin.

Although the major oroclinal folding appears to have been complete before Oligocene time, the axial trends of basins and upwarps involving middle and late Tertiary rocks conform, in general, to the configuration of the oroclines; many of the transcurrent faults remain active, displacing upper Tertiary and Quaternary sediments (Grim and McManus, 1970). Movement along transcurrent faults just north and south of the Bering Strait has probably played an important role in the development and destruction of intercontinental land connections in the area of the Bering Strait.

Bering Shelf Acoustic and Geologic Units

Three broad groups of rocks and sediments have been distinguished on the continental shelf in the Bering Sea: (1) folded rocks below a strongly reflecting horizon, termed the acoustic basement by Scholl et al., (1968); (2) the main layered sequence (MLS), a unit of gently deformed sediments of Tertiary age above the acoustic basement; and (3) the generally flat-lying sediments of Quaternary age at or near the sea floor.

Acoustic Basement

The acoustic basement is an erosion surface that bevels all rocks underlying the MLS. The petrology and age of the rocks below the acoustic basement are not well known. Speculations based on magnetic mapping, extrapolation from rocks exposed on the mainland and insular areas, deep profiling, and dredging at the continental margin suggest that three distinctive tectonic units may occupy successively more southern belts (Fig. 6).

The basement rocks beneath the Chirikov Basin and much of Norton Sound probably consist of strongly lithified rocks of Paleozoic age, metamorphosed in many places. Similar rocks are exposed in northern Chukotka, on the Seward Peninsula and on much of St. Lawrence Island.

The Okhotsk volcanic belt, a belt of late Mesozoic and early Cenozoic volcanogenic rocks (Tilman et al., 1969), extends across southern Chukotka and beneath Anadyr Strait to southwestern St. Lawrence Island, thence south of the

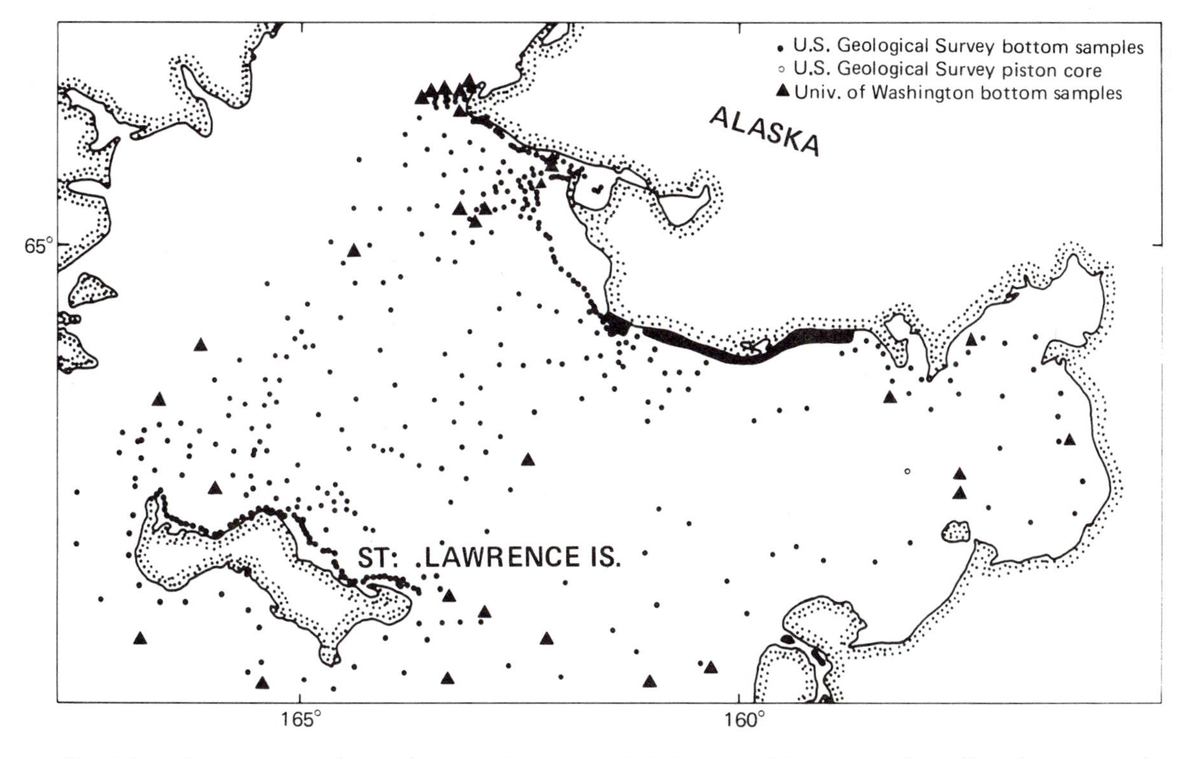

Fig. 4. Sample locations in the northeastern Bering Sea. U.S. Geological Survey samples collected in cooperation with the U.S. National Ocean Survey and the Department of Oceanography of the University of Washington.

central and eastern part of the island (Patton and Csejtey, 1970; Patton, 1970). The southern limit of the Okhotsk volcanic belt evidently lies a few tens of kilometers south of St. Matthew Island, as judged by the silicic volcanic rocks exposed on the island and the presence of many steep magnetic anomalies in the surrounding area (Scholl and Marlow, unpublished data). Broadly undulating reflectors detectable beneath the acoustic basement in this region (Scholl and Marlow, 1970) are suggestive of the open folding that characterizes the Okhotsk volcanic belt (Tilman et al., 1969). Patton (1970) suggests that the Okhotsk volcanic belt swings northeastward from St. Matthew Island, passing beneath Norton Sound to the Alaskan mainland (Fig. 6).

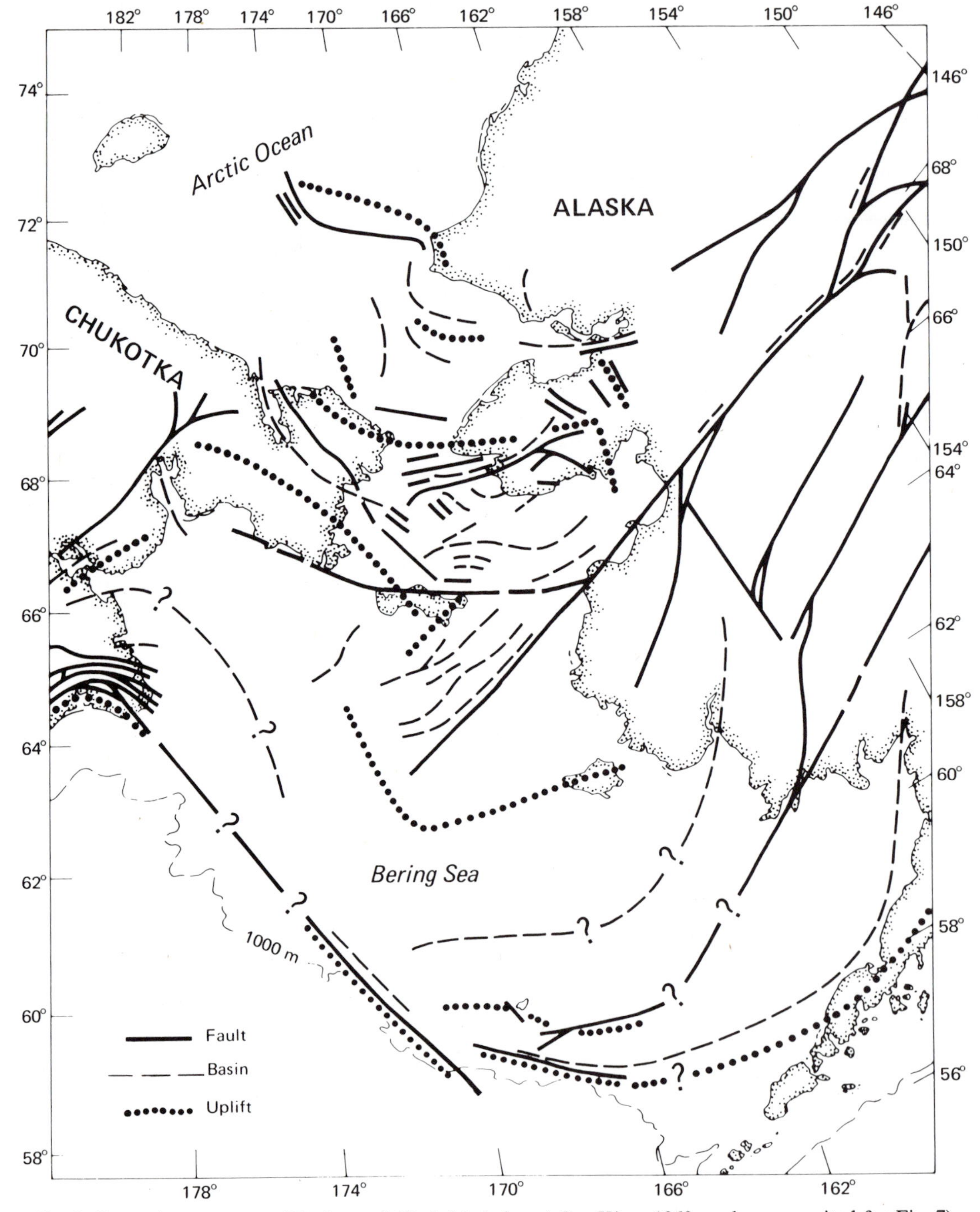

Fig. 5. Cenozoic structures of Bering and Chukchi shelves (after King, 1969, and sources cited for Fig. 7).

The southernmost tectonic unit (Fig. 6) seems to consist of intensely deformed Cretaceous flysch sediments intruded, in places, by serpentinite. A similar complex is exposed extensively in mainland areas of Alaska and Siberia (Scholl et al., 1966, 1968, in press); lithified turbidites of late Cretaceous (Campanian) age have been recovered from below the acoustic basement in the walls of the Pribilof Canyon (Hopkins et al., 1969); serpentinite lies beneath the Quaternary lavas that make up most of St. George Island (Barth, 1956).

Since its formation in early Tertiary time, the erosion surface represented by the acoustic basement has been considerably deformed. It has been flexed into upwarps in insular and peninsular areas where rocks of Paleocene and older age are now exposed and into downwarps that occur as numerous coalescent sedimentary basins on the continental shelf; these basins are now filled with strata of the main-layered sequence (Fig. 7). The acoustic basement is sharply downflexed at the continental margin and can be traced a considerable distance seaward beneath the continental rise (Scholl et al., 1966, 1968).

From these limited data, a late Mesozoic and early Cenozoic history of the Bering Sea area can be postulated. An episode of volcanism began during Cretaceous time and continued into the Paleocene Epoch in the Okhotsk volcanic belt, extending through the Bering Sea to the Alaskan mainland. Clastic and volcanic sediments were deposited at about the same time in several small narrow fault-bounded basins on the Alaskan mainland; for example, those filled with Chickaloon and Cantwell Formations (Wolfe and Wahrhaftig, 1966, 1970). Shortly thereafter uplift ended sedimentation, and a long period of erosion began throughout Alaska, Chukotka, and the continental shelves of the Bering and Chukchi Seas. Near the end of Oligocene time, most of Beringia was a surface of marine and subaerial planation. This surface was later warped and the basins filled with Tertiary sediments, the main-layered sequence.

Main-Layered Sequence

The main-layered sequence (MLS) is an acoustic unit that represents gently deformed sedimentary strata underlying most of the floor of the continental shelf of the Bering Sea. The unit is more than 500 m thick in about two-thirds of the continental shelf area (Fig. 7); locally it is some 4 km thick (Scholl et al., in press).

In a few places the MLS grades upward into Quaternary sediments with no visible unconformity. More commonly, however, the gently dipping strata of the MLS are truncated by a nearly horizontal erosion surface at or a short distance below the sea floor (Scholl and Hopkins, 1969; Grim and McManus, 1970). Outcrops of gently dipping beds within the MLS form cuestas on the sea floor south and east of St. Lawrence Island and probably elsewhere. The MLS forms a thick prograded sequence at the continental margin off Bristol Bay and the Gulf of Anadyr, and a thinner mass of prograded MLS evidently originally lay on the downwarped acoustic basement at the continental margin in the central Bering Sea. The prograded MLS is modified by submarine slumps off Bristol Bay and the Gulf of Anadyr; on the central continental slope, the MLS has been intensely dissected by submarine canyons and gullies that cut into and below the acoustic basement (Scholl et al., 1968, 1970).

The MLS is an offshore analogue to gently folded nonmarine sediments of middle and late Tertiary age that fill basins in mainland areas, for example, the Kenai Formation in the Cook Inlet Basin of southern Alaska (Wolfe and Wahrhaftig, 1966), the Tertiary coal-bearing formations of the central Alaska Range (Wahrhaftig et al., 1969), the Kougarok gravel of the Seward Peninsula (Hopkins, 1963), and the Koinatkhum Suite of the Chukotka Peninsula (Baranova et al., 1968; Biske, 1970).

The MLS has been sampled at many locations near the shore line and near the continental margin. Samples dredged from the continental margin consist of shallow-water detrital marine sediments rich in diatoms of Oligocene to late Pliocene or early Pleistocene age (Hopkins et al., 1969; Scholl et al., in press). Off Nome, several exploratory drill holes terminated in marine near-shore sands and clayey-silts of early Pliocene age (Hopkins and Nelson, unpublished data). Marine detrital sediments of Oligocene age, nonmarine sediments of late Oligocene or early Miocene age, and marine sediments of late Miocene age are exposed near the west coast of the Gulf of Anadyr (Petrov, 1966, and unpublished data; Baranova et al., 1968). Nonmarine sediments of late Oligocene age are exposed on northwestern St. Lawrence Island (Patton and Csejtey, 1970; Hopkins et al., 1972), and marine limestone containing diatoms of late Miocene or Pliocene age (I. Koizumi, written communication, 1971) has been dredged

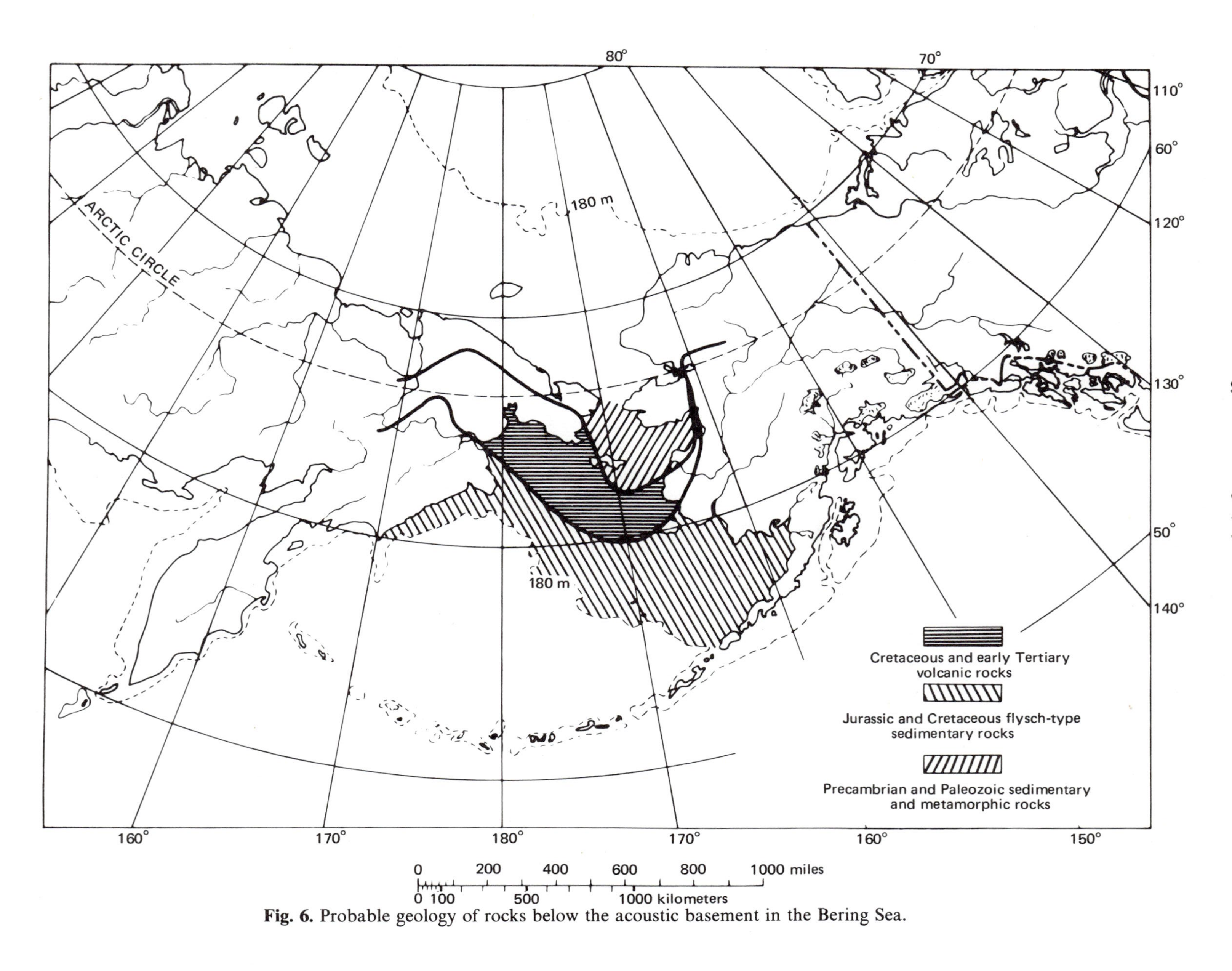

Fig. 6. Probable geology of rocks below the acoustic basement in the Bering Sea.

from the sea bottom about 30 km south of St. Lawrence Island (Nelson, unpublished data).

Stratigraphic studies of middle and late Tertiary beds in the Nenana and Cook Inlet basins of mainland Alaska (Wahrhaftig et al., 1969; Kirschner and Lyon, 1970) indicate major modification of the drainage pattern in Alaska that must have had pronounced effects upon sedimentation

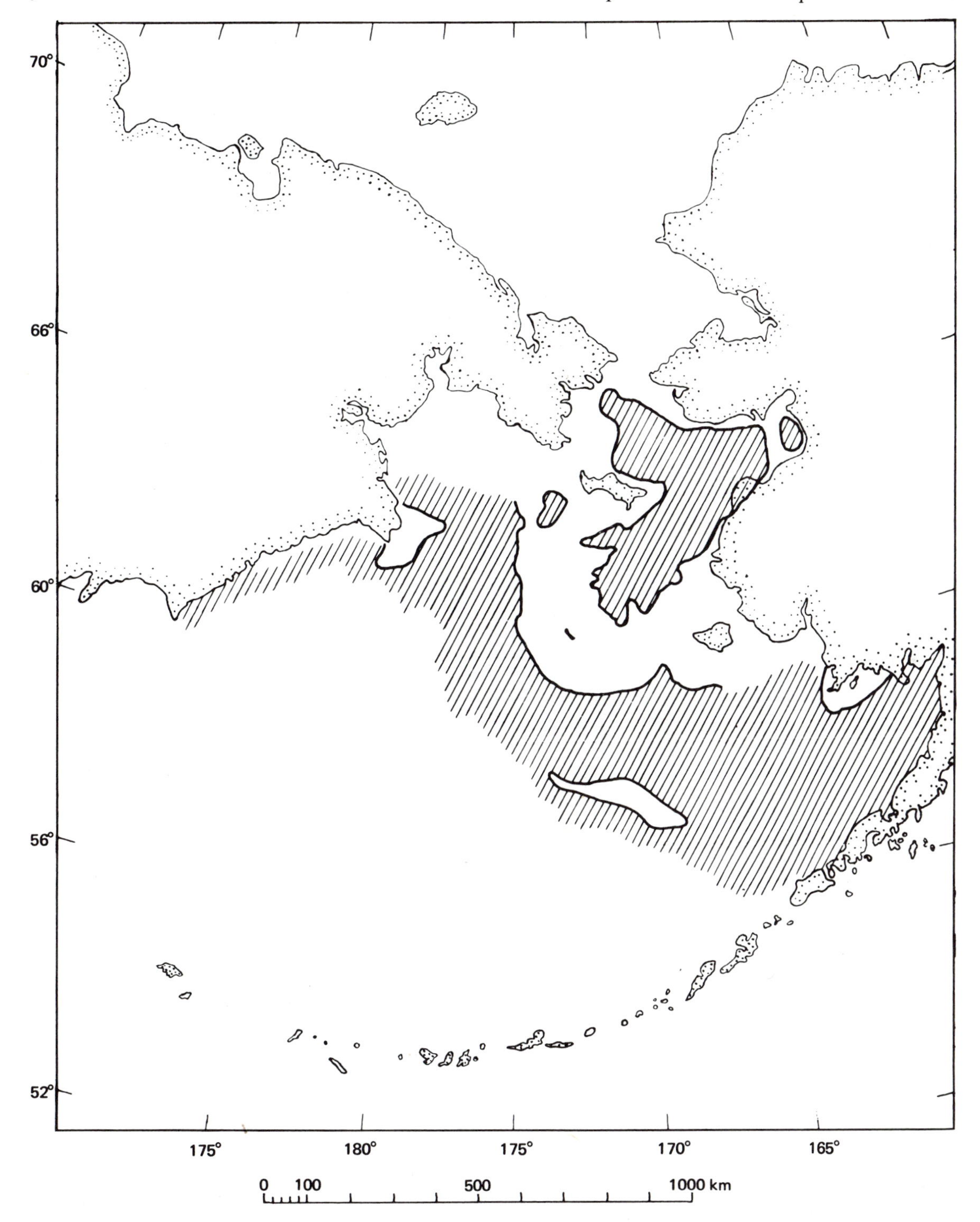

Fig. 7. Areas on the continental shelf of the Bering Sea that are underlain by 500 m or more of main-layered sequence (data from Scholl et al., 1968; Scholl and Hopkins, 1969; Granz et al., 1970a, 1970b; Scholl and Marlow, 1970; Verba et al., 1971; Kummer and Creager, 1971).

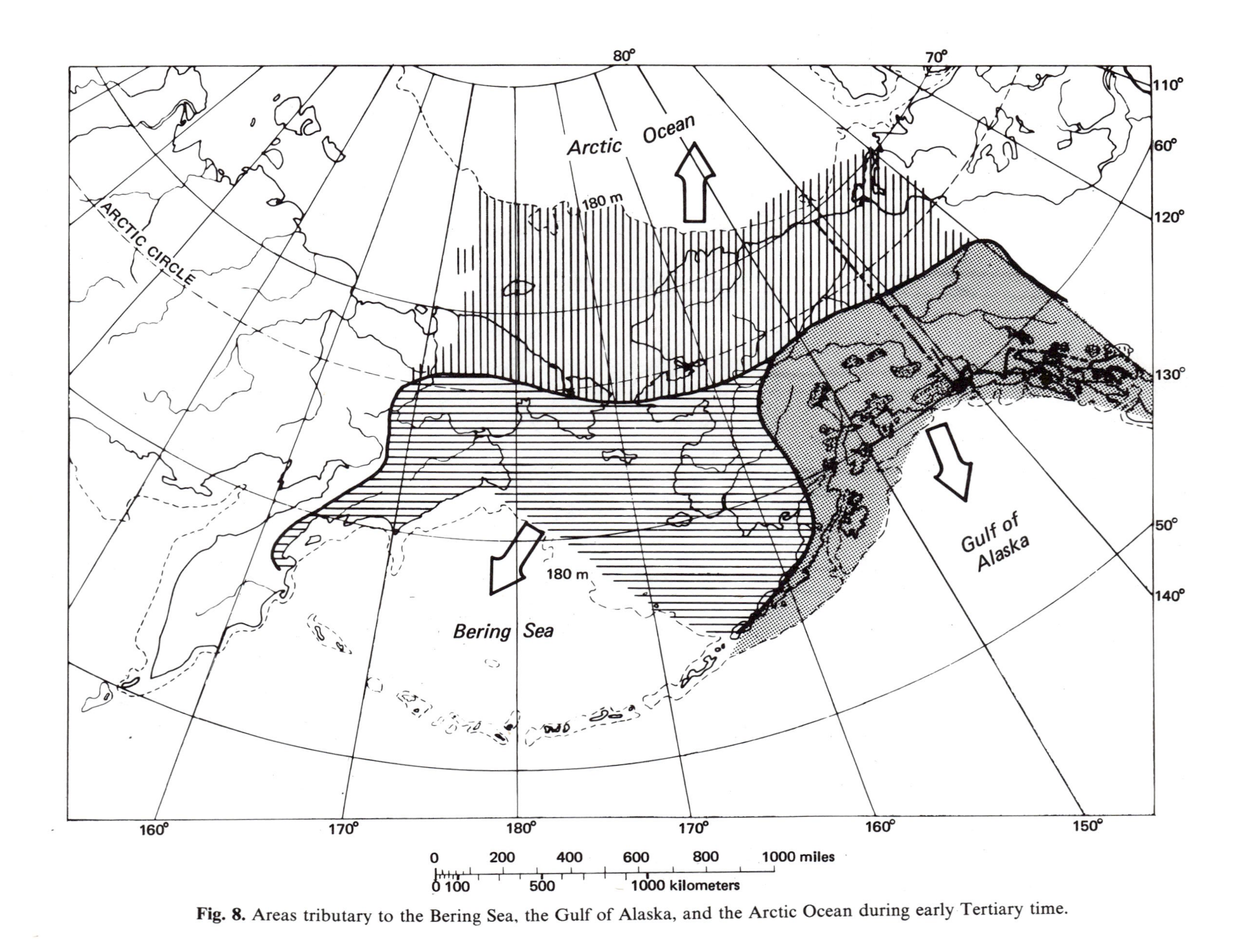

Fig. 8. Areas tributary to the Bering Sea, the Gulf of Alaska, and the Arctic Ocean during early Tertiary time.

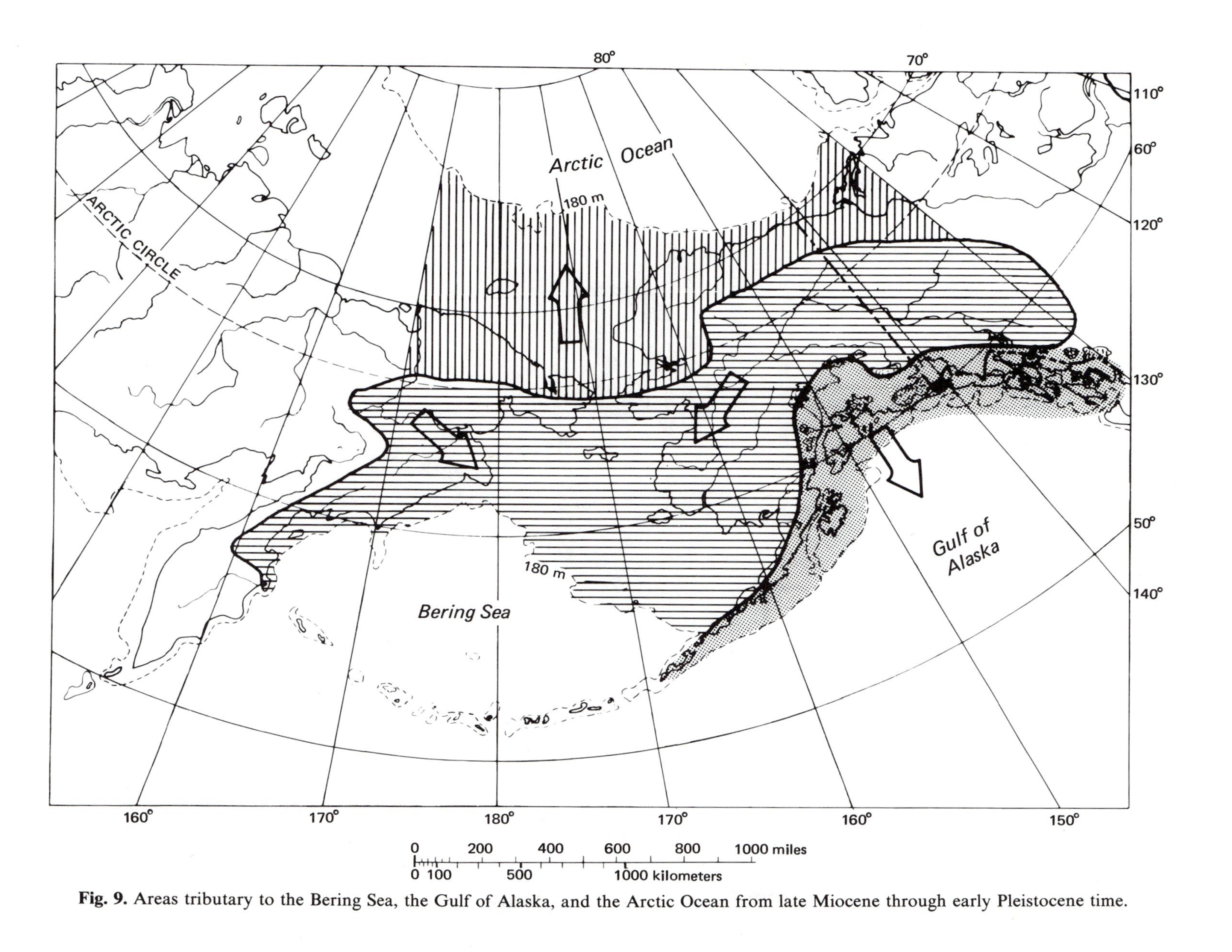

Fig. 9. Areas tributary to the Bering Sea, the Gulf of Alaska, and the Arctic Ocean from late Miocene through early Pleistocene time.

in the Bering Sea. Prior to late Miocene time, much of central Alaska was drained by the tributaries of a trunk stream that flowed southward across the present site of the Alaska Range to the area of Cook Inlet and thence to the Gulf of Alaska (Fig. 8). Uplift of the Alaska Range during late Miocene time defeated this trunk stream, diverting the drainage from central Alaska westward to the Bering Sea by way of the Yukon River and its tributaries. This drainage diversion must have resulted in an enormous increase in the volume of sediment reaching the Bering Sea and a change in its character; the event is probably recorded by changes in the petrology of the sediments making up the MLS.

The geologic history of the main-layered sequence in the Bering Sea begins, then, after a prolonged period of denudation. The erosional carving of the acoustic basement probably took place from Eocene into Oligocene time. The erosional detritus from the present Bering continental shelf, as well as from southwestern Alaska and southeastern Chukotka, presumably was deposited in the Aleutian and Kamchatka Basins of the Bering Sea (Fig. 8). Sediment from parts of Beringia north of the Arctic Circle was deposited in the Canadian Basin of the Arctic Ocean, and sediment from central, southern, and southeastern Alaska, as well as from the southern Yukon Territory, was dumped into the Gulf of Alaska (Fig. 8). By late Oligocene time, however, much of the sediment was being trapped in small basins on the Alaskan and Chukotkan mainlands and in a series of large shelf basins. Marine sediments were deposited on the actively subsiding continental margin and intermittently in the Anadyr Basin, but areas farther north were still emergent and receiving continental sediments.

During late Miocene time, uplift of the Alaska Range blocked at least one and possibly several drainage-ways that had been tributary to the Gulf of Alaska. These drainage ways became

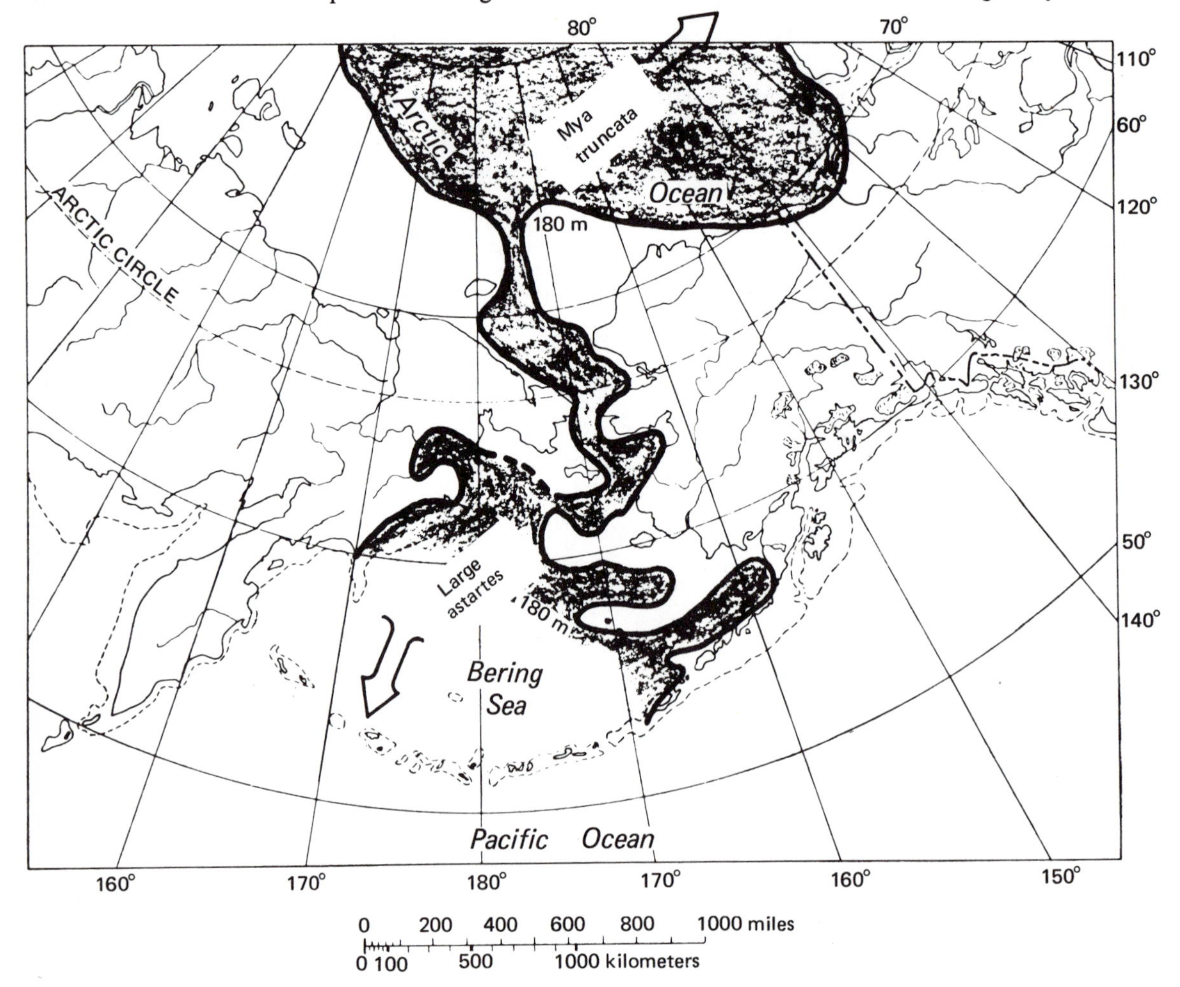

Fig. 10. Probable configuration of the Bering seaway 10 million years ago.

integrated to form the present Yukon River, resulting in an increase of nearly 50% in the land area shedding sediment to the Bering Sea (Fig. 9), and probably in a significant increase in sediment deposited in its offshore basins. Late in the Miocene Epoch, because of an eustatic rise in sea level or structural downwarping, the sea gradually invaded basins lying progressively northward on the continental shelf. Biogeographical evidence summarized by Hopkins (1967a,b) indicates that the Bering Strait was inundated and had, by the end of the Miocene Epoch (Fig. 10), formed a continuous marine connection between the Pacific and Arctic Oceans.

Most of Miocene and Pliocene time was a period of progradational sedimentation at the continental margin, and deposition of the MLS built the shelf outward a considerable distance off Bristol Bay and off the Gulf of Anadyr; the continental margin prograded somewhat less in the central part of the Bering Sea area. Indirect biogeographical evidence indicates that the Bering Strait seaway was blocked briefly about 5 million years ago and a narrow Siberian–Alaskan land bridge was formed (Fig. 11); the seaway was re-established about 3 million years ago, and the Bering Sea assumed approximately its present configuration (Hopkins, 1967a,b). In late Pliocene time clastic progradation ceased (Hopkins et al., 1969).

Quaternary Geologic History

Structural and Stratigraphic Framework

High-resolution seismic profiling on the continental shelf of the Bering Sea indicates that the Tertiary main-layered sequence is overlain, generally unconformably, by a more complex suite of deposits of probable Quaternary age (Walton et al., 1969; Grim and McManus, 1970; Nelson, 1970a; Kummer and Creager, 1971). The Quaternary sediments on the continental shelf are rarely thicker than 100 m and in many places are only a few meters thick; in some nearshore and insular areas they are lacking entirely and pre-Quaternary bedrock is exposed on the sea floor.

Quaternary sediments nearly 1 km thick mantle most of the deep basin of the Bering Sea; thicker deposits may underlie the continental rise (Scholl et al., 1968; Scholl et al., 1971). Several Quaternary units are recognized, including a slope-mantling unit, a continental-rise unit, and an abyssal-plain unit (Scholl et al., 1968). The slope-mantling unit, which is found only on the outer shelf, the continental slope, and on part of the Umnak Plateau in the southeastern sector, seems to consist of hemipelagic sediment draped over an undulating surface in areas of maximal organic productivity. The acoustic and topographic characters of the continental-rise unit indicate that it probably consists of turbidites from a complex system of coalescing fans radiating beyond mouths of submarine canyons. The abyssal-plain unit contains diatom tests as a large component of the surface sediments (Gershanovich, 1967), the major part is detrital sediment shed from Beringia during Quaternary time, carried through submarine canyons, and deposited in the deepest parts of the Bering Sea (Scholl et al., 1970a; Scholl et al., 1971).

The Quaternary deposits are relatively undisturbed, tectonically. However, fault scarps break the sea floor in northern Bering Sea (Grim and McManus, 1970) near the Pribilof Islands (D. M. Hopkins, unpublished data), and perhaps elsewhere. A large closed depression north of the middle part of St. Lawrence Island, shown on the new bathymetric map of the region (U.S. Coast and Geodetic Survey, 1969) is evidently of tectonic origin. Nearshore areas of Goodnews Bay seem to have undergone tectonic unwarping in late Quaternary time (Hopkins, 1959; P. Barnes and A. R. Tagg, unpublished data), and warped terraces indicate considerable tectonic activity along the coast of the Seward Peninsula (Sainsbury, 1967; D. M. Hopkins, unpublished data). Our unpublished studies show that lava flows and plugs of Quaternary age are present on the sea floor north of the central part of St. Lawrence Island, near the Pribilof Islands, and on the continental slope along the Aleutian Ridge.

Pleistocene Glacial and Fluvial Deposition

The Pleistocene history of the Bering continental shelf is characterized by repeated episodes of subaerial exposure, encroachment as well as deposition from large glaciers, and fluvial sedimentation. Evidence of Siberian glaciers on the continental shelf is highly deformed deposits observed in seismic profiles in the northern part of the Gulf of Anadyr (Kummer and Creager, 1971) and in the western part of the Chirikov Basin (Grim and McManus, 1970). From seismic records we interpret three distinct episodes of glaciation: The Siberian glaciers pushed as far as 150

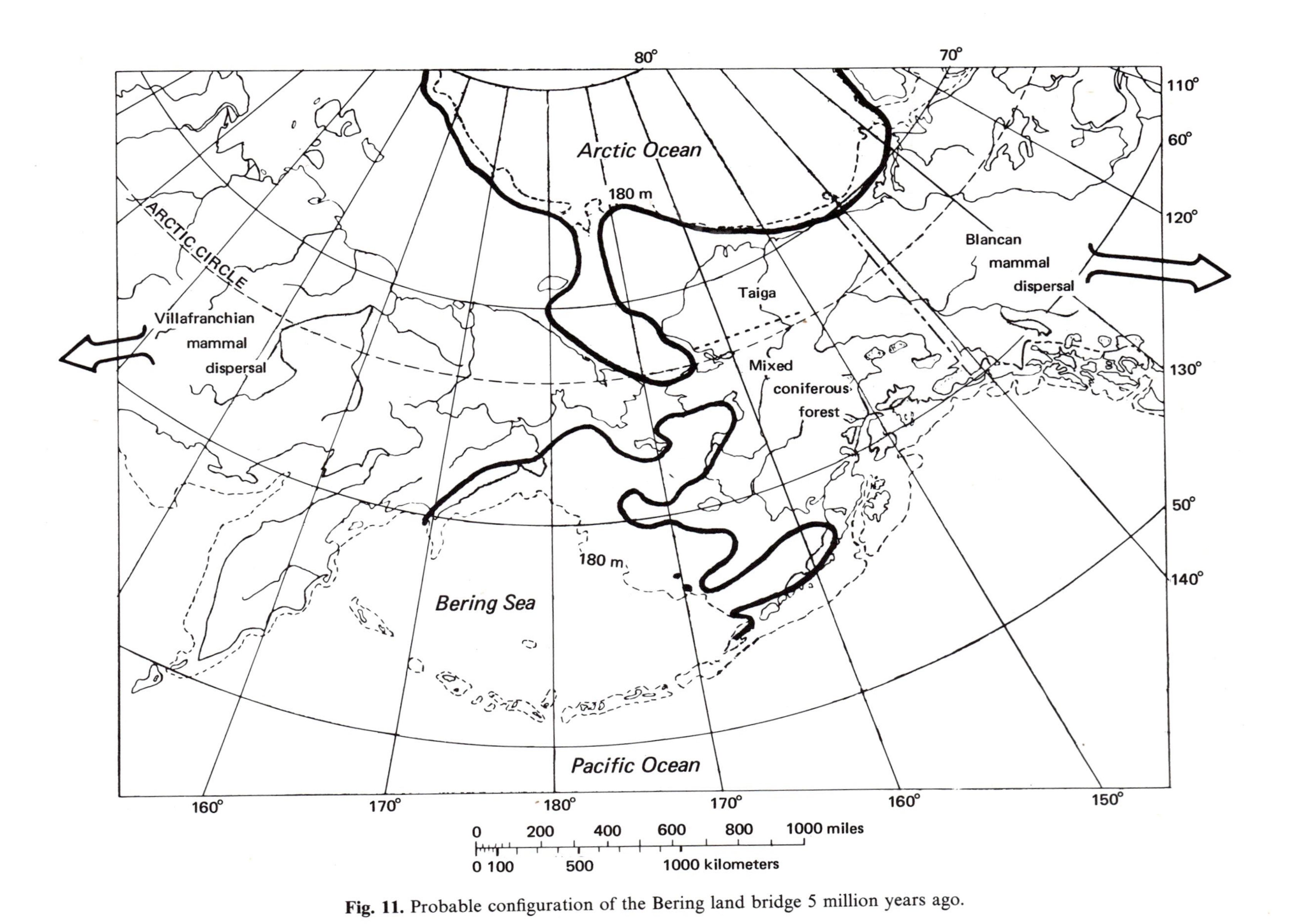

Fig. 11. Probable configuration of the Bering land bridge 5 million years ago.

km beyond the shore line of the Chukotka Peninsula (Fig. 1); they encroached on St. Lawrence Island and deposited a series of morainal ridges on the continental shelf that are now expressed as gravel bars extending northward from the island (Fig. 12) (Nelson and Hopkins, 1972; Hopkins et al., 1972). Other early and middle Pleistocene valley glaciers of the Seward Peninsula pushed debris a few kilometers seaward from the present coast. The distribution of glacial deposits in southwestern Alaska (Coulter et al., 1965) indicates that glaciers also extended far beyond the shore line into Bristol Bay.

Extensive migration of the Yukon River mouth and channels during Pleistocene periods of low sea level is indicated by reconnaissance seismic profiling. At least two major bodies of Yukon River deltaic sediment on the Bering shelf are apparent in high-resolution profiles, and presumably there are others. One, discovered by Kummer and Creager (1971), lies between St. Lawrence and St. Matthew Islands and probably formed during late Wisconsin time about 18,000 to 20,000 years ago. Another lies in northern Norton Sound, where buried prograded beds and distributary channels occur (Moore, 1964); it appears to extend as a buried surface far southwestward into Chirikov Basin (Grim and McManus, 1970). The presence of large partially buried channels suggests that the main Yukon drained to the southwest of St. Lawrence Island and through the central shelf. The Pribilof and Zhemchug submarine canyons probably were carved by turbidity currents when the Yukon River reached the continental margin during different times of lowered sea level (Scholl et al., 1970a).

Other trunk streams can also be traced across the continental shelf. The Kuskokwim River evidently flowed southward to central Bristol Bay and thence southwestward toward the abyssal Bering Sea (Hopkins, 1972); turbidites from an extended Kuskokwim River may have carved the Bering submarine canyon (Scholl et al., 1970a). The Anadyr River evidently flowed southward through the Gulf of Anadyr to the Pervenets canyon. Areas north of St. Lawrence Island were drained by a north-flowing trunk stream that passed through the Bering Strait to join the Hope Sea Valley (Creager and McManus, 1967; Hopkins, 1972). Faulting created two large depressions, one north of central St. Lawrence Island, the other south of the Bering Strait; these depressions seem to have contained large lakes when the continental shelf was last exposed.

Holocene Marine Deposition

In Holocene time the erosional and depositional features of the Pleistocene have been masked by sediments in parts of the Bering Sea. In the southern Bering Sea, offshore from local mud-filled embayments along the shore line, the Holocene deposits form a classical gradational sequence ranging from coarse sands nearshore to muds at the shelf edge (Sharma, 1970). Farther north, however, Holocene sediments on the Bering shelf do not form the nearly complete cover or show the gradation. Rather, beginning with the northern and western margins of the central area (Lisitsyn, 1966), the sediments are extremely heterogeneous and the distribution of relict and modern surface sediments is patchy; this sediment distribution is dependent upon positions of bedrock and glacial debris outcrops on the sea floor, locations of river sediment inflow, and velocity and patterns of water currents in Holocene time.

During the Holocene high sea level, strong currents affected the shape of the sea floor and the distribution of sediment. In the Bering and Anadyr Straits, relict gravels and hummocky topography of apparent glacial origin remain exposed (Grim and McManus, 1970; Nelson and Hopkins, 1972); on the lee side of such current-swept channels, weaker currents have deposited sediments to form shoals like those north of Cape Prince of Wales (Creager et al., 1970) and Northeast Cape (Nelson, 1970a). Most of the Chirikov Basin has a thin cover of relict fine marine sands from the Holocene transgression except for the Siberian morainal ridges, nearshore areas, and straits, where current scour preserves relict glacial and bedrock gravels (Nelson and Hopkins, 1972) (Fig. 12).

In the northern part of the Bering Sea, limited Holocene deposits, surface topographical features such as former beach ridges, outwash fans, stream valleys, and extensive relict sediments (Nelson and Hopkins, 1972) suggest that during the Holocene high sea level, sediments from the Yukon River and other streams have been swept from most of the northern Bering Sea into the Arctic Ocean by the strong northward-moving currents (Fleming and Heggarty, 1966; Husby, 1969, 1971). The present current regime and movement of Yukon–Kuskokwim sediment may explain the general lack of modern sediments in the northern

Bering Sea and the presence of Holocene deposits on the epicontinental shelf to the north (Creager et al., 1970; McManus et al., 1969) and south (Sharma, 1970).

Although nearly 90 million metric tons and 90% of the river sediment supplied to the Bering Sea is derived from the Yukon–Kuskokwim system (Lisitsyn, 1966), a layer less than 20 cm thick of Holocene Yukon silt apparently covers Norton Sound seaward of the river mouth (Fig. 12) (C. H. Nelson, unpublished data). A recent radiocarbon date of 10,500 B.P. (C. H. Nelson and M. Rubin, unpublished data) for well-defined subaerial peat layers immediately below the Yukon silt indicate rates of Holocene sedimentation as low as 2 cm/100 yr. Correlative paleontologic analyses in central and southern Norton Sound reveal the same low accumulation, yet the Yukon River input suggests that deposits here should be meters thick if all material of the past 10,500 years sedimented in the Bering Sea.

The low rate of Holocene sedimentation in Norton Sound and almost complete lack of Holocene sediment in the Chirikov Basin, but relatively high rate in the Chukchi Sea, suggests a major displacement and bypassing of Yukon–Kuskokwim sediment over the Bering epicontinental shelf. The several meters of Holocene silt in the Chukchi Sea (Creager et al., 1970) and high concentrations of suspended sediment in the easternmost Bering Sea (26 g/m^3; McManus and Smyth, 1970) indicates that major portions of Yukon sediment are being flushed out of Norton Sound and the Bering Sea into the Chukchi Sea. This flushing apparently has occurred throughout Holocene time, and could have taken place during other periods when the Bering Strait was submerged.

In contrast to the few measurements in the eastern Bering Sea, studies of suspended sediment transport as well as other modern sedimentary processes have been well documented in the western Bering Sea (Lisitsyn, 1966). Shore-line erosion, ice rafting, and organic production appear to be more important in the western Bering Sea, and measurements of suspended sediment indicate transport rates several orders of magnitude lower in the western (2 to 5 g/m^3; Lisitsyn, 1966) than in the eastern (5 to 25 g/m^3; McManus and Smyth, 1970) Bering Sea. Lisitsyn (1966) suggests that pack ice and shore-fast ice movements may be rafting pebbles as much as 100 km offshore. Gershanovich (1967) does not find evidence to support this in the eastern Bering Sea, nor do we (Fig. 12). Our data indicate that present sea-floor gravel patches (0 to 50% gravel) originated when the Holocene or earlier transgressions reworked glacial drift of the sea-floor surface. Grounding ice now plucks up pebbles from these exposed gravel patches, but drops scattered pebbles into finer-sized sediments within 10 to 20 km of the gravel-rich regions (Fig. 12).

Summary of the Sedimentary History

The present continental shelf of the Bering Sea is a mosaic of modern sediments and relict deposits formed in shallow water, at the strand, or in subaerial environments at times when sea level was lower than at present. Glacial deposits were formed on the Quaternary subaerial shelf far seaward of the present shore line; subaerial drainage systems and glacial scouring dissected previous sea floors; and shore-line transgressions, regressions, and late Quaternary still stands built shoals, partially filled in old stream valleys, and smoothed over the older topographic features. In other areas shore-line processes and current action reworked, winnowed, and prevented deposition through sediment bypassing.

The Quaternary deposits in the deep basin of the Bering Sea are dominated by two modes of deposition; turbidite deposits formed when sea level was low and when rivers delivered their sediment directly to the continental margin; diatomaceous hemipelagic deposits formed during interglacial epochs of high sea level. The hemipelagic sediments are especially thick in the area of high productivity north of the eastern passes through the Aleutian Islands and along the continental margin off the Gulf of Anadyr.

The Tertiary sediments have been drilled or developed in only a few places and thus are generally unavailable for direct study. We infer, however, that Tertiary patterns of sedimentation were similar to the modern interglacial pattern in the abyssal basin; they were quite different on the shelf, where large quantities of fluvial detritus were deposited in the early history of subsiding basins, and nearshore marine deposits may have been common during later history.

References

Baranova, Yu. P., S. F. Biske, V. F. Goncharov, I. A. Kul'kova, and A. S. Titkov. 1968. Cenozoic of the northeast of the U.S.S.R. *Akad. Sci., U.S.S.R.*,

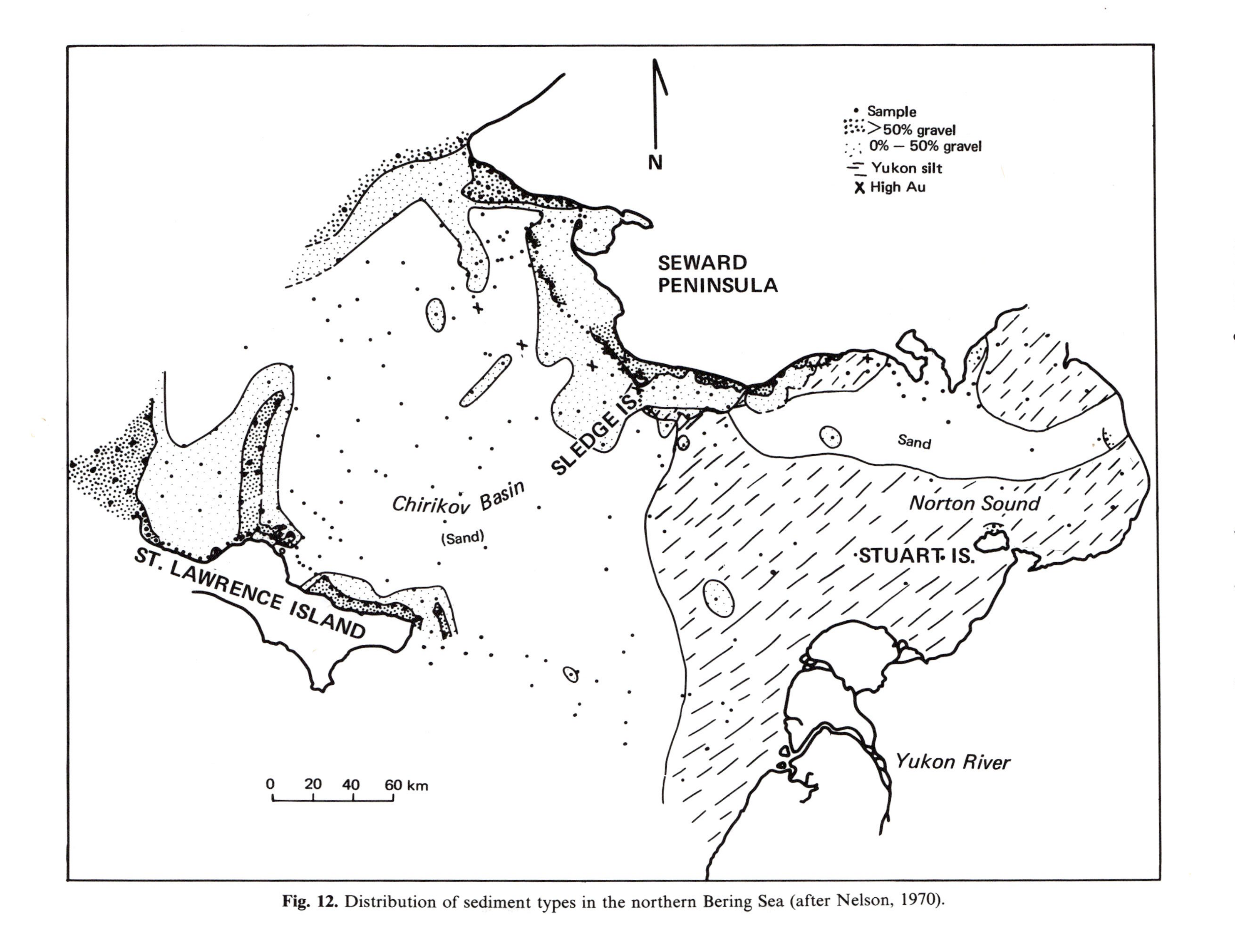

Fig. 12. Distribution of sediment types in the northern Bering Sea (after Nelson, 1970).

Siberian Division, Institute of Geology and Geophysics, Trudy. [In Russian. English translation U.S. National Translation Center, Chicago.] 125 pp.

Barth, T. F. W. 1956. Geology and petrology of the Pribilof Islands, Alaska. *U.S. Geol. Surv. Bull.* 1028-F:101–160.

Biske, S. F. 1970. Correlation of Tertiary nonmarine sediments in Alaska and northeastern Siberia (abstract). *Am. Assoc. Petroleum Geol.* 54:2469.

Buffington, E. C., A. J. Carsola, and R. S. Dietz. 1950. Oceanographic cruise to the Bering and Chukchi seas, summer, 1949. Part 1, sea floor studies. *U.S. Navy Electron. Lab. Rep.* 204 pp.

Burns, J. J. 1970. Remarks on the distribution and natural history of pagophilic pinnipeds in the Bering and Chukchi seas. *J. Mammalogy* 28:445–454.

Churkin, M. 1972. Western boundary of the North American continental plate in Asia. *Geol. Soc. Am. Bull.* 83:1027–1036.

Coulter, H. W., et al. 1965. Map showing extent of glaciation in Alaska. *U.S. Geol. Surv. Misc. Geol. Invest.* Map I-415.

Creager, J. S., and D. A. McManus. 1967. Geology of the floor of Bering and Chukchi seas—American studies. In: D. M. Hopkins, ed. *The Bering Land Bridge*. Stanford, California: Stanford University Press. pp. 32–46.

———, R. J. Echols, M. L. Holmes and D. A. McManus, 1970. Chukchi Sea continental shelf sedimentation (abstract). *Am. Assoc. Petroleum Geol. Bull.* 54(12):2475.

Dietz, R. A., A. J. Carsola, E. C. Buffington, and C. J. Shipek. 1964. Sediments and topography of the Alaska shelves. In: R. L. Miller, ed. *Papers in Marine Geology*. New York: Macmillan. pp. 241–256.

Ewing, M., W. J. Ludwig, and J. Ewing. 1965. Oceanic structural history of the Bering Sea. *J. Geophys. Res.* 70:4593–4600.

Fleming, R. H., and D. Heggarty. 1966. Oceanography of the southeastern Chukchi Sea. In: N. J. Willimovsky and J. N. Wolfe, eds. *Environment of the Cape Thompson region, Alaska*. U.S. Atomic Energy Commission. pp. 687–754.

Gershanovich, D. E. 1967. Late Quaternary sediments of the Bering Sea and the Gulf of Alaska. In: D. M. Hopkins, ed. *The Bering Land Bridge*. Stanford, California: Stanford University Press. pp. 32–46.

———. 1968. New data on geomorphology and recent sediments in the Bering Sea and the Gulf of Alaska. *Marine Geol.* 6:281–296.

Grantz, A., W. F. Hanna, and S. C. Wolfe. 1970a. Chukchi Sea seismic reflection and magnetic profiles, 1969, between northern Alaska and International Date Line. *U.S. Geol. Surv. Open-File Rep.* 26 pp.

———, S. C. Wolf, L. Breslau, T. C. Johnson, and W. F. Hanna. 1970b. Reconnaissance geology of the Chukchi Sea as determined by acoustic and magnetic profiling. In: W. L. Adkinson and W. W. Brosge, eds. *Proceedings of the Geology Seminar on the North Slope of Alaska*. American Association of Petroleum Geologists, Pacific Section, Los Angeles, F1-F28.

Greene, H. G. 1970. A portable refraction seismograph survey of gold placer areas near Nome, Alaska. *U.S. Geol. Surv. Bull.* 1312-B. 29 pp.

Grim, M. S., and D. A. McManus. 1970. A shallow-water seismic-profiling survey of the northern Bering Sea. *Marine Geol.* 8:293–320.

Grow, J. A., and T. Atwater. 1970. Mid-Tertiary tectonic transition in the Aleutian Arc. *Geol. Soc. Am. Bull.* 81:3715–3722.

Holmes, M. L., R. von Huene, and D. A. McManus. 1972. Seismic reflection evidence supporting underthrusting beneath the Aleutian Arc near Amchitka Island. *J. Geophys. Res.* 77:959–964.

Hopkins, D. M. 1959. Cenozoic history of the Bering land bridge. *Science* 129:1519–1528.

———. 1963. Geology of the Imuruk Lake area, Seward Peninsula, Alaska. *U.S. Geol. Surv. Bull.* 1141-C. 101 pp.

———. 1967a. The Cenozoic history of Beringia—a synthesis. In: D. M. Hopkins, ed. *The Bering Land Bridge*. Stanford, California: Stanford University Press. 495 pp.

———. 1967b. *The Bering Land Bridge.* Stanford, California: Stanford University Press. 495 pp.

———. 1972. The paleogeography and climatic history of Beringia during late Cenozoic time. *Internord* 12:121–150.

———, D. W. Scholl, W. O. Addicott, R. L. Pierce, J. A. Wolfe, D. Gershanovich, B. Kotenev, K. E. Lohman, J. E. Lipps, and J. Obradovich. 1969. Cretaceous, Tertiary, and early Pleistocene rocks from the continental margin in the Bering Sea. *Geol. Soc. Am. Bull.* 80:1471–1480.

———, and D. W. Scholl. 1970. Tectonic deveolopment of Beringia, late Mesozoic to Holocene (abstract). *Am. Assoc. Petroleum Geol. Bull.* 54:2486.

———, R. W. Rowland, and W. W. Patton. 1972. Middle Pleistocene mollusks from St. Lawrence Island and their significance for the paleo-oceanography of the Bering Sea. *Quat. Res.* 2:119–134..

Husby, D. M. 1969. Report oceanographic cruise U.S.C.G. *Northwind*, northern Bering Sea—Bering Strait—Chukchi Sea, 1967. *U.S.C.G. Oceanographic Rep.*, 24. Washington, D.C.: U.S. Coast Guard Oceanographic Unit. 75 pp.

———. 1971. Oceanographic investigations in the northern Bering Sea and Bering Strait, June–July, 1968. *U.S.C.G. Oceanographic Rep.*, 40. Washington, D.C.: U.S. Coast Guard Oceanographic Unit. 50 pp.

Kienle, J. 1970. Gravity and magnetic measurements over Bowers Ridge and Shirshov Ridge, Bering Sea. *J. Geophys. Res.* 76:7138–7153.

King, P. B., Jr. 1969. Tectonic map of North America. *U.S. Geol. Surv. Map.*

Kirschner, C. E., and C. A. Lyon. 1970. Stratigraphic and tectonic development of Cook Inlet petroleum province (abstract). *Am. Assoc. Petroleum Geol. Bull.* 54:2490.

Knebel, H. J., and J. S. Creager. 1970. Holocene sedimentary framework of east-central Bering Shelf: preliminary results (abstract). *Am. Assoc. Petroleum Geol. Bull.*, 54:2491.

Kummer, J. T., and J. S. Creager. 1971. Marine geology and Cenozoic history of Gulf of Anadyr. *Marine Geol.* 10:257–280.

Lamb, H. H. 1971. Climate-engineering schemes to meet a climatic emergency. *Earth Sci. Rev.* 7:87–96.

Lisitsyn, A. P. 1966. Recent sedimentation in the Bering Sea. *Akademiya Nauk SSSR, Institut Oceanologii* (translated) [Israel Program for Scientific Translations, publ., 1969]. 614 pp.

Ludwig, W. J., R. E. Houtz, and M. Ewing. 1971a. Sediment distribution in the Bering Sea: Bowers Ridge, Shirshov Ridge, and enclosed basins. *J. Geophys. Res.* 76:6367–6375.

———, S. Murauchi, N. Den, M. Ewing, H. Hotta, R. E. Houtz, T. Yoshii, T. Asanuma, K. Hagiwara, T. Saito, and S. Ando. 1971b. Structure of Bowers Ridge, Bering Sea. *J. Geophys. Res.* 76:6350–6366.

Marlow, M. S., D. W. Scholl, E. C. Buffington, R. E. Boyce, T. R. Alpha, P. J. Smith, and C. J. Shipek. 1970. Buldir depression—A late Tertiary graben on the Aleutian Ridge, Alaska. *Marine Geol.* 8:85–108.

McManus, D. A., J. C. Kelly, and J. S. Creager. 1969. Continental shelf sedimentation in an arctic environment. *Geol. Soc. Am. Bull.* 80:1961–1984.

———, and C. S. Smyth. 1970. Turbid bottom water on the continental shelf of northern Bering Sea. *J. Sedimentary Petrology* 40:869–877.

Menard, H. W. 1967. Transitional types of crust under small ocean basins. *J. Geophys. Res.* 72:3061–3073.

Moll, R. F. 1970. Clay mineralogy of the north Bering Sea shallows. Masters thesis, University of Southern California, Los Angeles. 101 pp.

Moore, D. G. 1964. Acoustic reflection reconnaissance of continental shelves: eastern Bering and Chukchi seas. In: R. L. Miller, ed. *Pap. Marine Geol.* New York: Macmillan Co. pp. 319–362.

Nelson, C. H. 1970a. Late Cenozoic history of deposition of northern Bering shelf (abstract). *Am. Assoc. Petroleum Geol. Bull.* 54:2498.

———, 1970b. Potential development of heavy metal resources in the northern Bering Sea—Science in Alaska. *Proceedings 20th Alaska Scientific Conference Alaska Division.* American Association for Advancement of Science, College, Alaska, 1969. pp. 366–376.

———, and D. M. Hopkins. 1972. Sedimentary processes and distribution of particulate gold in northern Bering Sea. *U.S. Geol. Surv. Prof. Pap.* 689. 27 pp.

———, D. L. Pierce, K. Leong, and F. Wang. 1972. Mercury distribution in ancient and modern sediment of northeastern Bering Sea. *U.S. Geol. Surv. Open-File Rep.* 25 pp.

Patton, W. W., Jr. 1970. Mesozoic tectonics and correlations in Yukon-Koyukuk Province, west-central Alaska (abstract). *Am. Assoc. Petroleum Geol. Bull.* 54:2500.

———, and B. Csejtey, Jr. 1970. Preliminary geologic investigations of western St. Lawrence Island, Alaska. *U.S. Geol. Surv. Prof. Pap. 684.* 15 pp.

Petrov, O. M. 1966. Stratigrafiya i fauna morskikh mollyuskov Chetvertichnikh otlozhenii Chukotskogo Poluostrova (stratigraphy and fauna of marine mollusks of Quaternary deposits of the Chukotka Peninsula). *Akademiya Nauk SSSR Institut Geologicheskogo Trudy*, 155. 257 pp.

———. 1967. Paleogeography of Chukotka during late Neogene and Quaternary time. In: D. M. Hopkins, ed. *The Bering Land Bridge.* Stanford, California: Stanford University Press. pp. 144–171.

Rowland, R. W., and D. M. Hopkins. 1971. Comments on the use of *Hiatella arctica* for determining Cenozoic sea temperatures. *Paleogeography, Paleoclimatology, Paleoecology* 9:59–64.

Sainsbury, C. L. 1967. Quaternary geology of western Seward Peninsula, Alaska. In: D. M. Hopkins, ed. *The Bering Land Bridge.* Stanford, California: Stanford University Press. pp. 121–143.

Scholl, D. W., E. C. Buffington, and D. M. Hopkins. 1966. Exposure of basement rock on the continental slope of Bering Sea. *Science* 153:992–994.

———, E. C. Buffington, and D. M. Hopkins. 1968. Geologic history of the continental margin of North America in Bering Sea. *Marine Geol.* 6:297–330.

———, and D. M. Hopkins. 1969. Newly discovered Cenozoic basins, Bering shelf, Alaska. *Am. Assoc. Petroleum Geol.* 53:2067–2078.

———, and M. S. Marlow. 1970. Bering Sea seismic profiles, 1969. *U.S. Geol. Surv. Open-File Rep.*

———, E. C. Buffington, D. M. Hopkins, and T. R. Alpha. 1970a. The structure and origin of the large submarine canyons of the Bering Sea. *Marine Geol.* 8:187–210.

———, H. G. Greene, and M. S. Marlow. 1970b. The Eocene age of the Adak "Paleozoic?" locality, Aleutian Islands, Alaska. *Geol. Soc. Am. Bull.* 81:3583–3592.

———, and E. C. Buffington. 1970. Structural evolution of Bering continental margin: Cretaceous to Holocene (abstract). *Am. Assoc. Petroleum Geol. Bull.* 54:2503.

———, J. S. Creager, R. E. Boyce, R. J. Echols, T. J.

Fullam, J. A. Grow, I. Koizumi, J. Homa Lee, H. Y. Ling, P. R. Supko, R. J. Stewart, T. R. Worsley, A. Ericson, J. Hess, G. Bryan, and R. Stoll. 1971. Deep-sea drilling project leg 19. *Geotimes* 16(11):12–15.

———, E. C. Buffington, and M. S. Marlow. in press. Plate tectonics and the structural evolution of the Aleutian-Bering Sea region: Solutions and complications. In: R. B. Forbes, ed. *The Geophysics and Geology of the Bering Sea Region*. Geological Society of America Special Paper.

Sharma, G. D. 1970. Recent sedimentation on southern Bering shelf (abstract). *Am. Assoc. Petroleum Geol. Bull.* 54:2503.

Sheth, M. 1971. A heavy mineral study of Pleistocene and Holocene sediments near Nome, Alaska. Masters thesis, San Jose State College, San Jose, California. 83 pp.

Shor, G. G., Jr. 1964. Structure of the Bering Sea and the Aleutian Ridge. *Marine Geol.* 11:213–219.

Silberman, M. L. 1969. Preliminary report on electron microscopic examination of surface texture of quartz sand grains from the Bering shelf. *U.S. Geol. Surv. Prof. Pap.* 650-C. pp. C33–C37.

Tagg, A. R., and H. G. Greene. 1971. High-resolution seismic survey of a nearshore area, Nome, Alaska. *U.S. Geol. Surv. Prof. Pap.* 759:A1–A23.

Tilman, S. M., V. F. Belyi, A. A. Nikolaevskii, and N. A. Shilo. 1969. Tektonika Severo-vostoka SSSR (Tectonics of northeastern USSR). *Akademiya Nauk SSSR, Sibirsk Otdeleniye, Severo-Vostochnogo Nauchno-Issledovatel'skiy Institut Trudy*, 33. 78 pp.

Udintsev, G. B., I. G. Biochenko, and V. F. Kanaev. 1959. Bottom relief of the Bering. In: P. L. Bezrukov, ed. *Geographical Description of the Bering Sea. Akademiya Nauk SSSR, Institut Oceanologii Trudy*, 29. [Israel Program for Sci. Translations, 1964]. pp. 14–16.

U.S. Coast and Geodetic Survey. 1969. St. Lawrence Island to Port Clarence, 1:250,000. *Environmental Science Services Administration Bathymetric Map* (Preliminary), C&GS PBM-1.

Venkatarathnam, K. 1971. Heavy minerals on the continental shelf of the northern Bering Sea. *U.S. Geol. Surv. Open-File Rep.* 93 pp.

Verba, M. L., G. I. Gaponenko, S. S. Ivanov, A. N. Orlov, V. I. Timofeev and Yu. F. Chernenkov. 1971. Glubinnoe stroenie i perspectivy neftegazonosnosti severo-zapadnoi chasti Beringia Morya (Deep structure and oil and gas prospects of the northwestern part of Bering Sea). [In Russian. English translation available from U.S. National Translation Center, Chicago.] *Nachno-Issled. Inst. Geologii Arktiki (NIIGA), Geofiz. Metody Razvedki v Arktike* 6:70–74.

Von Huene, R., and G. G. Shor. 1969. The structure and tectonic history of the eastern Aleutian Trench. *Geol. Soc. Am. Bull.* 80:1889–1902.

Wahrhaftig, C., J. A. Wolfe, E. B. Leopold, and M. A. Lanphere. 1969. The coal-bearing group in the Nenana coal field Alaska. *U.S. Geol. Surv. Bull.* 1274-D. 30 pp.

Walton, F. W., R. B. Perry, and H. G. Greene. 1969. Seismic reflection profiles, northern Bering Sea. *U.S. Dept. Commerce, ESSA, Operational Data Rep.*, C&GS DR-8.

Wolfe, J. A., and C. Wahrhaftig, 1966. Tertiary stratigraphy and paleobotany of the Cook Inlet region, Alaska, *U.S. Geol. Surv. Prof. Pap. 398-A*. 29 pp.

———, and C. Wahrhaftig. 1970. The Cantwell Formation of the central Alaska Range. *U.S. Geol. Surv. Bull. 1294-A*. pp. 41–46.

Chapter 5
Geological Oceanography of the Bering Shelf[1]

GHANSHYAM D. SHARMA[2]

Abstract

The Bering Sea is a unique and important subarctic sea. It has an area of 2.25 million km^2, almost half of which is continental shelf. This broad shelf is extremely smooth, with a gentle uniform gradient. The climate is dominated by extratropical cyclonic and arctic anticyclonic storms which tend to be brief but violent. During winter most of the Bering Sea shelf is covered by ice.

The Bering Sea exchanges waters with both the Pacific and the Arctic Oceans. The net flow is northward through the Aleutian Passes and also through the Bering Strait. Large-scale seasonal fresh water inflow from North America and severe winter conditions on the Bering shelf result in the formation of the principal water masses. The chemical and biological parameters are strongly influenced by upwelling along the Aleutian chain, by unusual conditions of river input near the coast, and by the seasonal ice cover over the shelf.

Sedimentation on the Bering shelf can be divided into five distinct depositional regimes: southeastern, central, northeastern, northwestern and outer shelves. The nature of sediments deposited in each regime is related to the provenance and the water dynamics in the area. In the northeastern Bering shelf the sediment cover consists of modern Yukon River mud and relict material. Other areas of the Bering Shelf are covered with graded recent sediments. Those sediments originating along the coast are primarily transported offshore in graded suspension by storm waves. Sands are predominant over the inner shelf, while silt and clay sediments are transported by storm waves and currents to the outer shelf and slope. This process results in a smoothly prograding shelf.

Introduction

The Bering Sea lies between 52 and 66°31′N latitude and 162°2′E, 157°W longitude, with an area of 2.25 million km^2. Almost half of this area (45%) is less than 200 m deep. Approximately 80% of the total shelf lies in the eastern and northeastern Bering Sea (Fig. 1). The shelf has a gentle uniform slope and an extreme degree of leveling over a large area. The shelf floor has an average depth of 60 m with a slope of less than 3×10^{-4}. The triangular shallow shelf is approximately 1335 km long and 1110 km wide and is bordered by an abrupt and very steep continental slope break at depths varying from 150 to 170 m. The slope is scarred with valleys and some of the world's largest submarine canyons.

The southeastern Bering Sea shelf is covered by a thin veneer of contemporary sediments, from 1.5 to over 6 m thick (Gershanovich, 1967). In the northeastern shelf recent Yukon River sediments filled old valleys and blanketed the floor of

[1] This is Contribution No. 30 from the Institute of Marine Science, University of Alaska, Fairbanks, Alaska.

[2] Institute of Marine Science, University of Alaska, Fairbanks, Alaska 99701, U.S.A.

Norton Sound (Nelson and Hopkins, 1972), but in spite of the large amounts of sediments supplied by the Yukon River, the overall thickness of recent sediments if negligible (Nelson, 1971). Relict sediments also cover large areas of northern region.

The purpose of this paper is to describe the nature and sources of sediments on the Bering Sea shelf and to elucidate the sediment transport and dispersal in terms of regional oceanography.

Climate

The Bering Sea is located in the subarctic climatic zone, with the exception of the southernmost part, which can be included in the temperate zone. The largest extratropical cyclone of the world, the winter Aleutian "low," predominates over the Bering Sea. This cyclonic circulation renders the eastern half warmer than the western half. During the annual cycle, climatic conditions are primarily controlled by the Arctic and the Honolulu highs, and the seasonal influences of the Siberian high; the Aleutian low and Asiatic depression are also significant. The dominant control, the Honolulu high, changes its position and influence throughout the year. During winter the Honolulu high occupies a southeastern position; in summer it becomes more vigorous and moves to a northwestern position. Throughout the year the Honolulu high produces strong southeasterly winds and is a source of heat advection for the Bering Sea. On the other hand the Arctic anticyclone is the source for cold advection and increased northern and northeastern winds.

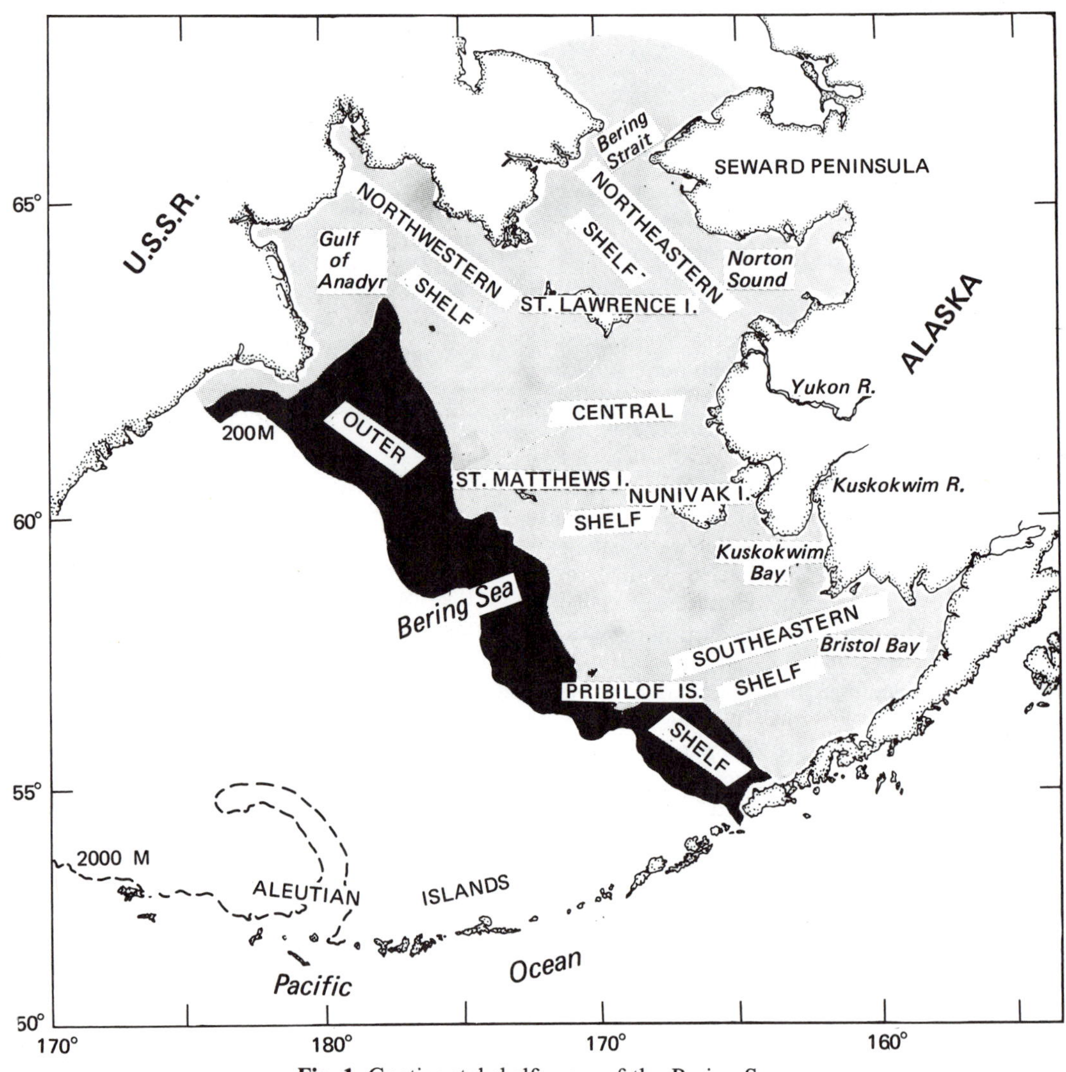

Fig. 1. Continental shelf areas of the Bering Sea.

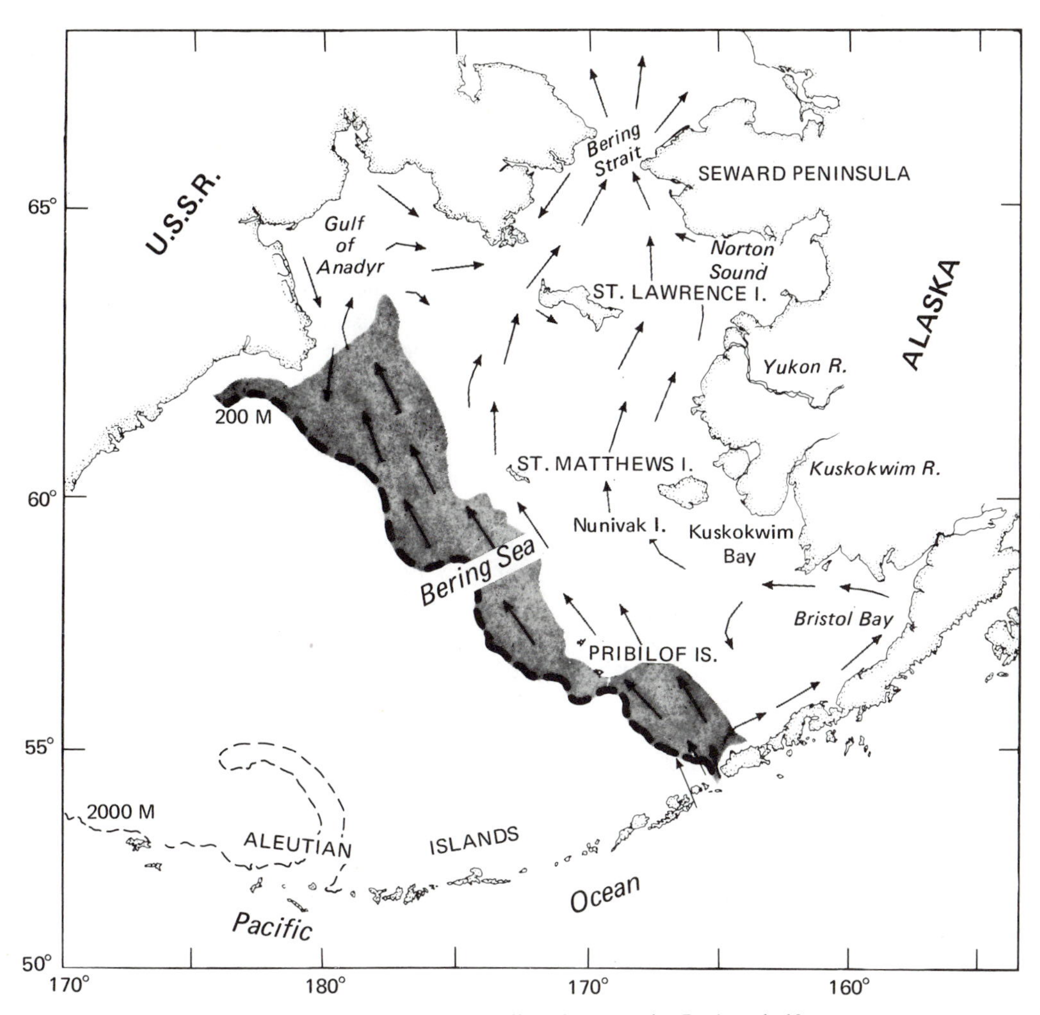

Fig. 2. Major current directions on the Bering shelf.

The summer shifting of the Honolulu high (anticyclone) to the west and northwest leads to intensification of cyclonic circulation and to an increase in frequency of south winds. The winter shifting of the Honolulu high eastward results in the simultaneous advance of the Arctic high (anticyclone) to the south, and thus to an increased frequency of north winds.

The Bering Sea is subjected to numerous storms because it lies in the path of extratropical cyclonic and the Arctic anticyclone storms. The general track of the migratory cyclonic storm is along the Japanese and Kurile Islands, and thence north and east over the Bering Sea.

During winter the shelf area of the Bering Sea is generally covered with sea ice. The southern and western deep parts of the Bering Sea are generally ice-free throughout the year due to the intrusion of warm Pacific water. Pack ice is common south of St. Lawrence Island and often extends as far as the Pribilofs. Formation of shore ice begins in December and it often lasts until April. Along the eastern coast the ice break-up occurs during June, and most of the Bering shelf including the Bering Strait becomes ice-free by the end of June.

Physical and Chemical Characteristics of the Bering Sea

Estimates of Pacific water volume moving into the Bering Sea have been made by various investigators and described by Lisitsyn (1969). It is estimated that approximately $1.5 \times 10^{14} m^3$ Pacific water enters the Bering Sea annually through various straits along the Aleutian Islands. Only $2.1 \times 10^{13} m^3$, or 14%, of the total remains in Bering Sea while most returns to the Pacific Ocean. The net gain from Pacific and surface flow is lost annually to the Arctic Ocean through the Bering Strait. A definite current pattern is set up in the Bering Sea deep and shallow waters. The net movement of water is northward. Cyclonic gyres are formed in deep basins (Fig. 2).

The dominant water movement of the eastern shelf originates in the vicinity of Unimak Islands. Pacific water entering these passes moves northward toward St. Matthews Island and eastward toward the head of Bristol Bay. The northward-moving stream bifurcates near St. Matthews Island into the Lawrence and the Anadyr Currents, both merging prior to passing through the Bering Strait (Fig. 2). The Lawrence Current flows along the Alaskan mainland and St. Lawrence Island. The Anadyr Current is the portion of the main stream flowing northwest which is deflected eastward at the mouth of Gulf of Anadyr to join the Lawrence Current south of the Bering Strait. The speed of these currents ranges up to 40 cm/sec (McManus and Smyth, 1970). The water movement may occasionally reverse.

The water moving eastward along the Alaskan Peninsula reaches the head of Bristol Bay and is deflected westward by the water of the Kvichak and Nushagak Rivers. Near the mouth of the Kuskokwim Bay the westward flowing water is mixed with Kuskokwim River flow and directed toward the south, thus forming a cyclonic gyre. Part of the Kuskokwim flow is carried northward.

The shallow eastern Bering Sea is characterized by pronounced seasonal layering of water masses, in contrast to the deep western basins (Ohotani, 1969). In summer the southern shelf has a well-defined thermocline and halocline reaching to depths of 20 to 30 m. Surface water temperatures may rise 8 to 12°C, while the bottom layer remains at 0 to -1.5°C. Salinity increases with increasing depth, and locally the Kuskokwim and Kvichak Rivers reduce the surface water salinity and increase water temperature. In the northern Bering shelf and Bering Strait during summer a warm (>5°C), low salinity (<32.5‰) water mass forms in the upper 10 m and overlies a layer with gradients between 10 and 15 m; the bottom layer consists of a colder (<3°C), and a more saline (>32.5‰) water mass (Husby, 1968). Large zonal temperature and salinity gradients are developed in this region and are attributed to dilution by the effluents of the Yukon and Kuskokwim Rivers. The northernmost part of the Bering Sea surface water is also locally influenced by intruding Arctic Ocean water, ice, and the flow of the Anadyr River of Siberia. The summer water stratification on the eastern shelf is occasionally destroyed by storms, but is subsequently re-established by the addition of surface flow.

Three water masses are formed on the Bering shelf: Alaskan coastal water, Gulf of Anadyr water, and northern St. Lawrence water. The Alaskan coastal water forms by mixing of sea water and the effluents of Yukon, Kuskokwim, and other rivers draining Alaska. This water mass is characterized by relatively high temperatures (8 to 12°C) and salinities less than 32‰.

The Gulf of Anadyr water, however, is colder (-1.5 to 2.0°C) and more saline (32.0 to 33.0‰). The northern St. Lawrence water is generally found in the region between St. Lawrence Island and the Bering Strait and is formed by the mixing of two other water masses. Saur et al. (1954) termed this water mass as modified shelf water, with temperatures between 1 and 4°C and salinities ranging from 32.0 to 33.0‰.

In winter intense mixing occurs on the eastern shelf due to the shallowness of the water, and the sea becomes isothermal and isohaline. In the northern area, sea ice generally forms in October, and by December the sea ice generally extends to the 150-m isobath, where the smooth shelf breaks off into the shelf margin. The thickness and areal extent of sea ice is greatly influenced by meteorological conditions prevailing over the area during this season. Maximum thickness occurs in shore ice formed in bays and straits, where ice is protected from winds and waves. During the spring melting, large amounts of sediments in the ice are carried offshore and deposited. Melting sometimes coincides with river floods, and river detritus is often carried with the sea ice. Because the pattern of ice movement is presently unknown, the sediment dispersal pathways by ice remains obscure.

On the shelf the water is saturated with oxygen during the summer season; in winter, however, when covered with ice the oxygen content decreases. The oxygen source during the ice-covered season is from the deep water. Atmospheric oxygen is available only in areas of open leads and polynya.

The pH in the eastern Bering Sea is controlled primarily by the CO_2 in solution. The water near the Unimak Island was found to be undersaturated with CO_2 (Kelley et al., 1971). The effect of coastal mixing on CO_2 is also indicated by salinities, which varied from 32.1‰ near Unimak

Island to 29.8‰ in the area of CO_2 supersaturation. Because of increased surface flow, it is evident that the pH of Bering Sea water is subject to seasonal variations as well. Other parameters which are closely associated with biological activity in the Bering Sea and vary seasonally are silica, phosphorus, nitrogen compounds, and organic carbon.

Productivity

The southeastern Bering Sea is well known for its high primary biological production which supports extensive fisheries, mammals, and birds. Cold, nutrient-rich water from depths of 100 m or more are brought to surface by tidal mixing along the northern Aleutian Island Chain. The upwelling and mixing of subsurface with surface water produces horizontal anomalies of physical and chemical properties. Phosphate ranges from 0.26 μg-at/1. at non-upwelling stations to 1.97 μg-at/1. near an upwelling region. Nitrate varied from 0.0 to 25.1 μg-at/1., ammonia from 0.0 to 3.5 μg-at/1., and silica-silicate from 0.0 to 60 μg-at/1. (Goering, personal communication, 1972). During midsummer the near-surface water (<100 m) just north of Unimak Pass contained 43 to 811 μg C/1. particulate organic carbon, while samples deeper than 100 m contained only 12 to 27 μg C/1. From the above data it appears that nitrogen and silicon are the two most important elements regulating productivity in the eastern Bering Sea.

Measurements of productivity during winter and summer months at various stations in the eastern Bering Sea have been made by McRoy et

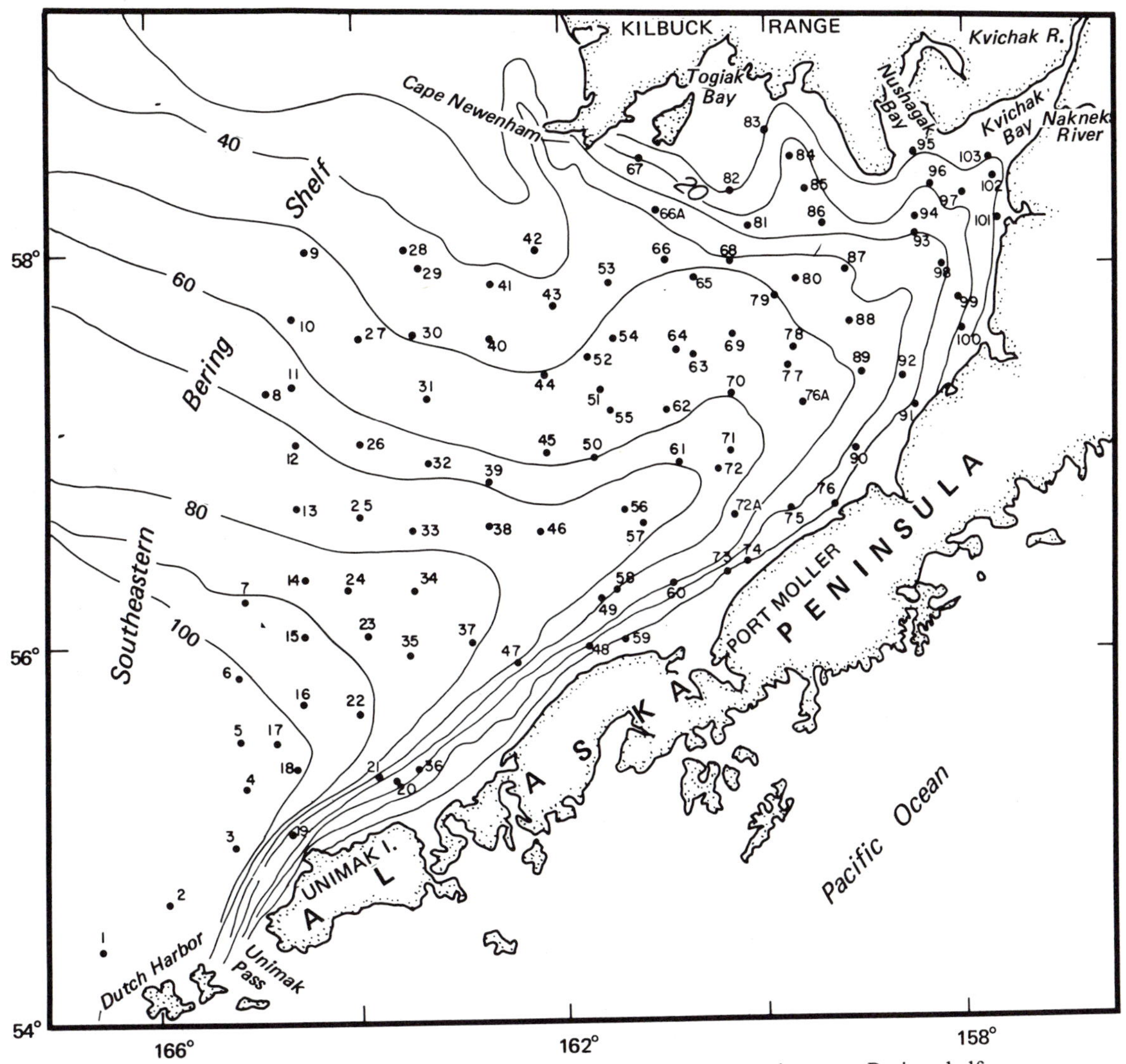

Fig. 3. Bathymetry (meters) and sample locations of the southeastern Bering shelf.

al. (1971). The highest surface productivity was noted during summer in the Bering Strait, where surface values as high as 410 mg C/m^3 day were encountered. The summer measurements in the vicinity of the Aleutian Islands varied from 2.2 to 165 mg C/m^3 day, and in lagoons located on the Alaskan coast the productivity ranged from 26.8 to 194 mg C/m^3 day. Winter productivity in ice-covered areas was significantly lower, and varied from 0.17 to 4.09 mg C/m^3 day. The Bering Sea annual cycle of primary production apparently begins with a development of microalgae below the sea ice surface. It is followed by phytoplankton bloom in the coastal zone that extends from the Aleutians to the Bering Strait and westward to the Gulf of Anadyr; production decreases toward the central Bering Sea. During the winter season production rates are highest in open water and lowest under ice.

Geology

The general geology of the Bering shelf has been described by Scholl and Hopkins (1969). Structural geology based on analyses of reflection profiles from this region suggest that the outer shelf is a narrow downwarped and downfaulted belt running parallel to the continental margin. A virtually undeformed inner shelf borders it. The outer shelf consists of the Pribilof and Zehmchug depressions, and the inner shelf includes the Bristol, Anadyr, and Norton Basins. The Bristol and Norton Basins are separated by the broad Nunivak Arch, which underlies Nunivak Island and possibly St. Matthews Island. These basins are filled with relatively undeformed thick Cenozoic sediments.

The distribution of surface sediments on the Bering shelf is primarily controlled by their source and secondarily by the water dynamics of the region. On the basis of these two factors the shelf can be divided into five regions. These regions follow closely the boundaries of the areas differentiated structurally by Scholl and Hopkins (1969). The regions include (1) the southeastern shelf, including Bristol Bay; (2) the central shelf, a broad region lying between St. Matthews and Nunivak islands; (3) the northeastern shelf, generally known as Chirikov Basin and Norton Basin, the region bounded by Seward Peninsula, the Yukon delta, St. Lawrence Island, and the Russian coast; (4) the northwestern shelf, essentially covering the Gulf of Anadyr; and (5) the outer shelf, a narrow elongated area running parallel to the continental margin (Fig. 1).

Southeastern Bering Shelf—Bristol Bay

The southeastern Bering shelf, commonly known as Bristol Bay, is a triangular embayment (Fig. 3). From the base of the Alaskan Peninsula the shelf extends about 740 km to the southwest, where a steep and canyon-scarred slope leads to the Aleutian Basin. The shelf is extremely flat with an average gradient of only 2.4×10^{-4} and with an average depth of 50 m.

The texture of sediments in the southeastern Bering shelf varies with depth and distance from shore. Nearshore the sediments consist of coarse sand and gravel, but they become progressively finer toward the open shelf as water depth increases (Fig. 4). The gravelly coarse sands of the coast and nearshore are extremely poorly sorted, the sands of the central shelf are moderately poorly sorted to moderately well sorted, and the muddy sands of the outer shelf are extremely poorly sorted. The sediments of the central shelf have symmetrical size distribution, but they grade progressively into strongly coarse-skewed sediments toward shore and strongly fine-skewed toward the outer shelf.

The suspended load contained both biogenic matter (mostly pelagic diatoms) and clay-sized sediments and varied laterally as well as vertically. The suspended load under calm sea conditions varied from 1.60 mg/l. to 12.0 mg/l. and decreased with increased distance from shore; in general most measurements lie in the range of 1.0 to 9.0 mg/l. Locally high values, however, off Port Moller and on the open shelf are caused by increased biogenic activity.

The mineralogy of sediments in Bristol Bay reflects weathering environment and provenance from adjacent shores and the hinterland. Predominance of finely divided primary silicates in the clay fraction suggests mechanical abrasion and lack of intense chemical weathering of soils. This is typical of the subarctic hinterland which provides the terrigenous sediments depositing in Bristol Bay (Sharma, 1970a, 1970b). In the southeastern Bering shelf the important sources of detrital sediments, as indicated by mineral assemblages, are the drainage from the east and north, the products of contemporary volcanism along the

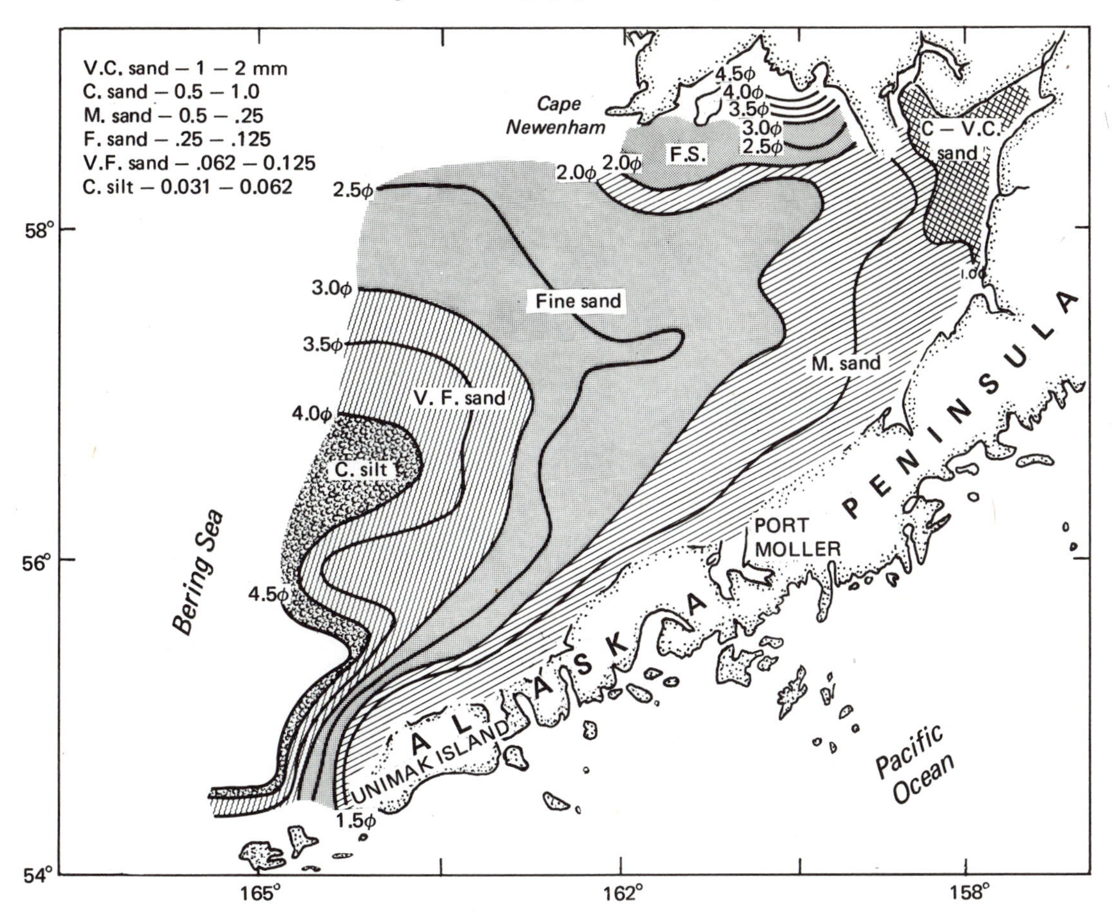

Fig. 4. Distribution of sediment size over the southeastern Bering shelf.

Alaskan Peninsula, and biogenic materials (Sharma et al., 1972). The predominance of euhedral and relatively unaltered hypersthene and the common occurrence of magnetite and ilmenite suggest that these sediments originate in nearby basic and ultrabasic rocks. The presence of small amounts of reddish-brown pleochroic hornblende in most samples suggests a source in the basalts or hornblende andesites that are extensively exposed in the Alaskan Peninsula. Sillimanite, garnet (almandite), staurolite, and epidote in the sediments indicate a source from high-grade metamorphic rocks that are exposed on the Alaskan mainland; regionally metamorphosed rocks are not exposed in the Alaskan Peninsula and in the Aleutian Islands.

Central Bering Shelf

The sediments from this area (Fig. 5) have been studied by the University of Washington, the U.S. Geological Survey, and a few samples have been investigated by Russian geologists. On the inner and central shelf the sediments are predominantly sand (50 to 90%), while along the outer shelf and the slope sands are replaced by silt (Fig. 5). As in Bristol Bay, the sediment mean size generally decreases with increasing depth and increasing distance from shore (3 to 6 ϕ). Unlike at the southeastern shelf, the sediment mean size does not grade evenly across the shelf. It is disrupted by large amounts of silt brought by various rivers. Sediments vary from moderately sorted on the central shelf to very poorly sorted toward the shore and the outer shelf. The size distribution is strongly fine-skewed.

McManus and Smyth (1970) found between 5 mg/l. in mid-shelf and 10 mg/l. of suspended sediments near the Alaskan mainland on a southeast transect between St. Lawrence Island and the Alaskan coast. The material consists of more than 85% mineral grains having a mode of 20 μ. The amount of suspended material increases with increasing depth.

The sources of surface sediments of the central shelf are the Quaternary alluvial, palus-

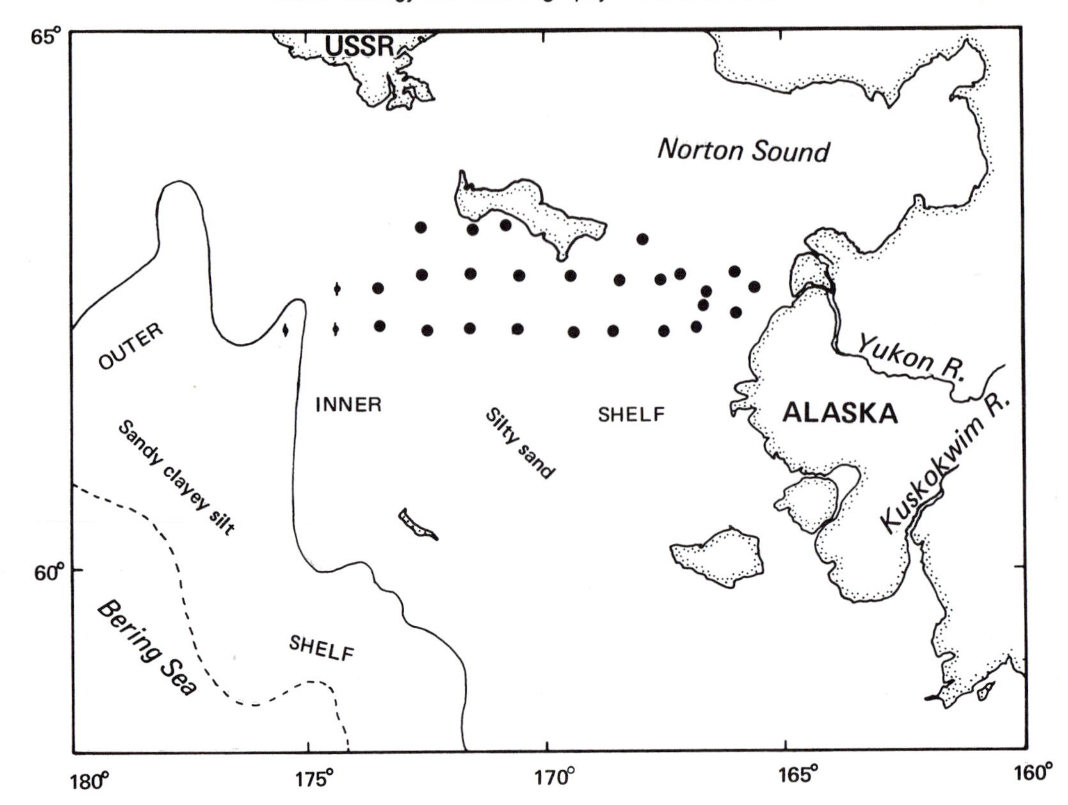

Fig. 5. Sample station locations and sediment texture in central Bering shelf (data obtained from McManus and Creager, 1965).

trine, lacustrine, and glacial deposits along the shore. Exposures of andesite, basalts, and intrusives of Cenozoic age along the coast provide local plutonic source. The Kuskokwim River provides debris of shales, sandstones, clays, lignite, and conglomerates of Tertiary age. Biogenic processes also contribute modern sediments.

Northeastern Bering Shelf

The northeastern Bering shelf is defined by the Bering Strait in the north, by the 46-m deep sill separating the Gulf of Anadyr in the southeast, and by the 32-m deep sill east of St. Lawrence Island. The eastern part, Norton Sound, is less than 30 m deep and receives the drainage of the Yukon River, the largest river in the area. The western part is slightly deeper, with a few channels and shallow sea valleys.

The sediments have been studied by Creager and McManus (1967), Venkatarathnam (1969), and Nelson and Hopkins (1972). The distribution of sediments in this region is complex because of the various modes of sediment transport. The basin floor is covered by glacially derived as well as by marine deposits (Fig. 6). Submarine exposures of relict gravel over glacial deposits and bedrock cover significant areas off the coast of Seward Peninsula and St. Lawrence Island in the west-central parts of the shelf and in the Bering Strait. These gravels are derived from morainal deposits which extend close to the present shore line of the Seward Peninsula. However, the gravel which extends southwest of the Bering Strait toward St. Lawrence Island, is probably a morainal ridge deposited by glaciers that originated in Siberia. The glacial debris has been reworked to gravels where moraines cover the sea floor, but elsewhere moraines are buried beneath younger deposits.

The marine sediments consist of silty-sand which covers much of the shelf (Fig. 6). In the southeastern basin the Yukon River has built a thin delta of clayey-silt, which encroaches on the silty-sand (Nelson, 1971).

Nine heavy mineral assemblages are recognized in the northeastern shelf by Venkatarathnam (1969). The entire shelf, however, can be divided into four regions on the basis of heavy mineral sources. Source areas for detrital sediments are: (1) the Yukon River depositing clayey-silt characterized by high clinopyroxene, hornblende, and hypersthene content which covers the region off the Yukon delta and most of the Norton Sound; (2) the Siberian coast, contributing relict glacial deposits characterized by hornblende, opaque minerals with sphene, zircon, and metallic copper, which covers the western and central region of Chirikov Basin; (3) the Seward sands and gravels, characterized by large quantities of garnet, opaque, and metamorphic minerals, which extends offshore to the central Chirikov and Norton Basins; and (4) St. Lawrence sandstone and gravel, characterized by diverse mineralogy, of igneous and sedimentary source rocks. The heavy mineral distribution and the sediment texture suggest that much of the silty-sand was deposited during lower sea levels, whereas the deposition of the clayey-silt in Norton Sound is controlled by the contemporary environment.

Northwestern Bering Shelf

This part of the Bering shelf, commonly known as the Gulf of Anadyr, is deeper than other shelf areas. Much of the shelf lies between 50 to 100 m isobath. The gentle slope is cut by numerous valleys, with relief averaging 20 m. The shelf displays broad ridges measuring 20 to 30 km at the base and up to 20 m high; these features are the manifestation of buried bedrock relief (Lisitsyn, 1969). The thickness of unconsolidated sediments in the Gulf of Anadyr varies between 15 and 20 m, and the upper 5 m is entirely of Postglacial origin (Lisitsyn, 1969). The Anadyr River, second in volume to the Yukon and Kuskokwim, empties into the Gulf of Anadyr.

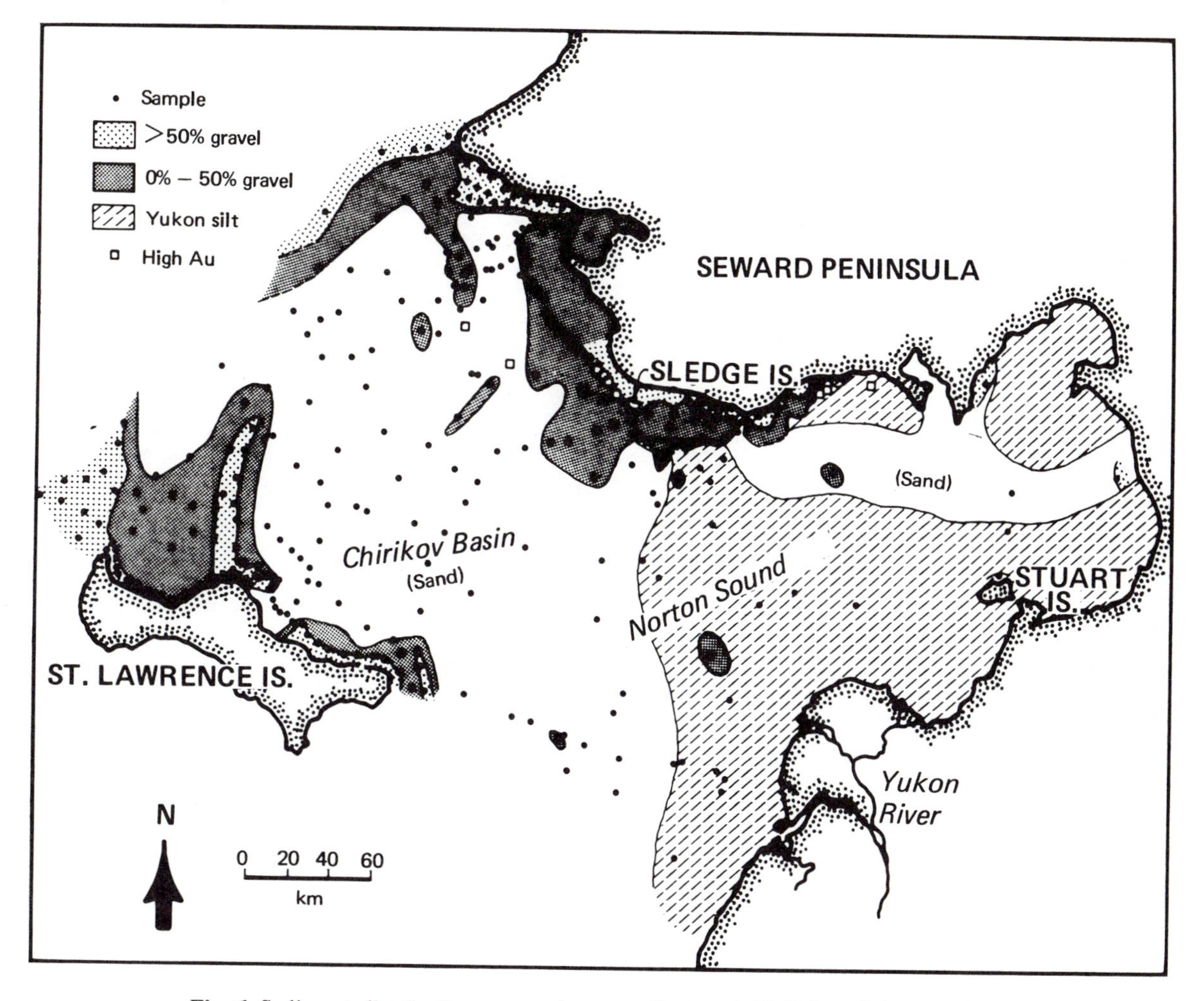

Fig. 6. Sediment distribution on northeastern Bering shelf. (After Nelson et al., 1972.)

The northern coast of the Gulf of Anadyr is characterized by Mesozoic sedimentary and igneous rocks. Thick Quaternary continental and marine deposits cover most of the western coast. Lacustrine and paludal deposits of silt constitute the continental beds. Marine Quaternary deposits are silty-clays and sands with pebbles and shells. Isolated exposures of volcanics are found throughout the coast.

The nearshore sediments in the Gulf of Anadyr consist of gravels which form a broad 50- to 60-km wide belt that extends from the shore to depths of 30 to 40 m (Lisitsyn, 1969). Greenish-gray coarse and medium sand occurs between 60 and 90 m depth. Fine sands are the most common constituents and are similar in distribution to the find sands in the southeastern Bering shelf, often containing large percentages of biogenic fragments. Fine sands with good sorting and rich in heavy minerals are widespread and cover extensive areas ranging 150 to 170 km in width.

Concentration of suspended matter in surface water reaches a peak in spring. Values up to 13 mg/l. have been measured during high flow of the Anadyr River. Normally the concentrations vary between 5 and 7 mg/l. in the gulf and 3 and 5 mg/l. on the adjacent shelf, and generally remain the same with increasing depth. Bottom turbid layers with high quantities of suspended (mineral and biogenous) material have been observed often (Lisitsyn, 1969).

The Gulf of Anadyr can be divided into an eastern and a western mineralogical province. The eastern sector minerals derived from metamorphic rocks are amphiboles, epidote, chlorite, muscovite, quartz, and potassium-feldspar; the percentages of pyroxene and magnetite are low.

The western sector sediments were transported by the Anadyr River. In the fine sand the heavy minerals exceed 40%, with highest concentrations occurring at the Anadyr River mouth and decreasing offshore. Typical heavy minerals are black ore minerals, hypersthene, augite-diopside, zircon, and tourmaline. These sediments are poor in quartz, hornblende, and pyroxene but contain large amounts of plagioclase and ilmenite.

The primary source of fine sediments is the Anadyr River, which drains the Anadyr Basin. The distribution of clay minerals (chlorite, illite, and montmorillonite) in the Gulf of Anadyr confirms this source. The coarse material originates in the shore and is generally transported along the coastline by ice (Lisitsyn, 1969).

Outer Bering Shelf

The outer Bering shelf lies between 100 and 175 m depth, and parallels the continental margin. At its southern end, north of Unimak Island, it is approximately 120 km wide, and it is steepest in the vicinity of the Pribilof Islands; in the north, near the Gulf of Anadyr, it becomes wider, reaching a width of about 350 km; here the gentlest gradients were measured.

The sediments consist of varying mixtures of sands, silts, and clays. Sand is predominant in shallow depths, whereas silts, clays, and biogenous materials are predominant in the deeper parts. The silt content at the outer edge of the shelf may reach over 50% by weight. The sediments are generally poorly sorted to extremely poorly sorted. They are characteristically fine-skewed to strongly fine-skewed and leptokurtic to extremely leptokurtic.

Seasonal and hydrodynamic conditions control the concentration of suspended material. The suspended load in surface water is low during sea-ice cover but may reach values as high as 3.0 mg/l. during phytoplankton bloom. The suspended load generally increases with depth to concentrations as high as 20 mg/l. (Lisitsyn, 1969). A significant increase occurs during and after severe storms and is attributed to roiling of bottom sediments by storm waves (Lisitsyn, 1969).

The mineralogy of sediments is complex: The southern portion is dominated by material originating in Bristol Bay and the Kuskokwim River; the central sector is covered by material originating in the adjacent coast; islands and the Kuskokwim River and the deposits in the Northwestern region are dominated by sediments carried by the Anadyr River.

Sediment Transport

The prominent feature of the Bering Sea shelf sediments is the exclusive distribution of sands over the entire area. The almost complete absence of silt and clay-size sediments in areas with water depths less than 100 m strongly suggest that these size grades are kept in suspension and are prevented from being deposited on the inner shelf. Fine sands are widely distributed and are well sorted. In general the principal factors controlling sand sorting on the shelf are the shelf width and gradient as well as wave intensity. Shelf gradient

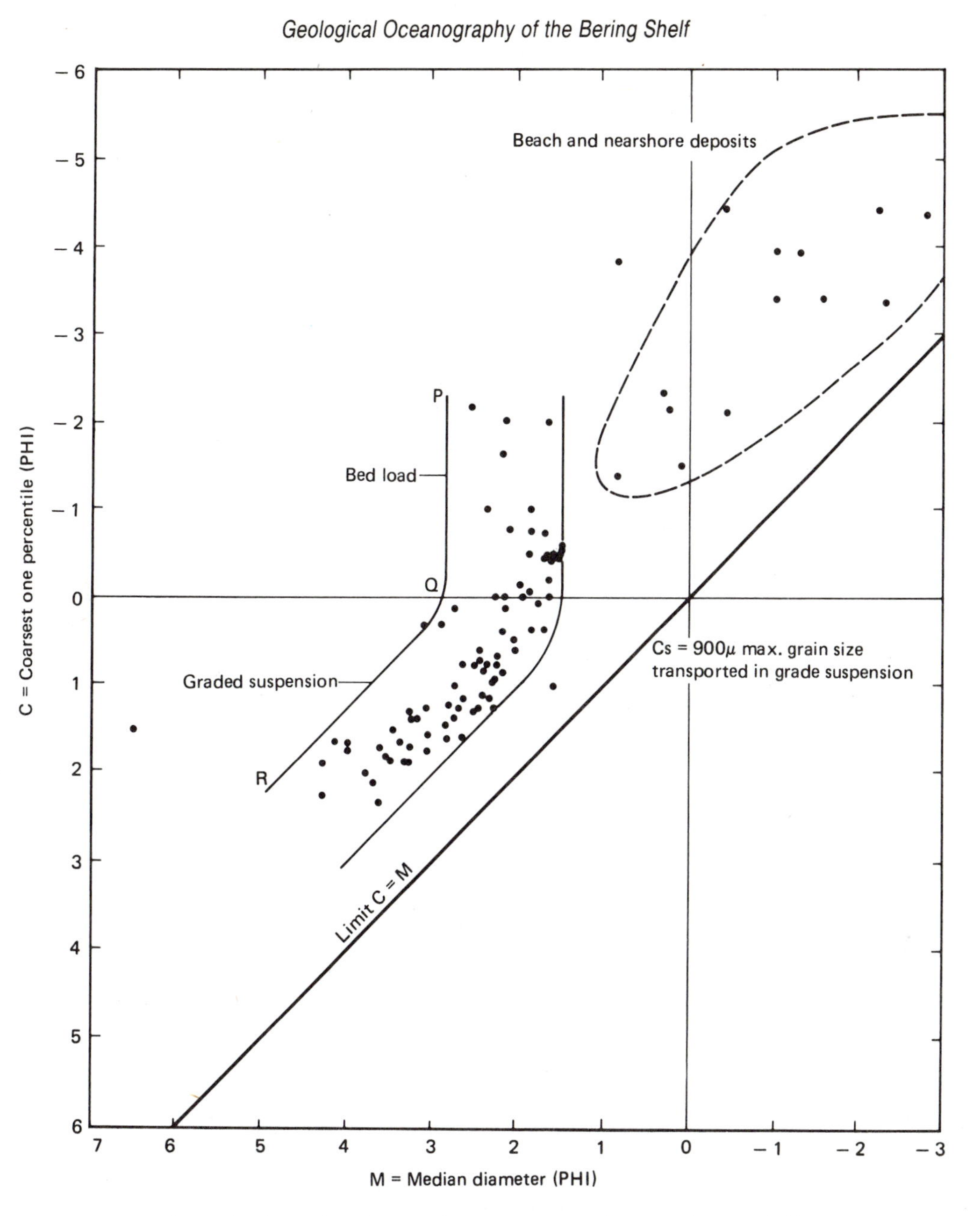

Fig. 7. C-M relationship of sediments from southeastern Bering shelf.

affects the hydrodynamic conditions, thereby confining sediment deposition to a certain particle size range. The wide distribution of fine sands on the inner Bering Shelf is attributed to its particle size that requires minimum water velocity for its translocation. Fine sands, therefore, are the first particles to be set in motion by wave resuspension.

The mode of sediment transport is emphasized by the C-M diagram of Passega (1964). Sediments deposited on the southeastern Bering shelf are characteristic of particles transported in graded suspension and as bed load (Fig. 7). Turbid layers near the bottom, and increasing concentration of suspended sediments with increasing depth in the northwestern Bering shelf have been reported by Lisitsyn (1969). He also demonstrated roiling of sediments by storms in the Gulf of Anadyr and suggested that roiling of sediments at depths less than 90 m occurs frequently and results in winnowing of fines and good sorting of sands.

To assess the influence of storm waves on the transport of sediments in the southern shelf, sporadic meteorological observations of the past 100 years have been compiled. Theoretical wave heights and periods for severe annual storms in this region also have been computed from synoptic surface wind charts. These considerations and wave forecast theory provide for the southern shelf a significant wave height (H) of approximately 10 m and a significant wave period (T) of about 11 sec. This results in a wavelength (L) of approximately 200 m. A wave of the above dimension at a depth (d) of 94 m would result in water particle velocity (u) of:

$$u = \frac{\pi H}{T \sinh \frac{2\pi d}{L}} = 30 \text{ cm/sec}$$

Consequently storm waves in this region at a depth of 94 m can be strong enough to cause incipient movement of fine sand due to water particle velocity resulting from the wave (Hjulstrom, 1939).

Scatterplots—phi mean size against sorting coefficient (Figs. 8 and 9) and skewness against kurtosis (Fig. 10) suggest two broad environments of deposition. These are the inner shelf, an area where wave deformation and winnowing of fine takes place; and the outer shelf, which lies below the 100-m isobath and is uninfluenced by storm waves. Waves generated by frequent severe storms are therefore the most important mode of sediment transport on the Bering shelf.

Sediments in the inner shelf are also influenced by tidal and permanent currents. The effects of these currents are generally superimposed on the sediment distribution, which is controlled by wave energy distribution. The extent of sediment transport by these currents, best illustrated by the heavy mineral distribution in the southeastern Bering shelf; this suggests that a current moving along the Alaska Peninsula forms a cyclonic gyre in Bristol Bay. In the central Bering shelf the major water movement is to the north; this water carries a fraction of Kuskokwim River sediments. The current velocity and suspended load measurements in central Bering shelf by McManus and Smyth (1970) also suggest that the currents are capable of transporting silt-size sediments. The sediments in the northeastern Bering shelf are influenced by the northward-moving water and the influx of the Yukon River. Deposition of river sediments occurs near the Yukon River delta and along the southern sector of the Seward Peninsula. Exposures of submarine relict sediments suggest areas of nondeposition in the Chirikov Basin. Most sediments brought by the Yukon River are carried into the Chukchi Sea by currents. The Gulf of Anadyr sediment distribution is influenced by southward-moving currents along the western coast and northward current along the eastern side.

During winter most of the Bering shelf is ice-covered, and therefore sediment transport by ice can be of significance. Sea ice samples were collected from offshore during January and February 1970 in order to study the nature of sea ice and sediments therein (Fig. 11). The amount of

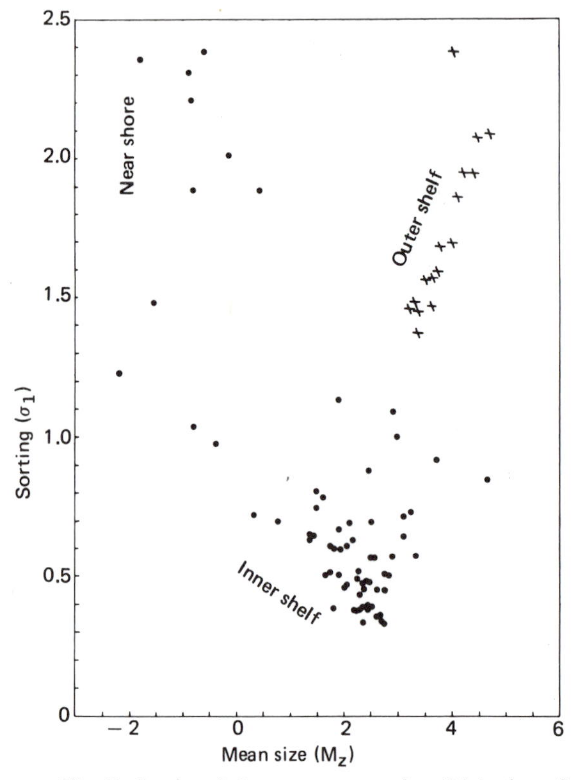

Fig. 8. Sorting (σ_1) versus mean size (M_Z) plot of sediments from southeastern Bering shelf.

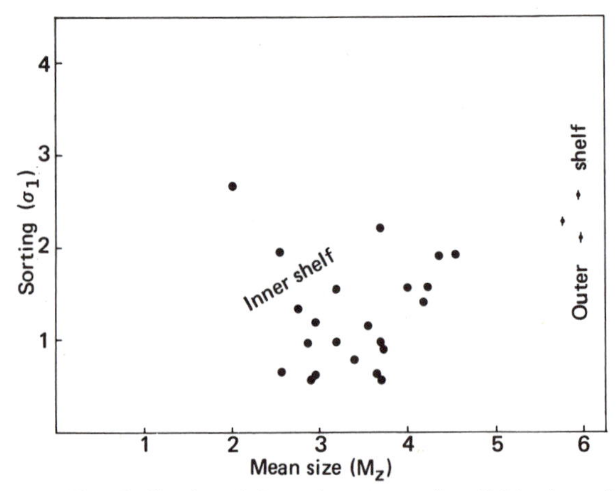

Fig. 9. Sorting (σ_1) versus mean size (M_Z) plot of sediments from central Bering shelf.

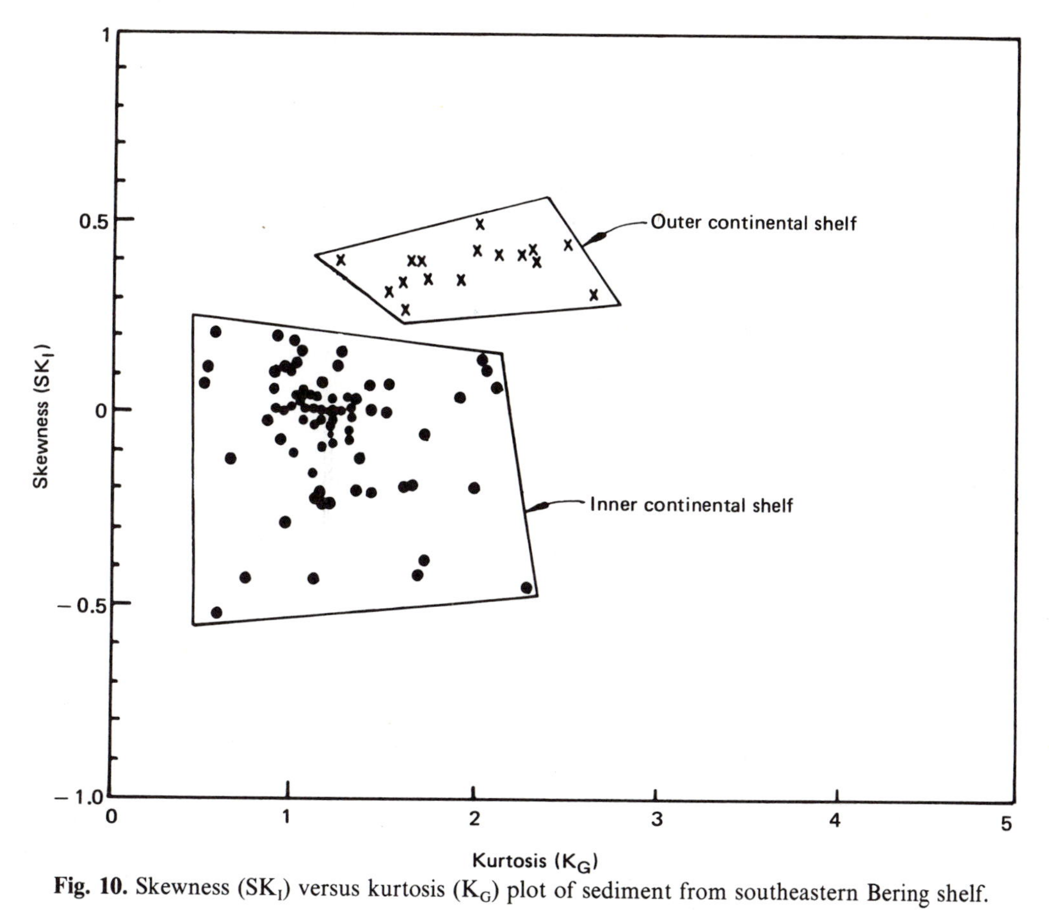

Fig. 10. Skewness (SK_I) versus kurtosis (K_G) plot of sediment from southeastern Bering shelf.

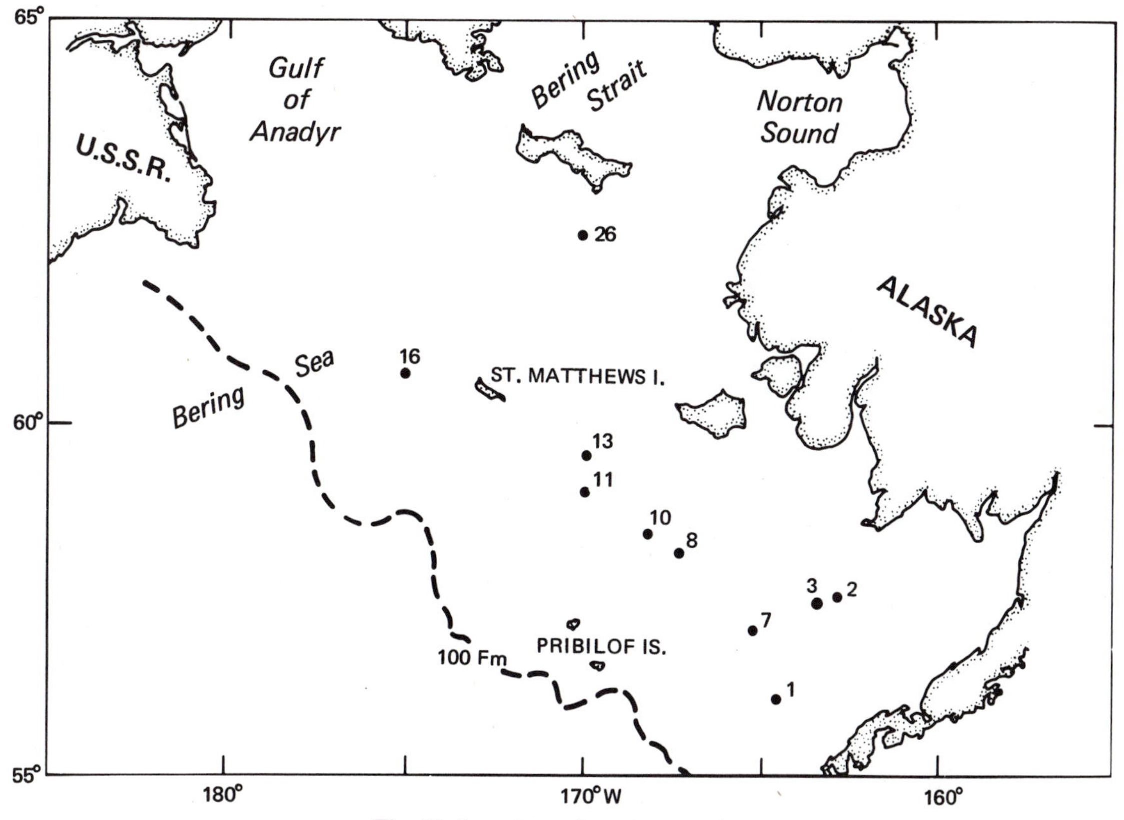

Fig. 11. Locations of sea ice samples.

sediments in sea ice varied regionally as well as temporally. Based on crystal orientation a maximum of five layers were observed in some sea-ice samples (Fig. 12). These layers were characterized also by size and shape of individual crystals, air bubbles, salt pockets, sediments, or phytoplankton content.

In the Bering shelf ice sediments were mostly found at the lower boundary of the middle ice layer with horizontal C-axis. The sediments consist of varying amounts of sand, silt, and clays. The sand varied from 1 to 8%, silt from 35 to 80%, and clays from 12 to 60% of the total weight (Fig. 13). The bottom sediments from the same stations were dominantly fine to medium sands (Fig. 13).

The clay mineralogy of the sea ice sediments, determined by X-ray diffraction, was found to be uniform. Sediment texture (grain size distribution) and the uniformity of clay minerals suggest that the sediments incorporated in the ice were in suspension during the freezing of the sea water and the aggradation of the sea ice. Fine sand and coarse silt found in sea ice in offshore areas can be brought in suspension by storms. The very fine-grained, dense, bubbly, milky ice associated with sediments in sea ice also indicates turbulent sea conditions. It is, however, not clear how the energy for suspension is supplied during ice cover.

Conclusion

The grading of sediments shown by spatial variation in the southeastern, central, and northwestern shelves indicate that coastal sediments are carried to the deeper parts of the Bering shelf and that the silts and clays are ultimately deposited on the outer Bering shelf, the slope, and the deep basins. In the northeastern Bering shelf the sedi-

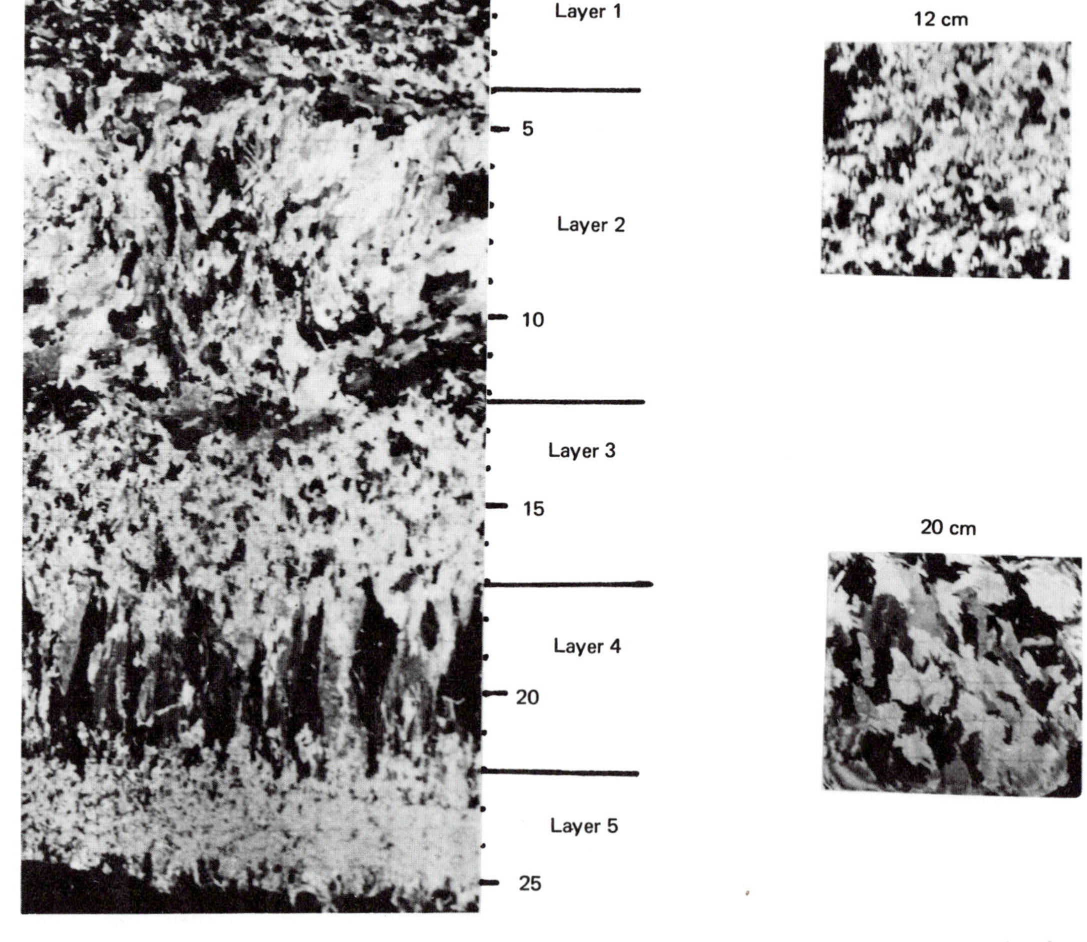

Fig. 12. Vertical and horizontal thin sections of ice under polarized light from station number 3.

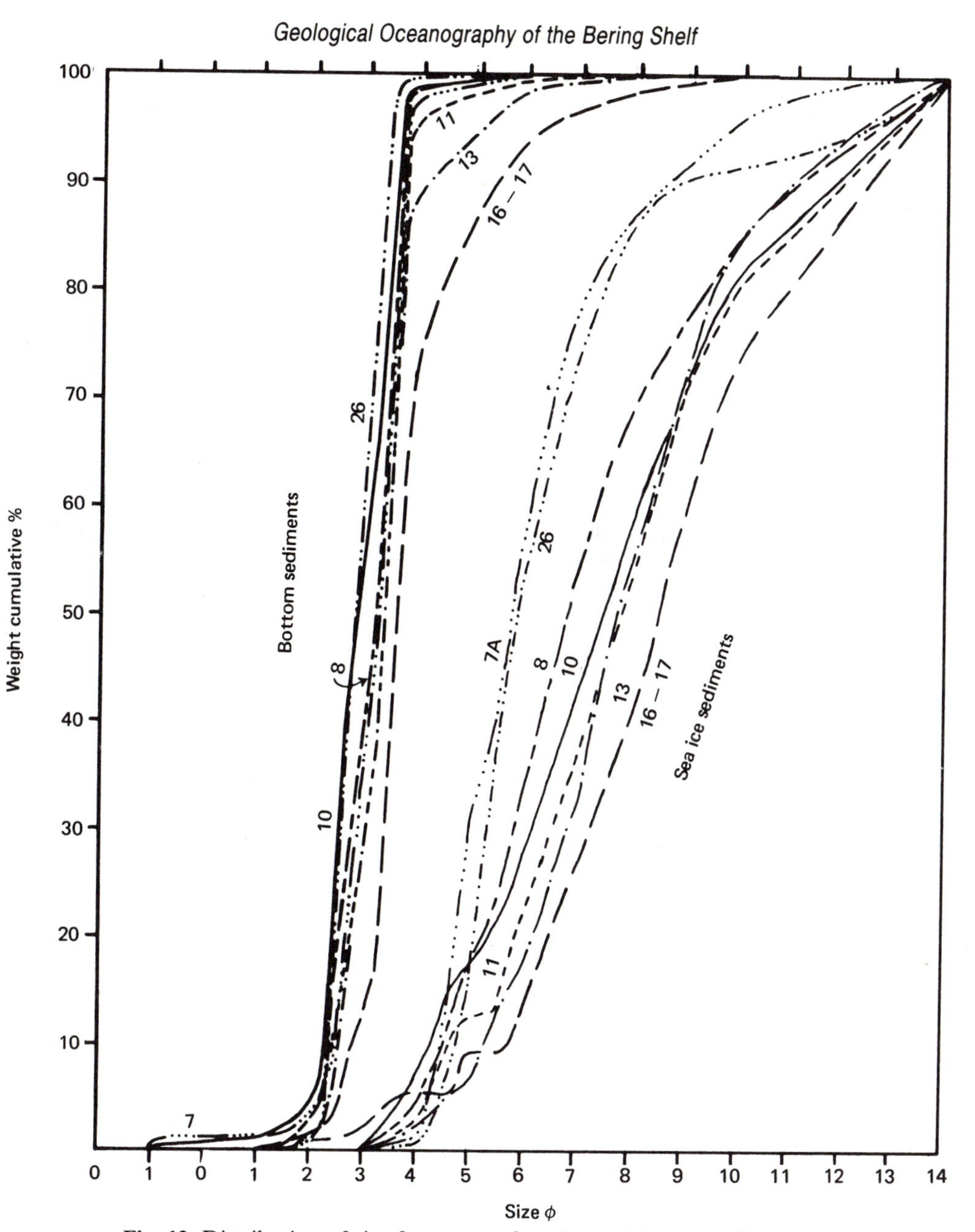

Fig. 13. Distribution of size frequency of sea ice and bottom sediments.

ments brought by the Yukon River are in part deposited in the Norton Basin and partly carried into the Chukchi Sea. Storm waves in shallow areas roil bottom sediments while currents move sediments laterally across the shelf. A smooth broad prograding shelf with graded sediments is formed in this manner.

Acknowledgment

The author is indebted to Dr. C. H. Nelson for reading the manuscript and for making suggestions.

References

Creager, J. S., and D. A. McManus. 1967. Geology of the floor of the Bering and Chukchi Seas; American studies. In: D. M. Hopkins, ed. *The Bering Land Bridge.* Stanford, California: Stanford University Press. pp. 7–31.

Gershanovich, D. E. 1967. Late Quaternary sediments of Bering Sea and the Gulf of Alaska. In: D. M. Hopkins, ed. *The Bering Land Bridge.* Stanford, California: Stanford University Press. pp. 32–46.

Hjulstrom, F. 1939. Transportation of detritus by running water. In: P. D. Trask, ed. *Recent Marine*

Sediments. Tulsa, Oklahoma: American Association of Petroleum Geologists. pp. 3–31.

Husby, D. M. 1968. Oceanographic investigations in the northern Bering Sea and Bering Strait. *Oceanographic Report, 40.* Washington, D.C.: U.S. Coast Guard Oceanographic Unit. p. 50.

Kelley, J. J., L. L. Longerich, and D. W. Hood. 1971. Carbon dioxide in the surface waters near the coast of southern Alaska and the eastern Aleutian Islands. In: *Oceanography of Bering Sea, Phase 1.* Institute of Marine Science, Report, R-71-9, Fairbanks, Alaska: University of Alaska. pp. 3–39.

Lisitsyn, A. P. 1969. Recent sedimentation in the Bering Sea. In: P. L. Bezrukov, ed. *Academy of Sciences of the U.S.S.R. Department of Earth Sciences, Commission of Sedimentary Rocks* (transl.). Institute of Oceanology, Moscow 1966. Washington, D.C.: U.S. Department of Commerce. p. 614.

McManus, D. A., and J. S. Creager. 1965. Bottom sediment data from the continental shelf of the Chukchi and Bering seas, 1959–1963. *Department of Oceanography, Technical Report, 135.* Seattle, Washington: University of Washington.

———, and C. S. Smyth. 1970. Turbid bottom water on the continental shelf of the northern Bering Sea. *J. Sed. Petrol.* 40(3):869–873.

McRoy, C. P., J. J. Goering, and W. E. Shiels. 1971. Studies of primary production in the eastern Bering Sea. In: *Oceanography of the Bering Sea. Phase 1.* Institute of Marine Science Report R-71-9. Fairbanks, Alaska: University of Alaska. p. 212.

Nelson, C. H. 1971. Trace metal content of surface relict sediments and displacement of northern Bering Sea Holocene sediments. In: *Second National Coastal and Shallow Water Research Conference* (abstract). University of Southern California Press. p. 269.

———, and D. M. Hopkins. 1972. Sedimentary processes and distribution of particulate gold in northern Bering Sea. *U.S. Geol. Surv. Prof. Pap. 689,* 27.

———, D. M. Hopkins, and D. W. Scholl. 1972. Cenozoic sedimentary and tectonic history of the Bering Sea. *International Symposium for Bering Sea Study*, Hakodate, Japan, January 31–February 4, 1972.

Ohotani, K. 1969. On the oceanographic structure and ice formation on the continental shelf in the eastern Bering Sea (in Japanese). *Bull. Fac. Fish. Hokkaido Univ.* 20(2):94–117.

Passega, R. 1964. Grain-size representation by C-M patterns as a geological tool. *J. Sed. Petrol.* 34:830–847.

Saur, J. F. T., J. P. Tully, and E. C. LaFond. 1954. Oceanographic cruise to the Bering and Chukchi seas, summer 1949. Part 4: Physical oceanographic studies Vol. 2. *U.S.N. Electron. Lab. Res. Rep. 416.* 31 pp.

Scholl, D. W., and D. M. Hopkins. 1969. Newly discovered Cenozoic basins, Bering Sea shelf, Alaska. *Am. Assoc. Petrol. Geol.* 53(10):2067–2078.

Sharma, G. D. 1970a. Sediment-sea water interaction in glaciomarine sediments of southeast Alaska. *Geol. Soc. Am. Bull.* 81(4):1097–1106.

———. 1970b. Evolution of interstitial waters in recent Alaskan marine sediments. *J. Sed. Petrol.* 40(2):722–733.

———, A. S. Naidu and D. W. Hood. 1972. Bristol Bay: A model contemporary graded shelf. *Am. Assoc. Petrol. Geol.* 56(10):2000–2012.

Venkatarathnam, K. 1969. Clastic sediments on the continental shelf of the northern Bering Sea. *Department of Oceanography Special Report 41.* Seattle, Washington: University of Washington: 40–61.

Chapter 6

Holocene Sedimentary Framework, East-Central Bering Sea Continental Shelf[1]

HARLEY J. KNEBEL,[2] JOE S. CREAGER,[3] AND RONALD J. ECHOLS[3]

Abstract

Granulometric, paleontologic, chronologic, and high-resolution sub-bottom data indicate considerable temporal and spatial variability in the Holocene stratigraphic sequence of the east-central Bering Sea continental shelf between St. Lawrence and St. Matthew Islands and the Alaskan mainland. Rates of sediment accumulation vary between 9 and 97 mg $cm^{-2}yr^{-1}$ (8 to 60 cm 10^3yr^{-1}) and indicate that the influence of fluvial and oceanographic processes have changed greatly during the last transgression of sea level across this part of the shelf. Older bathymetric features have also affected the sedimentary variability.

An isolated basin northeast of St. Matthew Island has been filled with varved lacustrine deposits which accumulated at about the time of the last major lowering of sea level. During the initial stage of the transgression, inner-shelf sediments were rapidly deposited within the northern sector; little modern sediment was deposited subsequently. In contrast, clayey silt has accumulated at an average rate of 60 mg $cm^{-2}yr^{-1}$ (70 cm 10^3yr^{-1}) since 10,000 to 11,000 years B.P. at one location in the south. A scheme for paleo-environmental interpretation based on foraminiferal fossils has been used in conjunction with radiocarbon ages and sedimentary characteristics to delineate the relative positions of sea level during the transgressive period. Evidence is presented suggesting that sea level did not fall below −120 m in this area.

[1] Contribution no. 659, Department of Oceanography, University of Washington.

[2] Marine Biological Laboratory, Woods Hole, Massachusetts 02543, U.S.A.

Introduction

The central part of the eastern Bering Sea continental shelf as used herein is bounded by latitude 60°N on the south, by the Alaskan mainland on the east, and by a semicircular line extending from the Yukon delta through St. Lawrence Island to St. Matthew Island on the north and west. The area includes Nunivak Island and comprises approximately 174,000 km^2 (Fig. 1).

Personnel of the Department of Oceanography, University of Washington, have been and are currently involved in a study of the bottom sediments of arctic continental shelves in an attempt to reveal variations and trends in the Holocene transgressive sea sediments.[4] Thus far primary attention has been given to the Chukchi (McManus et al., 1969; Creager and McManus, 1967), Laptev (Holmes, 1967; Holmes and Creager, this volume), East Siberian (Naugler, 1967; Naugler et al., this volume), and northern Bering Seas (Grim and McManus, 1970; McManus et al., 1970). This study adds a new area to this investigation.

The purposes of this paper are to outline the vertical and horizontal variation in the Holocene stratigraphic sequence within this section of the

[3] Department of Oceanography, University of Washington, Seattle, Washington 98105, U.S.A.

[4] In this paper Holocene sediments are those deposited since the last major sea level minimum.

Bering Sea continental shelf, to determine the cause of variation by evaluating the genesis of this sequence, and to determine the significance of variations in terms of changes in the physical processes of sedimentation and the postglacial rise of sea level.

Survey Results

Methods

Bathymetric, sedimentary, and sub-bottom data for this report were obtained along more than 5400 km of track line during cruise TT042 (1969) of the *RV Thomas G. Thompson*; supplemental depth information has been included from cruises TT020 (1967) and TT051 (1970) (Fig. 2). Bathymetric and high-resolution sub-bottom profiles were produced during TT042 by using a 12-kHz transceiver coupled with a precision depth recorder; no corrections were made for the deviation of sound velocity in water or sediment from 1464 m/sec. During TT042 the ship's track was controlled on a daily basis by the need to cover the area uniformly in the available time and by the findings of preceding days (Fig. 2). Whenever the sub-bottom reflectors indicated a thickened section of unconsolidated sediment, an attempt was made to obtain a core down to or below the first major reflector.

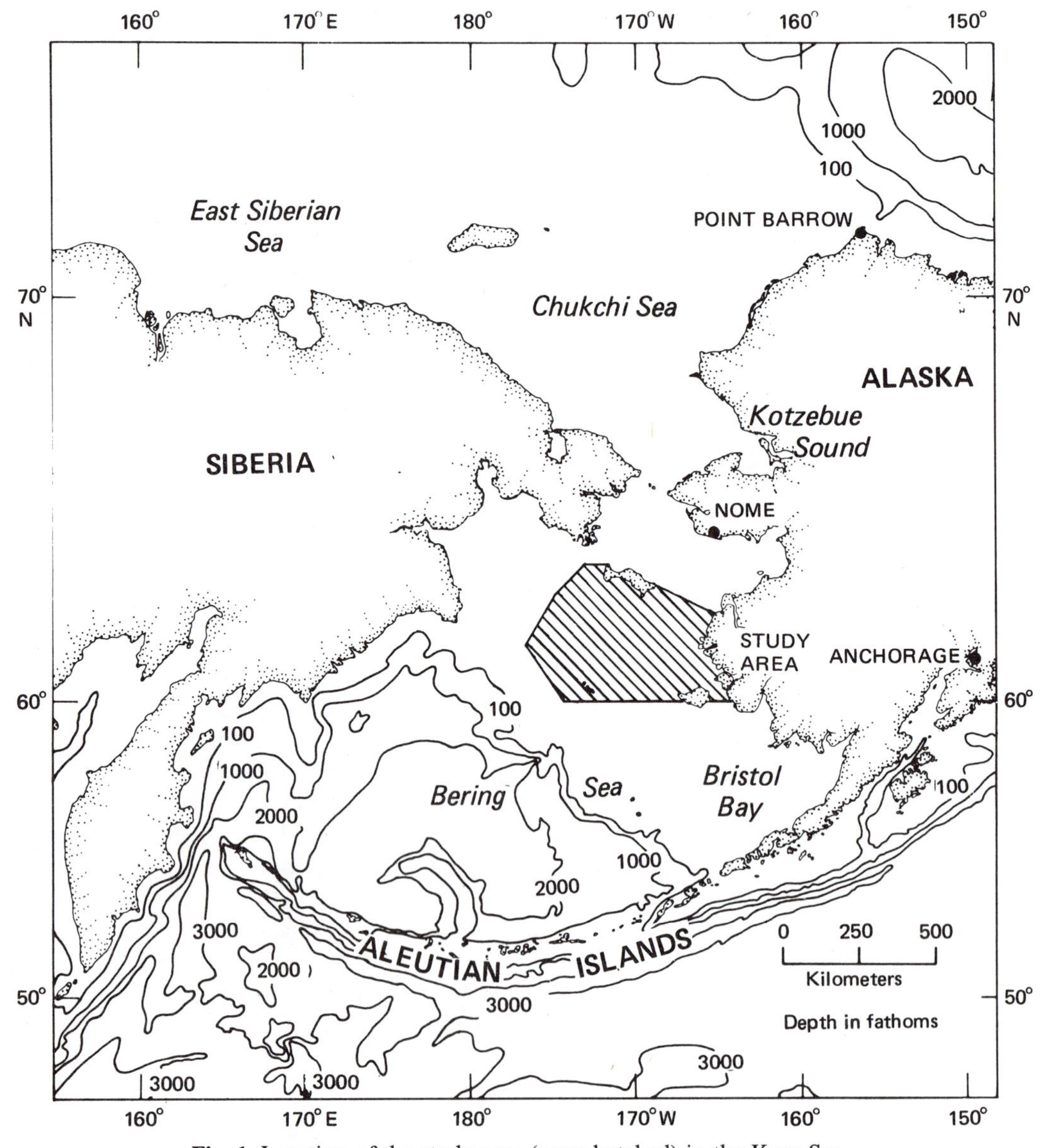

Fig. 1. Location of the study area (cross-hatched) in the Kara Sea.

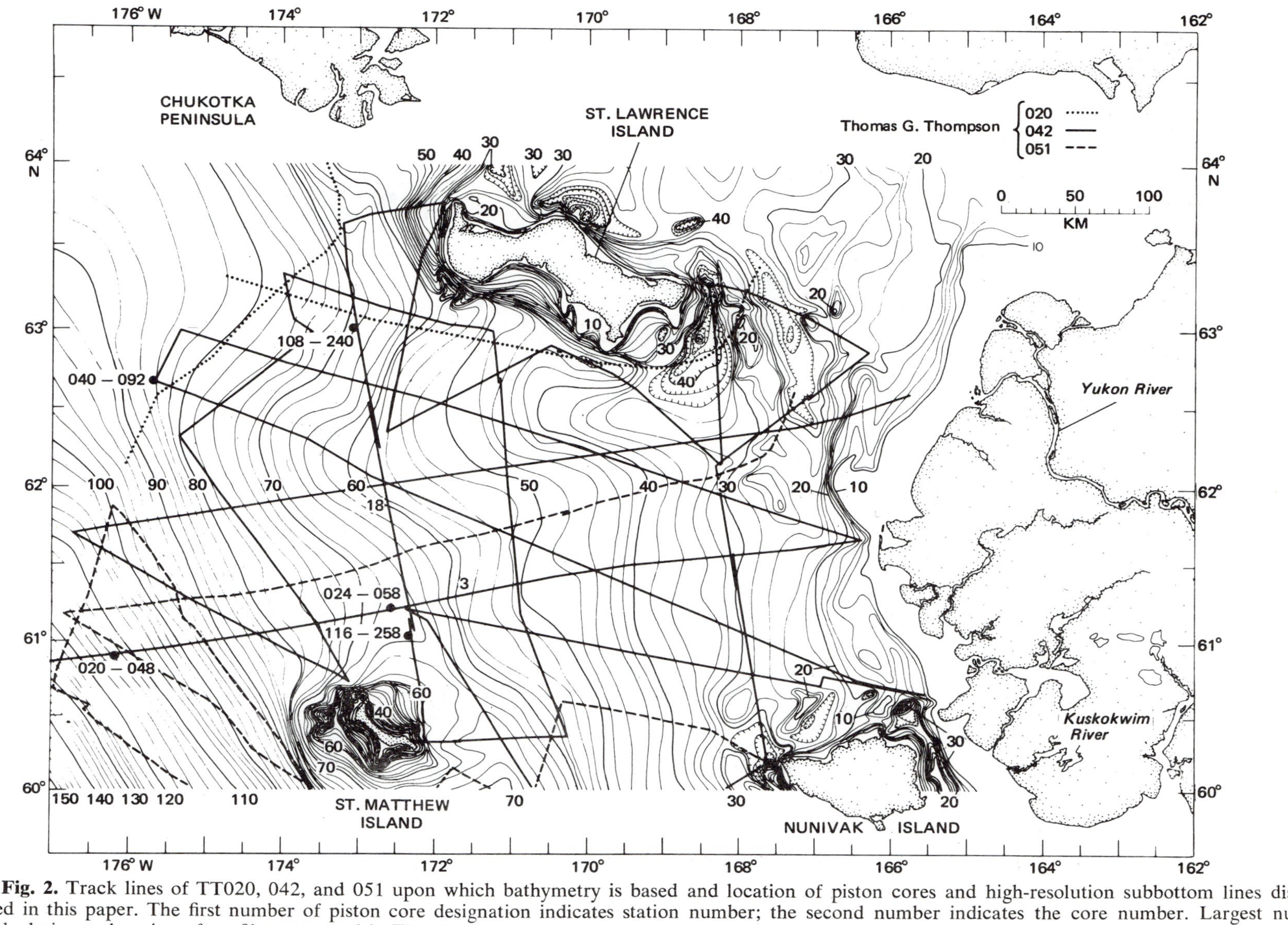

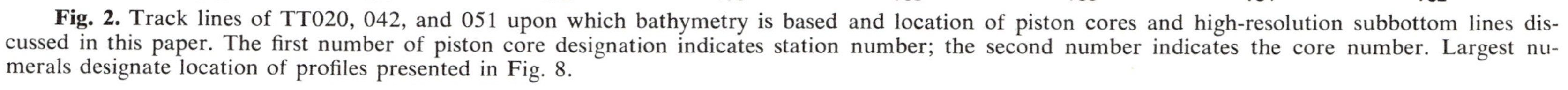
Fig. 2. Track lines of TT020, 042, and 051 upon which bathymetry is based and location of piston cores and high-resolution subbottom lines discussed in this paper. The first number of piston core designation indicates station number; the second number indicates the core number. Largest numerals designate location of profiles presented in Fig. 8.

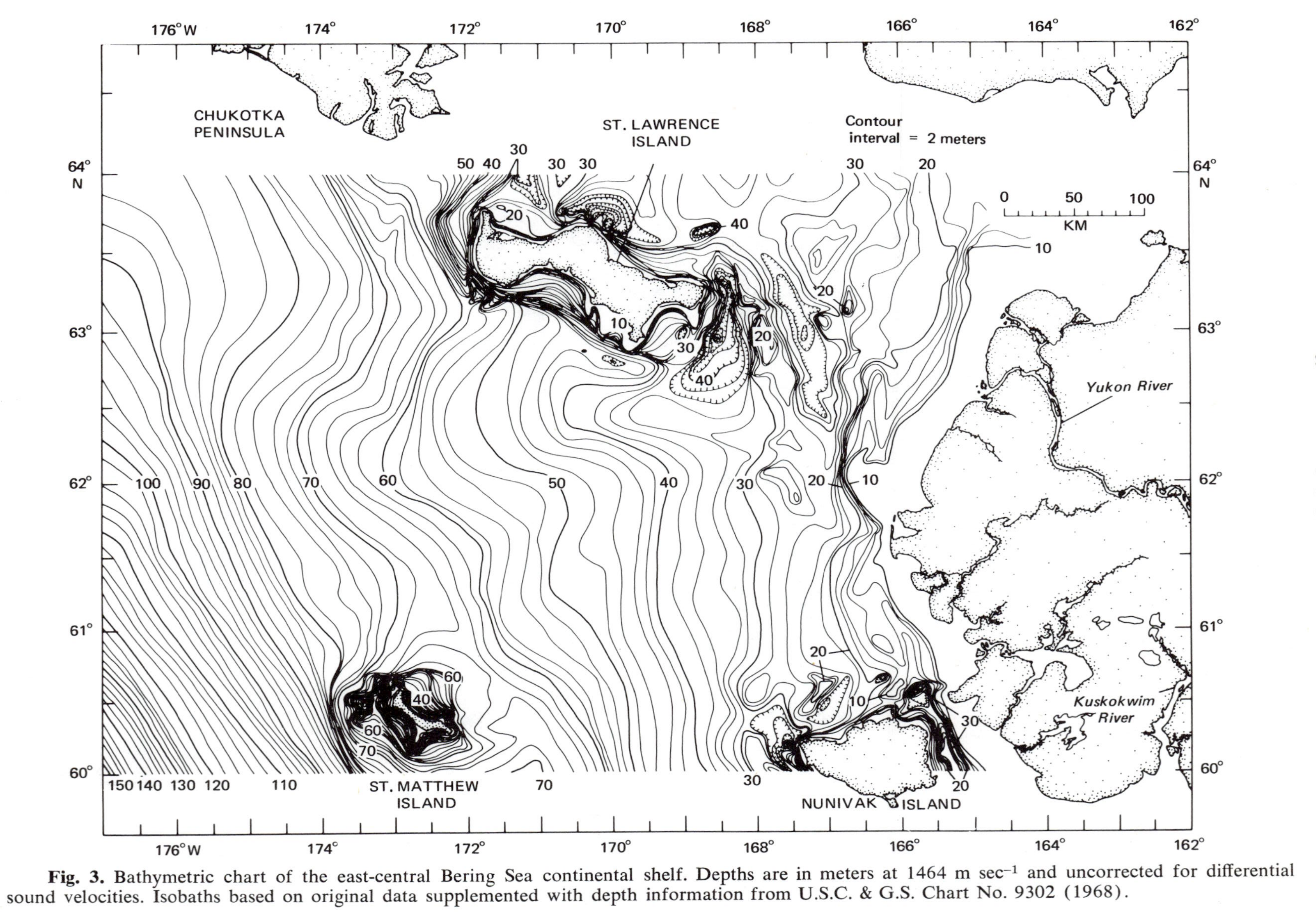

Fig. 3. Bathymetric chart of the east-central Bering Sea continental shelf. Depths are in meters at 1464 m sec^{-1} and uncorrected for differential sound velocities. Isobaths based on original data supplemented with depth information from U.S.C. & G.S. Chart No. 9302 (1968).

Granulometric, chronologic, and paleontologic information from five of the cores collected during this survey are presented in Figs. 5 through 7. The cores were split, logged visually, and photographed. Subsamples were taken at the top and bottom, at color (using the Rock Color Chart, Geological Society of America, 1963) and textural changes, and at regular intervals within homogeneous sections. Particle-size analyses were made with the standard sieve and pipette techniques; the appropriate three-end-member relationships of Shepard (1954) were determined subsequently with a computer program. The biogenic components that were retained on a 63-μ mesh sieve were examined microscopically after concentration by flotation on a bromoform-acetone solution of specific gravity 2.29. The abundance of fossils is expressed as the number of individuals in a gram of dry bulk sediment; the quantity of plant fragments is expressed as the volume (mm^3) of dry loose-packed material concentrated by bromoform-acetone flotation from a gram of dry bulk sediment. To determine the volume of plant fragments the total float volume is measured in a small-diameter calibrated glass tube, and then the percentage of the plant fragments in the float, spread upon a gridded tray, is estimated. All foraminiferal species discussed here are figured and referenced in Loeblich and Tappan (1953) and Cooper (1964).

Radiocarbon ages were determined by Teledyne Isotopes, Westwood, New Jersey, and were based on the carbonate-free carbon content of the sediments. In an attempt to correct the ages for contamination by inactive carbon, the "readily oxidizable organic material," as termed by Jackson (1958), was determined for the samples by using the Walkley-Black analysis and subtracted from the carbonate-corrected total carbon content. The remainder was attributed to inactive carbon. The corrections are those suggested by Broecker and Kulp (1956). Corrected ages appear without parentheses in the core diagrams (Figs. 5 through 7).

Bathymetry

The bathymetry for this area has been constructed from original data supplemented with depth information from U.S. Coast and Geodetic Survey Chart #9302 (1968) (Fig. 3). Bathymetric data beyond the northern and western limits of the area are from Grim and McManus (1970) and a revised chart of Kummer and Creager (1971).

This section of the Bering Sea continental shelf is very flat with gradients ranging from undetectable in some local areas to 8 m km^{-1} just off the southwestern tip of St. Lawrence Island. Other relatively pronounced sea-floor irregularities include several small knolls and depressions near Nunivak Island and a linear bathymetric indentation south of the bight of St. Lawrence Island. Two linear depressions with an intervening ridge occur between St. Lawrence Island and the Yukon delta (Fig. 3).

The most conspicuous bathymetric features of the open shelf are the sinuous bank (here called St. Lawrence Bank) that lies in the northern half of the area and the bathymetric discontinuity that occurs about 170 km northeast of St. Matthew Island (Fig. 3). St. Lawrence Bank begins at about 30 to 32 m depth and is still discernible at 86 to 90 m. Initially, it is well defined, but further offshore it becomes less distinct. The discontinuity in the south occurs at 48 to 50 m. Below this depth, the isobaths change orientation and are convex toward the southwest (Fig. 3).

Biogenic Components and Paleoenvironment

Criteria for paleoenvironmental interpretation of assemblages of fossil foraminifera in the cores have been developed from a study of assemblages in Bering Sea surface sediments by Anderson (1963) and unpublished data at the University of Washington. Of particular importance for paleoenvironmental interpretation is the bimodal depth distribution of *Eggerella advena* (Cushman) in surface sediments (Fig. 4). This species is dominant among arenaceous forms nearshore, but becomes rare on the central part of the shelf where *Reophax arctica* Brady is dominant. *E. advena* returns as a major species on the outer part of the shelf. From this distribution three zones are recognized. The inner shelf zone, extends outward to 40 to 50 m, where the frequencies of *E. advena* and *R. arctica* are equal. From there, the central shelf zone extends to about 100 m depth; across this part of the shelf *E. advena* becomes rare. Beyond 100 m in the outer shelf zone this species again occurs at frequencies consistently greater than 10%. The zonal limits proposed here apply only to the east-central Bering Sea shelf; *E. advena* is more abundant in the central shelf zone near Bristol Bay (Fig. 1).

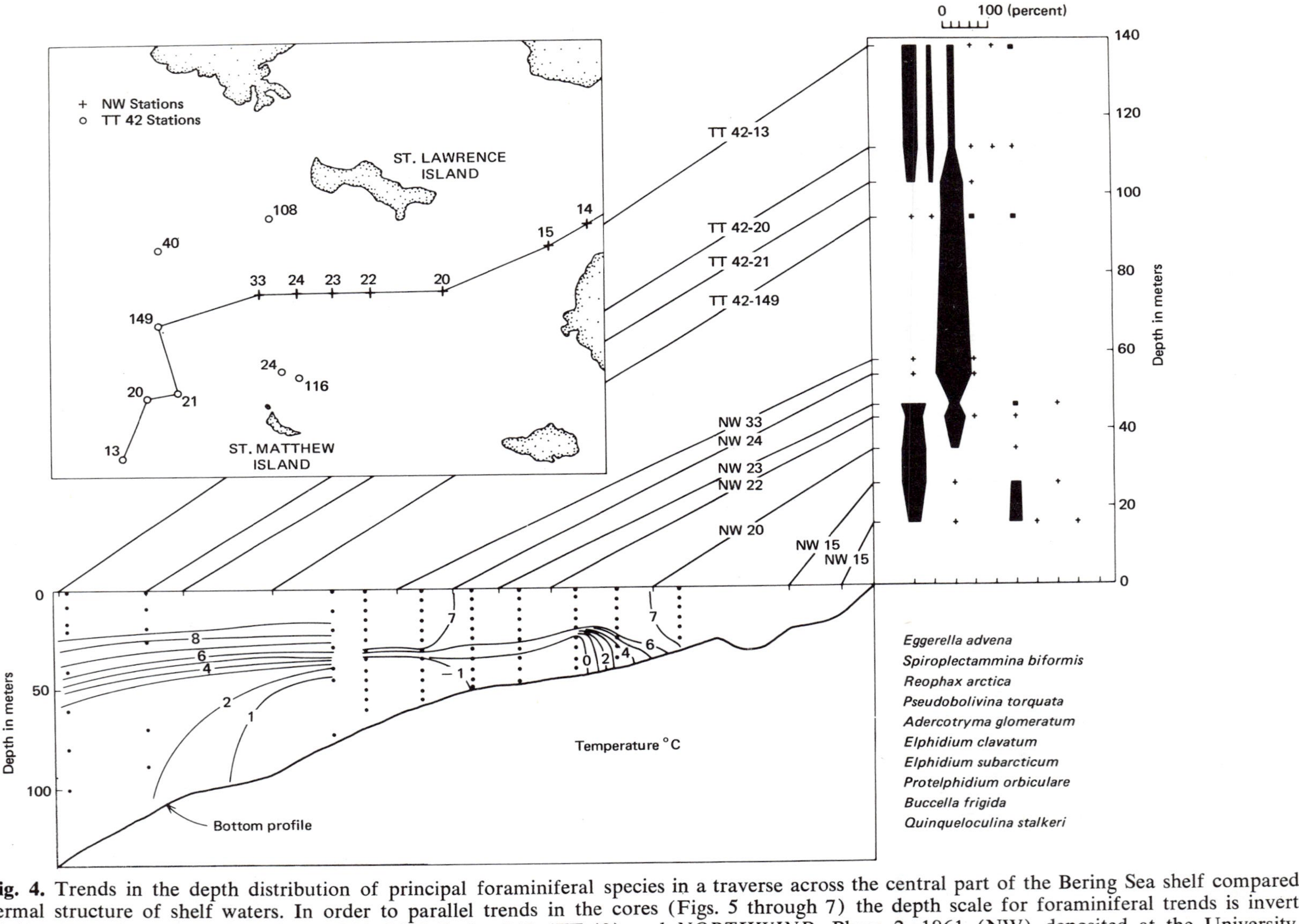

Fig. 4. Trends in the depth distribution of principal foraminiferal species in a traverse across the central part of the Bering Sea shelf compared to the thermal structure of shelf waters. In order to parallel trends in the cores (Figs. 5 through 7) the depth scale for foraminiferal trends is inverted. Foraminiferal samples are from *T. G. THOMPSON* Cruise 042 (TT-42) and *NORTHWIND*, Phase 2, 1961 (NW) deposited at the University of Washington. Only station numbers are shown for sample locations. Temperature data for the inner part of the profile is from *NORTHWIND* Cruise No. 31969 (Phase Bravo), stations 54, 55, 77, 78, 81, 82, and 89, September 1962 (U.S. Coast Guard, 1964). That for the outer part of the profile is from *OSHORO MARU* Cruise 24, stations 63, 67, and 69, August 1967 (Faculty of Fisheries, Hokkaido University, 1968). *NORTHWIND* stations have been transferred 15 minutes of latitude northward to the line of the profile of foraminiferal samples. *OSHORO MARU* stations have been transferred similar distances. Plus signs indicate 2 to 9% of foraminiferal assemblage.

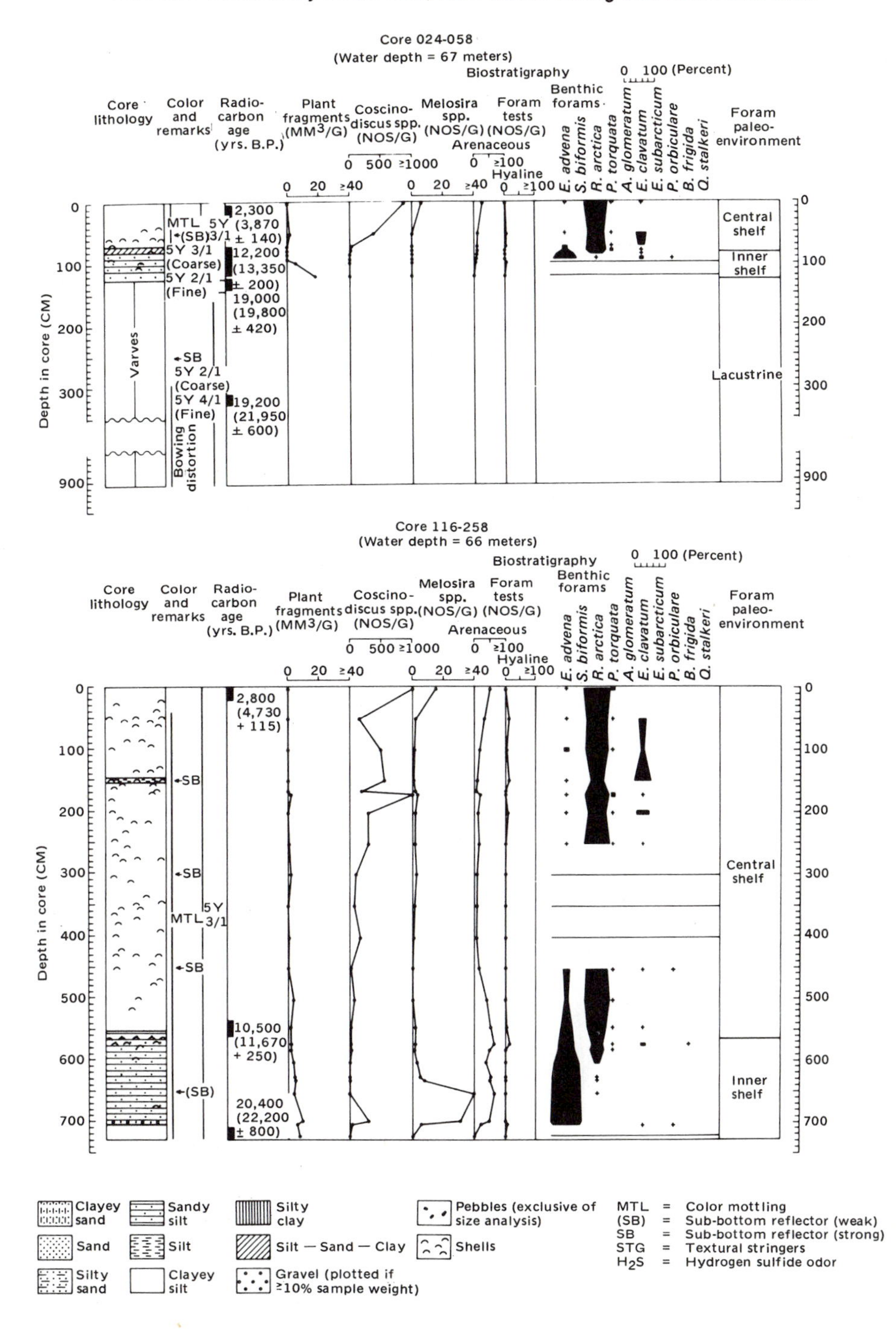

Fig. 5. Variation in sediment properties with depth in cores 024-058 and 116-258 from the east-central Bering Sea. Lithologies based on visual core logging and complete size analyses which determined appropriate three-end-member relationship of Shepard (1954). Color designations from the Rock Color Chart (Geological Society of America, 1963). Radiocarbon dates are based on carbonate-free carbon content of selected core intervals; ages corrected for inactive carbon appear without parentheses. Plus signs indicate 2 to 9% of foraminiferal assemblage; horizontal lines in the foraminifera column indicate a total count of less than 10 specimens per gram of bulk sample.

Assemblages of fossil foraminifera encountered downward in each core (Figs. 5 through 7) are generally similar to the sequence of assemblages in the surface sediments from the east-central Bering Sea shelf (Fig. 4). Inner and central shelf zone assemblages are present in all the cores.

Outer shelf zone assemblages occur only in core 020–048, because it is the only core from a water depth greater than 100 m (Fig. 6).

In cores 108–240 and 040–092 (Fig. 7) there are two kinds of inner shelf zone assemblages; one kind is dominated by *E. advena,* and the other by the calcareous species *Elphidium clavatum* (Cushman). In core 020–048 (Fig. 6) all inner shelf zone assemblages are dominated by *E. clavatum.* Assemblages in core 108–240 (Fig. 7) dominated by *E. clavatum* include significant frequencies (greater than 10%) of *Elphidium subarcticum* (Cushman), *Protelphidium orbiculare* (Brady), and *Buccella frigida* (Cushman). These assemblages may be fossil counterparts of a modern fluvial marine fauna that occurs to depths of about 20 m offshore from the Yukon River delta and the Kuskokwim River estuary (Anderson, 1963). The

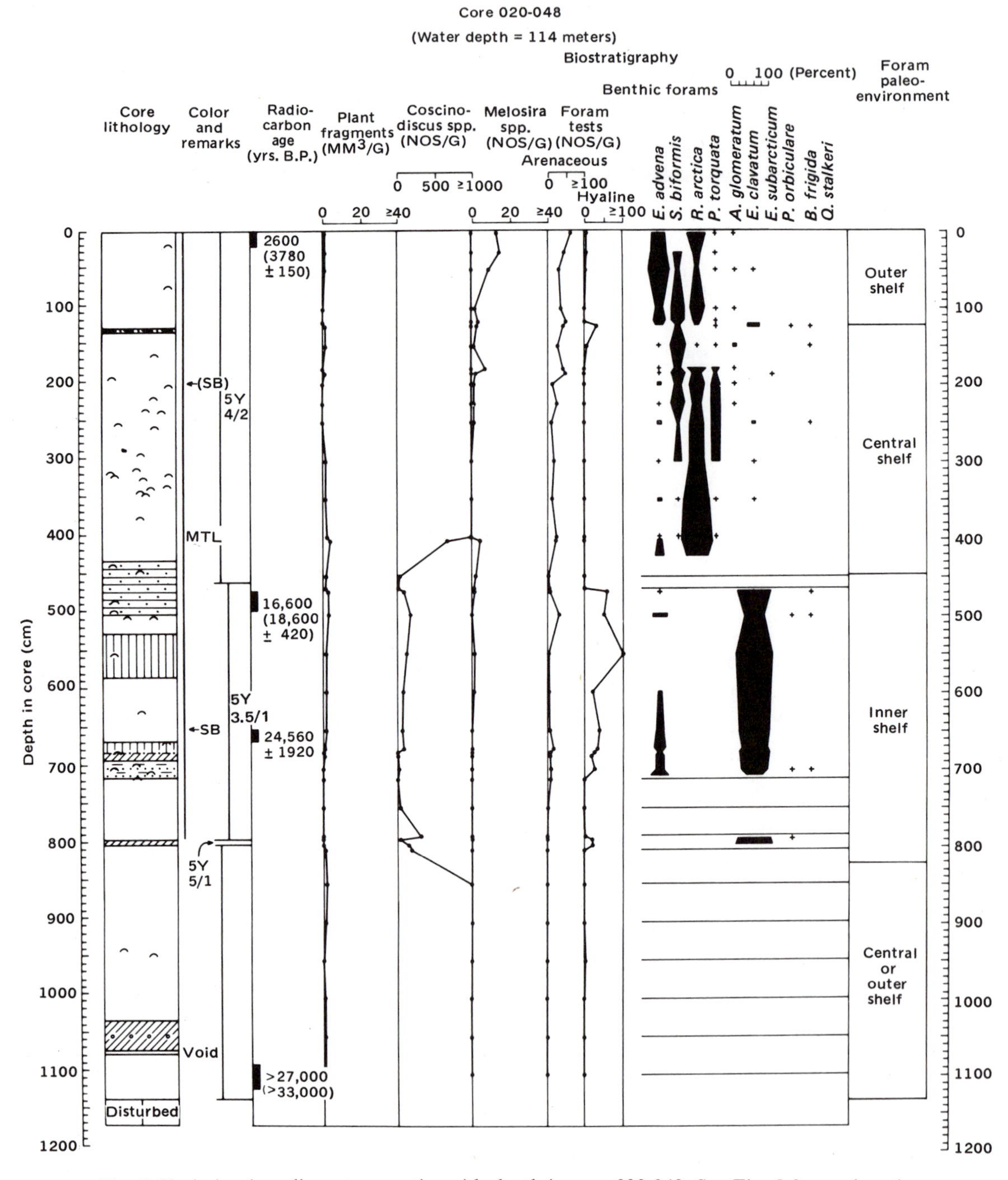

Fig. 6. Variation in sediment properties with depth in core 020-048. See Fig. 5 for explanation.

modern fluvial marine fauna, called the Deltaic Fauna by Anderson (1963) is dominated by *E. clavatum* and *E. subarcticum*. However, assemblages in cores 040–092 (Fig. 7) and 020–048 (Fig. 6) that are dominated by *E. clavatum* are different from the modern assemblages of the modern fluvial marine fauna. They differ because calcareous species other than *E. clavatum* (and *B. frigida* in some parts of core 040–092) do not occur at significant frequencies.

Because plant fragments are transported to the marine environment by rivers or streams, they may be indicative of fluvial marine environments. In core 108–240 (Fig. 7), however, plant fragments are most abundant in the section which contains *E. advena* assemblages rather than in that containing *E. clavatum* assemblages. This is probably because the *E. clavatum* assemblages occur in sand from which the plant fragments have been removed by winnowing. Thus, plant fragments may have accumulated with contemporary fine-grained sediment farther offshore. If this is true, then the abundance of plant fragments in the clayey silt of core 040–092 (Fig. 7) is not an indication that the *E. clavatum* assemblages in the same part of the core are counterparts of the fluvial marine fauna. Instead, they may indicate a paleoenvironment farther from the mouth of a river or stream and in water deeper than that in which the modern fluvial marine fauna occurs. Plant fragments are uncommon in core 020–048 (Fig. 6), and the sediments are too fine to have been winnowed. Thus, it is unlikely that the *E. clavatum* assemblages in this core are indicative of a fluvial marine environment. Cores 020–048 (Fig. 6) and 040–092 (Fig. 7) were collected at greater water depths than the other cores (114 and 84 m, respectively). Possibly, when inner shelf zone environments existed at these locations, *E. clavatum* dominated the foraminiferal assemblages. As sea level rose and inner shelf zone environments retreated farther from the shelf edge, *E. clavatum* was replaced by *E. advena* as the dominant inner shelf zone species.

Criteria for paleoenvironmental interpretation of assemblages in cores are also provided by the abundance of the planktonic diatom *Coscinodiscus* and the benthic diatom *Melosira. Coscinodiscus* valves are very abundant in most parts of cores having central or outer shelf zone foraminiferal assemblages, but much less abundant in parts of cores having inner shelf zone foraminiferal assemblages. Likewise, in modern inner shelf zone sediments all diatom valves are rare (less than 135 g^{-1}), including those of *Coscinodiscus* (Anderson, 1963). The scarcity of *Coscinodiscus* valves in sediments of the inner shelf zone may partially result from winnowing by strong bottom turbulence at shallow depths. However, this is an inadequate explanation where the sediment is silt-size or finer, such as in cores 040–092, 108–240, and 020–048 (Figs. 6 and 7). A more likely explanation is that *Coscinodiscus* is an oceanic form that is intolerant of coastal waters. According to Jousé (1960), *Coscinodiscus oculus-iridis* Ehrenberg and *Coscinodiscus marginatus* Ehrenberg[1] are important oceanic species that dominate the flora in bottom sediments of the deep basins of the Bering Sea. This oceanic complex of diatoms comprises only up to 40% of the total diatom assemblage in sediments of the Bering Sea shelf which are dominated by neritic diatoms. The important species of neritic diatoms are smaller than 63 μ and therefore are not present in the size fraction studied here.

Indigeneous populations of the benthic diatom *Melosira* are confined to shallow water areas, where the euphotic zone extends to the sea floor. Consequently, the abundance of fossil *Melosira* frustule chains that are associated with inner shelf zone assemblages in cores 116–258 (Fig. 5) and 108–240 (Fig. 7) may be further evidence that the sediments in which they occur were deposited in shallow water. The occurrence of *Melosira* frustule chains, however, must be interpreted with caution because they are also common in surface sediments of cores obtained at depths below the modern euphotic zone. The presence of *Melosira* in surface sediments from deeper parts of the shelf is probably due to their displacement from shallow water.

In four cores the section containing inner shelf zone microfossil assemblages either extends to the base or rests upon non-marine Wisconsin sediments. However, in core 020–048 (Fig. 6) this section overlies the sediments containing few fossils; these in turn overlie a bed containing microfossils indicating deposition on the central or outer part of the shelf during a pre-Late Wisconsin high stand of sea level. The latter contains abundant *Coscinodiscus* valves and rare individuals of calcareous foraminifera as well as

[1] Identification of diatom species during routine counting using a stereoscopic microscope is not possible. However, diatoms of a sample of surface sediment from the Bering Sea have been examined by transmitted light at high magnifications and *C. oculis-iridis* and *C. marginatus* were found to be the dominant large species.

the arenaceous species *Silicosigmoilina groenlandica* (Cushman). However, the arenaceous foraminifera that dominate modern assemblages of the outer and central shelf zones are completely absent. Probably central or outer shelf zone arenaceous foraminifera were originally present, but they have since been destroyed by a process that affected the organic cement of typical arenaceous foraminiferal tests, but not the siliceous (?) cement of *S. groenlandica* tests. Solution did not affect calcareous foraminifera tests or diatom valves. This destructive process and the environmental conditions necessary for it to act are not understood.

On the modern Bering Sea shelf the distribution of populations dominated by *R. arctica* accompanied by few *E. advena* is probably controlled by the perennial cold bottom water mass formed beneath the sea ice in winter (Fig. 4). Bottom water temperatures at any point may differ slightly from year to year (as evidenced by the discontinuity between isotherms of the two different data sources used to construct the inner and outer parts of the temperature profile of Fig. 4) and vary with time, generally becoming warmer as the summer progresses. Nevertheless, the general thermal structure is a recurrent feature of Bering Sea shelf hydrography (Kitano, 1970). In summer, warm surface water is mixed downward to the thermocline at depths of 30 to 35 m, primarily by wave turbulence. Nearshore, however, warm water extends to the bottom at depths greater than the offshore thermocline (Fig. 4), probably due to lateral mixing associated with north-flowing currents near the Alaskan mainland (Saur et al., 1954). Outer shelf bottom waters are warmed by subarctic water (Kitano, 1970).

The discussion above shows that the depth of the sea floor at which the transition from warm nearshore waters to cold central shelf waters occurs is a function of the intensity of summer heating, of wave turbulence, and of mixing associated with coastal currents. Marked long-term variations in any of these factors may have caused changes in the depth of the boundary between the inner and central shelf zones during the last transgression. Less is known about factors controlling the depth of the boundary across which the transition from cold central shelf water to warmer subarctic water occurs. However, if the depth of this transition changed during the last transgression, then the depth of the boundary between central and outer shelf zones also may have changed.

Sediments

Core 024–058 (at a water depth of 67 m, approximately 66 km northeast of St. Matthew Island) (Figs. 2 and 5), consists of a lower lacustrine sequence and an upper marine sequence. The lacustrine section, which begins at 124 cm and continues to the base at 905 cm, is composed of a series of distinct light and dark layers that are generally less than 1 cm thick; they show an alternate enrichment in sand and clay, contain no fossils, and have the appearance of varves. Below 293 cm the layers have been distorted by the action of the coring device. Two radiocarbon dates from this section (one at the top, the other just below the inception of bowing) differ by an amount of time consistent with an annual interpretation of the paired laminae. The lacustrine section is capped by a sandy-clayey silt sequence that contains a foraminiferal assemblage indicative of continuously deepening water during sediment accumulation. A radiocarbon date just above the lower boundary of this section indicates that the accumulation of marine sediment commenced prior to 12,200 years ago (Fig. 5). Since then, the average sedimentation intensity (Koczy, 1951) has been extremely slow, amounting to less than 9 mg $cm^{-2}yr^{-1}$; 8 cm 10^3yr^{-1})[1].

Core 116–258 was collected at a depth of 66 m northeast of St. Matthew Island and only 22 km from the site of 024–058 (Figs. 2 and 5). The upper 552 cm of the core is a shelly, clayey silt. Below, the sediment is sandier and fewer shells are present; gravel occurs near the base. Mottling is discernible throughout the core except for the top 40 cm (Fig. 5).

Foraminiferal and radiocarbon data indicate that the finer-grained section accumulated under central shelf conditions since 10,000 to 11,000 years ago. The average accumulation rate for this interval has been relatively high (60 mg $cm^{-2}yr^{-1}$; 70 cm 10^3yr^{-1}). In contrast, the sandy section contains an inner shelf fauna and accumulated sometime between 11,000 and 21,000 years B.P. (Fig. 5). The probability of a hiatus within this section precludes a definite conclusion regarding depositional rates during this time interval.

Three faunal assemblages can be recognized in core 020–048 which was obtained 165 km

[1] Rates of accumulation are based on an assumed sediment density of 2.70 g cm^{-3} (Hamilton, 1970) and an average water content for individual core intervals as determined by weight loss during desiccation.

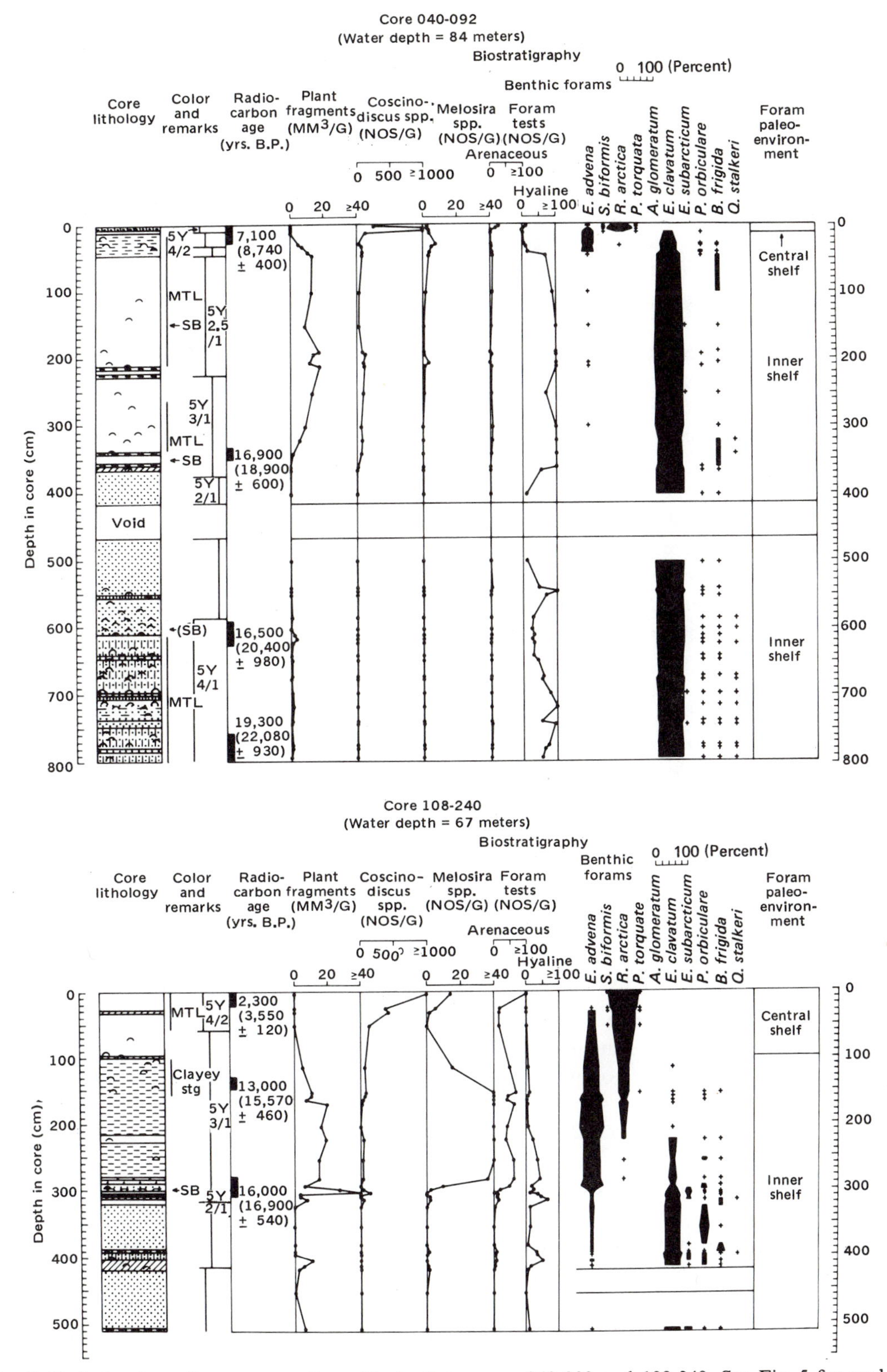

Fig. 7. Variation in sediment properties with depth in cores 040-092 and 108-240. See Fig. 5 for explanation.

northwest of St. Matthew Island at a depth of 114 m (Figs. 2 and 6). The upper 125 cm of the core is a homogeneous clayey silt that contains an outer shelf foraminiferal assemblage. From 125 to 450 cm the sediment remains dominantly a clayey silt, but a central shelf foraminiferal fauna is present. The interval 450 to 825 cm is texturally heterogeneous (sandy-clayey silts, silty clays, and sand-silt-clay) and paleontological data indicate deposition under inner shelf conditions. The lowermost section of the core consists of 3 m of clayey silt. The fauna in this interval, which is characterized by the paucity of foraminiferal tests and the abundance of *Coscinodiscus* frustules, suggests deposition at central or outer shelf depths. The core is disturbed below 1135 cm (Fig. 6).

Radiocarbon ages from core 020–048 indicate that the sediments within the uppermost intervals which contain central and outer shelf foraminifera probably began to accumulate shortly after 17,000 years B.P. An average rate of accumulation for this section based on dates at 10 and 483 cm is 30 mg $cm^{-2}yr^{-1}$ (34 cm 10^3yr^{-1}). The age of 1105 cm (>27,000 years B.P.) indicates that the sediments near the base of the core were laid down at a time greater than the range of radiocarbon dating. However, a date of 24,560 years B.P. has been obtained at 657 cm (Fig. 6). This date indicates that at least part of the sedimentary sequence which contains an inner shelf fauna accumulated prior to the last maximum lowering of sea level (generally postulated to have occurred about 20,000 years B.P.). In fact, the lack of a discernible nearshore lithofacies within this sequence suggests that a continuous late regression–Holocene stratigraphic section might be present. If this inference is correct, silty sediments accumulated at this location at an average rate of 24 mg $cm^{-2}yr^{-1}$ (22 cm 10^3yr^{-1}) during the time of the lowest position of sea level.

Core 040–092 was obtained near the northwestern margin of the survey area at a depth of 84 m (Figs. 2 and 7). One of the most remarkable features of this core is the short section with a distinct central shelf foraminiferal fauna. This section comprises only the upper 25 cm of the core. Below this unit the fauna indicate an inner shelf paleoenvironment. Above 366 cm the sediment size is dominantly clayey silt, but several silt layers are also present. The lower half of the core is principally sand and clayey sand. The sand unit was pulled apart while the core was being collected (Fig. 7).

The base of the core has been dated at 19,300 years B.P. From this time until about 16,500 years ago, 190 cm of clayey sand accumulated at this location. With no depositional hiatus, sediment could have accumulated during this time span at an average rate of 97 mg $cm^{-2}yr^{-1}$ (60 cm 10^3yr^{-1}). Thereafter, sediments continued to accumulate quite rapidly, as indicated by the similarity of radiocarbon ages at 341 and 607 cm (Fig. 7). Although it is possible that some of the sand within this interval might have been taken from a common level by the coring device, (lithologic, biostratigraphic, and chronologic differences between sediments) on either side of the section obviate extensive disarrangement of the core. The age of the top of the core is quite old relative to other surficial dates from this area. Even if one includes the possibility that the coring device failed to retrieve the upper few tens of centimeters of sediment (which is indicated in this case from auxiliary core data), undoubtedly there has been a substantial decrease in the rate of accumulation since the inner shelf sequence was deposited. Thus, the age discrepancy probably reflects the incorporation within the dated interval of older sediments which occur below the central-inner shelf faunal boundary.

Core 108–240 was collected 88 km southwest of St. Lawrence Island at a depth of 67 m (Figs. 2 and 7). This core is similar to core 040–092 in that it has a thin surface layer (75 cm) which accumulated under central shelf conditions. Below this section to 277 cm the sediments contain an inner shelf foraminiferal assemblage, are rich in plant fragments and *Melosira* frustules, and are dominantly silt-size. Penetration of the core was stopped in a well-sorted, very fine sand unit which contains a calcareous fauna dominated by *E. clavatum* suggestive of deposition in the inner part of the inner shelf zone (Fig. 7).

Radiocarbon ages show that deposition under inner shelf conditions commenced prior to 16,000 years B.P. and continued until sometime after 13,000 years B.P. (Fig. 7). An average rate of accumulation based on ages at 135 and 290 cm is 62 mg $cm^{-2}yr^{-1}$ (52 cm 10^3yr^{-1}). A change in the sedimentation intensity or a depositional hiatus (or both) is indicated subsequently by the 2300-year date at the top of the core within the central shelf sequence (Fig. 7).

Discussion

Evidence and implications concerning crustal stability in the Bering Sea are presented by Budanov et al. (1957), Hopkins (1959), and

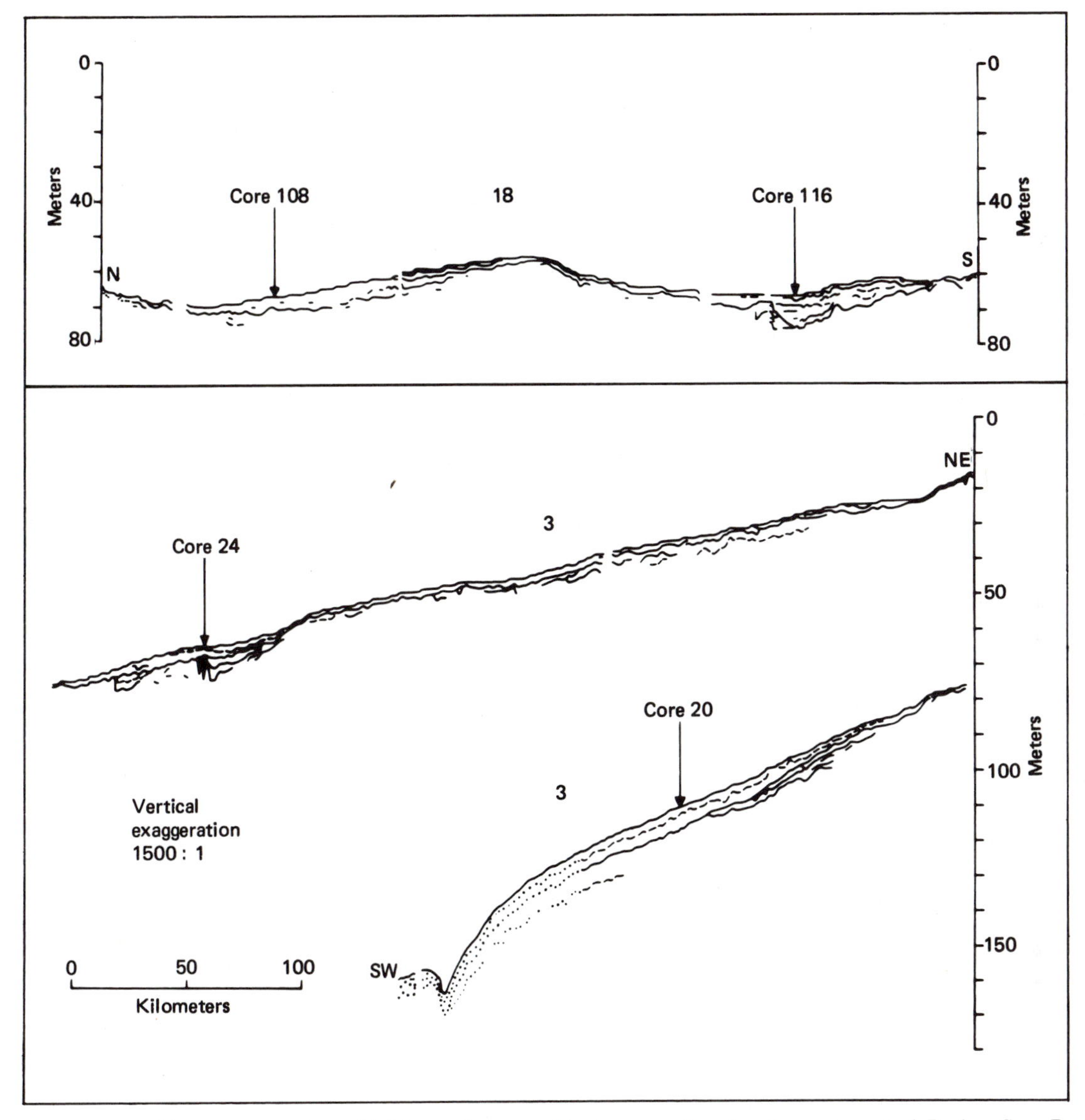

Fig. 8. Graphic representation of high-resolution subbottom profiles for the east-central Bering Sea. Profile locations are shown in Fig. 2. Data were obtained with a 12-kHz transceiver coupled with a PDR. Subbottom reflector depths are in meters and are uncorrected for deviations of sound velocity from 1464 m sec^{-1}. Dashed and dotted lines represent weak and discontinuous reflectors, respectively. Only the station numbers are presented for core locations.

Creager and McManus (1967). Although there is evidence of minor tectonic activity at present and of crustal instability during the early and middle Pleistocene in some areas of Siberia and Alaska, vertical movements during and since the Wisconsin have probably been small and localized.

Data from cores 040–092 and 108–240 suggest that prior to or during 16,000 to 17,000 years B.P., sandy inner shelf sediments were being laid down quite rapidly over a depth range of −70 to −90 m in the northern part of the area. If it is assumed that the well-sorted sand at the base of core 108–240 which contains an *E. clavatum*–dominated fauna was deposited at depths less than 20 m [where similar faunas occur today (Anderson, 1963)], sea level was probably somewhere between depths of −50 and −70 m at that time. After 16,000 years B.P. sea level was probably somewhat higher, but finer inner shelf sediments continued to accumulate at a fairly rapid rate (62 mg $cm^{-2}yr^{-1}$ for core 108–240) until at least 13,000 years B.P. Thereafter, the rate of accumulation decreased greatly, as indicated by the thin layers of modern sediments at the tops of these cores, and the anomalous age of the surficial sediments in core 040–092.

Inferences from cores 040–092 and 108–240 concerning sea-level positions are not contradicted by the characteristics and chronology of core 020–048. During the period 16,000 to 17,000 years B.P. inner shelf sediments were also being deposited at −119 m, but they were finer than contemporary sands at shallower depths to the north. Shortly thereafter, sediments containing a central shelf fauna began to accumulate. According to the scheme for paleoenvironmental interpretation, sea level was above −78 m when this sequence was initiated. If one applies the average rate of accumulation determined for the upper intervals of this core (30 mg $cm^{-2}yr^{-1}$) and assumes continuous sedimentation, then sea level stood at −14 m about 6000 years ago when sediments with an outer shelf foraminiferal assemblage began to accumulate at this site. That sea level did not fall below −120 m is supported by the fact that a basal transgressive deposit cannot be recognized texturally and that a continuous late-regressive Holocene stratigraphic sequence might be present. Furthermore, if no hiatus is present in the lowermost section of core 040–092, then sea level stood above −91 to −92 m at 19,300 years B.P. (Fig. 7).

Cores 024–058 and 116–258 show great disparity with regard to rates and times of sediment accumulation. Core 024–058 indicates that before 19,000 years B.P., sea level was below −68 m, and varved deposits accumulated northeast of St. Matthew Island. Although the shore line had passed over this location before 16,000 years B.P. (as indicated by core 108–240), only a thin inner shelf silt section had been preserved by about 12,000 years B.P. Less than 1 m of finer sediment with a modern fossil assemblage accumulated subsequently. In contrast, a very extensive section of marine sediments accumulated at the site of core 116–258 after about 11,000 years B.P. Sea level must have been above −32 m when these sediments were deposited because they contain fossils characteristic of central shelf depths.

In addition to sea level changes, the Holocene stratigraphic sequence is also a function of the temporal and spatial variability of sedimentary sources and of the oceanographic conditions responsible for the accumulation of sediment at a particular site. Core 108–240 was taken from a thickened lens of acoustically transparent silt north of St. Lawrence Bank (Figs. 2, 7, and 8). This lens of sediments decreases in thickness and appears to be draped over the northern flank of the bank, thus creating a more extensive bathymetric feature (Figs. 2, 3, and 8). The same relationship is observed along other lines normal to the crest of the bank. Although a comprehensive discussion of the genesis of St. Lawrence Bank is beyond the scope of this paper, deep penetration seismic records indicate that its western part is related to the deformation of the acoustic basement (Kummer and Creager, 1971; M. L. Holmes, personal communication), and, therefore, it may have been in existence prior to the last regression of sea level. Although the processes responsible for the accumulation of this lens of silt are uncertain at the present time, the stratigraphic relationship and the biogenic characteristics in core 108–240 suggest quiet water deposition on the north side of the bank. The thin surficial layers of modern sediment in cores 040–092 and 108–240 may reflect either a change in sediment source or the oceanographic regime (or both) after 13,000 years B.P. It is interesting to note that these changes took place at about the same time that the Bering Land Bridge was severed during the Holocene transgression (Creager and McManus, 1967).

The lower boundary of the acoustically transparent section coincides with the top of the inner shelf sand unit in core 108–240 (Figs. 7 and 8). Buried channels are observed below this reflector toward the north and east and suggest that the depression north of St. Lawrence Bank was a locus of deposition prior to 16,000 years B.P. The channels to the east and the proximity of St. Lawrence Island indicate that St. Lawrence Island supplied at least part of the sediments during this time.

The site of core 020–048 lies within the silt-rich zone of the central part of the Gulf of Anadyr, and the sediments reflect the combined influence of terrigenous and biogenic components (Lisitsyn, 1966, p. 301) (Figs. 2, 6, and 8). The average content of amorphous silica due to diatom frustules in the sediments of this area ranges between 16 and 25% (dry weight) (Lisitsyn, 1966, p. 451) and the concentrations of *Coscinodiscus* are greater than 1000 specimens per gram of sediment in the central and outer shelf intervals of core 020–048 (Fig. 6). Kummer and Creager (1971) attribute at least part of the Holocene sediments that mantle the central Anadyr shelf to fluvial detritus supplied during a stillstand of sea level at approximately −75 to −80 m. Modern silty sediments in the area are largely derived from the coasts of the Chukchi Peninsula and the Anadyr River (Lisitsyn, 1966, p. 430).

The shallow sub-bottom reflectors observed north and east of St. Matthew Island are presented in Fig. 8. These profiles show a "pock-marked" subsurface configuration which has been smoothed subsequently by sedimentary fill. Cores 024–058 and 116–258 indicate, however, that the infilling of these depressions has not been coeval (Figs. 5 and 8). The sub-bottom valley from which core 024–058 was collected, along with a contiguous depression to the east, were filled at least in part by varved lacustrine deposits which accumulated near the beginning of Holocene time when the area was subaerially exposed. Only a small marine sequence was deposited atop the varves subsequent to the inundation by the sea (Figs. 5 and 8). Conversely, the trough from which core 116–258 was collected received a large amount of finer sediment and was filled after the shore line had passed over the area during the last transgression. The attitude of the reflectors to the south of the trough and the surficial gradient suggest that most of the material has been derived from St. Matthew Island (Figs. 2, 5, and 8). Biogenic sediments are probably also important as a source. A suitable oceanographic environment exists for the accumulation of clayey silts, for the area is in the wave shadow of St. Matthew Island and a small current gyre exists over this location (Goodman et al., 1942).

Summary and Conclusions

1. A scheme for the paleoenvironmental interpretation of the foraminiferal fauna permits the identification of inner, central, and outer shelf sections within long cores of Holocene sediments collected from the east-central Bering Sea continental shelf.

2. Rates of accumulation of Holocene sediments based on corrected radiocarbon ages vary considerably over this section of the shelf, with sedimentation intensities ranging from 97 mg $cm^{-2}yr^{-1}$ (60 cm 10^3yr^{-1}) for core 040–092 collected near the northwestern corner of the survey area to less than 9 mg $cm^{-2}yr^{-1}$ (8 cm 10^3yr^{-1}) just northeast of St. Matthew Island.

3. The spatial and temporal variability within Holocene sediments as determined by the granulometric, paleontologic, and chronologic information in cores shows that sea level probably did not fall below −120 m, as indicated by the absence of a distinct nearshore lithofacies within the possible late regression–Holocene stratigraphic section of core 020–048; that varved lacustrine sediments were emplaced at one site northeast of St. Matthew Island at about the beginning of Holocene time; that the accumulation of inner shelf sediments was quite rapid within the northern sector during the initial stage of the transgression, but decreased markedly after 13,000 years B.P.; and that an extensive section of central shelf sediments accumulated after 11,000 years B.P. at one location in the southern sector, but only a surficial veneer of correlative sediments exist in the north.

4. High-resolution sub-bottom data indicate that the influence of fluvial and oceanographic processes have changed during the transgression of sea level across this part of the shelf and that older bathymetric features have probably contributed to the Holocene sediment variability.

Acknowledgments

The authors thank R. W. Roberts and S. S. Barnes for their assistance in the laboratory analyses, and D. R. Morrison for his assistance in obtaining the cores. The research was supported by Grants GA-11126 and GA-28002 from the National Science Foundation.

References

Anderson, G. J. 1963. Distribution patterns of Recent foraminifera of the Bering Sea. *Micropaleontology* 9(3):305–317.

Broecker, W. S., and J. L. Kulp. 1956. The radiocarbon method of age determination. *Amer. Antiquity* 22(1):1–11.

Budanov, V. I., A. T. Vladimirov, A. S. Ionin, P. A. Kaplin, and V. S. Medvedev. 1957. Modern vertical movements of the sea coast in the Far East (Consultants Bureau running translation). *Akademiya Nauk SSSR, Doklady Geologicheskogo* 116:829–832.

Cooper, S. C. 1964. Benthonic foraminifera of the Chukchi Sea. *Contrib. Cushman Found. Foram. Res.* 15:79–104.

Creager, J. S., and D. A. McManus. 1967. Geology of the floor of the Bering and Chukchi Seas—American studies. In: D. M. Hopkins, ed. *The Bering*

Land Bridge. Stanford, California: Stanford University Press. pp. 7–31.

Faculty of Fisheries. 1968. *Data Record of Oceanographic Observations and Exploratory Fishing No. 12*. Hakodate, Hokkaido, Japan: Hokkaido University. 421 pp.

Geological Society of America. 1963. *Rock color chart*. New York. 6 pp.

Goodman, J. R., J. H. Lincoln, T. G. Thompson, and F. A. Zeusler. 1942. Physical and chemical investigations: Bering Sea, Bering Strait, Chukchi Sea during the summers of 1937 and 1938. *Univ. Washington Publ. Oceanography* 3(4):105–169.

Grim, M. S., and D. A. McManus. 1970. A shallow seismic profiling survey of the northern Bering Sea. *Marine Geol.* 8:293–320.

Hamilton, E. L. 1970. Sound velocity and related properties of marine sediments, North Pacific. *J. Geophys. Res.* 75(23):4423–4446.

Holmes, M. L. 1967. Late Pleistocene and Holocene history of the Laptev Sea. Masters thesis, University of Washington, Seattle. 94 pp.

Hopkins, D. M. 1959. Cenozoic history of the Bering land bridge. *Science* 129(3362):1519–1528.

Jackson, M. L. 1958. *Soil Chemical Analysis*. Englewood Cliffs, N.J.: Prentice-Hall. 298 pp.

Jousé, A. P. 1960. Diatomovye v poverkhnostnom sloe osadkov Beringova morya (Diatoms in the upper layer of Bering Sea sediments). *Akademiya Nauk SSSR, Institut Okeanograficheskaya Trudy* 32:171–205.

Kitano, K. 1970. A note on the thermal structure of the Eastern Bering Sea. *J. Geophys. Res., Oceans Atmospheres* 75(6):1110–1115.

Koczy, F. F. 1951. Factors determining the element concentration in sediments. *Geochim Cosmochim Acta* 1:73–85.

Kummer, J. T., and J. S. Creager. 1971. Marine geology and Cenozoic history of the Gulf of Anadyr. *Marine Geol.* 10:257–280.

Lisitsyn, A. P. 1966. Recent sedimentation in the Bering Sea. *Israel Program Sci. Transl., Jerusalem*. Washington, D.C.: U.S. Department of Commerce, National Science Foundation. 614 pp.

Loeblich, A. R., Jr., and H. Tappan. 1953. Studies of Arctic foraminifera. Washington, D.C.: *Smithsonian Misc. Collections* 121(7):1–150.

McManus, D. A., J. C. Kelley, and J. S. Creager. 1969. Continental shelf sedimentation in an arctic environment. *Bull. Geol. Soc. Am.* 80:1961–1984.

———, K. Venkatarathnam, R. J. Echols, and M. L. Holmes, 1970. Bottom samples and seismic profiles of the northern Bering Sea, and associated studies. Menlo Park: U.S. Geological Survey. Unpublished Report. 181 pp.

Naugler, F. P. 1967. Recent sediments of the east Siberian Sea. Masters thesis, University of Washington, Seattle. 71 pp.

Saur, J. F. T., J. P. Tully, and E. C. Lafond. 1954. Oceanographic cruise to the Bering and Chukchi seas, summer 1949. Part 4: Physical oceanographic studies. *U.S. Navy Electron. Lab. Rep.* 416, 1:1–31.

Shepard, F. P. 1954. Nomenclature based on sand-silt-clay ratios. *J. Sedimentary Petrol.* 24(3):151–158.

U.S. Coast and Geodetic Survey. 1968. Washington, D.C.: Bathymetric chart 9302.

U.S. Coast Guard. 1964. Oceanography cruise, U.S.C.G.C. *Northwind*, Bering and Chukchi seas, July–September 1962. *U.S.C.G. Oceanographic Rep. 1*. 104 pp.

Chapter 7

Sedimentation in the Beaufort Sea: A Synthesis[1]

ANGI SATYANARAYAN NAIDU[2]

Abstract

Results of past and continuing studies on grain-size distributions, clay mineralogy, suspended loads, and geochemistry of bottom sediments and interstitial waters of the Beaufort Sea are presented.

Generally, bottom sediments are extremely poorly sorted muds or sandy muds. Gravels are frequently encountered on the shelf, sporadically on the upper slope, and never on the lower slope and abyssal basin. Invariably size distribution curves of shelf sediments are positive-skewed to nearly symmetrical and platykurtic, whereas those of the slope and basin sediments are mesokurtic and nearly symmetrical. These textural differences are the result of the relatively smaller amounts of mud depositing on the shelf. It is difficult to conjecture on the origin of shelf gravels; they could be all contemporaneously ice-rafted, or partly contemporaneously ice-rafted and partly relict sediments. Much of the bottom mud is probably transported by ice rafting from nearshore and deposited by ice melting.

Preliminary data suggest that the suspended sediment distribution is controlled by river discharge and that in the nearshore, suspensates are transported to the east. Differences between suspensate weights in the Chukchi Sea, Beaufort Sea, and Canada Basin are attributed to regional differences in planktonic productivity and supply of terrigenous suspensates.

In all sediments illite is the predominant clay mineral, chlorite and kaolinite occur in significant amounts, and smectite is the least abundant. It is suggested that the use of clay minerals in inferring paleoclimates should be done with great caution, since the generally accepted view is that kaolinite is a low-latitude mineral.

Beaufort Sea sediments are chemically different from lower-latitude marine sediments in being relatively deficient in Fe, Mn, K, Na, Ca, Mg, Ni, Zn, Co, and Cu, and having lower Na/K, Ca/Mg, and Fe/Mn ratios. The cause of this elemental deficiency is not known. The peculiar chemical character of Beaufort Sea sediments suggests that high-latitude marine paleosediments can be identified using geochemical criteria. The bulk of the alkali and heavy transition metals is presumably fixed in the argillaceous fraction. Except for cobalt, elemental scavenging by the ferrimanganic phase of the sediment appears to be of secondary importance.

Based on sediment interstitial water data and the presence of thick contemporary ferromanganic encrustations on gravels it is concluded that post-depositional dissolution, migration, and oxidative reprecipitation of Mn and Fe is taking place at the present.

Introduction

Knowledge of processes and products of contemporary marine sedimentation is vital to geologists for several reasons, such as for inferring

[1] Contribution no. 154 of the Institute of Marine Science, University of Alaska.

[2] Institute of Marine Science, University of Alaska, Fairbanks, Alaska 99701, U.S.A.

paleogeography and the genesis of various metal and nonmetal sedimentary deposits. A review of the literature shows that within the past 30 years considerable progress has been made in studies of marine sediments from several oceanographic provinces of the lower and middle latitudes. However, our knowledge of polar sea sediments, particularly those of the western Arctic Ocean, is limited. Studies conducted on the Arctic shelf sediments of the Chukchi Sea (Creager and McManus, 1967; McManus et al., 1969) make it one of the best known of shelf areas. A few investigations have been undertaken on deep-sea Canada Basin sediments—especially by the Lamont group, led by Hunkins, as well as by Herman and Clark from Washington State and Wisconsin Universities, respectively, but relatively little attention has been focused on sediments of the adjoining Beaufort Sea. Studies conducted by Carsola (1954a), are the earliest known extensive geological investigations of the Beaufort Sea concerned with sediment-size distribution and organic matter content. Additional data on these aspects were presented by Carsola et al. (1961), Hoskin et al. (1969), and Naidu et al. (1973, in press). Research was only recently initiated on the clay mineralogy and chemistry of the Beaufort Sea sediments by Naidu et al. (1971) and Naidu and Hood (1972). This paper summarizes the information available from the literature and from research projects, in progress at the Institute of Marine Science, University of Alaska, concerning Beaufort Sea sediments.

Geographic and Geologic Settings

Some confusion exists on the nomenclature of the Beaufort Sea (Beal et al., 1966). In this paper Beaufort Sea connotes the marginal sea constituting the southern extent of the Amerasian Basin of the western Arctic Ocean and arbitrarily includes the geographic unit that is limited in the south by the North Slope of Alaska and the Mackenzie Delta, in the north by the Canada Basin, in the east by the Canadian Archipelago, and in the west by the northeast Chukchi Sea (Fig. 1).

The bathymetry of the Beaufort Sea is now fairly well known from continuous echo soundings (Carsola, 1954b; and Carsola et al., 1961). These authors observed that the shelf edge is at the relatively shallow depth of about 64 m (35 fath) and that the shelf width is approximately 72 km (46 mi). However, the shelf and slope east of Herschel Island are relatively wider and less steep, respectively, than on the west of the island, presumably because of the prograding Mackenzie River Delta. With a few exceptions, such as the presence of submarine canyons off Point Barrow and a sea valley off the Mackenzie River Delta, the slope and deep basin of the Beaufort Sea are featureless (Carsola, 1954b; Carsola et al., 1961). On the inner shelf there are several barriers and elongated bars with orientations roughly parallel to the coast line. Some of these bars are probably submarine ice-pushed ridges. Recently Shearer et al. (1971) reported the presence of a number of submarine pingos in the inner shelf of the eastern Beaufort Sea, and concluded that they had formed in a marine environment.

The bulk of the terrigenous detritus supplied to the western and central Beaufort Sea is transported during spring by numerous Arctic rivers which rise from the Brooks Range and flow across the North Slope. The Brooks Range, a geanticline, is composed chiefly of Paleozoic and Mesozoic sedimentary and metasedimentary rocks. The North Slope of Alaska including the Foothill and Coastal Plain provinces, is underlain by a Mesozoic sedimentary trough (Committee on Polar Research, 1970). The Mackenzie River, which has an extensive drainage basin in central and northwestern Canada, transports enormous volumes of sediments (an average high of 538,000 c.f.s. in July, to an average low of 116,000 c.f.s. in April; Levinson et al., 1969) into the eastern Beaufort Sea. Soils of the Brooks Range and Foothill Provinces are chiefly glacially derived under low dynamic chemical weathering, whereas soils of the Coastal Plain are predominantly Tundra gleys.

For 9 to 11 months of the year extensive areas of the Beaufort Sea are covered either with heavy Arctic pack ice or broken sea ice. This ice cover limits the formation of waves because of the decrease in effective fetch. Storms are common over the Beaufort Sea, and they bring cataclysmic results to sedimentation in the nearshore and coast (Hume and Schalk, 1967). The mean annual air temperature over the Beaufort Sea is around 4°C (39°F), and the mean annual precipitation of 11.6 cm (4.5 in.) is comparable to that of arid or semi-arid regions (United States Weather Bureau, 1964).

Except for some preliminary results (Burrell et al., 1970), no extensive data are available on the physical and chemical oceanography of the Beau-

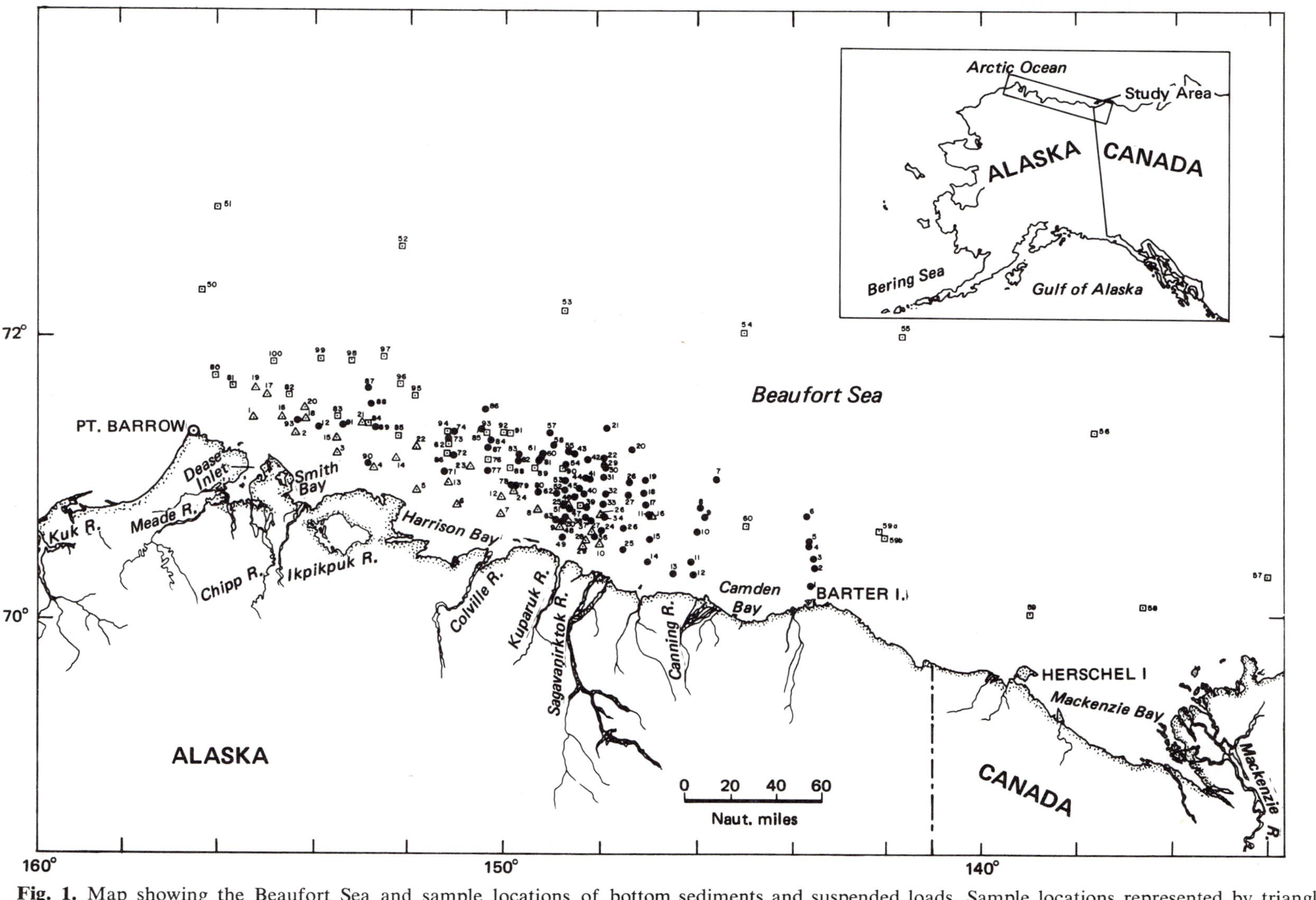

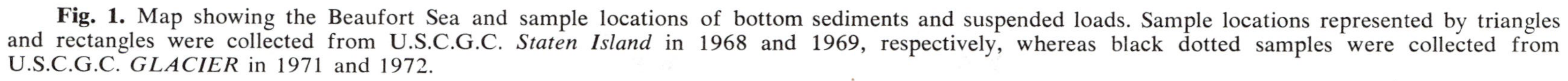

Fig. 1. Map showing the Beaufort Sea and sample locations of bottom sediments and suspended loads. Sample locations represented by triangles and rectangles were collected from U.S.C.G.C. *Staten Island* in 1968 and 1969, respectively, whereas black dotted samples were collected from U.S.C.G.C. *GLACIER* in 1971 and 1972.

fort Sea. During August and September 1971 a multidisciplinary cruise[1] was conducted by the U.S. Coast Guard to collect baseline ecological data from the central and western Beaufort Sea. The analyses of oceanographic data as well as the biological and geological samples collected during this cruise are underway.

Field and Laboratory Methods

Results presented in this paper are largely based on analyses of 62 Van Veen/Shipek grab or short gravity core samples collected from the shelf, slope, and abyssal basin of the Beaufort Sea on U.S. Coast Guard ice-breaker *Staten Island* during 1968 and 1969 (Fig. 1). Results of partial analyses of an additional suite of 84 Van Veen/Smith-Mcintyre grabs and 15 gravity cores, which were collected from the U.S.C.G.C. *Glacier* during the summer of 1971 (Fig. 1), are also included.

Grain-size analyses of sediments were conducted by the standard combined sieving-pipetting method (Krumbein and Pettijohn, 1938), and the conventional size parameters were calculated by using the formulae of Folk and Ward (1957).

Procedures for X-ray analyses of clay minerals in the >2-μ fraction and chemical analyses of the sediment fraction less than pebble size (>4 mm) have been described in earlier papers by Naidu et al. (1971) and Naidu and Hood (1972). To investigate elemental speciation in the iron and manganese oxide phases of sediments, ferrimanganic encrustations from 10 gravel surfaces were carefully scraped with an agate-tipped spatula. Fine powders of these encrustations were chemically analyzed by a direct reading emission spectroscopic method (Christensen et al., 1968). Randomly oriented mounts of these powders were subjected to X-ray analysis for mineral identification.

In order to determine weights of suspensates in the Beaufort Sea, water samples were collected in a Niskin bottle at several stations at various depths. One liter of these water samples was filtered through preweighed dry millipore filter papers with apertures of 0.45 μ.

The gravity core sediments were extruded on shipboard immediately after retrieval. The 2-mm surficial portions were immediately scraped and the temperature and pH of 10-cm sediment sections from the core top were measured. With the help of a squeezer (Reeburgh, 1967) interstitial water samples from the 10-cm core sections were pressed out. Iron and Mn contents from these samples were measured by atomic absorption spectrometry, and SiO_3 and PO_4 were determined colorimetrically in an autoanalyzer on board ship.

[1] U.S.C.G.C. *Glacier* WEBSEC-71 cruise.

Results and Discussion

Sediment Texture

Generally the surficial sediments of the Beaufort Sea are extremely poorly sorted silty clays or sandy muds. Gravels are frequently encountered on the shelf ($<$64m), sporadically on the upper slope (64 to 500 m), and not at all on the lower slope (500 to 1825 m) and abyssal basin[2] ($>$1825 m). A spatial variation in the size distribution of sediments is discernible from the shelf to the extrashelf region[3]. Invariably, the shelf is blanketed by positive to very positively skewed and platykurtic sediments and only occasionally by nearly symmetrically size distributed sediments. However, the extrashelf sediments commonly have mesokurtic and nearly symmetrical, to slightly positively skewed size distributions (Naidu et al., 1973, in press).

Presumably the slight regional differences observed in sediment skewnesses are a function of sediment mean size, as attested by the M_Z vs. SK_I scatterplots in Fig. 2. The sediment mean size in turn is determined by the amount of gravel, sand, silt, and clay in the entire sediment. On the shelf association of well-sorted coarse material, such as gravels and/or very coarse sands, with relatively poorly sorted muds (silty-clays or clayey-silts) results in a poorly sorted polymodal sediment. Such an association of different size modes also leads to a pronounced tailing in the fine end of the size distribution curves (positive skewness) on shelf sediments. On the other hand, consequent on the near or complete absence of the well-sorted gravel mode, considerable decrease in the sand mode and presence in almost equal amounts of the silt and clay mode, the extrashelf muds have size distribution curves that have either a slight tailing in the coarse end (slightly positively skewed) or no tailing at any end (nearly symmetrical). Both the shelf and extrashelf sediments are

[2] Includes abyssal plain and continental rise regions.
[3] Extrashelf includes the slope and abyssal basin regions.

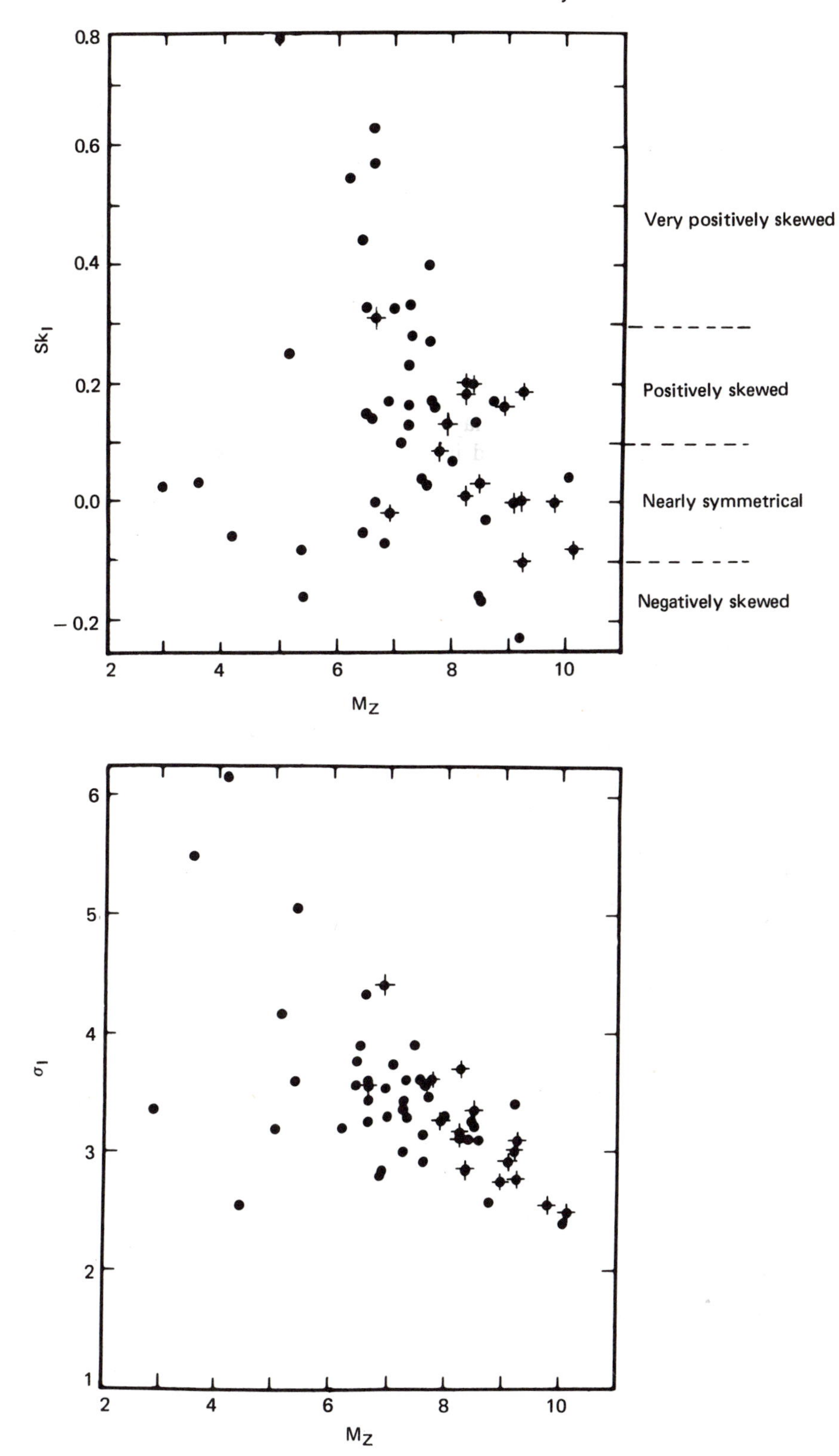

Fig. 2. Relationships of phi Mean size (M_Z) with sorting (σ_I) and skewness (Sk_I) of sediments. Closed dots represent shelf sediments and stars represent extrashelf sediments.

poorly sorted because of the presence of more than one size mode.

There are minor differences in the sediment skewness data presented by Carsola 1954a) and Naidu et al. (1973, in press). Carsola (1954a, p. 1566) observed that there is a progressive change in the skewness from negative to positive, from the shelf to the slope region. The data of Naidu et al. (1973, in press) do not concur with the above, inasmuch as they have commonly encountered positively skewed to nearly symmetrical shelf sediments and nearly symmetrically size-distributed slope and abyssal basin sediments. The conflicting results may be attributed to the different formulae used in the calculation of skewness by Carsola (1954a) and Naidu et al. (1973, in press).

The origin of the gravels on the shelf surface is difficult to ascertain and remains a matter of speculation. (See addendum at the end of this chapter.) Questions which immediately arise include: Are all the gravels (1) contemporary ice-rafted material or (2) relict, deposited during lowerings of sea level consequent to the Pleistocene glaciation, and/or ice rafted in Quaternary time, or (3) complex in origin, being partly relict and partly contemporaneously ice-rafted, or (4) residual deposits resting on submarine rock outcrops. At present it would seem that the gravels are either all contemporaneously ice-rafted, or both relict and contemporaneously ice-rafted components. The difficulty in arriving at a definite conclusion lies in the fact that there is as yet no specific criterion to distinguish satisfactorily, in a multi-origin gravel assemblage, gravels which were transported by ice rafting—either in Pleistocene or Holocene times—from gravels which were deposited in shallow marine waters under high dynamic wave and current conditions. Any speculation on the factors controlling the gravel distribution is difficult in the shallow region of the Beaufort Sea because the paleogeography, paleoclimatology, and tectonic history including eustatic changes in this polar sea shelf during and even after the most recent Wisconsin Glaciation are not known. In order to throw light on the origin of the shelf gravels it would be important to ascertain the extent of the lowering of sea level in Northern Alaska during the Last Glacial maximum and whether or not the Arctic pack ice had extended, as a shelf ice, further south to the present North Slope coast line just prior to the Wisconsin sea-level lowering. If it is assumed that the whole of the Beaufort Sea shelf had a perennial pack ice cover (not rigid ice cover) during the Wisconsin, then the only way by which gravels could have been transported to the shelf at that time would have been by ice rafting. Even if this were true, the problem of the origin of the shelf gravels would not be solved completely because it would be necessary to differentiate gravels that were ice-rafted during Wisconsin time and that have not been covered up by recent sedimentation, from those that are now being ice-rafted. Thus, the answer to the origin of gravels lies in proving whether or not in the recent geological past the Beaufort Sea shelf was completely free from ice cover. On the basis of researches carried out on long sediment core samples several geologists (Ericson et al., 1964; Herman, 1964, 1969; Hunkins and Kutschale, 1967; Hunkins et al., 1971; Olausson and Jonasson, 1969) have surmised that there has been no great change in the nature of ice cover in the Canada Basin and immediate surroundings since Wisconsin time. However, results of the Canada Basin studies probably cannot be extended to the Beaufort Sea shelf, which is nearer the coast and has relatively shallow depths.

There should be no doubt regarding the ice-rafted origin of the sporadic gravels observed in the upper slope beyond a depth of about 200 m, because Wisconsin lowering of sea level probably did not extend to such great depths. In the course of sampling, during the summer of 1971 a few large (up to 30 cm long), rounded, faceted, polished, and parallelly striated dolomite boulders were encountered on the outer shelf mud. It is inferred that these dolomite boulders are glacially derived and ice-rafted, but possibly not by the ice island T-3, which does not carry dolomite or limestone boulders (Stoiber et al., 1956). The author is making a detailed study of the lithology of the gravels which have been collected from the Beaufort Sea, and it is hoped that from a comparison of the mineral compositions of these gravels and associated heavy mineral sands with those of the ice islands (Shorey, 1953; Stoiber et al., 1956, 1960) and of Northern Alaska, the provenance of the gravels in the Beaufort Sea could be inferred.

Hundreds of observations made of ice-floe surfaces by the author and several observations made by divers under the ice-floes in the summer of 1971 in the Beaufort Sea revealed the presence of few gravels, although large volumes of mud and some fine sand were being ice-rafted. These observations, together with the relatively small amounts

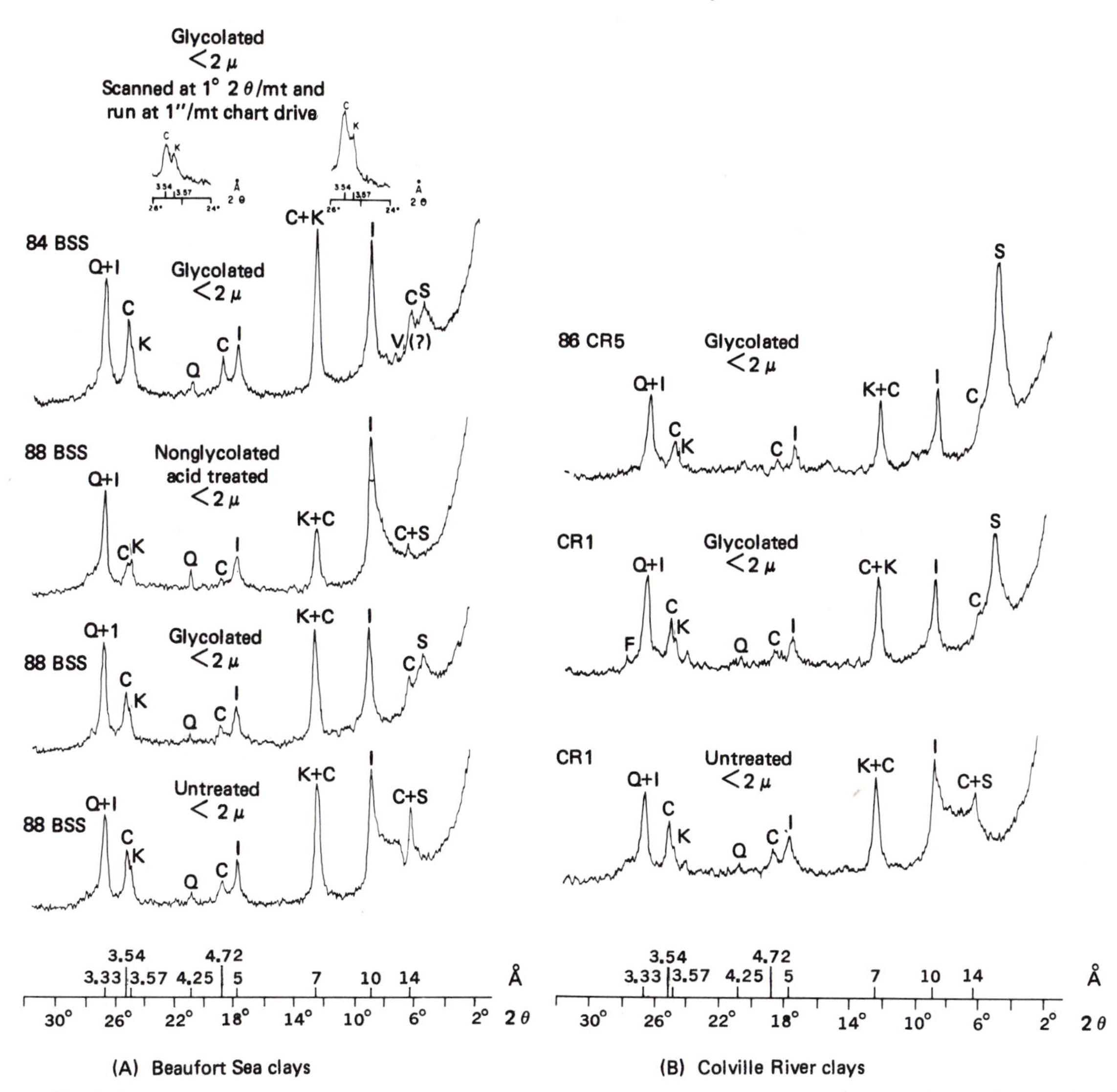

Fig. 3. Some typical X-ray diffraction traces of Beaufort Sea and Colville River clays. C: chlorite; S: smectite; I: illite; K: kaolinite; and Q: quartz.

of gravels generally noted in sediments by Naidu (in Burrell et al., 1970), suggest that their transportation by ice rafting to the Beaufort Sea is quantitatively not very significant. In fact, Naidu et al. (1973, in press) have pointed out that the 60 to 80% of mud observed in the Beaufort Sea shelf far outweighs the 15% of mud normally observed in polar shelves (Hayes, 1967).

The foregoing observations lead to another difficult question: How much and which mud fractions (silt and clay-size particles) of the Beaufort Sea sediments were ice-rafted and which were transported by wave and current actions? Ice rafting is generally accepted as mode of gravel transport in the polar seas, but ice rafting of fine-sized particles is rarely considered. However, in view of the common presence of mud-laden sea ice, ice rafting appears to be an important mode of transport of the silt and clay-size sediments to the Beaufort Sea, although aeolian transport could also be significant in some areas (Windom, 1969).

It would seem reasonable to assume that most of the gravels on the Beaufort Sea floor were ice-rafted. This statement is based on the findings of Carsola (1954a, p. 1567) and Naidu et al. (1973, in press, Fig. 3) that good correlation exists between the extent of broken sea ice and Arctic pack ice[1] and the distribution of gravels. The near-absence of gravels in the extrashelf region

[1] The demarcation line for the extent of broken sea ice and the Arctic pack ice oscillates roughly between the shelf edge and the upper slope.

may be due to the presence of Arctic pack ice, that provides an effective barrier for the movement of sediment-laden shore ice from the North Alaskan coast. In the shelf which has partial ice cover, in the form of broken sea ice, relatively higher contents of gravels are encountered because movement of broken shore ice is less restricted into that area.

Schwarzacher and Hunkins (1961), Clark (1969), Hunkins et al. (1970), and Hunkins et al. (1971) concluded that the gravels in the Canada Basin and adjacent deep basins might have been ice-rafted by ice islands such as T-3 and Arlis II. However, the author has not observed gravels in any of the 17 large grab samples collected from the lower slope and abyssal basin of the Beaufort Sea. The latter two regions are contiguous with the southern part of the Canada Basin. The absence of gravels in the Beaufort Sea deeps could be due either to the inadequate number of samples collected or to the possibility that the area over the upper slope and abyssal basin of the Beaufort Sea is bypassed by ice islands. This conclusion is supported by the drift tracks of ice islands (Schindler, 1968).

On the basis of sediment size distributions, Carsola (1954a, p. 1566) delineated two depositional regimes, the western and eastern Beaufort Seas. Carsola observed that at all depths west of 140°W meridian (or roughly west of Herschel Island) sediments are relatively coarser and more poorly sorted than in the corresponding depth zones east of the 140°W meridian. The regional differences in sediment texture are apparently caused by the relatively large deposition of fine-grained sediments in the eastern Beaufort Sea from the Mackenzie River discharge.

Distribution of Suspensates

Knowledge of the distribution, composition, and origin of the marine suspensates is vital to the understanding of several marine depositional processes. However, very little research has been carried out in this direction in the Arctic Ocean. Weights of suspensates reported by Burrell (1971) and those recently measured by the author (Table 1) are the only available suspended load data from the Beaufort Sea. On the basis of limited data, the following conclusions are tentatively drawn:

(1) The amounts of suspensates in the Beaufort Sea range from 0.03 to 12.19 mg/l.

(2) There are no systematic vertical variations in the suspended sediment contents at any station, nor are there any progressive variations from the coast outward. However, a regional variation in the offshore area is discernible. The amount of suspensates in the region between the Sagavanirktok River and Barter Island is relatively higher than in the region west of the Sagavanirktov River. Apparently, the areas of high concentration of suspended sediments are the areas to which the bulk of fluvial outflow or transportation of shore ice laden with fluvial suspensates is directed. The relatively low concentrations of suspensates observed in the shallow marine areas north and west of the Colville and Kuparuk Rivers suggest that much of the suspensate outfall of these rivers is ultimately directed eastward. The bulk of discharge of the North Slope rivers takes place in a few weeks during the spring breakup (Arnborg et al., 1967). Most of this sediment-laden outflow initially settles on fast ice which is situated off river mouths, and subsequent distribution of it in the offshore waters is regulated by the processes which determine the movement and melting of the broken nearshore ice. Reimnitz and Bruder (1972) have concluded that most of this fluvial discharge finally settles on the steeper slopes seaward of the 2-m depth contour, off river mouths, and that this area represents the delta front of Arctic rivers.

(3) When the quantities of suspensates collected from the upper 40 m depth of the Beaufort Sea (Table 1), northeastern Chukchi Sea, and the Canada Basin (Loder, 1971) are compared, some regional differences in these contiguous Arctic water masses are indicated. On the average the content of suspensates in the western Beaufort Sea are much higher than in the northeastern Chukchi Sea and Canada Basin. These differences are probably due to differences in the overall phytoplankton productivity in the three regions and to the supply of terrigenous suspensates. In view of the fact that Johnson (unpublished; in Carsola, 1954a, p. 1576) observed a larger standing crop of phytoplankton in the Chukchi Sea than in the Beaufort Sea, it appears that the larger quantities of suspensates in the latter is due to a greater supply of terrigenous detritus. In the Canada Basin the amount of suspensates is lowest, presumably because of the small supply of terrigenous detritus and the low planktonic production.

Clay Mineralogy

In a recent article Naidu et al. (1971) reported the clay mineral compositions of Beaufort Sea

Table 1 Weights of Suspended Particulate Matter in the Beaufort Sea
Samples were collected from U.S.C.G.C. *Glacier* by the author in August and September 1971.
Refer to Fig. 1 for sample locations.

Station	Depth (m)	Suspensates (mg/l.)	Station	Depth (m)	Suspensates (mg/l.)
GLA-71-1	1	2.06	GLA-71-15	1	11.46
	5	1.78		15	0.82
	10	1.51		34	1.60
	20	1.73	GLA-71-16	1	1.05
	30	1.66		18	1.63
GLA-71-2	1	1.05		36	1.24
	10	1.19	GLA-71-19	1	7.73
	20	0.99		125	9.50
	30	1.50		250	10.50
	40	1.40		375	11.70
GLA-71-4	1	0.97		500	12.19
	17	1.21	GLA-71-20	1	11.39
	24	0.99		450	2.30
GLA-71-6	1	0.16		900	2.64
	125	0.81		1350	2.15
	250	0.45		1800	9.95
	375	0.58	GLA-71-21	1	9.60
	500	0.79		375	9.63
GLA-71-7	1	0.78		750	0.68
	125	0.54		1152	0.75
	250	0.95		1500	0.96
	375	0.29	GLA-71-22	1	3.69
	500	3.43		200	3.00
GLA-71-8	1	0.41		450	6.00
	20	1.01		775	1.78
	40	1.13		900	0.85
GLA-71-10	1	1.34	GLA-71-23	1	0.85
	10	2.36		12	0.90
	20	1.21		22	3.22
	30	1.55	GLA-71-27	2	2.80
	40	1.20		25	2.26
GLA-71-11	1	1.56		47	2.10
	8	1.19	GLA-71-29	1	3.25
	16	1.15		125	1.05
	24	1.48		250	2.00
GLA-71-12	1	1.35		375	0.53
	5	1.37		500	1.63
	10	1.14	GLA-71-30	1	0.75
	15	1.49		22	0.26
	23	1.16		44	0.14
GLA-71-13	1	1.71		66	1.27
	3	0.71		88	0.80
	8	1.66	GLA-71-75	1	0.56
	13	0.47		45	1.33
	20	1.00		85	1.33
GLA-71-14	1	0.85	GLA-71-78	1	0.46
	12	7.46		22	1.44
	24	1.99	GLA-71-86	1	0.46
				1800	0.45

sediments. These authors showed that in all studied sediments illite is the predominant clay mineral (56 to 69%), chlorite (14 to 30%) and kaolinite (5 to 21%) occur in significant amounts and smectite is the least abundant (trace to 9%)[1]. No correlation has been noted between clay mineral-

[1] Percentages are expressed as weighted peak-area percentages, after Biscaye (1965).

ogy and water depth, distance from shore, or grain-size distributions of sediments.

The significance of the clay mineral data of the Beaufort Sea in the interpretation of paleoclimates has been discussed by Naidu et al. (1971), especially in view of the clay mineral distributions in the world ocean (Biscaye, 1965; Griffin et al., 1968) and the conclusion of Griffin et al. (1968) that kaolinite is a low-latitude clay mineral. The immediate sources of kaolinite in the Beaufort Sea are the North Slope rivers (Naidu et al., 1971) and the Mackenzie River (Dewis et al., 1972). These rivers cut across kaolinite-rich Mesozoic rocks in the north Alaskan-Canadian hinterland.

The East-Central Chukchi Sea, the western Beaufort Sea, and the Canada Basin have contiguous water masses. However, on the basis of relative abundances of different clay minerals (Carroll, 1970; Naidu, et al., 1971; and Naidu and Sharma, 1971) these water masses seem to have different physico-chemical depositional regimes. The identification of chamosite in the Canada Basin sediments (Carroll, 1970) and its absence from the east-central Chukchi Sea and western Beaufort Sea sediments (Naidu et al., 1971; Naidu and Sharma, 1971) suggest reducing environment in the deeper Canada Basin. However, positive distinction of chamosite from disordered kaolinite and chlorite is not always easy, and misinterpretation of the oxidation-reduction potentials of the environments in question could easily arise from misidentification of the above minerals. Relatively much higher smectite contents are observed in the far offshore areas of the east-central Chukchi Sea than in the adjacent western Beaufort Sea (Naidu and Sharma, 1971). This is attributable to a larger supply of smectite from the Chirikov Basin and Yukon Delta, probably supplied to the east-central Chukchi Sea by the turbid waters described by McManus and Smyth (1970). Clay-mineral analysis of the turbid suspensates could help to confirm the above conclusions. Apparently, the net drift of currents northward from the Bering Sea to the Chukchi Sea (Fleming and Heggarty, 1966) does not have a strong component moving into the western Beaufort Sea. This statement finds support in the fact that there is a general decrease in smectite content northeastward from the east-central Chukchi Sea (Naidu and Sharma, 1971). These conclusions are tentative, as they are based on clay-mineral analyses of only 60 sediment samples from the Chukchi and Beaufort Seas; they will be verified by the additional analyses of about 200 samples.

Studies initiated recently by Naidu (1972) and Naidu and Mowatt (1973, in press), concerning lateral variations of clay mineral assemblages from fresh to marine environments in the Arctic deltaic region of the Colville and adjacent rivers, have yielded results which may have bearing on the origin of clay minerals in the Beaufort Sea and may be potentially useful in interpreting the paleogeography of the nearshore environment of this sea. On the basis of field and laboratory evidence it appears that the smectite-rich fluvial clays of the Colville River (Fig. 3b), upon entering the marine environment, rapidly give way to illite-rich clays (Fig. 3a). Analyses of clay samples from the transitional environment along the north coast of Alaska, accompanied by laboratory experiments of these samples suggest that the increase in illite in the Beaufort Sea clays is the result of regeneration of degraded fresh-water illites by K^+ adsorption in marine environment.

Sediment and Interstitial Water Chemistry

The first detailed study on the geochemistry of Beaufort Sea sediments is that of Naidu and Hood (1972). Earlier studies (Carsola, 1954a; Hoskin et al., 1969) were restricted to determinations of organic matter and carbonate contents of sediments.

The concentrations of Na, K, Mg, Ca, Fe, Mn, Rb, Li, Zn, Ni, U, Co, Cu, organic carbon, and carbonate of the Beaufort Sea sediments (Naidu and Hood, 1972) are summarized in Table 2. Chemical composition of igneous rocks, crustal material, and contemporary low-latitude deep-sea clays are included in Table 2 for comparison. In the following paragraphs the important features of the Beaufort Sea sediment chemistry (Naidu and Hood, 1972) are presented.

The Beaufort Sea sediments have a chemical character that is distinctly different from low- and middle-latitude marine sediments. The points of difference are in the significant deficiency of all elements, except U, that have been analyzed and in the lower Na/K, Ca/Mg, and Fe/Mn ratios of the Beaufort Sea sediments. In fact, chemically the Beaufort Sea sediments (Table 2) are very similar to the Antarctic sediments which have been analyzed by Angino (1966). These results suggest that geochemical criteria such as impoverishments of

Fe, Mn, Ca, K, Na, Mg, Ni, Zn, Co, and Cu, and low Na/K, Ca/Mg, and Fe/Mn ratios may be used as potential tools in deciphering glacial-marine or high-latitude paleosediments.

Naidu and Hood (1972) have speculated that the relative deficiency of almost all these elements could be due to: (1) the impoverishment of all the elements analyzed in terrigenous source material and /or (2) low depositional rates of biogenous and chemogenous sediment fractions and retarded adsorption of ions. Presumably the near-freezing temperatures of the Beaufort Sea inhibit chemical and biochemical reactions. Preliminary results (Naidu, 1972; Dewis et al., 1972) support to a large extent the assumption that the terrigenous source materials are impoverished in the above-mentioned elements.

The average content of organic matter in the Beaufort Sea shelf and extrashelf is 1.71 and 2.14%, respectively. These values are notably lower than the average values of organic matter cited for corresponding lower-latitude marine environments (Trask, 1939). The factors which control the amount of organic matter in sediments have been discussed at length by Trask (1939). The chief factor in the Beaufort Sea seems to be the limited supply of organic matter to the bottom sediments, due to low organic productivity and the presence of relatively low biogenic suspensates in the overlying waters. This conclusion is based on observations made in the Arctic Basin adjacent to the Beaufort Sea, by Dietz and Shumway (1961) and Kinney et al. (1971). To date, however, no primary productivity studies have been made in the Beaufort Sea.

Shelf sediments are distinguishable from those of the extrashelf (Table 2) in being relatively enriched in carbonate, by having lower Na/K ratios and by their organic matter content. The relatively larger amounts of carbonate in shelf sediments has been attributed by Naidu and Hood (1972) to the higher content of calcareous bioclastics (molluscan shells and foraminiferal tests) and other lithogenous components. Higher

Table 2 Average Abundances of Some Elements and Ratios in the Beaufort Sea Sediments, Crust, Igneous Rock and Low-Latitude Deep-Sea Clays

Unless otherwise indicated, all abundances are in wt. %.

Chemical component	Beaufort Sea shelf	Beaufort Sea extrashelf[1]	Crust[2]	Igneous rock[3]	Low-latitude deep-sea clay[4]	Pacific pelagic clay[5]
Corg.	0.95	1.19				
$CO_3^{=}$	4.80	2.75				
Fe	3.57	3.52	5.60	5.00	6.50	5.07
Mn	0.03	0.09[a]	0.10	0.10	0.67	0.48
Ca	0.42	0.22	4.10	3.63	2.90	
Mg	2.22	1.73	2.30	1.26	2.10	
Na	1.59	1.97	2.40	2.83	4.00	
K	2.30	2.03	2.10	2.59	2.50	
Rb	0.0097	0.0084	0.0090	0.0090[b]	0.0110	
Li	0.0047	0.0043	0.0020	0.0065	0.0057	
Zn	0.0098	0.0082	0.0070	0.0075[b]	0.0165	
Ni	0.0047	0.0056	0.0075	0.0080	0.0225	0.0211
Co	0.0029	0.0028	0.0025	0.0023	0.0074	0.0101
Cu	0.0057	0.0059	0.0055	0.0070	0.0250	0.0323
U (ppm)	2.5	2.8	2.7	2.7[b]	1.3	
Mn/Fe	0.01	0.02	0.02	0.02	0.10	
Na/K	0.69	0.97	1.14	1.09	1.60	
Ca/Mg	0.19	0.13	1.78	2.88	1.38	

[1] Extrashelf includes slope and abyssal basin environments.
[2] After Taylor (1964).
[3] After Clarke and Washington (1924), Goldschmidt (1954), Rankama and Sahama (1950), and Sandell and Goldich (1943).
[4] After Turekian and Wedepohl (1961).
[5] After Cronan (1969). Averages exclude contributions by ferromanganese nodules.
[a] Average includes data of abyssal sediments which have much higher Mn.
[b] Average of granite and basalt values (Krauskopf, 1967).

content of organic matter and Na/K ratios in the extrashelf sediments are determined by the relatively larger amounts of mud in the extrashelf than in the shelf region.

Calculations of correlation coefficients (Naidu and Hood, 1972; Table 3) that consider elemental, granulometric, and clay mineral compositions of sediments have permitted discussion of the pathways of chemical elements. The bulk of the Ca^{++} seems to be fixed in carbonates, whereas alkalies are predominantly tied to adsorbed or exchangeable sites of clay minerals. From inter-element and clay-element relationships, it appears that except possibly for Co, the scavenging of other trace transition metals—Cu, Ni, and Zn—by hydrated oxides of either Mn or Fe is of secondary importance and that major portions of these transition metals are linked to clay minerals.

The partition patterns of Fe and Mn in the Beaufort Sea sediments are complex. Small portions of these elements are probably co-precipitating as ferrimanganic hydrate, while the rest of the portions and the bulk of the Mg^{++} are fixed in octahedral lattice sites of clay minerals and/or other silicates. Although no discrete ferromanganese nodules have been detected in any sediments, the presence of thick Fe-Mn encrustations on gravels has been commonly observed. In lower latitudes several trace elements are enriched in deep-sea sediments, compared to shallow marine sediments (Goldberg, 1965); this elemental enrichment is attributed to the scavenging of trace elements by ferromanganese nodules (Goldberg, 1965; Cronan, 1969). Naidu and Hood (1972) have surmised that the lack of enrichment of Ni, Co, Cu, Rb, and Zn in the extrashelf Beaufort Sea sediments (Table 2) is possibly related to the absence or lack of concentration of ferromanganese nodules in the extrashelf region.

The thick ferrimanganic encrustations on Beaufort Sea gravels are an interesting object of study. Preliminary investigations show that these encrustations are initially deposited as oxides and hydroxides in the form of a horizontal rim around gravel surfaces. These rims indicate the zone of demarcation between the top oxidizing and lower reducing environments within the subsurface regime of the Beaufort Sea floor. The accretion, if continued unchecked, completely envelops the gravel, giving it the appearance of a ferromanganese nodule.

In an attempt to elucidate both the processes involved in the origin of these encrustations and the geochemical role of ferrimanganic hydrate phase in the scavenging of trace elements in the Beaufort Sea, the mineralogy and chemistry of the ferrimanganic encrustations were analyzed. On the basis of X-ray analysis it is not possible to ascertain their mineralogy. The X-ray diffraction peaks at 3.27, 3.33, and 4.23 Å apparently indicate the presence of gamma iron oxide hydrate, beta iron oxide hydrate, and hydrated oxide of alpha iron (goethite) and/or manganese oxide, respectively. Chemical analysis (Table 3) indicates the following:

(1) There are two types of encrustations; one Fe-Mn-rich and another Al-rich.

(2) The quantities of P, K, Ca, Mg, Na, Ba, Sr, B, Cu, Zn, and Cr are much larger in the Fe-Mn-rich than in the Al-rich encrustations. On the basis of this observation it is concluded that ferrimanganic hydrate in the Beaufort Sea sediment is an important phase with which trace elements are associated. However, on the basis of

Table 3 Chemical Composition of the Ferrimanganic Encrustations of Gravels
Refer to Fig. 1 for sample locations.

Sample no.	Percent					Parts per million								
	P	K	Ca	Mg	Na	Al	Ba	Fe	Sr	B	Cu	Zn	Mn	Cr
GLA71-18	17	5.7	<1.0	3.9	2.4	30,500	870	174,000	543	1580	271	1030	118,000	136
GLA71-28	15	6.8	<1.0	3.2	1.8	34,900	1070	153,000	513	1600	271	800	79,200	148
GLA71-30	1.2	3.2	<1.0	2.7	0.93	60,000	760	62,000	139	690	31	100	436	<60
GLA71-31a	14	4.5	<1.0	3.2	1.5	39,000	1210	173,000	614	1560	237	530	80,800	<60
GLA71-31b	15	3.7	<1.0	2.8	1.5	35,000	1030	161,000	579	1600	184	390	56,400	94
GLA71-43	7.6	4.3	<1.0	1.8	1.2	62,000	690	134,000	165	1280	74	190	1910	94
GLA71-74a	1.4	3.2	<1.0	2.2	1.3	81,000	760	62,000	108	670	57	110	1350	99
GLA71-74b	15	6.3	<1.0	3.0	2.0	19,300	800	191,000	714	1720	207	420	92,000	148
GLA71-76	17	6.3	<1.0	2.8	1.8	35,000	1050	177,000	543	1760	309	590	100,000	142
GLA71-89	15	5.3	<1.0	2.8	1.7	21,800	850	171,000	564	1630	176	530	93,000	110

gross sediment chemistry the role of the ferrimanganic phase in elemental fixation in the Beaufort Sea is subordinate to that of clays.

(3) Some interesting differences are evident from a comparison of the chemistry of ferrimanganic encrustations of the Beaufort Sea gravels (Table 3) with that of a similar sediment phase from other oceans, e.g., ferromanganese nodules (Riley and Chester, 1971). The encrustations of the Beaufort Sea have much higher P contents and are relatively very deficient in all trace elements. The geochemical factors which cause these differences are not fully understood.

Some attempts have been made to determine whether these encrustations are relict or contemporary deposits. It was observed that living bryozoans and other sessile organisms are absent on gravel surfaces which have thick ferrimanganic encrustations on the top. This could be related to the possibility that a high rate of deposition/accretion of the ferrimanganic precipitate tends to quickly blanket the organisms. Such organic growths are, however, commonly observed on gravels which have barely any ferrimanganic encrustations on surfaces facing the sediment top. These observations suggest that the encrustations are contemporary materials; if they were relict features, then sessile organisms should grow on all gravel surfaces, irrespective of the presence or absence of the ferrimanganic encrustations. Variations of Mn and Fe in interstitial waters, discussed in more detail later, further attest to the contemporary origin of encrustations.

In Table 3 the pH, temperature, and concentrations of Fe, Mn, SiO_2, and PO_4 in interstitial waters of several Beaufort Sea core sections are given. Compared to the average composition of sea water (Goldberg, 1965, p. 164) the values of Fe, Mn, SiO_2, and PO_4 obtained in this study (Table 3) are much higher. It is yet to be determined whether the observed enrichments of the elements analyzed in the Beaufort Sea interstitial waters, over sea water, are real or apparent. This question might be elucidated by calculating the ratios of these elements, and some alakali and alkaline earth elements to the absolute content of Cl^- in the interstitial waters. If the enrichment of Fe, Mn, SiO_2, and PO_4 proves to be real, then we would have evidence for diagenetic changes in the Beaufort Sea sediments.

There are no progressive changes toward the top of all cores in Fe, Mn, SiO_2, and PO_4 contents in the interstitial waters (Table 3). However, in the uppermost 10 cm of a number of cores (GLA-71-6, 24, 45, 62, and 86; Table 3) there is a sharp decrease in Mn and Fe content, and in some cores (GLA-71-F, 19, 5F, and 86; Table 3) a net upward increasing trend in Mn is discernible. These variations in Mn and Fe concentrations of the interstitial waters are apparently related to post-depositional dissolution and migration from the subsurface reduced layers as well as to oxidative precipitation at or near surface of Fe and Mn from interstitial waters.

Summary and Conclusions

From data available it is possible to draw several conclusions regarding sedimentation in the Beaufort Sea. Almost all sediments are extremely poorly sorted muds or sandy-muds. Gravels are frequently encountered on the shelf, sporadically in the upper slope, and not at all in the upper slope and abyssal basin. The slight differences in the skewnesses and kurtosis of the size distribution curves of the shelf and extrashelf sediments are attributed to differences in contents of gravels and/or coarse sand in the muds of the two regions. The shelf sediments have overall positively skewed and platykurtic size distributions because of the presence of well-sorted gravels and/or coarse sand in poorly sorted muds. In the extrashelf the total or near absence of well-sorted coarse material and the presence of nearly equal amounts of silt and clay results in nearly symmetrically size-distributed sediments. Two depositional regimes the western and eastern Beaufort seas can be delineated, based on sediment size distributions. The eastern Beaufort Sea has relatively more fine grained and better sorted sediments and is greatly influenced by the Mackenzie River deltation.

The origin of gravels in the shelf remains a matter of speculation; they could be either all contemporaneously ice-rafted material or partly contemporaneously ice rafted and partly relict. The possibility that at least a part of them are residual deposits, derived from submarine rock weathering, is not completely ruled out. In order to explain the origin of the shelf gravels it would be important to ascertain the extent of sea-level lowering and that of the Arctic pack ice off the coast of Northern Alaska during the Last Glacial maximum. Because of the presence of large amounts of mud in the sea ice, it is concluded that

Table 4 Hydrogen Ion (pH) Concentrations, Temperatures and Concentrations (ppm) of Fe, Mn, SiO_2, and PO_4 in Interstitial Waters of Core Sediments from the Beaufort Sea

Sample no.	Depth in core (cm)	Environment and water depth	pH	Temp, °C	Fe	Mn	SiO_2	PO_4
GLA-71-3	0–10	Outer shelf; 48 m	7.13	7.0	0.15	0.05		
GLA-71-6	0–5	Upper slope; 496 m	7.35	3.5	0.20	0.03		
	5–10		7.14	3.5	0.75	6.42		
	10–20		7.03	2.5	1.65	6.57		
	20–30		7.14	2.5	2.35	7.15		
	30–40		7.14	3.0	1.35	7.43		
	40–50		7.02	3.0	1.05	6.73		
	50–60		7.04	3.0	1.65	6.75	10.9	1.2
	60–70		7.15	3.0	0.35	6.00		
	70–80		7.15	3.0	1.15	6.80	10.8	1.1
	80–90		7.18	3.5	0.35	7.27	10.8	2.4
	90–105		7.14	4.0	3.60	7.67		
GLA-71-7	0–10	Upper slope; 406 m	7.38	4.0	0.70	3.15	34.9	3.0
	10–20		7.54	3.0	0.45	3.00	33.6	1.0
	20–30		7.59	4.0	0.55	3.10	34.9	1.5
	30–40		7.28	7.00	0.15	2.14	24.8	1.0
	40–50		7.39	7.00	0.20	2.28	28.2	1.6
	50–60		7.44	7.50	0.30	2.40	28.2	2.0
	60–70		7.52	8.00	0.20	2.35	23.8	1.0
	70–80		7.43	8.50	0.30	2.48	31.8	3.9
	80–90		7.52	9.00	0.25	2.42	30.6	2.9
	90–100		7.50	9.00	0.10	1.47	21.9	0.9
GLA-71-19	30–40	Lower slope; 980 m	7.56	3.0	3.10	6.75	32.3	4.1
	40–50		7.34	3.0	2.55	6.85	35.2	5.3
	50–60		7.37	4.0	2.10	6.72	34.6	4.8
	60–70		7.48	2.5	1.15	6.35	33.8	3.5
	70–80		7.43	3.5	0.25	5.55	29.1	1.9
	80–90		7.43	3.0	0.40	5.95	29.0	2.3
	90–100			3.5	0.35	6.05	31.5	2.7
	100–110			3.0	0.25	4.65	24.3	1.8
	110–120			3.0	0.80	5.45	31.2	3.1
	120–135			3.0	1.75	5.75	36.3	7.1
GLA-71-21	0–3	Abyssal basin; 2560 m			0.15	6.02	20.0	4.3
GLA-71-24	0–10	Outer shelf; 29 m	7.20	0.0	4.65	2.93	17.2	0.8
	10–20		7.15	0.0		4.45	19.6	1.1
	20–30		7.04	0.5		5.00	16.0	2.0
	30–40		7.02	0.5		4.80	15.7	1.0
	40–50		6.99	1.0		5.70	15.5	1.1
	50–60		7.06	2.0	2.25	4.30	12.2	0.5
	60–70		7.03	3.0	5.30	4.40	12.4	0.4
	70–75		7.00	5.0		6.25	12.0	0.3
GLA-71-45	0–2	Outer shelf; 38 m	7.17	4.0	0.15	0.20	14.4	0.6
	2–11		7.20	4.0	0.20	0.75	19.2	0.6
	11–18		7.29	4.0	1.00	1.22	20.9	0.6
	18–23		7.29	4.0	1.25	0.73	10.9	0.5
	23–30		7.16	4.5	0.50	2.17	25.5	0.4
GLA-71-57	0–4	Outer slope; 1649 m			0.15	12.90	24.8	1.7
	4–10				0.30	10.15	31.6	2.5
	10–20		7.78	2.0	0.70	9.70	38.1	3.9
	20–30		7.61	1.5	0.40	10.15	46.9	6.7
	30–40		8.20	1.7	0.35	9.60	44.6	6.0
	40–50		7.83	1.5	0.90	7.78	40.6	5.0
	50–60		7.35	1.5	0.25	8.15	40.9	4.2
	60–70		7.52	1.5	1.60	9.15	49.5	10.1
	70–80		7.64	2.0	0.15	8.40	42.4	5.2
	80–90		7.61	2.5	0.25	8.20	48.3	6.3

Table 4 (*Continued*)

Sample no.	Depth in core (cm)	Environment and water depth	pH	Temp, °C	Fe	Mn	SiO_2	PO_4
	90–100		7.60	3.0	0.80	8.20	49.5	8.3
	100–110		7.70	3.0	0.20	6.98	37.3	3.4
GLA-71-62	0–10	Outer shelf; 36 m	7.55	5.0	0.25	0.41	20.5	0.6
	10–15		7.00	5.0	0.30	0.79		
	15–20		6.83	5.5	0.55	0.68		
	20–25		7.06	5.0	0.20	1.76		
	25–30		6.94	5.5	0.25	1.43		
	30–35		6.54	5.5		1.65		
	35–40		6.69	6.0				
GLA-71-75	0–10	Upper slope; 142 m	7.63	4.0	0.25	1.75	23.3	1.5
	10–20		7.22	2.0	0.52	0.20		
	20–30		7.53	2.0	0.25	0.17		
	30–40		7.22	1.0	0.25	0.23		
	40–50		7.62	1.5	0.25	0.33	19.4	0.5
	50–60		7.65	2.0	0.30	0.36	29.0	0.9
	60–70		7.57	2.5	0.50	0.25		
	70–80		7.35	2.5	0.30	0.26		
	80–90		7.51	3.5	0.25	0.30		
	90–100		7.39	4.5	0.15	0.32		
GLA-71-86	0–5	Abyssal plain; 2136 m			0.55	0.40	23.6	1.2
	30–40		7.69	1.0	3.45	12.77	38.7	8.7
	40–50		7.63	1.0	2.90	10.85	40.6	7.6
	50–60		7.64	1.2	0.90	10.15	34.4	3.3
	60–70		7.54	1.7	2.60	9.22	35.8	5.8
	70–80		7.53	1.0	0.35	7.02	27.4	
	80–90		7.52	1.5	0.65	7.55	31.5	21.8
	90–100		7.41	2.0	0.25	4.95	26.2	12.3
	100–110		7.38	2.0	3.30	4.94	33.9	4.5
	110–120		7.54	2.0	1.75	4.45	27.4	2.2
	120–130		7.46	2.5	1.25	2.42	21.2	1.3
	130–140		7.42	4.0	1.75	3.25	26.3	0.6

a large proportion of the silt and clay in the bottom sediment has been transported by ice and deposited subsequent to its melting. However, there are no means by which to distinguish between ice-rafted silt and clay from those transported by waves and currents. Higher percentages of suspensates are present at and near the surface water in the Beaufort Sea than in the adjacent east-central Chukchi Sea and Canada Basin. These differences are attributed to greater supply of terrigenous detritus to the Beaufort Sea.

Clay mineral analysis of bottom sediments shows predominance of illite (56 to 69%), significant amounts of chlorite and kaolinite (14 to 30% and 5 to 20%, respectively) and small amounts of smectite (trace to 9%). The presence of kaolinite in high-latitude sediments is notable because it is generally considered to be a typical low-latitude mineral. As such, interpretations of paleoclimate based on clay mineral data must be made with great caution. Beaufort Sea clay minerals are essentially detrital, having as a source the North Alaskan–Canadian hinterland. Presumably some of the illite has resulted from regeneration of degraded fresh-water illite by cation exchange/adsorption of K^+ in open marine environment. From a comparative study of the clay mineral distributions in the contiguous marine regime of the east-central Chukchi and Beaufort Seas, movement of currents in that region has been speculated. It is believed that the net northward drift of currents from the Bering Sea into the Chukchi Sea does not have a major component moving into the Beaufort Sea.

The low organic carbon contents in sediments is attributed to the small supply of organic detritus from overlying water, which in turn is probably due to the limited biogenic productivity. Compared with low-latitude deep-sea sediments, the Beaufort Sea sediments are deficient in Na, K,

Mg, Ca, Fe, Mn, Rb, Li, Zn, Ni, Co, and Cu, and have lower Na/K, Ca/Mg, and Fe/Mn ratios. On the basis of these data it is suggested that geochemical criteria may have potential use in deciphering high-latitude paleomarine sediments. Sediments from the eastern and western Arctic Ocean are being chemically analyzed to check the validity of this assumption. The deficiency of elements in sediments analyzed to date is attributed to terrigenous source material impoverished in all of these elements and/or to low rates of deposition of biogenous and chemogenous fractions and retarded ion adsorption by sediments. It would seem that the bulk of the alkali and trace transition metals are fixed in argillaceous fractions and the calcium is fixed in carbonates. The chemistry of ferrimanganic encrustations of gravels attest to the coprecipitation of trace elements with the hydroxides of iron and manganese. However, on the basis of gross sediment chemistry it is suggested that, except possibly for Co, fixation of minor and trace elements in sediments is predominantly brought about by clays rather than by the ferrimanganic hydrate phase.

Compared to the average composition of sea water, the amount of Fe, Mn, SiO_2, and PO_4 in interstitial waters of sediment cores are much higher. It is yet to be determined whether these higher elemental concentrations are real or apparent. Although there are no progressive elemental variations toward the core top, in a few cores a net upward increase in Mn concentration, up to 10 cm below the core top, and an abrupt decrease in Mn and Fe in the core surfaces are evident. On the basis of these data and based on the assumption of the contemporary nature of the ferrimanganic encrustations on gravels, it is concluded that post-depositional dissolution and migration from lower reduced layers and oxidative precipitation at surfaces of Fe and Mn is taking place. What these diagenetic changes are and what the geologic implications of these changes could signify form an important part of the author's continuing studies of the geochemistry of Arctic Ocean sediments.

Addendum

In a recent paper Naidu and Mowatt (1973) have discussed at length the origin of gravels on the floor of the Beaufort Sea and have concluded that most of the gravels on the shelf are relict deposits. This contention is based on results of extensive study by Naidu and Mowatt on the size distributions and heavy mineral contents of continental margin and marine sediments of north Arctic Alaska, petrography and ferromanganese coatings of gravels, glacial history and marine transgressions in the North Slope of Alaska, and insignificant contemporary transport of gravels by ice-rafting in the Beaufort Sea.

Acknowledgments

Thanks are due to Dr. P. D. Rao, Mr. R. Hadley, and Miss Joanne Groves for the help in chemical analyses. Silica and phosphate in interstitial waters were kindly analyzed by Messrs. Gary Hufford and Bruce Rutland of the Oceanographic Unit, U.S. Coast Guard.

Ice-breaker ship support was provided by the U.S. Coast Guard. The help given by Dr. Peter Barnes and Jim Trumbull of the U.S. Geological Survey; Tom Furgatsch, Dave Mountain, officers and crew of the U.S.C.G.C. *Glacier* in the collection of samples is gratefully acknowledged. Chemical analysis of the gravel encrustations were kindly provided by Professor J. R. Moore of the Marine Research Laboratory, University of Wisconsin. The author is greatly indebted to him for this help. Appreciation for suggestions in manuscript are due Professor F. F. Wright.

This work was supported in part by the U.S. Atomic Energy Commission (Contract No. AT (04-3)-310), U.S. Geological Survey (Contract No. 14-09-001-12559), the NOAA-Sea Grant (Contract 36109), and the U.S. Environmental Protection Agency (Contract 16100 EOM).

References

Angino, E. E. 1966. Geochemistry of Antarctic pelagic sediments. *Geochim Cosmochim Acta* 30:939–961.

Arnborg, L., H. J. Walker, and J. Peippo. 1967. Suspended load in the Colville River, Alaska, 1962. *Geografiska Annaler* 49A:131–144.

Beal, M. A., F. Edvalson, K. Hunkins, A. Molloy, and N. Ostenso. 1966. The floor of the Arctic Ocean: geographic names. *Arctic* 19:215–219. Biscaye, P. E. 1965. Mineralogy and sedimentation of recent deep-sea clays in the Atlantic Ocean and adjacent seas and oceans. *Bull. Geol. Soc. Am.* 76:803–832.

Burrell, D. C. 1971. Trace metal associations in Subarctic and Arctic marine environments. *Institute of*

Marine Science, Report No. R71-12. University of Alaska. 120 pp.

———, P. J. Kinney, R. S. Hadley, and M. E. Arhelger. 1970. Beaufort Sea environmental data: 1968–1969. *Institute of Marine Science, Report No. R70-20.* University of Alaska. 274 pp.

Carroll, D. 1970. Clay minerals in Arctic Ocean seafloor sediments. *J. Sediment. Petrol.* 40:814–821.

Carsola, A. J. 1954a. Recent marine sediments from Alaskan and northwest Canadian Arctic. *Bull. Am. Assoc. Petrol. Geol.* 38:1552–1586.

———. 1954b. Submarine canyons on the Arctic slope. *J. Geol.* 62:605–610.

———, R. L. Fisher, C. J. Shipek, and G. Shumway. 1961. Bathymetry of the Beaufort Sea. In: G. O. Raasch, ed. *Geology of the Arctic.* Toronto: University of Toronto Press. pp. 678–689.

Christensen, R. E., R. M. Beckman, and J. J. Birdsall. 1968. Some mineral elements of commercial spices and herbs as determined by direct reading emission spectroscopy. *J.A.O.A.C.* 51:1003–1010.

Clark, D. L. 1969. Paleoecology and sedimentation in part of the Arctic Basin. *Arctic* 22:233–245.

Clarke, F. W., and H. S. Washington. 1924. The composition of the Earth's crust. *U.S. Geol. Surv. Prof. Pap. 127.*

The Committee on Polar Research. 1970. *Polar research: a survey.* Washington, D.C.: National Academy of Science. 204 pp.

Creager, J. S., and D. A. McManus. 1967. Geology of the floor of Bering and Chukchi Seas—American studies. In: D. M. Hopkins, ed. *The Bering Land Bridge.* Stanford, California: Stanford University Press. pp. 7–30.

Cronan, D. S. 1969. Average abundances of Mn, Fe, Ni, Co, Cu, Pb, Mo, V, Cr, Ti, and P in Pacific pelagic clays. *Geochim Cosmochim Acta* 33:1562–1565.

Dewis, F. J., A. A. Levinson, and P. Bayliss. 1972. Hydrogeochemistry of the surface waters of the Mackenzie River drainage basin, Canada. IV. Boron-salinity-clay mineralogy relationships in modern deltas. *Geochim Cosmochim Acta* 36:1359–1375.

Dietz, R. S., and G. Shumway. 1961. Arctic Basin geomorphology. *Geol. Soc. Am. Bull.* 72:1319–1319.

Ericson, D. B., M. Ewing, and G. Wollin. 1964. Sediment cores from the Arctic and Subarctic seas. *Science* 144:1183–1192.

Fleming, R. H., and D. E. Heggarty. 1966. Physical and chemical oceanography of the southeastern Chukchi Sea. In: N. J. Wilimovsky, ed. *Environment of the Cape Thompson Region, Alaska.* U.S. Atomic Energy Commission. pp. 697–754.

Folk, R. L., and W. C. Ward. 1957. Brazos River bar—a study in the significance of grain size parameters. *J. Sediment. Petrol.* 27:2–26.

Goldberg, E. D. 1954. Chemical scavengers of the sea. *J. Geol.* 62:249–265.

———. 1965. Minor elements in sea water. In: J. P. Riley and G. Skirrow, eds. *Chemical Oceanography,* Vol. 1. London: Academic Press. pp. 163–196.

Goldschmidt, V. M. 1954. *Geochemistry.* Oxford: Clarendon Press. 730 pp.

Griffin, J. J., H. Windom, and E. D. Goldberg. 1968. The distribution of clay minerals in the world ocean. *Deep-Sea Res.* 15:433–459.

Hayes, M. O. 1967. Relationship between coastal climate and bottom sediment type on the inner continental shelf. *Marine Geol.* 5:111–132.

Herman, Y. 1964. Temperate water foraminifera in Quaternary sediments of the Arctic Ocean. *Nature* 201:386–387.

———. 1969. Arctic Ocean Quaternary microfauna and its relation to paleoclimatology. *Paleogeography Paleoclimatology Paleoecology* 5:251–276.

Hoskin, C. M., D. C. Burrell, and P. J. Kinney. 1969. Size fractionation data for shelf sediments of the Alaskan Arctic shelf. *Institute of Marine Science, Report No. R69-12.* University of Alaska. 11 pp.

Hume, J. D., and M. Schalk. 1967. Shoreline processes near Barrow, Alaska: a comparison of the normal and catastrophic. *Arctic* 20:86–103.

Hunkins, D., and H. Kutschale. 1967. Quaternary sedimentation in the Arctic Ocean. In: M. Sears, ed. *Progress in Oceanography, Vol. 4.* Oxford: Pergamon Press, 4:89–94.

———, G. Mathieu, S. Teeter, and A. Gill. 1970. The floor of the Arctic Ocean in photographs. *Arctic* 23:175–189.

———, A. W. H. Bé, N. D. Opdyke, and G. Mathieu. 1971. The Late Cenozoic history of the Arctic Ocean. In: K. K. Turekian, ed. *Late Cenozoic Glacial Ages*, New Haven-London: Yale University Press. pp. 215–237.

Kinney, P. J., T. C. Loder, and J. Groves. 1971. Particulate and dissolved organic carbon matter in the Amerasian Basin of the Arctic Ocean. *Limnology and Oceanography* 16:132–137.

Krauskopf, K. B. 1967. *Introduction to Geochemistry.* New York, New York: McGraw-Hill Book Co. 721 pp.

Krumbein, W. C., and F. J. Pettijohn. 1938. *Manual of Sedimentary Petrography.* New York, New York: Appleton-Century-Crofts, Inc. 549 pp.

Levinson, A. A., B. Hitchon, and S. W. Reeder. 1969. Major element composition of the Mackenzie River at Normal Wells, N.W.T., Canada. *Geochim Cosmochim Acta* 33:133–138.

Loder, T. C. 1971. Distribution of dissolved and particulate organic carbon in Alaskan polar and subpolar and estuarine waters. Ph.D. Dissertation, *Institute of Marine Science, Report No. R71-15.* University of Alaska. 236 pp.

McManus, D. A., J. C. Kelley, and J. S. Creager. 1969. Continental shelf sedimentation in an Arctic environment. *Geol. Soc. Am. Bull.* 80:1961–1984.

———, and C. S. Smyth. 1970. Turbid bottom water on the continental shelf of the northern Bering Sea. *J. Sediment. Petrol.* 40:869–873.

Naidu, A. S. 1972. Clay mineralogy and heavy-metal geochemistry of deltaic sediments of the Colville and adjacent rivers. In: *Baseline data study of the Alaskan Arctic aquatic environment. Institute of Marine Science, Report No. R72-3.* University of Alaska. pp. 123–138.

———, D. C. Burrell, and C. M. Hoskin. 1970. Sediments of the south Beaufort Sea. *Trans. Am. Geophys. Union Meetings, April 1970* (abstract). Washington, D.C. p. 332.

———, D. C. Burrell, and D. W. Hood. 1971. Clay mineral composition and geologic significance of some Beaufort Sea sediments. *J. Sediment. Petrol.* 41:691–694.

———, and G. D. Sharma. 1971. Texture, mineralogy and chemistry of Arctic Ocean sediments. *Institute of Marine Science, Report No. R71-16.* University of Alaska. 17 pp.

———, D. C. Burrell, D. W. Hood, and J. A. Dygas. 1973. Some aspects of texture, clay mineralogy and chemistry of bottom sediments, west Beaufort Sea, Arctic Ocean. *Geol. Soc. Am. Spec. Pap. 151.*

———, and D. W. Hood. 1972. Chemical composition of bottom sediments of the Beaufort Sea, Arctic Ocean. *Proceedings 24th International Geological Congress.* Montreal, Canada. Section 10:307–317.

———, and T. C. Mowatt. 1973. Aspects of size distributions, mineralogy, and geochemistry of deltaic and adjacent shallow marine sediments, north Arctic, Alaska, *U.S. Coast Guard C. G. C. Report.* Washington, D.C. in press.

Olausson, E., and U.C. Jonasson. 1969. The Arctic Ocean during the Würm and Early Flandrian. *Geologiska Föreningens i Stockholm Forhandlingar* 91:185–200.

Rankama, K., and T. G. Sahama. 1950. *Geochemistry.* Chicago, Illinois: University of Chicago Press. 921 pp.

Reeburgh, W. S. 1967. An improved interstitial water sampler. *Limnology, Oceanography* 12:163–165.

Reimnitz, E., and K. F. Bruder. 1972. River discharge into an ice-covered ocean and related sediment dispersal, Beaufort Sea, Coast of Alaska. *Geol. Soc. Am. Bull.* 83:861–866.

Riley, J. P., and R. Chester. 1971. *Introduction to Marine Chemistry.* London-New York: Academic Press. 465 pp.

Sandell, E. B., and S. S. Goldich, 1943a. The rarer metallic constituents of some American igneous rocks, Vol. 1. *Intern. J. Geol.* 51:99–115.

———, and S. S. Goldich. 1943b. The rarer metallic constituents of some American igneous rocks, Vol. 2. *Intern J. Geol.* 52:167–189.

Schindler, J. F., 1968. The impact of ice island—the story of Arlis II and Fletcher's Ice-Island, T-3, since 1962. In: J. E. Sater, coordinator. *Arctic Drifting Stations.* Arctic Institute of North America. pp. 49–78.

Schwarzacher, W., and K. Hunkins, 1961. Dredged gravels from the central Arctic Ocean. In: G. O. Raasch, ed. *Geology of the Arctic.* Toronto: Toronto University Press. pp. 666–677.

Shearer, J. M., R. F. Macnab, B. R. Pelletier, and T. B. Smith, 1971. Submarine pingos in the Beaufort Sea. *Science* 174:816–818.

Shorey, R. R., 1953. Fletcher's Island rocks. Unpublished report. Terrestrial Sciences Laboratory, Geophysics Research Directorate, Air Force Cambridge Research Centre.

Stoiber, R. E., J. B. Lyons, W. T. Elberty, and R. H. McCrehan, 1956. The source area and age of Ice-Island T-3. Final report under Contract No. AF19(604)-1075, Dartmouth College, Department of Geology. 41 pp.

———, J. B. Lyons, W. T. Elberty, and R. H. McCrehan. 1960. Petrographic evidence on the source area and age of T-3: Scientific studies on Fletcher's Ice Island, T-3 (1952–1955). *Geophys. Res. Papers.*

Taylor, S. R. 1964. Abundance of chemical elements in the continental crust: a new table. *Geochim Cosmochim Acta* 28:1273–1285.

Trask, P. D. 1939. Organic content in recent marine sediments. In: P. D. Trask, ed. *Recent marine Sediments.* London: Thomas Murby and Co. pp. 428–453.

Turekian, K. K., and K. H. Wedephol. 1961. Distribution of the elements in some major units of the Earth's crust. *Geol. Soc. Am. Bull.* 72:175–192.

United States Weather Bureau. 1964. Local climatological data for Barrow, Alaska. Ashville, N.C.: U.S. Department of Commerce.

Windom, H. L. 1969. Atmospheric dust records in permanent snowfields: implications to marine sedimentation. *Geol. Soc. Am. Bull.* 80:761–782.

Chapter 8
Recent Sediments of the East Siberian Sea

FREDERIC P. NAUGLER [1], NORMAN SILVERBERG [2], AND JOE S. CREAGER [3]

Abstract

The East Siberian Sea, one of the large epicontinental arctic seas off northeastern Siberia, is shallow and covered with ice most of the year. The sea bottom is monotonously flat except where intersected by two drowned river valleys. The surficial sediments are fine grained, often containing over 15% colloidal (finer than 11ø) material. This is attributed to mechanical weathering in the arctic permafrost region, the low gradients of incoming rivers, and the low energy conditions that exist in the East Siberian Sea. The primary sources of sediment are the Indigirka and Kolyma Rivers and the New Siberian Island region. Sediment transport is generally easterly to northeasterly. On the basis of factor analysis of grain-size data and heavy mineral analysis of the 4ø sand fraction, three distinctive sediment groups have been defined within regions corresponding roughly to the western, central, and eastern portions of the East Siberian Sea. Sediment derived from the New Siberian Islands dominates the shallow (10 to 15 m) western region. Currents here are stronger than average and silt is the dominant sediment type. The central plateau is dominated by material introduced by the Indigirka River, characterized by low concentrations of heavy minerals. Currents here are generally weaker than in the New Siberian Shoal region and sediments are typically clayey-silts. Zones of cleaner silts indicate locally more intense currents. The eastern third of the East Siberian Sea is characterized by relatively deep (30 to 50 m) irregular topography, and variable sediment texture and mineralogy. Winnowed sandy sediments and gravelly-sandy-muds possibly associated with ice rafting break the pattern of silty-clays and clayey-silts. The Kolyma River has introduced most of the sediment into this region, although local shore line sources are indicated clearly by the mineralogy. Ice rafting generally appears to be insignificant in the East Siberian Sea sediments.

[1] Pacific Oceanographic Laboratories. NOAA, University of Washington, Seattle, Washington 98105, U.S.A. Deceased August 6, 1973.

[2] Centre d'Etudes Universitaires de Rimouski, Université du Quebec, Rimouski, Quebec, Canada.

[3] Department of Oceanography, University of Washington, Seattle, Washington 98105, U.S.A.

Introduction

General

A broad shallow-water platform extends from the Taimyr Peninsula of eastern Siberia to Alaska and includes the Laptev, East Siberian, and Chukchi epicontinental seas (Fig. 1). The East Siberian is the widest and shallowest of these, having a width of more than 800 km in its western part and narrowing to about 550 km near Wrangel Island. Depths generally are less than 40 m.

During the summer of 1962 the *U.S.C.G.S. Northwind* occupied several oceanographic stations in the eastern East Siberian Sea. The following summer the *Northwind* ventured into the western portion and then into the Laptev Sea, occupying 140 stations and providing the United States with its first oceanographic samples from these regions of the Siberian shelf. In the summer of 1964 the *U.S.S. Burton Island* occupied 80

Fig. 1. The shallow Arctic seas off northeastern Siberia. Currents after Sverdrup (1929). Physiography after Suslov (1961).

oceanographic stations in the East Siberian Sea. A total of 147 sediment samples, including Van Veen grab samples and short gravity cores, were taken in the East Siberian Sea during these cruises (Fig. 2).

The objectives of this paper are to describe the physical and mineral properties of the recent sediments, relate these properties to the depositional environments, and try to define the sources and modes of transport of the sediments.

Environmental Factors

The East Siberian Sea is subject to a severe continental climate, with air temperatures in January averaging about −40°C (Suslov, 1961, p. 124) and in July about 2 to 4°C (Suslov, 1961, p. 127). Runoff and precipitation greatly exceed evaporation. This produces intense stratification, with a layer of relatively fresh water overlying more saline and slightly colder water. Convective circulation arising from atmospheric cooling is limited to this thin low-density surface layer, facilitating the formation of ice.

Pack ice is most intense during the winter, when it averages between 3 and 3½ m in thickness (U.S. Navy Hydrographic Office, 1954). In the summer melting reduces the ice thickness to about 1 m and open areas are common near the coast where relatively warm water is being introduced by rivers. North of Ayon Island an extensive region of pack ice known as the Ayon Ice Massif exists throughout the year. This had proved a formidable obstruction to navigation and few data have been obtained from this area. There are no icebergs in the East Siberian Sea but small bits of ice are continually calving from the De Long Islands glaciers north of the New Siberian Islands.

Salinity varies from less than 10‰ near the Kolyma River to about 33‰ in the bottom waters near Wrangel Island and sea-water temperatures are generally below 0°C (Codispoti and Richards, 1968).

Currents are sluggish, with speeds generally less than 15 cm/sec. They are not well developed and vary seasonally as well as locally under the influence of the prevailing winds. According to Sverdrup et al. (1942, p. 576), "The most striking example of the influence of friction on tides is found on the wide shelf along the arctic coast of eastern Siberia." Tidal currents generally do not exceed 20 cm/sec and the tidal range is less than 20 cm along the entire mainland coast (U.S. Navy Hydrographic Office, 1954).

Physiographic Setting

The Siberian coast consists of low tundra interrupted in places by higher ground and cliffs with low hills inland (Fig. 1). Further inland there are mountain ranges, rolling plateaus, and vast swampy lowlands (Suslov, 1961, p. 208). East of the Kolyma River the mountain ranges of the Chukchi Peninsula extend to the coast line, where the shore is cliffed. West of the Kolyma is a vast lowland area that is poorly drained and dotted with many small lakes and marshy areas separated by tundra-covered ground. Northeastern Siberia is covered by continuous permafrost that reaches depths of up to 180 m.

The sea coast consists of frozen strata of loose material containing lenses of fossil ice. This material is eroded rapidly by waves, producing sharp and swift changes in the configuration of the shores. Fossil ice extends below sea level and forms an icy bottom beneath the western part of the East Siberian Sea (Suslov, 1961, p. 172).

Discharge of rivers emptying into the East Siberian Sea is markedly seasonal, with over 90% taking place in the warm spring and summer months. The largest of these rivers, the Kolyma, drains an area of about 644,000 km^2 and has an annual discharge of 120 km^3 (Samoilov, 1952, p. 319). Flooding occurs in the spring because of the rapid thawing of snow and ice. Sediment that has collected in the channels of the delta during the previous year is swept out into the sea during this period of intense outflow.

The second largest river, the Indigirka, drains an area of about 360,400 km^2 and has an annual discharge of 58 km^3. It originates in the Cherski Mountains and meanders through several hundred kilometers of the poorly drained lowland before reaching the sea. Its average turbidity is 150 gm/m^3, which is about five times that of most other rivers in eastern Siberia (Samoilov, 1952, p. 315). Thus, although the discharge of the Indigirka is about one-half that of the Kolyma, its yearly contribution of silt (8.5×10^6 tons) to the East Siberian Sea is nearly twice that of the Kolyma (4.7×10^6 tons).

Geologic Setting

The region lying roughly between the Lena River and the Pacific Ocean to the east is termed the Pacific Ocean Geosyncline (Nalivkin, 1960, p. 131). This region contained a geosynclinal basin

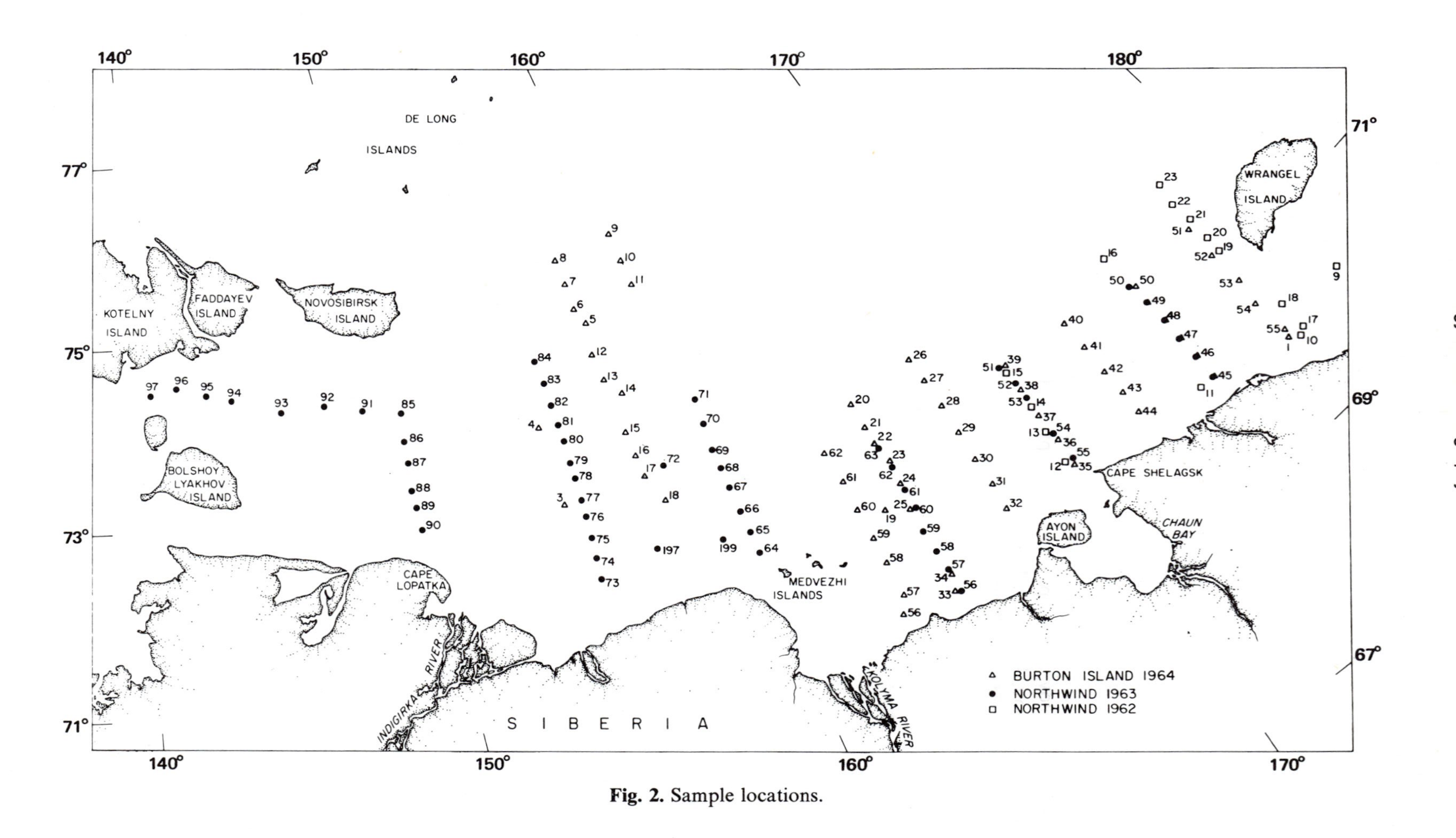

Fig. 2. Sample locations.

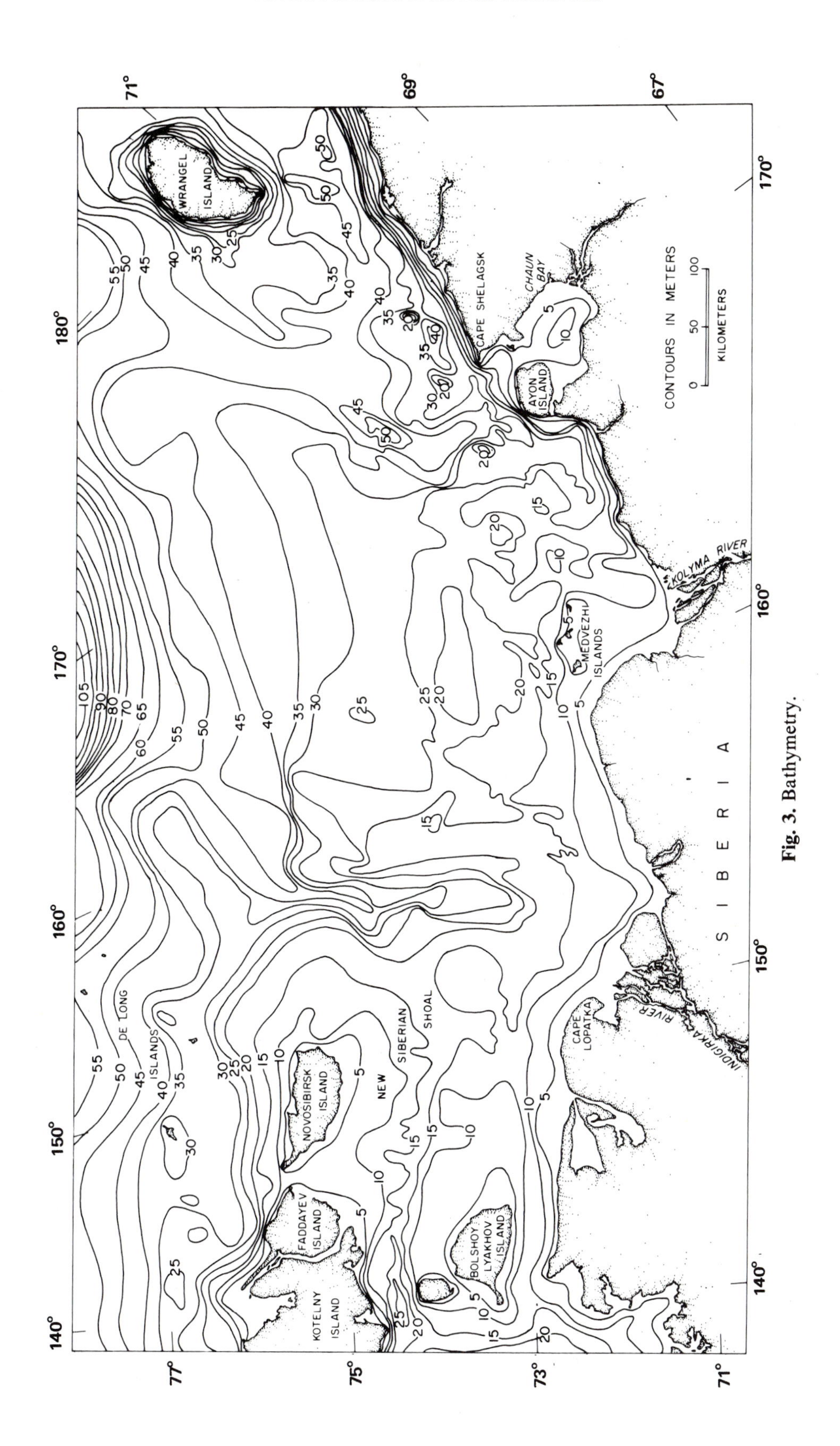

Fig. 3. Bathymetry.

during Paleozoic and Mesozoic times but became part of the continental mass during the lower Cretaceous. It is divided into the outer Alpine subregion and the inner Kimmerian subregion. The latter contains the drainage basins of the Indigirka and Kolyma Rivers. The geologic systems of this area can be traced from the Cambrian to the Quarternary Periods. Especially notable are the upper Paleozoic and Mesozoic marine deposits, with many Mesozoic and Tertiary igneous rocks.

The upper reaches of the Indigirka and Kolyma Rivers intersect Permian through Cretaceous limestones, shales, and sandstones, with small masses of granites and syenites scattered throughout. East of the Kolyma River, extensive areas are covered by Cretaceous effusive series of various compositions. These crop out along the coast east of Chaun Bay. Undifferentiated Quaternary deposits occupy extensive areas in the lowlands west of the Kolyma River. Clays and silts containing lenses and layers of fossil ice predominate (Nalivkin, 1960, p. 142). These are tundra deposits that were once covered by numerous lakes and bogs. Some recent marine sediments occur along the shore. Quaternary sediments cover the New Siberian Islands except for Kotelny, where Miocene marine deposits crop out over large areas. Several extinct volcanoes that produced basaltic flows occupy the Indigirka and Kolyma valleys. The geology of Wrangel Island is similar to that of the adjacent mainland, with the addition of small patches of Quaternary sediments along the coast.

Continental glaciation during the Pleistocene did not extend into the Indigirka–Kolyma lowland (Flint and Dorsey, 1945). The shallow and flat nature of the continental shelf from Taimyr Peninsula to the northwestern tip of Alaska suggests that this entire area was free of glaciers during the Pleistocene. Wrangel Island and several of the New Siberian Islands show local evidence of recent glaciation.

Bathymetry

The East Siberian Sea (Fig. 3) occupies a broad submerged continental margin that was subaerially exposed numerous times during glacial lowerings of sea level. It is one of the flatest areas of comparable size on earth, including the vast abyssal plain regions of the deep ocean. Slopes throughout the area vary between 1:1000 and 1:10,000. Contouring at 5-m intervals was necessary to reveal even some of the large-scale features. The extreme flatness is attributed to several factors: A subaerial history as a vast marshy lowland; numerous and extensive horizontal migrations of the beach zone related to sea-level variations; and also the lack of Pleistocene glaciation.

A northerly trending extension of the Indigirka River valley dominates the topography on the western half of the shelf. A somewhat less well-defined submarine valley is associated with the Kolyma River. This feature hugs the coast line near Ayon Island and then swings northward. The limited amount of available soundings makes it difficult to determine whether it continues north or south of Wrangel Island. It is possible that the ancient Kolyma River used both of these alternate routes.

West of the Kolyma submarine valley the continental shelf out to about 250 km from the mainland is shoaler than 30 m. West of the Indigirka valley, in the region we shall refer to as the New Siberian Shoal, depths are extremely shallow, averaging between 10 and 15 m, whereas east of the valley depths range around 20 to 25 m. The region including and east of the Kolyma valley is on the average 15 to 20 m deeper and is characterized by small knolls and depressions.

The shelf break occurs somewhere between depths of 60 and 70 m. Lack of soundings prevents its precise location.

Observations and Results

Methods

A total of 270 samples from cores and grabs were analyzed for grain size using standard sieve-pipette techniques (Krumbein and Pettijohn, 1938). The data were processed by computer to provide listings of the grain-size distribution in phi sizes and the various Inman (1952) parameters. Only the results for the 147 surface samples of cores and the grabs are presented here.

With an aim to better describing the data, factor analysis was undertaken using the following variables: % gravel, % sand, % silt, % clay, % 12ø, sand/mud ratio, median phi, major mode, first percentile, and water depth. Methods used were essentially similar to those of McManus et al. (1969) save for the inclusion of % 12ø, major

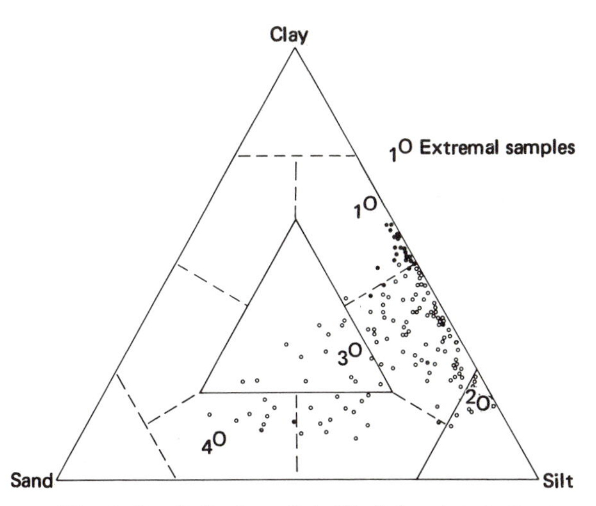

Fig. 4. Sand-silt-clay plot. Circled points indicate extremal samples.

mode, and water depth and exclusion of several percentile positions. Percent 12ø is an artifact of the size-analysis procedure which groups all sizes smaller than the 11ø size group together. Examination of samples which have not been disturbed for over a year indicates that this material extends well down into the colloid range. Many samples show 12ø values over 20%, which is a high percentage for continental shelf sediments. Because of this, many of the samples did not provide effective measures of sorting, as the tails of the size distribution were not detailed. Although sorting was not included in the 10 variables, its influence on the factors studied is apparent indirectly. Major mode and water depth were included as meaningful criteria because of the polymodal character of some samples and possible correlation of texture with bathymetry.

Only the results of factor analysis of 100 samples are included in this report[1]. Four factors, with eigenvalues greater than 1, were extracted from the data and account for 97.5% of the variability among samples. For a description of this type of analysis see Imbrie and Van Andel (1964).

To obtain a sufficient amount of sand-size material for heavy mineral analyses it was necessary to extract the coarse fraction from several hundred grams of sediment. Four-phi material was used for the analysis because this was the only size which consistently contained sufficient amounts. The 4ø fraction was split into heavy and light fractions with bromoform (S.G. = 2.89). The heavy minerals were mounted in Hyrax (R.I. = 1.71), and 400 grains from each sample were identified under the petrographic microscope. The light minerals were mounted in Canada Balsam and scanned.

Clay-mineral, carbonate, and organic-carbon analyses, as well as foraminifera identifications, were conducted and are presented elsewhere (Naugler, 1967). To summarize these briefly, illite and chlorite are the predominant clay minerals; calcium carbonate was undetectable in most of the samples; organic carbon varied from 0.4 to 1.5% of the sediment; and there were few foraminifera—generally less than 3/g of sediment.

[1] These are all surface samples except for three subsurface samples, included because they were extremals in a larger study of 270 surface and subsurface samples. When choosing the samples for this report, gravity core surface samples were used in preference to grabs, as they probably are less disturbed.

Grain Size Analysis

Basic presentation is based upon a sand-silt-clay diagram (Fig. 4). Using nomenclature derived from Shepard (1954) the sediments can be described as predominantly clayey-silts, some silty-clays, and sand-silt-clays, silty-sands, sandy-silts, and a few silts. Notably absent are sands and clays or mixtures of these two.

Examination of this diagram indicates that no obvious breaks occur at the arbitrary boundaries set by Shepard, the distinction between silty-clays and clayey-silts being nonexistent, with a strong concentration of samples all along the silt-clay axis. A diffuse array trends away from this concentration toward the sand apex. Actually a much smaller triangle is formed by the sediment distribution with apices found at silt, silty-clay, and silty-sand positions on the diagram.

This pattern is strongly emphasized in the results of the factor analysis, as can be seen by the positions of the extremal samples (circled in Fig. 4). These samples embody the characteristics of the mathematically derived end-member sediments, or factors. The loadings on the rest of the samples indicate the degree of similarity or representation of each of the four end-member sediments. The characteristics of each factor may be inferred by examining the attributes of the highest loaded members. The six highest loaded members of each factor (except for factor 4, which included only four samples) are described in Table 1. Size frequency diagrams of the four highest loaded members are shown in Fig. 5, the areal distribution of factor loadings on all samples in Fig. 6, and the areal distribution of 12ø and finer material in Fig. 7.

Factor I. This is the strongest factor, with an eigenvalue of 86.20 accounting for most of the variability among all the samples. Many samples have factor loadings greater than 0.50 on this factor (a value indicating 50% contribution and the arbitrary cut off used here to indicate inclusion in a particular factor group).

An examination of the highest loaded mem-

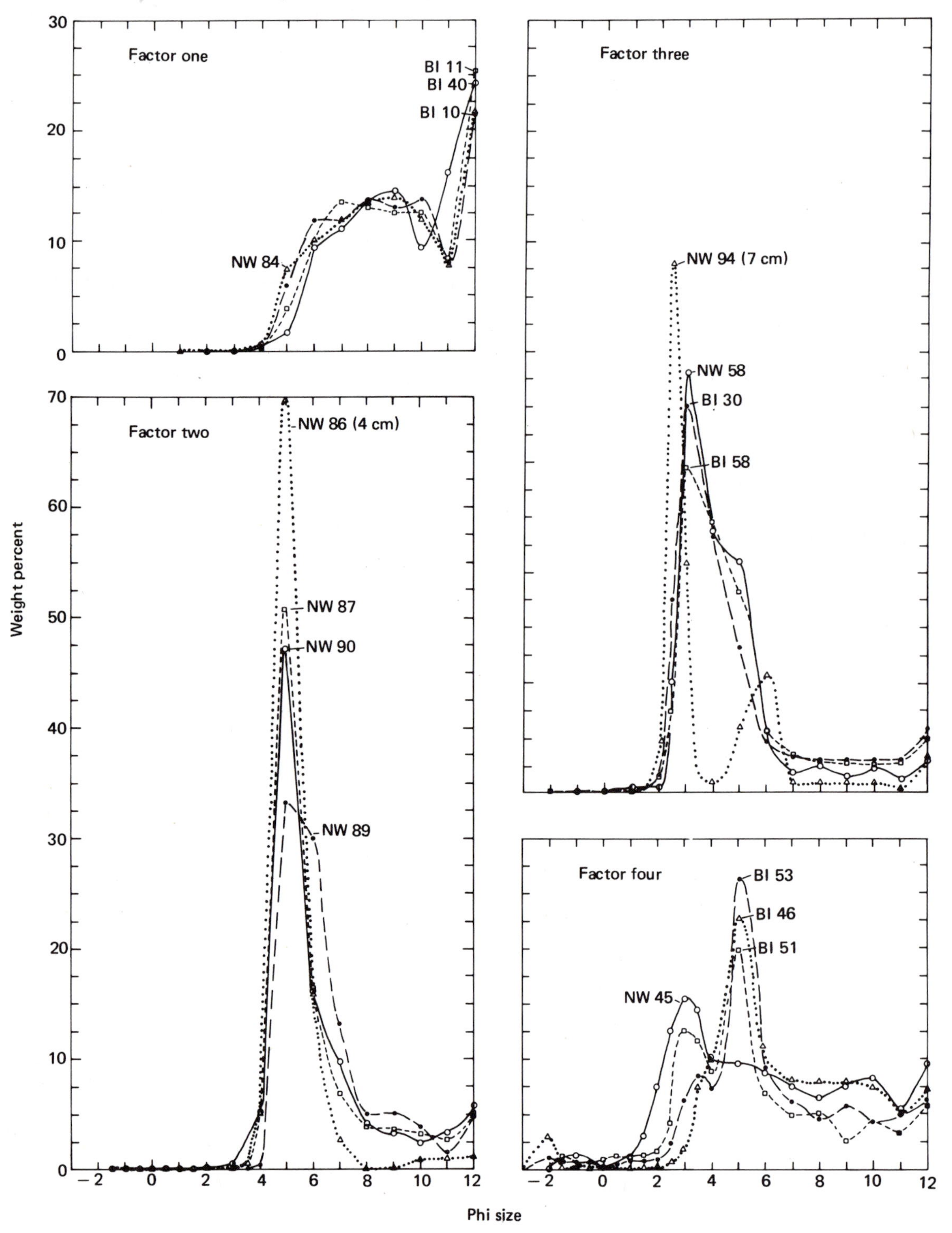

Fig. 5. Frequency curves of highest loaded samples in each factor.

Table 1 Characteristics of the Factors: Surface samples except where indicated.

Loading	Sample no.	% Gravel	% Sand	% Silt	% Clay	% 12φ	Sand/mud ratio	Median	Mode	First %	Depth (m)
FACTOR 1											
.915	BI-38	0.0	1.6	39.4	59.0	16.7	0.02φ	8.45	9.00	3.46	48
.919	BI-41	0.0	0.7	42.4	56.9	19.0	0.01φ	8.43	9.00	4.19	42
.928	BI-10	0.0	1.4	42.6	56.0	21.7	0.01φ	7.99	9.00	3.70	46
.931	NW-84	0.0	0.0	43.6	56.4	21.7	0.00φ	8.17	10.00	4.19	39
.971	BI-11	0.0	0.5	40.6	58.9	25.6	0.00φ	7.99	7.00	4.30	42
1.000	BI-40	0.0	0.6	35.7	63.7	23.8	0.01φ	6.58	11.00	4.34	44
FACTOR 2											
.768	NW-88	1.5	10.4	76.0	12.1	4.6	0.14φ	4.88	5.00	−1.67	15
.810	NW-86	0.0	12.4	70.9	16.7	6.3	0.14φ	4.84	5.00	2.96	15
.871	NW-87	0.1	6.3	78.4	15.2	5.2	0.07φ	4.73	5.00	3.41	16
.885	NW-89	0.1	1.0	82.3	16.6	5.6	0.01φ	5.38	5.00	3.92	15
.922	NW-90	0.0	6.3	77.9	15.8	5.8	0.07φ	4.85	5.00	2.92	10
1.000	NW-86 (04 cm)	0.0	6.7	89.8	3.5	1.2	0.07φ	4.49	5.00	3.62	15
FACTOR 3											
.695	NW-95	0.6	49.9	36.6	12.9	4.4	1.02φ	3.95	3.00	1.97	16
.705	BI-44	0.0	49.6	27.9	22.5	6.2	0.98φ	4.08	2.50	1.32	30
.809	BI-58	0.1	55.8	30.6	13.6	5.3	1.26φ	3.74	3.00	1.67	15
.845	BI-30	0.2	60.2	24.6	15.0	6.0	1.52φ	3.48	3.00	1.73	26
.873	NW-58	0.0	61.7	30.1	8.2	3.0	1.61φ	3.50	3.00	2.20	17
1.000	NW-94 (07 cm)	0.0	76.2	18.2	5.6	3.4	3.21φ	2.46	2.50	1.64	17
FACTOR 4											
.664	NW-45	1.2	35.2	32.4	31.3	9.9	0.57φ	5.48	3.00	−1.13	40
.843	BI-51	1.8	44.4	37.1	16.7	6.0	0.86φ	4.14	5.00	−1.59	35
.983	BI-53	2.7	28.6	46.7	22.1	6.5	0.45φ	4.59	5.00	−2.07	42
1.000	BI-46	2.9	19.9	48.9	28.3	7.5	0.30φ	5.42	5.00	−2.63	46

bers of this group shows that clay contents higher than 50% with 12ø fractions contributing over 15% characterize this group. Median phi values and major modes lie close to or finer than the clay-silt boundary. Other characteristics include a dearth of gravel and sand and water deeper than 39 m. In general this factor represents the finest-grained sediments encountered in the submarine river channels and extending offshore in deep water.

Factor II. The six highest loaded members of this group consist of more than 70% silt (the extremal is 90% silt), and have median phi and major mode values very close to 5ø. Water depth tends to be shallow, less than 22 m. This group occupies most of the bank regions in the East Siberian Sea.

Factor III. This factor and factor IV do not include many samples but help explain the distribution of the remaining data. The extremal sample, NW 094, is from 7 cm below the surface but shows the highest sand content in the East Siberian Sea, 76.2%. The other samples range down to 50% sand. The major mode lies in the sand range and the median in the fine sand range. Silt is secondary and clay subordinate in this group. The samples in this group are widely spaced but are associated with bank regions.

Factor IV. Only four samples comprise this group but their characteristics and associations are unusual. The presence of small amounts of gravel, moderate sand, and sizes continuous through the colloid range produces rather poor sorting (Inman deviations 2.93 to 3.81). They are polymodal and are associated with water depths of 35 to 46 m, in scattered locations off the Chukotskiy coast.

Heavy Minerals

The percentage of heavy minerals in the 4ø fraction (Fig. 8) varies from 0.15 to 11.6. The

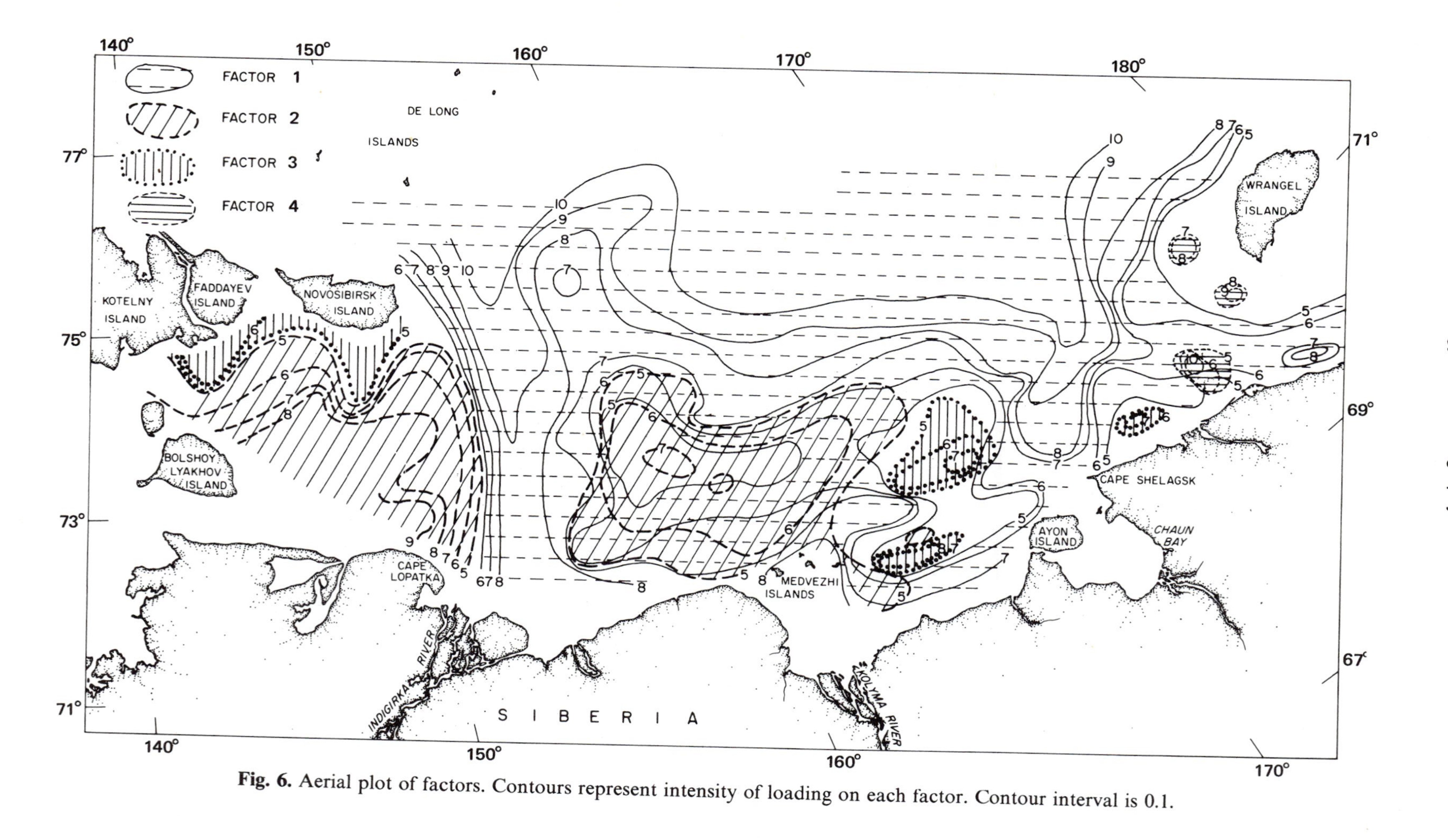

Fig. 6. Aerial plot of factors. Contours represent intensity of loading on each factor. Contour interval is 0.1.

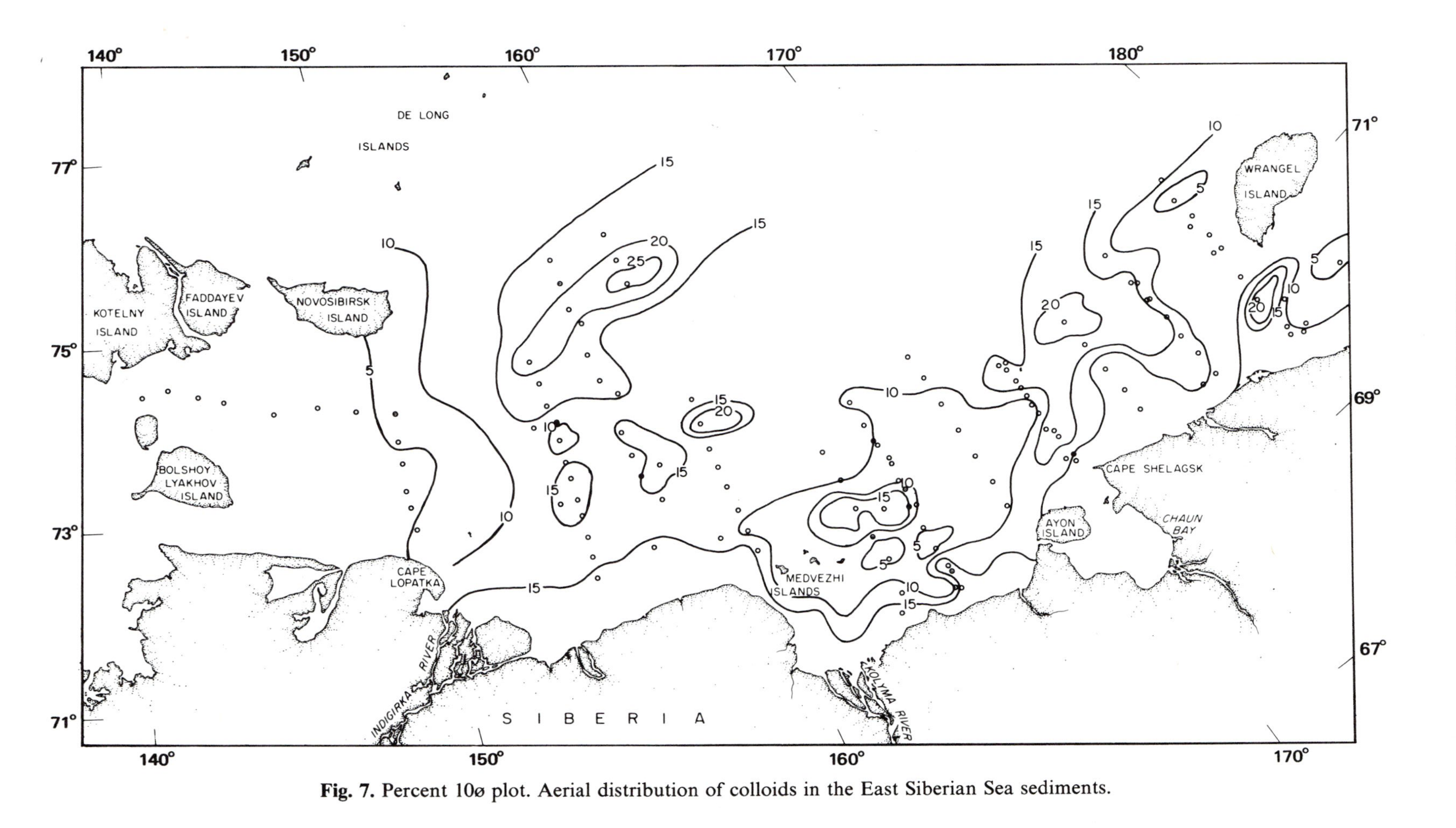

Fig. 7. Percent 10ø plot. Aerial distribution of colloids in the East Siberian Sea sediments.

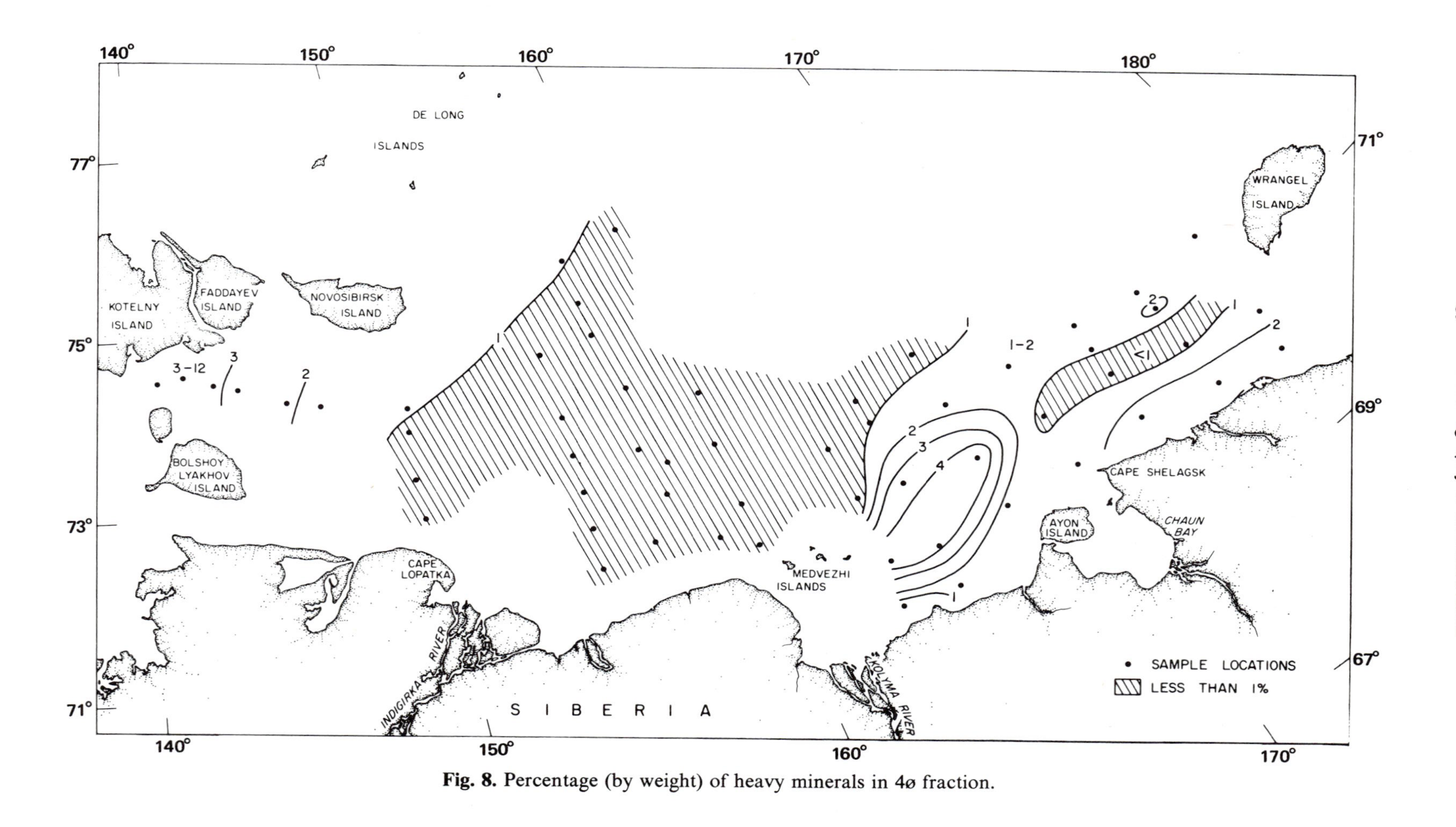

Fig. 8. Percentage (by weight) of heavy minerals in 4ø fraction.

typical assemblage throughout the area (Table 2 and Fig. 9) consists of amphibole, clinopyroxene, and epidote, with appreciable quantities of the following accessory minerals: garnet, orthopyroxene, muscovite, chlorite, biotite, apatite, zircon, tourmaline, sphene, and opaque minerals (illmenite, magnetite, hematite). Traces of pyrite, glaucophane, and carbonate (siderite?) occur locally.

Amphiboles. Common hornblende comprises about 40% of the 4ø heavy mineral assemblage. The maximum concentration, 68%, occurs off the mouth of the Indigirka River just north of Cape Lopatka. There are relatively low concentrations in the shoal area east of the Medvezhi Islands. Hornblende occurs largely as green to brown elongate cleavage fragments of rather strong birefringence, moderate pleochroism and inclined extinction. Traces of basaltic hornblende and glaucophane occur throughout the area.

Clinopyroxenes. Augite is the most abundant of the clinopyroxenes. It varies from about 10% of the heavy minerals from the western regions to over 35% north of Ayon Island. It occurs chiefly as greenish to greenish-brown somewhat-rounded elongate cleavage fragments. It is characterized by high birefringence, oblique extinction, and refractive index that changes from higher to lower than that of Hyrax (1.71) upon rotation in plane polarized light. Diopside comprises from 2 to 5% of the clinopyroxenes.

Epidote. Epidote comprises about 10 to 20% of the heavy minerals, with as much as 30% west of Wrangel Island. Epidote is distinguished by high refractive index, yellowish tint, inclined extinction, weak pleochroism, and "compass needle" interference figure. Pistacite and clinozoisite make up most of the epidote assemblage, with rare traces of zoisite. Various degrees of alteration are exhibited by these minerals. Some are exceptionally clear with no inclusions, while others are highly altered and difficult to distinguish from rock fragments.

Garnet. Garnet is a major constituent of the heavy minerals in the western regions, with concentrations of about 15 to 25% near the New Siberian Islands. These values decrease toward the east, with as little as 1.1% (NW 64) just off the Kolyma River mouth. Relatively high concentrations also occur off Wrangel Island.

Orthopyroxene. Orthopyroxenes are generally less than 5% of the heavy minerals but were major fractions of several samples from the eastern regions. The maximum concentrations are in the three samples from along the coast east of Chaun Bay (Fig. 9). The orthopyroxenes occur as slightly rounded elongate prisms with varying degrees of pleochroism and refractive index. The most common of these show faint pleochroism (green to pinkish-orange), a refractive index above that of Hyrax, and generally give a good biaxial negative interference figure. These properties, particularly the high refractive index, are characteristic of the iron-rich ferrohypersthene. True hypersthene is present in much smaller amounts and is distinguished by its lower refractive index and more intense pleochroism.

Apatite. The chief minerals included under "Others" (Fig. 9) are: apatite, zircon, tourmaline, and sphene. Apatite is generally from 1 to 5% of the heavy minerals, but up to 10% is found in a strip extending northward off the Indigirka River mouth. Apatite generally occurs as colorless, unaltered, rounded grains. It is characterized by low birefringence and a refractive index below that of Hyrax. Several very fresh euhedral prisms were noted.

Zircon. Zircon is a trace constituent of almost every sample but makes up about 5% of the heavy minerals near the New Siberian Islands. It occurs as colorless, elongate, euhedral, or slightly rounded prismatic crystals. It is identified by its high refractive index, high birefringence, and straight extinction, and inclusions are almost always present.

Tourmaline. Tourmaline is most abundant (up to 2.5%) in the broad bank region northeast of the Indigirka River mouth and rare or absent in the adjacent regions. It occurs as subrounded prismatic grains with high birefringence, straight

Sphene. Sphene comprised 2 to 4% of the heavy minerals near the Indigirka submarine valley and to the west. Elsewhere, it is generally less than 1% of the heavy minerals. It occurs mainly as colorless to brownish-yellow well-rounded grains of very strong birefringence and extreme dispersion of the optic axes, with absence of total extinction.

Mica. The micas (muscovite, chlorite, and biotite) are the most variable in concentration of all the heavy minerals present and have been excluded from the results. They make up from 1 to 35% of the heavy minerals assemblage, generally occurring in larger amounts in the finer sediments.

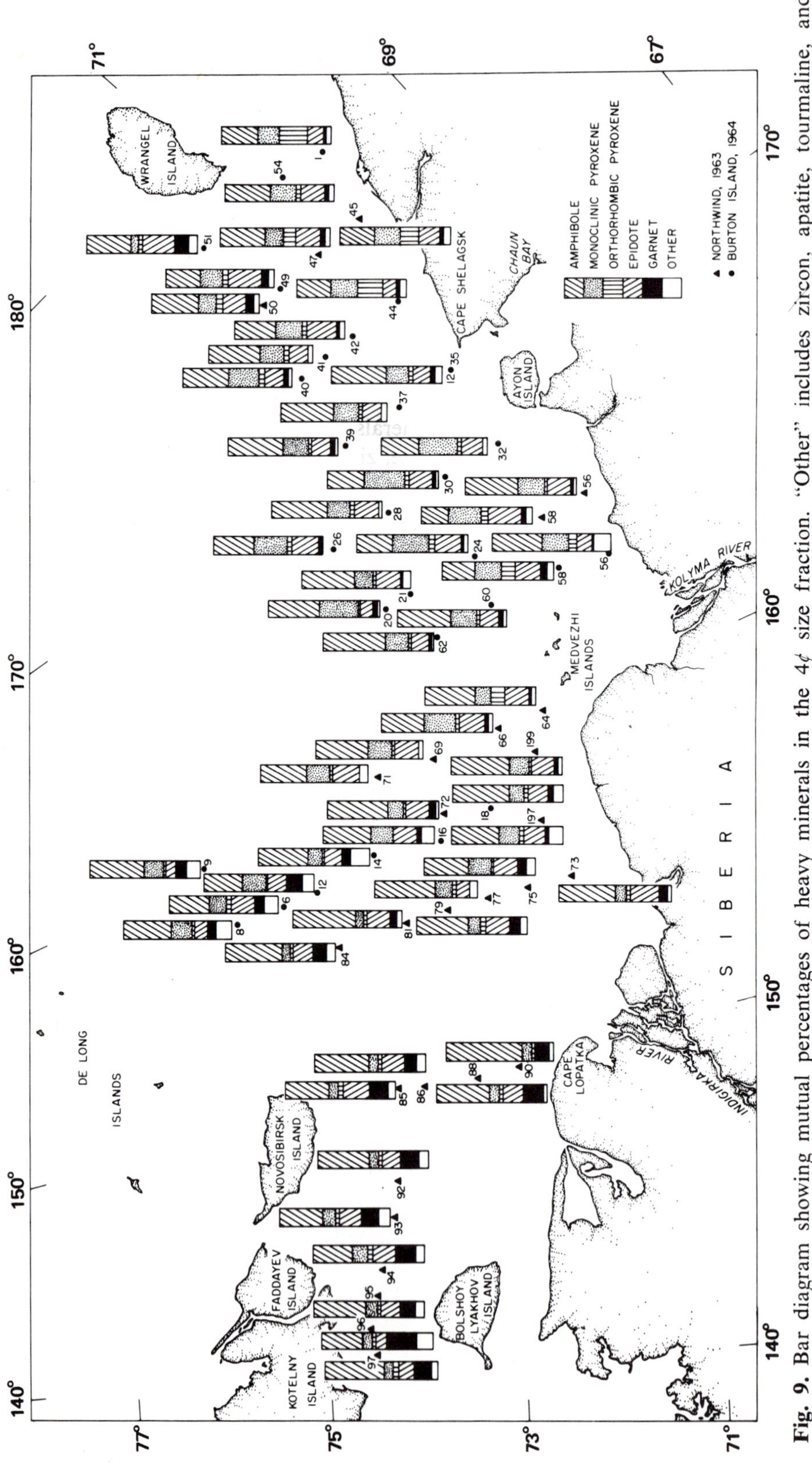

Fig. 9. Bar diagram showing mutual percentages of heavy minerals in the 4φ size fraction. "Other" includes zircon, apatite, tourmaline, and sphene. Lithic fragments, opaques, and micas are excluded.

Table 2 Heavy Minerals from 4ø Size Fraction Given in Mutal Percentages. Micas, opaques, and lithic fragments are excluded.

Sample no.	Clino-pyroxene	Amphibole	Epidote	Apatite	Sphene	Ortho-pyroxene	Zircon	Tourma-line	Garnet	Other
BI-1	18.6	33.8	15.7	02.6	01.1	25.8	01.9	00.0	02.6	00.0
BI-6	16.9	37.2	20.3	04.8	02.8	04.8	00.7	01.7	08.6	02.1
BI-8	18.7	44.2	12.0	05.6	03.0	03.0	02.6	02.2	07.9	00.7
BI-9	18.4	48.5	09.8	05.5	03.1	01.8	01.2	02.5	09.2	00.0
BI-12	21.9	35.5	16.7	06.2	01.9	00.6	01.9	01.2	14.2	00.0
BI-14	14.0	45.0	15.5	10.0	02.1	02.7	02.1	07.3	00.0	06.5
BI-16	20.2	42.2	21.4	08.6	01.9	00.0	00.8	00.8	03.9	00.0
BI-18	13.9	51.6	17.4	00.0	00.0	04.0	00.0	00.0	04.0	00.0
BI-20	35.6	46.3	11.1	02.3	00.0	00.3	01.0	00.0	03.4	00.0
BI-21	16.7	48.1	22.2	03.7	00.5	02.8	00.5	00.0	01.4	04.2
BI-24	33.1	32.8	22.8	01.9	01.6	04.8	01.0	00.0	02.6	00.3
BI-26	29.5	37.3	25.4	01.1	00.4	03.7	00.0	00.0	02.6	00.0
BI-28	20.5	51.1	21.8	01.3	00.3	03.2	00.9	00.0	00.9	00.0
BI-30	36.7	33.9	21.2	01.3	00.6	04.1	00.3	00.0	01.9	00.0
BI-32	34.4	34.7	19.3	01.5	00.8	03.9	01.2	00.0	04.2	00.0
BI-35	21.1	49.8	15.2	04.2	00.8	04.2	00.8	00.0	03.8	00.0
BI-37	22.3	48.3	17.8	00.0	02.2	03.7	00.7	00.4	04.5	00.0
BI-39	24.6	49.1	17.5	02.5	00.4	01.8	00.4	00.0	03.9	00.0
BI-40	28.0	42.7	16.6	00.6	00.6	05.1	01.3	00.0	03.8	01.3
BI-41	18.0	47.4	16.6	02.4	00.9	04.7	00.5	00.0	04.3	05.2
BI-42	24.4	36.3	27.5	02.0	01.4	04.3	01.1	00.7	02.0	00.2
BI-44	23.2	31.1	11.6	02.3	03.0	23.5	02.0	00.3	03.0	00.0
BI-49	20.6	32.4	30.1	02.0	01.4	05.2	01.7	00.3	06.3	00.0
BI-51	07.2	38.2	30.4	02.8	00.6	03.1	06.6	00.0	11.0	00.0
BI-54	23.0	41.5	23.0	00.5	00.9	02.7	01.2	03.3	00.9	03.0
BI-56	20.9	45.3	13.7	05.8	01.8	07.2	00.4	00.7	01.4	02.6
BI-58	24.0	28.3	22.7	02.1	01.3	12.9	02.6	00.4	05.6	00.0
BI-60	23.4	48.5	16.5	03.0	00.6	03.0	00.8	00.6	03.6	00.3
BI-62	21.9	56.2	16.2	01.3	00.3	01.3	00.3	00,3	01.9	00.0
NW-45	22.0	31.4	19.3	01.8	02.1	16.5	03.0	00.0	03.9	00.0
NW-47	16.4	40.7	22.0	01.6	01.0	10.5	02.0	00.3	04.6	00.0
NW-50	17.7	42.6	20.4	02.4	01.9	06.1	01.3	00.5	06.3	00.8
NW-56	24.5	47.2	21.9	00.0	00.4	01.9	01.9	00.4	01.5	00.4
NW-58	30.1	23.7	29.2	01.4	00.5	06.2	05.0	00.0	03.9	00.0
NW-64	14.7	44.3	22.0	02.7	00.0	12.5	00.0	01.6	01.1	00.0
NW-66	27.2	38.6	22.8	03.6	01.0	03.3	00.3	00.3	03.0	00.0
NW-69	21.6	46.3	21.3	02.8	00.6	02.2	00.9	00.3	0.40	00.0
NW-71	21.5	41.4	23.9	02.8	00.6	02.1	00.3	00.3	07.1	00.0
NW-72	15.2	54.1	20.3	02.1	00.3	01.4	01.0	00.3	05.2	00.0
NW-73	09.3	50.5	25.8	03.1	00.0	03.1	00.0	00,0	07.2	01.0
NW-75	20.1	39.5	22.0	05.4	01.0	01.3	02.2	01.3	07.3	00.0
NW-77	15.3	54.0	11.9	04.5	02.0	03.5	01.5	00.5	06.9	00.0
NW-79	10.5	46.0	21.9	01.3	03.4	05.1	00.8	00.4	09.3	01.3
NW-81	07.8	56.5	20.1	01.9	03.0	03.0	01.1	00.4	06.3	00.0
NW-84	07.8	50.2	18.9	02.3	02.6	00.7	11.1	00.7	11.1	00.0
NW-85	06.3	39.6	23.8	01.6	01.8	04.2	04.4	00.4	16.4	01.6
NW-86	05.7	47.6	21.7	02.4	00.7	02.4	06.9	00.0	10.2	02.4
NW-88	03.7	54.1	13.7	00.0	00.9	02.8	00.9	00.9	22.9	00.0
NW-90	07.2	69.3	12.6	01.4	01.0	02.7	01.4	00.0	03.4	02.0
NW-92	08.3	46.1	16.4	02.3	02.8	02.5	04.8	00.2	16.6	00.0
NW-93	11.4	39.4	18.3	03.4	01.8	03.7	05.3	00.2	15.4	00.0
NW-94	12.4	35.4	21.0	02.0	02.8	04.3	03.0	00.4	18.4	00.4
NW-95	10.2	47.0	16.7	01.5	02.8	03.2	04.2	00.2	14.0	00.2
NW-96	08.1	36.7	09.9	01.6	02.9	03.7	09.4	00.0	27.5	00.0
NW-97	07.7	52.0	13.1	02.9	02.3	05.0	02.3	00.0	14.8	00.0
NW-197	16.6	43.4	18.3	07.6	04.4	03.4	01.4	02.4	03.4	02.1
NW-199	17.9	53.2	20.6	02.4	00.0	01.2	00.4	00.8	03.6	00.0

Light Minerals

A cursory examination of the light minerals (S.G. less than 2.89) of the 4ø fraction shows they are composed primarily of quartz, orthoclase, and plagioclase. Plagioclase is only a small fraction of the feldspars in the western regions, but reaches major proportions east of the Kolyma River mouth.

Lithic fragments are a minor portion of the sediments south of the New Siberian Islands, increasing considerably toward the east, and reaching their highest concentrations along the coast east of Chaun Bay.

Discussion

The broad distribution of factor I sediments reflects the general low energy conditions in the East Siberian Sea. The regions of highest loaded factor I sediments correlate well with regions of high percent 12ø (compare Figs. 6 and 7). It is apparent that factor I represents quiet water conditions and deposition of mostly clays and colloids. The Kolyma and Indigirka sediment sources are definitely factor I but of lower loading than the deeper areas of the submarine valleys, indicating that considerable amounts of silt accompany the clay and colloid in the river effluent. The dominance of clayey-silts, as noted in the ternary diagram (Fig. 4), and the widespread pattern of factor I sediments with low loading (Fig. 6), reflect the characteristic mechanical weathering process, congelifraction, of the Siberian permafrost region. The intense stratification developed by the river effluent during summer and the general eastward current patterns assures transport of the clays and colloids across the entire sea. The plateau regions, being quite shallow, apparently experience stronger currents and are bypassed by clays and colloids with only silts deposited. Factor II sediments typify these regions.

Coarse sediments are rare but are included in factors III and IV. Factor III sandy sediments occur close to shore off the Chukotskiy coast and the New Siberian Islands. The bank region north of the Kolyma submarine valley locally shows considerable amounts of sand, but the presence of a small patch of factor II sediment and broad background of mixed sediment where no factor is dominant, suggests a complex environment. Local winnowing probably is occurring on this bank area.

Factor IV sediments, gravelly, poorly sorted, and polymodal, are found sporadically off the Chukotskiy coast and Wrangel Island in depths of 35 to 46 m. These gravelly sandy muds may be related to ice rafting (see discussion below). No evidence for ice rafting was found elsewhere in the East Siberian Sea sediments.

Heavy mineral analysis provides a direct means of determining sediment sources and is especially valuable when there are several sources whose mineralogic make-up varies considerably. From a qualitative standpoint heavy minerals in the East Siberian Sea sediments are similar throughout the area, consisting primarily of amphibole, clinopyroxene, and epidote, with an assortment of accessory minerals. Distinctive sources, however, are revealed by the varying percentages of certain specific minerals as well as by the total heavy mineral contribution.

In the central portion of the East Siberian Sea, between the Indigirka and Kolyma submarine valleys, sediments are characterized by notably low concentrations of heavy minerals (Fig. 8). The predominant source of this material is undoubtedly the Indigirka River. This river has an extremely low gradient and meanders through several hundred kilometers of marshy lowland before entering the sea. The lower part of the Indigirka drainage basin is covered primarily by Quaternary alluvium derived for the most part from marine sediments and acid igneous rocks which inherently are low in heavy mineral concentrations. This central region is further characterized by relatively high apatite and tourmaline concentrations.

Sediments covering the New Siberian Shoal are characterized by relatively high concentrations of garnet and zircon and low concentrations of clinopyroxene, suggesting a western source for these sediments. These distinctive mineral characteristics extend into the central offshore region, suggesting a certain degree of mixing between the Indigirka River and the New Siberian Shoal sediments (see Fig. 10).

The eastern third of the East Siberian Sea is the most complex mineralogically. It is characterized by high percentages of heavy minerals and a general increase in clinopyroxene content. These attributes appear to characterize sediments introduced by the Kolyma River, although other sources are contributing to this region. The main tributaries of the Kolyma River flow in from the highlands to the east, intersecting regions with

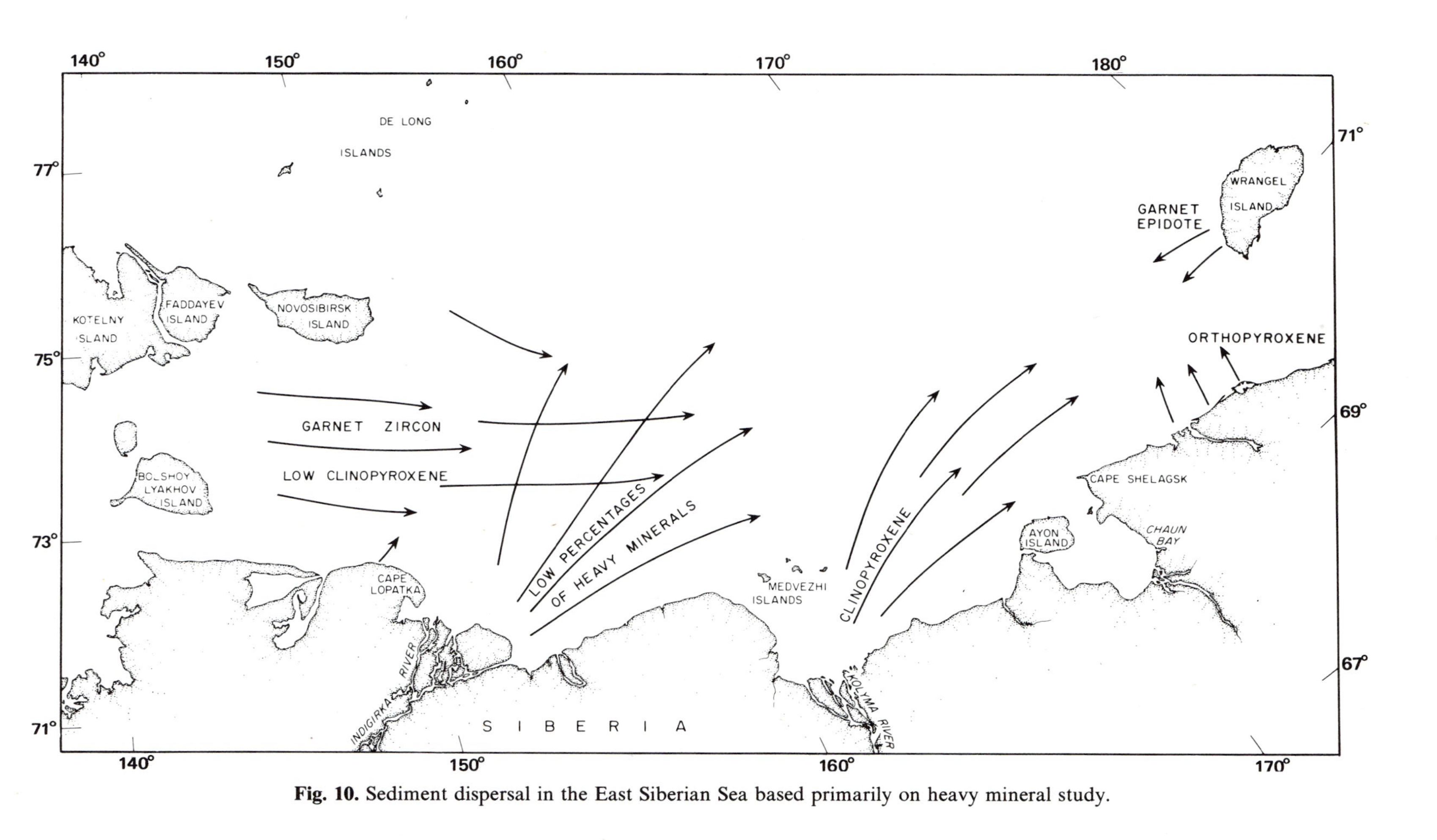

Fig. 10. Sediment dispersal in the East Siberian Sea based primarily on heavy mineral study.

numerous outcrops of basic igneous material along with the more acidic granites and syenites. Cretaceous volcanics occur in patches and crop out along the coast east of Cape Shelagsk. The highest concentrations of clinopyroxene occur in the area immediately northeast of the Medvezhi Islands. Off the Chukotskiy coast in the extreme eastern portion of the East Siberian Sea the heavy minerals show rather large local variations reflecting sediments from several sources. Unusually high concentrations of orthopyroxene are found in sediments just off the coast east of Cape Shelagsk. These undoubtedly are contributed by the Cretaceous volcanic rocks cropping out along the coast. West of Wrangel Island sediments rich in garnet and epidote and poor in clinopyroxene suggest that this island also provides a local source.

On the basis of the grain size and mineral analyses, the East Siberian Sea sediments can be divided into three regional groups, each with more or less distinctive characteristics: the New Siberian Shoal; the central plateau; and the region including and east of the Kolyma submarine valley. The New Siberian Shoal region is very shallow (5 to 15 m). Currents here are stronger than in most areas of the East Siberian Sea, as reflected by a dominant silt sediment, sand close to shore and in the channel immediately south of Kotelny Island, and moderate to high heavy mineral concentrations. Relatively high concentrations of zircon and garnet, typically resistant minerals, suggest reworking of these sediments. Low percentage of lithics and low relative amounts of plagioclase in this region are further evidence of reworking. Even though the 5ø silts characterizing this region would likely be transported during reworking, it is felt that this material is simply redistributed locally.

The large central region is dominated by sediments supplied by the Indigirka River although some mixing with sediment derived from the west apparently occurs in the offshore region. Sediments here are silty-clays and clayey-silts with very low heavy mineral concentrations. The smooth plateau between the Indigirka and Kolyma submarine valleys is 10 to 15 m deeper than the New Siberian Shoal. Currents here are slower but still capable of transporting fines out of the region. The central portion of the central plateau is covered with factor II silty sediments but finer-grained clayey-silts cover most of the region, and particularly the Indigirka submarine valley where currents are probably lowest. It is possible that the fines are maintained in suspension during the summer open-water season and settle to the bottom only during the winter ice cover when currents become negligible and the density stratification in the water column breaks down.

The Kolyma–Chukotskiy region, east of the Kolyma Delta, is characterized by relatively deep irregular topography and variable sediment texture and mineralogy. A generally high clinopyroxene content reflects the strong influence of the Kolyma River, whereas local occurrences of orthopyroxene, and garnet and epidote reveal contributions from the Chukotskiy coast and Wrangel Island, respectively. Winnowed sandy sediments and gravelly-sandy-muds break the pattern of silty-clay and clayey-silts carried into the region from the Kolyma and further west. The gravelly-sandy-muds occur in the regions where a local shore-line source is clearly indicated in the mineralogy. Ice rafting is strongly indicated by this association. All of the gravel found is associated with coarse sand and, thus, does not reflect deposition into a muddy sediment characteristic of the classical case of ice rafting. The classical case, however, involves glacial ice. The only mechanism for ice rafting in the East Siberian Sea is sea ice—either grounded ice or fast ice which is connected directly to shore in winter. Larionova (1959) classifies the East Siberian Sea platform as a transgressive continental shelf with a coast line of thermal abrasion, and ice undoubtedly plays an important role in altering the coast line. Coarse material from shore-line cliffs may fall onto this ice (see Holmes and Creager, this volume, Fig. 10). However, a significant amount of sand along with finer material is also incorporated into the bottom (Holmes and Creager, this volume, Fig. 9), especially in the case of fast ice, which actually freezes to the bottom within beach zones. This ice is thin and melts rapidly during the spring breakup. Thus, its entire load may be unloaded in one place. Material incorporated into thick glacial ice, on the other hand, is often winnowed to a coarse gravel (perhaps containing cobbles and boulders), through the process of surface melting and removal of fines by surface streamlets, before it is unloaded. If this unloading occurs in mud, the classical ice-rafting relationship occurs. There is no evidence for this mechanism in the samples studied.

An alternate explanation for the patches of gravelly sandy muds is that it is material related to deltaic deposition associated with lower sea-level stands. The depths at which these samples occur (35 to 46 m) are consistent with the stillstand described from the Chukchi Sea (Creager and McManus, 1965) and the Laptev Sea (Holmes and Creager, this volume). The Kolyma submarine valley is distinct only in its central portion and may have continued along several alternate routes—north or south of Wrangel Island or possibly even along the mainland coast where closed basins are now found. The position of the mouth of the Kolyma during lower sea-level stands is thus unclear.

Further complicating the picture in the far eastern portion of the East Siberian Sea are seasonal current patterns in Long Strait between Wrangel Island and the mainland (Coachman and Rankin, 1968) which permit exchange of sediment with the Chukchi Sea.

Summary and Conclusions

The East Siberian Sea occupies a broad shallow-water platform off northeastern Siberia. For the most part it represents a drowned extension of the vast Indigirka–Kolyma lowland which is covered by marshy tundra deposits that have developed in an arctic environment.

It is extremely shallow and monotonously flat except where interrupted by two broad submarine valleys related to the Indigirka and Kolyma Rivers. Depths in the western and central regions average between 10 and 25 m. This portion of the shelf appears to be an extension of the marshy coastal plain of the mainland, with very gradual slopes from land out to about 250 km. East of the Kolyma River the nearshore slopes are noticeably steeper, and within about 30 km of the shore, depths exceed 30 m and in places 50 m. This region is characterized by knolls and depressions that may be related to erosional and depositional processes operating during lower stands of sea level.

Sediments are extremely fine grained for a shallow shelf environment, often containing over 15% colloidal material (finer than 11ø). This is attributed to the fine-grained nature of the source material (which is transported by low gradient rivers and developed under permafrost conditions) and to the low energy conditions in the East Siberian Sea imposed by lengthy ice cover, sluggish currents, and negligible tides.

Although ice conditions suggest that ice rafting might be an important dispersing agent, this mechanism appears to be insignificant, except perhaps in the nearshore regions.

On the basis of textural and mineral analyses the East Siberian Sea sediments can be divided into three regional groups: the New Siberian Shoal, the central plateau, and the region including and east of the Kolyma submarine valley. The New Siberian Shoal is characterized by silt-rich sediment high in garnet and zircon and low in clinopyroxene. This material is derived from the New Siberian Islands and is subjected to appreciably greater currents than occur throughout most of the East Siberian Sea. The relatively strong currents are attributed to shallow depths and also the current accelerations through the straits to the west. The mineralogy suggests that these sediments have undergone reworking.

The central plateau is dominated by sediments supplied by the Indigirka River. Sediments here are silty-clays containing low amounts of heavy minerals. In the central portion of this plateau some cleaner silty sediments occur, indicating that current activity is sufficient to keep most of the fines from settling except perhaps during periods of ice cover.

The third regional group, lying east of the Kolyma River, is characterized by relatively deep irregular topography and variable sediment texture and mineralogy. Winnowed sandy sediments and gravelly sandy muds break the pattern of silty-clay and clayey-silts. The Kolyma River has introduced most of the sediment to this region, although local shore-line sources are indicated clearly by the mineralogy. The patches of gravelly sandy muds appear to be related to ice rafting, although there is the alternate possibility that they are the result of deltaic deposition during lower stands of sea level.

Acknowledgments

This paper is Contribution No. 733 of the Department of Oceanography, University of Washington, Seattle. This research was supported by National Science Foundation Grants GA-808 and GA-28002 to the Department of Oceanography.

References

Coachman, L. K., and D. A. Rankin. 1968. Currents in Long Strait, Arctic Ocean. *Arctic*, 21(1):27–38.

Codispoti, L. A., and F. A. Richards. 1968. Micronutrient distributions in the East Siberian and Laptev Seas during summer 1963. *Arctic*, 21:67–83.

Creager, J. S., and D. A. McManus. 1965. Pleistocene drainage patterns on the floor of the Chukchi Sea. *Marine Geol.*, 3:279–290.

Flint, R. F., and H. G. Dorsey. 1945. Glaciation of Siberia. *Geol. Soc. Am. Bull.*, 56:89–106.

Imbrie, J., and T. H. Van Andel, 1964. Vector analysis of heavy-mineral data. *Geol. Soc. Am. Bull.*, 75:1131–1156.

Inman, D. L. 1952. Measures for describing size of sediments. *J. Sedimentary Petrol.*, 22:125–145.

Krumbein, W. C., and F. J. Pettijohn. 1938. *Manual of Sedimentary Petrography*. New York: Appleton-Century-Crofts, 549 pp.

Larionova, A. N. 1959. Attempts to classify elevations and depressions of the ocean bottom. *Traveling on the Sea Bottom.* Leningrad: The Hydrometeorological Publishers, 20–70.

McManus, D. A., J. C. Kelley, and J. S. Creager. 1969. Continental shelf sedimentation in an arctic environment. *Geol. Soc. Am. Bull.*, 80:1961–1984.

Nalivkin, D. V. 1960. *The Geology of the U.S.S.R.* New York: Pergamon Press, 170 pp.

Naugler, F. P. 1967. *Recent sediments of the East Siberian Sea.* Masters thesis. Seattle: University of Washington, 71 pp.

Samoilov, I. V. 1952. Ust'ia rek. Gosudarstvennoe Izdatel'stvo Geograficheskoi Literatury, Moskva, 526 pp.

Shepard, F. P. 1954. Nomenclature based on sand-silt-clay ratios. *J. of Sedimentary Petrol.*, 24(3):151–158.

Suslov, S. P. 1961. *Physical Geography of Asiatic Russia.* San Francisco: W. H. Freeman and Co., 594 pp.

Sverdrup, H. U. 1929. The waters on the north Siberian Shelf. *Scientific Results Norwegian North Polar Expedition, MAUD* (1918–1925), 4, 131 pp.

———, M. W. Johnson, and R. H. Fleming. 1942. *The Oceans: Their Physics, Chemistry, and General Biology.* New York: Prentice-Hall, 1087 pp.

U.S. Navy Hydrographic Office. 1954. Sailing directions for the northern U.S.S.R. Vol. 3. Hydrographic Office Publication 137c, 346 pp.

Chapter 9
Holocene History of the Laptev Sea Continental Shelf [1]

M. L. HOLMES [2] AND J. S. CREAGER [2]

Abstract

The 400-km-wide, low gradient Laptev Sea continental shelf consists of flat terrace-like features at regular depth intervals from 10 to 40 m below present sea level. The five large submarine valleys traversing the shelf do not continuously grade seaward, but contain elongated, closed basins. These terraces and closed basins plus deltaic sediments associated with the submarine valleys quite possibly mark sea level stillstands, and enable reconstruction of the paleogeography of the Laptev Sea shore line at five periods during post-Wisconsin (Holocene) time.

Radiocarbon dates on the silty-clay to clayey-silt sediments from cores of the northeastern Laptev Sea indicate average sedimentation intensity of 2 to 15 mg/cm^2/yr. The presence of manganese nodules and crusts in surface samples from less than 55 m depths and a general decrease in total foraminiferal abundances with depth in the cores suggest that the present deposition rate is less than when sea level was lower. The main components of the shelf deposits are nearshore sediments which were spread over the shelf as Holocene sea level fluctuated and marine currents distributed modern fine sediment. Rare silty-sand layers and the coarser nuclei of the manganese crusts and nodules indicate ice rafting. However, this mechanism is probably only locally important as a significant transporting agent.

[1] Contribution no. 734, Department of Oceanography, University of Washington.

[2] Department of Oceanography, University of Washington, Seattle, Washington 98105, U.S.A.

Introduction

The Laptev Sea is one of three epicontinental seas lying along the northern coast of central Asiatic Russia (see Fig. 1, Naugler et al., this volume). Field work was carried out during July and August 1963, aboard the U.S. Coast Guard ice-breaker *Northwind.* Ninety-nine stations were occupied along approximately 4000 km of track, and bottom samples collected include 58 gravity cores and 62 grab samples (Fig. 1). This paper is concerned with the analysis of bathymetric data and the study of 14 cores from the northeastern Laptev Sea, and discusses sea-level history and paleogeography of coast lines of the Laptev Sea during Holocene (post-Wisconsin) time.

The climatic and physiographic features of the surrounding area are described by Suslov (1961, p. 125), and Naugler et al. (this volume). Sverdrup (1927), Suslov (1961, p. 162), and Codispoti (1965) discuss the physical oceanography of the Laptev Sea. The extent of various continental glaciations in Siberia are discussed by Flint and Dorsey (1945), Donn et al. (1962), and Flint (1971). At various times during the Pleistocene, continental ice sheets covered only the western shores and sea floor of the Laptev Sea, and did not extend east of the present-day Olenek River. A small thin ice sheet, approximately 250 km in diameter, spread out from the New Siberian Islands (Flint, 1971, p. 665). Saks (Samoilov, 1952,

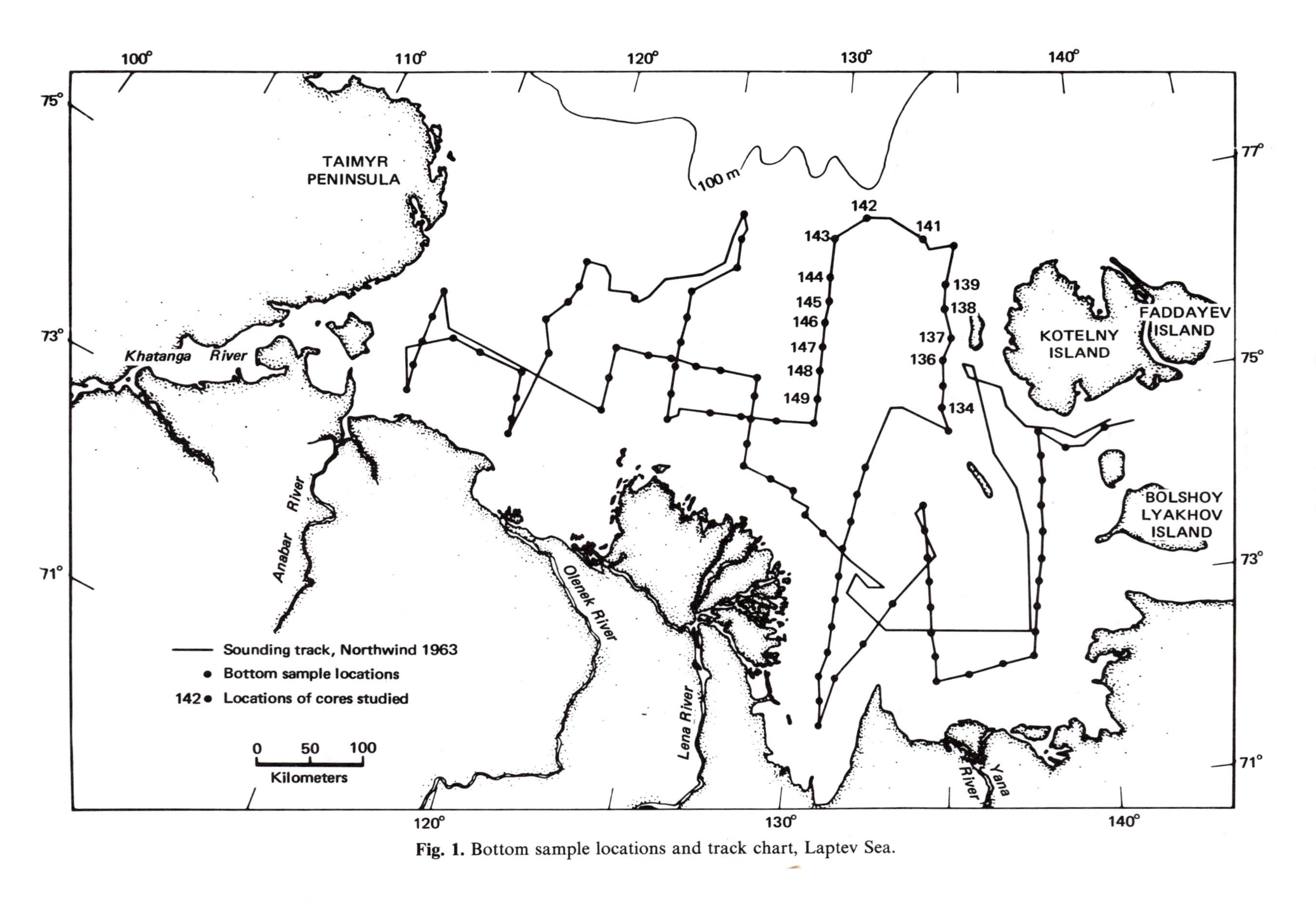

Fig. 1. Bottom sample locations and track chart, Laptev Sea.

p. 307) briefly describes the mouths of the Lena and Yana Rivers (Fig. 1) during the Quaternary. He states that during the Karginsky transgression (about 11,500 years B.P.) the Lena formed two branches near the area of its present delta. The mouth of one branch was 300 km northwest of the present delta, and a separate branch flowed northward. Saks (Samioilov, 1952, p. 307) also states that the mouth of the Yana River during this period was slightly more than 300 km north of its present position. Gusev (1959) discusses the evolution of the coastal plain during the Quaternary, but gives coast-line positions only for Wisconsin time and for an unspecified period during Neogene time.

Results and Discussion

Bathymetry

Sounding data from the *Northwind* cruise, supplemented by nearly 500 Soviet spot soundings, were used to compile the bathymetry of the Laptev Sea (Fig. 2). The bathymetry of the continental slope is from a chart by Gakkel' (1958). The continental shelf covers approximately 460,000 km^2, and varies in width from 300 km in the western part to over 500 km in the east. The shelf break occurs at 50 to 60 m depth, much shallower than the 200-m break mentioned by Suslov (1961, p. 261) and Eardley (1964) for this area. The shelf is very flat, with gradients ranging from undetectable, in some areas, to about 5 m/km. The bathymetry is more irregular in the northwestern area, probably due in part to the action of the continental ice sheet which covered portions of this region during part of the Pleistocene. An extensive shoal (called Stolbovoy Shoal in this paper), with depths less than 15 m, occupies the central area of the eastern Laptev Sea. The shoal is probably a primary structural feature representing the northward extension of the foothills of the Verkhoyansk Range. Numerous small banks, less than 5 m in depth, occurring on Stolbovoy Shoal were probably formed as a result of the melting of ice cores in small islands. This process has been documented in the case of Vasilevsky Bank by Grigorov (1946).

Five large submarine valleys cross the Laptev Sea shelf and are named after the rivers with which they seem to be associated. A conspicuous feature of all of these valleys is that they are not continuously graded seaward, but are marked by a series of linear depressions. The depressions in the Yana and eastern Lena valleys occur at depths of 25 and 40 m, while those of the Olenek, western Yana, and Anabar-Khatanga valleys exhibit closures at depths of 30 and 45 m. Similar depressions have been found along Hope valley in the Chukchi Sea (Creager and McManus, 1965) and in the Kolyma and Indigirka valleys of the East Siberian Sea (Naugler et al., this volume).

Extensive flat areas occur at remarkably regular depth intervals on the shelf. The greater part of Stolbovoy Shoal is between 10 and 15 m in depth, and in the western Laptev Sea between the Anabar River and the Lena Delta the 10- and 15-m isobaths are also very widely spaced. Northeast of Kotelny Island, the 20- and 25-m contours may delineate another large flat area, but the bathymetry here is based only on Soviet spot soundings. In the central area of the shelf north of the Lena Delta, two large plains are well developed at depths of 30 to 35 m and 40 to 45 m.

The subaerial portions of the Lena Delta do not extend very far seaward (Fig. 1). The Gulf of Bourkhaya and Oleneksky Bay probably represent delta flank depressions (Bates, 1953) which have been formed by a combination of wave and current action, local crustal downwarping, and compaction of fine-grained sediments with time.

The larger size of the Gulf of Bourkhaya, the distribution of delta channels, and the lack of barrier islands along the eastern edge of the delta (and their presence to the west) seem to indicate that most of the Lena River's sediment load is being deposited along the eastern flank of the delta at present.

The general circulation pattern of surface currents in the Laptev Sea is described by Suslov (1961, p. 162). A cold current sets southward along the eastern coast of the Taimyr Peninsula toward the Khatanga estuary, where it mixes with the waters of the Khatanga and Anabar Rivers and turns eastward. This current, reinforced by the outflow from the Lena River, then sets northeastward and bifurcates, one branch setting northward and the other eastward between Kotelny Island and the mainland (Fig. 2). The northern branch again divides, with the main portion proceeding northwestward into the Arctic drift, while the other branch sets eastward along the northern side of Kotelny Island (U.S. Navy Hy-

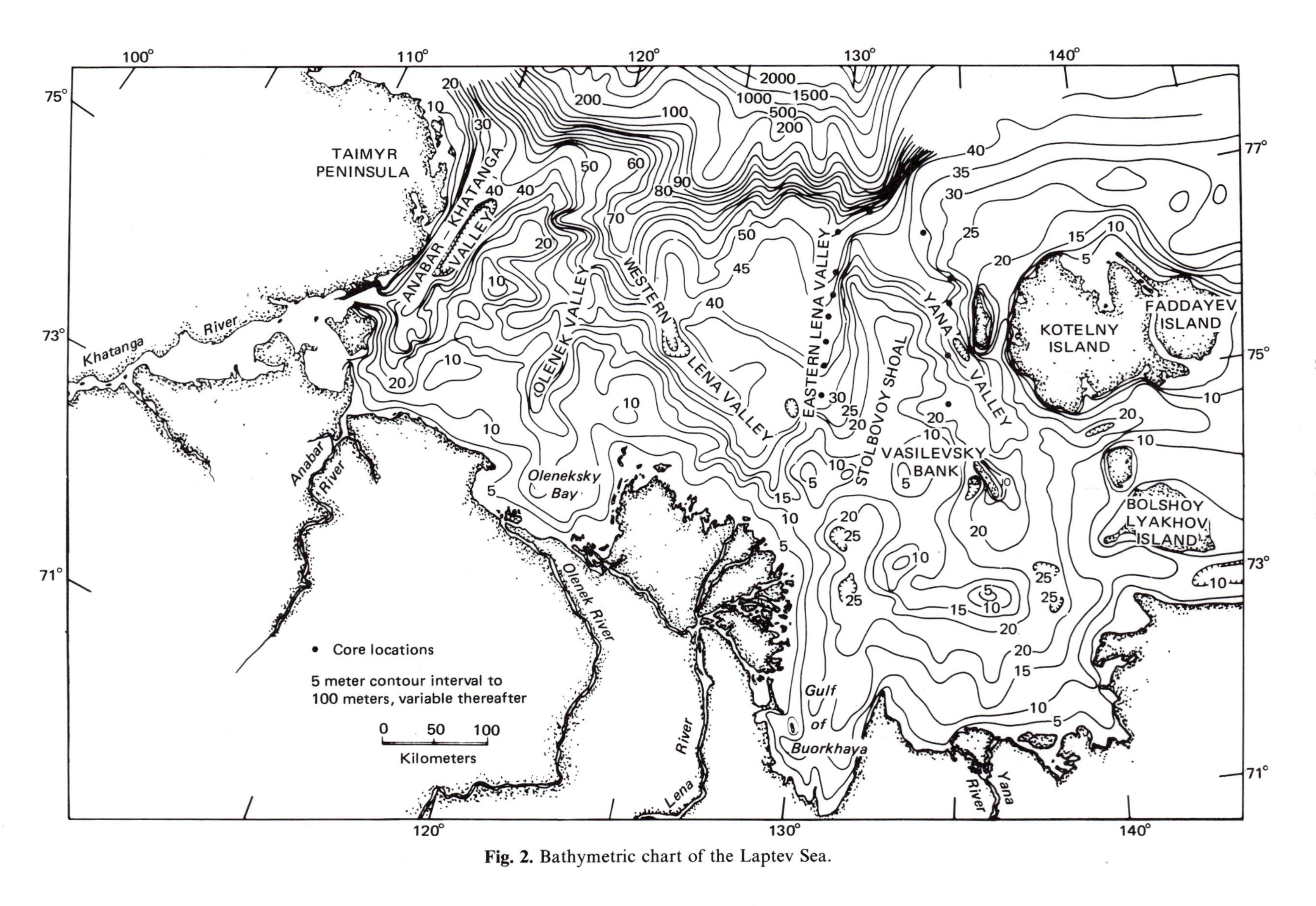

Fig. 2. Bathymetric chart of the Laptev Sea.

drographic Office, 1944). Currents generally do not exceed 10 cm/sec, although speeds of 25 to 50 cm/sec have been observed for the northward-flowing current in the eastern-central portion of the Laptev Sea.

Stolbovoy Shoal would have acted to control the location and magnitude of currents as the sea level rose and fell over the ridge during late Pleistocene and Holocene times. The configuration of the northwestern part of the shoal suggests that the bathymetry in this area may have been modified by current action. During times of lowered sea level, the shore line forming the northwestern bank of Stolbovoy Shoal may have acted to deflect the northward and northeastward flowing currents slightly to the northwest, resulting in a piling up of water and an increase in current speed. This current could have either prevented deposition or permitted the bypassing of finer sediments. The extreme northern tip (40-m isobath) of Stolbovoy Shoal may represent an extension of the shoal by deposition of suspended sediment from this current as it slowed while passing beyond the confining part of the bathymetric ridge. A similar process has apparently been responsible for the formation of Cape Prince of Wales Shoal and other smaller shoals downstream from promontories in the Chukchi Sea (McManus and Creager, 1963; Fleming and Heggarty, 1966; Creager and McManus, 1966; Holmes et al. 1968).

The continental slope north of the Laptev Sea exhibits gradients as steep as 95 m/km, and has several troughs and canyons incised into it. The largest of these, the Sadko Trough (Gakkel', 1958), appears to merge with the Yana and eastern Lena valleys on the shelf (Fig. 2).

Sediments

Fourteen cores (Fig. 1) from the vicinity of the Yana and eastern Lena valleys were selected for detailed study (Holmes, 1967). Where grab samples had been collected with the cores, they were examined for the presence of material coarser than sand size. The 80 subsamples collected were subjected to mechanical analysis to determine textural relations. The carbon content, average grain density, and mineralogy of selected samples were analyzed. Foraminiferal species were identified and six ^{14}C dates were obtained from four cores using the total organic carbon method.

The predominant sediment types are silty-clays and clayey-silts, with mean grain sizes from 7.5 to 8.5ø. With the exception of cores 136 and 143 (Fig. 1), they are quite uniform with respect to sediment type, mineralogy, and chemistry. Sediments below sharp unconformities, at 25 cm in core 136 and 18 cm in core 143, show many of the characteristics of deltaic sediments as described by Shepard (1960) from the Mississippi Delta. The silts and silty-sands in these zones are coarser than the overlying material, and plant fragments and fibers are abundant, as are ferruginous aggregates. No foraminifera were found in samples taken below the unconformities. Ferruginous aggregates also occur in cores 137, 138, and 139. High concentrations of organic matter are not present in these cores, but the presence of ferruginous aggregates below about 40 cm suggests that these sediments were deposited close to sea level, where they may have been subjected periodically to subaerial oxidation.

Manganese nodules were found in the surface grab samples from two stations in the eastern Lena valley (Fig. 2). One nodule, a small quartzite pebble (−4ø) with an iron-manganese crust, was found at station 145, while three smaller crusts were found at station 143 (Fig. 1). These latter crusts appear to have been formed around nuclei (−2 to −3ø) of highly altered basalts. The crusts contained about 10% manganese and 22% iron, and the total composition is very similar to that of manganese nodules from the North Pacific Ocean (Y. R. Nayudu, 1967, personal communication).

With the exception of the manganese nodule and crusts from cores 143 and 145, the eastern Laptev Sea sediments are very fine grained. Although silty-sands with mean grain diameter of 4.5 to 5.5ø occur in cores 143 and 146 (Fig. 1), over 70% of the samples have mean grain diameters finer than 7.0ø (very fine silt). Ninety-five percent of the samples are very poorly sorted, having sorting coefficients greater than 2.0ø units (Folk and Ward, 1957).

These general sediment characteristics are the result of the combined effects of the soil forming processes, transporting agents, and depositional environment. The extensive permafrost areas of north-central Asia favor the development and subsequent erosion of fine-grained soils. The major rivers flowing into the Laptev Sea have very low gradients where they cross the normally wide coastal plain, and the coarsest materials transported from their upper reaches would be deposited here as the rivers rapidly lose competency. The deltas and nearshore areas then trap most of the sandy sediments (Suslov, 1961, p. 177; Samoilov, 1952, p. 309). Since wave heights rarely exceed 1 m and current speeds are generally less than 10 cm/sec (Suslov, 1961, p. 162), the wide shallow Laptev Sea represents a very low energy environment in which the suspended material carried to sea in the plumes of the rivers is allowed to settle with very little modification and reworking.

The manganese nodules and crusts from the surface sediment of stations 143 and 145 represent one of the shallowest known occurrences of such material (45 to 55 m). The quartzite and altered basalt fragments forming the nuclei of these nodules and crusts were probably transported into the area by ice rafting from the southern or southwestern shores of the Laptev Sea prior to the formation of the rather fragile iron-manganese encrustations. The average rate of nodule growth is thought by Mero to be about 1 mm/1000 yr (Shepard, 1963, p. 405), and the 2 to 3 mm thickness of the Laptev Sea crusts would represent a time of accumulation of 2000 to 3000 years. Similar manganese crusts recovered from British Columbia fjords in water 70 m deep appear to have been deposited more rapidly than this, however (J. Murray, 1967, personal communication).

Sedimentation Intensities and Depositional Rates

Six radiocarbon dates from four cores (137, 143, 144, and 147; Fig. 1) were determined, using the total organic carbon method. Table 1 lists the dates, sampling intervals, and average sedimentation intensities (Koczy, 1951) for each of the cores. In discussing these dates, the ages will be applied to the mid-point of the sampling interval. In the cases of cores 143 and 147, a surface interval date of 6000 years B.P. was assumed. This is the age of the upper interval from core 144, which is located between these cores along the axis of the eastern Lena valley, and is probably a satisfactory approximation. The sedimentation intensities range from 2 to 15 $mg/cm^2/yr$, and values from cores in the eastern Lena valley show a distinct decrease northward.

The apparently anomalous old age of the surface sediments in core 137 is due to contamination by inactive carbon. Small black macroscopic particles, believed to be coal fragments, were observed in cores 137 and 141 (Fig. 1). In core 137 these particles were well dispersed above 40 cm, but increased noticeably and became more patchy below that level. In core 141 the fragments occurred in two separate zones at 21 and 38 cm and in much lower concentrations than in core 137. As a test for coal, two samples (at 12 and 71 cm) from core 137 were subjected to the Wakely-Black analysis (Jackson, 1958, p. 213). The test was positive, and indicated that 17 and 52% of the total carbon at 12 and 71 cm, respectively, was due to the presence of coal. Allowing for the 17% coal contamination, which was found in the surface interval of core 137, would result in a corrected age of 6500 years B.P., using the curves of Olson and Broecker (1958).

This is very close to the surface interval date from core 144, in which there was no apparent coal contamination. It is therefore probably safe to assume that the date for the surface interval of core 144 is actually 6050 years B.P. This is considerably older than the surface sediments (4350 years B.P.) in the Chukchi Sea (Creager and McManus, 1965). This can be explained by lower depositional rates in the Laptev Sea and by the fact that a large interval and thus a wide span of ages was covered in the dated sample.

Of the ages for subsurface intervals of the cores, only that from core 137 appears to be in error. The coal content in the section of the core from which the radiocarbon date was obtained appeared to be the same as the sample at 71 cm. If one assumes a 50 to 60% contamination for the dated interval, this would result in an actual age of 11,500 to 12,000 years B.P., and a corrected

Table 1 Carbon-14 Dates and Sedimentation Intensities

Core	Depth in core (cm)	Radiocarbon age (years B.P.)	Average sedimentation intensity ($mg/cm^2/yr$)[1]
137	3–17	8410 ± 230	
			9
	103–117	18,400 ± 540	
143	24–32	14,200 ± 370	2[2]
144	3–20	6050 ± 200	
			5
	44–64	15,000 ± 460	
147	89–102	11,040 ± 310	15[2]

[1] Based on elapsed time between dates.

[2] Assumed surface interval (3 to 20 cm) date 6000 yr B.P. See text.

sedimentation intensity since that time of about 25 mg/cm²/yr.

The dated interval from core 143 was taken from a section (17 to 36 cm) containing large amounts of wood fragments and fibers. Because of the unique and uniform lithology of this zone, the date can probably be applied to the entire section. This, coupled with the sharp boundary between this portion and the surface interval, would indicate that the core may not represent continuous deposition, and the calculated average sedimentation intensity should be regarded as a maximum.

The combined annual discharge of suspended sediment from the Lena and Yana Rivers is 15×10^6 metric tons (Samoilov, 1952, pp. 310, 312). The area of that portion of the Laptev Sea east of 125°E is approximately 200,000 km². It is characterized in the summer by less than $\frac{7}{10}$ ice cover, and probably represents the greatest area over which the plumes of the Lena and Yana spread. If it is assumed that all of the suspended sediment discharged annually by these two rivers is deposited uniformly over this area, the present sedimentation intensity would be about 7 mg/cm²/yr. To a first approximation, this agrees well with the average intensities determined from the radiocarbon dates.

The small manganese nodule and manganese crusts found in the surface grab samples from stations 143 and 145 indicated that the present rate of sediment accumulation is less than the average sedimentation intensities (2 and 5 mg/cm²/yr, respectively) would indicate (Y. R. Nayudu, 1967, personal communication). Even if these shallow-water crusts were growing much more rapidly than deep-sea crusts, their presence would still indicate a very slow accumulation of detrital material (J. Murray, 1967, personal communication). The foraminifera number (test/gram of sediment) shows a general decrease with depth in all studied cores. These data, with the old surface interval dates, support the interpretation that present depositional rates are actually lower than those inferred from lapsed time between dates in the cores.

Sea-level Fluctuations

A potential difficulty in recognizing ancient stillstands of sea level in the Laptev Sea and correlating them with stillstands documented from other areas is the fact that the region is currently tectonically active. The only definite quantitative evidence cited in the literature (Suslov, 1961, p. 173) of the amount of possible post-Pleistocene uplift are the 3- to 5-m terraces on Bolshoy Lyakhov Island (Fig. 2). The change in river mouth configuration from estuaries in the west to deltas in the east indicates that the region may have undergone tilting during late Pleistocene or Holocene time, although this apparent regional tilt could also be the result of crustal depression by ice loading (Flint, 1971, p. 366, 662).

Certain bathymetric features also suggest an uplift of the eastern portion of the Laptev Sea relative to the western part. The series of depressions along the axes of the submarine valleys occur at two different depths. The basins in the Yana and eastern Lena valleys exhibit closures at 25 and 40 m, whereas those in the western Lena, Olenek, and Anabar–Khatanga valleys occur at 30 and 45 m. Although this supports other evidence for the relative emergence of the eastern Laptev Sea, it would be very difficult to determine if the uplift occurred during one brief period or rather slowly and uniformly throughout most of late Pleistocene and Holocene times. Therefore, no effort will be made to correct features discussed in the following stillstand discussion for the amount of uplift which may have occurred after they were formed. Most of these features would require a water depth of 2 to 4 m for their formation, and a correction for this would merely cancel that for subsequent tectonic effects.

The well-developed shoals and benches with depths from 10 to 15 m (Fig. 2) were probably formed by thermal erosion (Larionova, 1959) and other nearshore processes at a time of lowered sea level. A bathymetric chart of the East Siberian Sea (Naugler et al., this volume) also shows extensive flat areas between 10 and 15 m. Fairbridge (1961, p. 147) indicates a stillstand at −10 m during the period from 7500 to 6700 years B.P. (Fig. 3). No cores from these flat areas were studied or dated, and so any correlation with Fairbridge's −10-m stillstand must be based solely on the similarity of depths.

The configuration of the 20- and 25-m isobaths northeast of Kotelny Island (Fig. 2) may give similar evidence for a stillstand of sea level at about −20 m. This bench-like feature is more weakly developed than the one at −10 m, and its existence is based only on Soviet soundings. Fairbridge (1961, p. 147) lists limits of −15 to −24 m for a stillstand during the period 10,300 to 8900 years B.P. (Fig. 3). The sea-level curves of

Curray (1960, 1961, 1965) show two stillstands within these same depth limits, one at about 9600 years B.P. and a later one at 8100 years B.P. Mörner (1971) also shows a stillstand at −20 m during the period from 8700 to 7900 years B.P. (Fig. 3). Again, because no actual dates are available from the Laptev Sea feature, correlation must be on the basis of bathymetry only, and it could thus be the result of two −15 to −24 m stillstands at different times (9600 and 8100 years B.P.).

Another well-developed terrace with depths from −30 to −35 m occurs north of the Lena River Delta. This large flat area is cut by both the eastern and western Lena valleys, but can be easily recognized in spite of its dissected nature. Naugler et al. (this volume) also found a well-developed terrace at these depths in the East Siberian Sea.

The linear depressions or basins which interrupt the grade of the submarine valleys on the Laptev Sea shelf could have been formed at times of lowered sea level. Similar depressions are found in present-day estuaries and may be the result of tidal and river current scour or nondeposition. If this is the case, the depressions indicated by closures of the 30-m isobath in the Olenek valley (Fig. 2) and the 25-m isobaths in branches of the Yana valley would have been formed when sea level was approximately 25 to 30 m below its present position.

Shepard (1960) lists criteria for recognition of ancient deltas. He stated that deltaic sediments would include abundant land plant remains and ferruginous aggregates. Invertebrate remains would be scarce and if found, should show little diversity of species. These characteristics are well shown by the sediments in the lower portion (below 25 cm) of core 136 (Fig. 1). Plant matter, wood fragments, and ferruginous aggregates are present in the sediments, and only one foraminifera test was found in the two samples from this zone. It is therefore possible that these sediments represent an ancient delta of the Yana River. The present depth of water at station 136 is 29 m, indicating that the stillstand during which the delta might have formed was at this depth or only slightly shallower.

The sea-level curve of Curray (1960, 1961) in Fig. 3 shows a low stillstand of sea level at −27 to −37 m at about 8700 years B.P. The bathymetric and sedimentary features from the Laptev Sea

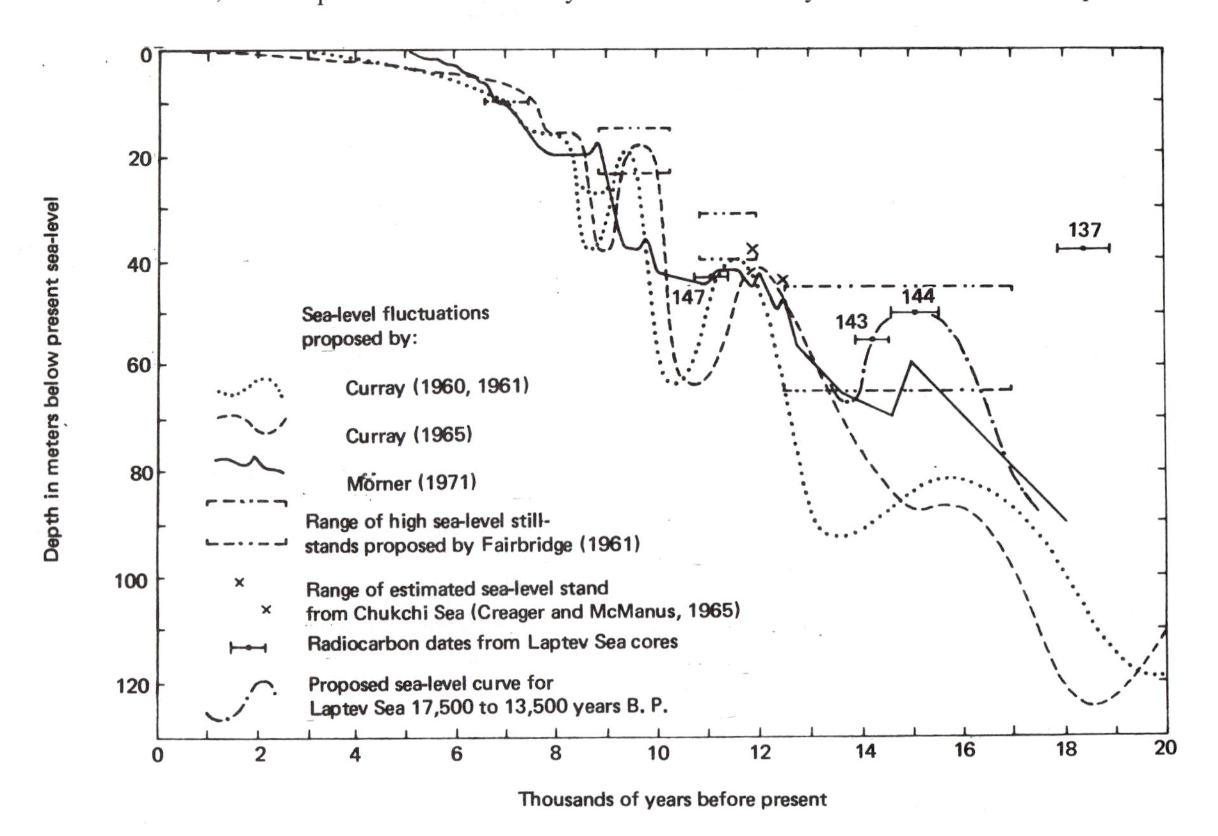

Fig. 3. Sea-level stillstands and fluctuations during late Pleistocene and Holocene times.

indicate a stillstand at −25 to −35 m, and although no dates are available, this could represent a low sea-level stand which was contemporary with Curray's.

An extensive terrace or bench 40 to 45 m deep is situated in the central portion of the Laptev Sea. Flat areas occurring at the same depths were cited by Naugler et al. (this volume) as possible evidence for a stillstand of sea level. Linear basins with closures at 45 m occur in the western Lena and Anabar–Khatanga valleys (Fig. 2). In the Yana valley a depression formed by a 40-m isobath closure may be seen. Following the assumptions concerning the origin of similar basins at 25 and 30 m, these might indicate a sea-level stillstand at −40 to −45 m.

Ferruginous aggregates occur in cores 137 to 139 (Fig. 1). High concentrations of organic matter are not present in these cores, and the foraminiferal population is not significantly smaller than that from the other cores. This excludes the possibility of these sediments being classified as deltaic using Shepard's (1960) criteria, but the presence of the ferruginous aggregates would still seem to indicate that the sediments from the lower portions of the cores were deposited very close to sea level, where they may have been subjected periodically to subaerial oxidation.

The radiocarbon dates (Table 1) from the lower intervals of cores 137, 143, 144, and 147 are plotted in Fig. 3. The occurrence of marine sediments with these ages places a lower limit on the position of sea level during these times. Curray's (1960, 1961, 1965) sea-level curves (Fig. 3) show a high sea-level stillstand at −40 to −41 m 12,000 to 11,600 years B.P., and Fairbridge (1961, p. 147) indicates a stillstand at −32 to −40 m during the period from 12,000 to 10,800 years B.P. The sea-level curve of Mörner (1971) also shows a high stand at −42 m during this same time interval. Figure 3 shows the dates (corrected for possible coal contamination) for a section of ancient deltaic sediments from the Chukchi Sea (Creager and McManus, 1965). The date from core 147 falls very close to the position of this previously documented stillstand, and strongly supports evidence from bathymetry and the sediments that sea level in the Laptev Sea stood at −37 to −45 m between 12,000 and 11,000 years B.P.

In the preceding section on sedimentation intensities it was stated that the average age of the lower interval of core 137 appeared to be in error due to contamination by coal. In Fig. 3 this date is well displaced from the sea-level curves and limits of Curray (1965) and Fairbridge (1961). If the contamination by inactive carbon is 50 to 60 percent, as indicated by the single coal analysis at 71 cm, the corrected age for that interval would be about 12,000 years B.P. This would place the date very close to those for the stillstand recognized by Curray (1960, 1961, 1965), Fairbridge (1961, p. 147), Creager and McManus (1965), and Mörner (1971).

Evidence for a stillstand at −50 to −55 m depths is provided by the subsurface dates of cores 143 and 144 (Fig. 3), and the character of the sediments in the lower half of core 143 (18 to 36 cm). These sediments contain large amounts of wood material and plant fibers, and no foraminifera were detected in the samples from this interval. Ferruginous material such as that found in cores 136 through 139 does not occur, but instead large concentrations of pyrite aggregates and pyritized wood fragments are present. The pyrite aggregates were probably produced by a reduction of original ferruginous granules under the anoxic conditions which apparently existed during or after deposition of this part of the core. Thus the sediments of this interval, like those in the lower portion of core 136, probably represent ancient deltaic deposits and give evidence for a sea-level stillstand at about −55 m.

The dates from this interval in core 143 and the lower portion of core 144 fall within the limits which Fairbridge (1961) proposed for a stillstand at −45 to −65 m from 17,000 to 12,500 years B.P. (Fig. 3). Creager and McManus (1965) found bathymetric evidence for a stillstand at −53 m in the Chukchi Sea, and concluded that it had probably occurred between 17,000 and 12,000 years B.P. Mörner (1971) also cites evidence for a high stand of sea level at about −60 m during this time (15,000 years B.P.). It would therefore appear that the stillstand of sea level during this period was considerably higher than the −81 to −86 m shown by Curray (1960, 1961, 1965). The stillstand date from the Laptev and Chukchi Seas (Creager and McManus, 1965) would modify the curve as shown in Fig. 3, supporting the view of Fairbridge (1961, p. 147) that there was a stillstand at this depth.

Paleogeography

Using the sea-level stillstand data from the preceding sections, a series of paleogeographic

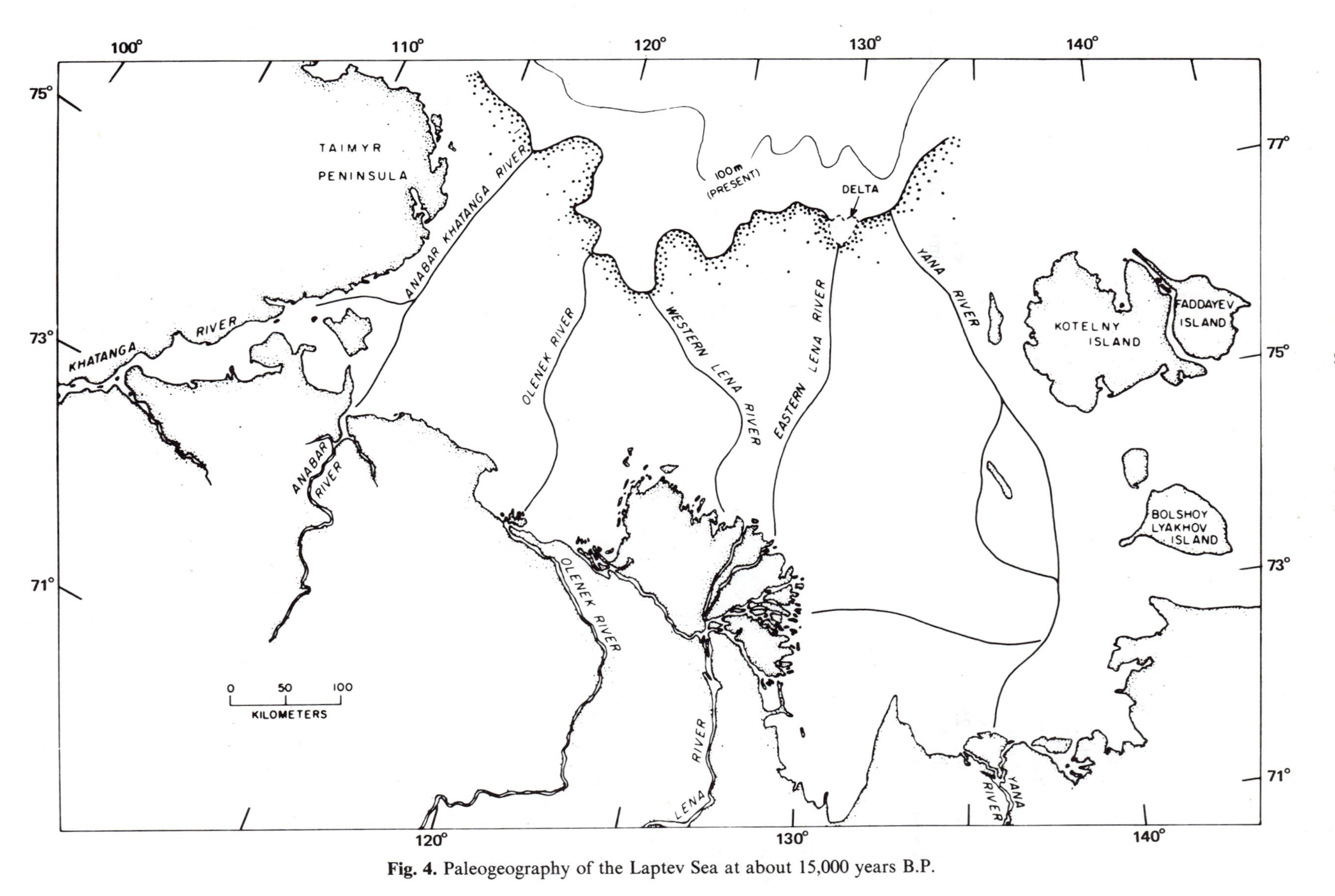

Fig. 4. Paleogeography of the Laptev Sea at about 15,000 years B.P.

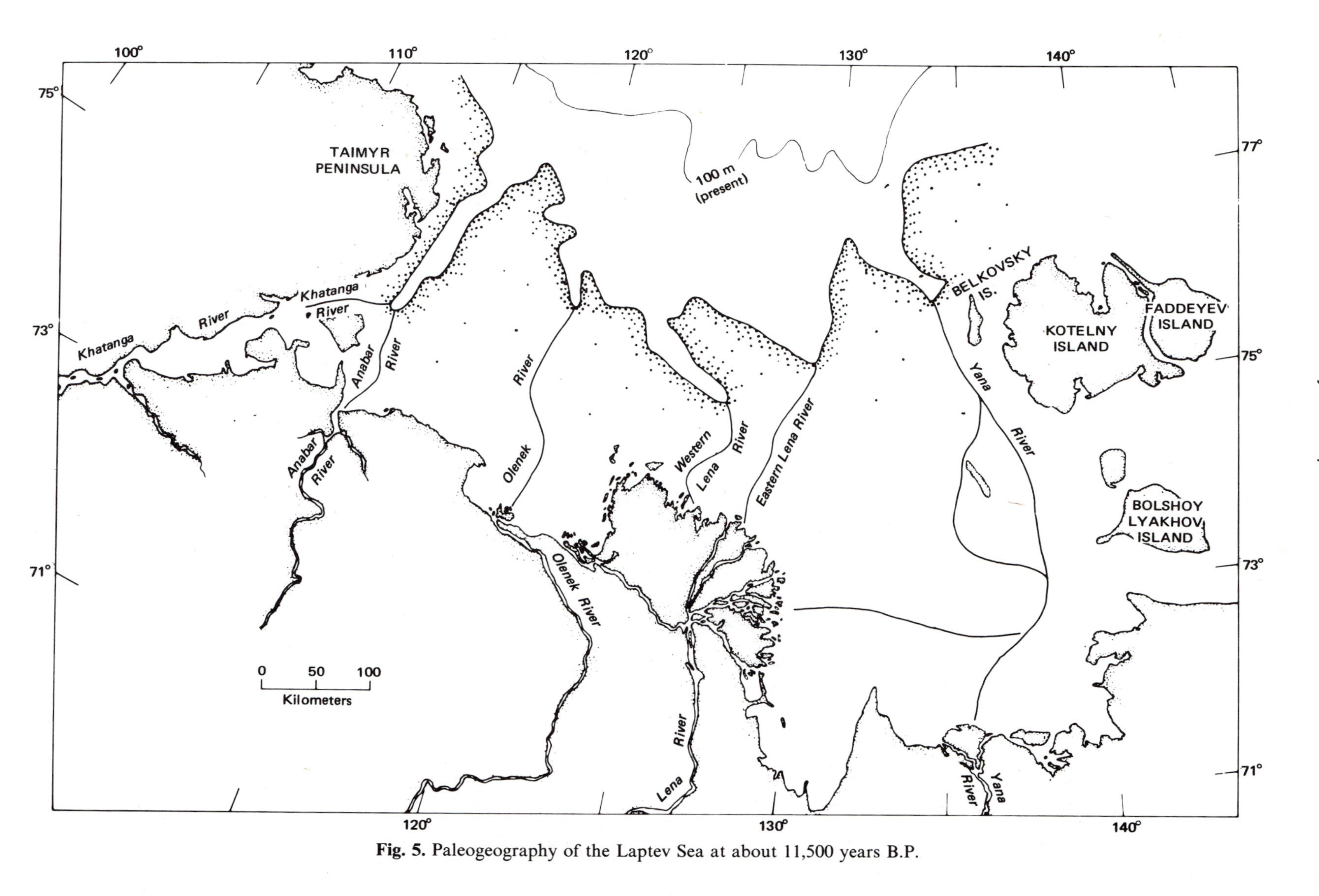

Fig. 5. Paleogeography of the Laptev Sea at about 11,500 years B.P.

charts were prepared (Figs. 4 through 8). They are numbered in order of decreasing age, and show the fluctuations of shore line and river valley locations on the Laptev Sea shelf during the late Pleistocene and Holocene transgressions and regressions (Fig. 3).

The high stillstand at −50 to −55 m would correspond to the Masurian Submergence about 15,000 years B.P. (Fairbridge, 1961). Figure 4 shows that the shore line of the Laptev Sea at this time was very close to the present shelf break. Gusev (1959) described an ancient coast occupying the same position, but dated it only as Neogene. The Lena River probably sent three distributaries across the shelf. The Western Lena and Olenek Rivers entered a large embayment about 200 km northwest of the present Lena Delta, and the mouths of the Yana and eastern Lena Rivers were within 100 km of each other northwest of the New Siberian Islands. The Khatanga and Anabar Rivers entered the Arctic Ocean through a common river valley.

Saks (Samoilov, 1952, p. 307) stated that during the Karginsky Submergence about 11,500 years B.P. (Fig. 5) the Lena River branched near the present delta and that the mouth of the western branch was 300 km to the northwest. He also stated that the mouth of the Yana River was a little more than 300 km north of its present position. The paleogeographic chart for the −37 to −45 m stillstand (Fig. 5) shows that although Saks' figure for the Yana River mouth was quite good, the locations of the mouths of the Lena River branches were approximately 200 km north of the present delta. The Anabar and Khatanga Rivers entered a long narrow estuary along the eastern coast of the Taimyr Peninsula, and the Yana flowed into a similar feature near present-day Belkovsky Island.

Fig. 6 shows the paleogeography for the −20-m-high stands of sea level during the Yoldia Submergence at about 9600 yr B.P. and the Ancylus Emergence about 8100 yr B.P. (Fairbridge, 1961). The configuration of the shore line is quite complex, with many embayments, peninsulas, and well-developed estuaries. The main flow of the Lena had probably shifted from the northward-flowing channels to those flowing east into the system of straits formed by present-day Stolbovoy Shoal and Stolbovoy Island.

During the Post-Yoldia Emergence (8700 years B.P.) sea level fell to approximately −25 to −30 m (Fig. 7). The rivers again flowed out across the shelf, and the mouths of all but the Lena entered the heads of long narrow estuaries. The Lena River, whose main outflow had probably shifted to the north once again, flowed into a large bay north of its present delta. The shallow channel trending eastward from the Lena Delta may have carried some of the Lena's discharge into the Yana River valley. This coastal configuration closely resembles one which Gusev (1959) described as lying along the present 25-m isobath, but he dates the shore line only as second half of the upper Quaternary.

Gusev (1959) also states that at the end of the Age of Mammoths the coast of the Laptev Sea followed the 15-m isobath. This may correspond to the shore line during the Hydrobia Submergence (−10 m) approximately 7000 years B.P., as shown in Fig. 8. The river mouths were very close to their present locations, and active modern delta formation probably began at this time. Many large islands were produced as a result of the partial flooding of the large shoal areas in the southeastern and western portions of the Laptev Sea.

Ice As a Geological Agent

Several writers (Tarr, 1897; Kindle, 1924; and Lisitsyn, 1957; and Hume and Schalk, 1967) discussed the importance of Arctic sea ice as an agent of erosion, transportation, and deposition of sediments. There seems to be little agreement as to the criteria for recognizing ice-borne sediments unless the material in question is noticeably coarser than the associated sediments. Opinions as to the actual importance of rafting by floe ice vary considerably also.

One of the most widely accepted methods by which floe ice acquires sediment is that of bottom accretion and progressive upward migration. Ice floes which scrape or freeze to the bottom pick up sediment which then makes its way upward as the surface ice melts and new ice forms on the bottom of the floe. This process undoubtedly occurs, and it may be illustrated in Fig. 9. The dark band near the sea surface in the ice floes may represent a layer of sediment slowly working its way upward. Another process which would be more effective in transporting large amounts of material involves fast ice which has frozen to the bottom near the river mouths and high coastal cliffs. The river ice tends to be broken up by the spring floods prior to the melting of the fast ice near shore. The rivers

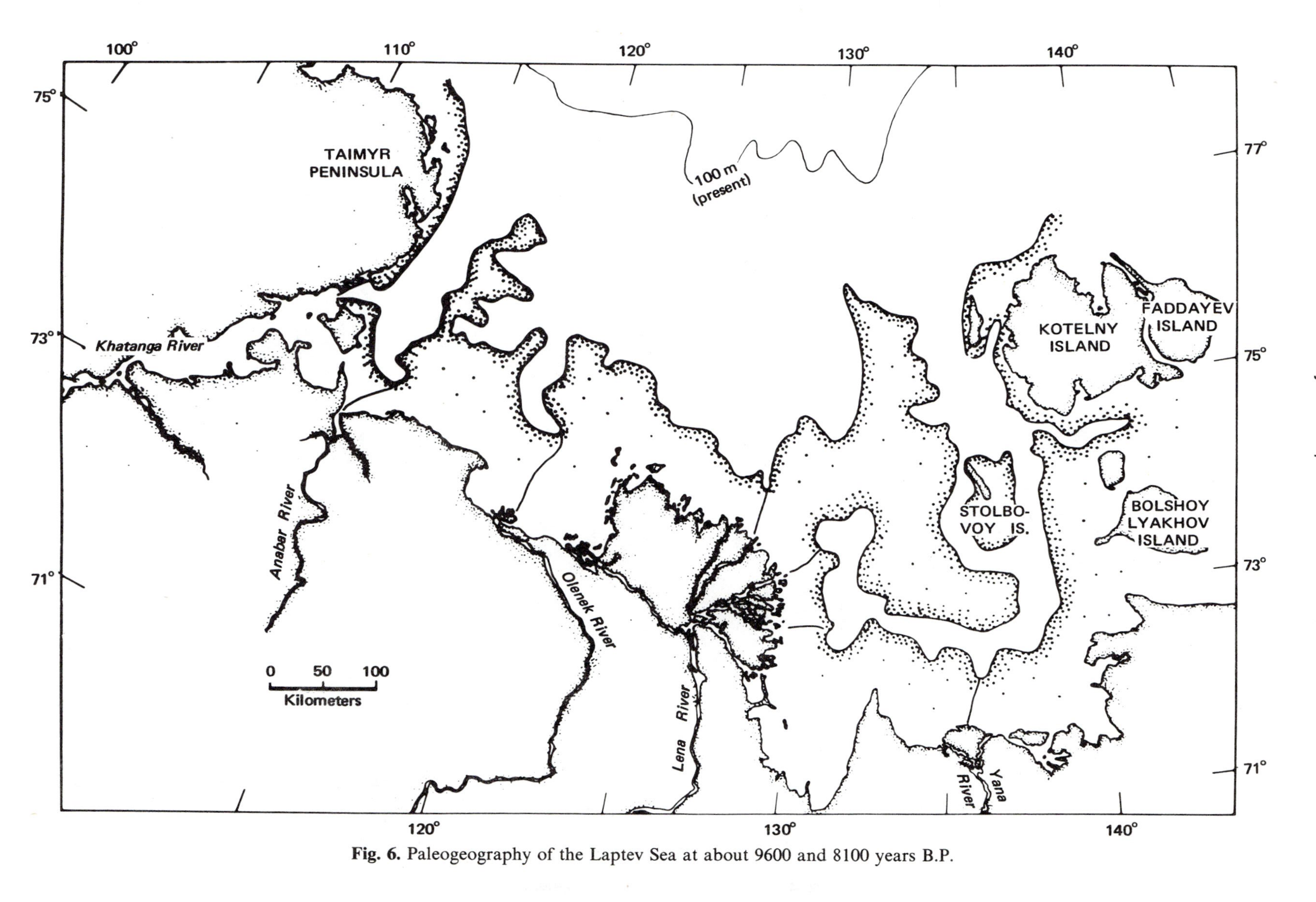

Fig. 6. Paleogeography of the Laptev Sea at about 9600 and 8100 years B.P.

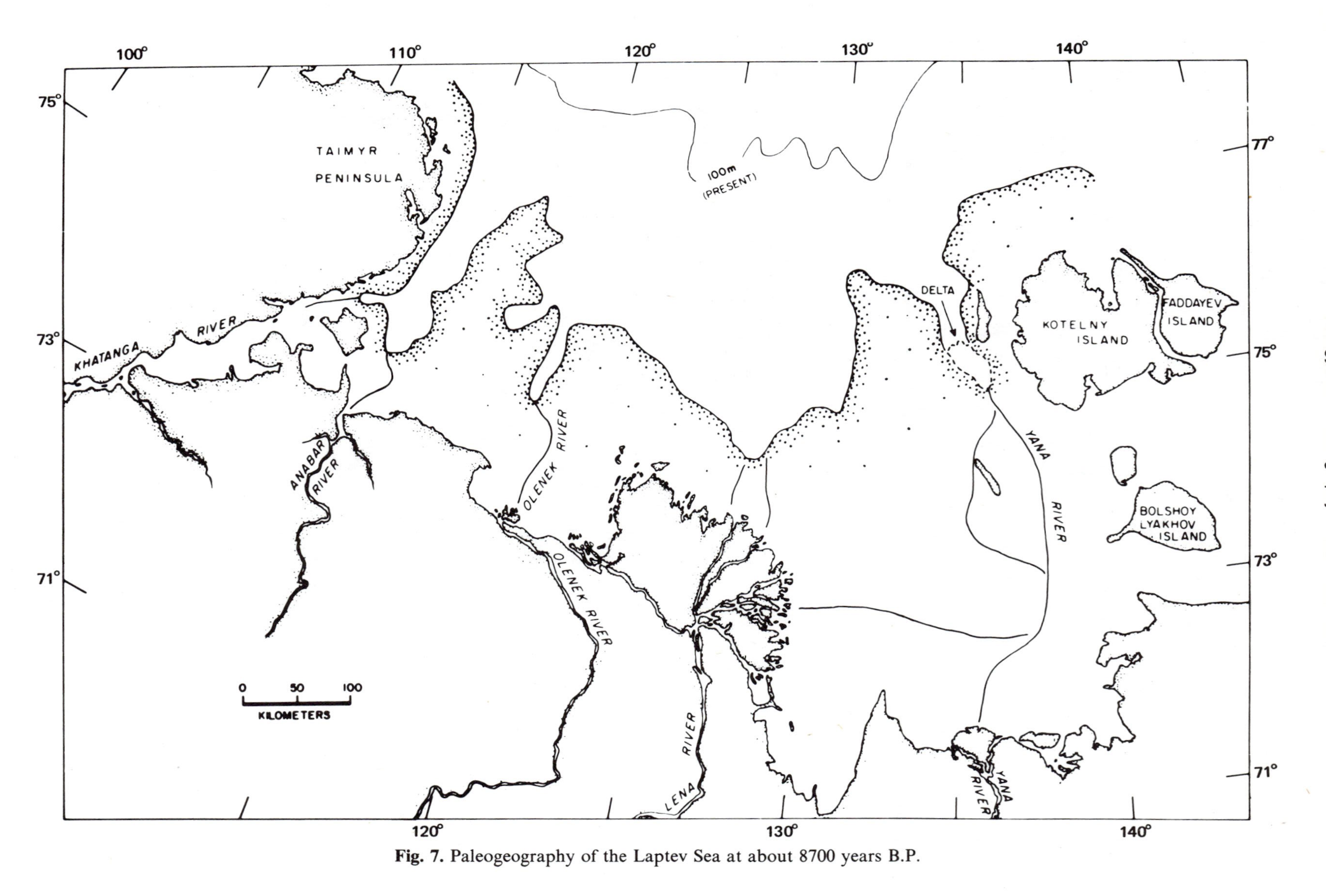

Fig. 7. Paleogeography of the Laptev Sea at about 8700 years B.P.

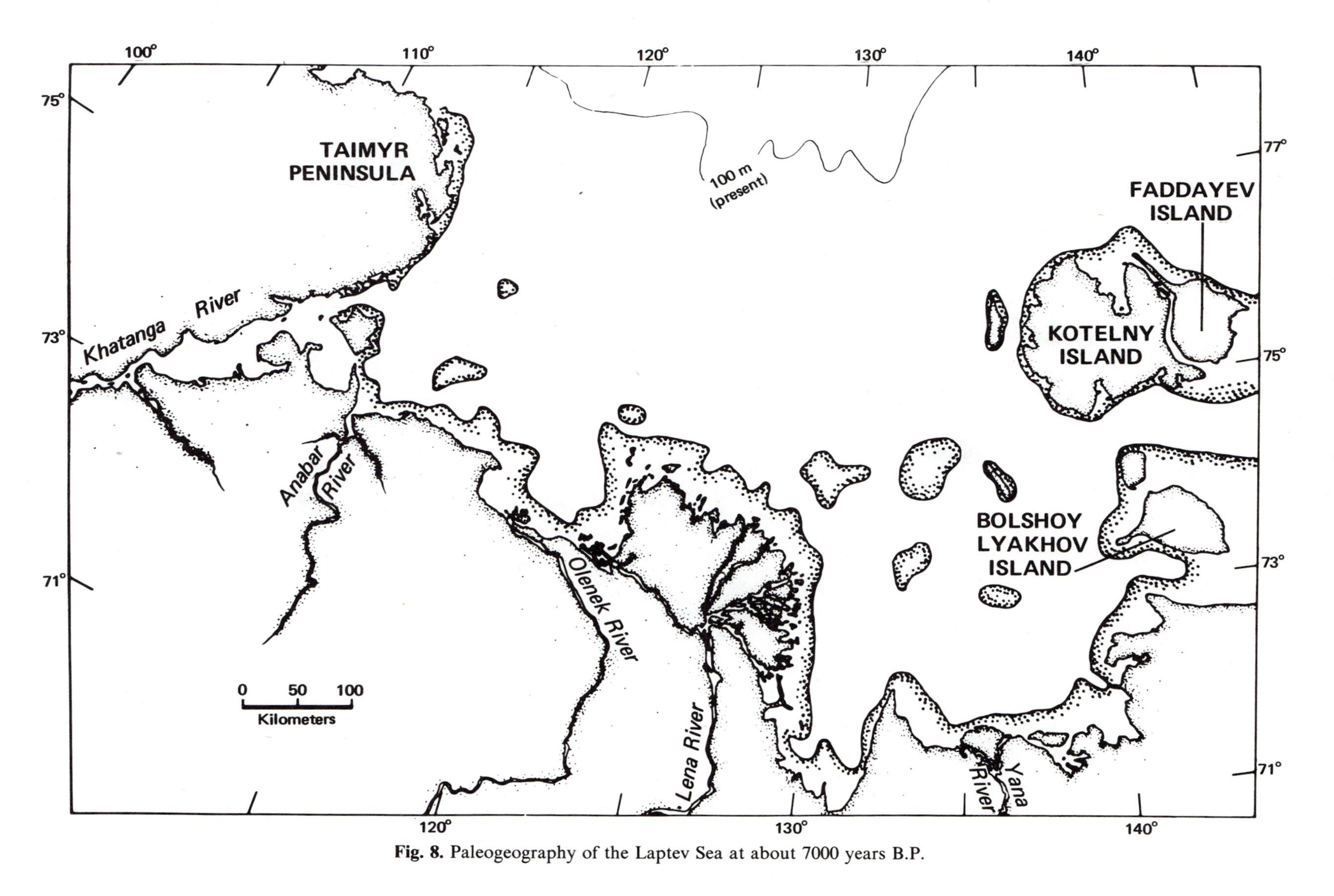

Fig. 8. Paleogeography of the Laptev Sea at about 7000 years B.P.

Fig. 9. Large pieces of floe ice containing a single dark stratum of sediment just above the water line (arrows).

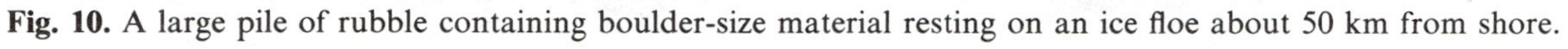

Fig. 10. A large pile of rubble containing boulder-size material resting on an ice floe about 50 km from shore.

would thus carry large loads of sediment out onto the pack ice, and when the ice near shore breaks up it would transport this sediment load with it until the floes melted or were overturned. Spring and summer warming also results in very active slumping of the high banks containing large amounts of buried ground ice (Suslov, 1961, p. 172). Any fast ice still attached to the shore below these cliffs could acquire a large surface load of material as a result of this mass movement, as may be illustrated in Fig. 10.

The only evidence for ice rafting in the sediments of the cores comes from those in the outer parts of the Eastern Lena valley. The nuclear material of the manganese nodules and crusts was undoubtedly brought into the area by ice because of their very large size (-2 to -4ø) compared to the fine-grained (7.5 to 8.5ø) sediments. The original source of these small rock fragments would be very difficult to identify, but the general current and ice drift patterns (Suslov, 1961, p. 162) would indicate that they were probably carried to sea from the Lena Delta area or the southwestern shores of the Laptev Sea. Cores 144 through 146 also contain small layers of silty-sand which are coarser than the other sediments in the cores. If these do represent ice-borne detritus, it is odd that no similar layers were observed in any of the other cores. A possible explanation is provided by the extent of ice cover during mid-summer. The area from which the cores were taken is located at the northern edge of the ice-free region caused by the plume of the Lena River. The offshore surface currents may carry masses of pack ice, some with appreciable concentrations of sediment acquired in the near-shore areas, northward to the edge of the open water zone where they would melt and discharge their sediment load.

Ice rafting of material for short distances probably occurs quite regularly in the areas near the coast and the mouths of the rivers. Offshore, however, this process is only locally important, and ice rafting does not appear to be a prominent factor in the distribution of sediments in the Laptev Sea.

Summary and Conclusions

From the data presented and the interpretations made, the following summarizing statements may be set forth:

1. The sediments in 14 cores from the eastern Laptev Sea are fine-grained and poorly to very poorly sorted. This is the result of: (a) fine-grained soils produced by the permafrost weathering processes; (b) low river gradients; (c) trapping of sandy sediments by the river deltas; and (d) the very low energy environment of the Laptev Sea.

2. Manganese nodules and crusts were recovered from the surface at two of the coring stations and represent the shallowest known occurrence of these materials. The manganese encrustations were probably formed after the nuclear material (rock fragments) had been rafted into place by ice.

3. Average sedimentation intensities were calculated from four cores using radiocarbon dates. Values ranged from 2 to 15 mg/cm^2/year. An approximate sedimentation intensity based on the annual silt discharge of the Yana and Lena Rivers (7 mg/cm^2/yr) agrees well with these values, although the presence of the manganese crusts indicates that the present rate of deposition in the outer shelf area may be considerably slower.

4. Data from the bathymetric features, certain sediment characteristics and subsurface carbon-14 dates suggest that at least five stillstands of sea level may have occurred since about 17,000 yr B.P. The depths below present sea level and dates for these stillstands agree quite well with those which have been documented from other parts of the world.

5. Ice rafting is probably not as important a geological agent as the severity of ice conditions in the Laptev Sea would indicate. The nuclear material of the manganese nodules and possibly some silty-sand zones in some of the cores indicate that floe ice is capable of transporting material into the outer shelf regions, but these seem to represent only local occurrences.

Acknowledgments

The work was supported by grants from the National Science Foundation (GA808, 11126, 28002).

References

Bates, C. C. 1953. Rational theory of delta formation. *Bull. Am. Assoc. Petroleum Geol.*, 37(9):2119–2162.

Codispoti, L. A. 1965. *Physical and chemical features of the East Siberian and Laptev Seas in the summer.* Unpublished Thesis, University of Washington. 51 pp.

Creager, J. S., and D. A. McManus. 1965. Pleistocene drainage patterns on the floor of the Chukchi Sea. *Marine Geol.*, 3:279–290.

———, and D. A. McManus. 1966. Geology of the Southeastern Chukchi Sea. In: N. J. Wilimovsky and J. M. Wolfe, eds. *Environment of the Cape Thompson Region, Alaska.* Oak Ridge, Tennessee: U.S. Atomic Energy Commission. pp. 755–786.

Curray, J. R. 1960. Sediments and history of Holocene transgressions, continental shelf, northwest Gulf of Mexico. In: F. P. Shepard, F. B. Phleger, and T. H. van Andel, eds. *Recent Sediments, Northwest Gulf of Mexico.* Tulsa, Oklahoma: American Association of Petroleum Geologists. pp. 221–266.

———. 1961. Late Quaternary sea level: a discussion. *Geol. Soc. Am. Bull.*, 72:1707–1712.

———. 1965. Late Quaternary history, continental shelves of the United States. In: H. E. Wright, Jr. and D. G. Frey, eds. *The Quaternary of the United States.* Princeton, N.J.: Princeton University Press. pp. 723–735.

Donn, W. L, W. R. Farrand, and M. Ewing. 1962. Pleistocene ice volumes and sea-level lowering. *J. Geol.*, 70(2):206–214.

Eardley, A. J. 1964. Polar rise and equatorial fall of sea level since the Cretaceous. *J. Geol. Educ.*, 12:1–11.

Fairbridge, R. W. 1961. Eustatic changes in sea level. *Phys. Chem. Earth*, 4:99–185.

Fleming, R. H., and D. Heggarty. 1966. Oceanography of the Southeastern Chukchi Sea. In: N. J. Wilimovsky and J. N. Wolfe, eds. *Environment of the Cape Thompson Region, Alaska.* Oak Ridge, Tennessee: U.S. Atomic Energy Commission. pp. 697–754.

Flint, R. F. 1971. *Glacial and Quaternary Geology.* New York, N.Y.: John Wiley and Sons. 892 pp.

———, and H. G. Dorsey, Jr. 1945. Glaciation of Siberia. *Geol. Soc. Am. Bull.*, 56:89–106.

Folk, R. L., and W. C. Ward. 1957. Brazos River bar: a study in the significance of grain size parameters. *J. Sedimentary Petrol.*, 27:3–26.

Gakkel', Ya. Ya. 1958. Signs of recent volcanic activity in the Lomonosov Range (transl.). *Priroda*, 4:87–90.

Grigorov, I. P. 1946. Ischezayuschie Ostrova. *Priroda*, 10:65–68.

Gusev, A. I. 1959. K. Istorii Razvitia Primorskoi Ravniny v Chetvertichnoe Vremya. In: F. G. Markova, ed. *Sbornik Statei Geologii Arktiki.* Nauchno-Issledovatel'skog Instituta Geologii Arktiki, Leningrad. Trudy, 102(10). pp. 160–165.

Holmes, M. L. 1967. *Late Pleistocene and Holocene History of the Laptev Sea.* Unpublished thesis, University of Washington. 176 pp.

———, J. S. Creager, and D. A. McManus, 1968. High frequency acoustic profiles on the Chukchi Sea continental shelf. *Trans. Am. Geophys. Union* (abstract), 49(1):207.

Hume, J. D., and Schalk, M. 1967. Shoreline processes near Barrow, Alaska—a comparison of the normal and the catastrophic. *Arctic*, 20(2):86–103.

Jackson, M. L. 1958. *Soil Chemical Analysis.* Englewood Cliffs, N.J.: Prentice-Hall. 498 pp.

Kindle, E. M. 1924. Observations in ice-borne sediments by the Canadian and other Arctic expeditions. *Am. J. Sci.*, 7(40):251–286.

Koczy, F. F. 1951. Factors determining element concentration in sediments. *Geochem. Cosmochim. Acta*, 1:73–85.

Larionova, A. N. 1959. Attempts to classify elevations and depressions of the ocean bottom. In: *Traveling on the Sea Bottom.* Leningrad: Hydrometeorological Publishers. pp. 20–70.

Lisitsyn, A. P. 1957. Types of marine sediments connected with the activity of ice. *Akademia Nauk S.S.S.R., Geological Science Section*, 118(1):45–47.

McManus, D. A., and J. S. Creager. 1963. Physical and sedimentary environment on a large spitlike shoal. *J. Geol.*, 71(4):498–512.

Mörner, N. A. 1971. Eustatic changes during the last 20,000 years and a method of separating the isostatic and eustatic factors in an uplifted area. *Paleogeography, Paleoclimatology, Paleoecology,* 9(3):153–181.

Olson, E. A., and W. S. Broecker. 1958. Sample contamination and reliability of radiocarbon dates. *Trans., New York Academy Sci.*, Ser. 2, 20(7):593–604.

Samoilov, I. V. 1952. *Ust'ya Rek.* Moskava: Gosudarstvennoe Izdatel'stvo Geograficheskoi Literatury. 526 pp.

Shepard, F. P. 1960. Mississippi delta: marginal environments, sediments, and growth. In: F. P. Shepard, F. B. Phlegar, and T. H. van Andel, eds. *Recent Sediments, Northwest Gulf of Mexico.* Tulsa, Oklahoma: American Association of Petroleum Geologists. pp. 56–81.

———. 1963. *Marine Geology.* New York: Harper and Row. 557 pp.

Suslov, S. P. 1961. *Physical Geography of Asiatic Russia.* San Francisco, California: W. H. Freeman and Co. 594 pp.

Sverdrup, H. U. 1927. *The Waters of the North Siberian Shelf.* Scientific Results Norwegian North Polar Expedition Maud (1918–1925), Vol. 4. 131 pp.

Tarr, R. S. 1897. The Arctic Sea ice as a geological agent. *Am. J. Sci.*, 3(15):223–229.

U.S. Navy Hydrographic Office. 1954. Sailing directions for the northern U.S.S.R. Vol. 3. *Hydrographic Office Publication 137c.* 346 pp.

Chapter 10
Sediment Distribution in Deep Areas of the Northern Kara Sea

JOHN A. ANDREW [1] AND JOSEPH H. KRAVITZ [2]

Abstract

The northern margin of the Asiatic Arctic continental shelf, underlying the Kara Sea, is cut by two deep troughs. The Svyataya Anna (or St. Ann) Trough to the west is the larger (140,000 square kilometers at the 200-m isobath) and deeper (exceeding 640 m) of the two troughs with a general U-shape. The Central Kara Plateau, a shallow area which contains two islands and several shoal areas, separates the St. Ann Trough from the Voronin Trough (20,000 square kilometers at the 200-m isobath, and over 400 m deep).

The areal distributions of sedimentary variables such as grain-size, organic carbon content, water content, sand content of sediment cores and clay mineral ratios show a consistent north-south zonation in the St. Ann Trough and a less clear axial trend in the Voronin Trough. A persistent set of bottom currents may best explain the origin of the sediment dispersal pattern in the St. Ann Trough, and an intermittent bottom current may best explain the less clear dispersal pattern observed in the Voronin Trough.

Bottom currents on the west side of the St. Ann Trough flow toward the south, introducing deep-ocean fauna and warmer waters into the trough. This current does not transport terrigenous sediment, although it may rework existing sediments. A north-flowing bottom current best explains the distribution of sediment variables on the eastern side of the St. Ann Trough. This postulated current must owe its existence to the near proximity of two of the world's largest rivers, the Ob and the Yenisey, and to the different behavior of north-flowing rivers above the Arctic Circle.

The southern reaches of the rivers open first during the spring, then breakup proceeds seaward. There is a sudden and dramatic flushing of the downstream portions of the rivers. The first slug of flood water would mix with the cold saline waters filling the estuaries and rivers as a salt water wedge. The now-saline slug of flood water would erode and transport the last deposited sediment from the previous year. The result would be the formation of a density current which would continue seaward along the drainage system existing on the submerged portions of the Ob–Yenisey Delta, and eventually flow into the St. Ann Trough. Once in the U-shaped trough, such currents would tend to flow along the eastern side as a tractive current under the influence of the Coriolis effect.

The current enters the Arctic Ocean at a depth greater than 600 m and deposits considerable terrigenous sediment on the Arctic sea floor (the Svyataya Anna Cone on the Heezen-Tharp physiographic map) over 600 km from the Ob–Yenisey Delta. The common genetic relationship of river, delta, submarine canyon, and abyssal fan is not definitive for the Ob–Yenisey system, since the St. Ann Trough is neither a true delta nor canyon. It functions as a 600-km-long pipeline

[1] Shell Oil Company, P.O. Box 127, Metairie, Louisiana 70004, U.S.A.

[2] U.S. Naval Oceanographic Office, Washington, D.C. 20390, U.S.A.

passing terrigenous sediments from the delta to the Arctic floor.

The surface patterns of sediment variables persist at depth in cores from the two troughs, suggesting that the processes now active have been active for some time.

Introduction

During the summer of 1965 the U.S.C.G.C. *Northwind* made the first reconnaissance survey conducted by a United States ship in the Kara Sea (Wells, 1966; Petrow, 1967). The scientific program aboard the *Northwind* included standard Nansen casts for water chemistry, bottom sampling and coring, bathymetric profiling, and a geophysical program consisting of gravity, magnetic, seismic refraction and reflection, and hydroacoustic studies. The bottom sampling program was limited to depths greater than 200 m by the Continental Shelf Treaty of 1961, which gives a coastal state exclusive rights to its continental shelf.

Several studies using data collected from the *Northwind* have been completed. They include the works of Johnson and Milligan (1967), Stoll (1967), Vogt (1968), Milligan (1969), Turner and Harriss (1970), and Turner (1972). The last two studies were based mainly on data collected during the U.S.C.G.C. *Eastwind*–U.S.C.G.C. *Edisto* cruise in 1967. Several Russian articles have been translated by the U.S. Naval Oceanographic Office (NAVOCEANO) as background material for the 1965 expedition. Among them are the articles by Saks (1948), Yermolayev (1948a, 1948b) Gorshkova (1957), Kulikov (1961), Rikhter (1963), and Strelkov (1965). Other important available translations include the works of Suslov (1947), Dibner (1957), and Zenkevitch (1963). Volume 1 of the *Geology of the Arctic* contains papers by Atlasov and Sokolov (1961), Markov and Tkachenko (1961), Rabkin and Ravich (1961), and Saks and Strelkov (1961) on the geologic history and tectonic development of the area.

Regional Geography

The Kara Sea is an epicontinental sea which lies over the western parts of the Asiatic Arctic

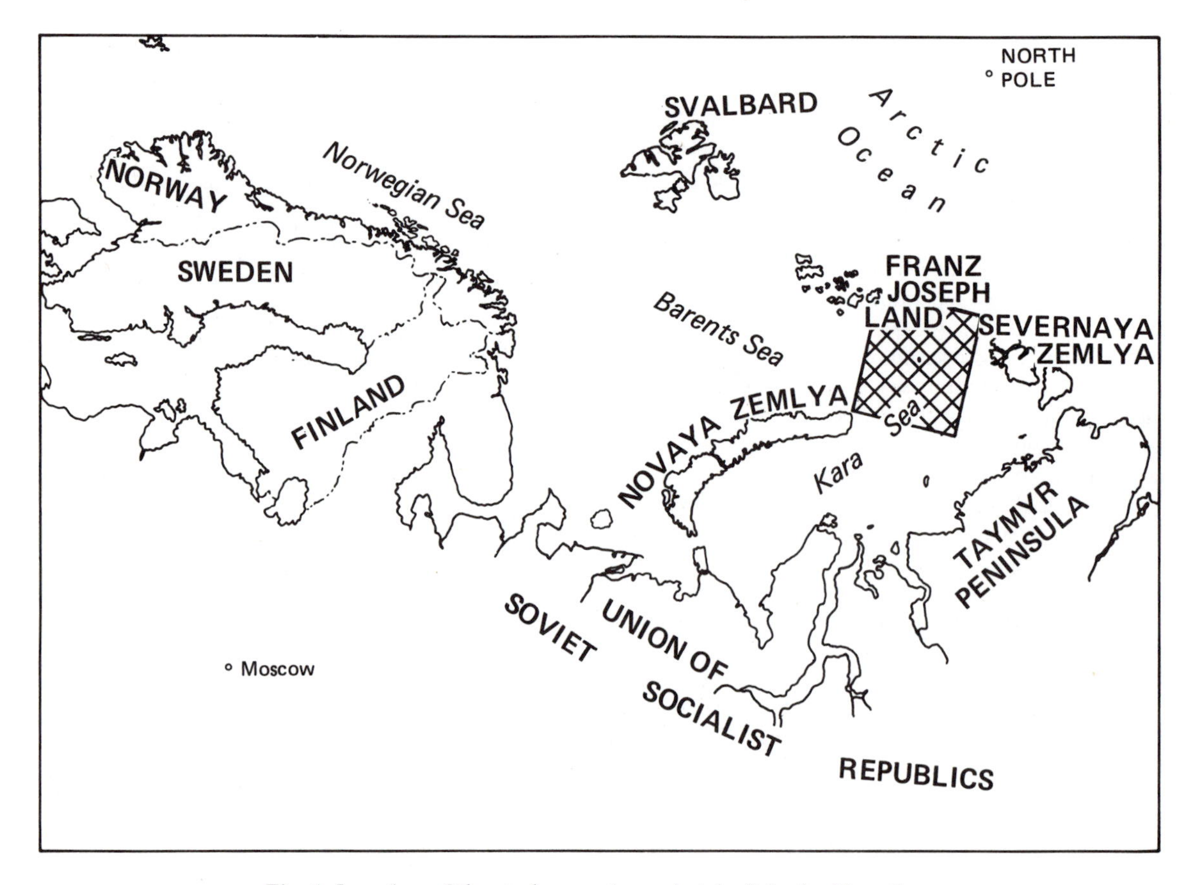

Fig. 1. Location of the study area (cross-hatched) in the Kara Sea.

continental shelf (Fig. 1). The islands of Franz Joseph Land and Novaya Zemlya and the imaginary line connecting them separate the Barents Sea from the Kara Sea. The southern limits are the West Siberian Lowlands and the Taymur (Taymyr) Peninsula. The eastern limits are the islands of Severnaya Zemlya. The northern limits are defined by the imaginary line connecting the northernmost islands of Franz Joseph Land and Severnaya Zemlya, at approximately 81°30′ north latitude.

The area of the Kara Sea is 851,000 km² with

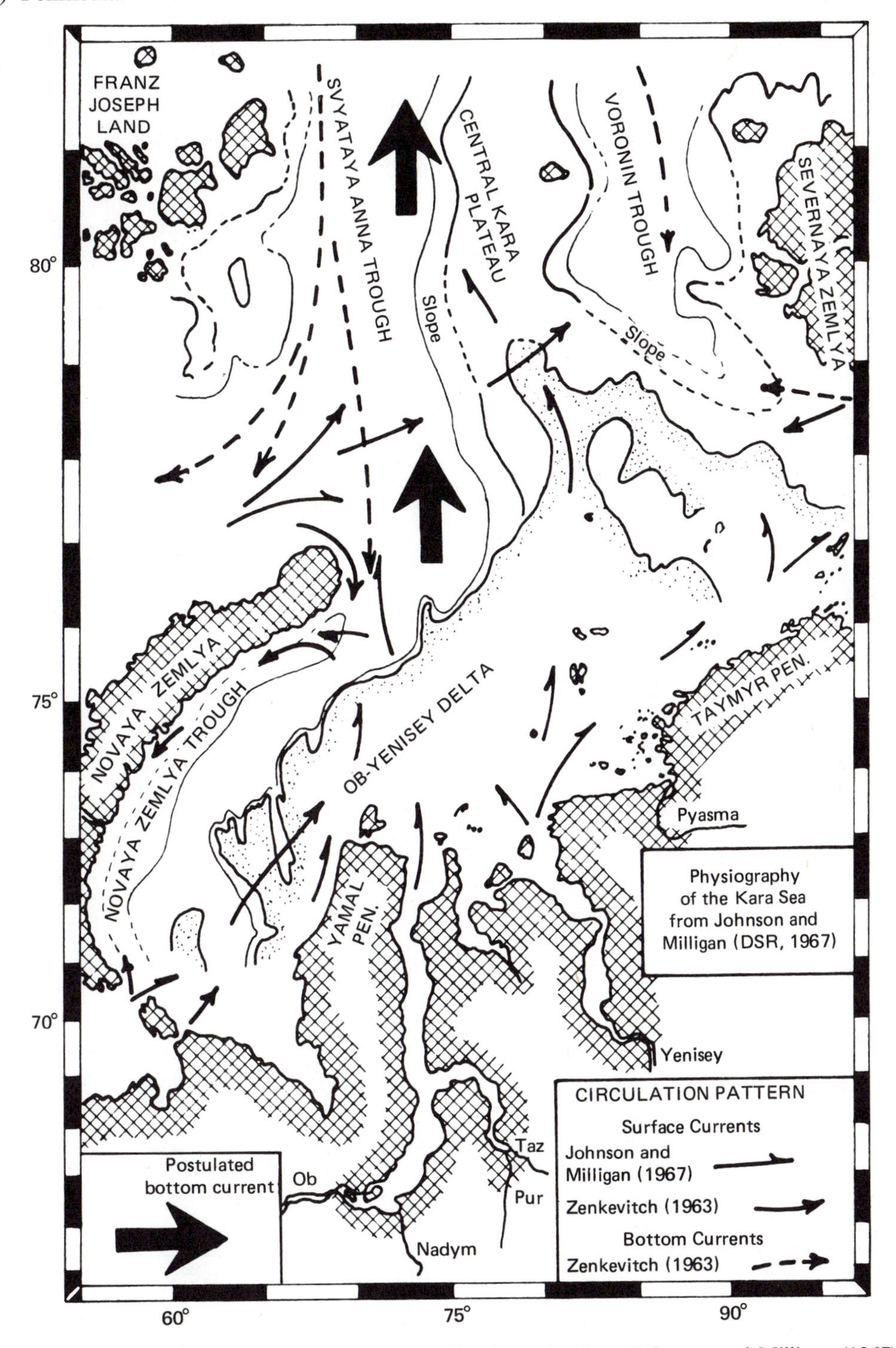

Fig. 2. Physiography and currents of the Kara Sea. Physiography from Johnson and Milligan (1967) based on data collected from the *Northwind*. The large arrow indicates the position of the north-flowing current postulated from data contained in this study.

a volume of 111,000 km³. The mean water depth is 90 m with a maximum exceeding 640 m. Approximately 15% of the total area exceeds 200 m in depth. The deeper water in the northern Kara Sea (north of 76° north latitude) is confined to the two north-facing, open-ended troughs shown in Fig. 2, the Svyataya Anna (or St. Ann) Trough to the west, and the Voronin Trough to the east.

The St. Ann Trough, at the 200-m isobath, is 550 km long and 150 to 200 km wide in the north and opens to the west in the south (Figs. 3 and 4). The trough floor is generally broad and flat but is slightly rougher on the western side. The width of the Voronin Trough varies from 175 km in the north to 65 km in the south, and it is 175 km long.

Johnson and Milligan (1967) report slopes of up to 1:20 on the eastern side of the St. Ann Trough and slopes greater than 1:90 on the west. The steep slopes on the eastern side probably reflect sea-cropping Mesozoic sedimentary rocks (Dibner, 1957). The Mesozoic age was determined by Soviet geologists using palynologic and other paleobotanical data on detrital coal found in nearby sediments.

Gentle slopes (1:100) and a smooth but undulating floor are present in the Voronin Trough. Johnson and Milligan (1967) attribute the swale topography to the blanketing effect of sediments on an uneven bottom. Despite the proximity of a terrigenous sediment supply, they report that there is no development of depositional plains on the trough floors.

The area of the St. Ann Trough, including a portion in the Barents Sea, is 140,000 km²; and that of the Voronin Trough is approximately 20,000 km². The two troughs are separated by the Central Kara Plateau, a shallow area which contains two islands and several shoal areas. The

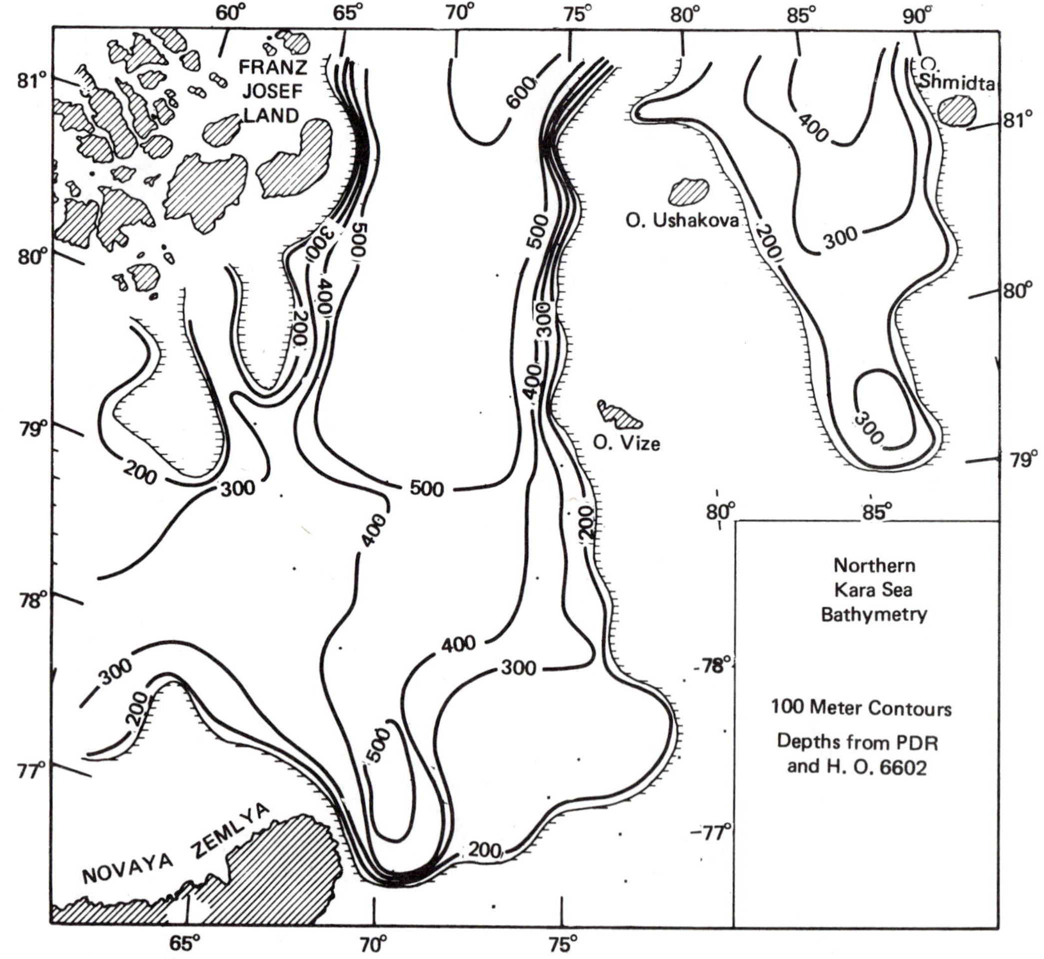

Fig. 3. Bathymetry of the northern Kara Sea based on data collected from the *Northwind*, supplemented by data from H.O. Chart 6602. For this and the following similar figures, the hatched line represents the approximate limit of legal samples allowed by the Continental Shelf Treaty of 1961. Land areas are shown with diagonal lines. For scale, one degree latitude equals 60 nautical miles or approximately 110 kilometers.

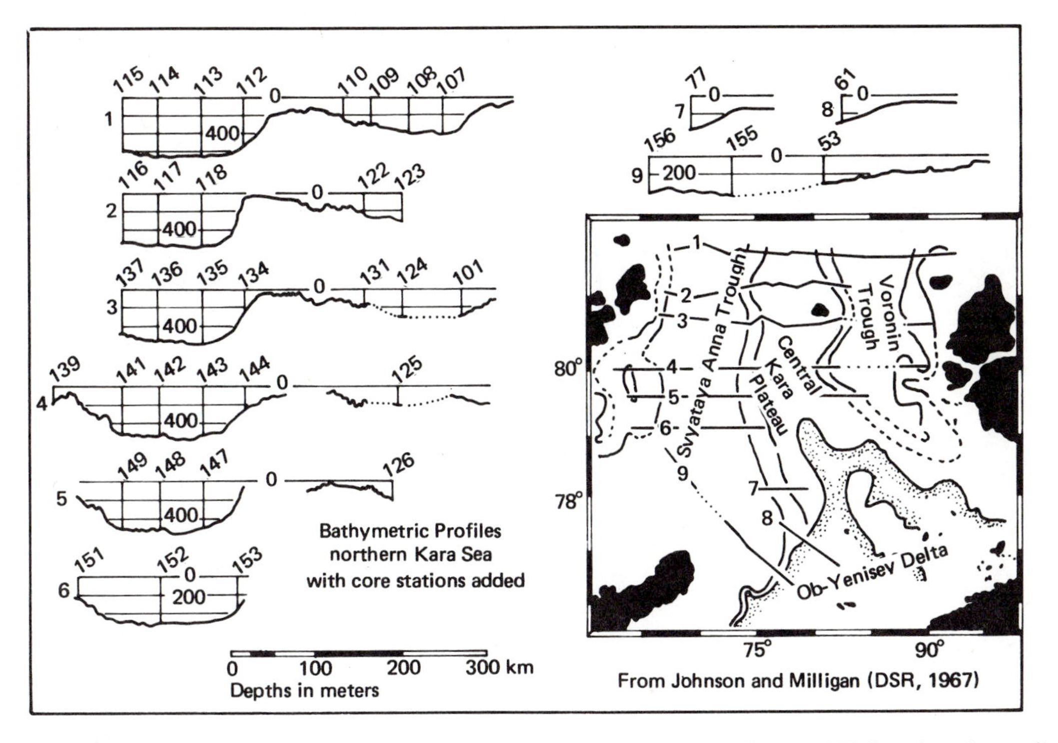

Fig. 4. Bathymetric profiles of the Northern Kara Sea from Johnson and Milligan (1967) based on data collected from the *Northwind.* The station numbers refer to core locations.

dominant feature of the southern reaches of the Kara Sea is the submerged portion of the Ob–Yenisey Delta (Johnson and Milligan, 1967). The area of the delta, at the 50-m isobath, is approximately 230,000 km^2.

Regional Oceanography

The main oceanic water masses affecting the Northern Kara Sea enter from the west through the opening between Novaya Zemlya and Franz Joseph Land, from the north at depth in the two troughs, and from the east through the straits separating the islands of Severnaya Zemlya from the Asiatic mainland (Milligan, 1969). The dominant water affecting the surface of the Kara Sea comes from the fresh-water discharge of the main Siberian rivers. Their combined annual discharge is over 1200 km^3 (Table 1), approximately 1% of the volume of the Kara Sea.

Kulikov (1961) distinguishes three main surface currents dominated by the discharge of the Siberian rivers (shown on maps by Zenkevitch, 1963; and Johnson and Milligan, 1967). The westernmost current issues from the mouth of the Ob Gulf (Fig. 2), moves northwestward toward

Table 1 Rivers of the Kara Sea Area.

River	Length, km	Basin area, 1000 sq km	Discharge: Annual, cu km	Discharge: Mean, cu m/sec	Sediment load*, millions of tons: Sus. sed.	Sediment load*, millions of tons: Dis. sed.
Yenisey	3354	2599	548	17400	14.90	121.0
Ob	3676	2485	394	12500	13.40	56.1
Pechora	1790	327	129	4100	—	—
Pyasima	820	192	80	2550	1.70	5.5
Taz	779	108	47	1500	—	—
Pur	256	67	29	930	0.95	—
Taymur	600	72	20	650	—	—

L'vovich (1953) in Johnson and Milligan (1967).
* From Kulikov (1961).

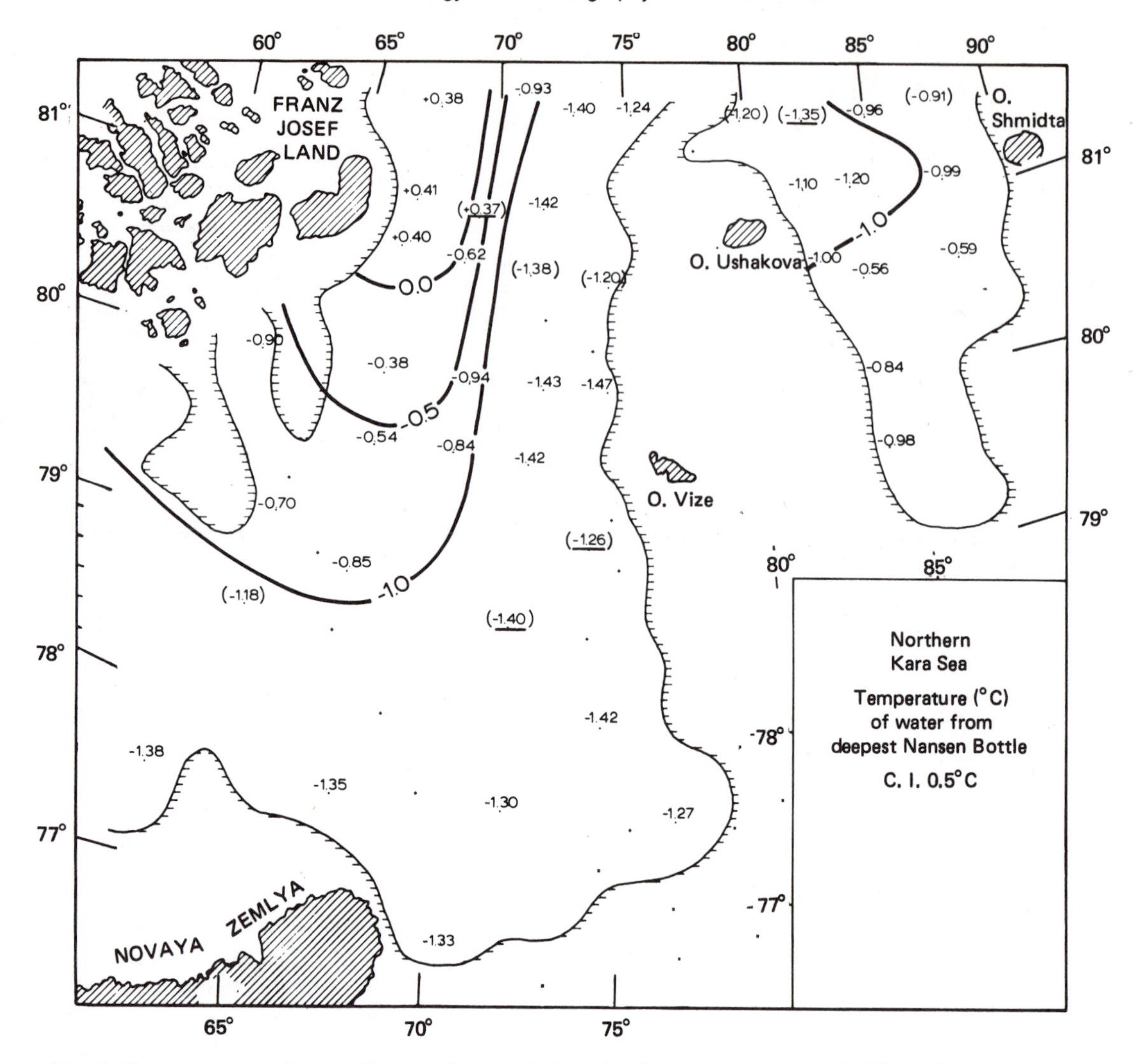

Fig. 5. Temperature in degrees Centigrade recorded by the deepest Nansen bottle. Those figures in parentheses are eighteen or more meters off bottom. Those also underlined are more than 30 meters off bottom.

the island of Novaya Zemlya, continues south along the east coast of the islands, then completes the circle by moving northward along the west coast of the Yamal Peninsula as the Yamal Current. This current predominates early in the summer season following the dramatic spring discharge of the rivers.

The central current flows northward, along the western side of the Central Kara Plateau, into the Arctic Ocean. The easternmost current consists mainly of Yenisey River water. The current moves northeastward along the Taymur coast and passes through the strait between Severnaya Zemlya and the Asiatic mainland. However, our data, and those of Zenkevitch (1963) and Milligan (1969), do not support Kulikov's description. The *Northwind* detected a bottom current in the strait at a depth of 82 m which sets slightly south of west with an average velocity of 10 cm/sec (range 0 to 21 cm/sec). By using radar on floating icebergs, a surface current was plotted which sets slightly north of west with an average velocity of 26 cm/sec. Zenkevitch (1963) shows both surface and bottom currents flowing toward the west. His current chart is based on the penetration and distribution of benthonic organisms into the Kara Sea. In addition, westward-flowing currents would conform to the model experiments conducted by Gudkovich and Nikiforov (1964).

Kulikov (1961) and Coachman (1962) mention the existence of a south-flowing counter-current at depth in the St. Ann Trough. Data collected from the *Northwind* confirm the presence of this current and locates its position on the western side of the trough. The temperature recorded by the deepest Nansen bottle (Fig. 5) clearly outlines the incursion of the relatively warmer water from intermediate depths in the Arctic Ocean (Sater et al., 1963).

The warm water at depth in the Arctic Ocean has its origin in the cooled waters of the North Atlantic Drift. The warm but saline waters are cooled while passing through the Norwegian Sea. Near Spitsbergen the water sinks below the cold but fresher Arctic water and spreads out over the Arctic Basin. Panov and Shpaiker (1963) report that Dobrovol'skii and Timofeyev established, by different methods, almost identical times for the progress of the Atlantic water from Spitsbergen to the Kara Sea. Dobrovol'skii determined that the Atlantic water reaches the Kara Sea in two years, while Timofeyev lists a time of a year and a half. This would indicate a flow rate on the order of 2 cm/sec.

Nature and Scope of This Study

The data presented in this study were obtained from a total of 42 samples collected from the St. Ann Trough and 13 from the Voronin Trough (Fig. 6). The total includes 10 small volume surface samples collected during the U.S.C.G.C. *Eastwind*–U.S.C.G.C. *Edisto* cruise in 1967. Some additional data, for depths less than 200 m, is available in the literature (Gorshkova, 1957).

The purpose of this paper is to discuss the distribution of surface sediments in deep areas of the Northern Kara Sea and to put forth some ideas concerning the observed distribution pattern. Only those data germaine to the discussion are displayed in Figs. 5 through 16. Additional data, collected during the study, are presented in

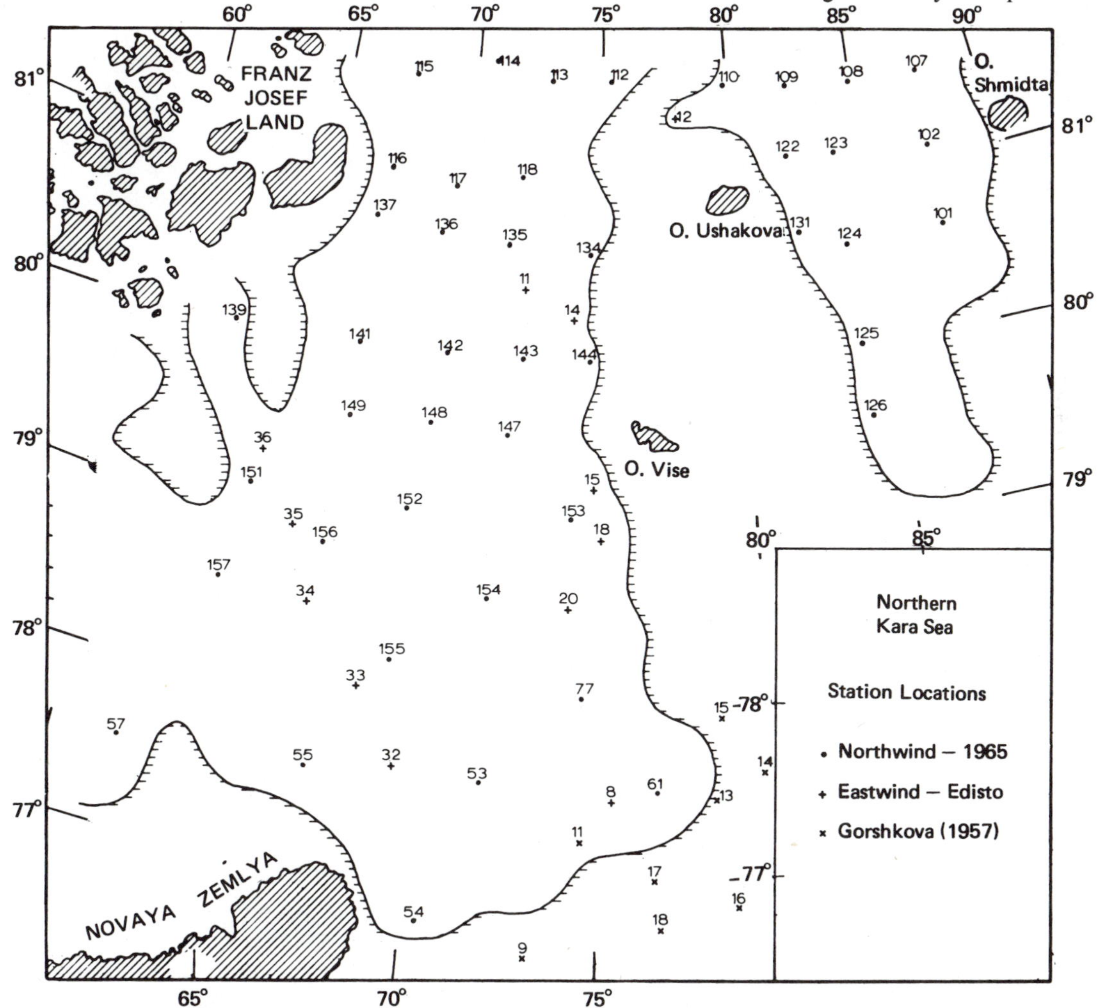

Fig. 6. Station locations in the Northern Kara Sea. Only those *Northwind* stations where sediment samples were collected are shown. The U.S.C.G.C. *Eastwind*—U.S.C.G.C. *Edisto* stations indicate the location of small volume surface sediment samples used for clay analysis furnished by Ralph Turner of Florida State University. The remaining stations refer to the location of grain-size data published by Gorshkova (1957).

Table 2 Location, Depth, Grain-Size, and Clay Data.

A	B	C	D	E	F	G	H	I	J	K	L	M	N	O	P	Q	R	S	T	A
53	291	77	32.5	71	46	3.6	1.8	23.0	71.6	100.0	5.4	43.0	8.2	2.3	−2.0	9.7	18.7	1.1	1.0	53
54	443	76	44.5	70	24	4.3	43.1	20.4	33.0	100.8	47.4	25.0	5.9	4.3	−2.5	12.7	12.7	.9	2.0	54
55	356	77	32.0	67	00	29.5	31.3	20.1	19.2	100.2	60.9	13.0	4.0	4.0	−3.0	5.0	5.0	1.2	4.0	55
57	265	77	31.0	61	50	2.9	39.9	29.5	27.7	100.0	42.8	19.0	5.8	4.2	−2.0	2.0	1.7	1.1	5.0	57
61	280	77	31.8	76	42	.1	1.3	35.0	64.0	100.4	1.4	35.0	8.6	2.0	4.0	8.5	10.0	.8	1.0	61
77	362	78	03.4	74	39	.0	.4	31.2	69.2	100.8	.4	44.0	8.0	1.7	4.2	8.0	5.7	.7	2.0	77
101	310	80	36.0	87	39	.0	.9	35.8	63.3	100.0	.9	42.0	8.4	1.9	4.0	1.7	1.2	.9	****	101
102	340	81	04.0	87	32	.2	26.3	34.6	38.5	99.6	26.5	26.0	6.9	3.6	.7	2.1	1.2	1.0	7.0	102
107	420	81	30.5	87	39	.0	11.3	21.3	67.4	100.0	11.3	38.0	7.8	3.0	2.0	2.2	1.7	2.0	4.5	107
108	410	81	30.5	84	54	.0	.2	48.9	51.1	100.2	.2	36.0	8.1	2.6	4.5	2.2	2.0	2.3	5.0	108
109	423	81	30.6	82	15	.0	12.4	61.6	26.0	100.0	12.4	15.0	6.9	2.7	2.6	1.7	1.3	1.2	7.0	109
110	203	81	34.8	79	52	****	****	****	****	*****	****	****	***	***	****	2.7	1.8	2.0	5.5	110
112	421	81	37.0	75	20	.1	40.2	35.5	24.5	100.3	40.3	13.0	6.0	3.0	1.1	7.5	4.0	1.4	3.0	112
113	640	81	36.0	73	00	.0	.1	42.7	58.5	101.3	.1	30.0	8.8	1.6	4.6	2.4	3.3	1.6	3.0	113
114	631	81	42.3	70	46	.1	2.5	34.9	62.5	100.0	2.6	33.0	8.3	2.0	2.7	2.0	1.4	2.3	5.0	114
115	567	81	35.5	67	32	.0	5.8	31.1	63.5	100.4	5.8	40.0	8.2	2.3	2.2	.9	1.0	2.0	6.5	115
116	475	81	02.2	67	08	.0	1.8	34.4	64.0	100.2	1.8	47.0	8.7	2.0	2.8	.7	1.2	2.9	5.0	116
117	566	80	58.1	69	34	.0	9.3	26.9	63.8	100.0	9.3	39.0	7.9	2.6	2.1	1.6	1.2	1.7	7.5	117
118	588	81	03.0	72	00	.0	.1	38.2	62.3	100.6	.1	45.0	8.8	1.8	4.3	2.1	2.7	1.4	3.5	118
122	268	81	08.0	82	05	****	****	****	****	*****	****	****	***	***	****	3.2	2.4	1.8	4.0	122
123	298	81	07.0	83	58	.1	1.5	42.2	56.2	100.0	1.6	39.0	7.7	3.1	3.6	1.7	1.8	1.5	5.0	123
124	315	80	35.8	83	59	.0	.9	34.6	64.5	100.0	.9	44.0	8.7	2.1	4.0	2.3	2.3	1.4	6.0	124
125	204	80	00.5	84	01	.0	4.3	28.0	67.7	100.0	4.3	49.0	8.0	2.8	2.0	1.0	2.0	1.3	4.5	125
126	217	79	35.5	84	02	.0	2.9	16.8	80.3	100.0	2.9	44.0	8.7	1.9	2.3	2.3	2.0	1.5	4.0	126
131	202	80	41.1	82	13	****	****	****	****	*****	****	****	***	***	****	1.5	1.5	1.7	7.0	131
134	243	80	36.8	74	32	.2	18.8	41.7	39.4	100.1	19.0	25.0	7.0	3.2	1.4	3.3	3.3	1.4	3.0	134
135	593	80	39.9	71	43	.0	.2	39.2	60.4	99.8	.2	31.0	8.7	1.8	4.4	3.7	2.3	1.6	6.0	135
136	549	80	42.9	69	10	.0	5.3	30.2	65.0	100.5	5.3	38.0	8.3	2.3	2.2	1.6	1.9	1.9	4.0	136
137	498	80	45.1	66	48	.0	3.4	31.8	66.6	101.8	3.4	47.0	8.5	2.2	2.3	.6	1.0	1.5	10.0	137
139	228	80	00.0	64	15	.1	12.4	37.8	50.3	100.6	12.5	31.0	7.5	2.9	−2.0	.3	.4	.9	14.0	139
141	520	80	00.0	66	55	.1	1.9	24.3	73.8	100.1	2.0	57.0	8.9	1.9	3.0	.8	.7	1.0	11.0	141
142	564	80	00.0	69	47	.0	1.9	24.8	73.3	100.0	1.9	42.0	8.7	1.8	3.7	1.5	1.8	1.5	5.5	142
143	538	80	00.0	72	11	.0	.2	39.8	60.3	100.3	.2	33.0	8.6	1.9	4.1	2.0	2.2	1.4	4.5	143
144	215	80	00.0	74	36	.0	5.8	42.3	52.2	100.3	5.8	31.0	8.1	2.3	1.8	4.3	6.7	1.8	1.5	144
147	521	79	34.9	72	00	.0	5.7	16.0	78.3	100.0	5.7	50.0	8.8	1.8	2.0	1.5	2.2	1.2	5.0	147
148	532	79	35.0	69	27	.0	14.1	29.6	56.3	100.0	14.1	35.0	7.4	3.1	2.5	1.4	1.0	1.7	6.0	148
149	526	79	35.2	66	57	.0	3.2	39.5	57.3	100.0	3.2	44.0	8.0	2.3	3.4	1.2	.8	1.6	10.0	149
151	260	79	06.2	64	06	.0	4.6	41.8	53.8	100.2	4.6	40.0	8.1	2.4	3.0	.6	.5	1.2	12.0	151
152	525	79	05.5	69	02	.0	3.3	36.8	60.0	100.1	3.3	37.0	8.2	2.2	2.9	1.3	****	1.5	9.5	152
153	366	79	05.0	74	09	****	****	****	****	*****	****	****	***	***	****	5.0	3.8	1.4	4.0	153
154	485	78	37.3	71	41	.0	.1	30.5	69.4	100.0	.1	17.0	8.6	1.4	4.2	2.0	1.6	1.5	5.0	154
155	443	78	12.6	69	00	****	****	****	****	*****	****	****	***	***	****	2.3	2.9	1.6	4.0	155
156	374	78	50.7	66	39	.1	2.3	44.4	53.2	100.0	2.4	35.0	8.0	2.5	3.0	1.2	1.2	1.4	6.0	156
157	346	78	33.0	63	38	.0	6.2	38.0	55.8	100.0	6.2	40.0	6.1	2.5	2.6	1.0	1.1	1.5	8.0	157

A. Station number (Fig. 6)
B. Water depth (m) (Fig. 3)
C. Degrees north latitude
D. Minutes north latitude
E. Degrees east longitude
F. Minutes east longitude
G. Weight percent gravel
H. Weight percent sand
I. Weight percent silt
J. Weight percent clay
K. Sum of weight percents
L. Weight percent sand plus gravel (Fig. 11)
M. Weight percent fine clay
N. Inman mean $M_\phi = (84\phi + 16\phi)/2$
O. Inman sorting $\sigma_\phi = (84\phi - 16\phi)/2$
P. Coarse one percentile (Fig. 16)
Q. Chlorite (004)/kaolinite (002) ratio—surface samples (Fig. 15)
R. Chlorite (004)/kaolinite (002) ratio—grab samples
S. Illite (002)/chlorite (002) + kaolinite (001) ratio—surface samples
T. Kaolinite (002) peak height (Fig. 14)
*** Indicates no data.
Data displayed on figures indicated.

Table 3 Sediment and Water Chemistry, Core, and Heavy Mineral Data.

A	B	C	D	E	F	G	H	I	J	K	L	M	N	O	P	Q	R	S	A
53	82	1.46	10	11	7.2	1.08	−1.30	***	**	**	****	****	549.1	5.5	1.0	.08	1.45	.01	53
54	***	.99	4	49	7.3	.80	−1.33	***	**	**	****	****	*****	*****	****	****	****	***	54
55	47	****	4	6	6.8	1.06	−1.35	***	**	**	****	****	*****	*****	****	****	****	***	55
57	43	1.07	4	7	6.9	.97	−1.38	***	**	**	****	****	*****	*****	****	****	****	***	57
61	93	1.67	11	12	7.6	.86	−1.27	108	5	90	2.8	.5	*****	*****	****	****	****	***	61
77	150	1.75	12	12	7.2	.84	−1.42	118	4	95	9.9	7.2	314.7	11.7	3.7	.22	1.87	.07	77
101	107	1.17	9	13	6.8	.82	−.59	84	11	58	4.2	4.2	*****	*****	****	****	****	***	101
102	81	1.20	10	26	7.7	.44	−.99	77	77	50	7.4	7.1	*****	*****	****	****	****	***	102
107	86	1.01	7	60	7.3	.70	−.91	104	16	**	4.1	5.2	*****	*****	****	****	****	***	107
108	103	1.13	11	50	6.8	.73	−.96	67	21	44	2.0	1.1	*****	*****	****	****	****	***	108
109	53	.66	3	182	7.4	.69	−1.35	***	**	**	****	****	*****	*****	****	****	****	***	109
110	***	****	**	55	5.9	.71	−1.20	83	15	72	6.5	6.0	*****	*****	****	****	****	***	110
112	51	1.11	5	21	7.3	.70	−1.24	75	21	75	18.5	17.9	121.0	10.6	8.7	.12	1.13	.10	112
113	135	1.46	11	16	7.8	.75	−1.40	96	39	73	.9	.9	304.3	1.3	.4	.00	.00	.00	113
114	108	1.06	11	31	7.3	.72	−.93	91	29	63	10.8	8.2	130.5	2.6	2.0	.06	2.33	.05	114
115	100	.96	11	17	6.6	.83	.38	92	34	74	10.2	6.8	160.3	7.2	4.5	.10	1.38	.06	115
116	98	.92	8	5	7.1	.83	.41	***	**	**	****	****	*****	*****	****	****	****	***	116
117	99	.89	8	108	7.4	.80	.37	83	28	64	8.5	6.0	185.7	13.7	7.4	.00	.00	.00	117
118	148	1.82	13	46	7.7	.76	−1.42	91	12	77	2.0	.5	*****	*****	****	****	****	***	119
122	***	****	**	14	7.2	.71	−1.10	72	19	61	14.4	8.6	*****	*****	****	****	****	***	122
123	82	1.59	6	8	7.6	.78	−1.20	65	17	39	2.5	.9	*****	*****	****	****	****	***	123
124	111	1.25	10	18	7.2	.78	−.56	75	19	55	1.6	.9	*****	*****	****	****	****	***	124
125	112	1.48	14	20	6.9	.75	−.84	103	15	98	1.9	.3	*****	*****	****	****	****	***	125
126	140	1.62	12	29	7.1	.79	−.98	111	17	85	1.5	.1	*****	*****	****	****	****	***	126
131	***	****	**	23	7.0	.38	−1.00	58	9	42	11.0	7.9	*****	*****	****	****	****	***	131
134	39	1.12	5	75	7.2	.72	−1.20	23	8	**	41.5	13.7	201.1	107.8	53.6	1.56	1.45	.78	134
135	164	1.60	15	73	7.7	.79	−1.38	81	13	58	.7	.3	301.9	1.3	.4	.00	.00	.00	135
136	115	1.05	10	51	7.1	.87	−.62	87	21	77	8.7	5.0	167.5	14.4	8.6	.31	2.15	.19	136
137	117	1.22	11	20	6.9	.88	.40	83	23	78	11.3	7.4	71.9	2.8	3.9	.10	3.55	.14	137
139	76	1.50	5	11	7.0	.75	−.90	19	3	**	17.8	7.6	166.3	15.3	9.2	.34	2.22	.20	139
141	114	1.04	12	34	6.8	.74	−.38	***	**	**	****	****	150.4	3.0	2.0	.10	3.37	.07	141
142	118	1.08	15	45	7.3	.88	−.94	76	28	64	5.7	5.6	158.6	2.7	1.7	.04	1.45	.03	142
143	97	1.64	4	38	7.3	.81	−1.43	112	9	82	5.8	3.5	102.9	3.5	3.4	.02	.57	.02	143
144	77	1.15	8	12	7.2	.68	−1.47	84	8	84	8.5	2.1	134.0	10.7	8.0	.13	1.21	.10	144
147	101	1.59	16	26	7.9	.78	−1.42	70	10	52	8.0	4.8	186.0	11.6	6.2	.22	1.90	.12	147
148	81	.86	7	40	6.9	1.23	−.84	84	20	62	9.5	5.3	172.8	23.2	13.4	.45	1.94	.26	149
149	106	1.17	9	43	6.6	.56	−.54	98	25	50	2.2	.9	134.5	2.4	1.7	.06	2.54	.04	149
151	130	1.65	11	29	7.5	.75	−.70	***	**	**	****	****	194.9	35.9	18.4	.40	1.11	.21	151
152	117	1.41	10	225	7.1	.76	*****	83	21	59	6.6	5.1	*****	*****	****	****	****	***	152
153	51	****	6	102	8.0	.70	−1.26	***	**	**	****	****	192.4	82.0	42.6	1.18	1.44	.61	153
154	156	1.77	15	221	7.6	.80	−1.40	70	12	70	1.5	.4	*****	*****	****	****	****	***	154
155	146	****	16	57	7.8	.77	*****	100	6	85	3.8	1.0	*****	*****	****	****	****	***	155
156	113	1.07	9	57	7.0	.75	−.85	66	17	57	13.5	5.0	273.1	7.2	2.7	.00	.00	.00	156
157	109	1.19	8	68	7.7	.71	−1.18	***	**	**	****	****	198.5	10.6	5.3	.18	1.70	.09	157

A. Station number (Fig. 6)
B. Water content (percent dry weight) (Fig. 13)
C. Organic carbon content (Fig. 12)
D. Weight percent soluble in HCl
E. Height off bottom of deepest Nansen bottle (ft)
F. Oxygen content—water from deepest Nansen bottle
G. Inorganic phosphate content—water from deepest Nansen bottle
H. Temperature (Centigrade)—water from deepest Nansen bottle (Fig. 5)
I. Core length (cm)
J. Depth to oxidized layer (cm)
K. Depth to coarse layer (cm)
L. Mean weight percent sand in cores (Fig. 9)
M. Standard deviation of weight percent sand in cores (Fig. 10)
N. Weight total grab sample
O. Weight sand from total grab sample
P. Weight percent sand from total grab sample
Q. Weight heavy minerals in sand from total grab sample
R. Weight percent of sand
S. Weight percent of sample
*** Indicates no data.
Data displayed on figure indicated.

Table 4 Foraminifera Data—Number of Individuals Counted (from Stoll, 1967).

A	B	C	D	E	F	G	H	I	J	K	L	M	N	O	P	Q	R	S	T	U	V	W	X	Y	X+Y	Z1	Z2	Z	A
54	11	0	0	9	0	0	5	0	0	1	2	1	0	0	0	0	0	0	0	0	0	1	7	0	7	29	0	30	54
77	14	23	4	2	26	5	158	37	190	110	4	2	0	0	0	0	0	0	0	10	2	2	13	4	17	575	12	589	77
102	6	1	0	3	54	5	70	10	61	104	8	2	0	1	1	2	3	4	0	1	5	0	11	7	18	324	17	341	102
107	0	1	2	0	12	0	24	18	21	35	0	0	0	0	1	0	0	0	0	0	0	0	7	1	8	113	1	114	107
108	4	2	8	8	11	3	24	74	70	64	2	15	0	0	0	1	0	1	0	6	2	2	12	6	18	285	10	297	108
109	1	0	2	1	19	11	44	1	17	34	2	1	0	0	2	1	0	0	0	0	0	1	11	3	14	133	3	137	109
110	0	1	0	0	90	5	10	6	47	10	0	1	0	0	0	0	0	0	0	0	0	0	8	0	8	170	0	170	110
112	16	0	0	25	0	0	7	14	0	0	0	0	12	27	13	75	14	108	5	28	219	36	4	18	22	62	501	599	112
113	2	0	13	0	0	0	28	6	57	4	0	0	0	0	0	0	0	1	0	0	0	0	6	1	7	110	1	101	113
114	3	0	2	1	0	1	50	46	61	7	3	3	0	1	2	40	0	12	0	30	46	0	10	7	17	177	131	318	114
115	0	0	0	3	0	0	8	28	0	3	2	1	3	20	15	95	0	18	1	10	574	1	6	10	16	45	736	782	151
116	2	0	74	0	19	19	47	19	61	2	0	0	0	0	0	0	0	0	0	0	0	1	9	0	9	243	0	244	116
117	2	0	0	2	1	3	51	42	43	57	1	0	11	10	6	57	0	1	10	8	247	1	9	9	18	202	350	553	117
118	0	0	0	0	0	0	0	0	0	0	0	0	0	0	0	0	0	0	0	0	0	0	0	0	0	0	0	0	118
122	2	0	6	0	30	11	79	31	199	180	1	3	0	0	0	0	0	0	0	1	0	0	10	1	11	542	1	543	122
123	10	3	1	0	3	1	85	30	93	52	0	0	0	0	0	0	0	0	0	0	0	0	9	0	9	278	0	278	123
124	29	25	6	9	12	3	20	33	85	27	3	3	0	0	0	0	0	1	0	3	0	0	12	2	14	255	4	259	124
125	3	30	3	6	2	1	15	9	69	12	0	3	0	1	0	0	0	0	0	2	0	2	12	3	15	153	3	158	125
126	1	1	1	0	1	0	3	2	5	0	0	2	0	0	0	0	0	0	0	0	0	0	8	0	8	16	0	16	126
131	0	1	0	0	0	0	3	0	12	0	0	0	0	0	0	0	0	0	0	0	0	0	3	0	3	16	0	16	131
134	1	0	0	2	0	2	19	1	18	5	0	0	0	0	0	0	0	1	0	0	0	1	7	2	9	48	1	50	134
135	56	0	7	2	2	1	68	40	124	8	9	19	0	0	0	2	0	1	0	5	2	6	11	5	16	336	10	352	135
136	2	0	14	1	11	4	89	41	115	8	0	0	0	0	0	2	0	0	0	0	9	0	9	2	11	285	11	296	136
137	4	2	111	1	13	15	41	42	75	96	1	8	0	0	0	0	0	0	0	0	0	0	12	0	12	409	0	409	137
139	3	5	8	0	5	13	40	2	37	13	2	12	0	0	0	0	0	0	0	0	0	0	11	0	11	140	0	140	139
141	5	1	143	1	0	12	90	48	159	2	1	3	0	0	0	0	0	0	0	1	0	0	11	1	12	465	1	466	141
142	11	0	80	0	17	20	52	53	133	31	0	0	0	0	0	0	0	0	0	0	0	2	9	0	9	397	0	399	142
143	3	0	0	1	3	1	70	15	146	15	0	9	0	0	0	0	0	0	0	0	0	1	10	0	10	263	0	264	143
144	1	3	2	5	32	1	29	4	2	8	1	1	0	1	0	1	0	0	0	1	0	3	12	3	15	89	3	95	144
148	2	0	47	0	0	18	101	42	199	29	0	0	0	0	0	0	0	0	0	0	0	0	7	0	7	438	0	438	148
149	7	0	242	1	5	5	95	59	219	9	3	1	0	0	0	0	0	0	0	0	0	0	11	0	11	646	0	646	149
151	1	0	8	3	0	0	33	7	2	13	2	0	0	0	0	0	0	0	0	0	0	2	9	0	9	69	0	71	151
153	0	0	0	1	1	0	38	1	7	14	0	0	0	0	0	0	0	0	0	0	0	0	6	0	6	62	0	62	153
156	2	2	40	0	0	6	109	33	165	21	1	0	0	0	0	2	0	0	0	0	1	0	9	2	11	379	3	382	156
157	1	0	21	3	0	3	61	35	45	48	0	0	0	0	0	0	0	0	0	0	0	0	8	1	9	217	0	217	157

Index to Tables 4, 5, and 6
(Sarah Stoll, unpublished)

A. Station number (Fig. 6)
B. *Adercotryma glomeratum* (Brady)
C. *Alveolophragmium crassimargo* (Norman)
D. *Alveolophragmium subglobosum* (G. O. Sars)
E. *Haplophragmoides canariensis* (d'Orbigny)
F. *Hormosina globulifera* Brady
G. *Psammosphaera fusca* Schulze
H. *Reophax atlantica* (Cushman)
I. *Reophax nodulosus* Brady
J. *Reophax scorpiurus* Montfort
K. *Saccorhiza ramosa* (Brady)
L. *Trochammina nana* (Brady)
M. Unidentified extremely fine arenaceous formanimifera
N. *Astrononion stelligerum* (d'Orbigny)
O. *Buccella frigida* (Cushman)
P. *Cassidulina islandica* Norvang var. *minuta* Norvang
Q. *Cassidulina laevigata* d'Orbigny
R. *Cibicides lobatulus* (Walker and Jacob)
S. *Cibicides lobatulus* (Walker and Jacob) var. *A*
T. *Elphidium incertum* (Williamson) var. *clavatum* Cushman
U. *Gavelinonion barleeanum* (Williamson)
V. *Globigerina pachyderma* (Ehrenberg)
W. Other
X. Arenaceous species
Y. Calcareous species
X + Y. Total species
Z1. Total arenaceous counted
Z2. Total calcareous counted
Z. Total counted

Table 4—Number of individuals counted
Table 5—Percentage of each species
Table 6—Percentage of arenaceous and calcareous species

Tables 2 and 3; since we have no plans to publish these data elsewhere, they are included here for completeness. The same applies to the unpublished foraminiferal population counts of Stoll (1967) in Table 4 and our computer recalculations of her data in Tables 5 and 6.

Analytical Techniques

Sand Content and Sediment Color of Cores

The sediment color was determined by comparing the moist newly opened core (Kravitz, 1968) to the Standard Geological Society of America color chart. The sand content of the cores is the weight percent of material retained on a 61-μ sieve. A statistical mean and standard deviation of the sand content were computed for each core. The statistics are based on a minimum of a 10-cm sample increment or samples from adjacent sedimentological units.

Grain-Size Data of Surface Samples

Samples from the upper 10 cm of the cores were mechanically dispersed using a sonic disrupter (Kravitz, 1966). The sand-size distribution was determined by wet sieving. The silt and clay fractions were determined by the standard pipette technique outlined by Krumbein and Pettijohn (1938).

Organic Carbon Content

Organic carbon in the sediment from the upper 10 cm of the cores was determined by the Allison (1935) titration method, which is based on the reduction of chromic acid.

Water Content

The wet sediment from the upper 10 cm of the cores was weighed, dried at 105°C for 24 hours, and reweighed. The water content is expressed as a percentage of dry weight (Taylor, 1948).

Clay Mineral Content

Samples for clay mineral analysis were collected from three sources: grab samples, the top 2 cm of cores (here called surface samples), and small volume surface samples furnished by Ralph Turner. These samples were prepared by washing, disaggregating, and separating the clay fraction (by decanting the aliquot containing the less than 0.002-mm size fraction to a glass slide and air drying).

Two sets of X-ray analyses were obtained from both the grab and surface samples from all available sample locations. A fast scan was made at a goniometer scan speed of one 2ø per minute. A slow scan was made across the chlorite (004)–kaolinite (002) peak doublet at a scan speed of $\frac{1}{4}$ 2ø per minute (Biscaye, 1964). The results from the two sets of analyses across the chlorite (004)–kaolinite (002) peak doublet show close agreement between the data from the grab samples and the surface samples (Table 2).

Discussion of Data

Sediment Color and Sand Content of Cores

The colors of the sediment in the cores from the Kara Sea are, in general, similar to those described by Yermolayev (1948a); a thin (to 20 or 30 cm) upper, oxidized zone and a lower, gray, reduced zone, each with several subzones (Turner and Harriss, 1970; Turner, 1972). Regional patterns of sediment color (Figs. 7a, 7b, and 7c) and sand content (Figs. 8a, 8b, and 8c) can be defined. The same pattern is defined by the areal distribution of the mean weight percent sand in cores (Fig. 9) and the standard deviation of weight percent sand in cores (Fig. 10).

The cores from the western St. Ann Trough and extreme western Voronin Trough contain sediment with all of the zones and subzones described by Yermolayev (1948a). These cores contain sediment with a higher mean weight percent sand (greater than 6%) and a greater variation in sand content (standard deviation greater than 5) than the sediment in the cores from the northeastern side and the southeastern end of the St. Ann Trough (stations 113, 118, 135, 61, 144, 154, and 155). The sediment here is much more uniform in color in the lower portions of the cores. The sand content of these cores is low (less than 4%) and the standard deviation is low (less than 1). The same is true for the sediments collected in the axial portions of the Voronin Trough.

Cores from the margins of the two troughs contain a marked increase in sand content at core tops (Stations 112, 134, and 139 in the St. Ann

Table 5 Foraminifera Data—Percentage of Each Species (from Stoll, 1967).

A	%B	%C	%D	%E	%F	%G	%H	%I	%J	%K	%L	%M	%N	%O	%P	%Q	%R	%S	%T	%U	%V	%W	%X	%Y	%Z1	%Z2	Z	A
54	36.7	0.0	0.0	30.0	0.0	0.0	16.7	0.0	0.0	3.3	6.7	3.3	0.0	0.0	0.0	0.0	0.0	0.0	0.0	0.0	0.0	3.3	100.0	0.0	96.7	0.0	30	54
77	2.4	3.9	0.7	0.3	4.4	0.8	26.8	6.3	32.3	18.7	0.7	0.3	0.0	0.0	0.0	0.0	0.0	0.0	0.0	1.7	0.3	0.3	76.5	23.5	97.6	2.0	589	77
102	1.8	0.3	0.0	0.9	15.8	1.5	20.5	2.9	17.9	30.5	2.3	0.6	0.0	0.3	0.3	0.6	0.9	1.2	0.0	0.3	1.5	0.0	61.1	38.9	95.0	5.0	341	102
107	0.0	0.9	1.8	0.0	10.5	0.0	21.1	15.8	18.4	30.7	0.0	0.0	0.0	0.0	0.9	0.0	0.0	0.0	0.0	0.0	0.0	0.0	87.5	12.5	99.1	0.9	114	107
108	1.3	0.7	2.7	2.7	3.7	1.0	8.1	24.9	23.6	21.5	0.7	5.1	0.0	0.0	0.0	0.3	0.0	0.3	0.0	2.0	0.7	0.7	66.7	33.3	96.0	3.4	297	108
109	0.7	0.0	1.5	0.7	13.9	8.0	32.1	0.7	12.4	24.8	1.5	0.7	0.0	0.0	1.5	0.7	0.0	0.0	0.0	0.0	0.0	0.7	78.6	21.4	97.1	2.2	137	109
110	0.0	0.6	0.0	0.0	52.9	2.9	5.9	3.5	27.6	5.9	0.0	0.6	0.0	0.0	0.0	0.0	0.0	0.0	0.0	0.0	0.0	0.0	100.0	0.0	100.0	0.0	170	110
112	2.7	0.0	0.0	4.2	0.0	0.0	1.2	2.3	0.0	0.0	0.0	0.0	2.0	4.5	2.2	12.5	2.3	18.0	0.8	4.7	36.6	6.0	18.2	81.8	10.4	83.6	599	112
113	1.8	0.0	11.7	0.0	0.0	0.0	25.2	5.4	51.4	3.6	0.0	0.0	0.0	0.0	0.0	0.0	0.0	0.9	0.0	0.0	0.0	0.0	85.7	14.3	99.1	0.9	111	113
114	1.0	0.0	0.6	0.3	0.0	0.3	16.2	14.9	19.8	2.3	1.0	1.0	0.0	0.3	0.6	13.0	0.0	3.9	0.0	9.7	14.9	0.0	58.8	41.2	57.5	42.5	308	114
115	0.0	0.0	0.0	0.4	0.0	0.0	1.0	3.6	0.0	0.4	0.3	0.1	0.4	2.6	1.9	12.1	0.0	2.3	0.1	1.3	73.4	0.1	37.5	62.5	5.8	94.1	782	115
116	0.8	0.0	30.3	0.0	7.8	7.8	19.3	7.8	25.0	0.8	0.0	0.0	0.0	0.0	0.0	0.0	0.0	0.0	0.0	0.0	0.0	0.4	100.0	0.0	99.6	0.0	244	116
117	0.4	0.0	0.0	0.4	0.2	0.5	9.2	7.6	7.8	10.3	0.2	0.0	2.0	1.8	1.1	10.3	0.0	0.2	1.8	1.4	44.7	0.2	50.0	50.0	36.5	63.3	553	117
118	0.0	0.0	0.0	0.0	0.0	0.0	0.0	0.0	0.0	0.0	0.0	0.0	0.0	0.0	0.0	0.0	0.0	0.0	0.0	0.0	0.0	0.0	0.0	0.0	0.0	0.0	0	118
122	0.4	0.0	1.1	0.0	5.5	2.0	14.5	5.7	36.6	33.1	0.2	0.6	0.0	0.0	0.0	0.0	0.0	0.0	0.0	0.2	0.0	0.0	90.9	9.1	99.8	0.2	543	122
123	3.6	1.1	0.4	0.0	1.1	0.4	30.6	10.8	33.5	18.7	0.0	0.0	0.0	0.0	0.0	0.0	0.0	0.0	0.0	0.0	0.0	0.0	100.0	0.0	100.0	0.0	278	123
124	11.2	9.7	2.3	3.5	4.6	1.2	7.7	12.7	32.8	10.4	1.2	1.2	0.0	0.0	0.0	0.0	0.0	0.4	0.0	1.2	0.0	0.0	85.7	14.3	98.5	1.5	259	124
125	1.9	19.0	1.9	3.8	1.3	0.6	9.5	5.7	43.7	7.6	0.0	1.9	0.0	0.6	0.0	0.0	0.0	0.0	0.0	1.3	0.0	1.3	80.0	20.0	96.8	1.9	158	125
126	6.3	6.3	6.3	0.0	6.3	0.0	18.8	12.5	31.3	0.0	0.0	12.5	0.0	0.0	0.0	0.0	0.0	0.0	0.0	0.0	0.0	0.0	100.0	0.0	100.0	0.0	16	126
131	0.0	6.3	0.0	0.0	0.0	0.0	18.8	0.0	75.0	0.0	0.0	0.0	0.0	0.0	0.0	0.0	0.0	0.0	0.0	0.0	0.0	0.0	100.0	0.0	100.0	0.0	16	131
134	2.0	0.0	0.0	4.0	0.0	4.0	38.0	2.0	36.0	10.0	0.0	0.0	0.0	0.0	0.0	0.0	0.0	2.0	0.0	0.0	0.0	2.0	77.8	22.2	96.0	2.0	50	134
135	15.9	0.0	2.0	0.6	0.6	0.3	19.3	11.4	35.2	2.3	2.6	5.4	0.0	0.0	0.0	0.6	0.0	0.3	0.0	1.4	0.6	1.7	68.8	31.3	95.5	2.8	352	135
136	0.7	0.0	4.7	0.3	3.7	1.4	30.1	13.9	38.9	2.7	0.0	0.0	0.0	0.0	0.0	0.7	0.0	0.0	0.0	0.0	3.0	0.0	81.8	18.2	96.3	3.7	296	136
137	1.0	0.5	27.1	0.2	3.2	3.7	10.0	10.3	18.3	23.5	0.2	2.0	0.0	0.0	0.0	0.0	0.0	0.0	0.0	0.0	0.0	0.0	100.0	0.0	100.0	0.0	409	137
139	2.1	3.6	5.7	0.0	3.6	9.3	28.6	1.4	26.4	9.3	1.4	8.6	0.0	0.0	0.0	0.0	0.0	0.0	0.0	0.0	0.0	0.0	100.0	0.0	100.0	0.0	140	139
141	1.1	0.2	30.7	0.2	0.0	2.6	19.3	10.3	34.1	0.4	0.2	0.6	0.0	0.0	0.0	0.0	0.0	0.0	0.0	0.2	0.0	0.0	91.7	8.3	99.8	0.2	466	141
142	2.8	0.0	20.1	0.0	4.3	5.0	13.0	13.3	33.3	7.8	0.0	0.0	0.0	0.0	0.0	0.0	0.0	0.0	0.0	0.0	0.0	0.5	100.0	0.0	99.5	0.0	399	142
143	1.1	0.0	0.0	0.4	1.1	0.4	26.5	5.7	55.3	5.7	0.0	3.4	0.0	0.0	0.0	0.0	0.0	0.0	0.0	0.0	0.0	0.4	100.0	0.0	99.6	0.0	264	143
144	1.1	3.2	2.1	5.3	33.7	1.1	30.5	4.2	2.1	8.4	1.1	1.1	0.0	1.1	0.0	1.1	0.0	0.0	0.0	1.1	0.0	3.2	80.0	20.0	93.7	3.2	95	144
148	0.5	0.0	10.7	0.0	0.0	4.1	23.1	9.6	45.4	6.6	0.0	0.0	0.0	0.0	0.0	0.0	0.0	0.0	0.0	0.0	0.0	0.0	100.0	0.0	100.0	0.0	438	148
149	1.1	0.0	37.5	0.2	0.8	0.8	14.7	9.1	33.9	1.4	0.5	0.2	0.0	0.0	0.0	0.0	0.0	0.0	0.0	0.0	0.0	0.0	100.0	0.0	100.0	0.0	646	149
151	1.4	0.0	11.3	4.2	0.0	0.0	46.5	9.9	2.8	18.3	2.8	0.0	0.0	0.0	0.0	0.0	0.0	0.0	0.0	0.0	0.0	2.8	100.0	0.0	97.2	0.0	71	151
153	0.0	0.0	0.0	1.6	1.6	0.0	61.3	1.6	11.3	22.6	0.0	0.0	0.0	0.0	0.0	0.0	0.0	0.0	0.0	0.0	0.0	0.0	100.0	0.0	100.0	0.0	62	153
156	0.5	0.5	10.5	0.0	0.0	1.6	28.5	8.6	43.2	5.5	0.3	0.0	0.0	0.0	0.0	0.5	0.0	0.0	0.0	0.0	0.3	0.0	81.8	18.2	99.2	0.8	382	156
157	0.5	0.0	9.7	1.4	0.0	1.4	28.1	16.1	20.7	22.1	0.0	0.0	0.0	0.0	0.0	0.0	0.0	0.0	0.0	0.0	0.0	0.0	88.9	11.1	100.0	0.0	217	157

See index at end of Table 4.

Table 6 Foraminifera Data—Percentage of Arenaceous and Calcereous Species (from Stoll, 1967).

A	%B*	%C*	%D*	%E*	%F*	%G*	%H*	%I*	%J*	%K*	%L*	%M*	%N*	%O*	%P*	%Q*	%R*	%S*	%T*	%U*	%V*	%W	%X	%Y	%Z1	%Z2	W	X	Y	X+Y	Z1	Z2	Z	A
54	37.9	0.0	0.0	31.0	0.0	0.0	17.2	0.0	0.0	3.4	6.9	3.4	—	—	—	—	—	—	—	—	—	3.3	100.0	0.0	96.7	0.0	1	7	0	7	29	0	30	54
77	2.4	4.0	0.7	0.3	4.5	0.9	27.5	6.4	33.0	19.1	0.7	0.3	0.0	0.0	0.0	0.0	0.0	0.0	0.0	83.3	16.7	0.3	76.5	23.5	97.6	2.0	2	13	4	17	575	12	589	77
102	1.9	0.3	0.0	0.9	16.7	1.5	21.6	3.1	18.8	32.1	2.5	0.6	0.0	5.9	5.9	11.8	17.6	23.5	0.0	5.9	29.4	0.0	61.1	38.9	95.0	5.0	0	11	7	18	324	17	341	102
107	0.0	0.9	1.8	0.0	10.6	0.0	21.2	15.9	18.6	31.0	0.0	0.0	0.0	0.0	100.0	0.0	0.0	0.0	0.0	0.0	0.0	0.0	87.5	12.5	99.1	0.9	0	7	1	8	113	1	114	107
108	1.4	0.7	2.8	2.8	3.9	1.1	8.4	26.0	24.6	22.5	0.7	5.3	0.0	0.0	0.0	10.0	0.0	10.0	0.0	60.0	20.0	0.7	66.7	33.3	96.0	3.4	2	12	6	18	285	10	297	108
109	0.8	0.0	1.5	0.8	14.3	8.3	33.1	0.8	12.8	25.6	1.5	0.8	0.0	0.0	66.7	33.3	0.0	0.0	0.0	0.0	0.0	0.7	78.6	21.4	97.1	2.2	1	11	3	14	133	3	137	109
110	0.0	0.6	0.0	0.0	52.9	2.9	5.9	3.5	27.6	5.9	0.0	0.6	—	—	—	—	—	—	—	—	—	0.0	100.0	0.0	100.0	0.0	0	8	0	8	170	0	170	110
112	25.8	0.0	0.0	40.3	0.0	0.0	11.3	22.6	0.0	0.0	0.0	0.0	2.4	5.4	2.6	15.0	2.8	21.6	1.0	5.6	43.7	6.0	18.2	81.8	10.4	83.6	36	4	18	22	62	501	599	112
113	1.8	0.0	11.8	0.0	0.0	0.0	25.5	5.5	51.8	3.6	0.0	0.0	0.0	0.0	0.0	0.0	0.0	100.0	0.0	0.0	0.0	0.0	85.7	14.3	99.1	0.9	0	6	1	7	110	1	111	113
114	1.7	0.0	1.1	0.6	0.0	0.6	28.2	26.0	34.5	4.0	1.7	1.7	0.0	0.8	1.5	30.5	0.0	9.2	0.0	22.9	35.1	0.0	58.8	41.2	57.5	42.5	0	10	7	17	177	131	308	114
115	0.0	0.0	0.0	6.7	0.0	0.0	17.8	62.2	0.0	6.7	4.4	2.2	0.4	2.7	2.0	12.9	0.0	2.4	0.1	1.4	78.0	0.1	37.5	62.5	5.8	94.1	1	6	10	16	45	736	782	115
116	0.8	0.0	30.5	0.0	7.8	7.8	19.3	7.8	25.1	0.8	0.0	0.0	—	—	—	—	—	—	—	—	—	0.4	100.0	0.0	99.6	0.0	1	9	0	9	243	0	244	116
117	1.0	0.0	0.0	1.0	0.5	1.5	25.2	20.8	21.3	28.2	0.5	0.0	3.1	2.9	1.7	16.3	0.0	0.3	2.9	2.3	70.6	0.2	50.0	50.0	36.5	63.3	1	9	9	18	202	350	553	117
118	—	—	—	—	—	—	—	—	—	—	—	—	—	—	—	—	—	—	—	—	—	0.0	0.0	0.0	0.0	0.0	0	0	0	0	0	0	0	118
122	0.4	0.0	1.1	0.0	5.5	2.0	14.6	5.7	36.7	33.2	0.2	0.6	0.0	0.0	0.0	0.0	0.0	0.0	0.0	100.0	0.0	0.0	90.9	9.1	99.8	0.2	0	10	1	11	542	1	543	122
123	3.6	1.1	0.4	0.0	1.1	0.4	30.6	10.8	33.5	18.7	0.0	0.0	—	—	—	—	—	—	—	—	—	0.0	100.0	0.0	100.0	0.0	0	9	0	9	278	0	278	123
124	11.4	9.8	2.4	3.5	4.7	1.2	7.8	12.9	33.3	10.6	1.2	1.2	0.0	0.0	0.0	0.0	0.0	25.0	0.0	75.0	0.0	0.0	85.7	14.3	98.5	1.5	0	12	2	14	255	4	259	124
125	2.0	19.6	2.0	3.9	1.3	0.7	9.8	5.9	45.1	7.8	0.0	2.0	0.0	33.3	0.0	0.0	0.0	0.0	0.0	66.7	0.0	1.3	80.0	20.0	96.8	1.9	2	12	3	15	153	3	158	125
126	6.3	6.3	6.3	0.0	6.3	0.0	18.8	12.5	31.3	0.0	0.0	12.5	—	—	—	—	—	—	—	—	—	0.0	100.0	0.0	100.0	0.0	0	8	0	8	16	0	16	126
131	0.0	6.3	0.0	0.0	0.0	0.0	18.8	0.0	75.0	0.0	0.0	0.0	—	—	—	—	—	—	—	—	—	0.0	100.0	0.0	100.0	0.0	0	3	0	3	16	0	16	131
134	2.1	0.0	0.0	4.2	0.0	4.2	39.6	2.1	37.5	10.4	0.0	0.0	0.0	0.0	0.0	0.0	0.0	100.0	0.0	0.0	0.0	2.0	77.8	22.2	96.0	2.0	1	7	2	9	48	1	50	134
135	16.7	0.0	2.1	0.6	0.6	0.3	20.2	11.9	36.9	2.4	2.7	5.7	0.0	0.0	0.0	20.0	0.0	10.0	0.0	50.0	20.0	1.7	68.8	31.3	95.5	2.8	6	11	5	16	336	10	352	135
136	0.7	0.0	4.9	0.4	3.9	1.4	31.2	14.4	40.4	2.8	0.0	0.0	0.0	0.0	0.0	18.2	0.0	0.0	0.0	0.0	81.8	0.0	81.8	18.2	96.3	3.7	0	9	2	11	285	11	296	136
137	1.0	0.5	27.1	0.2	3.2	3.7	10.0	10.3	18.3	23.5	0.2	2.0	—	—	—	—	—	—	—	—	—	0.0	100.0	0.0	100.0	0.0	0	12	0	12	409	0	409	137
139	2.1	3.4	5.7	0.0	3.6	9.3	28.6	1.4	26.4	9.3	1.4	8.6	—	—	—	—	—	—	—	—	—	0.0	100.0	0.0	100.0	0.0	0	11	0	11	140	0	140	139
141	1.1	0.2	30.8	0.2	0.0	2.6	19.4	10.3	34.2	0.4	0.2	0.6	0.0	0.0	0.0	0.0	0.0	0.0	0.0	100.0	0.0	0.0	91.7	8.3	99.8	0.2	0	11	1	12	465	1	466	141
142	2.8	0.0	20.2	0.0	4.3	5.0	13.1	13.4	33.5	7.8	0.0	0.0	—	—	—	—	—	—	—	—	—	0.5	100.0	0.0	99.5	0.0	2	9	0	9	397	0	399	142
143	1.1	0.0	0.0	0.4	1.1	0.4	26.6	5.7	55.5	5.7	0.0	3.4	—	—	—	—	—	—	—	—	—	0.4	100.0	0.0	99.6	0.0	1	10	0	10	263	0	264	143
144	1.1	3.4	2.2	5.6	36.0	1.1	32.6	4.5	2.2	9.0	1.1	1.1	0.0	33.3	0.0	33.3	0.0	0.0	0.0	33.3	0.0	3.2	80.0	20.0	93.7	3.2	3	12	3	15	89	3	95	144
148	0.5	0.0	10.7	0.0	0.0	4.1	23.1	9.6	45.4	6.6	0.0	0.0	—	—	—	—	—	—	—	—	—	0.0	100.0	0.0	100.0	0.0	0	7	0	7	438	0	438	148
149	1.1	0.0	37.5	0.2	0.8	0.8	14.7	9.1	33.9	1.4	0.5	0.2	—	—	—	—	—	—	—	—	—	0.0	100.0	0.0	100.0	0.0	0	11	0	11	646	0	646	149
151	1.4	0.0	11.6	4.3	0.0	0.0	47.8	10.1	2.9	18.8	2.9	0.0	—	—	—	—	—	—	—	—	—	2.8	100.0	0.0	97.2	0.0	2	9	0	9	69	0	71	151
153	0.0	0.0	0.0	1.6	1.6	0.0	61.3	1.6	11.3	22.6	0.0	0.0	—	—	—	—	—	—	—	—	—	0.0	100.0	0.0	100.0	0.0	0	6	0	6	62	0	62	153
156	0.5	0.5	10.6	0.0	0.0	1.6	28.8	8.7	43.5	5.5	0.3	0.0	0.0	0.0	0.0	66.7	0.0	0.0	0.0	0.0	33.3	0.0	81.8	18.2	99.2	0.8	0	9	2	11	379	3	382	156
157	0.5	0.0	9.7	1.4	0.0	1.4	28.1	16.1	20.7	22.1	0.0	0.0	—	—	—	—	—	—	—	—	—	0.0	88.9	11.1	100.0	0.0	0	8	1	9	217	0	217	157

See index at end of Table 4.

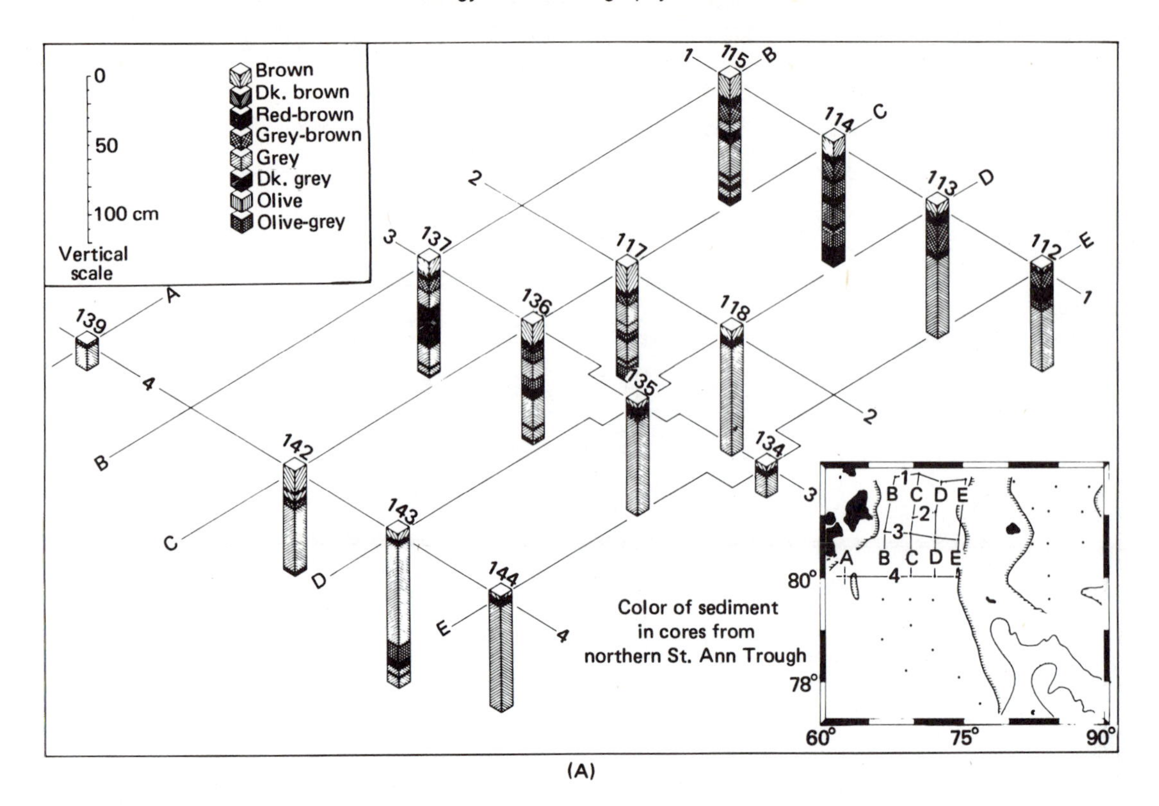

(A)

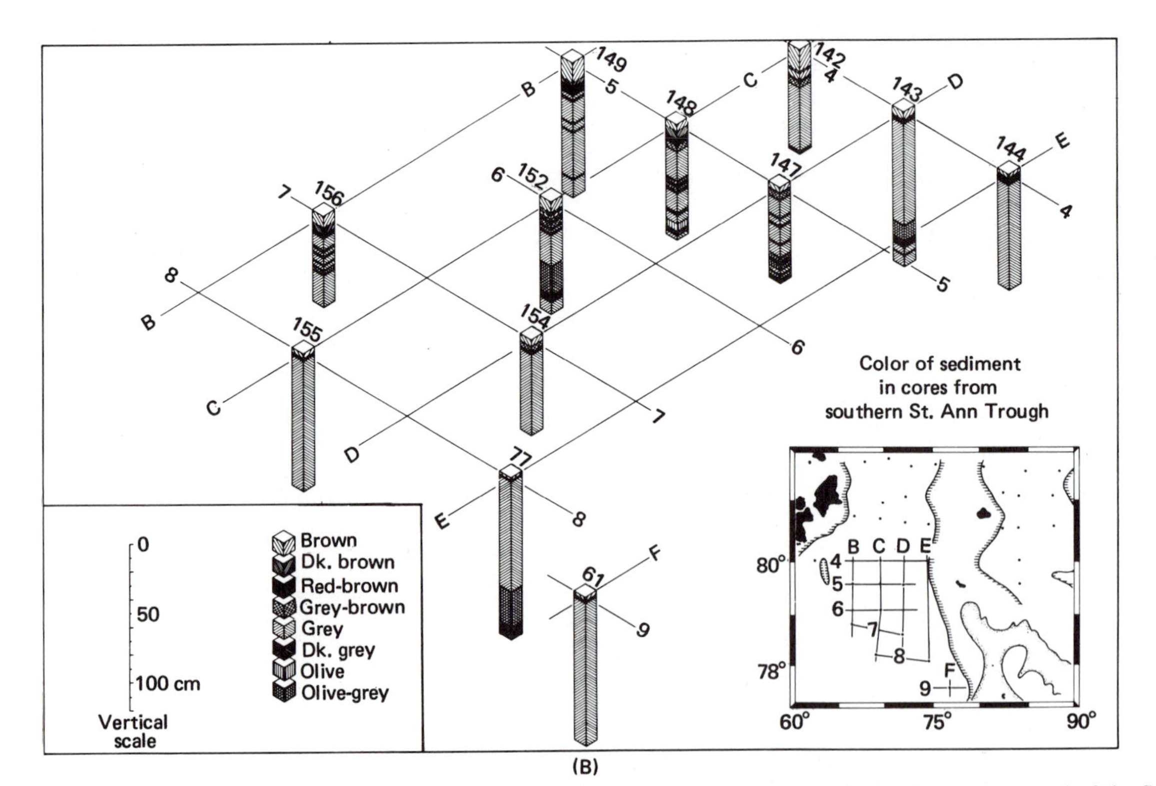

(B)

Fig. 7. Sediment color in cores from: (a) the northern end of the St. Ann Trough, (b) the southern end of the St. Ann Trough, and (c) the Voronin Trough. The northernmost line on (b) duplicates the southernmost line on (a).

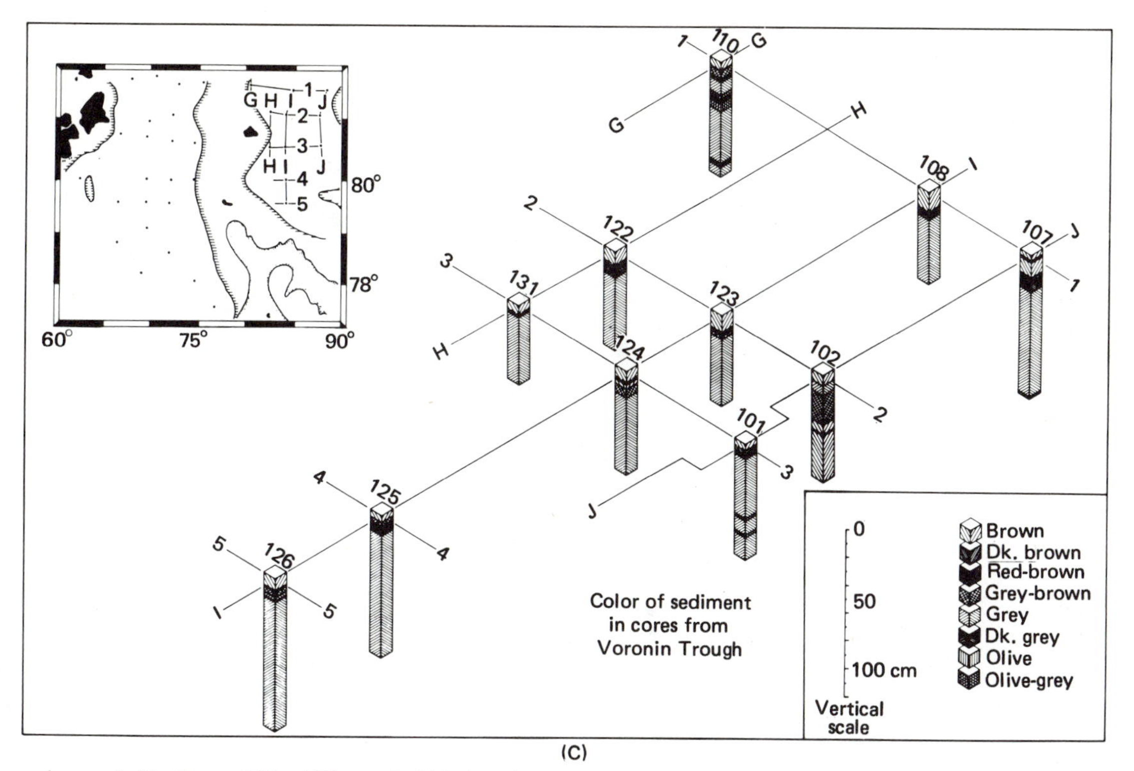

(C)

Trough, and Stations 102, 107, and 110 in the Voronin Trough). The mean weight percent sand in these cores is high (approximately 18% or greater for the stations in the St. Ann Trough and greater than 4% for the stations in the Voronin Trough). The standard deviation is high (greater than 5 in the Voronin Trough and up to 18 in the St. Ann Trough).

One core contained sediments with a color not consistent with other sediments from the two troughs. The core from station 102 contained brown-colored sediments throughout its length (77 cm). It can be inferred, from east-west traverses both north and south of the location, that the core was collected from a slope. The sand content is, however, similar to other cores from the trough margins.

Grain-Size Data from Surface Samples

The regional distribution of the sand and gravel content of the upper 10 cm of the cores (Fig. 11) is similar to the areal distribution of the sand content of the whole cores. Sandy sediment is found at the trough margins and in the southwestern St. Ann Trough. Greater than 2% gravel is found only in the southwestern St. Ann Trough. The presence of gravel in this area is probably the result of ice rafting (Latkinow, 1936; Armstrong, 1958). We sighted numerous icebergs in this area, presumably coming from the western side of the northern island of Novaya Zemlya. An alternate possibility is that the area contains relict morainal material. Less than 2% sand occurs in the northeastern St. Ann Trough, in the southeastern St. Ann Trough, and in the Voronin Trough.

Organic Carbon Content

High concentrations of organic carbon (greater than 1.6%) in surface sediments occur along the eastern side of the St. Ann Trough (Fig. 12). The lowest values (less than 1%) occur in the northwest St. Ann Trough. The source of the organic carbon is assumed to be river water which flows into the southern reaches of the Kara Sea (Gorshkova, 1957). Stoll (1967) identified cellulose in sediments from the southern ends of both troughs. We noted numerous large floating logs during the cruise. These logs had been stripped of their bark by the turbulent flood waters which occur following the spring breakup of the Siberian rivers (Vendrox, 1965).

Water Content

The highest values of the water content of the surface sediments (greater than 120% dry weight) occur in the same areas as the highest organic carbon content (Fig. 13).

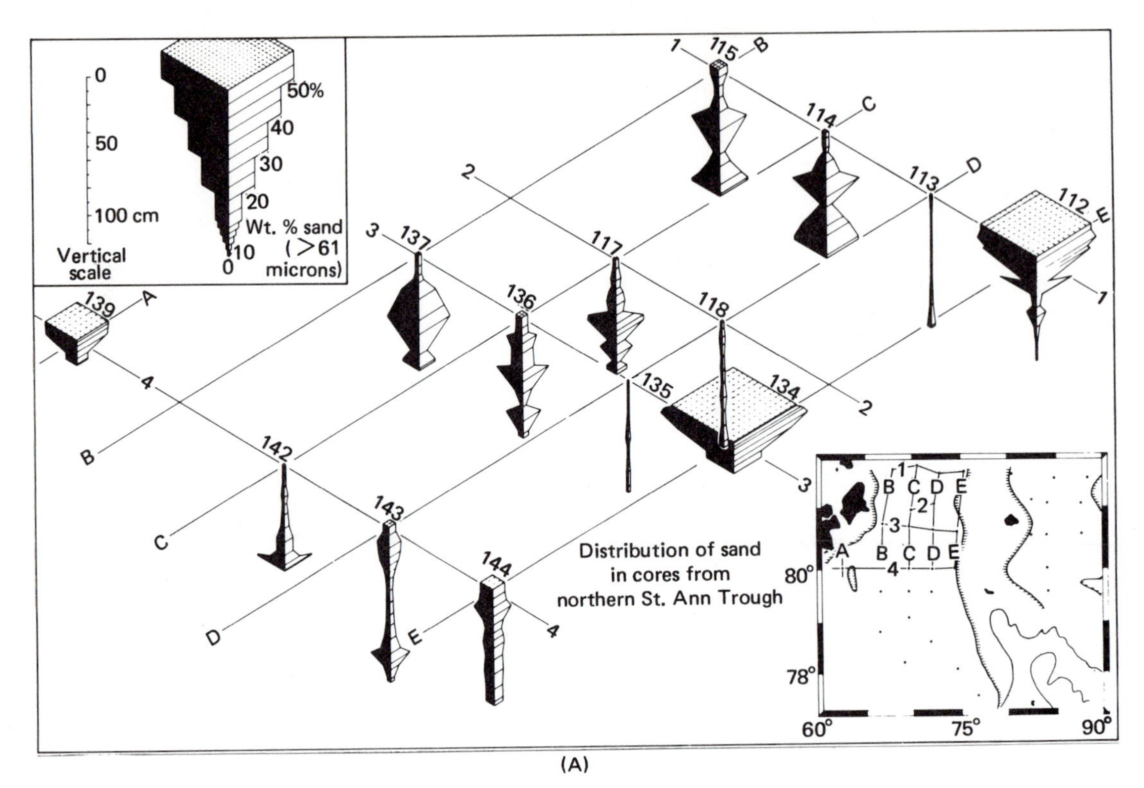

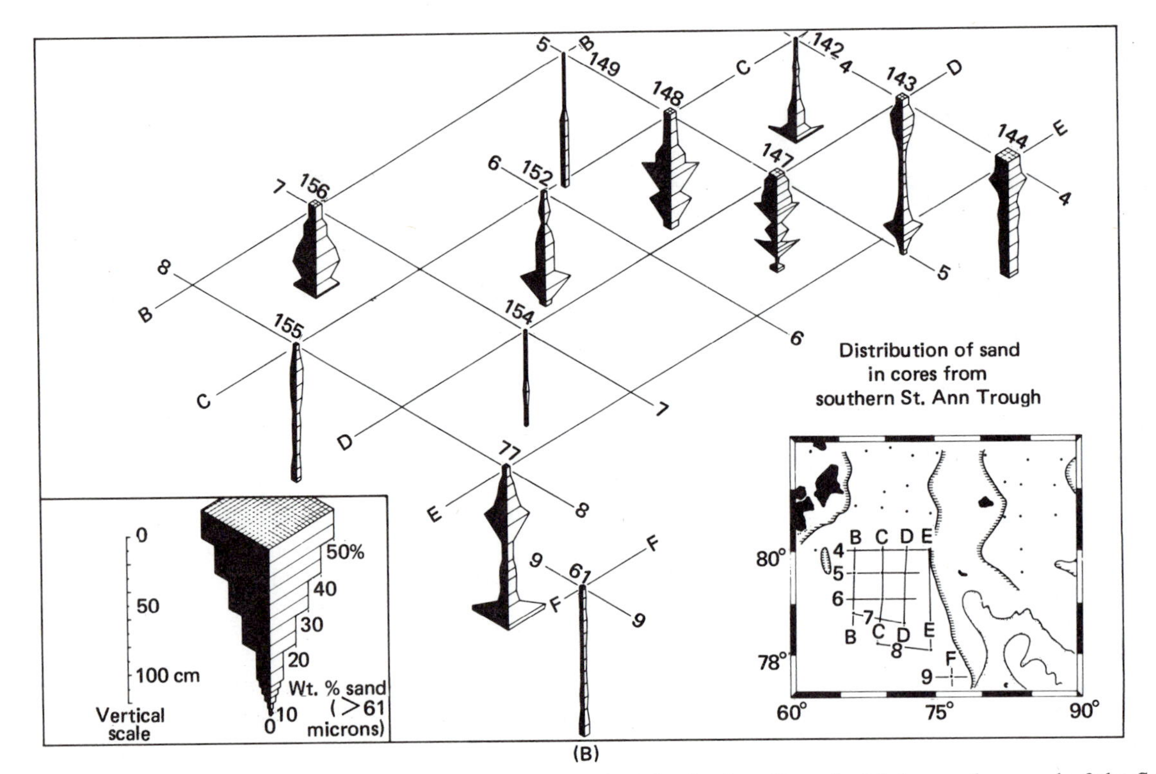

Fig. 8. Sand content of cores from: (a) the northern end of the St. Ann Trough, (b) the southern end of the St. Ann Trough, and (c) the Voronin Trough. The northernmost line on (b) duplicates the southernmost line on (a).

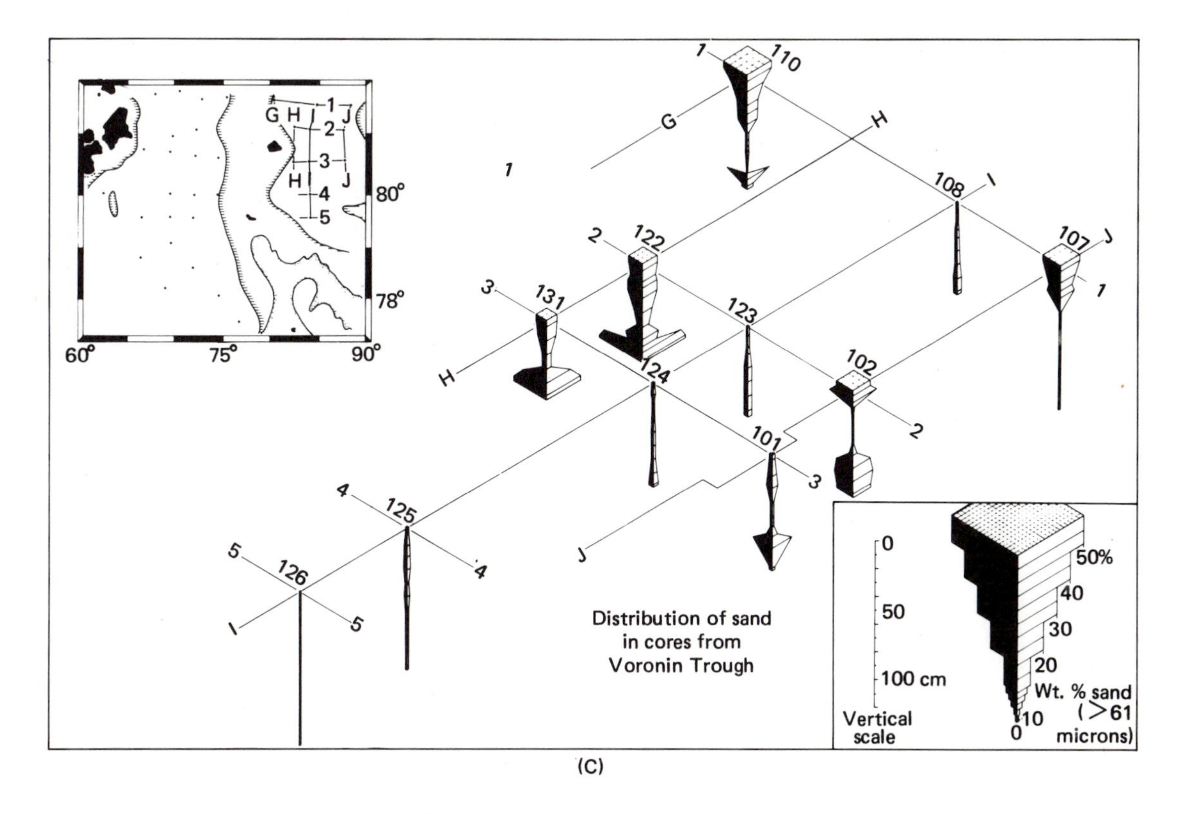

(C)

Clay Mineral Content

The clay mineral content of the less than 0.002-mm size fraction of the upper parts of the cores and of the grab samples consists predominantly of illite and mixed-layer clays with significant amounts of chlorite and kaolinite. Harry Giles (personal communication) found a similar composition while conducting an independent study on samples collected from the *Northwind.* Kulikov (1961) also reported large amounts of illite (hydromica) often interspersed with kaolinite (beidellite).

The highest values of the kaolinite (002) peak height occur near Franz Joseph Land (Fig. 14). The highest values of the chlorite (004) to kaolinite (002) peak height ratio occur in the southeastern portions of the St. Ann Trough (Fig. 15).

Foraminifera

Stoll (1967) identified 15 genera and 19 species of foraminifera in the Kara Sea sediment samples. Arenaceous foraminifera predominate, except for the northernmost stations where calcareous forms predominate. The only planktonic foraminifera identified is *Globigerina pachyderma,* and it is also associated with the northernmost stations and the direct influence of the Arctic Ocean.

The assemblages are characterized by the high dominance of a single species. The maximum number of species to comprise 95% of the fauna is 10, while the minimum is four (Stoll, 1967). The dominance, or percent of the most abundant species, ranges from 27 to 70%. The maximum number of species is found in the northwestern St. Ann Trough, along the eastern slope of the St. Ann Trough, and along the axis of the Voronin Trough (Fig. 16). The minimum number of species occur along the eastern side of the St. Ann Trough. Stoll did not find any foraminifera in the pint of sediment collected at Station 118.

Summary and Results

All the maps displaying the data discussed in this study show a consistent north-south zonation of variables in the St. Ann Trough and a less clear axial trend of variables in the Voronin Trough. The zonation in the St. Ann Trough is best explained by a persistent set of bottom currents. The patterns in the Voronin Trough are best explained by an intermittent axial bottom current, compounded by the effects of lateral sediment

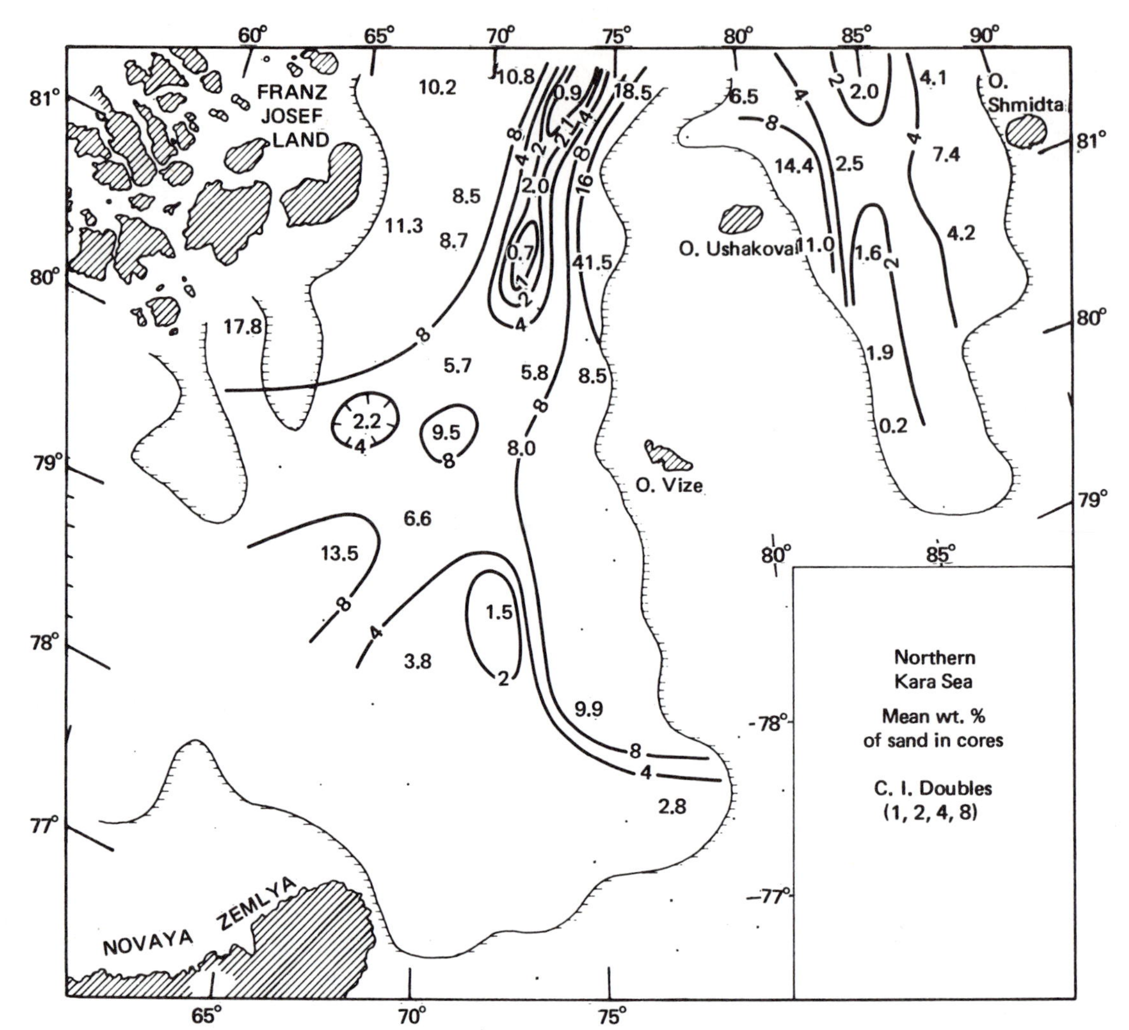

Fig. 9. Mean weight percent sand in cores from the Northern Kara Sea. This figure and several following utilizes an exponential contouring system in order to eliminate overcrowding of contour lines.

transport on the slopes into the trough.

The St. Ann Trough can be divided into four areas containing sediments displaying similar characteristics. The northwestern area contains sediment that has a larger and more varied grain-size, and lesser amounts of organic carbon and water than sediments from the eastern and southeastern regions, and more kaolinite than chlorite. The kaolinite has an apparently higher degree of crystallinity as indicated by the greater peak height on the diffractograms. The near-bottom water temperature is higher than any other area. The foraminiferal population is predominantly calcareous and the only planktonic species found occurs in this area. The sediment contained in cores collected in the northwestern area exhibits the greatest variation in color and sand content. This area is affected by the previously mentioned south-flowing bottom current coming from the Arctic Ocean. It is presently an area of nondeposition (except from possible density currents or gravity transport from the west) and, perhaps, active erosion.

The southeastern area exhibits opposite characteristics. The sediment from this area is much finer, and contains more organic carbon and water than sediments from other areas, and chlorite predominates over kaolinite. With the exception of the core collected at Station 77, the sand content of the cores is uniformly low, and there is less variation in sediment color.

An elongate belt of sediment with similar characteristics exists along the eastern side of the St. Ann Trough. This belt most probably marks the path of a persistent, presently active, north-flowing bottom current, not documented previ-

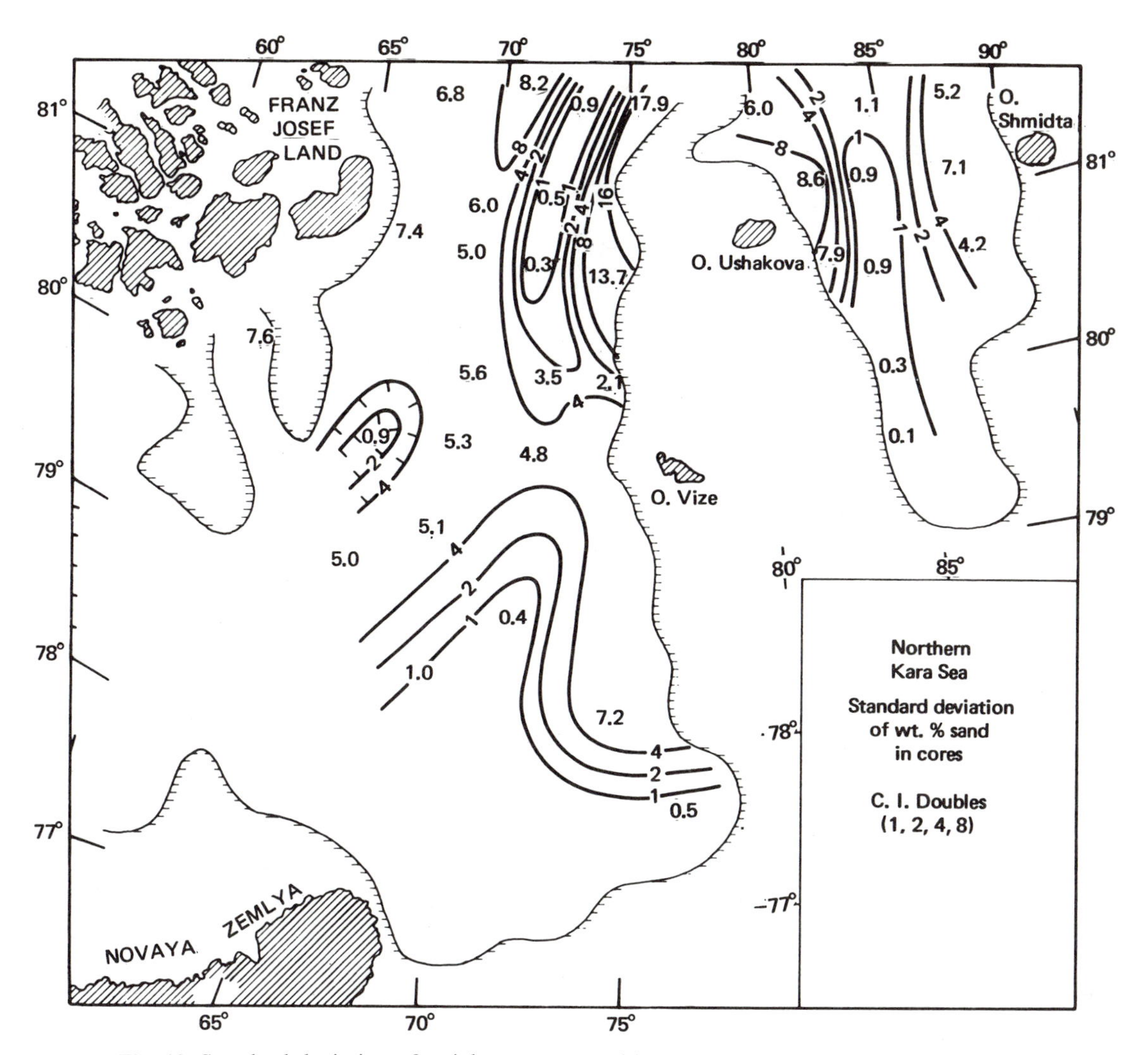

Fig. 10. Standard deviation of weight percent sand in cores from the Northern Kara Sea.

ously in the literature. Its existence has been implied on maps of Arctic Ocean bathymetry which show a cone of sediments on the Arctic floor at the northeastern end of the St. Ann Trough (the Svyataya Anna Cone on the Heezen-Tharp map of the Arctic Floor, *National Geographic Magazine*, October 1971).

The postulated north-flowing current must owe its existence to the proximity of two of the world's largest rivers, the Ob and the Yenisey (their combined flow of approximately 942 km^3 is fourth greatest after the Amazon, Congo, and Ganges-Brahmaputra Rivers), and to the peculiar behavior of some north-flowing rivers beyond the Arctic Circle (Walker, 1969).

The two rivers cannot be considered true Arctic rivers; they are too large and too long, with their headwaters in more temperate regions. In addition, there are differences in flow regime and drainage area topography between the two rivers which must be considered. The Angara River, the major tributary to the Yenisey, is the outlet for Lake Baykal, the largest lake in Asia. Other tributaries to the Yenisey drain the Angara Highlands to the east. As a result, the Yenisey is the largest river in Russia and has the steadiest flow.

The Ob and its major tributary, the Irtysh, drain the West Siberian Lowlands, which contain up to 70% swamps, lakes, and waterlogged areas in the northern area (Vendrox, 1965). As a result, spring breakup in the Ob is more dramatic than breakup in the Yenisey, though there is a marked effect in both rivers.

The breakup of the Ob travels north at a rate of approximately 0.35 m/sec and has a high water stage of 5 to 6 m over the low winter stage near the

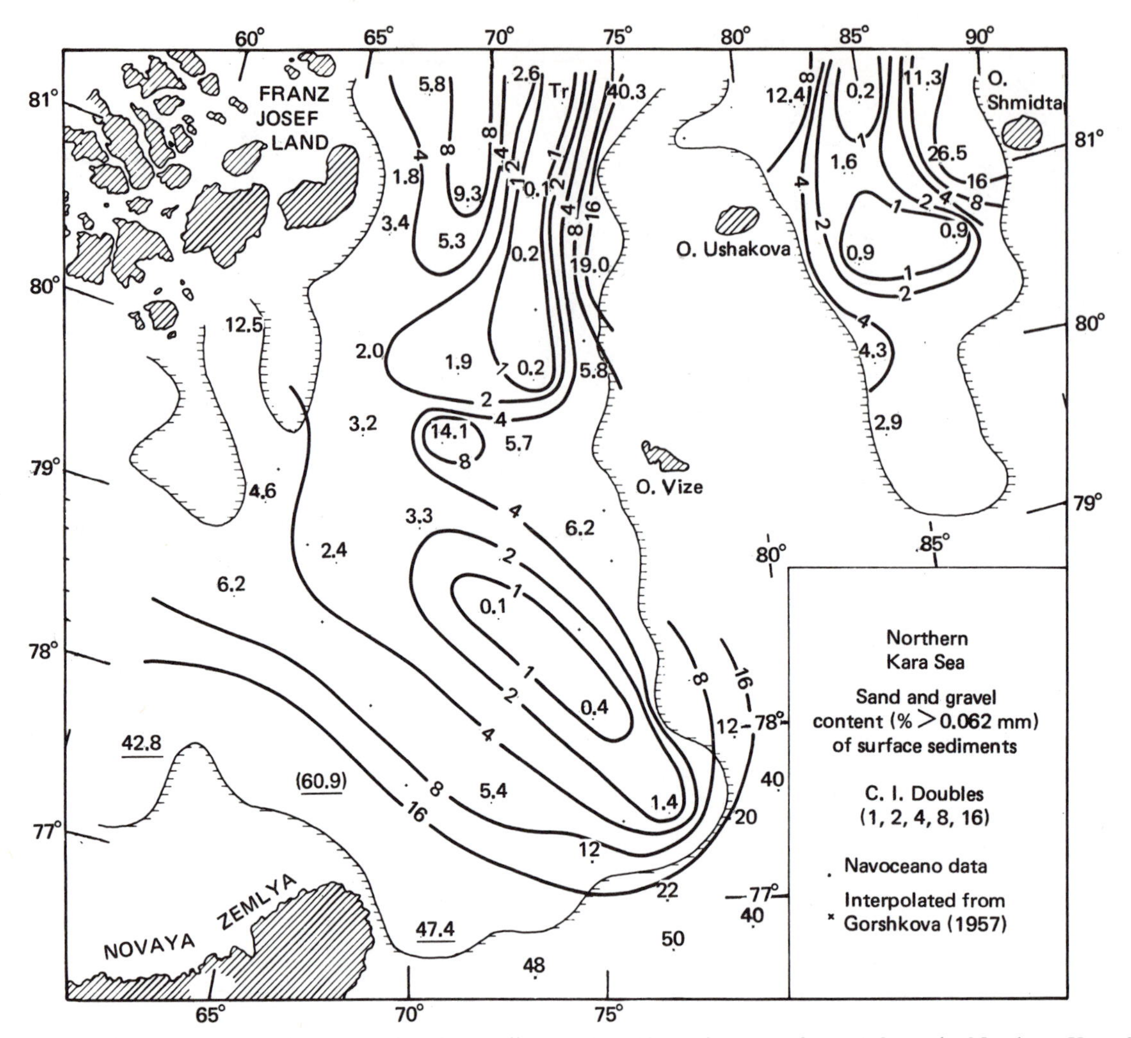

Fig. 11. Sand and gravel content of surface sediments (top 10 centimeters of cores) from the Northern Kara Sea. Those numbers underlined contain 2 to 5% gravel-sized material. The number also in parentheses contains 30% gravel-sized material. Tr indicates trace.

mouth (Vendrox, 1965). The spring floods cause large-scale ice jams and wide-scale flooding in the downstream portion of the Ob. The main channel may reach a width of 30 to 50 km, with hundreds of kilometers on either side being flooded to some degree (Lydolph, 1964). It takes several weeks for the flood waters to subside, due to the poorly developed drainage network (Vendrox, 1965).

Both rivers flow into estuaries. The Gulf of Ob is 800 km long, 25 to 65 km wide, and up to 24 m deep, with an area of 39,000 km². The Yenisey Gulf is about half as long and is slightly narrower and shallower. The area is 14,000 km². During the winter, when river discharge is at a minimum, cold saline water fills the estuaries and probably extends up the rivers as a salt water wedge.

The first slug of flood water would mix with the cold saline waters, first in the rivers, then in the estuaries. The now-saline slug would erode and transport the last deposited sediment from the previous year. The result would be the formation of a density current which would continue flowing seaward, following the submerged drainage net existing on the delta (visible on *Northwind* bathymetric surveys and reported by Kulikov, 1961, and Johnson and Milligan, 1967). The current would continue seaward and flow over the southern lip of the St. Ann Trough. Once in the U-shaped trough, it would tend to flow along the eastern side as a bottom current, held against the eastern slope under the influence of the Coriolis effect.

The current would flow into the Artic Ocean at a depth >600 m and deposit sediment on the

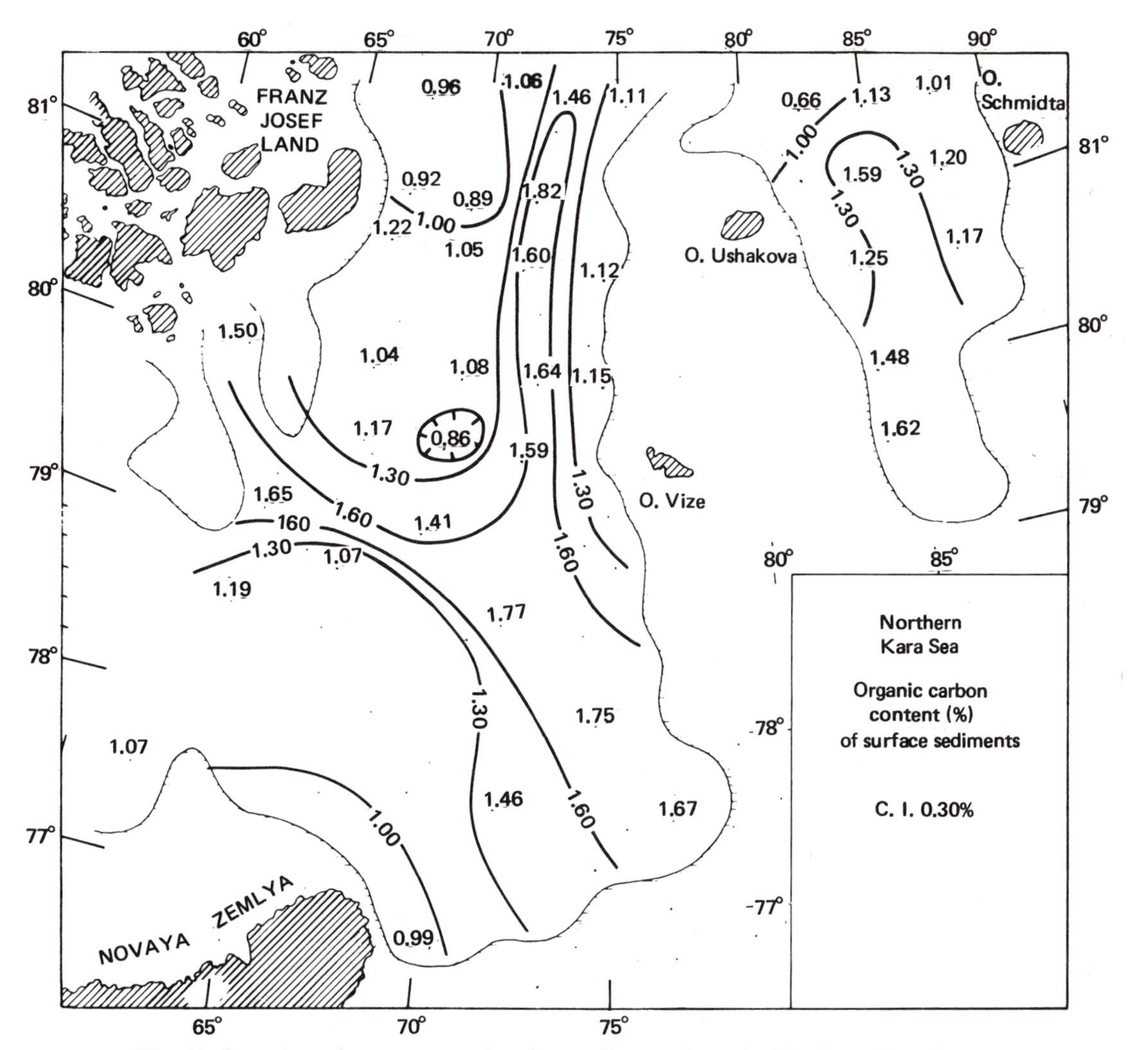

Fig. 12. Organic carbon content of surface sediments from the Northern Kara Sea.

sea floor at a distance greater than 600 km from the delta. The common genetic relationship of river, delta, submarine canyon, and abyssal fan found in many parts of the world (Moore, 1969) does not strictly hold true for the Ob–Yenisey system, since the St. Ann Trough can be called neither delta nor canyon. It must be considered as a 600-km-long pipeline allowing terrigerious sediments to be transported from the delta to the Arctic floor.

The marginal slope into the St. Ann Trough is the fourth area mentioned above. The data from samples taken on the slopes are more varied than from other areas. That is, the data on the individual parameters suggest that there are several different processes or sources of detritus active on the slopes. Gravity-initiated sediment transport must be the dominant process, moving material from the shoal areas surrounding the trough. The only appreciable amounts of gravel-size material found in surface samples came from locations in the south end of the St. Ann Trough.

The Voronin Trough can be divided into two principal areas, the slopes (as in the St. Ann Trough) and the trough axis. Cores collected along the trough axis consistently exhibit a low sand content, similar to the cores from the path of the postulated north-flowing bottom current in the St. Ann Trough. This also applies to the grain-size data from the surface sediments. Data on other parameters from the two areas do not show the same relationship. This may indicate the absence of a persistent axial north-flowing current, or the simple mixing of materials from the shoal areas by gravity-initiated sediment transport.

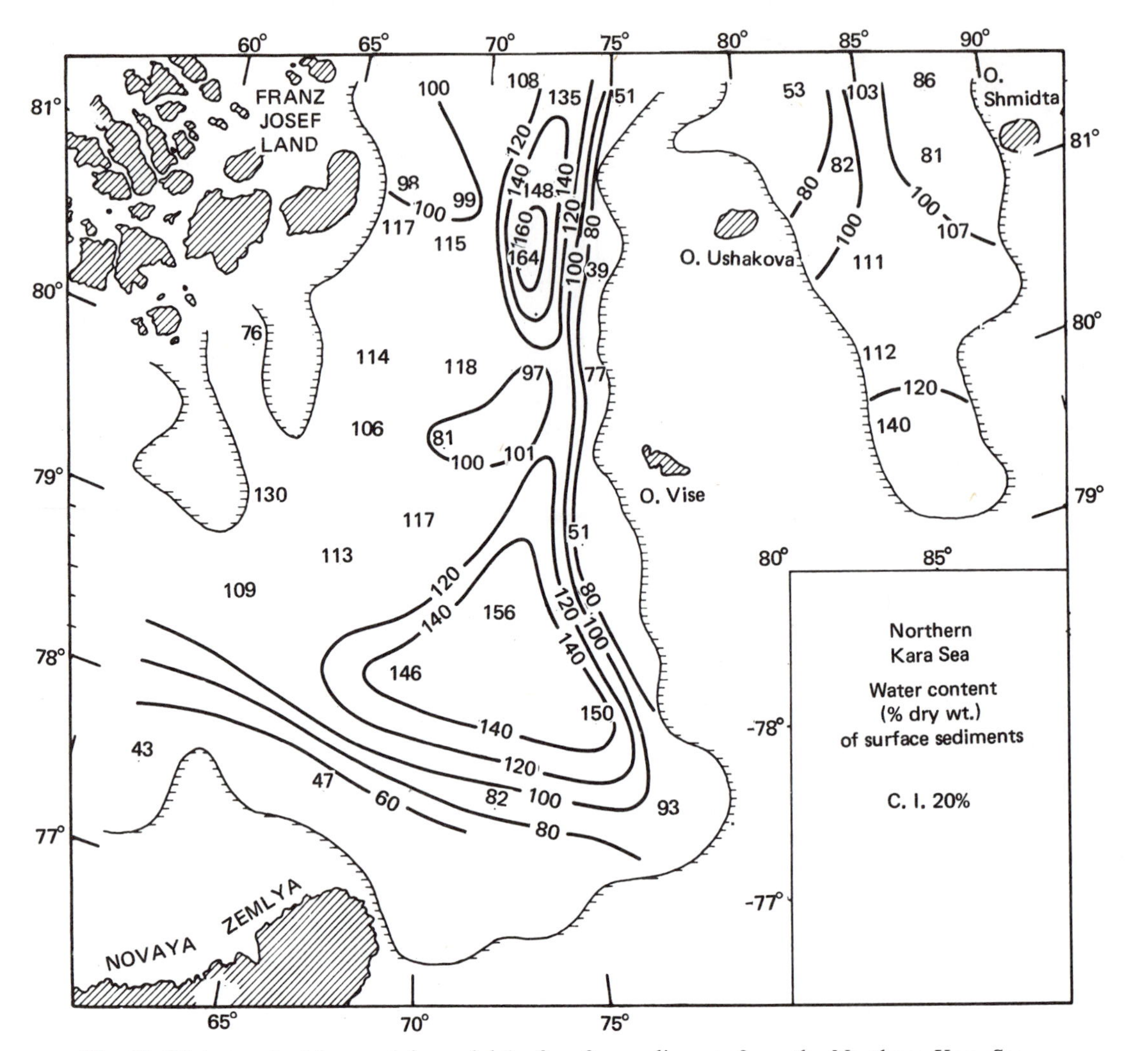

Fig. 13. Water content (percent dry weight) of surface sediments from the Northern Kara Sea.

Acknowledgments

This study is a condensed version of the senior author's doctoral dissertation directed at the University of Wisconsin by Professor J. R. Moore. Financial support was provided by Office of Naval Research Contract 1202 (29), Geophysical Studies from Icebreakers. Dr. Ned A. Ostenso and Dr. Richard J. Wold were the principal investigators. We wish to acknowledge the able assistance of the scientific personnel, officers, and crew of the U.S.C.G.C. *Northwind* involved in the basic data collection, and the thoughts and comments of numerous individuals involved in discussions on the project during the past several years. Messrs. R. K. McCormack and M. A. Meylan read this manuscript and offered valuable suggestions for improvement. Special thanks are due Sarah Stoll, who allowed us to present her unpublished foraminiferal data.

Contribution No. 219 of the University of Wisconsin Geophysical and Polar Research Center.

References

Allison, L. E. 1935. Organic soil carbon by reduction of chromic acid. *Soil Sci.*, 40:311–320.

Armstrong, T. 1958. *The Russians in the Arctic*. London: Methuen and Co., Ltd. 175 pp.

Atlasov, I. P., and V. N. Sokolov. 1961. Main features of the tectonic development of the central Soviet Arctic. In: G. O. Raasch, ed. *Geology of the Arctic*. Toronto: University of Toronto Press, 1. pp. 5–17.

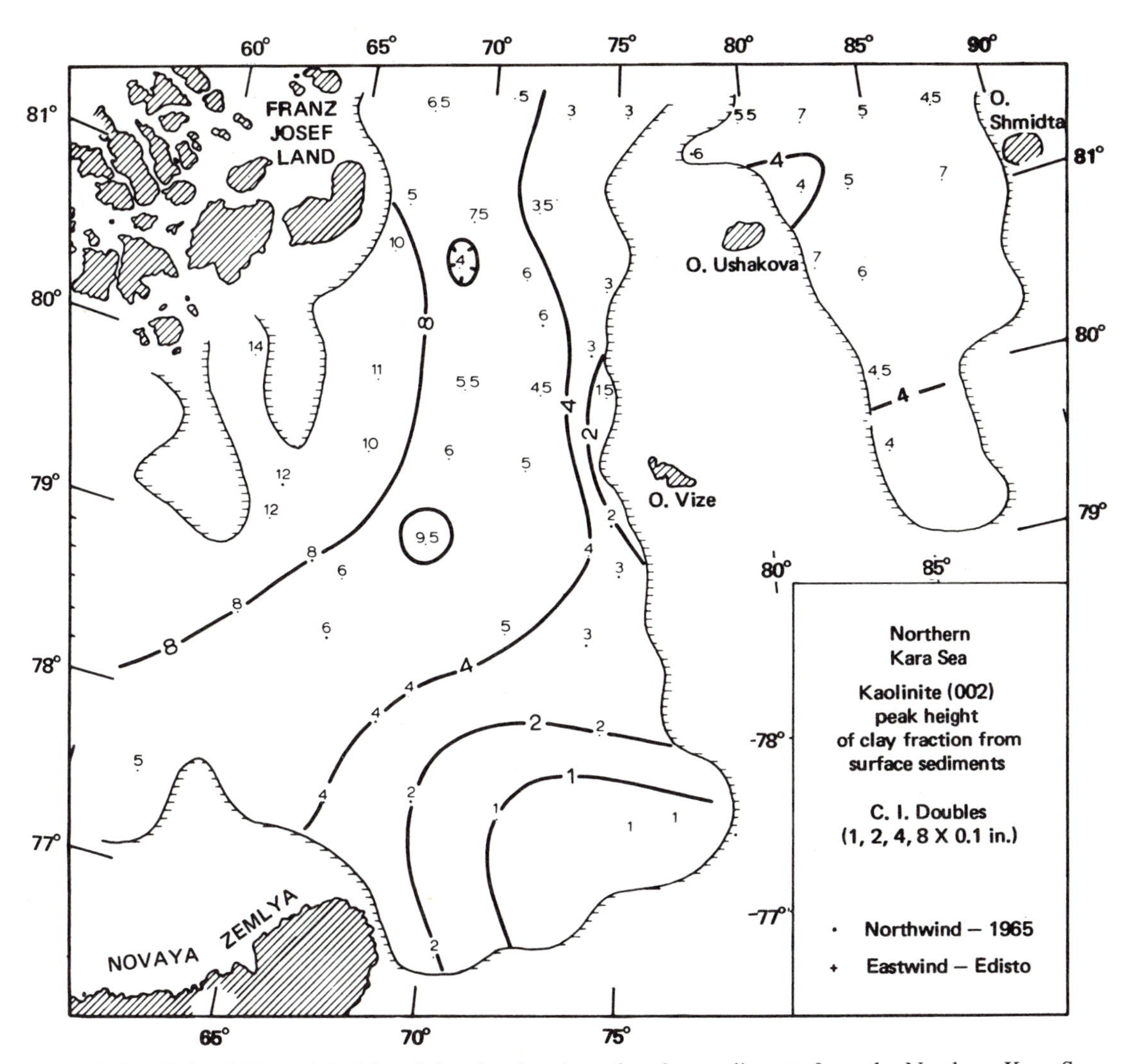

Fig. 14. Kaolinite (002) peak height of the clay fraction of surface sediments from the Northern Kara Sea.

Belov, N. A., and N. N. Lapina. 1962. Geologic studies of the floor of the Arctic Ocean during the last twenty-five years. In: N. A. Ostenso, ed. *Problems of the Arctic and Antarctic, 11*. Arctic Institute of North America. 16 pp.

Biscaye, P. E., 1964. Distinction between kaolinite and chlorite in recent sediments by x-ray diffraction. *Am. Mineral.*, 49:1281–1289.

Brujewicz, S. W., 1938a. Oxidation-reduction potential and the pH of sediments of the Barents and Kara Seas. *DAN SSSR*, 19:637–640.

———. 1938b. Oxidation-reduction potentials and the pH of sea bottom deposits. *Limnologie*, 8:35–49.

Burkhanov, V., 1956. *New Soviet Discoveries in the Arctic*. Moscow: Foreign Languages Publishing House.

Coachman, L. K., 1962. On the water masses of the Arctic Ocean. Doctoral Thesis. Seattle: University of Washington. 94 pp.

Dibner, V. D., 1957. The geological structure of the central part of the Kara Sea. In: *Geology of the Soviet Arctic* (translated by K. Price and W. A. Kneller). Eastern Michigan University. 16 pp.

Fairbridge, R. W., 1966. Kara Sea. In: R. W. Fairbridge, ed. *Encyclopedia of Oceanography*. New York, New York: Reinhold Publishing Co. pp. 430–432.

Garcia, A. W., 1969. Oceanographic observations in the Kara Sea and eastern Barents Sea. *U.S. Coast Guard Oceanographic Report, 373-26*. 99 pp.

Gorshkova, T. I. 1957. Sediments of the Kara Sea. *Trudy Vsesoyuznogo Gidrobiologicheskogo Obschestva*, 8:68–99. In: J. H. Kravitz, ed. 1967. *U.S.N.O.O. Translations*, 330 (transl. by M. Slessers). Washington, D.C. 35 pp.

———. 1966. Manganese in the sediments of the northern seas. In: A. P. Vinogradov, ed. *Abstracts of Papers, Second International Oceanographic Con-*

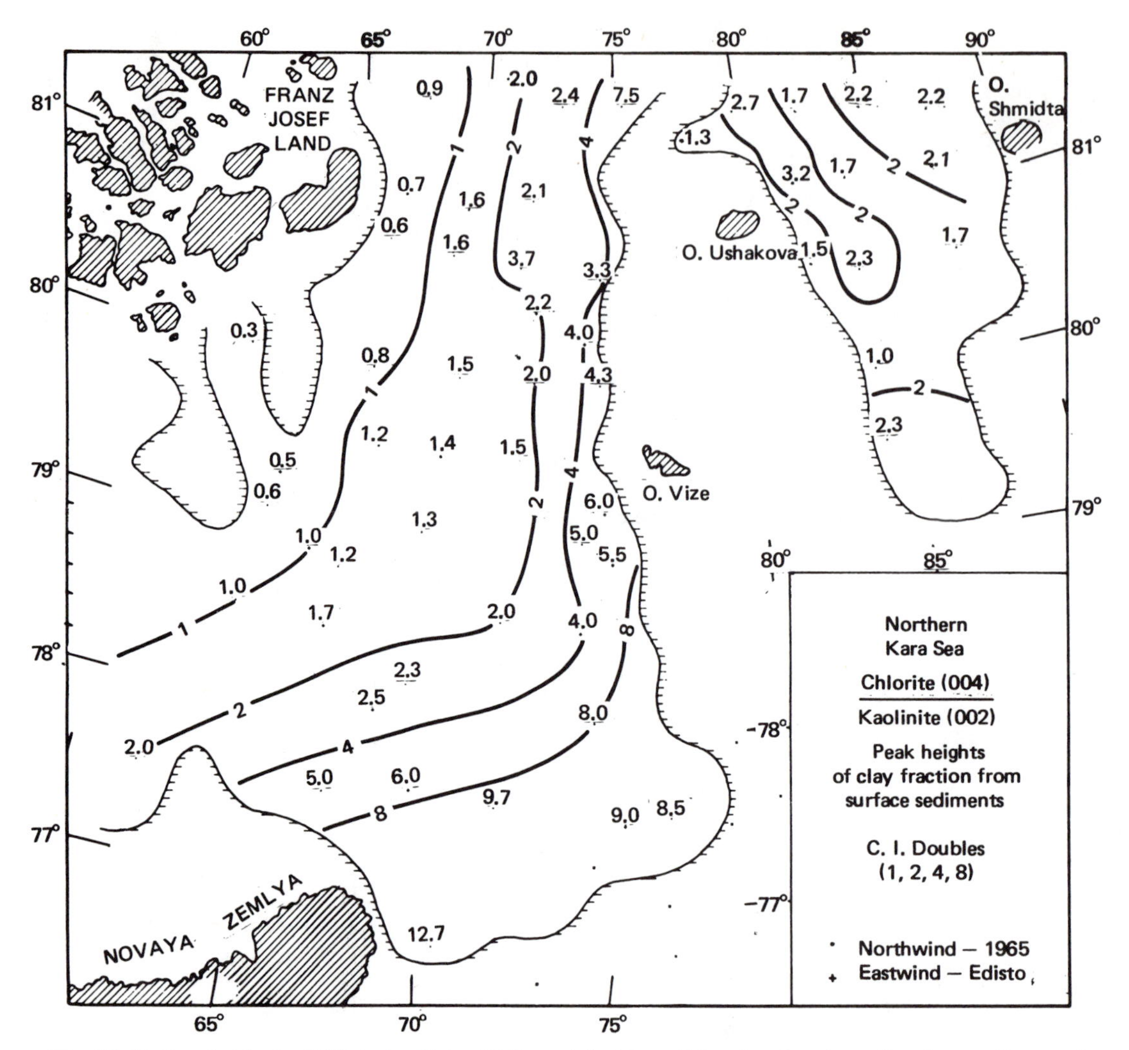

Fig. 15. Ratio of chlorite (004) to kaolinite (002) peak heights of the clay fraction of surface sediments from the Northern Kara Sea.

ference. Moscow: Publishing House NAUKA. pp. 140–141.

Gudkovich, Z. M., and Ye. G. Nikiforov. 1964. A study of the nature of circulation in the Arctic Basin using a model. *Oceanology* (AGU transl.), 6:52–60.

Inman, D. L. 1952. Measures for describing the size distribution of sediments. *J. Sed. Petrol.*, 22(3): 125–145.

Johnson, G. L., and D. B. Milligan. 1967. Some geomorphological observations in the Kara Sea. *Deep-Sea Res.*, 14:19–28.

Klenova, M. V. 1936. Sediments of the Kara Sea. *DAN SSSR*, 4(13):4.

———. 1938. Colouring of polar sea sediments. *DAN SSSR*, 19:8.

———, and M. L. Budianskaya. 1940. Phosphorus in the sediments of polar seas. *DAN SSSR*, 28:1.

———, and A. S. Pakhoma. 1940. Manganese in the sediments of polar seas. *DAN SSSR*, 28:1.

———. 1966. Barents Sea and White Sea. In: R. W. Fairbridge, ed. *Encyclopedia of Oceanography.* New York, New York: Reinhold Publishing Company. pp. 95–101.

Kravitz, J. H. 1966. Using an ultrasonic disrupter as an aid to wet sieving. *J. Sed. Petrol.*, 36:811–812.

———. 1968. Splitting core liners with a mica undercutter. *J. Sed. Petrol.*, 38:1358–1361.

Krumbein, W. C., and F. J. Pettijohn. 1938. *Manual of Sedimentary Petrography.* New York: Appleton-Century-Crofts, Inc. 549 pp.

Kulikov, N. N. 1961. Sedimentation in the Kara Sea. *Sovremennyye Osadki Morey I Okeanov*: 437–447. In: J. H. Kravitz, ed. 1967. *U.S.N.O.O. Translations*, 317 (transl. by M. Slessers). Washington, D.C. 16 pp.

Latkinov, A. F. 1936. Expedition of the icebreaker SEDOV to the northeast part of the Kara Sea in 1934. *Transactions of the Arctic Institute*, 64. Leningrad: Hydrology and Meteorology.

Lydolph, P. E. 1964. *Geography of the USSR.* New

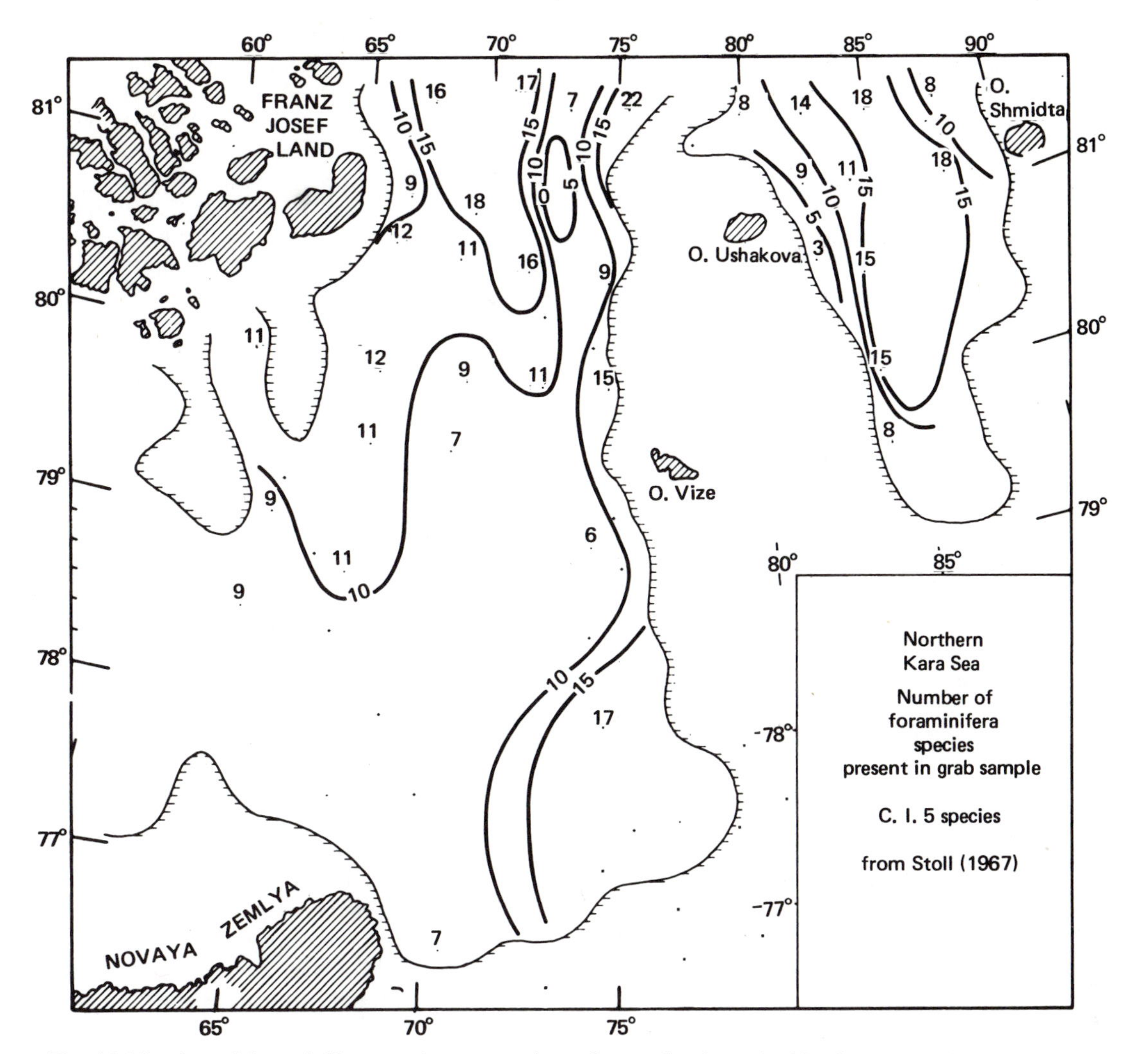

Fig. 16. Number of foraminifera species present in grab samples from the Northern Kara Sea. Identification by Stoll (1967).

York, New York: John Wiley and Sons, Inc. 683 pp.

Markov, F. G., and B. V. Tkachenko. 1961. The Paleozoic of the Soviet Arctic. In: G. O. Raasch, ed. *Geology of the Arctic.* Toronto: University of Toronto Press, 1. pp. 31–47.

Milligan, D. B., 1969. *Oceanographic survey results of the Kara Sea—summer and fall 1965.* Washington, D.C.: U.S. Naval Oceanographic Office Technical Report TR-217. 64 pp.

Moore, G. T., 1969. Interaction of rivers and oceans—Pleistocene petroleum potential. *Am. Assoc. Petrol. Geol. Bull.*, 53(12):2421–2430.

Mullen, R. E., D. A. Darby, and D. L. Clark. 1971. Sedimentary processes in the Arctic Ocean. *Second International Symposium on Arctic Geology.* San Francisco (abstract):39.

Naidu, A. S., D. C. Burrell, and J. Dygas. 1971a. Texture, organic carbon, and clay mineralogy of Western Beaufort Sea sediments. *Second International Symposium on Arctic Geology.* San Francisco (abstract):39–40.

———, D. C. Burrell, and D. W. Hood. 1971b. Geological significance of some Beaufort Sea sediments. *J. Sed. Petrol.*, 41(3):691–694.

Nalivkin, D. V. 1960. *The Geology of the USSR, A Short Outline.* New York, New York: Pergamon Press. 179 pp.

Nordenskjold, O., and L. Mecking. 1928. The geography of the polar regions. *American Geographical Society Special Publication, 8.* New York.

Ostenso, N. A., 1962. Geophysical investigations of the Arctic Ocean Basin. *Research Report 62–4.*

———. 1966. The Arctic Ocean. In: R. W. Fairbridge ed. *Encyclopedia of Oceanography.* New York, New York: Reinhold Publishing Company. pp. 49–55.

Panov, V. V., and A. O. Shpaiker. 1963. Influence of Atlantic waters upon some features of the hydrology of the Arctic Basin and adjacent seas. *Okeanologiya*, 3(4):579–590. *Deep-Sea Res.*, (transl.) 11:275–285.

Petrow, R. 1967. *Across the Top of Russia*. New York, New York: David McKay Co., Inc. 373 pp.

Rabkin, M. I., and M. I. Ravich. 1961. The Precambrian of the Soviet Arctic. In: G. O. Raasch, ed. *Geology of the Arctic*. Toronto: University of Toronto Press, 1. pp. 18–30.

Rikhter, G. D. (editor). 1963. Western Siberia. *Akademiya Nauk SSR Institute Geografii*: 129–134. In: J. H. Kravitz, ed. 1967. *U.S.N.O.O. Translations*, 347 (transl. by M. Slessers). Washington, D.C. 7 pp.

Saks, V. N. 1948. Quaternary period in the Soviet Arctic. *Trudy Arkicheskogo Nauchno-Issled Instituta*, 201:12–17 and 62–74. In: J. H. Kravitz, ed. 1967. *U.S.N.O.O. Translations*, 338 (transl. by M. Slessers). Washington, D.C. 25 pp.

———, and S. A. Strelkov. 1961. Mesozoic and Cenozoic of the Soviet Arctic. In: G. O. Raasch, ed. *Geology of the Arctic*. Toronto: University of Toronto Press, 1. pp. 48–67.

Sater, J. E. (Coordinator). 1963. *The Arctic Basin*. Arctic Institute of North America Publication. Centerville, Maryland: Tidewater Publishing Corp. 319 pp.

Stoll, S. J. 1967. A foraminiferal study of the Kara Sea north of 76 degrees north latitude. Masters Thesis, Madison, Wisconsin: University of Wisconsin. 86 pp.

Strelkov, S. A. 1965. North Siberia. *Istoriya Razvitia Rel'yefa Sibiri I Dal'nego Vostoka*: 12–15, 26–30, 49–52, 261–266, and 283–288. In: J. H. Kravitz, ed. 1967. *U.S.N.O.O. Translations*, 352 (transl. by M. Slessers). Washington, D.C. 32 pp.

Suslov, S. P. 1947. *Physical Geography of Asiatic Russia* (transl. N. O. Gershevsky). J. E. Williams, ed. San Francisco, California: W. H. Freeman Co. 594 pp.

Taylor, D. W. 1948. *Fundamentals of Soil Mechanics*. New York, New York: John Wiley and Sons, Inc. 700 pp.

Timofeyev, V. T. 1963. Interaction of waters from the Arctic Ocean with those from the Atlantic and Pacific. *Okeanologiya*, 3(4):569–578. *Deep-Sea Res.*, 9(transl): 358–361.

Turner, R. R. 1972. The significance of color banding in the upper layers of Kara Sea sediments. *U.S. Coast Guard Oceanographic Report*, 36:1–36.

———, and W. C. Harriss. 1970. The distribution of nondetrital iron and manganese in two cores from the Kara Sea. *Deep-Sea Res.*, 17:633–636.

Twenhofel, W. H., and S. A. Tyler. 1941. *Methods of Study of Sediments*. New York, New York: McGraw-Hill Book Co., Inc. 183 pp.

U.S. Coast Guard Cutter *Northwind*. 1965. *1965 Arctic cruise report*, unpubl.

U.S. Navy Hydrographic Office. 1958. Oceanographic Atlas of the Polar Seas. *Hydrographic Office Publication* 705. 149 pp.

Vendrox, S. L. 1965. A forecast of changes in natural conditions in the northern Ob Basin in case of construction of the Lower Ob hydro project. Soviet Geography, *Review and Translation*. pp. 3–18.

Vogt, P. R. 1968. A reconnaissance geophysical survey of the North, Norwegian, Greenland, Kara, and Barents seas and the Arctic Ocean. Doctoral Thesis, Madison, Wisconsin: University of Wisconsin. 133 pp.

Walker, H. J. 1969. Some aspects of erosion and sedimentation in an Arctic delta during breakup. *Geoscience Reprint Series*, 70-S. School of Geoscience, Louisiana State University, Baton Rouge, Louisiana. Reprinted from *Hydrologie des Deltas*. Association International d'Hydrologie Scientifique, Actes du Colloque de Bucarest, mai 1969.

Wells, R. D. 1966. Surveying the Eurasian Arctic. *U.S. Naval Institute Proceedings*, 92(10):79–85.

Yermolayev, M. M. 1948a. Lithogenesis of plastic clay sediments in seas. *Isvestiya Akademii Nauk SSR Seriya Geologicheskaya*. In: J. H. Kravitz, ed. 1967. *U.S.N.O.O. Translations*, 355 (transl. by M. Slessers). Washington, D.C. 1:121–138.

———. 1948b. The problem of historical hydrology of seas and oceans. *Voprosy Geografii Klimatologiya I Gidrologiya*, 7:27–36. In: J. H. Kravitz, ed., 1967. *U.S.N.O.O. Translations*, 346 (transl. by M. Slessers). Washington, D.C. 11 pp.

Zenkevitch, L. A. 1963. *Biology of the Seas of the USSR* (transl. by S. Botcharskaya). New York, New York: Interscience Publishers. 955 pp.

Chapter 11
Subarctic Pleistocene Molluscan Fauna[1]

SERGEI L. TROITSKIY[2]

Abstract

In the coastal zone of the Holsteinian (Mindel-Riss) seas the northeastern limit of the Atlantic middle-boreal subregion was near southwestern Jylland in Denmark, the high-boreal subregion extended up to the foot of the Polar Ural Range, and the boundary between the low-arctic and high-arctic subregions was east of the Khatanga River estuary. The Pacific boreal region reached the west coast of the Bering Strait. In the Eemian (Riss-Würm) seas the eastern boundaries of the Atlantic middle-boreal and the high-boreal subregions lay in the North Dvina River basin and the Pyassina River basin, respectively. The Atlantic low-arctic subregion was connected directly with the same subregion in the eastern subarctic. In eastern Fennoscandia a specific Baltic-Kola high-boreal province existed. In Bölling-Alleröd time the middle-boreal molluscs spread as far as Bergen and the high-boreal forms reached Tromsö and the Skaggerrak Strait. The western boundary of the Pacific low-arctic subregion was probably displaced to long. 175°W. The Mindel, Riss, and Würm seas were inhabited by typical high-arctic faunas along the entire coast.

[1] This chapter has been edited and abbreviated with the author's permission by the editor (Yvonne Herman).

Introduction

The North Eurasian lowlands and the valleys of the coastal mountain regions between the Netherlands to the west and the Gulf of Anadyr to the east have been inundated repeatedly by Pleistocene marine transgressions. The dry land between the recent shore line and the Pleistocene marine limit occupies about 1.6 million km^2 (617,600 mi^2). In the western European part of the region marine sediments occupy small and discontinuous areas, but the northern part of the Russian plain may be called a true kingdom of marine Pleistocene with an estimated area of 560,000 km^2 (216,150 mi^2). In the coastal zone of western and northern Siberia and in the Taimyr Lowland, marine deposits cover an area of 930,000 km^2 (358,950 mi^2). Near Cape Svyatoi Nos and Cape Schmidt, east of the Anabar River and in the New Siberian Islands, marine beds are exposed in a few places. Small patches of marine Pleistocene reappear in the valleys and intermontane depressions of the Chukotskiy Peninsula (Fig. 1).

[2] Institute of Geology and Geophysics, Siberian Division, Academy of Sciences of the U.S.S.R., Novosibirsk, U.S.S.R.

The great areal extent of the northern Eurasian coast (spanning 186° of longitude or 8000 km/4970 mi/estimated between the projections of the terminal points to the Arctic Circle) suggests *a priori* that a faunal differentiation existed in the Pleistocene as it does today. Based on marine molluscan associations two zoogeographic regions, six subregions, at least eight provinces, and many districts may be distinguished (Filatova, 1957b; Figs. 2 and 3).

The main objective of this chapter is to determine the geographic location of the boundaries at which the most pronounced marine molluscan faunal changes occurred.

To the best of my knowledge this is the first attempt to reconstruct the late Pleistocene paleo-

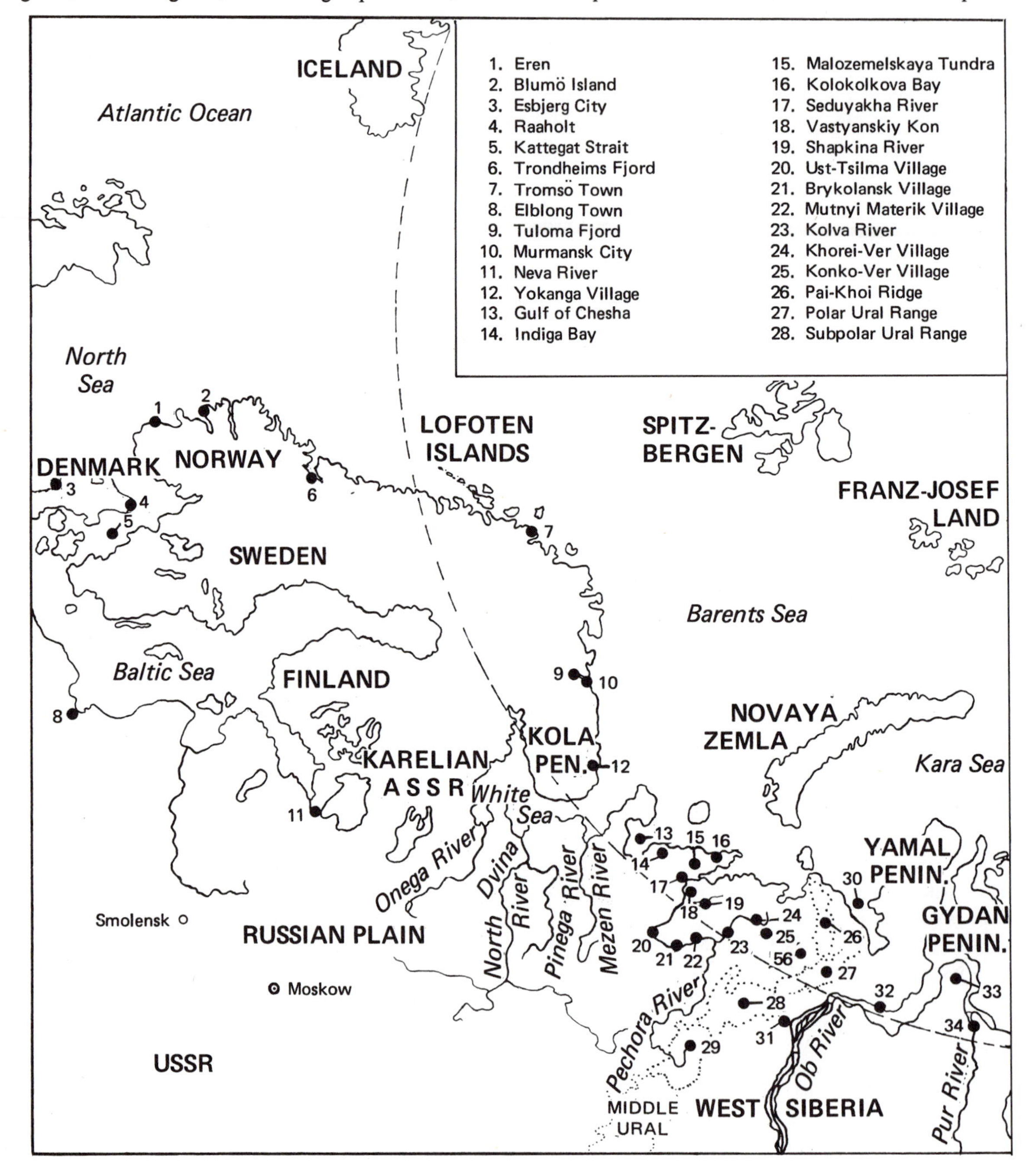

Fig. 1. Map of northern Eurasia showing the locations of important marine fossiliferous sites.

zoogeographic provinces of northern Eurasia. These reconstructions are based principally on a publication by Lavrova and Troitskiy (1960), as well as on the author's current research.

Methods

The present-day zoogeographic divisions may be used as a model for Pleistocene paleozoogeographic interpretations, inasmuch as 98% of Pleistocene molluscan fauna consists of species that live in neighboring water masses. The morphologic stability of these species suggests that their ecology was similar to that of the present (Table 1).

Paleozoogeographic divisions of Pleistocene sea basins are established in the same manner as the present-day zoogeographic provinces: They are based on the assumed relationship of faunas to their environment. Sympatric species, or species with similar areal distribution, are united in zoogeographic groups. The relationship of different

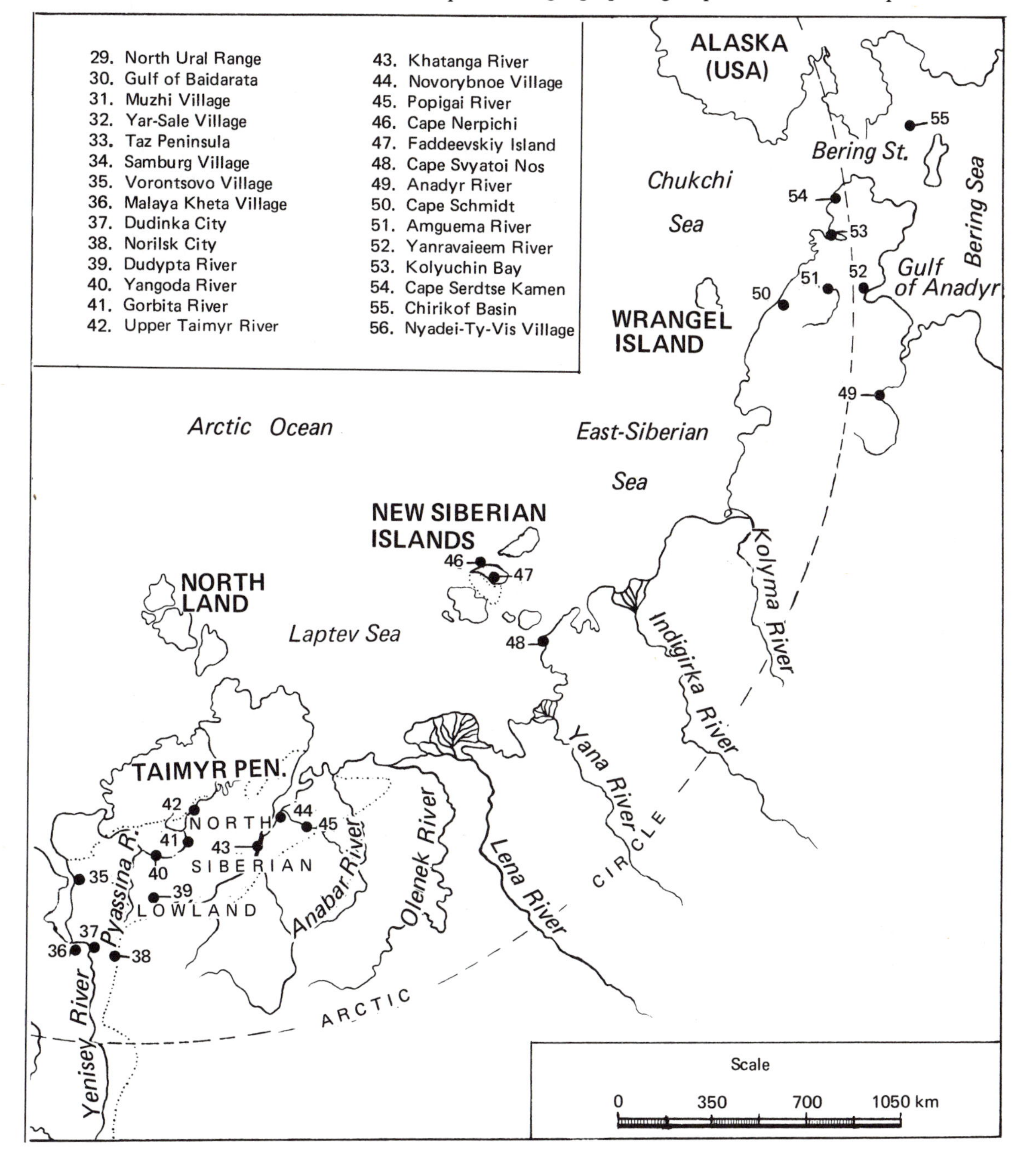

groups within the local fauna is called its zoogeographic structure. Stenotherm species are considered particularly important. If these species appear only once in the regional Quaternary sequence and at a definite stratigraphic level they may be named locally significant species. If they appear more than once they may be named characteristic species.

The paleozoogeographic boundaries are reconstructed principally for coastal water (littoral to mid sublittoral) because temperature changes are reflected most conspicuously in the fauna of this zone. Conclusions regarding the simultaneous existence of different local sea basins is based on biostratigraphic correlations (Table 2).

Division of Marine Pleistocene Basins

Paleontologic and stratigraphic data permit the reconstruction of zoogeographic regions and subregions during the Penultimate, Last Interglacial and the interstadials of the Last Glacial.

Table 1 Recent Zoogeographical Division of the Northern Eurasian Nearshore Sea Waters Based on the Distribution of *Bivalvia* (After Filatova, 1957b; Lavrova, 1960)

Region	Subregion	Average yearly temperature of water, C°	Position of boundary
Boreal	Low-Boreal (including Luzitanian)	20–8	Trondheims Fjord
	Mid-Boreal	8–5	Lofoten Islands
	High-Boreal	5–0	Jokanga (sublittoral) Gulf of Chesha (littoral)
Arctic	Low-Arctic (Atlantic)	Low positive and negative	Gulf of Baidarata
	High-Arctic	Constantly negative	Cape Serdtse Kamen—Wrangel Island
	Low-Arctic (Pacific)	Low positive and negative	

In the *Holstein* (*Mindel-Riss*) *Sea* the northern limit of the mid-boreal subregion was situated near southwest Jylland, Denmark, where shells of *Barnea candida* were found in the Vognsbol Bed near Esbjerg (Hansen, 1965). This species was considered to be Lusitanian until it was found in Trondheims Fjord, i.e., at the northern limit of mid-boreal water (Holtedal, 1958). As *B. candida* was the only Lusitanian, low-mid-boreal species found with an otherwise boreal and arctic assemblage and because it is absent from the boreal-arctic fauna of the Netherlands and Schleswig-Holstein (Heide, 1957), the author assumes that a narrow tongue of mid-boreal water penetrated from the west and reached the coast of Denmark. The faunal evidence suggests that the average annual water temperature was about +8°C, whereas at present it is +9.5°C. Layers with the warmest Holstein fauna have evidently not been preserved (or have not been found) in northwestern Europe.

Stratigraphic correlations between different marine strata in the northern and northeastern sectors of the Russian plain are not clear; however, in many places we can distinguish a thick unit of gray marine clays with a *Propeamussium* fauna including *Propeamussium greonlandicum, Yoldiella lenticula,* and *Yoldiella fraterna* (Lavrova, 1946; Lavrova and Troitskiy, 1960; Troitskiy, 1964; Zarkhidze, 1963; Pavlenko, 1963; Vollosovich, 1966; Zagorskaya et al., 1969). In numerous borings on the coast of the Gulf of Chesha, Indiga Bay, Kolokolkova Bay, in the Pechora River valley near the villages of Khangurei, Ust-Tsilma, Brykolansk, and Mutnyi Materik, in the Seduyakha, Shapkina, and Kolva River basins near the villages of Khorei-Ver and Konko-Ver, and in the Malozemelskaya tundra the above-mentioned marine unit occurs below the present-day sea level (Fig. 1). However, in the famous "Vastyanskiy Kon" outcrop on the Pechora River it is situated above the river water level. These clays can be partly grouped together into the informal *P. groenlandicum–Y. lenticula* zone (Troitskiy, 1964). Zarkhidze and Vollosovich grouped them into the Kolva Formation, Pavlenko named them the Seduyakha Formation, and Zagorskaya included them in the Kamensk Horizon (excluding the clay of the Vastyanskiy Kon). Zagorskaya concluded that the entire section of Vastyanskiy Kon is represented by the intermediate Budrinskiy Horizon; however, the molluscs from the lower clay are close to the Kolva Formation fauna and the upper sand fauna is identical to the local Riss-Würm deposits.

The molluscan fauna of these clays generally consist of arctic mid- and low-sublittoral species, but in some localities *Mytilus edulis, Arctica islandica,* and *Zirphaea crispata* are also present (core 35, Seduyakha River area). Shell fragments belonging to these species occur in the Vastyanskiy clays; they are thought to have drifted in from the nearshore zone. In core 305 (Nyadei-ty-Vis) the Kolva Formation is overlain by marine sands and silts with *M. edulis, Macoma baltica,* and *A. islandica* (Padymei Formation). As the Padymei Formation lies under the Rogovskaya Formation (the author considers the Rogovskaya Formation to be a continental till deposited during the Riss Glacial; this till contains fragments of marine shells including tests of foraminifera, incorporated from the underlying marine deposits), the Kolva-Padymei sequence has been deposited during the

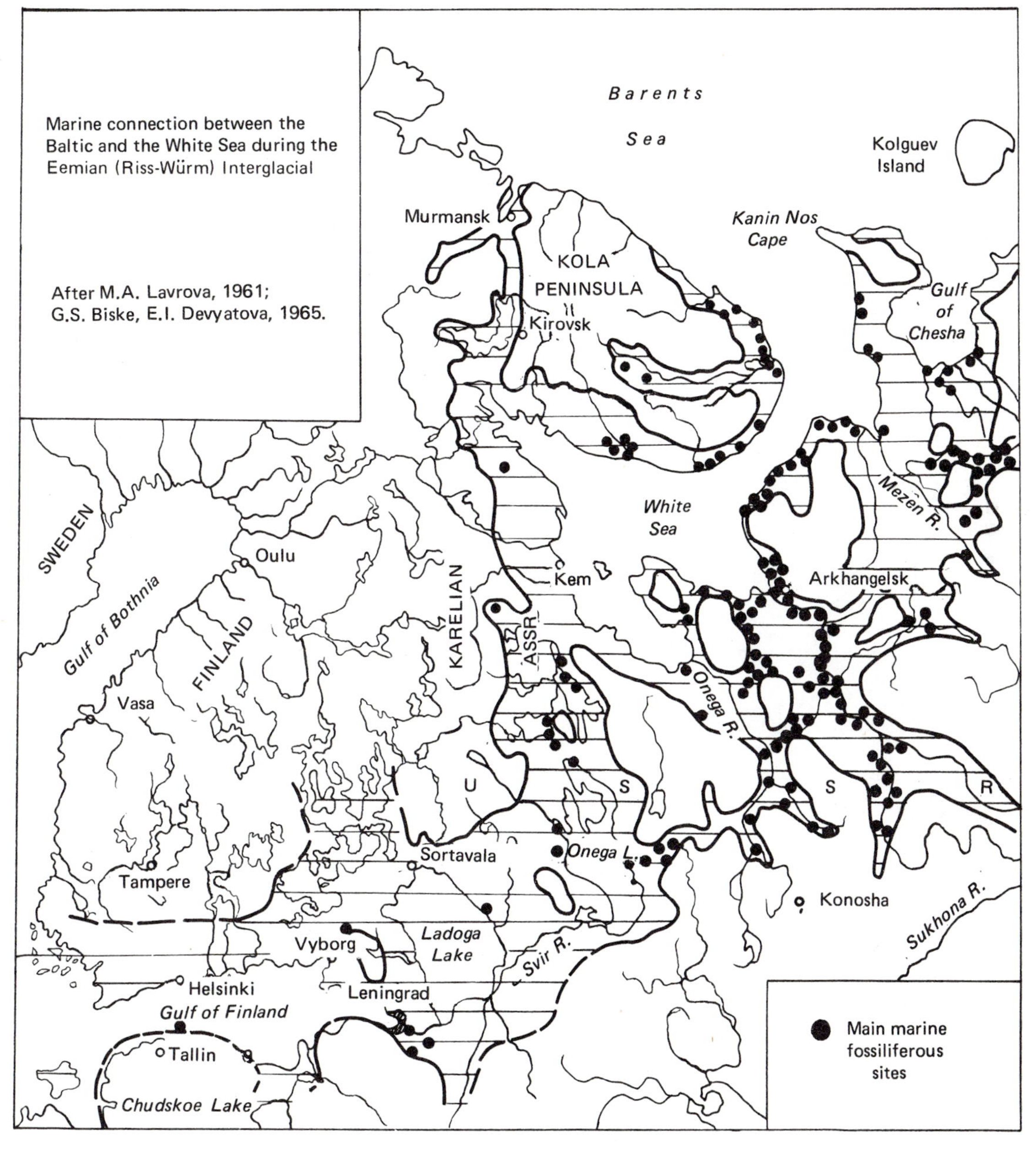

Fig. 2. Gulfs and straits in eastern Fennoscandia and in the northern sector of the Russian plain during the Riss-Würm Interglacial.

Table 2 Correlation of Marine Pleistocene

Years B.P.		Netherlands, Denmark, and West Germany		Southern Norway and southern Sweden	Western and northern Norway	Kola Peninsula	East Baltic
10,270 10,830		Upper Dryas		*Yoldia* clay	*Yoldiella* clay	*Portlandia* clay	
		Alleröd	*Arctica islandica* bed	*Littorina* and *Chlamys islandicus* bed	*Arctica islandica* bed	*Mytilus edulis* bed	*Portlandia* clay
11,800	WÜRM	Middle Dryas					
12,100		Bölling	*Littorina rudis* bed				
12,400		Lower Dryas	*Hiatella arctica* sand and *Portlandia* clay				
14,250 ?		*Portlandia arctica* bed					
70,000	EEMIAN	*Abra nitida* bed *Turritella terebra* bed Olander bed			Shells of boreal species in Eren till	White Sea bed with *Chlamys*	*Portlandia* clay
		Typical Eemian				Boreal bed with *Venus*, *Spisula elliptica*	*Cardium edule* clay
		Continental deposits, erosion					*Portlandia* clay
140,000 ?	RISS	*Portlandia* clay (reworked)					
	HOLSTEIN	Marine Needian, Holstein formation: Vognsböl bed Upper Esbjerg bed *Macoma calcarea* clay Lower Esbjerg clay (*Portlandia* clay)					

Horizons and Beds Along the North Eurasian Coast

	North of the Russian Plain		Northeast of the Russian Plain		Western Siberia	Taimyr Lowland	New Siberian Islands		Chukotsky Peninsula
	Portlandia clay			Zyryanka horizon	*Portlandia* varved clay (Dyuryussa bed)		*Portlandia* clay		Amguema bed with *Macoma baltica* (?)
									Portlandia–Cyrtodaria kurriana sand
					Portlandia clay (?)				
Mikulino horizon	White Sea bed (?)	Vashutkino horizon	White Sea bed (?)	Kazantsevo horizon	River sand	River sand	*Venericardia* clay (?)		Continental deposits
	Boreal bed with *Nassa*, *Cardium edule*		Boreal bed with *Spisula elliptica*, *Cardium edule*		*Mytilus edulis*, *Macoma baltica* sand, *Astarte* sand, *Yoldia hyperborea* clay			Valkatlen horizon	Valkatlen bed with *Buccinum baeri*, *Astarte rollandi* clay
					Sand and silt with *Arctica islandica*, *Zirphaea crispata*, *Cardium edule*				
	Macoma calcarea clay				*Macoma baltica*, *Mytilus edulis* sand				Lower Valkatlen bed with *Portlandia*
	Propeamussium clay	Kamensk horizon	Padymei formation: sand with *Arctica islandica*, *Zirphaea crispata*	Sanchugovka horizon		Nikitinskiy sand with *Hiatella*, *Macoma calcarea*, *Mya*	*Portlandia* clay (?)	Pinakul and Kresta Formations	The sequence of beds needs some revision
			Kolva clay with *Propeamussium*, *Yoldiella*, *Mytilus*		*Propeamussium-Yoldiella* clay with rare *Mytilus*, *Macoma baltica*				
						Portlandia varved clay			

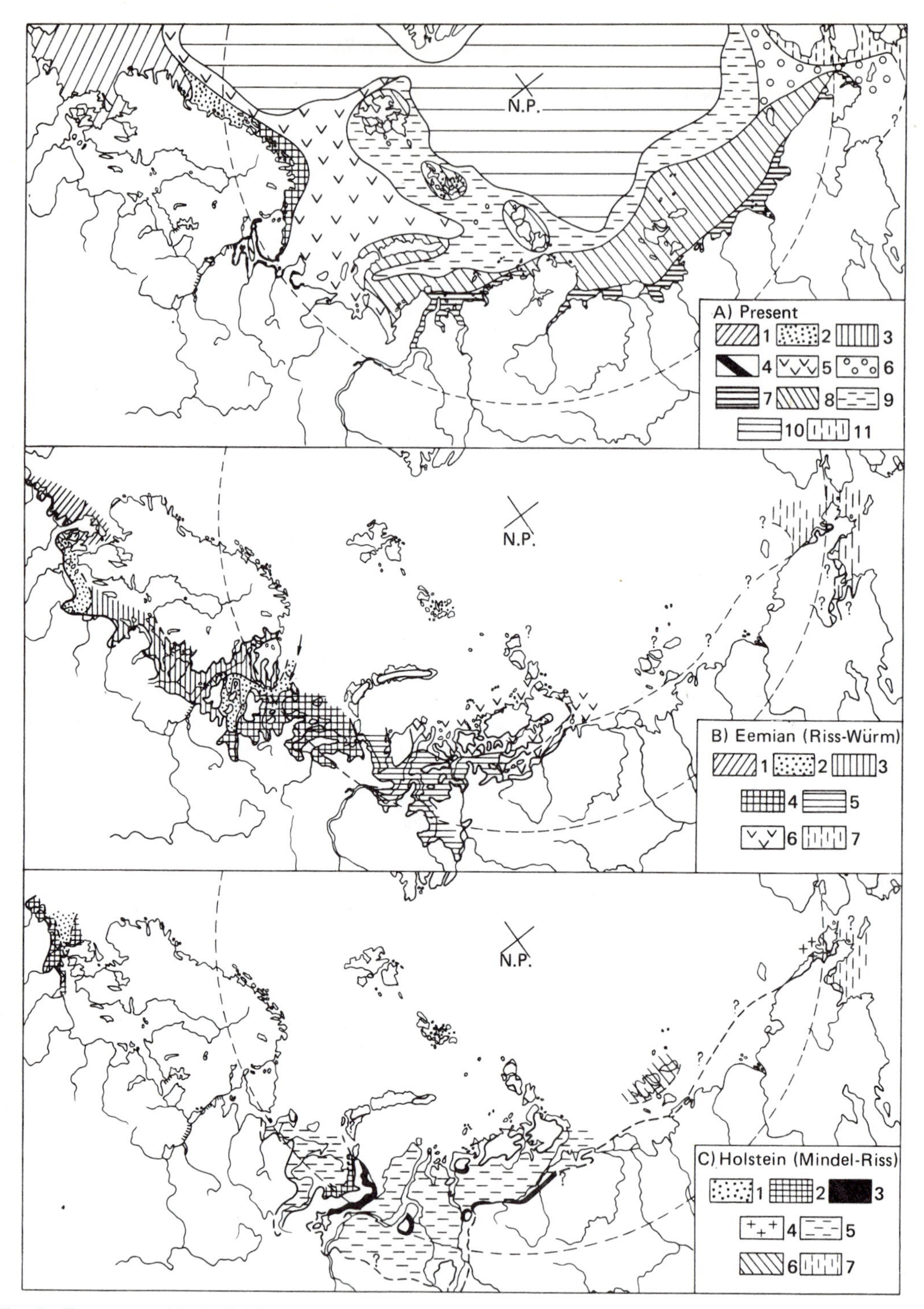

Fig. 3. Zoogeographical division of the present seas and paleozoogeographical division of the Eemian and Holsteinian marine basins.

A. Present: 1–4 = Atlantic boreal region: 1 = low-boreal subregion, 2 = middle-boreal subregion, 3 = high-boreal subregion, 4 = Murman-White Sea littoral province; 5–10 = Arctic region: 5, 6 = low-arctic sublittoral subregion: 5 = Barents Sea province, 6 = Chukchi Sea province; 7–10 = high-arctic subregion: 7 = Siberian brackish water province, 8 = Siberian-White Sea sublittoral province, 9 = Eurasian open sea province, 10 = abyssal province; 11 = Pacific boreal region: 11 = high-boreal subregion: Alaskan littoral and sublittoral provinces (Z. A. Filatova, 1957b).

B. Eemian: 1–5 = Atlantic boreal region: 1 = low-boreal subregion, 2 = middle-boreal subregion; 3–5 = high-boreal subregion: 3 = Baltic-Kola province, 4 = Mezen-Pechora province, 5 = Siberian province; 6 = Arctic region: low-arctic subregion; 7 = Pacific boreal region: high-boreal subregion.

C. Holsteinian: 1–2 = Atlantic boreal region: 1 = middle-boreal subregion, 2 = high-boreal subregion; 3–6 = arctic region: 3–4 = low-arctic subregion: 3 = Siberian upper-sublittoral province, 4 = Chukchi sublittoral province; 5–6 = high-arctic subregion: 5 = Eurasian open sea province, 6 = Eastern Siberian sublittoral province; 7 = Pacific boreal region: high-boreal subregion.

Likhvin (Holstein, Mindel-Riss) Interglacial.

Thus the nearshore fauna of the Kolva-Padymei basin was high-boreal and the average annual temperature in shallow water was somewhat higher than today. The floor of the open sea was inhabited by an arctic fauna. Such contrast is typical in recent high-boreal seas (Filatova, 1957a) as well as in Quaternary western European seas. The Pechora River basin is the easternmost area with marine deposits overlain by Riss till that contains boreal species. These species are distributed today in the extreme borders of the Boreal region. This fact, in addition to the absence of other thermophilic species, suggests that the boundary between the boreal and arctic regions was at the foot of the Polar Ural Range and the Pai-Khoi Ridge (Fig. 3).

The homogeneous and characteristic Sanchugovka open-sea arctic molluscan assemblage was widely distributed in the western Siberian and Taimyr lowlands during the maximum marine transgression (the Sanchugovka transgression of Sachs). Sachs, who was the first to establish and to describe this assemblage, showed that it contains the characteristic *Y. lenticula* associated with frequent occurrences of *Bathyarca glacialis* and *Portlandia arctica* (Sachs and Antonov, 1945; Sachs, 1945, 1953). The coefficients of species occurrence (the author uses the coefficients which are commonly used in hydrobiology, i.e., the ratio between site number containing one species divided by the total number of sites) calculated for 84 sites are *P. arctica* (55%), *Y. lenticula* (50%), *Macoma calcarea* (48%), *Hiatella arctica* (32%), *Astarte montagui* (24%), *Nucula tenuis* (24%), *Astarte compressa* (20%), *B. glacialis* (17%), *P. groenlandicum* (17%), *Sipho curtus* (12%), and *Cylichna alba* (11%). These percentages confirm the high stability of assemblages from the lower part of the Ob River basin up to the Popigai River and in the eastern coast of the Taimyr Peninsula. The importance of *Y. lenticula* as a characteristic or a locally significant species is also confirmed by these percentages.

The Sanchugovka fauna is dominated by arctic elements (54%) followed by arctic-boreal species, which constitute 38% of the fauna; however, in some sites single shells of boreal forms occur. Broken valves of *Mytilus* (tentatively identified as *M. edulis*) that have been carried from the littoral zone by currents were found together with typical Sanchugovka molluscs in core 11 near the town of Muzhi on the Ob River (Zaionts and Krapivner, 1967) and in core 3, located in the vicinity of Yar-Sale, in the Yamal Peninsula (Fig. 1; Arkhipov et al., 1968).

Single valves of *M. baltica* were found in the *Propeamussium* clays near the village of Vorontsovo. In core 57 (near the town of Malaya Kheta) and in core B7 (between Dudinka and Norilsk; Fig. 1) as well as in an outcrop near Novorybnoe, in clays with the same fauna, shells of the relatively deep-water boreal scaphopod *Dentalium entalis* were found. These findings indicate the penetration of Atlantic water into the deeper parts of the Sanchugovka Basin. Furthermore, they suggest that this basin was within the limits of a province similar to the recent Eurasian Open Sea Province (Filatova, 1957b). The shallow coastal water was inhabited by arctic and arctic-boreal species as well as by three boreal species: *M. edulis, M. baltica*, and *B. hameri*, which today inhabit only the southwestern sector of the Kara Sea.

Thus the low-arctic subregion extended not only along the banks of the Yenisey River (Troitskiy, 1970b) but also up to long. 105°E., as indicated by the new findings in the Khatanga River estuary (Fig. 11).

The assumption of simultaneous transgression in the Holstein, Kolva-Padymei, and Sanchugovka Basins is based on the following stratigraphic criteria: (1) they underlie middle Pleistocene (Riss) glaciofluvial deposits; (2) they are found below present sea level within an ancient depression; and (3) their fauna is similar. The aforementioned criteria permit the correlation of the Kamensk Horizon of the northern Russian plain, with the Sanchugovka Horizon. The author believes that the Sanchugovka Horizon of northern Siberia includes the Kazym Formation and the lower part of the Salekhard Formation (Lazukov, 1970), the Ob Bed (Gudina, 1966), the Sanchugovka Formation (Sachs, 1953, 1959), the Turukhan Marine Formation (Zagorskaya et al., 1965), and the Ust-Solenaya Bed (Slobodin and Suzdalskiy, 1969). These deposits contain in many sites Sanchugovka molluscan faunal assemblages and foraminiferal assemblages belonging to the informal *Miliolinella pyriformis* zone (Gudina and Saidova, 1969). This horizon is separated from the Kazantsevo (Riss-Würm) Horizon by tills, glaciofluvial sediments, pebble-boulder beds (residual moraine), or by a regional erosive surface (Troitskiy, 1969).

It is difficult to trace the middle Pleistocene transgression east of the Popigai River. The marine deposits are widespread only in the New Siberian Islands and their stratigraphy is not clear. If the "lower clays" with *P. arctica* underlying the main horizon of fossil ice was formed before middle Pleistocene glaciations, then we may assume that this part of the ancient Arctic Ocean belonged to the high-arctic subregion. Sachs (1953), as well as Ivanov and Yashin (1959) correlated the *Portlandia* clay with the Sanchugovka Horizon (but assumed that it belongs to the upper Pleistocene).

The lower and middle marine Pleistocene stratigraphy of the Chukotskiy Peninsula is poorly understood in spite of the excellent investigation of the molluscan fauna (Petrov, 1966). The age of some marine beds was determined mainly from their stratigraphic position. This method is not sufficiently reliable when applied to ancient marine deposits, particularly in an area which has undergone alternating cycles of transgressions and glaciations. The age of the Pinakul and Kresta Formations is probably middle Pleistocene; differences in their faunas is either due to facies or is insignificant chronologically. The rich high-boreal fauna including *Natica russa, Neptunea vinosa, Neptunea rugosa, Lora schantarica,* and *Mya arenaria* suggests that the boreal boundary was displaced toward the west coast of the Chiricof Basin (northern Bering Sea) during the climatic optimum of the Penultimate Interglacial.

The *Eemian* (*Riss-Würm*) offshore water in the Netherlands and Denmark was inhabited by boreo-Lusitanian and Lusitanian taxa including Mediterranean species. The open sea fauna had a moderate boreal character (Lavrova, 1961; Hansen, 1965). Average annual water temperatures approached +12°C (3°C higher than present day temperatures). It is impossible to assess the northern limit of the Eemian boreo-Lusitanian (low-boreal) fauna as there are no Eemian sites on the coast between northern Denmark and the Kola Peninsula.

In the Baltic during this period temperatures decreased sharply eastward, as indicated by the disappearance of typical Eemian forms between the Danish Islands and the town of Elblong near the mouth of the Vistula River, Poland (Fig. 1). Between the Vistula and Neva River basins a few boreal and arctic-boreal species are scattered along with relict populations of *P. arctica* (Lavrova, 1961, 1962; Fig. 1). Identical paleozoogeographic provinces existed in the gulfs and straits of Karelia and in the Kola Peninsula (Fig. 1); these regions were inhabited by a depauperate high-boreal fauna. Some species were displaced 150 to 200 km (90 to 125 mi) eastward from their present range and in several localities relict *Portlandia* associations were found (Lavrova, 1960; Biske and Devyatova, 1965). It is assumed that the main North Atlantic warm current flowed northeastward, then bent around Scandinavia, and that there was no appreciable increase in regional temperature. This part of the Eemian–Boreal Sea, inhabited by a characteristic fauna, may be regarded as a special Baltic–Kola high-boreal province (Fig. 2).

In contrast with the Baltic–Kola area, the southern nearshore zone of the White Sea was very warm and was inhabited by a diversified boreal fauna with eight boreo-Lusitanian species dominated by *Nassa reticulata, Bittium reticulatum, Spisula subtruncata,* and *Corbula gibba* (Lavrova, 1961; Devyatova, 1961; Biske and Devyatova, 1965). The distribution of these taxa along the Onega River and along the tributaries of the North Dvina River indicates that warm marine currents penetrated into the basin from the north. The North Dvina valley was the eastern boundary of the mid-boreal subregion. The Pinega, Mezen, and Pechora River basins (Fig. 1) were inhabited by boreal and high-boreal species (Lavrova and Troitskiy, 1960; Bylinskiy, 1962; Zarkhidze, 1963; Biske and Devyatova, 1965). *Spisula elliptica, Astarte sulcata, Cardium edule,* and *Nucella lappilus* were the warmest-water forms in this assemblage.

The open sector of the Northern Russian Boreal (Eemian) Sea was inhabited by a low-arctic fauna as well as by rare boreal species (Lavrova, 1961).

In the Siberian Kazantsevo (Eemian) basins, boreal fauna was present in the southern sector of the Yamal Peninsula, in the main part of the southern Gydan Peninsula, in the southeastern part of the Taimyr Lowland, in the tributaries of the Pyassina River (Dudypta and Yangoda Rivers), and in the Gorbita River (Upper Taimyra River basin) (Fig. 1). Sachs, who was the first to describe the Kazantsevo fauna, suggested that *A. islandica* was locally the most significant species. This region was inhabited by a diversified fauna composed of more than 118 species, comparable to the northwestern European Eemian fauna (of the Netherlands, Denmark, and West Germany). However, the northwestern European fauna is composed of 120 species that belong to a different zoogeographic province, inasmuch as it contains 20% boreal, 37% arctic-boreal, and 43% arctic species. According to new data, there are 23 moderately warm-water taxa in the Siberian Kazantsevo. This number is considerably higher than that given in earlier publications (Troitskiy, 1966). The new information indicates that we deal with a rather unique, high-boreal fauna which has no counterpart in present seas. Boreal species constitute 14% of the assemblage making the character of this fauna transitional between the recent East Murman boreal fauna, in which boreal elements constitute 22% of the fauna, and the southeastern Barents Sea low-arctic fauna in which boreal taxa make up 10.5% of the preserved assemblage (Filatova, 1957b; Fig. 3). The existence of this unique province was due to the penetration of boreal fauna into the "precursor" of the Arctic Ocean.

The Valkatlen fauna inhabited the Riss-Würm sea along the coastal zones and in the valleys of the Chukotskiy Peninsula; this fauna has a high-boreal character. It contains 23% boreal and only 14% arctic forms. The boreal species *Sipho martensi, N. vinosa,* and *Cyngula martyni* were distributed one to five degrees of latitude farther north than at present. *S. martensi* and *N. vinosa* reached the Kolyuchin Bay. The high-boreal Pacific molluscs were dispersed not only in the nearshore zone of the Bering Sea, but also along the southern coast of the Chukchi Sea (Petrov, 1966). It is probable that the North Pacific species *Venericardia crebricostata* found in Cape Nerpichiy (Faddeevshky Island, New Siberian Islands; Fig. 1) reached this part of sea, although Ivanov and Yashin suggest that this dispersal took place in middle Pleistocene time (Volnov et al., 1970).

In spite of the deep penetration of boreal water into the Arctic Ocean during the Riss-Würm Interglacial there was probably a wide belt of low-arctic water between the Atlantic and Pacific boreal regions. This prevented free faunal exchange along the Siberian coast of Eurasia but did not preclude the occasional penetration of some boreal species. The assumption of such episodes is based on the presence of Pacific boreal species *Liomesus cf. nassula* (identified by A. N. Golikov) on the Taz Peninsula. This occurrence is the only evidence of westward migration along the Siberian shore.

Late-Würm interstadial transgressions have left few shell-bearing deposits and our reconstructions are inevitably schematic.

The Bölling fauna of North Jylland (Raaholt) is low arctic and contains two boreal forms, *Littorina rudis* and *M. edulis* (Hansen, 1965). The

Table 3 Displacement of the Boundaries of the Marine Zoogeographical Subregions during Pleistocene (km)

Boundaries	Atlantic side			Pacific side	
Age	MB[1]–HB[2]	HB–LA[3]	LA–HA	HB–LA	LA–HA[4]
Alleröd	−600 (−370)*	−600 (−370)	−900? (−560?)		
Bölling			−2700 (−1670)		−750 (−460)
Würm (maximum)			> −3600 (−2250)		
Riss-Würm (optimum)	+2000 (+1250)	+2200 (+1370)	—	+400 (+250)	
Riss (early)			> −3600 (−2250)		
Mindel-Riss (optimum)		+800 (+500)	+1500 (+930)	+400? (+250?)	
Mindel-Riss (early)	−1000 (−620)				
Mindel (late)			> −3600 (−2250)		

[1] MB—Middle-Boreal subregion.
[2] HB—High-Boreal subregion.
[3] LA—Low-Arctic subregion.
[4] HA—High-Arctic subregion.
* The mileage is placed in parentheses (1 km = 0.621 mi).

boundary between this fauna and the high-arctic *Portlandia* assemblage of southern Sweden and Finland lies along the Kattegat and the Oresunn Straits (Lundqvist, 1965; Donner, 1965). It may have extended to Iceland, as indicated by the occurrence of *L. rudis* together with arctic forms in the Bölling Bed of the Skorradallur Valley (Ashwell, 1967). The moderate Bölling stage was not traced in marine deposits east of Denmark, although it was distinguished in continental facies up to Smolensk (Markov et al., 1965).

The Alleröd sea and its fauna are better known. The rapid water warming, which commenced approximately 12,000 years ago, resulted in a wide migration of warm-water molluscs. Judging from the Blumö Island site near Bergen, Norway, the molluscan associations of the Western Norwegian coast were similar to the recent molluscan faunas found near the Lofoten Islands (Undas, 1942; Andersen, 1965). The radiocarbon age of driftwood from this site is 12,100 ± 100 yr B.P. (i.e., latest Bölling or earliest Middle (or Older) Dryas); however, the faunal composition led to different conclusions concerning the climates than did deductions based on other criteria. The average annual water temperature was +6°C, which is comparable with present-day temperatures near the mid-high-boreal boundary. It seems that this warming was caused not only by increased insolation but also by warm North Atlantic currents. The Danish and Northern Norwegian Alleröd fauna was dominated by *A. islandica, Z. crispata,* and *Chlamys islandicus* (Hansen, 1965; Holtedal, 1958; Andersen, 1965). The eastern boundary of the Alleröd boreal region lay between Tromsö, where *A. islandica* was found under the youngest till (Holtedal, 1958) and Tuloma Fjord near the city of Murmansk, where the warmest species was *M. edulis* (Markov et al., 1965). The southern part of that boundary can be traced along the Skaggerrak and Kattegat Straits, which separated the low-arctic fauna of southern Scandinavia with *Littorina littorea* and *C. islandicus* (Lundqvist, 1965), from the Danish high-boreal fauna of northern Vendsyssel (Hansen, 1965).

The eastern part of the Baltic Sea and the ocean coast east of Murmansk were inhabited by high-arctic *Portlandia* associations. This fauna was homogeneous and consisted mainly of *P. arctica* (Donner, 1965; Lavrova, 1960, 1968; Troitskiy, 1966; Gudina et al., 1968). It is likely that the *Portlandia* clays found north of the Russian plain, in the western Siberian and Taimyr lowlands, as well as in the New Siberian Islands, were deposited simultaneously with the *P. arctica* and *Cyrtodaria kurriana* sands in the Yanramaiveem River valley (southwestern Chukotskiy). Petrov (1966), who described these deposits, combined them with the underlying Kresta Formation (middle Pleistocene) but the author assumes that these sands form a low late-glacial terrace and that the overlying sediments are Holocene wind-blown sands.

It appears that my previous assumption concerning the Alleröd age of the Amguema marine terraces on the northern coast of the Chukotskiy Peninsula, containing *M. edulis* and *M. baltica* (Troitskiy, 1970b), is no longer tenable, as the peat underlying alluvial Amguema beds has a radiocarbon age of 9530 ± 230 yr B.P. (Serebranyi, 1965). These terraces may be younger (Holocene) than Troitskiy and Petrov have suggested. The latter author assigned them to the Karginskiy interstage. If the discrepancy does not lie in the age of the terrace but is due to incorrect correlation between marine and river deposits, then we are able to determine the northern boundary of the Alleröd Pacific low-arctic subregion near 175°W. long.

The Mindel, Riss, and Würm periglacial marine faunas are known only from a few sites and they have a typical arctic character, being composed mainly of *P. arctica* populations. During glacial maxima the northern Eurasian land and ice sheets were bordered by the high-arctic subregion along the entire coast.

Discussion

The paleozoogeographic division of northern Eurasian seas, during late Pleistocene time, is based on molluscan assemblages. The information available allows the distinction of several zoogeographic provinces and the reconstruction of the temporal displacement of their boundaries (Fig. 3; Table 3). The data presented in this chapter, concerning the timing and amplitude of temperature fluctuations are in agreement with the results of other investigations which are based on continental faunal and floral evidence.

The divisions and correlations proposed by the author may become the subject of future discussion and it is anticipated that different correlations and interpretations will be proposed.

Acknowledgments

I should like to thank Dr. W. O. Kupsh (Saskatoon, Canada), Dr. Yvonne Herman (Pullman, Washington), and Dr. D. M. Hopkins (Menlo Park, California) for their editorial comments and revisions, as well as V. P. Kobkov, A. M. Bezobrazova, and A. N. Stroganov (Novosibirsk, U.S.S.R.) for their kind and friendly help in preparing the English text.

References

Andersen, B. 1965. The Quaternary of Norway. In: K. Rankama, ed. *The Geologic Systems: The Quaternary*, Vol. 1, New York: Interscience Publishers, pp. 91–138.

Arkhipov, S. A., V. I. Gudina, and S. L. Troitskiy. 1968. Raspredelenie paleontologicheskikh ostatkov v chetvertichnykh valunnykh otlozheniyakh Zapadnoi Sibiri v svyasi s voprosom ob ikh proiskhozhdenii. In: V. N. Sachs, ed. *Neogenovye i chetvertichnye otlozheniya Zapadnoi Sibiri.* Moskva: Izdatelstvo Nauka, pp. 98–112.

Ashwell, I. Y. 1967. Radiocarbon ages of shells in the glaciomarine deposits of western Iceland. *Geograph. J.*, 133(1):48–50.

Biske, G. S. and E. I. Devyatova. 1965. Pleistotsenovye transgressii na severe Evropy. *Ministerstvo Geologii S.S.S.R., Nauchno-Issledovatelskiy Institut Geologii Arktiki Trudy*, 143:155–176.

Bylinskiy, E. N. 1962. Novye dannye po stratigrafii chetvertichnykh otlozheniy i paleogeografii basseina r. Mezeni. *Dokl. Akad. Nauk S.S.S.R.*, 147(6):1421–1424.

Devyatova, E. I. 1961. *Stratigrafiya chetvertichnykh otlozheniy i paleogeografiya chetvertichnogo perioda v basseine r. Onegi.* Moskva–Leningrad: Izdatelstvo Akademii Nauk S.S.S.R., 89 pp.

Donner, J. J. 1965. The Quaternary of Finland. In: K. Rankama, ed. *The Geologic Systems: The Quaternary*, Vol. 1. New York: Interscience Publishers, pp. 199–272.

Filatova, Z. A. 1957a. Obshchiy obzor fauny dvustvorchatykh molluskov severnykh morei S.S.S.R. *Akad. Nauk S.S.S.R., Inst. Okeanologii Tr.*, 20:195–212.

———. 1957b. Zoogeograficheskoe raionirovanie severnykh morei S.S.S.R. po rasprostraneniyu dvustvorchatykh molluskov. *Akad. Nauk S.S.S.R., Inst. Okeanologii Tr.*, 23:195–215.

Gudina, V. I. 1966. *Foraminifery i stratigrafiya chetvertichnykh otlozheniy severo-zapada Sibiri.* Moskva: Izdatelstvo Nauka, 131 pp.

———, N. I. Nuzhdina, and S. L. Troitskiy. 1968. Novye dannye o morskom pleistotsene zapadnoi chasti Taimyrskoi nizmennosti. *Geol. Geofiz.*, 1:40–48.

———, and Kh. M. Saidova. 1969. Biostratigraficheskaya zona *Miliolinella pyriformis* v chetvertichnykh otlozheniyakh Arktiki. *Dokl. Adad. Nauk S.S.S.R.*, 185(4):1109–1111.

Hansen, S. 1965. The Quaternary of Denmark. In: K. Rankama, ed. *The Geologic Systems: The Quaternary*, Vol. 1. New York: Interscience Publishers, pp. 1–90.

Heide, van der, S. 1957. Correlations of marine horizons in the middle and upper Pleistocene of the Netherlands. *Geol. Mijnbow* (nieuwe serie), 7:272–276.

Holtedal, U. 1958. *Geol. Norvegii*, Vol. 2. Moskva: Izdatelstvo Inostrannoi literatury, 395 pp.

Ivanov, O. A., and D. S. Yashin. 1959. Novye dannye o geologocheskom stroenii ostrova Novaya Sibir. *Ministerstvo geologii S.S.S.R., Nauchno—Issledovatelskiy Institut geologii Arktiki Trudy*, 96:61–78.

Lavrova, M. A. 1946. O geograficheskikh predelakh rasprostraneniya borealnogo morya i ego fisikogeograficheskom rezhime. *Akademiya Nauk S.S.S.R., Institut Geografii Trudy*, 37:64–79.

———. 1960. *Chetvertichnaya geologiya Kolskogo poluostrova.* Moskva: Izdatelstvo Akademii Nauk S.S.S.R., 233 pp.

———. 1961. Sootnoshenie mezhlednikovoi borealnoi transgressii na severe S.S.S.R. s eemskoi v Zapadnoi Evrope. *Akademiya Nauk Estonskoy S.S.R., Institut Geologii Trudy*, 8:65–88.

———. 1962. Osnovnoy razrez verkhnego pleistotsena Leningradskogo rayona. In: M. A. Lavrova, A. P. Faddeeva, and A. T. Zhingarev-Dobroselskiy, eds. *Voprosy stratigrafii chetvertichnykh otlozheniy severo-zapada Evropeyskoy chasti S.S.S.R.* Leningrad: Gosudarstvennoe Nauchno-Tekhnicheskoe Izdatelstvo neftyanoy i gorno-toplivnoy literatury, pp. 125–139.

———. 1968. Pozdnelednikovaya i poslelednikovaya istoriya Belogo morya. In: V. N. Sachs, ed. *Neogenovye i chetvertichnye otlozheniya Zapadnoi Sibiri.* Moskva: Izdatelstvo Nauka, pp. 140–163.

———, and S. L. Troitskiy. 1960. Mezhlednikovye transgressii na severe Evropy i Sibiri. In: V. I. Gromov, ed. *Mezhdunarodnyi Geologicheskiy Kongress, 21 sessiya, Doklady sovetskikh geologov, sektsiya 4, Khronologiya i klimaty chetvertichnogo perioda.* Moskva: Izdatelstvo Akademii Nauk S.S.S.R., pp. 124–136.

Lazukov, G. I. 1970. *Antropogen severnoi poloviny Zapadnoi Sibiri.* Moskva: Izdatelstvo Moskovskogo Universiteta, 331 pp.

Lundqvist, J. 1965. The Quaternary of Sweden. In: K. Rankama, ed. *The Geologic Systems: The Quaternary,* Vol. 1. New York: Interscience Publishers, pp. 139–198.

Markov, K. K., G. I. Lazukov, and V. A. Nikolaev. 1965. *Chetvertichnyi period*, Vol. 1. Moskva: Izdatelstvo Moskovskogo Universiteta, 351 pp.

Pavlenko, V. V. 1963. K voprosu o stratigrafii kainozoyskikh otlozheniy na Seduyakhinskom podnyatii v predelakh Malozemelskoy tundry. In: A. I. Popov, ed. *Kainozoiskiy pokrov Bolshezemelskoy tundry.* Moskva: Izdatelstvo Moskovskogo Universiteta, pp. 12–16.

Petrov, O. M. 1966. Stratigrafiya i fauna morskikh molluskov chetvertichnykh otlozheniy Chukotskogo poluostrova. *Akademiya Nauk S.S.S.R., Geologicheskiy Institut Trudy, 155,* 258 pp.

Sachs, V. N. 1945. Novye dannye o geologicheskom stroenii basseina r. Pyasiny. *Glavsevmorput, Gorno-geologicheskoe upravlenie Trudy,* 16:3–64.

———. 1953. Chevertichnyi period v Sovetskoy Arktike. *Glavsevmorput, Nauchno-Issledovatelskiy Institut geologii Arktiki Trudy, 77,* 627 pp.

——— 1959. Nekotorye spornye voprosy istorii chetvertichnogo perioda v Sibiri. *Ministerstvo geologii S.S.S.R., Nauchno-Issledovatelskiy Institut geologii Arktiki Trudy, 96,* 151–163.

———, and K. V. Antonov. 1945. Chetvertichnye otlozheniya i geomorfologiya raiona Ust-Yeniseyskogo porta. *Glavsevmorput, Gorno-geologicheskoe upravlenie Trudy,* 16:65–117.

Serebryannyi, L. R. 1965. *Primenenie radiouglerodnogo metoda v chetvertichnoy geologii.* Moskva: Izdatelstvo Nauka, 269 pp.

Slobodin, V. YA., and O. V. Suzdalskiy. 1969. Stratigrafiya pliotsena i pleistotsena severo-vostoka Zapadnoi Sibiri. In: N. G. Zagorskaya, ed. *Materialy k problemam geologii pozdnego kainozoya.* Leningrad: Izdanie Nauchno-Issledovatelskogo Instituta geologii Arktiki, 115–130.

Troitskiy, S. L. 1964. Osnovnye zakonomernosti izmeneniya sostava fauny po razrezam morskikh mezhmorennykh otlozheniy Ust-Yeniseyskoi vpadiny i Nizhne-Pechorskoi depressii. *Akademiya Nauk S.S.S.R. Sibirskoe otdelenie, Institut Geologii i geofiziki Trudy, 9,* 48–65.

———. 1966. *Chetvertichnye otlozheniya i relief ravninnykh poberezhiy Yeniseyskogo zaliva i prilegayushchei chasti gor Byrranga.* Moskva: Izdatelstvo Nauka, 207 pp.

———. 1969. Obshchiy obzor morskogo pleistotsena Sibiri. In: V. N. Sachs, ed. *Problemy chetvertichnoi geologii Sibiri.* Moskva: Izdatelstvo Nauka, 32–43.

———. 1970a. Obshchiy obzor pleistotsenovykh morskikh faun severnogo poberezhya Evrasii. In: A. I. Tolmachev, ed. *Severnyi Ledovityi okean i ego poberezhje v kainozoe.* Leningrad: Gidrometeorologicheskoe Izdatelstvo, 179–185.

———. 1970b. Paleozoological division of Pleistocene marine basins on Arctic Coast of Eurasia (abstract). *Am. Assoc. Petroleum Geol. Bull.,* 54(12):2510.

Undås, I. 1942. On the Late Quaternary history of Møre and Trøndelag (Norway). *Kungliga Norske Videnskab Selskab Skrifter,* 2.

Vollosovich, K. K. 1966. Materialy dlya poznaniya osnovnykh etapov geologicheskoy istorii Evropeyskogo severo-vostoka v pliotsene–srednem pleistitsene. In: A. I. Popov and V. S. Enokyan, eds. *Geologiya pozdnego kainozoya severa Evropeiskoy chasti S.S.S.R.* Moskva: Izdatelstvo Moskovskogo Universiteta, 3–37.

Volnov, D. A., V. N. Voitsekhovskiy, O. A. Ivanov, D. S. Sorokov, and D. S. Yashin. 1970. Novosibirskie ostrova. In: *Geologiya S.S.S.R., 26,* 324–374.

Zagorskaya, N. G., Z. I. Yashina, V. Ya. Slobodin, F. M. Levina, and A. M. Belevich. 1965. Morskie neogen(?)–chetvertichnye otlozheniya nizhnego techeniya r. Yeniseya. *Ministerstvo geologii S.S.S.R., Nauchno-Issledovatelskiy Institut geologii Arktiki Trudy, 144,* 91 pp.

———, O. F. Baranovskaya, G. N. Berdovskaya, I. G. Gladkova, O. M. Lev, and I. I. Ryumina. 1969. Kratkiy ocherk stratigrafii i paleogeografii pozdnego kainozoya Pechorskoy nizmennosti. In: N. G. Zagorskaya, ed. *Materialy k problemam geologii pozdnego kainozoya.* Leningrad: Izdanie Nauchno-Issledovatelskiy Institut geologii Arktiki, 6–29.

Zayonts, I. L., and R. B. Krapivner. 1967. Stratigraficheskoe razdelenie yamalskoy serii v svete novykh dannykh. *Ministerstvo geologii S.S.S.R., Gidrogeologicheskoe upravlenie. Sbornik statey po geologii i inzhenernoy geologii, 6,* 11–20.

Zarkhidze, V. S. 1963. K istorii rasvitiya yugo-vostochnoi chasti Barentsova morya i ego fauny s verkhnechetvertichnogo vremeni. In: A. I. Popov, ed. *Kainozoiskiy pokrov Bolshezemelskoi tundry.* Moskva: Izdatelstvo Moskovskogo Universiteta, 91–99.

Chapter 12
The Neogene Period in the Subarctic Sector of the Pacific[1]

YURI B. GLADENKOV[2]

Abstract

World-wide correlations of Tertiary strata in tropical and subtropical areas are generally based on planktonic foraminiferal zones. These zones are based on phylogenetic successions of rapidly evolving taxa. Current studies show that many of these zones can be traced from Japan to Sakhalin, Kamchatka, and North America.

For the first time the stratigraphic scales of the North Pacific area can be tied in with the global scale. Studies of nannoplankton, mammals, and radioisotope and paleomagnetic determinations supplemented by paleoclimatic investigations have helped make world-wide correlations more reliable.

Since the planktonic fauna is very sparse in Tertiary North Pacific sediments, the establishment of regional scales must be based mainly on benthonic faunas.

Changes in composition of certain molluscs (*Yoldia, Acila, Arcidae, Pectenidae,* and *Turritellidae*), as well as changes in the number of species and the first appearance of species in certain periods may be used to subdivide the sections into several zones and to correlate them. Establishing these zones independently of facies composition and correlating them with planktonic zones are the basis for applying the international scale to the boreal Pacific area. Seven zones have been proposed for the Neogene of the U.S.S.R. based on *Yoldia* and other groups; many of these zones have analogues in adjacent areas. Other fossil groups may also be used. The establishment of evolutionary stages accompanied by studies of faunal assemblages of various facies of contemporaneous and successive levels are of great importance in working out regional stratigraphic scales.

Introduction

To date the solution of many problems regarding the correlation of Neogene deposits in the North Pacific have remained elusive. This concerns various stage units of the North Pacific and their correlatives in Europe. Recent biostratigraphic, paleomagnetic, and radiometric determinations have resulted in a revision of world-wide correlations of Neogene strata and in the formulation of new, more objective long-range correlations. Studies of benthonic molluscan fauna from Kamchatka and Sakhalin allow for the division of the Neogene into seven units, based on *Yoldia* (Fig. 1). These units can be traced from Japan to North America. The new data have allowed revision of earlier correlations between certain formations and reassessment of their ages. Independent confirmation from other fossil groups including diatoms, echinoids, and mammals support these revisions.

[1] This chapter has been edited and abbreviated with the author's permission by the editor (Yvonne Herman).

[2] Geological Institute of the Academy of Sciences of the U.S.S.R., Pyzhevsky per 7, Moscow, U.S.S.R.

Studies presently underway should help elucidate some of the unresolved problems concerning further subdivisions and correlations between the Arctic and North Pacific Paleogene and Neogene sections on one hand and their correlation with the European stratotypes on the other.

Methods of Correlation of Stratigraphic Subdivisions in the North Pacific and their Relation to the International Scale

Publications concerning the stratigraphic subdivision of the North Pacific with well-preserved Tertiary strata suggest that age assignments to the various "stages" and "horizons" were tentative. Consequently, it is not certain whether the "middle Eocene" of Sakhalin and the "lower Oligocene" of Kamchatka or the "upper Miocene" of North America are analogous to corresponding European subdivisions.

Stage subdivisions of Paleogene and Neogene deposits are known to be in need of clarification, mainly because the faunal characterization of certain stages in their stratotype is inadequate. Most stages have been distinguished in semi-closed European basins, for example, the Paleogene stages in England, France, and Belgium, as well as the Neogene stages in the Mediterranean region. In these regions stage definition was based primarily upon benthonic faunas. However, temporal alterations in the character of these organisms are believed to have been caused by environmental rather than by evolutionary changes. Further, the distribution of benthonic assemblages in European sections is restricted, and consequently these assemblages are thought to be unreliable when correlations are made over large areas.

The first attempt to establish a stratigraphic scale in the North Pacific based on benthonic faunas was that of Weaver et al. (1944). Despite certain shortcomings, the American correlations played an important role in the unification of stratigraphic subdivisions in the Pacific region. The North American scale consists of various units taken from sections that are separated from one another; for this reason these correlations should be considered provisional. They are based on benthonic molluscan and foraminiferal assemblages.

Five molluscan assemblages have been distinguished in the Kamchatka "Pliocene" (Kakert, Etolon, Erman, Enemten, and Bering), whereas the European molluscan assemblages do not allow the subdivision of the Pliocene into smaller units. Hence the correlation of the North Pacific strata with the European standard sections still remains unresolved.

In recent years an attempt has been made to correlate Paleogene and Neogene sequences on a global scale by using planktonic fauna, paleomagnetic stratigraphy, and absolute age determinations (Berggren, 1969). For example, in Japan the analogues of the Danian, upper Paleocene, middle Eocene, the lower part of the Oligocene, Aquitanian, Burdigalian, Tortonian, Helvetian, Messinian, and Pliocene have been recognized (Asano and Hatai, 1967; Ikebe et al., 1969). In Kamchatka various Paleogene and middle Miocene sediments were recognized (Serova, 1969), and in North America the Paleogene, Aquitanian, and the middle and upper Miocene have been identified (Bandy et al., 1969).

Not all the problems of zonal correlation have been resolved to date; the planktonic fauna is impoverished in certain North Pacific strata. These zones with few faunal remains are interlayered with layers entirely devoid of fossils.

Correlations based on both benthonic and planktonic faunas were found to be more reliable than correlations based on a single group of organisms. Biostratigraphy, supplemented by paleomagnetic and radiometric determinations afford the best means for correlations on a global scale.

Stratigraphic Correlations in the North Pacific

Correlations of Tertiary Pacific strata extending from Japan to Koryak upland and America has attracted the attention of scientists for a long time. In the Soviet Union a provincial scale based mainly on molluscs was compiled by L. V. Krishtofovich (1961, Fig. 2).

However, the provincial correlations based on benthonic faunas frequently meet with difficulties. Temperature-dependent faunal differences make correlations difficult between Japanese and Kamchatka faunas. Superimposed upon latitudinal variations are changes induced by differing substrate and water depth.

In Kamchatka as well as in other regions, a new method for stratigraphic subdivision is being

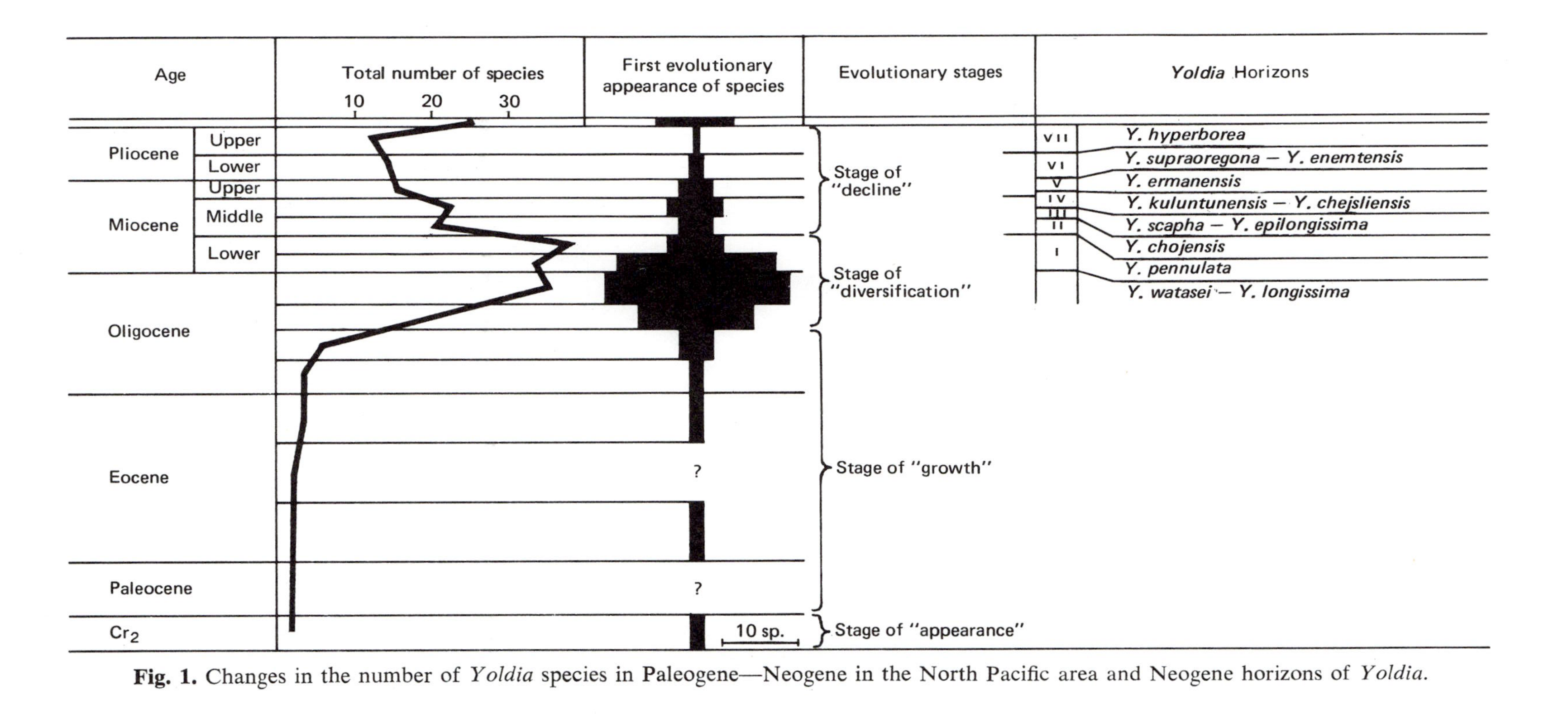

Fig. 1. Changes in the number of *Yoldia* species in Paleogene—Neogene in the North Pacific area and Neogene horizons of *Yoldia*.

Table 1 Neogene in Pacific Region,

Age		Standard stages of Japan	N. Japan	Sackalin		W. Kamchatka	E. Kamchatka
Q							Tusatuvajam beds
Pliocene		Shibikawa	Setana A	Pomir. horiz.	4 horizon of marujam. stage A		Ust-limimtenvajam A
Pliocene		Wakimoto	Takikawa P	Pomir. horiz.	3 horizon of marujam. stage P	Enemten P	P
Miocene	upper	Kitaura	Wakkanai	Takoi		Erman	Limimtenvajam
Miocene	upper	Funakawa		Okobikai		Etolon D	
Miocene	upper	Onnagawa	D	D		E	?
Miocene	middle	Nishikurosawa	Kawabata	Sertunai		Kakert	Yunjunvajam Cape Plosky
Miocene	middle			?		Iljin	
Miocene	middle					Kuluwen	
Miocene	lower	Daishima	Takinoue	Sackalin	Upper-duja	Wiwentec	Pestrocvet
Miocene	lower	Nisioga	Asahi	Sackalin	Tshechov	Utcholok	
Oligocene		Nishisonogi	Poronai X ?	Cholm Machigar X		Gakch Amanin X	Sandstone with *Laternula besshoensis* Ilchatun X
Oligocene		Madze / Funadsu	Ishikari	Lesogor		Kovatshin	Cape Tons

A Assemblage with *Astarte*.
P Assemblage with *Fortipecten takahashii*, *Chlamys cosibensis*, *C cosibensis heteroglypta*, *Anadara trilineata trilineata*.
XXX Assemblage with *Papyridea harrimani*, *Yoldia watasei*, *Yoldia longissima*.
D Upper stratigraphic limit of *Desmostylidae*.
H Upper stratigraphic limit of upper Miocene *Hipparion*.
E Echinoidea with *Astrodapsis antiselli*.

employed; it is based on evolutionary changes of molluscs (e.g., Kotaka, 1959, on turritellids; Masuda, 1962, on pectinids; Noda, 1966, on arcids; MacNeil, 1966, on *Mya*; and Uozumi, 1962, on *Yoldia*).

In Kamchatka changes in *Yoldia* occur and are found to be independent of facies changes. Seven zones have been distinguished in the Neogene, commencing with the lower Miocene (Gladenkov, 1970, 1971, 1972). They are shown in Fig. 1. Similar faunal changes have been observed in Sakhalin, Koryak, Japan, and North America (Table 1). In Japan the number of other molluscan species (turritellids, arcids, and pectinids) increases considerably in early Miocene and in early Pliocene, simultaneously with the first evolutionary appearance of many genera and subgenera.

The evolution of *Yoldia* and the *Acila* is somewhat different. The burst of speciation in these groups occurs in late Paleogene (upper Oligocene), as indicated in Fig. 1; the lower Pliocene boundary based on taxodonts is not sharp.

Stages and Horizons in the Pacific Region

N. America	Neogene horizons of *Yoldia*		Horizons usage in 1959 scale			Age by author/1970	
Anvilian beds						Q	
Beringian beds A	VII	*Y. hyperborea*	?	Pleistocene		Pliocene	"upp."
Etchegoin Jacalitos P	VI	*Y. enemtensis–Y. supraoregona*					"low."
- - - - H - - -							
Neroly Cierbo	V	*Y. ermanensis*	Erman	upp.	Pliocene	Miocene	upper
D	IV	*Y. chejsliensis–Y. kuluntunensis*	Etolon	middle			
- E - - - - - - -							
Briones	III	*Y. scapha–Y. epilongissima*	Kakert	low.			middle
- - - - - - - -							
Temblor	II		Kamtchat	upper	Miocene		
		- - - - - - - - - - - - - - - -					
		Y. chojensis	Sertunai				
Vaqueros	I	*Y. pennulata*	Sackalin	middle			lower
- - - - - - - -							
Blakeley "Lincoln" X		*Y. watasei–Y. longissima*	Machigar	lower		Oligocene	
			Kovatshin	Oligocene			

Age Revision of the Kamchatka Horizons

Early correlations between Kamchatka, North America, Sakhalin, and Japanese deposits, based on molluscan faunal assemblages had to be revised because many of the faunal identifications appear to have been faulty (Table 2) (Resolutions of the Interdepartmental Conference . . . ,1961). The American Etchegoin Formation (middle Pliocene), characterized by a rich molluscan fauna, was thought to be equivalent in age with the Kamchatka Etolon Suite; the Monterey shale was correlated with the diatomaceous-argillitic Kakert Suite. The maximum Pliocene transgression was considered to have taken place in Etcheginian time in the Pacific region. With the subsequent regression (in San Joaquinian time) continental deposits were laid down. The Erman Suite in Kamchatka with its lignites and flora-bearing horizons was thought to correspond to this regression. However, subsequent detailed faunal analyses showed that these correlations were rather broad.

The correlation between the Etchegoin Formation and the Etolon Suite was based principally on the following molluscs: *Chlamys etchegoini, Chlamys nutteri, Pecten caurinus oregonensis, Securella securis, Anadara trilineata,* and *Macoma nasuta.* The Etolon fauna was first described by Slodkevich (1938). His determinations were adopted by later authors without revision. However, subsequent detailed studies of molluscs showed that some of Slodkevich's determinations required revision. Thus, Rheinhart, studying the *Arcidae* concluded that the typical *Anadosa trilineata* differs considerably from the Kamchatka *Anadara* and that the latter is probably an ancestral form of *A. trilineata* (Rheinhart, 1943). Parker,

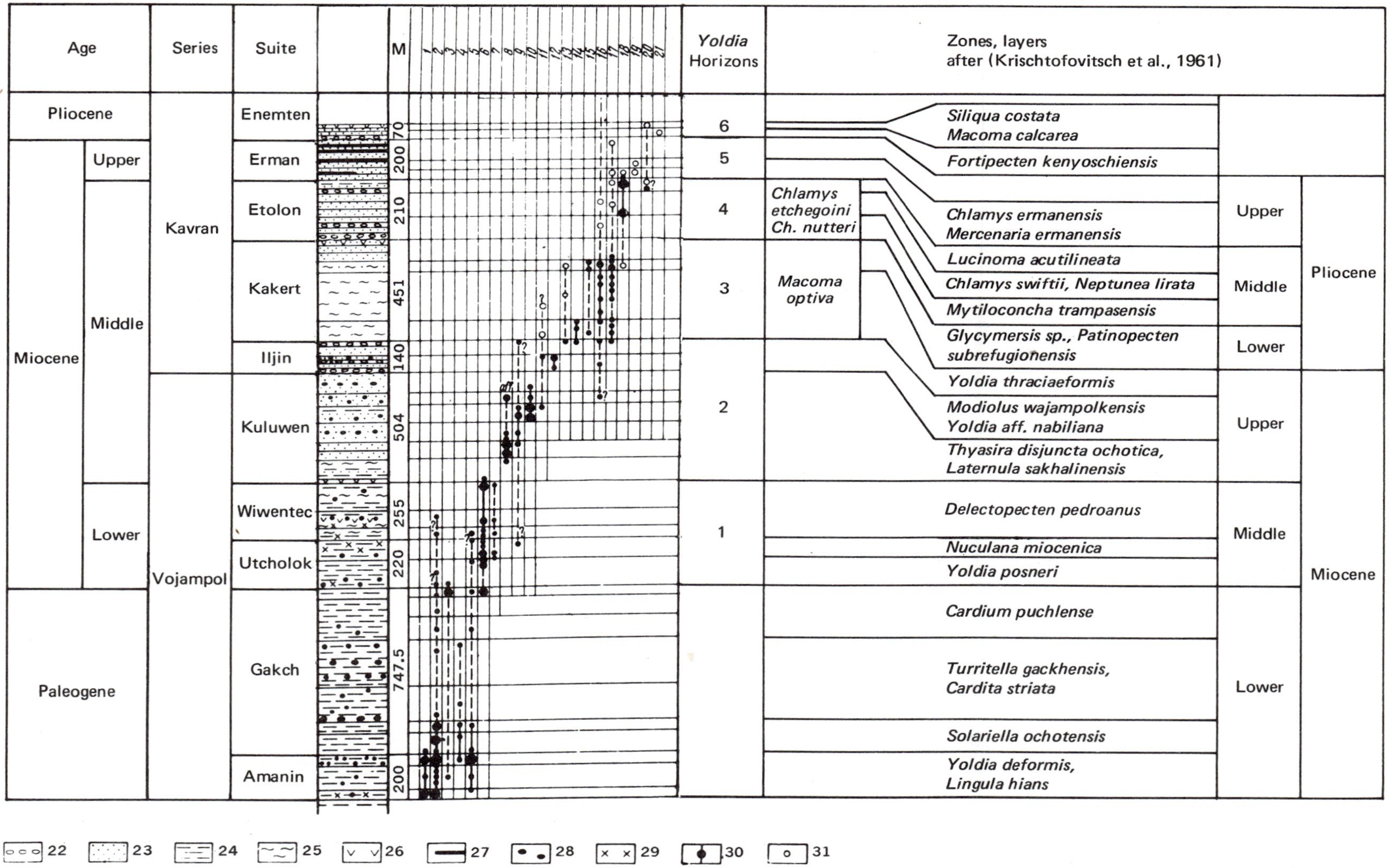

Fig. 2. Stratigrafic distribution of *Yoldia* in Neogene stratotypes of western Kamchatka (Tochilin region) 1. *Y. kovatschensis;* 2. *Y. longissima;* 3. *Y. laudabilis*; 4. *Y. watasei*; 5. *Y. nitida*; 6. *Y. pennulata*; 7. *Y. posneri*; 8. *Y. makarovi*; 9. *Y.* sp. nov. (aff. *Y. orientalis*); 10. *Y. chojensis*; 11. *Y. nabiliana*; 12. *Y.* sp. nov. (aff. *Y. chojensis*); 13. *Y. epilongissima*; 14. *Y.* sp. nov.; 15. *Y.* aff. *scapha*; 16. Y. *thraciaeformis*; 17. *Y.* sp. nov. (*Y hyperborea*); 18. *Y. kuluntunensis*; 19. *Y. ermanensis*; 20. *Y. chejsliensis*, *Y. enemtensis*; 21. *Y. supraoregona*; 22. conglomerate; 23. sandstone; 24. silt rock; 25. diatomite; 26. tuffs; 27. lignite; 28. carbonate concretion; 29. gennojshies; 30. remaines of *Yoldia*; 31. *Yoldia* in Hejsli region.

who studied the Pacific *Securella* species, suggested the similarity of *Securella ensifera chehalisensis* to the Etolon form. This species has been identified as *S. securis* (Parker, 1949). According to Scarlato (personal communication), among the Kamchatka species of *Macoma, M. nasuta* is not represented. The Etolon forms previously identified as *M. nasuta* are probably *Macoma optiva.*

The work of Masuda and Sinelnikova showed that the Kamchatka forms identified as *P. caurinus oregonensis* belong to the genus *Mizuhopecten* (Miocene-Recent) and are representatives of a new species related to the *Macoma kumurai–Macoma tokyoensis* branch (Miocene). Species belonging to the genus *Chlamys,* which were considered subspecies of the American *Swiftopecten swiftii,* are well-known Miocene subspecies of the Japanese *Chlamys cosibensis* and *Chlamys daishakaensis.* Recent work has shown the close similarity between the Kamchatka and the Japanese faunas of equivalent age (Gladenkov and Sinelnikova, 1972).

Re-examination of the Kamchatka suites suggests that the Etolon taxa are principally of Miocene age; they contain the following species: *Yoldia kuluntunensis, C. cosibensis, C. diashakaensis, Nanochlamys anapleus, S. ensifera chehalisensis, Pitar gretschischkini, Mytiloconcha kewi, M. optiva,* and *Neptunea lirata;* consequently the correlation of the Etolon Suite with the Etchegoin is likely to be erroneous.

The Kavrano–Utkholok Bay (Kamchatka) Neogene fauna was also re-examined. Here the facies is different from that in the stratotype section of western Kamchatka. In the section studied by Sinelnikova shallow-water species belonging to the Etolon Suite were found (Gladenkov and Sinelnikova, 1972).

The Kakert deposits underlying the Etolon Suite were considered upper Miocene–lower Pliocene, or lower Pliocene. In the Tochilin section the clayey lower part of the suite was thought to be Miocene and was described as being characterized by the presence of cold-water, thin-shelled species, whereas the sandy upper part, deposited in Pliocene time, was thought to contain a more diversified fauna, including shallow-water thermophilic elements. These data were interpreted to indicate that climates were colder in Miocene than during the Pliocene. Our new information, however, suggests that both the upper and lower Kakert fauna contains warm-water tolerant elements. The fauna includes numerous tropical pelecypods and echinoids. The presence of *Acila conradi, Anadara watanabei, Mytilus tichanovitchi, Mya cuneiformis, Dosinia ausiensis, Mizuhopecten matschiense, Limatula pilvoensis, Securella panzana, Securella ensifera, Yoldia epilongissima, Yoldia scapha, Kotorapecten subrefugionensis,* and other species throughout the suite suggests that the Kakert Formation is correlative with the Okobykaisk-Sertunai Horizon of Sakhalin, the Astoria Formation of North America, and the Kawabato-Wakkanai Formation of Japan, and is probably of middle Miocene age (Table 2).

The Erman deposits were considered upper Pliocene. Recently a detailed study of marine and fresh-water molluscs from Trohilin, the Kavrano-Utkholk Bay, and the Ycha River was undertaken; the following taxa were identified: *C. cosibensis, M. kewi, Mytiloconcha coalingensis, Modiolus tenuistriatus, Septifer margaritanus, Protothaca staleyi, Acila blancoensis, Yoldia ermanensis, Anadara obispoana, Glycymeris coalingensis, Macoma affinis, Mulinia densata,* and *Turritella fortilirata habei.* The large percentage of extinct species in the Erman assemblage suggests that these deposits are of Miocene age. A number of common forms with the fauna of the Santa Margarita formation (e.g., *S. margaritanus, M. kewi,* and *M. coalingensis*) suggests that they were deposited synchronously (Adegoke, 1969).

The Enemten Suite had been attributed to the marine post-Pliocene. However, a study of molluscs collected from this suite by Sinelnikova suggests that a revision of its age is pertinent (Gladenkov and Sinelnikova, 1972). The dominant taxa in these assemblages are: *Fortipecten konyoshiensis* (forms that are similar to *Fortipecten takanashii*), *Chlamys cosibensis heteroglypta, Swiftopecten swiftii kindlei, S. ensifera, Anadara trilineata trilineata, Yoldia supraoregona, Yoldia enemtensis,* and *Acila marujamensis.* All the above-listed species are extinct. Their prevalence and the presence among them of *F. kenyoshiensis,* and *A. trilineata,* which also occur in the early Pliocene sections of Takikawa, Japan, and in the third horizon of the Maruyama Suite in Sakhalin, suggest that the Enemten strata were deposited in early Pliocene time (Table 2). The presence of *Anadara, Securella, P. staleyi,* and *Turritella vanvlecki* indicates that these deposits are correlative with the Etchegoin Formation of California, which contains similar faunas. This correlation is confirmed by the microfauna; at the base of the Enemten Suite *Elphidiella retzens* (*E. pliocenica*) has been recently found. This taxon is present in both the third horizon of the Maruyama Suite and in the Etchegoin Formation.

Until recently fossiliferous material from the above-mentioned suites consisted mostly of molluscs. We now also have data on desmostylids and echinoids. Oligocene and Miocene desmostylids are found in middle Miocene Japanese and Californian deposits; however, they are less common in upper Miocene deposits (Rheinhart, 1959). Teeth and bones of desmostylids were found in the Kamchatka Kakert Suite (in the Etolon Suite *Desmostylius hesperus* was identified; Dubrovo and Sinelnikova, 1971). These findings confirm the Miocene age of the above-mentioned suites. The abundant remains of desmostylids in the Kakert Suite allow correlation with the Sertunai—the lower Kobykai Horizon of Sakhalin. These deposits are believed to have been deposited in middle Miocene time.

Age assignments based on desmostylids should be considered provisional and subject to modification as new information becomes available. Nevertheless, the findings of mammal remains are of a great interest because this group has evolved rapidly and can be used for a detailed subdivision of certain Neogene horizons.

Among echinoids collected in these suites (identified by Schmidt, in Schmidt and Sinelnikova, 1971) representatives of the genera *Astrodapsis* and *Kewia* predominate and are followed by *Echinarachnius* and *Vaquerosella. Kewia, Vaquerosella,* and *Astrodapsis* are known only as fossils; some species of *Astrodapsis* become extinct in lower Pliocene and species of *Vaquerosella* have been described only from the Miocene. In all deposits, up to the uppermost parts of the Etolon Suite, representatives of the Pliocene–Recent genus *Dendraster* and of Pliocene species of *Astrodapsis* (or *Pseudoastrodapsis*) are absent. Miocene age is further suggested by the presence of *Astrodapsis antiselli* (this species is also present in the Santa Margarita Formation, corresponding to the lowermost Delmontian) in the uppermost parts of the Kakert Suite; this is the basis for the correlation of the upper Kakert subsuite with the lowermost parts of the Cierbo or the upper part of the Briones "stages" of the American standard.

There are other arguments for "lowering" the above suites. The first is "an increase in age" of underlying suites. Thus the Iljin, underlying the Kakert Suite, contains middle Miocene fossils (e.g., *Modiolus wajampolkensis, Macoma astoria, Macoma secta, Glycymeris hitanii, Taras gravis, Mya grewingki, M. tichanovichi,* and *Buccinum haromaiacum*). Stratigraphically the lower Gakkhin and Amanin Suites, that were previously considered lower Miocene, now have been placed into the Paleogene, since assemblage with *Yoldia watasei, Yoldia longissima,* and *Papyridea matschigarica* are considered upper Paleogene in Japan.

This age is further suggested by the presence of strata younger than the Enemten (upper Pliocene, according to the American and Japanese scales). The Beringian beds of Alaska (Hopkins, 1965), Ustlimimten stratum of East Kamchatka and Setana of Japan can be assigned to the Beringian Horizon. In contrast with the Enemten, the Beringian Horizon is characterized by a colder-water assemblage with various species of *Astarte.* These deposits are overlain by Tusatuvayam (Kamchatka) layers; their fauna has close affinities with the Recent fauna, but contains some extinct species (Gladenkov et al., 1972). These deposits are overlain by Pleistocene Pinakul-Karagin Formations with Arctic-boreal faunas (with *Portlandia arctica*).

In recent years correlations of Pacific deposits, particularly those of upper Neogene age, have been based on diatoms and spore-pollen spectra. For example, analyses of diatoms carried out recently showed that in the middle Miocene–Pliocene sections of Japan (Koizumi, 1968), Sakhalin, and East Kamchatka (Gladenkov and Muzylev, 1972) four assemblages based on phylogenetic successions may be distinguished: The first (lowermost) is characterized by *Actinocyclus ingens* s.l, *Stephanopyxis schenckii,* and *Pterotheca kittoniana kamtschatica*; the second by *Thalassiosira nativa, Thalassiosira nidulus,* and *Thalassiosira zabelinae*; the third by *Bacterosira fragilis, Thalassiosira nordenskioldii,* and *Thalassiosira gravida*; and the fourth by *Melosira albicans.* The correlations based on diatoms (Table 2) also support the older age of the Sakhalin and Kamchatka strata.

It is noteworthy that the stratigraphic correlations mentioned above, based mainly on paleontologic considerations are in agreement with the data obtained by other methods. Paleoclimatic evidence supports the scheme suggested herein (Durham and MacNeil, 1967; Gladenkov, 1972). It is known that in northern Europe the thermophilic early Pliocene benthonic fauna (Coralline Crag of England and its analogues) are replaced by cold-water fauna of late Pliocene age (Red Crag of England), and these are replaced in turn by colder-water associations of Pleistocene age. The North Pacific horizons are similar to formations of equivalent age in other regions. A successive change from the "warm" Takikawa Horizon (with analogues in the Enemten and Etchegoin) to the cold-water horizons of Setara (Beringian beds) and finally to the colder Pleistocene is observed.

Radiometric age determinations confirm the biostratigraphic correlations between the Pacific sections and the European horizons.

With the aid of paleomagnetic stratigraphy it is now possible to relate the ranges of stratigraphically important fossils to an independent time scale (e.g., Einarsson et al., 1967; Hopkins, 1965; Niitsuma, 1970; Menner et al., 1972; Pevzner, 1972; Fig. 3).

In conclusion, the data summarized in this chapter indicate that biostratigraphic correlations over wide areas are based principally on planktonic microfossils, whereas provincial correlations rely mainly on benthonic faunas. The most satisfactory stratigraphic division is through erection of zones based on phylogenetic successions of rapidly evolving taxa. By simultaneous faunal, paleomagnetic, and radiometric determinations a more objective and precise correlation over wide areas may be achieved.

References

Adegoke, O. S. 1969. Stratigraphy and paleontology of the marine Neogene formations of the Coalinga region, California. *University of California Publications in Geological Sciences,* 80:241.

Asano, K., and K. Hatai. 1967. Micro- and macropaleontological Tertiary correlations with Japanese islands and with planktonic foraminiferal sequences of foreign countries. Tertiary correlations and climatic changes in Pacific. *XI Pacific Science Congress,* Tokyo: 77–87.

Bandy, O. L., R. W. Morin, and R. C. Wright. 1969. Definition of the *Catapsydrax stainforthi* Zone in the Saucesian Stage, California, *Nature,* 222:468–469.

Berggren, W. A. 1969. Cenozoic chronostratigraphy, planktonic foraminiferal zonation, and the radiometric time scale. *Nature,* 224:1072–1075.

Cooke, C. W., T. A. Gardner, and W. P. Woodring. 1943. Correlation of the Cenozoic formations of the Atlantic and Gulf Coastal plain and the Caribbean region. *Geol. Soc. Am. Bull.,* 54:1713–1723.

Cox, A. 1969. Geomagnetic reversals. *Science,* 163 (3864): 237–245.

Dubrovo, J. A., and V. N. Sinelnikova. 1971. Neogene *Desmostylidae* of Kamchatka (in Russian). *Dokl. Akad. Nauk, Geol.* 199, 3:670–673.

Durham, J. W. and F. S. MacNeil. 1967. Cenozoic migrations of marine invertebrates through the

Table 2 Correlation of Northwest Pacific Neogene Deposits Based on *Diatomaceae*

Age	Assemblages of *Diatomaceae*	Japan (Oga Peninsula)	N. Sakhalin (Schmidt Peninsula)	E. Kamchatka (Karaginsky Island)
Pliocene		?	?	Ust-limimtenvajam
	IV			Limimtenvajam
Upper Miocene	III	Kitaura		?
	II	Funakava	Majamraf* ?	Yunjunvajam
Middle Miocene	I	Onnagawa	Vengerian**	
				Cape Plosky

* Upper Miocene in 1959 scale.
** Middle-upper Miocene in 1959 scale.

Bering Strait region. In: D. M. Hopkins, ed. *The Bering Land Bridge.* Stanford: Stanford University Press, pp. 326–349.

Einarsson, T., D. Hopkins, and R. Doell. 1967. The stratigraphy of Tjornes, northern Iceland and the history of the Bering land bridge. In: D. M. Hopkins, ed. *The Bering Land Bridge.* Stanford: Stanford University Press. pp. 312–325.

Gladenkov, Yu. B. 1970. *Yoldia* in the Paleogene and Neogene of the North Pacific area (in Russian). *Izvestia Akad. Nauk, Seria Geologicheskaya* 2:112–122.

——— 1971. On development of the Neogene stratigraphy of the northern part of the Pacific area (in Russian). *Sovetskaya Geologia,* 4:23–44.

———. 1972. Neogene of Kamchatka (in Russian).

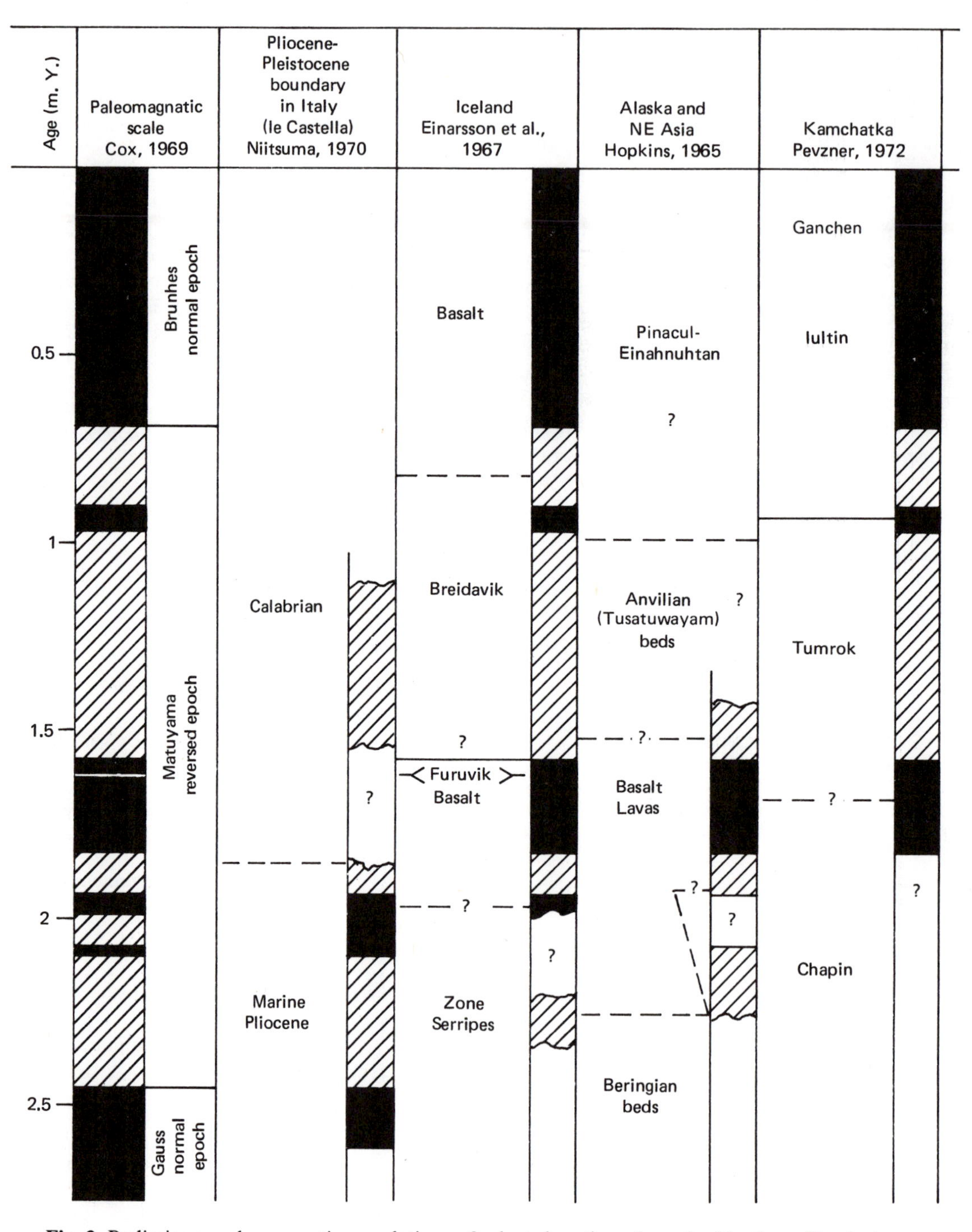

Fig. 3. Preliminary paleomagnetic correlations of selected sections from the Northern Hemisphere.

Transactions of the Geological Institute. Nauka, 214:1–251.

———, and N. G. Muzylev. 1972. Marine Neogene diatoms of East Kamchatka and North Sakhalin. (in Russian). *Izvestia Akad. Nauk, Seria Geologicheskaya,* 8:92–104.

———, and V. N. Sinelnikova. 1972. Neogene stratigraphy of the Far East in light of new paleontological data. (in Russian). *Bull. MOIP, Geol.,* 4:82–92.

———, O. M. Petrov, and V. N. Sinelnikova. 1972. Pliocene-Pleistocene boundary in the Northwest Pacific (in Russian). *International Colloquium on the Boundary between Neogene and Quaternary,* 3:58–73.

Hopkins, D. M. 1965. Quarternary marine transgressions in the Alaskan Anthropogene Period in the Arctic and Subarctic (in Russian). *Tr. NIIGA,* 143:47–90.

Ikebe, N., M. Chiji, and K. Makaseko. 1969. Problems on the terminal Miocene in Japan. *Committee on Mediterranean Neogene Stratigraphy, Proc. 4th Session Bologna, Giornale di Giologia,* 35(4):11–25.

Koizumi, I. 1968. Tertiary diatom flora of Oga Peninsula, Akita Prefecture, Northeast Japan, *Tohoku Scientific Reports,* 2nd seria (Geologia) 40:171–240.

Kotaka, T. 1959. The Cenozoic *Turritellidae* of Japan. *Scientific Reports, Tohoku University.* Sendai, Japan, 31(2):1–135.

Krishtofovich, L. V. 1961. Correlation of Tertiary deposits of the northern Circum Pacific of the Cenozoic folding. Unified stratigraphical schemes of the northeast of the USSR (in Russian). *Gostoptekhizdat,* 83–90.

MacNeil, F. S. 1965. Evolution and distribution of the genus *Mya* and Tertiary migrations of Mollusca. *U.S. Geol. Surv.,* Prof. Pap. 483–g:1–51.

Masuda, K. 1962. Tertiary *Pectinidae* of Japan. *Scientific Reports, Tohoku University,* 2 seria (Geologia). Sendai, Japan, 33(2):117–238.

Menner, V. V., K. V. Nikiforova, M. A. Pevzner, M. N. Alekseev, Y. B. Gladenkov, G. Z. Gurarii and V. M. Trubichin. 1972. Paleomagnetism in detailed stratigraphy of the Upper Cenozoic. (in Russian). *Uzvestia Akad. Nauk, Seria Geologicheskaya,* 6:3–17.

Minato, M., M. Corai, and M. Hwnahashi. 1965. Geological development of the Japanese islands. Tokyo: Tsukiji Shokan Co. Ltd. 442 pp.

Niitsuma, N. 1970. Some geomagnetic stratigraphical problems in Japan and Italy. *J. of Marine Geol.,* 6(2):99–112.

Noda, H. 1966. The Cenozoic *Arcidae* of Japan. *Scientific Reports, Tohoku University.* Sendai, Japan, 38(1):1–161.

Parker, P. 1949. Fossils and recent species of the pelecypod genera *Chione* and *Securella* from the Pacific Coast. *J. Paleontology,* 23(6):577–594.

Pevzner, M. A. 1972. Paleomagnetism and stratigraphy of Pliocene-Quaternary deposits of Kamchatka (in Russian). *Trans. Geol. Inst., Nauka,* 235:1–66.

Resolutions of the Interdepartmental Conference on the Compilation of Unified Stratigraphic Schemes for Sakhalin, Kamchatka, Kuril, and Komandor Islands (in Russian). 1961. *Gostoptekhizdat,* 1–21.

Rheinhart, P. W. 1943. Mesozoic and Cenozoic *Arcidae* from the Pacific slope of North America. *Geol. Soc. Am. Spec. Pap.,* 47:1–117.

Rheinhart, R. H. 1959. A review of the *Sirenia* and *Desmostylia. University of California Publications in Geological Sciences,* 35(1):1–146.

Schmidt, O. J. and V. N. Sinelnikova. 1971. Marine *Echinoidea* of the Kavranian series of western Kamchatka (in Russian). *Dokl. Akad. Nauk, Geologia,* 199(4):909–912.

Serova, M. Y. 1969. Zonal subdivision and correlation of the Paleogene deposits of the northwestern part of the Pacific Province (in Russian). Biogeography, fauna, flora of the Cenozoic of the northwestern part of the Pacific mobile belt. *Nauka,* 101–113.

Slodkevich, V. S. 1938. Tertiary pelecypods of the Far East (in Russian). *Paleontology USSR,* 10, 3, 18:1–508.

Uozumi, S. 1962. Neogene molluscan faunas in Hokkaido. *Faculty of Sciences, Hokkaido University,* seria 4, 11(3):539–596.

Weaver, C. E., S. T. Beck, M. N. Bramlette, S. Carison, B. L. Clark, T. W. Dibblee, J. M. Durham, G. C. Ferguson, L. C. Forest, U. S. Grant, M. Hill, F. R. Kelley, R. M. Kleinpell, W. D. Kleinpell, J. Marks, W. C. Putnam, H. G. Schenck, N. L. Tallaferro, R. L. Thorup, E. Watson, and R. T. White. 1944. Correlation of the marine Cenozoic formations of western North America. *Geol. Soc. Am. Bull.,* 55:569–598.

Chapter 13

Arctic Ocean Sediments, Microfauna, and the Climatic Record in Late Cenozoic Time

YVONNE HERMAN [1]

Abstract

Deep-sea cores from the Central Arctic Basin yield significant faunal and lithologic evidence of normal and low salinity cycles superimposed upon temperature fluctuations in late Cenozoic time. Lowest temperatures correspond to the upper Pleistocene (the Brunhes normal polarity epoch), whereas higher temperatures and lower salinities were recorded by planktonic foraminifera during the Matuyama reversed polarity epoch.

Biostratigraphic and lithologic correlations between cores, some with established paleomagnetic stratigraphy, supplemented by radiometric dating and oxygen isotope measurements, were used to estimate ages and sedimentation rates as well as to reconstruct the climatic and oceanographic history of the Arctic.

Ice-rafted detritus throughout the cores indicates that high latitude glaciation commenced prior to 3 million years ago. Three major climatic units may be distinguished. The sediments of unit III were deposited earlier than 2.4 million years B.P., probably during the Gauss normal polarity epoch. Lower-than-present sedimentation rates and/or corrosive deep water may account for the selective solution of the less resistant limy tests and the impoverished character of the fauna. The paucity of the fauna precludes definitive paleoclimatic reconstruction of this period; it is tentatively suggested that environments were similar to those that prevailed during the deposition of the foraminifera-rich layers of unit I. Sediments of unit II, deposited between approximately 2.4 and 0.7 million years ago, during the Matuyama epoch, are poor in both Fe and Mn oxides and in foraminifera but contain one foraminifera-rich layer. Surface water temperatures were generally higher and salinities were lower during this period than in the preceding and following epochs. It is assumed that the Arctic was free of permanent pack-ice in Matuyama time. The Brunhes cold-"warm" temperature fluctuations are represented by 4 to 6 foraminifera-rich, foraminifera-poor sequences, possibly correlative with the classic Donau, Günz, Mindel, Riss, and Würm Glacials and intervening interglacials, respectively. The former were deposited during pack-ice-covered, the latter in seasonally pack-ice-free periods.

An apparent correspondence between geomagnetic polarity reversals and climatic changes exists.

The record of climatic changes based on paleontologic data, oxygen isotope measurements, magnetic stratigraphy, and radiometric dating indicates that the late Cenozoic major glacial-interglacial cycles were broadly synchronous throughout the world.

[1] Department of Geology, Washington State University, Pullman, Washington 99163, U.S.A.

Introduction

With the systematic large-scale sampling of the sea floor which started about two decades ago, thousands of sediment cores have been raised from major oceans as well as from marginal and inland seas. Studies of these sediments have added to our knowledge of Quaternary climates, ocean-

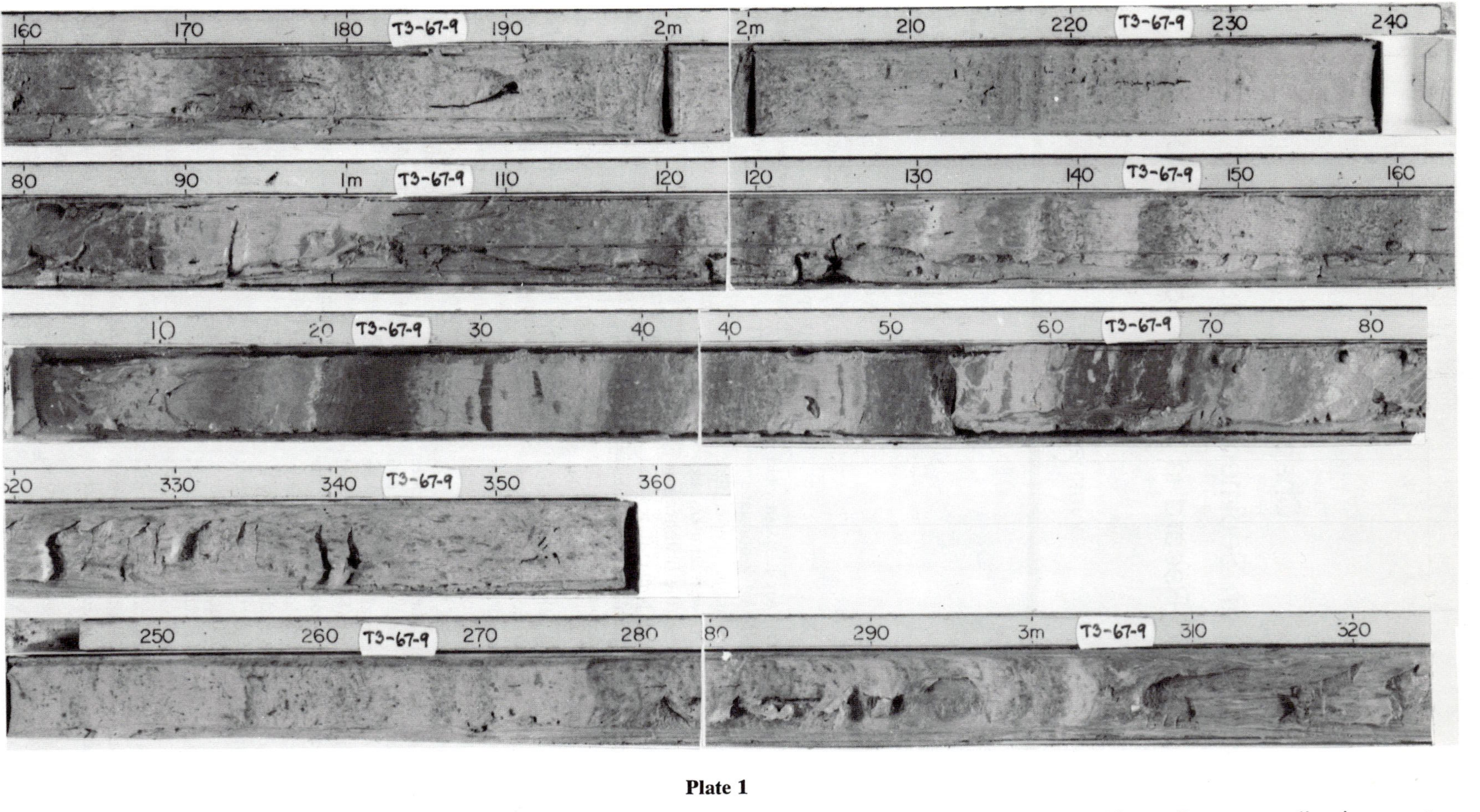

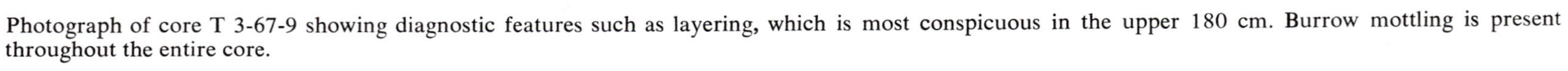

Plate 1

Photograph of core T 3-67-9 showing diagnostic features such as layering, which is most conspicuous in the upper 180 cm. Burrow mottling is present throughout the entire core.

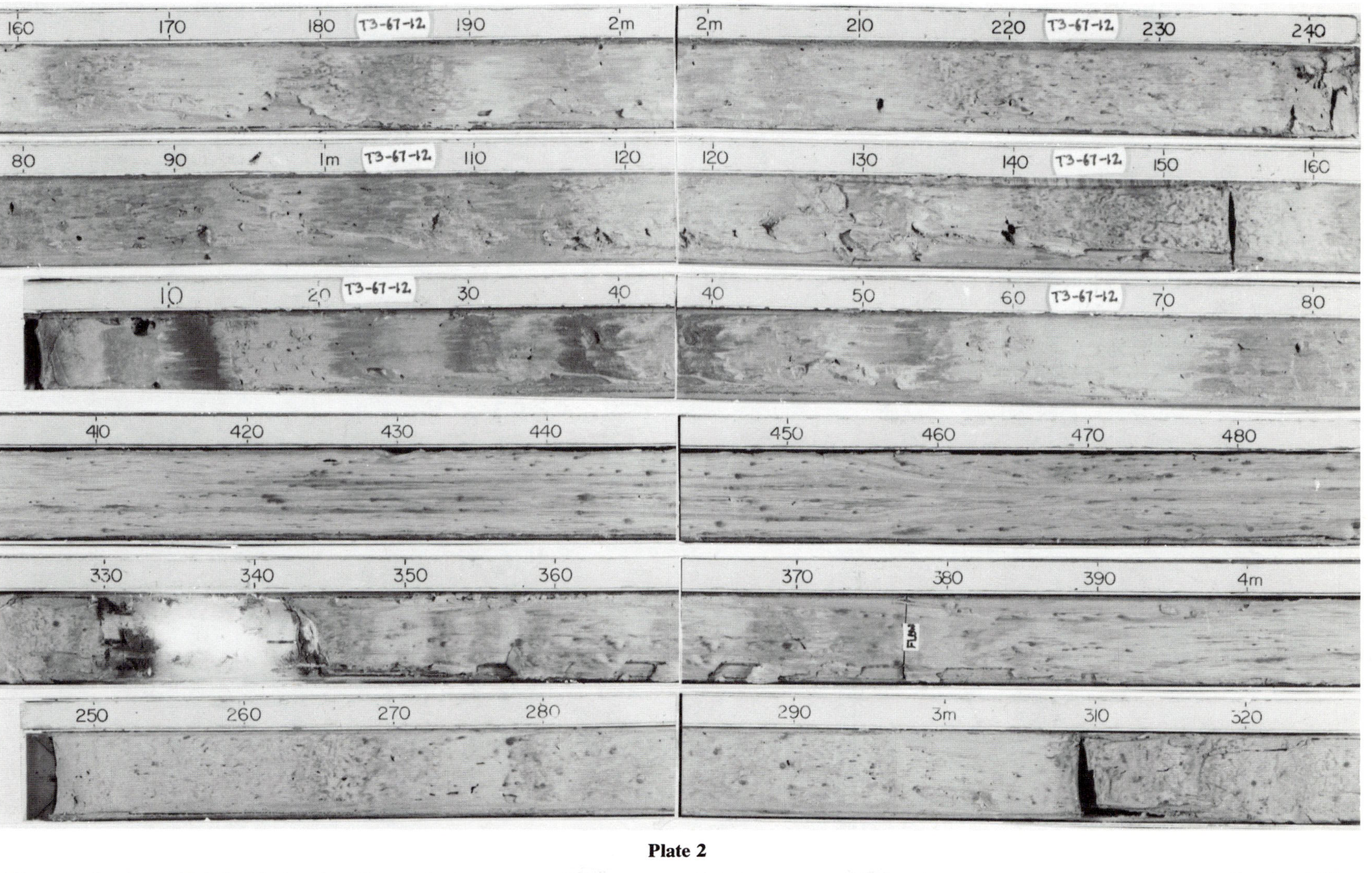

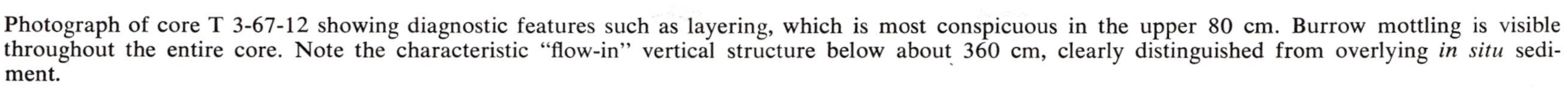

Plate 2

Photograph of core T 3-67-12 showing diagnostic features such as layering, which is most conspicuous in the upper 80 cm. Burrow mottling is visible throughout the entire core. Note the characteristic "flow-in" vertical structure below about 360 cm, clearly distinguished from overlying *in situ* sediment.

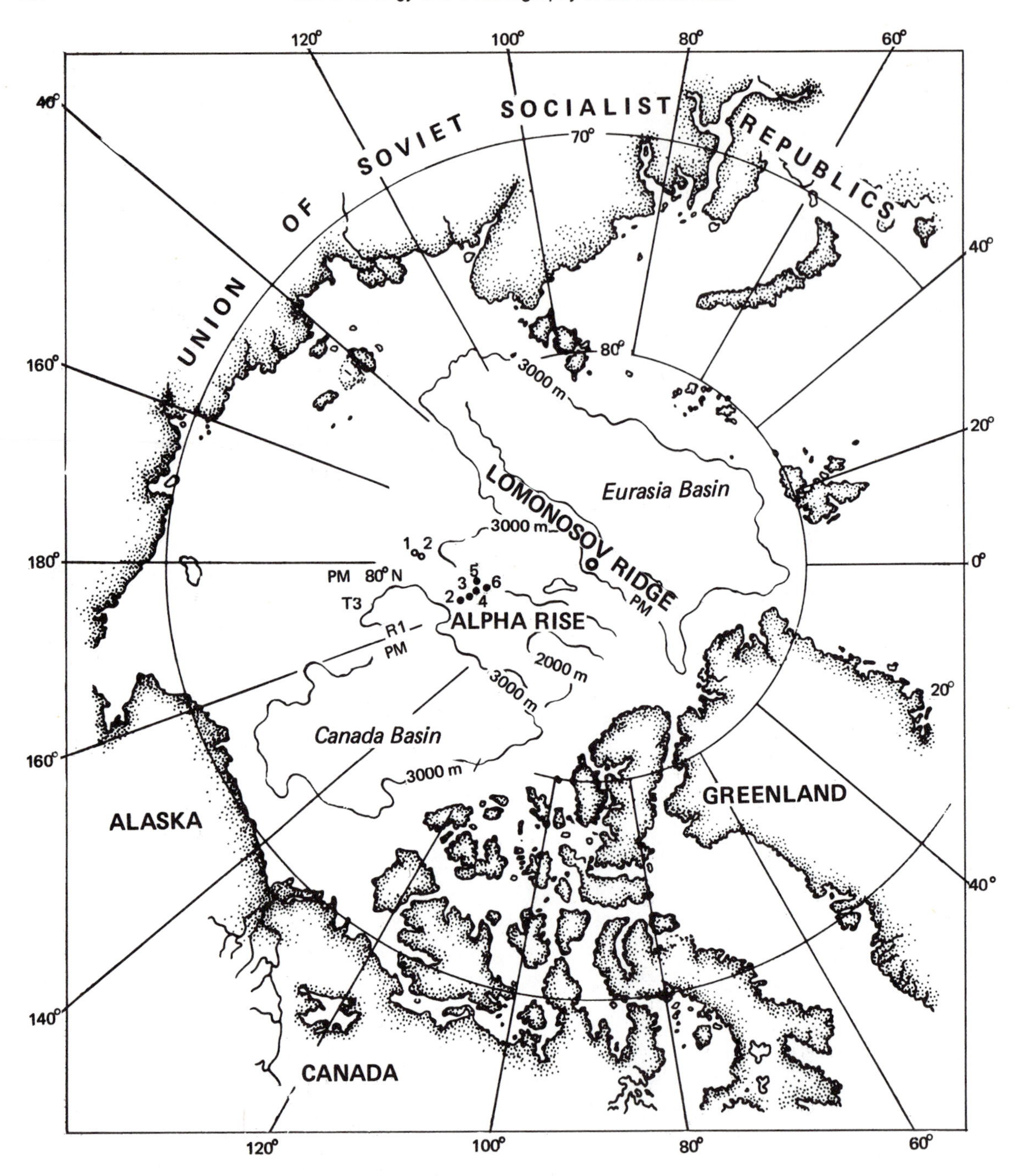

Fig. 1. Bathymetric map of the Arctic Ocean based on the Geological Map of the Arctic (1960).
T 3 = Location of cores T 3-67 4, 9, 11, and 12.
PM = Important cores with known paleomagnetic stratigraphy.
RI = Radioisotopically dated core.
Dark and light circles show the location of important previously studied cores (Herman, 1964, 1969, 1970a). For a complete list of core locations see Table 1.

ography and processes of sedimentation and have led to modifications of traditional concepts concerning the late Cenozoic Ice Age. Since major efforts were directed toward the study of readily accessible marine basins, the investigation of the Arctic Ocean was largely neglected. In recent years progress in polar geology has been achieved through studies of sediments from the Central

Arctic Basin (e.g. Green, 1960; Herman, 1964–1971; Ericson et al., 1964).

The main objectives of this chapter are to review data on lithology, fauna, and oxygen isotope measurements on foraminiferal tests and to present new results based on the author's research. Furthermore, an attempt has been made, however tentative, to reconstruct the late Cenozoic climatic and oceanographic history of the North Polar Basin and to compare the amplitude, timing, and the duration of Arctic temperature fluctuations with those from other marine regions.

Arctic Environment

General Characteristics

The polar ice-cap is the most distinctive feature of the Arctic Ocean determining biological processes and hydrologic conditions. The central part of the basin is covered all year round by sea ice, which attains thicknesses of 3 to 4 m at the end of winter, decreasing to about 2 to 3 m in summer (Charlesworth, 1957). The most pronounced variations in ice conditions occur over the broad continental shelves, particularly near land areas, where the ice melts almost completely in summer (Charlesworth, 1957).

Hydrology

Inasmuch as a detailed discussion of water masses and circulation patterns is presented elsewhere in this volume (Coachman and Aagaard), only the major hydrologic characteristics are summarized in the following paragraph:

Three main water masses can be distinguished in the basin: (1) A surface water with low temperatures (−1.6 to 0°C) and low salinites (<29 to ~34‰) formed in the Arctic basin itself and mixed with Pacific and river water (Coachman and Aagaard, this volume). In this layer plankton of Pacific origin is found almost to the North Pole (Treshnikov, 1959); (2) Between 200 and 900 m the intermediate water of Atlantic origin is found; this layer is characterized by relatively high temperature, salinity, and oxygen. When it enters the Arctic Ocean between Spitzbergen and Greenland, this water has a maximum temperature of 4°C and salinity slightly higher than 35‰. As it spreads out in the basin and mixes with polar water, both temperature and salinity decrease. However, the temperature remains at least 0.8°C higher than that of the surface and bottom layers (Coachman and Barnes, 1962); (3) The Arctic deep-water layer formed in the Norwegian Sea and mixed with Arctic water (Nansen, 1902) has temperatures of −0.4°C and salinities ranging between 34.90 and 34.94‰ in the Pacific sector, and slightly lower temperatures in the Atlantic sector (Treshnikov, 1959; Coachman and Aagaard, this volume).

Biological Production

In the central basin phyto- and zooplankton production is limited by the perennial sea ice and the long lightless winters. With the late July–August ice-melting, leads and cracks form and phytoplankton development occurs in the sunlit surface water (Zenkevitch, 1963). Inorganic nutrients which are vital for plant production are replenished at the surface by upward mixing of the nutrient-rich deep water. However, vertical mixing is probably reduced in summer, when increased temperatures coupled with decreased salinities lower surface water density and produce horizontal stratification. The most abundant phytoplankton groups are the diatoms, followed by peridineans, flagellates, silicoflagellates, and green algae (Zenkevitch, 1963).

According to Zenkevitch (1963), the zooplankton presently populating the Arctic consists of endemic forms which survived the Ice Age by adaptating to the extreme environment, of thermophilic Atlantic species carried by the relatively warm Atlantic current, and lastly of Pacific elements.

Material and Methods

Piston, gravity, and multiple-barrel cores raised from the Alpha Rise, Northwind Escarpment, Canada Basin, and Wrangel Plain were examined megascopically for details of texture and structure; visual inspection of cores is of particular value for detecting anomalous layers such as slump, turbidite, and ice-rafted deposits as well as "flow-in" sections (Plates 1 and 2). The locations, water depths, and core lengths are given in Table 1 and Fig. 1.

Equal volume samples weighing 8 to 14 g were taken at 10- to-50-cm intervals or in places

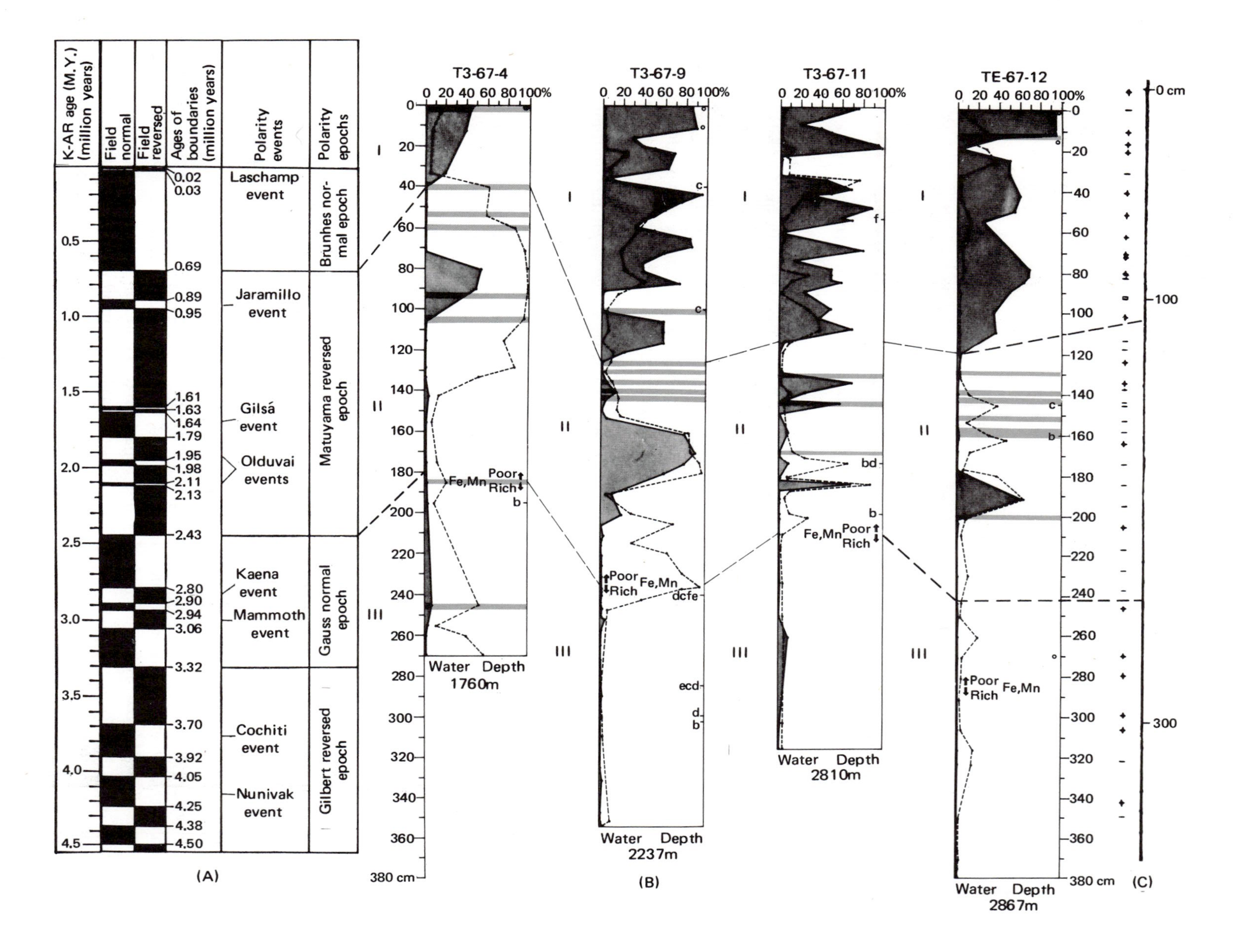
K-AR age (M.Y.) (million years)
Field normal
Field reversed
Ages of boundaries (million years)
Polarity events
Polarity epochs
Laschamp event
Jaramillo event
Gilsá event
Olduvai events
Kaena event
Mammoth event
Cochiti event
Nunivak event
Brunhes normal epoch
Matuyama reversed epoch
Gauss normal epoch
Gilbert reversed epoch
0.02
0.03
0.69
0.89
0.95
1.61
1.63
1.64
1.79
1.95
1.98
2.11
2.13
2.43
2.80
2.90
2.94
3.06
3.32
3.70
3.92
4.05
4.25
4.38
4.50
(A)
T3-67-4
Water Depth 1760m
T3-67-9
Water Depth 2237m
T3-67-11
Water Depth 2810m
TE-67-12
Water Depth 2867m
Fe,Mn
Poor
Rich
I
II
III
380 cm
(B)
0 cm
100
300
(C)

Table 1 Locations, Depths, and Lengths of Cores

Core	Latitude	Longitude	Depth (m)	Length (cm)
D.st.A.2	83°52′N	168°12′W	1,521	206
D.st.A.3	84°12′N	168°33′W	2,409	95
D.st.A.4	84°21′N	168°49′W	2,041	116
D.st.A.5	84°28′N	169°04′W	1,934	125
D.st.A.6	85°15′N	167°54′W	1,842	88
Arlis II,1	81°57′N	168°07′E	2,816	44
Arlis II,2	82°00′N	168°04′E	2,864	80
T3-63-1/S	82°56.2′N	155°54′W	3,548	18
T3-63-3/SA	82°57.8′N	162°46′W	3,400	24.5
T3-63-4/SA	82°20.7′N	161°49′W	3,795	11
T3-63-4/SB	82°20.7′N	161°49′W	3,795	13
T3-63-4/SC	82°20.7′N	161°49′W	3,795	13.5
T3-63-4/SD	82°20.7′N	161°49′W	3,795	14.5
T3-66-S-8	75°41.4′N	158°00′W	864	108
T3-66-S-28	75°42.5′N	161°01′W	2,097	243
T3-66-S-30	75°46.3′N	162°02′W	2,097	287
T3-66-S-31	75°47.4′N	162°20′W	1,741	279
T3-66-S-32	75°47′N	162°21′W	2,059	288
T3-66-S-33	75°50.4′N	162°59′W	2,047	225
T3-67-3	79°11′N	175°09′W	2,285	380
T3-67-4	79°22.7′N	174°46′W	1,760	272
T3-67-9	79°37.9′N	172°07′W	2,237	356
T3-67-11	79°34.9′N	172°30′W	2,810	250
T3-67-12	80°21.9′N	173°33′W	2,867	374

where lithologic changes were observed. However, the cores were sampled at closer intervals in sections of particular interest, and continuous sequences of samples were taken in the D.st.A. cores. Generally the dried and weighed material was washed through a 250-mesh sieve which retains particles >62 μ. Several D.st.A. cores originally wet-sieved through a 74-μ sieve were rewashed through 38-μ and 53-μ screens. After drying at 100°C the coarse fraction was weighed for a second time and recorded as a percentage of the total. Curves constructed from these percentages provide valuable evidence as to processes of sedimentation and yield means for cross-correlation between cores. For faunal analyses the coarse fraction was split into subsamples and counts totaling 500 to 800 specimens were made; the entire population within a sample was counted when total specimens numbered less than 500. The coiling direction of planktonic and benthonic foraminifera was recorded. Tests were separated from clastics in samples having low concentration of shells by flotation on carbon tetrachloride and stained using the method described by Herman and Metz (1972). When consecutive samples from the same faunal zone were examined, counts were made at 10 to 40 cm intervals only and in the intervening samples frequencies were estimated according to the following scale: Rare (R), 1 to 5 tests; frequent (F), 6 to 11 tests; common (C), 12 to 25 tests; abundant (A), 26 to 100 tests; very abundant (VA), >100 tests.

It was found that each sample is characterized by (1) the co-occurrence of a number of species and the relative abundance of one species to the rest of the population; (2) the ratio of two species whose abundance is negatively correlated (with different ecologic requirements); (3) the absolute number of species.

Climatic Units

Climatic units are used here informally to denote important environmentally controlled depositional sequences. Accordingly, based on faunal characteristics and lithology three major units were recognized.

The oldest, unit III (Fig. 2), represented by sediments deposited earlier than 2.4 m.y.b.p.,

Fig. 2.
(A) Paleomagnetic time scale after Cox (1969).
(B) Ratio of microfauna to clastics in the coarse fraction of cores T 3-67-4, 9, 11, and 12.
Shaded areas represent the microfauna, white areas the clastics.
Occurrence of low latitude foraminifera:
a = *Globorotalia menardii—tumida*
b = *Globorotalia scitula*
c = *Globorotalia* sp.
d = *Globigerinoides* sp.
e = *Globigerinoides ruber*
f = *Sphaeroidinella dehiscens*
Circles: Pteropods
Dark horizontal lines are zones in which benthonic foraminifera are in excess of 10%.
Dashed line represents the percentage of *G. quinqueloba* complex (including *G. exumbilicata*) out of the total planktonic foraminiferal population.
(C) Magnetic polarity changes in core T 3-67-12, from Opdyke (in Hunkins et al., 1971a). Modified after Herman (1970a).

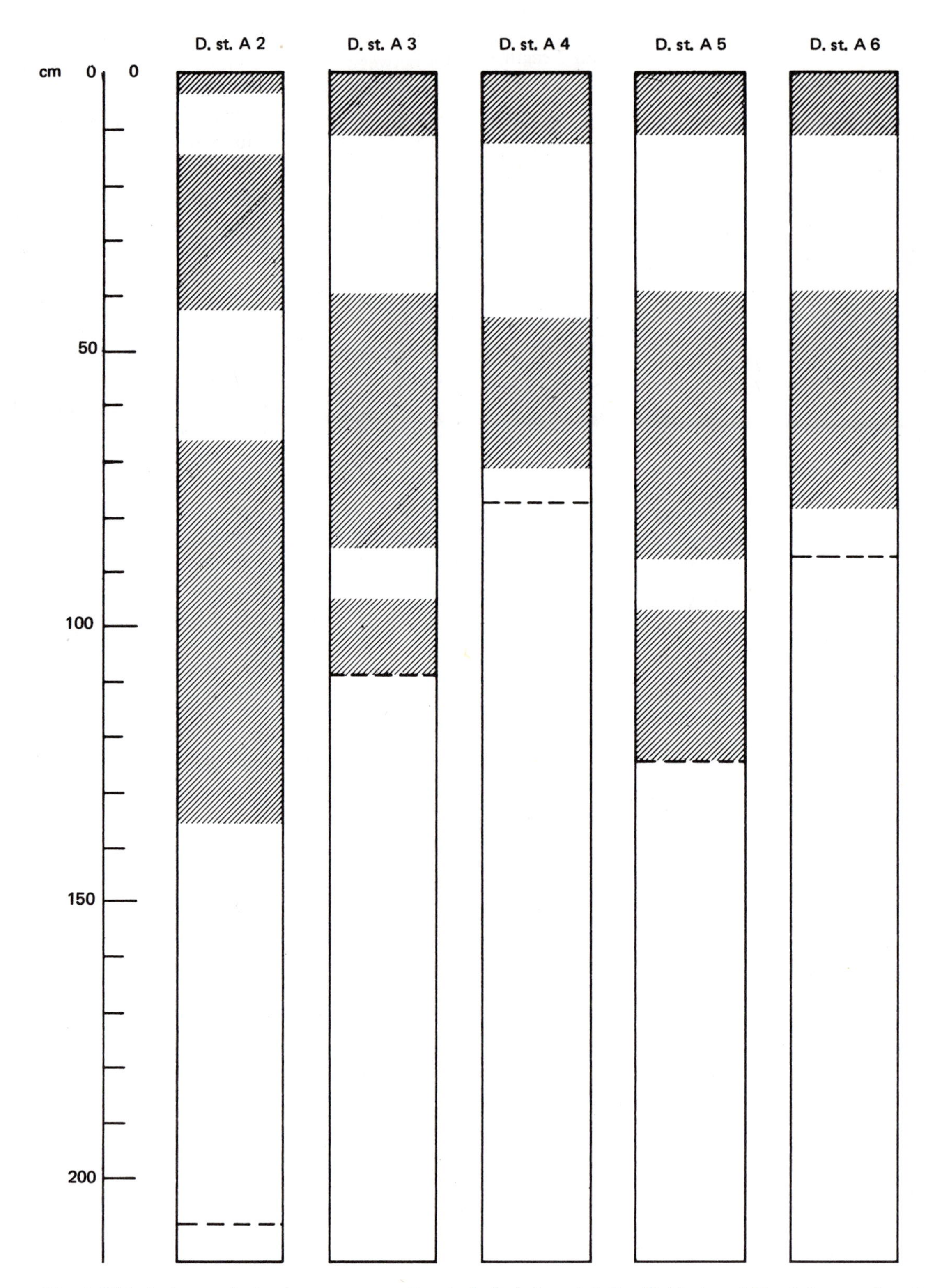

Fig. 3. Lithologic correlation between cores D.st.A. 2, 3, 4, 5, and 6. See Table 1 and Fig. 1 for core locations. The shaded zones indicate foraminifera-rich beds; the white zones indicate foraminifera-poor beds. Modified after Herman (1964).

probably during the Gauss epoch, consists of Mn and Fe oxide rich, foraminifera-poor silty-lutites. Planktonic foraminifera exhibit evidence of partial solution. Embryonic specimens constitute generally 1 to 2% of the fauna in this unit but are much more abundant in units II and I, and arenaceous foraminifera dominate the benthonic assemblages.

Sediments of unit II, deposited between about 2.4 and 0.7 m.y.b.p. are poor in both Fe and Mn oxides and in foraminifera but contain one foraminifera-rich layer (Fig. 2). Coarse ice-rafted detritus, including pebble-size rock fragments, accompanied by shallow-water *Elphidium* spp. and occasionally by high percentages of calcareous deep-water benthonic foraminifera occur sporadically and are concentrated in zones dominated or followed by *Globigerina quinqueloba* and *Globigerina exumbilicata*.

The sediments of unit I were laid down during the last 0.7 m.y. in the Brunhes epoch, a time of conspicuous climatic fluctuations, as indicated by temporal variations in the faunal composition and in the fauna/mineral ratio. Foraminifera-rich and foraminifera-poor layers alternate in these sediments (Herman, 1969, 1971a; Fig. 2).

Sediments

General

Processes of sedimentation in the Arctic Ocean differ from those in other marine regions because transport of terrigenous and shallow-water sediments by surface currents and winds is inhibited during the greater part of the year by the polar ice cap. In summer, when the sea-ice breaks up, the debris-laden ice islands, icebergs and ice-floes drift across the Arctic under the influence of prevailing currents and winds, gradually disintegrating and releasing to the sea floor the incorporated unsorted detritus. While drift-ice is an important sediment transporting agent, winds and surface currents play only a secondary role in carrying and distributing clasts to the central basin. Faunal productivity is limited, and dissolution of calcareous tests is enhanced by low temperatures and long exposure to the cold corrosive bottom water before burial.

Stratification

Color change is the most obvious physical characteristic determining stratification in Arctic sediments. In cores raised from topographic highs such as the crest and upper flank provinces of the Alpha Rise and the Northwind Escarpment, where rates of sedimentation are low, cyclic layering of brown (5 YR 5/2), yellow-brown (10 YR 6/2) and grayish (5 Y 5/2) silty-clays occurs (Fig. 3; Plates 1 and 2).

Changes in sediment color are generally accompanied by alterations in faunal composition and occasionally by variations in the percentage of coarse fraction (e.g. Figs. 4 through 10).

Sediments deposited during the Brunhes epoch (unit I) consist of (1) brown foraminifera-rich silty-lutites, the coarse fraction of which contains generally 5 to 90% microfaunal remains and is characterized by the high planktonic/benthonic foraminifera ratio, the dominance of solution-resistant limy foraminifera, and the relatively large amounts of Mn and Fe oxides (Saks et al., 1955; Herman, 1969; Bostrom, 1971; Figs. 2 and 4); (2) grayish foraminifera-poor silty-lutites containing less than 5% faunal remains and distinguished by the variable planktonic/benthonic foraminifera ratio which is generally lower than in foraminifera-rich layers, and by the absence or low percentage of Mn and Fe oxides (Saks et al., 1955; Herman, 1964, 1969, 1970a; Bostrom, 1971; Figs. 2 and 4). Sediments laid down in Matuyama time (unit II) are grayish silty-lutites poor in both foraminifera and in Fe or Mn oxides but containing one foraminifera-rich layer. The fauna is dominated by thin-shelled benthonic and planktonic foraminifera (Herman, 1969, 1970a; Fig. 2). The brown silty-lutites deposited during the Gauss epoch (unit III) are rich in Mn and Fe oxides but poor in foraminifera (Herman, 1970a; Fig. 2).

The deep Wrangel and Canada Plain cores (Fig. 1; Table 1) raised from areas where rates of sedimentation are relatively high, penetrate only one foraminifera-rich bed (Herman, 1969).

Rates of Sedimentation

Among the important factors controlling rates of sediment accumulation are topographic setting, supply of biogenic and clastic materials, deep current scour and post-depositional solution of minerals and organic remains. Seismic refractions (Beal, 1969), supplemented by ^{14}C, uranium series isotope dating (Ku and Broecker, 1967), and paleomagnetic determinations (Linkova, 1965;

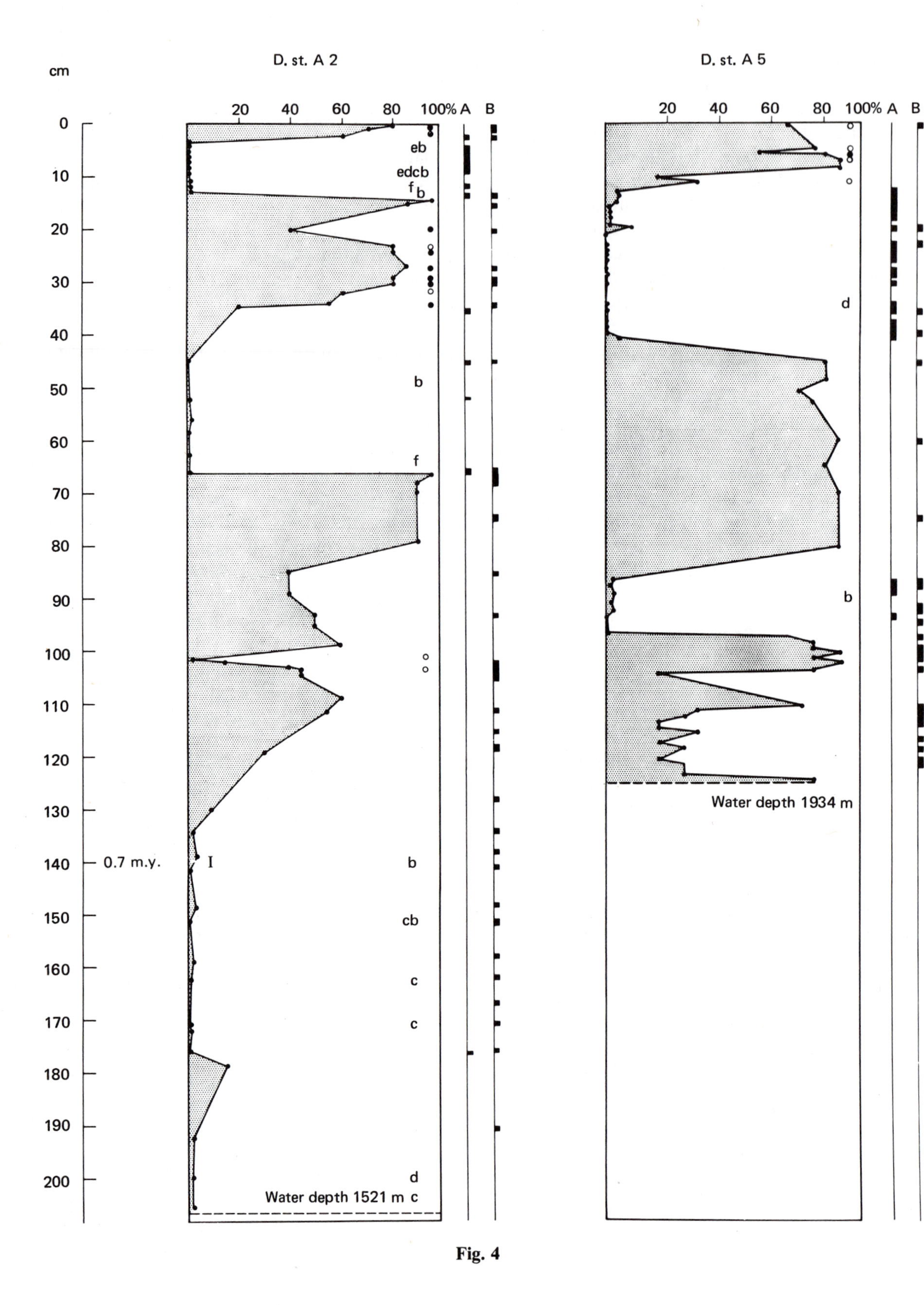
D. st. A 2
D. st. A 5
cm
20 40 60 80 100% A B
0 10 20 30 40 50 60 70 80 90 100 110 120 130 140 150 160 170 180 190 200
0.7 m.y.
I
eb
edcb
f b
b
f
b
cb
c
c
d
c
d
b
Water depth 1521 m
Water depth 1934 m

Fig. 4

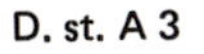

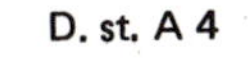

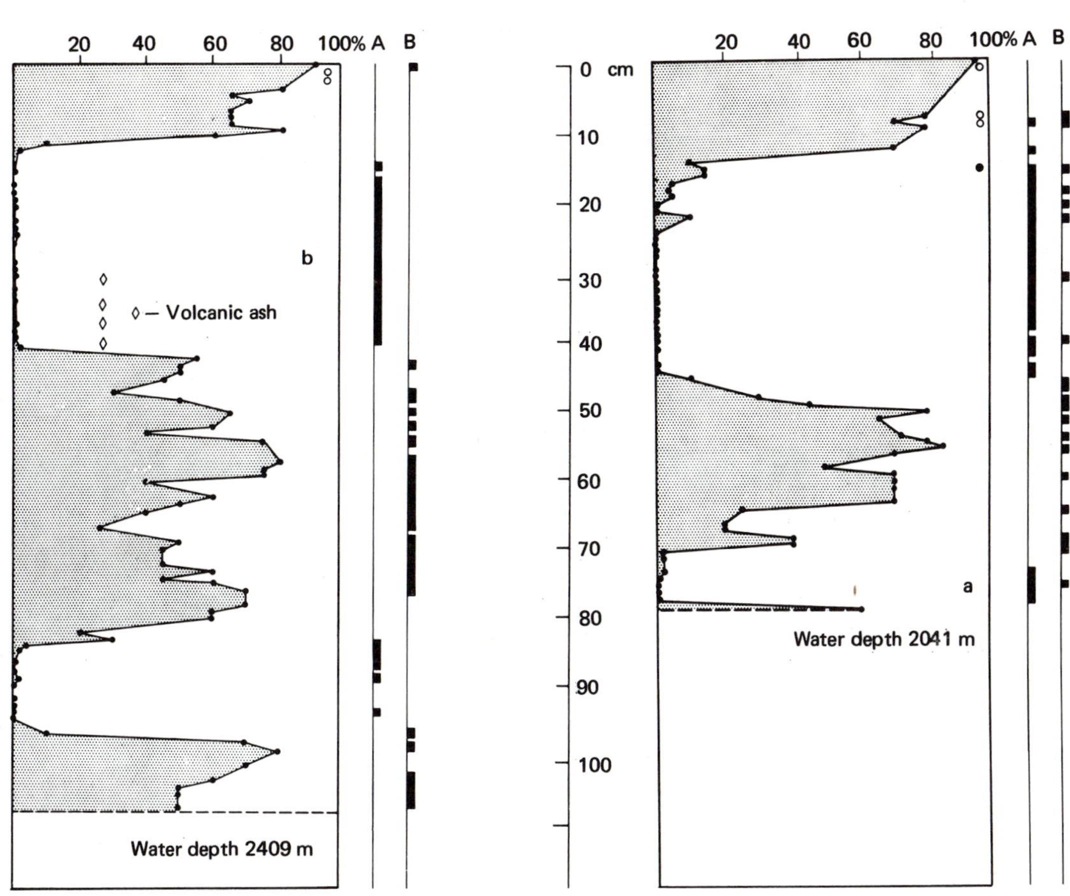

Fig. 4. Ratio of microfauna to clastics in the coarse fraction of cores D.st.A. 2, 3, 4, and 5.
Occurrence of low latitude foraminifera:

a = *Globigerinoides* cf. *ruber*
b = *Globigerinoides* sp.
c = *Globorotalia crassaformis*
d = *Globorotalia inflata*
e = *Globorotalia* sp.
f = *Globigeriniodes* cf. *sacculifer*

Open circles: *Limacina* cf. *helicina* = 1% of the total fauna.
Dark circles: *Limacina* cf. *helicina* = >1 to 5%.
First column to the right of each core indicates occurrence of plant debris.
Second column to the right of each core indicates occurrence of granules and pebbles. Modified after Herman (1969).

Opdyke, in Hunkins et al., 1971a) indicate that sedimentation rates are very low, ranging from 1 to 3 mm/1000 yr on the crest and upper flank provinces of the Alpha Rise and the Lomonosov Ridge but are higher in adjacent plains. Rates of sedimentation higher by one order of magnitude than those mentioned above on the Alpha Rise are suggested by Hunkins and Kutschale (in Demenitskaya and Hunkins, 1970); however, these values should be regarded with caution.

Geophysical data suggest that the thickness of unconsolidated sediments in the Central Arctic Basin ranges from 0 to about 1000 m (Beal, 1969). Sediment thickness is greatest near continents and in plains of intermediate depth, such as the Wrangel Plain (Kutschale, 1966), and thins toward the submarine mountain ranges.

Composition

Six sediment types were recognized in the studied cores: (1) ice-rafted detritus, (2) pelagic sediments settling from the water column and composed principally of tests of planktonic organ-

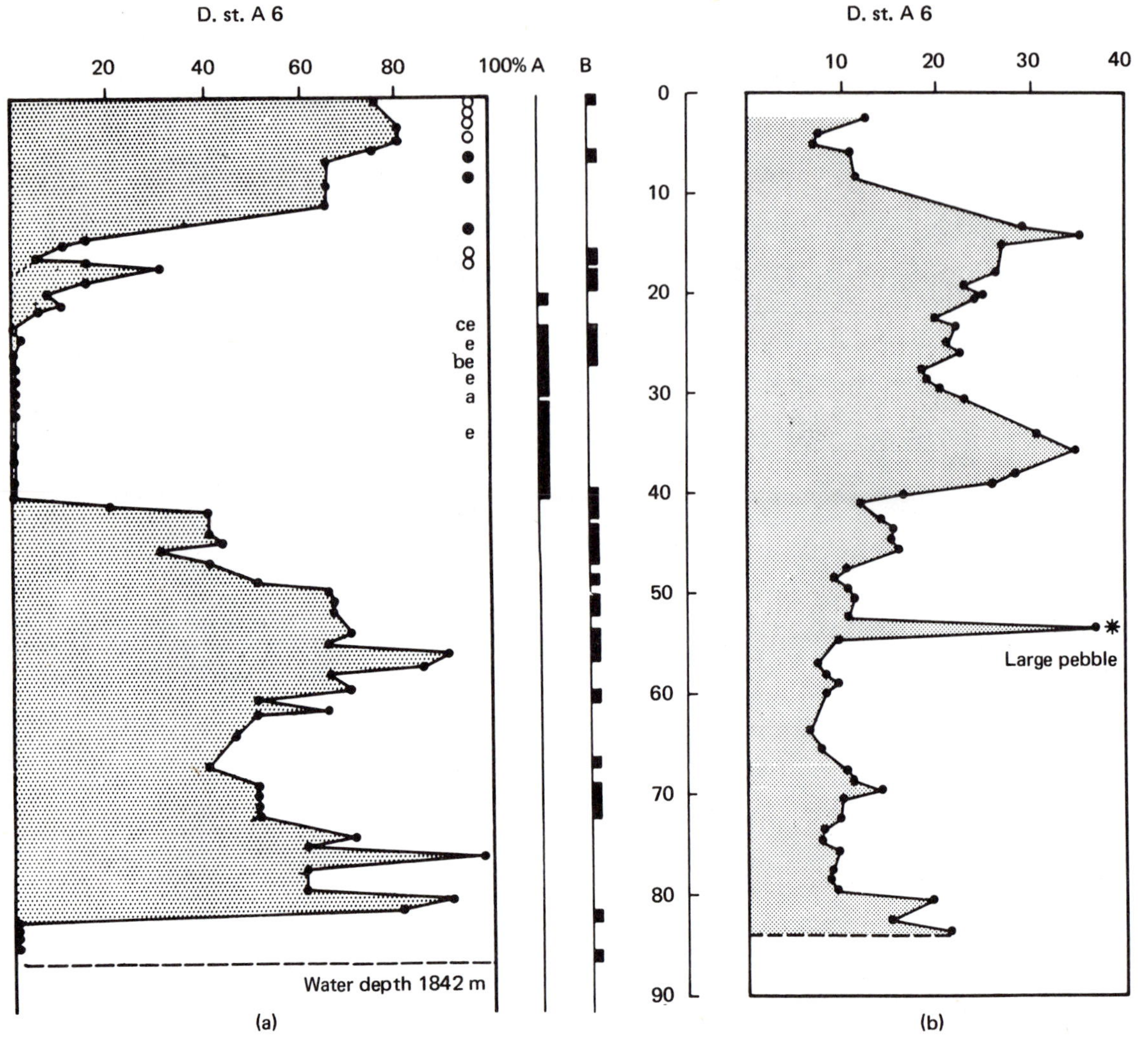

Fig. 5. Core D.st.A. 6.
(a) Ratio of microfauna to clastics. See the legend for Fig. 2. Modified after Herman (1969).
(b) Percentage of coarse fraction (>74 μ).

isms and of detrital minerals, (3) skeletal elements of benthonic organisms, (4) inorganic precipitates, (5) displaced sediments transported by turbidity and other deep currents, and (6) volcanic ash.

Ice-rafted Detritus

Land-derived clasts carried to the sea are deposited near shore and on drift-ice; moreover, by basal freezing both faunal remains and lithic fragments are incorporated in shelf-ice (Charlesworth, 1957; Hattersley-Smith, 1960, 1963; Hattersley-Smith et al., 1955).

As mentioned in a preceding paragraph, in summer when the sea-ice breaks up, the debris-laden ice islands, icebergs, and ice-floes drift across the Arctic, gradually disintegrating and releasing to the sea floor the entrapped unsorted sediments (Charlesworth, 1957; Cromie, 1961). It appears that drift-ice is an important sediment transporting agent carrying both terrigenous rock fragments and skeletal elements of shallow-water benthonic molluscs and foraminifera to the heart of the Arctic Basin (Charlesworth, 1957; Hattersley-Smith, 1960, 1963; Hattersley-Smith et al., 1955; Herman, 1969). Sediments transported from land by ice are poorly sorted rock and mineral fragments, ranging in size from boulder to clay and of variable lithologies including dolomites, metamorphic and igneous rocks (Hattersley-Smith, 1960, 1963). Sand-size particles consist predominantly of angular to subangular quartz (Herman, 1969). The distribution of coarse (granule and pebble size) ice-rafted detritus in sediments is random, as shown in Fig. 4.

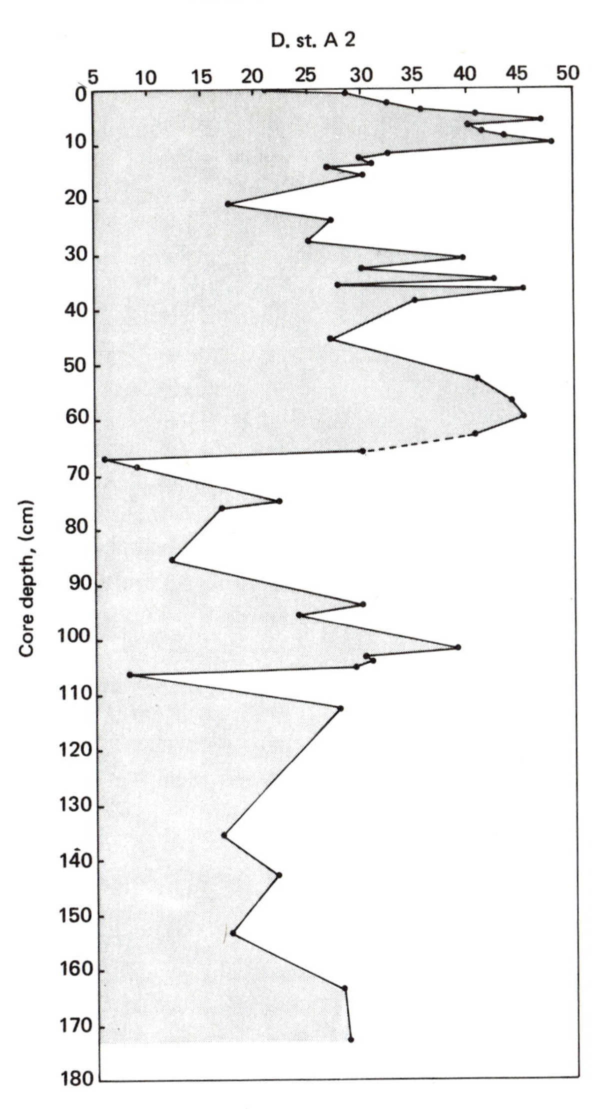

Fig. 6. Core D.st.A. 2.
Variation in the percentage of coarse fraction (>38 μ).

Pelagic Sediments

Biogenic Constituents. Although radiolarians and diatoms are common in the overlying water masses (Zenkevitch, 1963; Herman, unpublished), their skeletons are rarely preserved in sediments. Calcareous tests of planktonic foraminifera constitute the bulk of faunal remains in most late Cenozoic deposits and are accompanied by pteropods; the latter are preserved in the uppermost 10 to 15 cm of cores raised from depths <2500 m (Figs. 2 and 4).

Detrital Minerals. Clay-size particles have a long residence time in water and are carried in suspension over great distances before settling to the sea floor (Griffin et al., 1968). It is likely that these minerals have been transported continuously to the Arctic from lower latitudes by subsurface currents, and during pack-ice free periods by surface currents also.

Knowledge of the distribution and composition of clay-size minerals in the Arctic and subarctic seas is due to the work of Berry and Johns (1966), Carroll (1970), Naidu et al., (1971), and Naidu (this volume). According to Carroll (1970), the average mineralogic composition of the clay fraction in cores raised in the vicinity of those discussed in this chapter is about 60% mica, 20% chlorite, 10% mixed-layer mica and vermiculite, 10% non-clay minerals (quartz, 1 to 5%; feldspar, <1%; and dolomite, >5%). In some samples amorphous clay is present; dolomite is concentrated in the silt fraction. The time interval represented by Carroll's cores exceeds one million years, and the clay mineralogy does not change with core depth.

Benthonic Skeletal Remains

Calcareous foraminifera generally dominate the benthonic assemblages, however, arenaceous tests occur sporadically in the Alpha Rise and the Northwind Escarpment core tops and predominate in the lower sections of these cores, where most calcareous tests appear to have been dissolved. Ostracods, pelecypods, tubeworms, echi-

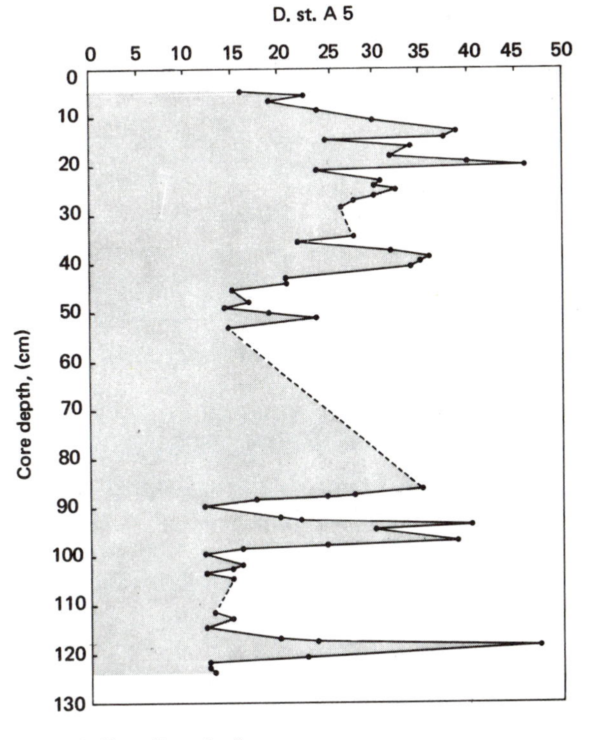

Fig. 7. Core D.st.A. 5.
Percentage of coarse fraction (>53 μ).

noid spines, and sponge spicules are minor components.

Inorganic Precipitates

Manganese and iron oxide mineral aggregates form thin crusts, coating foraminiferal tests and rock fragments in the brown foraminifera-rich zones deposited in Brunhes time (unit I) and in the brown silty-lutites deposited during the Gauss epoch (unit III); however, they are absent or rare in the Brunhes and Matuyama grayish and tan foraminifera-poor sediments. Bostrom (1971), who studied core T 3-67-12, suggests a possible correlation between the Mn oxide-rich horizons and glacial temperature minima.

A thick barite bed associated with Mn- and Fe-bearing oxides was found in an Alpha Rise T3 core; since the core terminates in this bed, its total thickness is unknown. A detailed study of the fauna and mineralogy is underway (Herman, in preparation). The occurrence of barite in deep-sea sediments has been described by Arrhenius (1963) and Degens (1965).

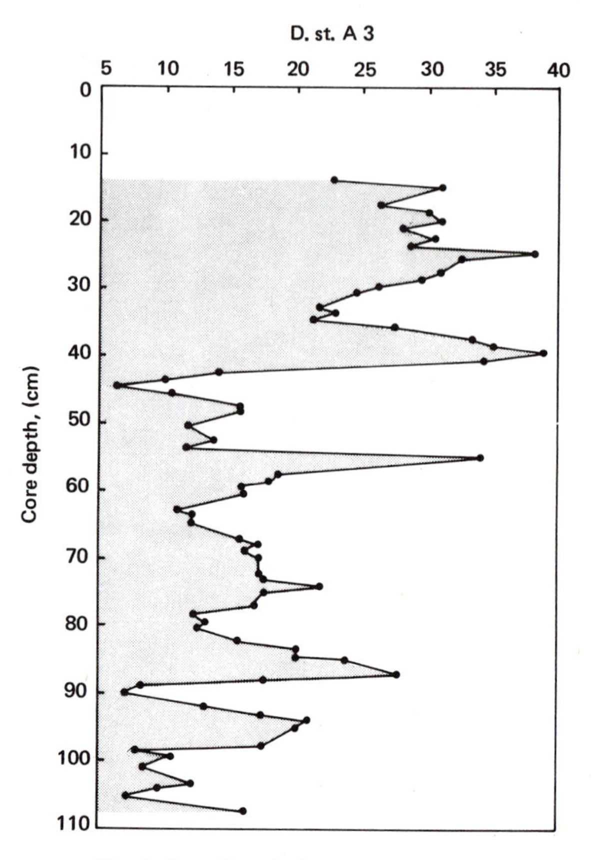

Fig. 9. Core D.st.A. 3.
Percentage of coarse fraction (>53 μ).

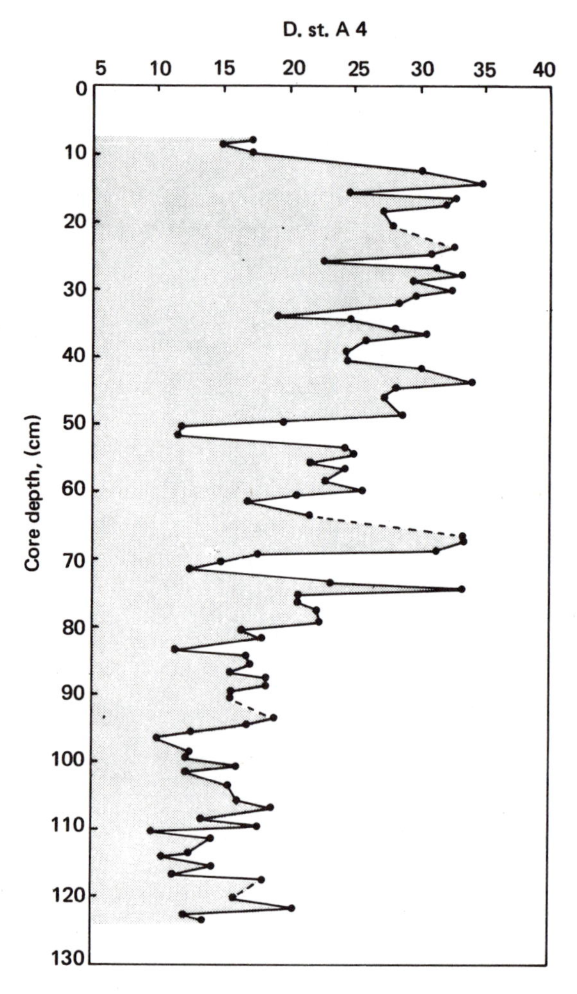

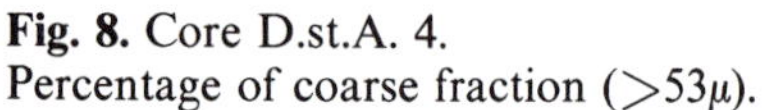

Fig. 8. Core D.st.A. 4.
Percentage of coarse fraction (>53μ).

Displaced Sediments

Deep currents erode and redistribute sediments on the sea floor. In cores, turbidite and slump deposits are generally detectable with the naked eye by their texture and structure. On the other hand stratigraphic hiatuses resulting from removal of sediments by various submarine processes are difficult to ascertain and close-interval intrabasinal biostratigraphic and lithologic correlation between cores are needed to establish such depositional lacunae. An example of missing sediments is provided by core T 3-67-4, where several decimeters from the core "top" are missing (Figs. 2 and 11).

Volcanic Ash

Two cores (D.st.A 3 and 6) contain colorless volcanic glass shards (Herman, 1969; Fig. 4). The origin and age of these shards is unknown; however, in both cores they occur in the upper-

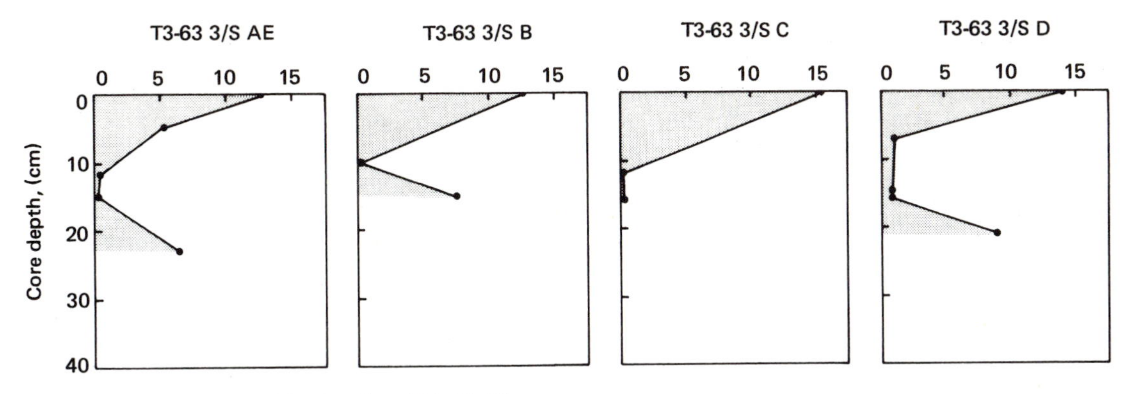

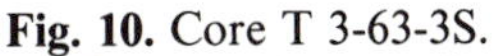

Fig. 10. Core T 3-63-3S.
Percentage of coarse fraction (>62 μ) in a multiple-barrel core.

most foraminifera-poor layer.

Correlation between Cores

Biostratigraphic and lithologic correlations supplemented by radiometric and paleomagnetic determinations have been employed to assess the continuity of sedimentary sequences and the equivalency of stratigraphic units. The cyclic and noncyclic criteria used include: (1) Duration and amplitude of successive normal salinity-low salinity and cold-cool temperature cycles, based on faunal composition and oxygen-isotope measurements. (2) Limited stratigraphic occurrence of *G. quinqueloba, G. exumbilicata,* and of several benthonic arenaceous and calcareous foraminifera. (3) Gross lithologic character of sediments, mainly color (Fig. 3; Plates 1 and 2) and variations in the percentage of coarse fraction (Figs. 5 thru 11). (4) Radiometric age dating. (5) Paleomagnetic measurements; the magnetic reversals were correlated with the dated standard geomagnetic polarity reversal scale of Cox (1969). In addition to the aforementioned criteria knowledge of topographic setting, sedimentation rates and post-depositional history are of considerable value in establishing correlations between various climatic units.

Micropaleontology

General

Distributional patterns of living planktonic foraminifera and pteropods indicate that many species have a limited tolerance to changes in temperature and salinity; food, light, and oxygen are also known to determine their distribution and abundance. Accordingly, variations in planktonic faunal composition in consecutive sediment layers are considered to reflect alterations in production rates as well as changes in climatic and hydrologic conditions at the time and shortly after their burial. In addition to the above-mentioned factors other variables determine the composition of faunal remains in sediments. Important among these are redistribution by currents and burrowers, changes in accumulation rates of detrital sediments and the solution of limy tests.

The chemical composition of shells is also thought to play an important role in determining their preservation on the sea floor. Preliminary electron-probe analyses of *Globigerina pachyderma, G. quinqueloba,* and *G. exumbilicata* indicate that their tests are composed of three discrete layers; the inner and outer layers are Mg-rich whereas the intermediate zone is Ca-rich. (Plate 10, Figs. 1 and 2). Among the analyzed specimens a few lack the outer Mg-rich layer; the absence of Mg coating is believed to be due to its dissolution. The zonation of juvenile chambers is similar to the adult; however, the Mg-content is twice that of the adult. The spines of *G. quinqueloba* are composed of an inner Mg-rich core surrounded by a Ca-rich envelope. The amount of Mg in the spine core is about the same as in the Mg-rich layers of the adult chambers (Herman and Knowles, in preparation). These data suggest that the low percentages of juvenile tests in core sediments raised from deep water and in core levels exhibiting evidence of differential solution is partially due to their higher Mg-content as well as to shell thickness.

Distribution of Planktonic Microfauna in Plankton Tows

Information on the composition of Arctic Ocean fauna is based on two publications (Meisenheimer, 1905; Bé, 1960). These authors report the occurrence of one pteropod species (*Limacina helicina*) and one planktonic foraminifer (*G. pachyderma*), respectively. The author has examined several plankton tows collected from the Beaufort Sea in late August 1971 by P. Barnes. Preliminary results indicate the occurrence of *G. pachyderma* and *G. quinqueloba* in the studied samples.

Distribution of Microfauna in Cores

Planktonic foraminifera and calcareous benthonic foraminiferal tests generally constitute over 90% of the faunal remains except in unit III sediments, where arenaceous foraminifers dominate the preserved benthonic foraminiferal assemblages. Pteropods, ostracods, pelecypods, tubeworms, and echinoid spines are minor components. Occasionally siliceous sponge spicules, radiolarians, and bryozoan fragments were observed.

Planktonic Foraminifera

G. pachyderma, *Globigerina* sp., cf. *G. pachyderma* A, *Globigerina occlusa*, *Globigerina paraobesa*, *G. quinqueloba*, and *G. exumbilicata* are the

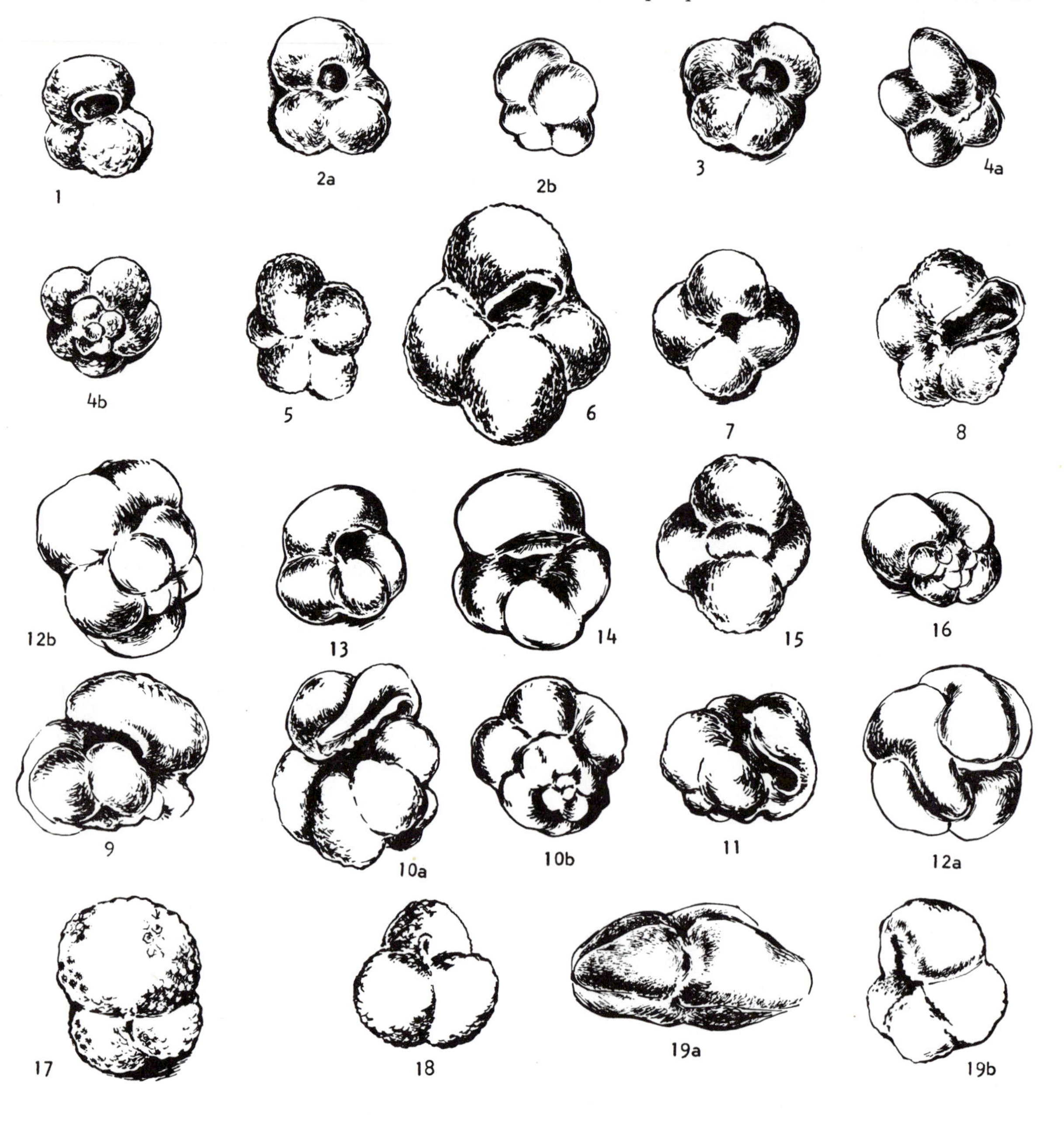

dominant taxa, generally forming 98% of the planktonic foraminiferal population; the coiling direction of these forms is given in Figs. 11 through 13 and in Tables 2 and 3. Low-latitude species occur sporadically and at low frequencies throughout the cores, but are more frequent in the foraminifera-poor zones (Herman 1964, 1969, 1970a; Figs. 2 and 4). A detailed synonymy and taxonomic discussion has not been attempted here; the nomenclature used by Herman (1969) has been followed in most cases.

Globigerina bulloides d'Orbigny
(Plate 3, Figs. 1 through 3)
Globigerina bulloides d'Orbigny, 1826. *Ann. Sci. Nat.* ser. 1,7:277.

Included in the counts of *G. bulloides* may be juvenile specimens of *G. pachyderma.* Forms possessing four spherical chambers in the final whorl and a large umbilical aperture were present sporadically in both foraminifera-rich and foraminifera-poor beds, usually constituting <0.1% of the planktonic foraminiferal population. The coiling direction of this taxon is predominantly sinistral. Typical forms are illustrated in Plate 1, Figs. 1 through 3. The lowest Antarctic surface water temperatures in which this species was found living was −0.48°C (Boltovskoy, 1966).

Globigerina exumbilicata, new species, Herman
Plate 17, Fig. 1; Plate 18, Figs. 1 through 5; Plate 19, Figs. 1 through 5)

Description: The small trochospiral test has a rounded lobate periphery; chambers are spherical, enlarging gradually, arranged in 3 to $3\frac{1}{2}$ whorls; there are generally $4\frac{1}{2}$ to 5 chambers in the last whorl. The final chamber is spherical and possesses a thin but distinct lip. The umbilicus is deep, and is surrounded by umbilical teeth. The aperture of the last chamber is a low arch, umbilical-extraumbilical. Sutures are distinct, narrow, and depressed. The wall is perforate, thin, and spinose; the coiling direction is random. The main difference between this taxon and *G. quinqueloba* is in the shape of the last chamber, which is spherical in the newly named species, and its well-developed umbilicus, higher spire, and larger size. The wall texture of the new taxon also differs from that of *G. quinqueloba,* as illustrated in Plates 13 through 19.

Electron-probe analyses indicate that the test is composed of three discrete layers; an outer and inner Mg-rich layer and an intermediate Ca-rich layer.

Distribution: G. exumbilicata is abundant in the shallow Northwind Escarpment and in Alpha Rise sediments deposited during the Matuyama epoch.

Plate 3

Fig. 1. *Globigerina bulloides,* ventral view; X 55. D.st.A. 6, 59.5 to 60.5 cm.
Fig. 2a. *G. bulloides,* ventral view; X 75. D.st.A. 5, 12.5 to 13.5 cm.
Fig. 2b. *G. bulloides,* side view; X 75. D.st.A. 5, 12.5 to 13.5 cm.
Fig. 3. *G. bulloides,* ventral view; X 75. D.st.A. 6, 59.5 to 60.5 cm.
Fig. 4a. *G. quinqueloba,* ventral view; X 105. D.st.A. 3, 10 cm.
Fig. 4b. *G. quinqueloba,* dorsal view; X 105. D.st.A. 3, 10 cm.
Fig. 5. *G. quinqueloba,* ventral view; X 105. D.st.A. 5, 34.3 to 35.3 cm.
Fig. 6. *G. pachyderma,* ventral view; X 105. D.st.A. 6, 59.5 to 60.5 cm.
Fig. 7. *G. pachyderma,* ventral view; X 90. D.st.A. 6, 59.5 to 60.5 cm.
Fig. 8. *G. paraobesa* n.sp., ventral view; X 53. Note that the last grotesque chamber is broken. D.st.A. 6, 54.1 to 55.6 cm.
Fig. 9. *G. paraobesa* n.sp., ventral view; X 75. D.st.A. 3, 10 cm.
Fig. 10a. *G. paraobesa* n.sp., holotype, ventral view; X 75. D.st.A. 3, 10 cm. U.S.N.M. No. 186535.
Fig. 10b. *G. paraobesa* n.sp., holotype, dorsal view; X 75. D.st.A. 3, 10 cm.
Fig. 11. *G. paraobesa* n.sp., ventral view; X 75. D.st.A. 6, 52.7 to 53.7 cm.
Fig. 12a. *G. paraobesa* n.sp., ventral view; X 83. D.st.A. 3, 8 to 9 cm.
Fig. 12b. *G. paraobesa* n.sp., side view; X 94. D.st.A. 3, 8 to 9 cm.
Fig. 13. *G. occlusa* n.sp., ventral view; X 70. D.st.A. 6, 42.2 to 43.2 cm.
Fig. 14. *G. occlusa* n.sp., ventral view; X 72. D.st.A. 3, 4 to 5 cm.
Fig. 15. *G. occlusa* n.sp., holotype, ventral view; X 75. D.st.A. 6, 58.5 to 59.5 cm. U.S.N.M. No. 186537.
Fig. 16. *G. occlusa* n.sp., dorsal view; X 65. D.st.A. 6, 55 to 56 cm.
Fig. 17. *Globigerinoides* sp., cf. *G. sacculifer,* ventral view; X 75. D.st.A. 2, 65.5 to 66.5 cm.
Fig. 18. *G.* sp., cf. *G. ruber,* ventral view; X 135. D.st.A. 4, 75 to 76 cm.
Fig. 19a. *Globorotalia crassaformis,* side view; X 53. D.st.A. 2, 172 to 173 cm.
Fig. 19b. *G. crassaformis,* ventral view; X 75. D.st.A. 2, 172 to 173 cm.

In earlier publications (Herman 1970a; Herman et al., 1971) and in Figs. 2 and 14 this species was included with the counts of *G. quinqueloba* and was also referred to as *Globigerina* sp., cf. *G. quinqueloba*.

Holotype: (figured) dimension: Diameter 0.26 mm. The holotype specimen is from sample D.st.A. 2 II, 144 cm. U.S.N.M. No. 186540 (Catalog No. 36).

Paratypes: (figured) dimensions: Diameter range 0.18 to 0.33 mm. The paratype specimens are from sample D.st.A. 2 II, 144 cm. U.S.N.M. No. 186539 (Catalog No. 36).

Globigerina occlusa Herman, new species (Plate 3, Figs. 13 through 16; Plate 10, Figs. 3 and 4)

Description: Test trochospiral with subrounded compact periphery; early whorls obscured by the last whorl which is composed of 4 to 4½ subspherical chambers gradually increasing in size; occasionally a diminutive last chamber is present. The aperture is small umbilical-extraumbilical, generally obscured by a wide apertural flap or lip. The wall is thick, perforate; the sutures are radial and narrow. The average length of this species is 0.32 mm and its width is 0.26 mm. The coiling direc-

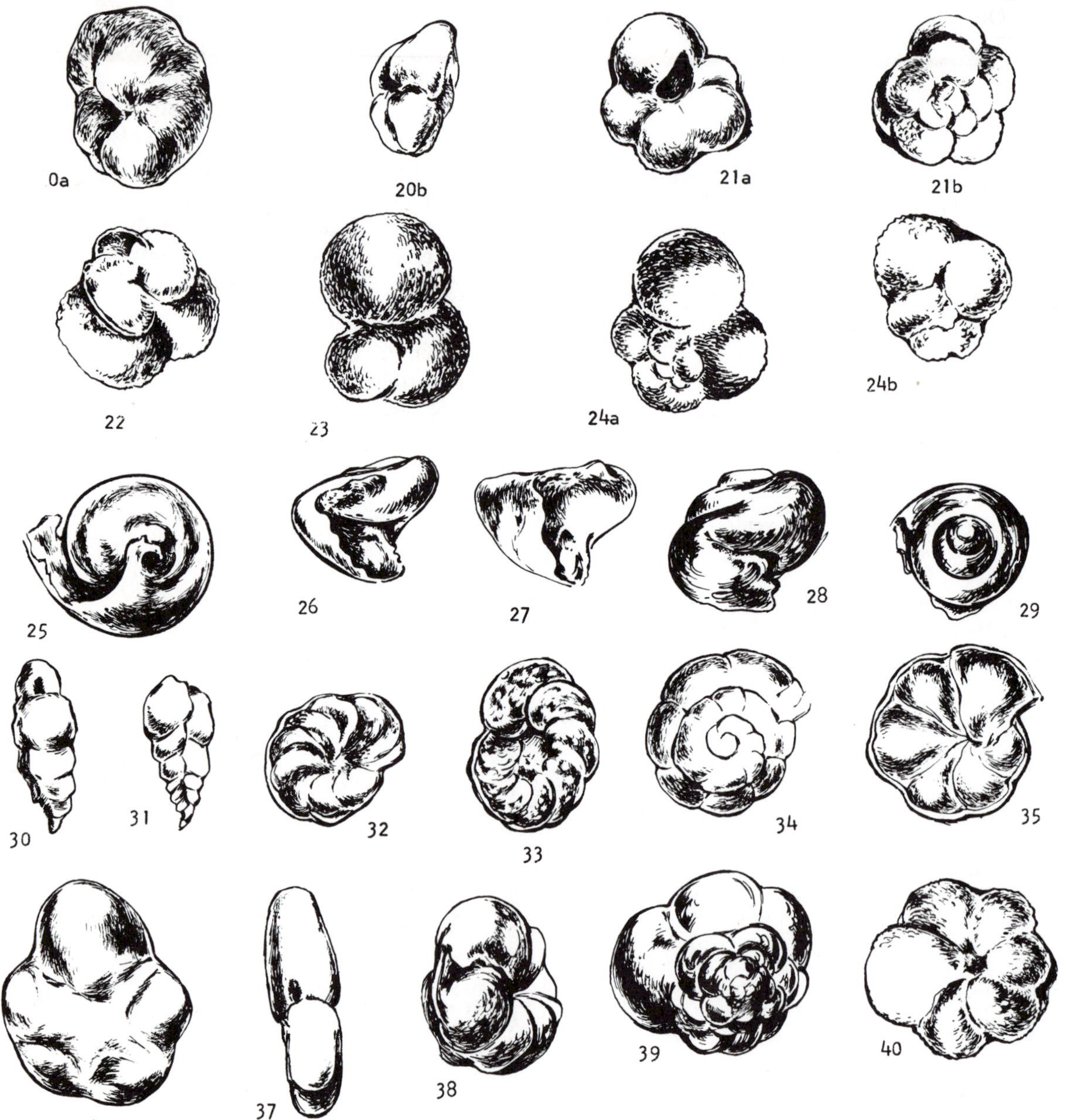

tion of *G. occlusa* is predominantly sinistral in unit I sediments; the percentage of right-coiling specimens is highest in the foraminifera-poor sediments of unit II (Fig. 12). The new species differs from *G. pachyderma* by the elongate last chamber, position of aperture, and its wider lip.

Distribution: G. occlusa is common in late Cenozoic Arctic Ocean sediments (in units I, II, and III; Fig. 14), particularly in the foraminifera-rich zones of unit I.

Holotype (figured): dimensions: Width 0.25 mm; length 0.30 mm. The holotype specimen is from sample D.st.A. 6, 58.5 to 59.5 cm. U.S.N.M. No. 186537 (Catalog No. 36).

Paratypes (unfigured): Length range is 0.21 to 0.30 mm; width range is 0.15 to 0.25 mm. The paratypes are from sample D.st.A. 6, 58.5 to 59.5 cm. U.S.N.M. No. 186538 (Catalog No. 36).

Globigerina pachyderma (Ehrenberg) (Plate 3, Figs. 6 and 7; Plate 5, Fig. 1; Plate 6, Figs. 1 through 6; Plate 7, Figs. 1 through 5; Plate 8, Figs. 1 through 4; Plate 9, Figs. 1 through 5; Plate 10, Figs. 1 and 2)
Aristerospira pachyderma Ehrenberg, 1861, *K. Preuss. Akad. Wiss. Berlin, Monatsber.*, pp. 276, 277, 303.

Table 2 Coiling Direction* of *Globigerina pachyderma* in Unit 3 Sediments

Core T3-67-4:	Average based on counts from twelve levels: 90% sinistral.
Core T3-67-9:	Average based on counts from seven levels: 95% sinistral.
Core T3-67-11:	Average based on counts from six levels: 95% sinistral.
Core T3-67-12:	Average based on counts from eight levels: 96% sinistral.

Total average: 94% sinistral

* Range: 82 to 100% sinistral. Counts include levels with 2 to 5 specimens; thus the range as well as the averages should be considered with caution.

This species constitutes a significant proportion of the planktonic population. Only typical forms characterized by their low trochospired test with 4 to 4½ chambers in the last whorl and the narrow low aperture usually possessing a lip were included in the counts. The coiling direction of this taxon is predominantly sinistral in unit III and I sediments (Table 2); the percentage of right-coiling forms is highest in the foraminifera-poor sediments of unit II (Herman, 1969; Figs. 11 and 12). The test is relatively thick and a calcitic cortex obscures the pores in specimens from foraminifera-rich zones. However, the shell is thinner and the pores are visible in individuals from foraminifera-poor zones (Plates 5 through 10). Electron-

Plate 4

Fig. 20a. *G. crassaformis,* ventral view; X 56. D.st.A. 2, 206 to 207 cm.
Fig. 20b. *G. crassaformis,* side view; X 41. D.st.A. 2, 206 to 207 cm.
Fig. 21a. *Globoquadrina dutertrei,* ventral view; X 53. D.st.A. 6, 51.7 to 52.7 cm.
Fig. 21b. *G. dutertrei,* dorsal view; X 53. D.st.A. 6, 51.7 to 52.7 cm.
Fig. 22. *Globigerinita glutinata,* ventral view; X 120. Note that the bulla has been broken. D.st.A. 5, 37.3 to 38.3 cm.
Fig. 23. *G. glutinata,* ventral view; X 84. D.st.A. 3, 2 to 3 cm.
Fig. 24a. *G. glutinata,* dorsal view; X 105. D.st.A. 6, 29.2 to 30.2 cm.
Fig. 24b. *G. glutinata,* ventral view; X 105. D.st.A. 6, 29.2 to 30.2 cm.
Fig. 25. *Limacina helicina,* ventral view; X 40. D.st.A. 2, 34 to 35 cm.
Fig. 26. *L. helicina,* apertural view; X 45. D.st.A. 2, 0.5 to 1.5 cm.
Fig. 27. *L. helicina,* apertural view; X 45. D.st.A. 2, 0.5 to 1.5 cm.
Fig. 28. *L. helicina,* side view; X 60. D.st.A. 2, 0.5 to 1.5 cm.
Fig. 29. *L. helicina,* apical view; X 60. D.st.A. 2, 0.5 to 1.5 cm.
Fig. 30. *Bolivina arctica,* side view; X 109. D.st.A. 5, 104 to 104.8 cm.
Fig. 31. *B. arctica,* side view; X 102. D.st.A. 5, 104 to 104.8 cm.
Fig. 32. *Cibicides wüellerstorfi,* ventral view; X 50. D.st.A. 2, 0.5 to 1.5 cm.
Fig. 33. *C. wüellerstorfi,* dorsal view; X 47. D.st.A. 2, 0.5 to 1.5 cm.
Fig. 34. *Pseudoeponides umbonatus,* dorsal view; X 75. D.st.A. 6, 5.5 to 6.5 cm.
Fig. 35. *P. umbonatus,* ventral view; X 75. D.st.A. 5, 5 to 6 cm.
Fig. 36. *Stetsonia horvathi,* X 270. D.st.A. 3, 91.6 to 92.6 cm.
Fig. 37. *S. horvathi,* side view; X 270. D.st.A. 3, 91.6 to 92.6 cm.
Fig. 38. *Eponides tumidulus horvathi,* edge view; X 161. D.st.A. 6, 5.5 to 6.5 cm.
Fig. 39. *E. tumidulus horvathi,* dorsal view; X 180. D.st.A. 5, 5 to 6 cm.
Fig. 40. *E. tumidulus horvathi,* ventral view; X 139. D.st.A. 6, 5.5 to 6.5 cm.

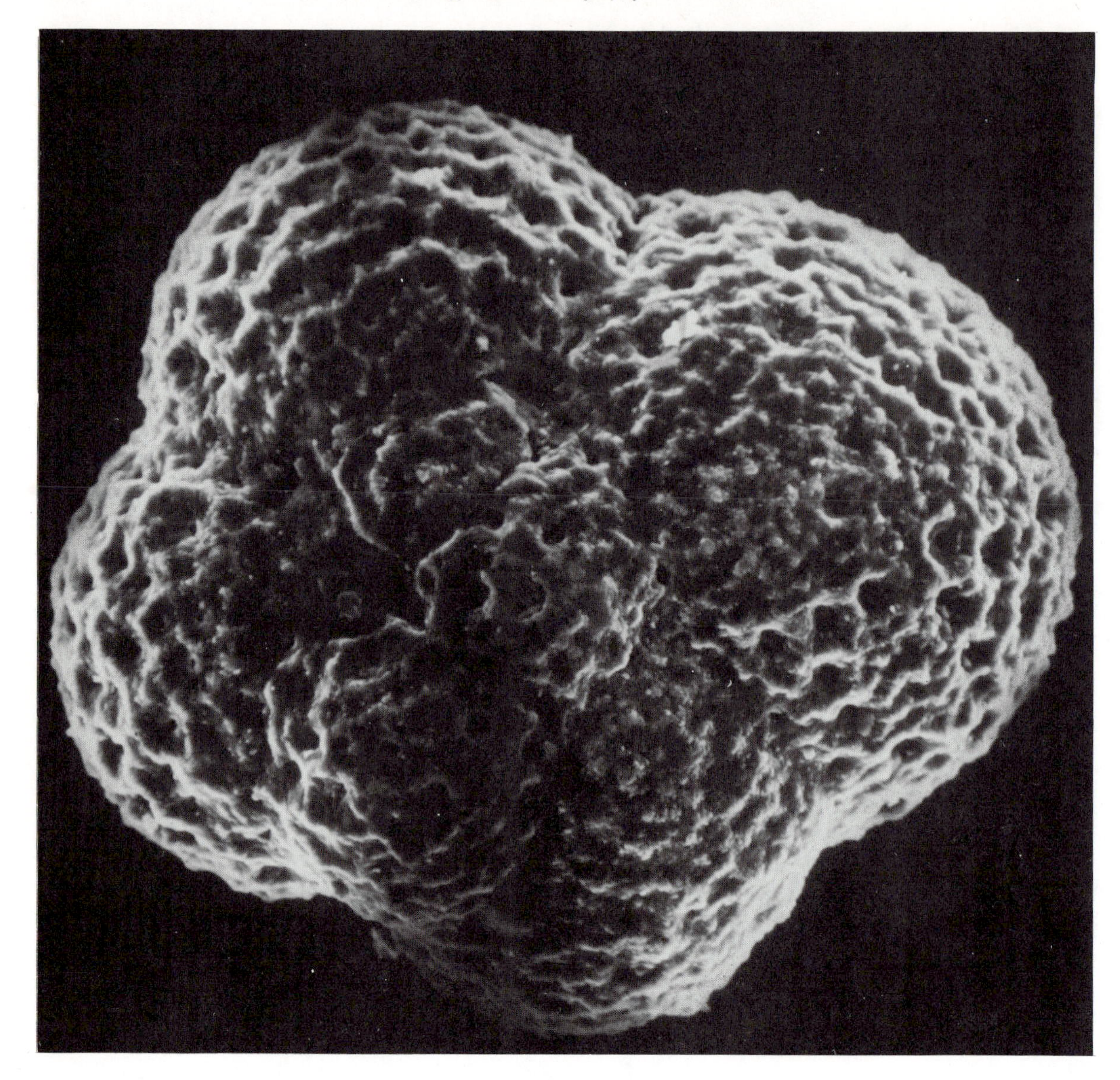

Plate 5
(SEM Illustration)

Globigerina pachyderma, dorsal view; about X 600. This specimen is from a foraminifera-poor zone. Note presence of pores and interpore ridges. T 3-67-9, 220 cm.

Plate 6
(SEM Illustrations)

Fig. 1. *Globigerina pachyderma,* ventral view; X 361. This specimen is from a foraminifera-poor zone; note the presence of large pores. T 3-67-9, 220 cm.

Fig. 2. *G. pachyderma,* ventral view; X 510. This specimen is from a foraminifera-rich zone. The test is covered with a thick calcitic cortex; note the distinct calcite crystals. The pores are not visible in this specimen. T 3-67-11, top.

Fig. 3. *G. pachyderma,* dorsal view; X 272. Note the presence of pores. T 3-67-12, 192 cm.

Fig. 4. *G. pachyderma,* dorsal view; about X 255. T 3-67-9, 220 cm.

Fig. 5. *G. pachyderma,* dorsal view; X 340. T 3-66-S.8, 8 to 10 cm.

Fig. 6. *G. pachyderma,* part of the last and penultimate whorls have been broken to show juvenile portion of test, wall thickness, and internal wall surface and pores; X 298. T 3-66, S.8, 5 cm.

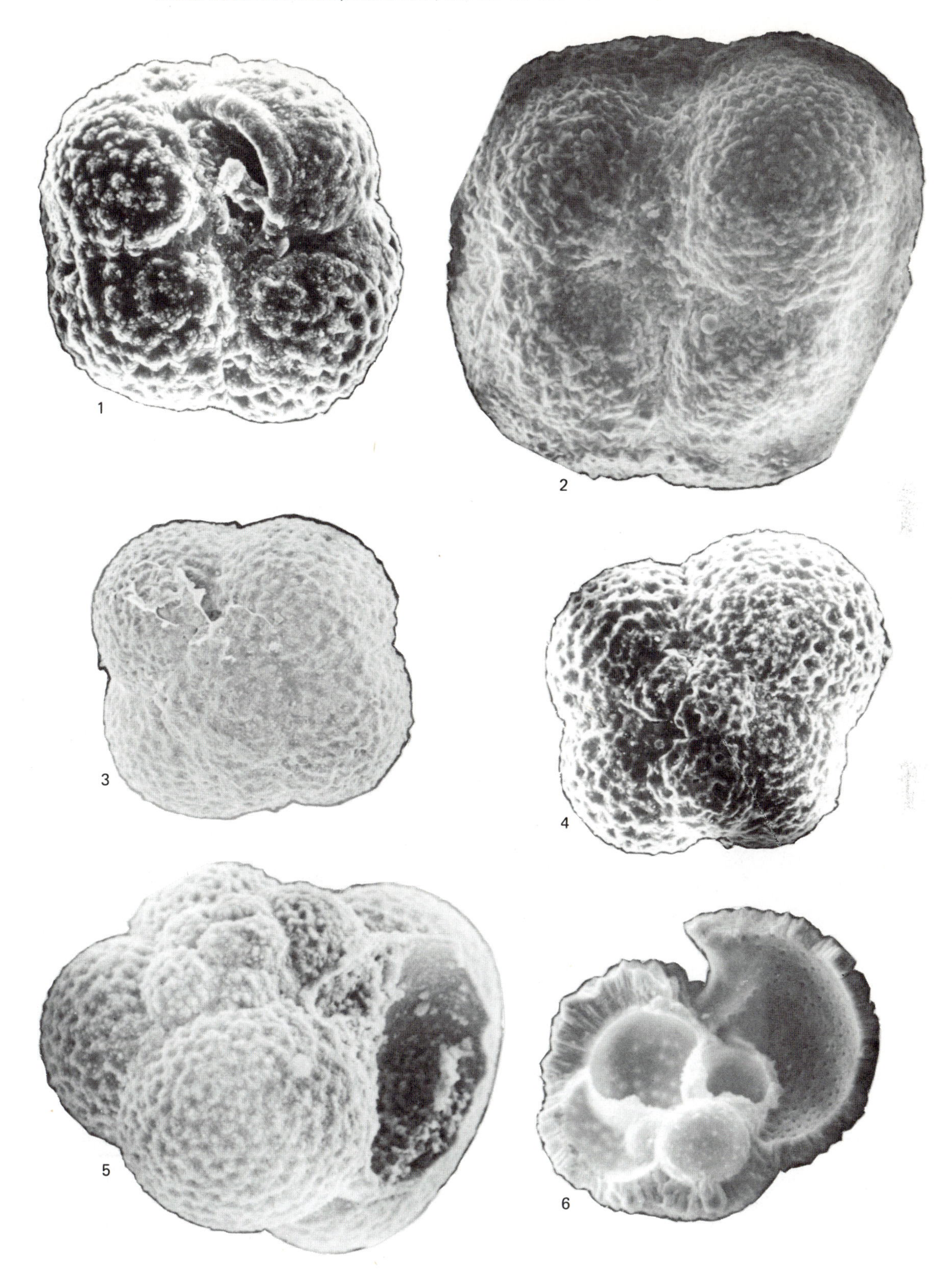
1
2
3
4
5
6

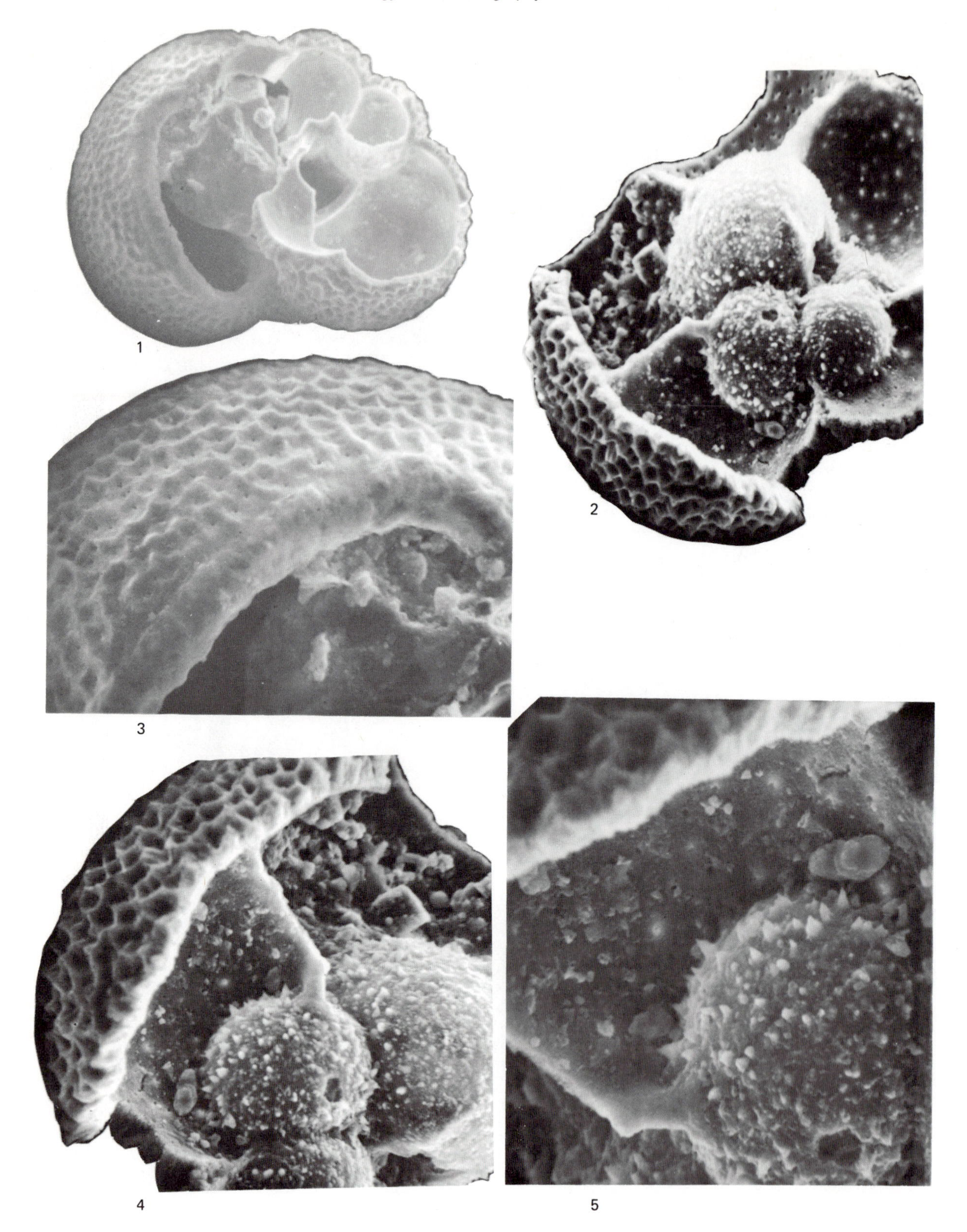
1
2
3
4
5

probe analyses show that the test is composed of three discrete zones; an outer and inner Mg-rich and an intermediate Ca-rich layer (Plate 10, Figs. 1 and 2). The lowest Antarctic surface water temperatures in which this species was found living was −1.38°C (Boltovskoy, 1966).

Globigerina sp., cf. *G. pachyderma* A is a variant of *G. pachyderma* possessing 5 to 5½ chambers in the final whorl. The relative abundance and the coiling direction of this form were calculated and are presented in Figs. 13 and 14. Counts of this taxon may include *Globoquadrina dutertrei.*

Globigerina paraobesa, Herman, new species
(Plate 3, Figs. 8 through 12; Plate 11, Figs. 1 through 4; Plate 12, Figs. 1 through 4)

Description: The fairly large and robust test (0.40 to 0.50 mm average diameter) consists of five chambers per whorl, with 5 to 6 chambers in the last whorl visible ventrally, arranged in 2½ to 3 whorls, in a high loosely coiled trochospire. The chambers are globular; a modified larger, grotesque last chamber is generally present. The aperture possesses a well-developed lip extending from umbilicus to periphery. The wall is thick, perforate, and the sutures are radial and narrow.

The new species resembles *G. pachyderma* but can be distinguished from the latter by its larger size, number of chambers, outline of test, which is circular in *G. paraobesa* and subquadrate in *G. pachyderma,* and its modified larger last chamber. The coiling direction of this taxon is predominantly sinistral (average: 98% sinistral, based on 19 sample counts in core T 3-67-3).

Distribution: G. paraobesa is present throughout the studied Arctic cores in units I, II, and III sediments at low frequencies, constituting 0.1 to 1.5% of the planktonic foraminiferal population.

Holotype: Diameter 0.50 mm (Plate 3, Figs. 10a and 10b); the holotype specimen is from sample D.st.A. 3, 10 cm (Pleistocene). U.S.N.M. No. 186535 (Catalog No. 36).

Paratypes (unfigured): Diameter range, 0.37 to 0.45 mm. U.S.N.M. No. 186536 (Catalog No. 36). The paratypes are from sample D.st.A. 3, 10 cm.

Globigerina quinqueloba Natland
(Plate 3, Figs. 4 and 5; Plate 13, Fig. 1; Plate 14; Plate 15; Plate 16, Figs. 1 through 5)
Globigerina quinqueloba Natland, 1938, University of California, Scripps Institute of Oceanography, *Bull., Tech. Ser.,* 4(5):149, pl. 6, Fig. 7.

This spinose species has an elongate last chamber which usually extends across the umbilical area. There are 4½ to 5 chambers in the last whorl. The coiling direction of *G. quinqueloba* is random (Herman, 1969; Table 3). This taxon is present in all cores, attaining highest frequencies (up to 99% of the planktonic foraminiferal population in unit II sediments and at the commencement and end of several foraminifera-rich layers deposited during the Brunhes epoch; Figs. 2, 11, 14 and 15). *G. quinqueloba* is more abundant in shallow cores (<2000 m), suggesting that its absence or low frequency in cores raised from greater depths is due to its dissolution.

Electron-probe analyses indicate that the test is composed of three distinct layers: the inner and outer layers are Mg-rich, whereas the intermediate zone is Ca-rich.

Globigerinita glutinata (Egger)
(Plate 4, Figs. 22 through 24)
Globigerina glutinata Egger, 1893, *Abhandl. K. Bayer. Akad. Wiss. München,* CL 2, 18:371, pl. 13, Figs. 19 through 21.

Plate 7
(SEM Illustrations)

Fig. 1. *Globigerina pachyderma,* internal view, also shows the thick apertural lip; X 204. T 3-66-S.8, 5 cm.

Fig. 2. *G. pachyderma,* the last whorl was broken to show juvenile test; note the rapid increase in chambers and position of aperture. This illustration shows the difference between the juvenile *G. pachyderma* and the adult *G. quinqueloba;* the latter is illustrated elsewhere; X 340. T 3-66-S.8, 5 cm.

Fig. 3. *G. pachyderma,* same as Fig. 1; X 468. Note the presence of pores and interpore ridges. T 3-66-S.8, 5 cm.

Fig. 4. *G. pachyderma,* same as Fig. 2. Note wall texture and structure as well as presence of pustules in juvenile portion of shell; X 468. T 3-66-S.8, 5 cm.

Fig. 5. *G. pachyderma,* same as Fig. 2. Note the presence of pustules in the juvenile portion of the shell and compare the wall structure and texture with that of the adult; X 850. T 3-66-S.8, 5 cm.

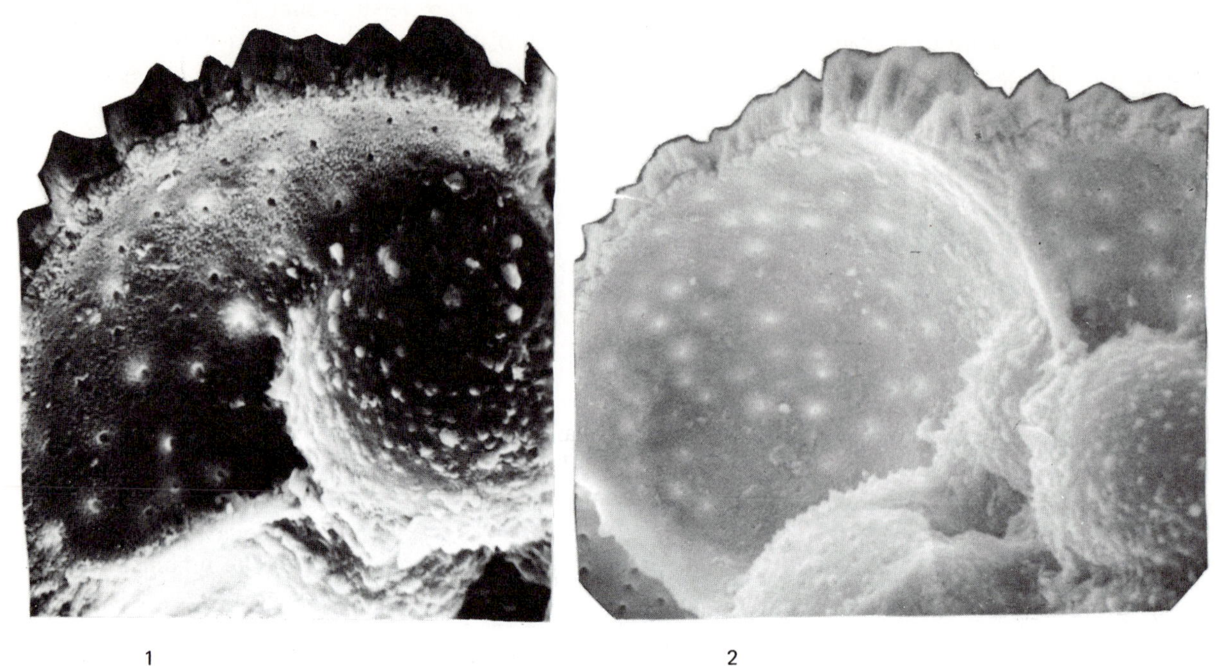

1 2

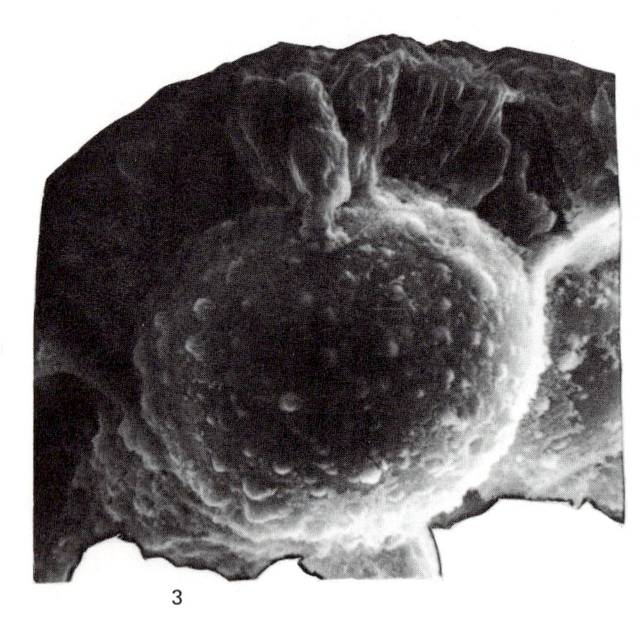

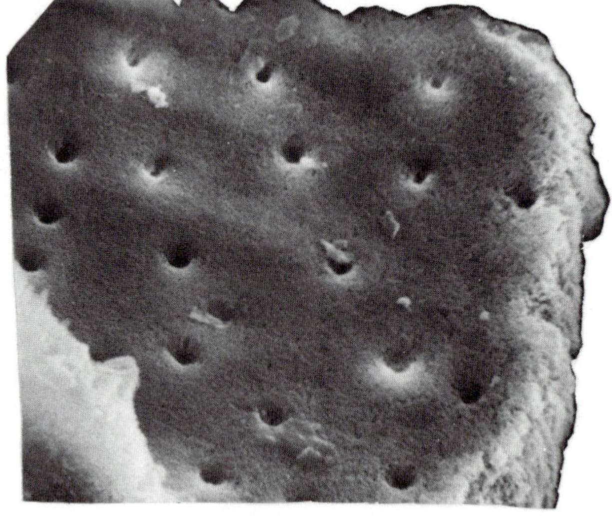

3 4

Plate 8
(SEM Illustrations)

Fig. 1. *Globigerina pachyderma,* same as Fig. 2 of Plate 6. This illustration shows wall pores of the last whorl internally and the cross-section of the wall. Note the dark median layer separating the "internal" portion of test from the "outer" cortex; X 808. T 3-66-S.8, 5 cm.

Fig. 2. *G. pachyderma,* same as Fig. 2 of Plate 6, showing the junction between the inner and outer wall marked by a dark median layer; X 850. T 3-66-S.8, 5 cm.

Fig. 3. *G. pachyderma;* the last whorl was broken to show the juvenile chamber. Note the presence of pustules on early chamber; X 1020. T 3-66-S.8, 5 cm.

Fig. 4. *G. pachyderma,* same as Fig. 2 of Plate 6. This illustration shows wall pores internally; X 1700. T 3-66-S.8, 5 cm.

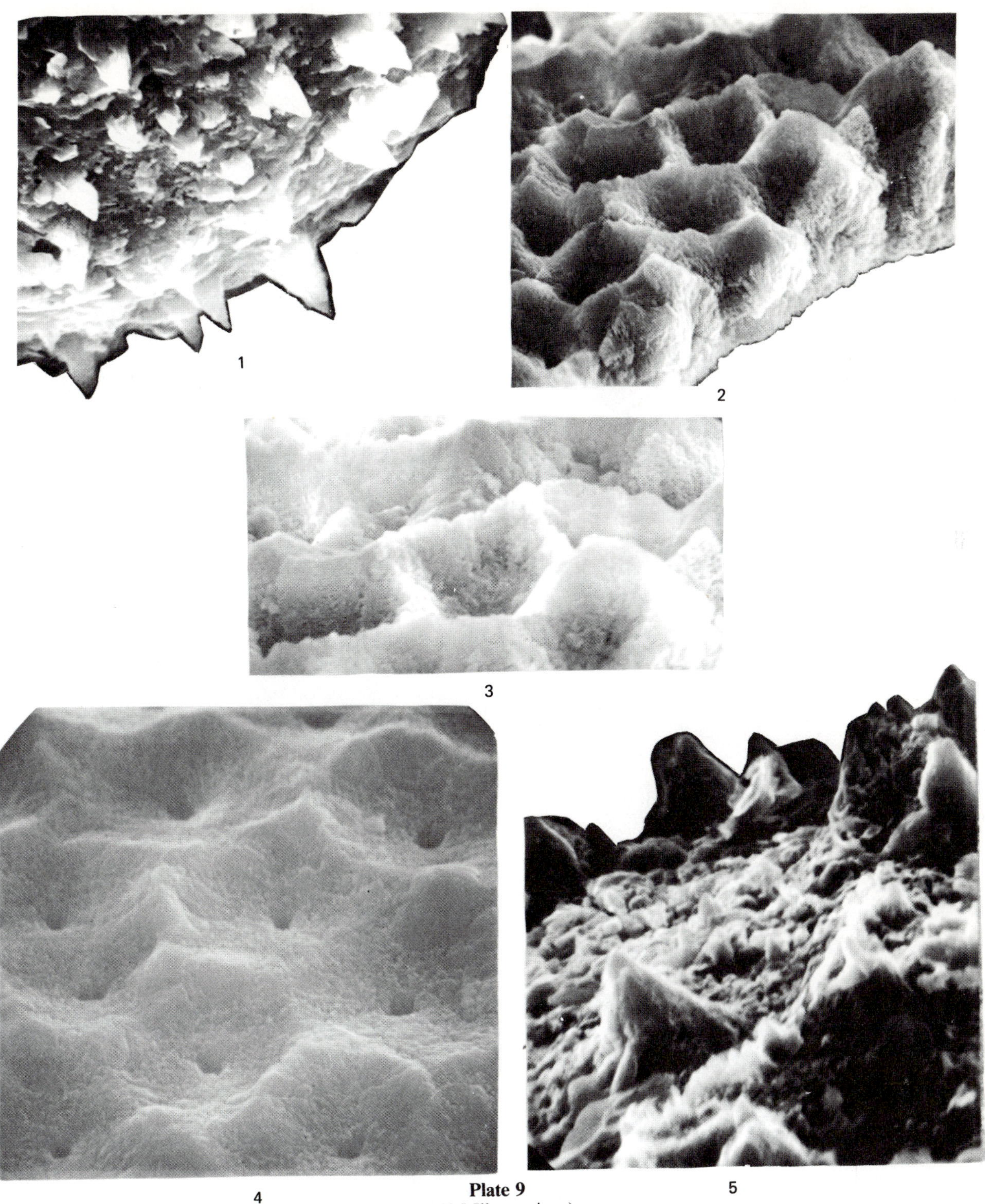

Plate 9
(SEM Illustrations)

Fig. 1. *Globigerina pachyderma,* same as Fig. 2 of Plate 6, showing portion of juvenile test. Note the short pointed pustules; X 3400. T 3-66-S.8, 5 cm.

Fig. 2. *G. pachyderma,* same as Fig. 2 of Plate 6; the outer surface of wall in the last whorl, showing pores and high interpore ridges; X 1700. T 3-66-S.8, 5 cm.

Fig. 3. *G. pachyderma,* same as Fig. 2 of Plate 6; the outer surface of the wall in the last whorl showing the high interpore ridges surrounding the evenly distributed pores; X 2550. T 3-66-S.8, 5 cm.

Fig. 4. *G. pachyderma,* same as Figs. 1 and 3 of Plate 6; the outer surface of the wall in the last chamber showing regularly spaced pores and interpore ridges; X 3400. T 3-66-S.8, 5 cm.

Fig. 5. *G. pachyderma,* portion of juvenile test showing pointed pustules; X 8500. T 3-66-S.8, 5 cm.

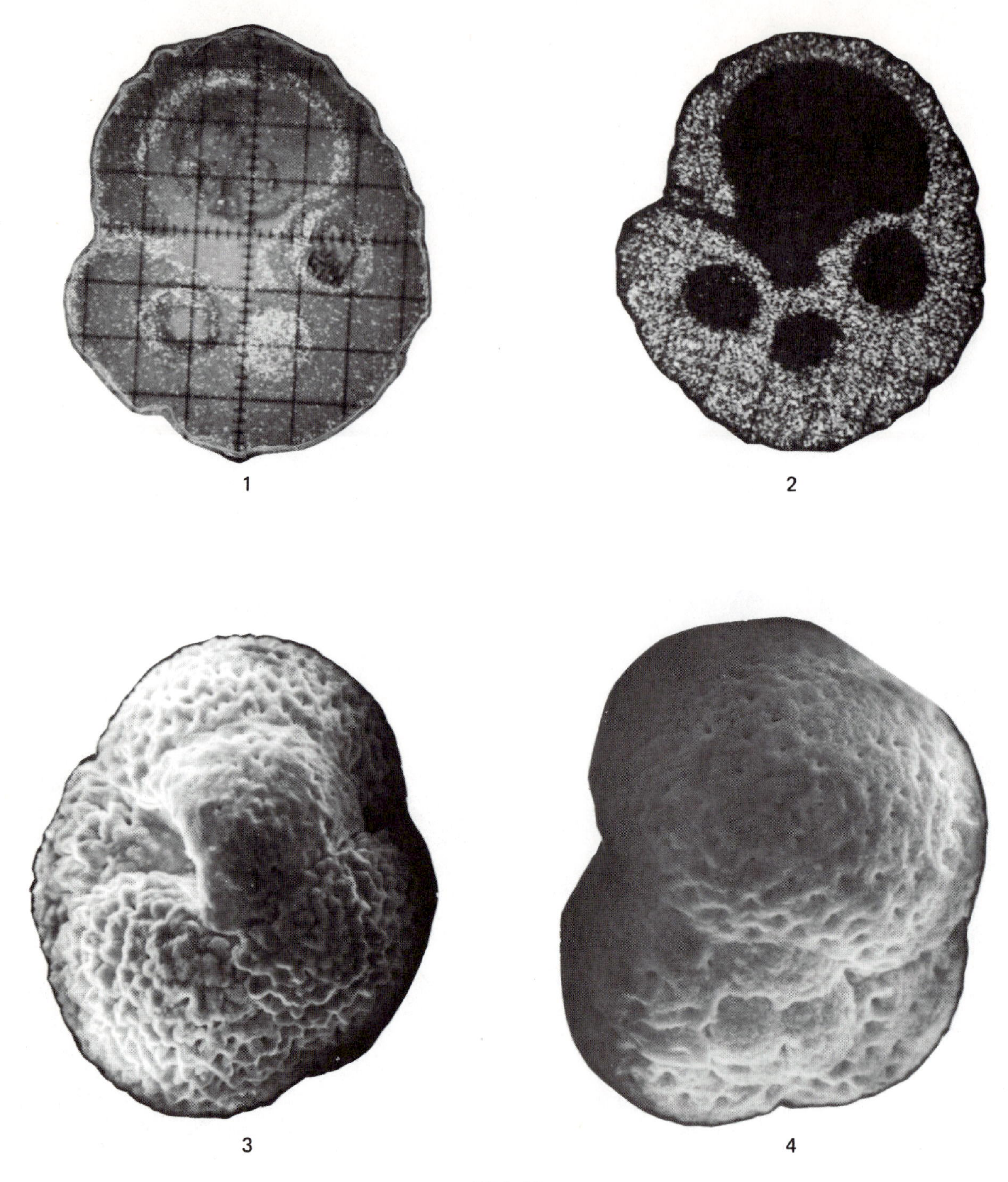

Plate 10
(SEM Illustrations)

Fig. 1. *Globigerina pachyderma,* polished cross section; about X 400. Electron-probe analyses show that the test is composed of three distinct layers. In this photograph the outer and inner Mg-rich layers are light gray, the intermediate Ca-rich zone is darker gray.

Fig. 2. *G. pachyderma,* same as Fig. 1, polished cross section; about X 400. The intermediate Ca-rich zone is indicated by the light layer.

Fig. 3. *G. occlusa,* n.sp., ventral view; X 360. Note the diminutive last chamber and coarse wall texture. T 3-66-S.8, 5 cm.

Fig. 4. *G. occlusa* n.sp., dorsal view; X 420. Note the presence of numerous minute pores. T 3-66-S.8, 5 cm.

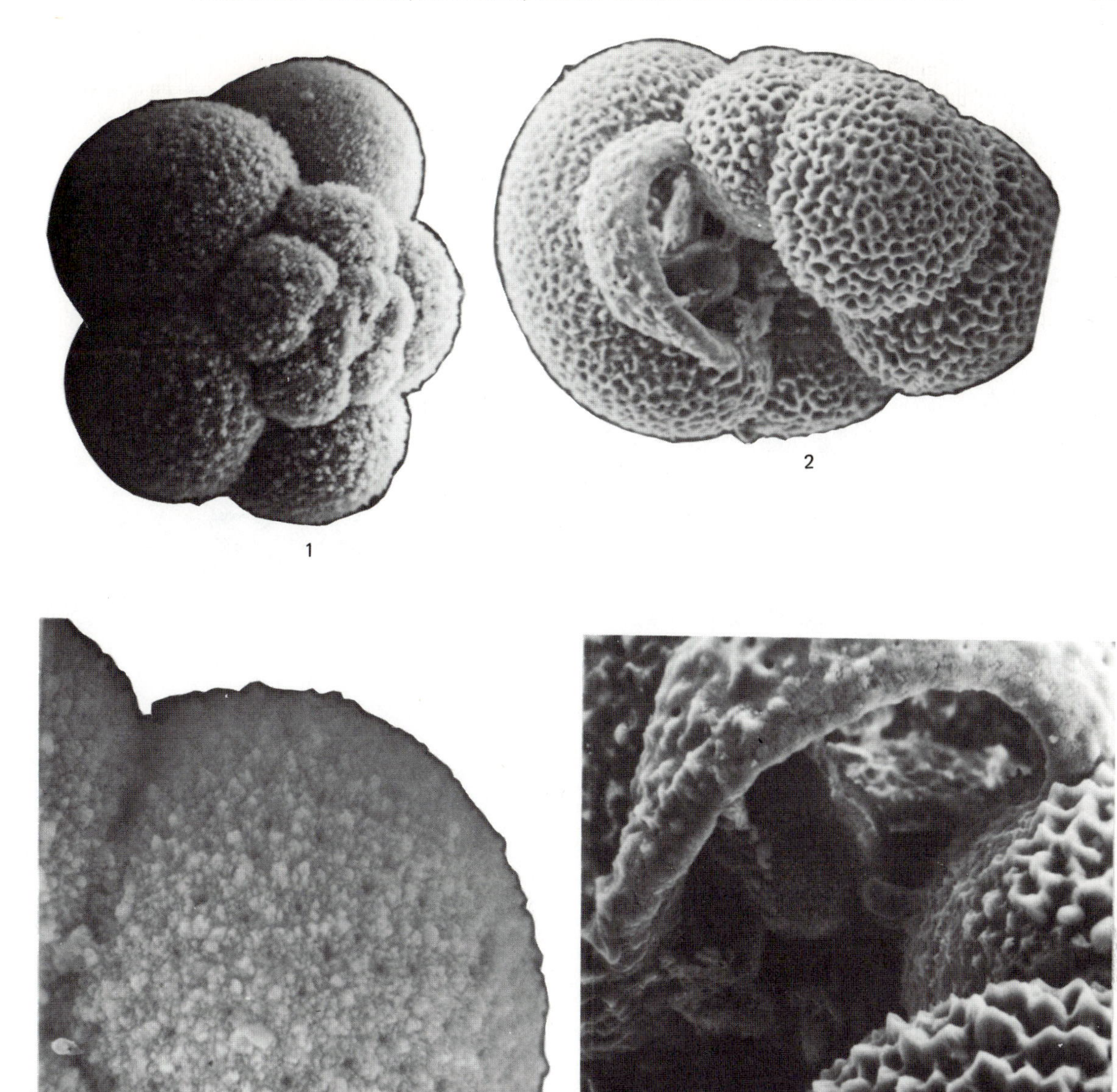

Plate 11
(SEM Illustrations)

Fig. 1. *Globigerina paraobesa* n.sp., dorsal view; X 340. T 3-66-S.8, 5 cm.

Fig. 2. *G. paraobesa* n.sp., ventral view; X 204. Note the diminutive last chamber and the thick apertural lip. T 3-66-S.8, 5 cm.

Fig. 3. *G. paraobesa* n.sp., same as Fig. 1. Illustration of chamber in the last whorl showing pores and pustules; X 850. T 3-66-S.8, 5 cm.

Fig. 4. *G. paraobesa* n.sp., same as Fig. 2. Note the presence of minute pores on the apertural lip; X 459. T 3-66-S.8, 5 cm.

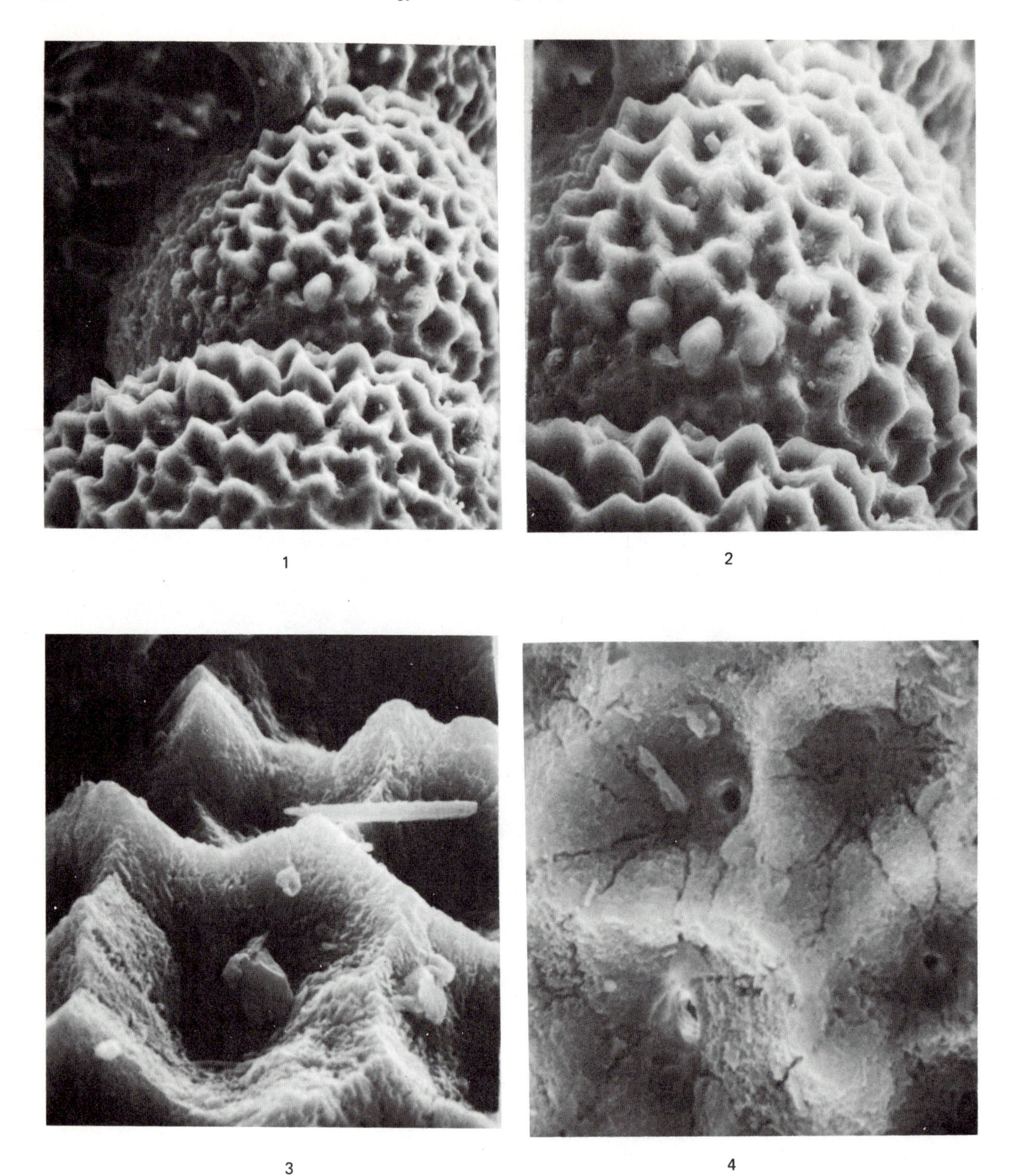

Plate 12
(SEM Illustrations)

Fig. 1. *Globigerina paraobesa* n.sp., same as Fig. 2 of Plate 11. The illustration shows the close relationship of this taxon to *G. pachyderma* in regard to wall texture and structure. Note the well-developed interpore ridges; X 553. T 3-66-S.8, 5 cm.

Fig. 2. *G. paraobesa* n.sp., same as Fig. 2 of Plate 11; X 850. T 3-66-S.8, 5 cm.

Fig. 3. *G. paraobesa* n.sp., same as Fig. 2 of Plate 11; X 3400. T 3-66-S.8, 5 cm.

Fig. 4. *G. paraobesa* n.sp., same as Fig. 2 of Plate 11. Note the circular pores surrounded by interpore ridges; X 3400. T 3-66-S.8, 5 cm.

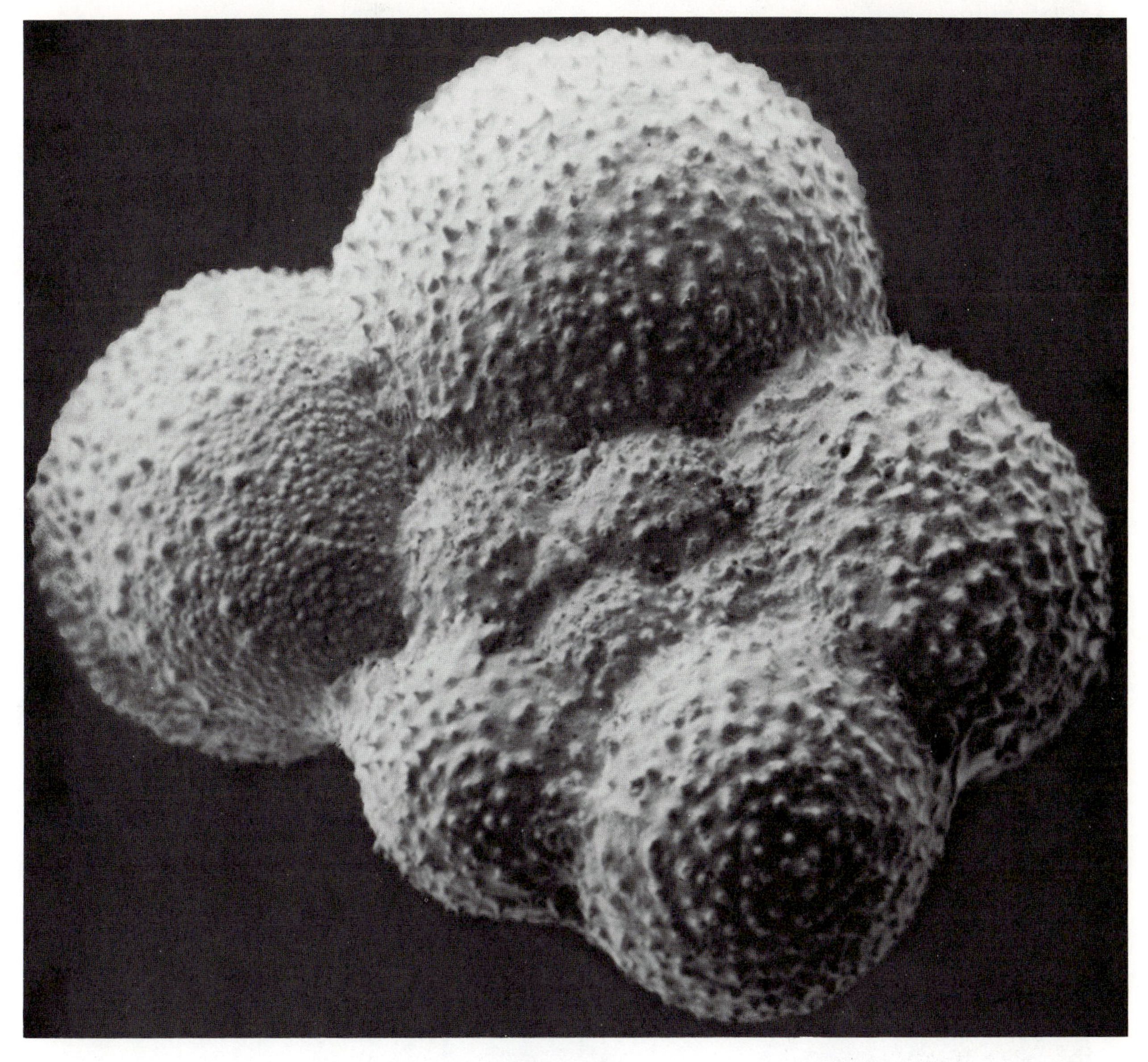

Plate 13
(SEM Illustration)

Globigerina quinqueloba, dorsal view; about X 808. This specimen is from a foraminifera-poor zone. Note the presence of numerous spine bases. T 3-67-9, 220 cm.

Most Arctic specimens lack bulla. This species was not observed in the two deep Arlis cores (Herman, 1969); it is present sporadically and at low frequencies in cores D.st.A. 2 to 5, T 3, and it is common in D.st.A. 6 between 25 and 84.5 cm. Its absence from deep cores may be due to its dissolution (Herman, 1969; Fig. 4). The lowest Antarctic surface water temperatures in which *G. glutinata* was found living was +0.30°C (Boltovskoy, 1969a).

Globigerinita uvula (Ehrenberg)
Pylodexia uvula Ehrenberg, 1861, *K. Preuss. Akad. Wiss. Berlin, Monatsber.*, pp. 276, 277, 308.

Rare specimens were observed in the Alpha Rise and the Wrangel Plain core samples. The lowest Antarctic surface water temperatures in which this species was found living was −0.58°C (Boltovskoy, 1966).

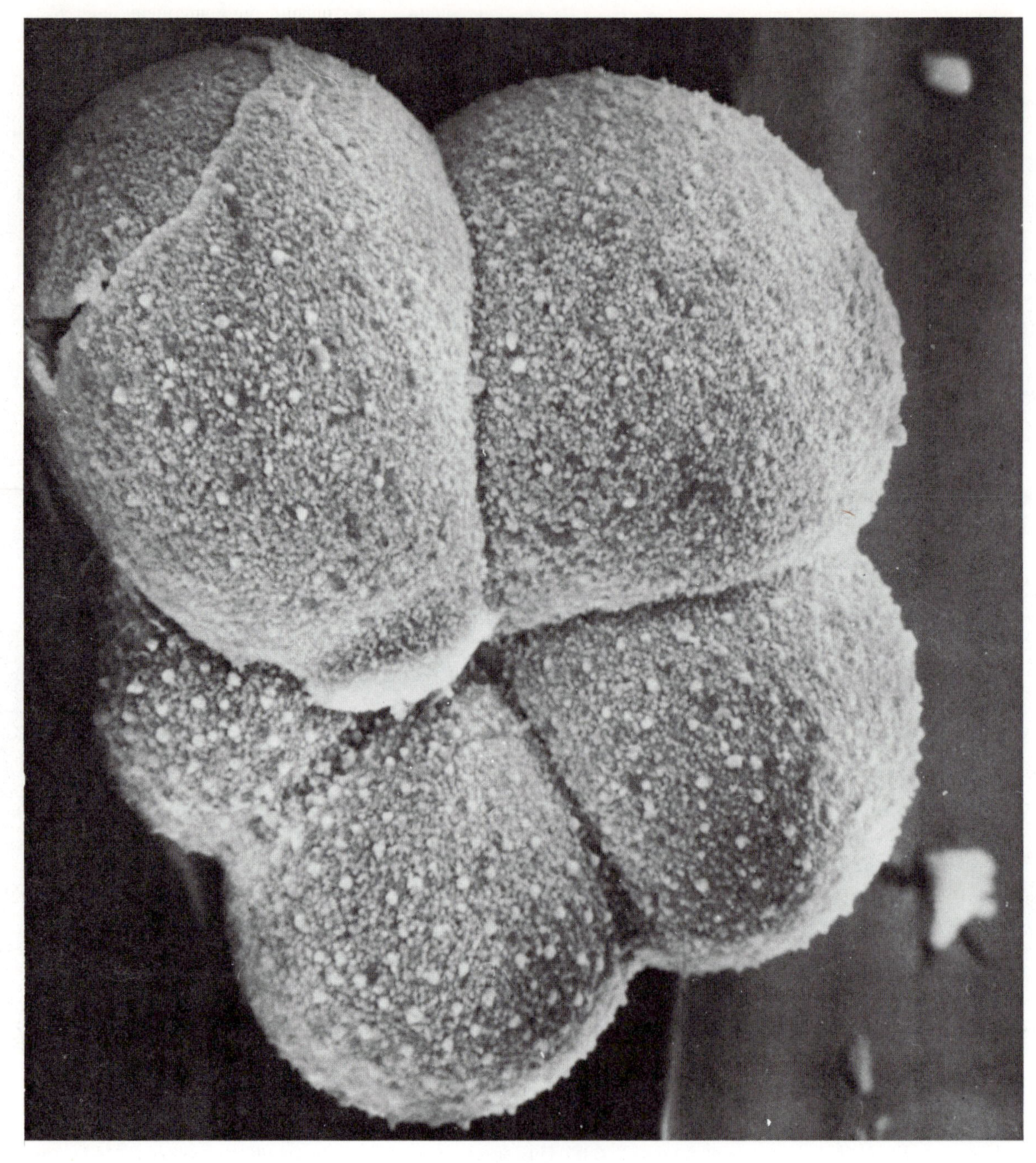

Plate 14
(SEM Illustration)

Globigerina quinqueloba, ventral view; about X 850. The illustration shows the differences between this taxon and *G. exumbilicata* n.sp. Note the elongate last chamber in *G. quinqueloba.* The wall texture and structure are distinctly different from those shown by *G. exumbilicata.* T 3-66-S.8, 8–10 cm.

Globigerinoides sp., cf. *G. ruber* (d'Orbigny)
(Plate 3, Fig. 18)

Globigerina rubra d'Orbigny, 1839, in *de la Sagra, Histoire physique, politique et naturelle de l'Ile de Cuba, "Foraminifères,"* 8:82, pl. 4, Figs. 12 through 14.

This species differs from *G. ruber* by its outline, which is compressed, and its thicker and rougher wall texture. Also, the main aperture of the last chamber is small and elongate and the secondary apertures are smaller than in the typical *G. ruber,* resembling *Globigerinoides ruber gomitulus.* Rare specimens were observed in foraminifera-poor

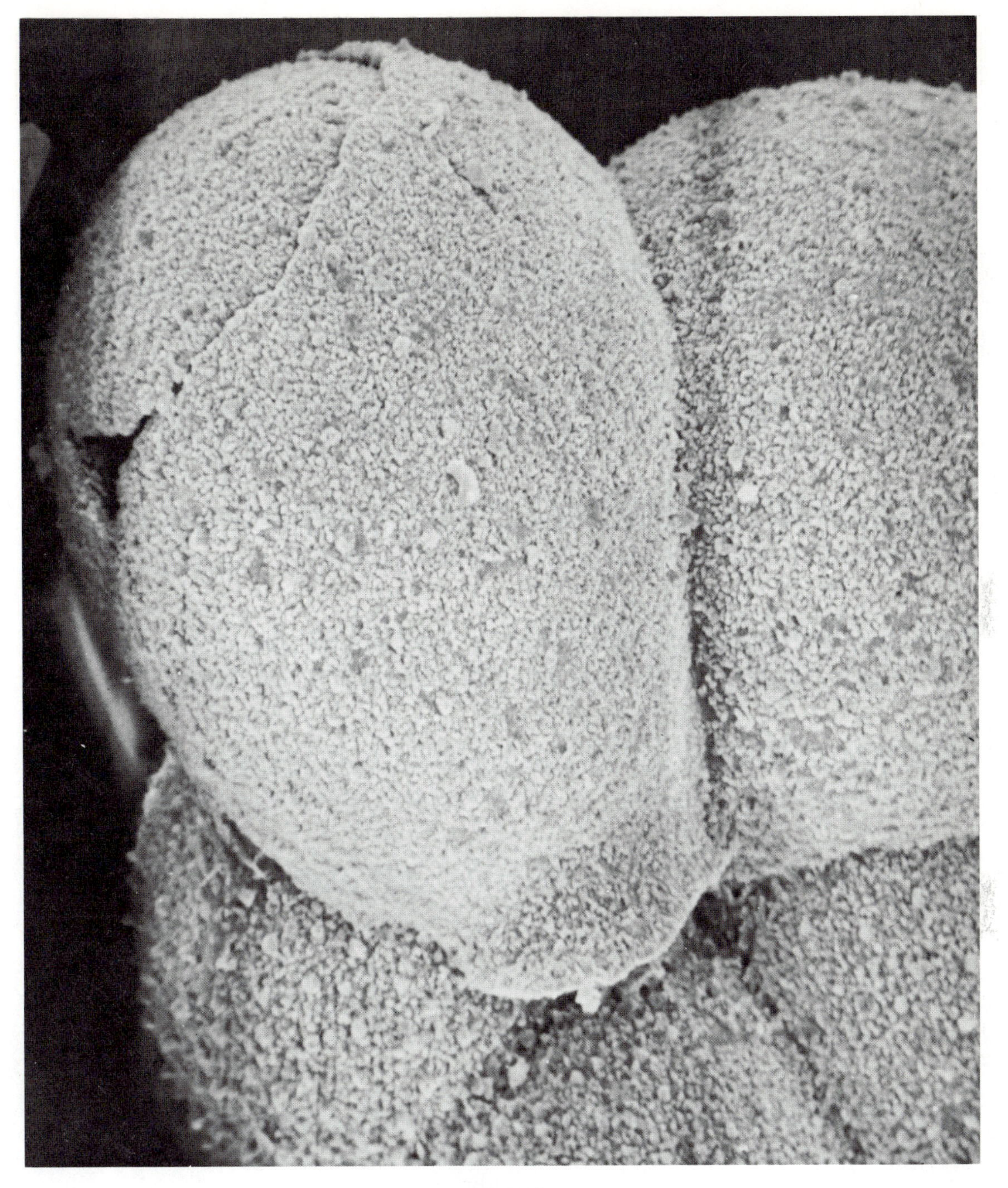

Plate 15
(SEM Illustration)

Globigerina quinqueloba, ventral view; about X 1100. Note the shape of the last chamber and the lack of apertural lip. T 3-66-S.8, 8–10 cm.

beds in Alpha Rise cores (Herman, 1969; Figs. 2, 4, and 12).

Globigerinoides sp., cf. *G. sacculifer* (Brady)
(Plate 3, Fig. 17)
Globigerina sacculifera Brady, 1877, *Geol. Mag.*, new ser., Dec. 2, 4(12):535.

A few immature specimens, tentatively identified as *G.* sp., cf. *G. sacculifer*, were observed in foraminifera-poor layers from the Alpha Rise (Herman, 1969, 1970a; Fig. 4).

Globorotalia crassaformis (Galloway and Wissler)
(Plate 3, Fig. 19; Plate 4, Fig. 20)
Globigerina crassaformis Galloway and Wissler 1927, *J. Pal.*, 1:41, pl. 7, Fig. 12.

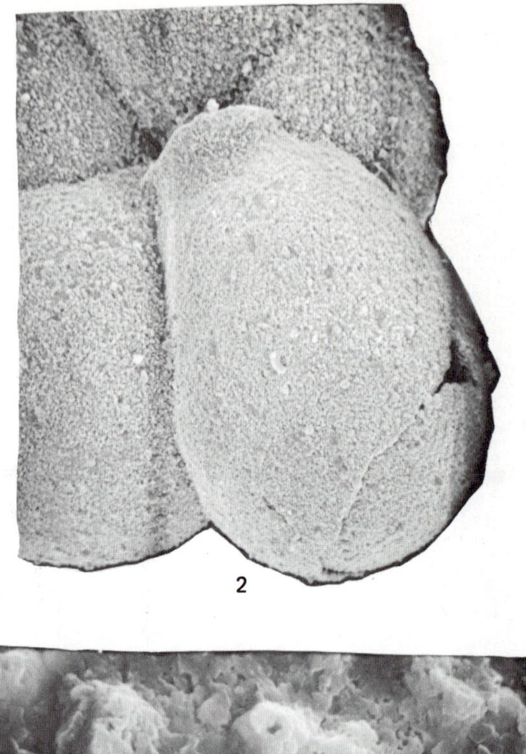

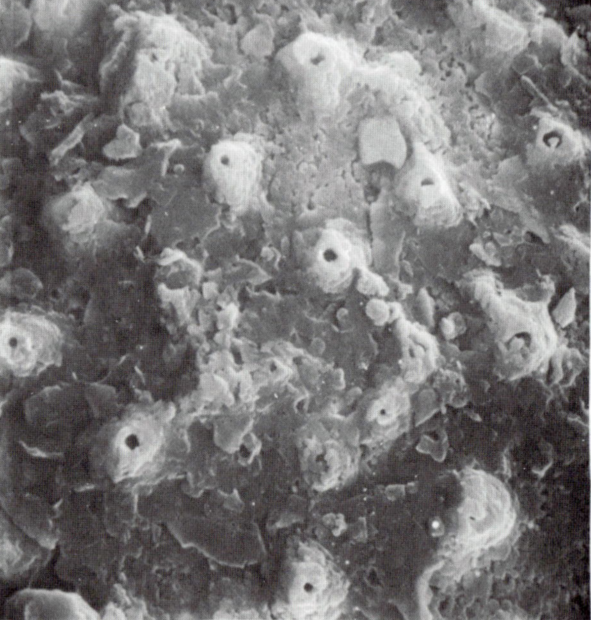

Plate 16
(SEM Illustration)

Fig. 1. *Globigerina quinqueloba*, ventral view; about X 351. Note the shape of the last chamber and the absence of the apertural lip. T 3-66-S.8, 8 to 10 cm.

Fig. 2. *G. quinqueloba*, ventral view; about X 486. Note the shape of the last chamber and the absence of the apertural lip. T 3-66-S.8, 8 to 10 cm.

Fig. 3. *G. quinqueloba*, dorsal view; X 396. T 3-66-S.8, 8 to 10 cm.

Fig. 4. *G. quinqueloba*, equatorial view; X 585. T 3-66-S.8, 8 to 10 cm.

Fig. 5. *G. quinqueloba*, last whorl, penultimate chamber, showing broken spine bases; about X 1800. T 3-67-12, 192 cm.

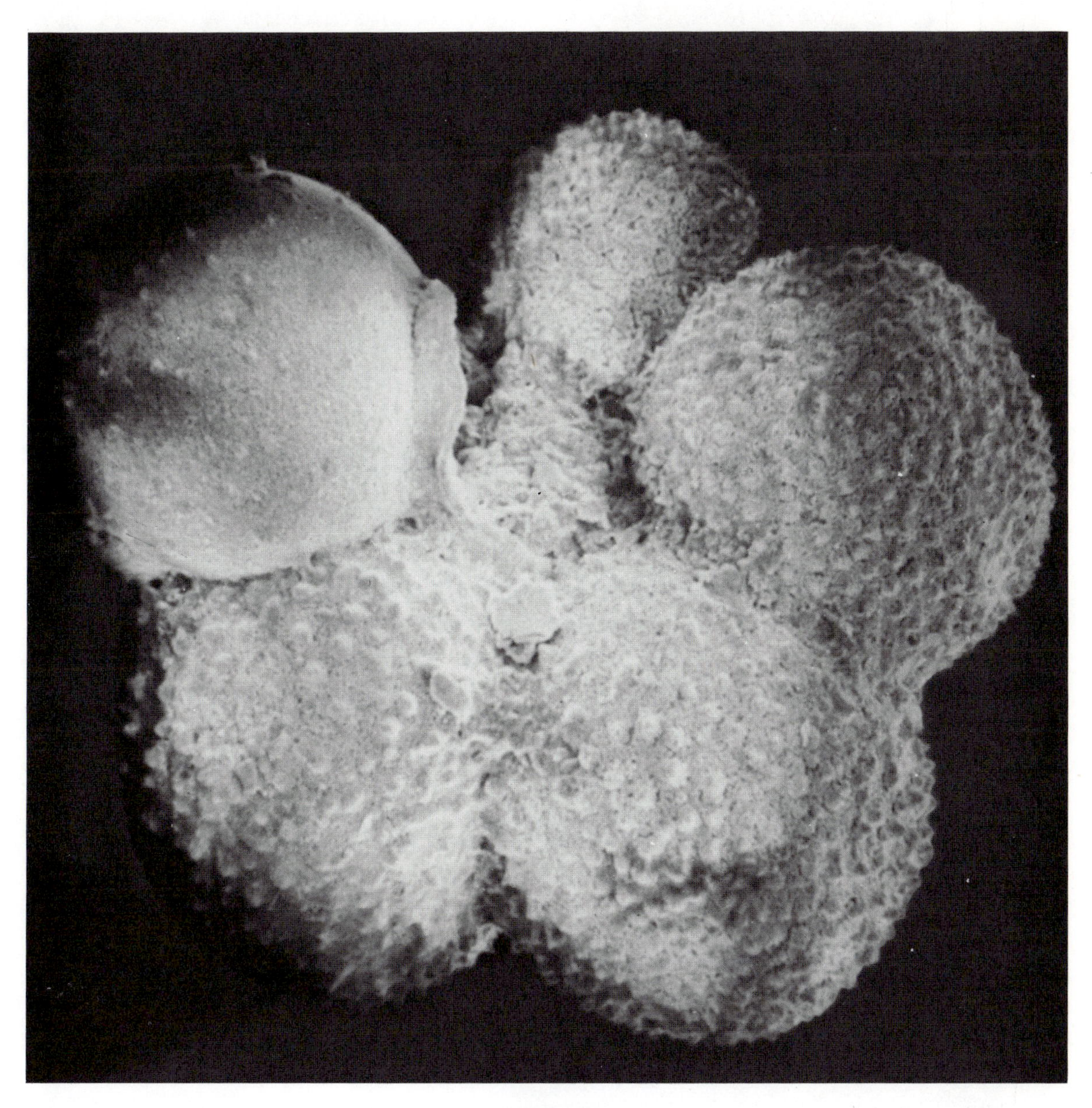

Plate 17
(SEM Illustration)

Globigerina exumbilicata n.sp., ventral view; about X 900. Note the shape of the last chamber, and the presence of a distinct apertural lip and a deep umbilicus. The wall texture and structure are different from those of *G. quinqueloba*; the wall of the newly named taxon is smoother and the spines are spaced farther apart.

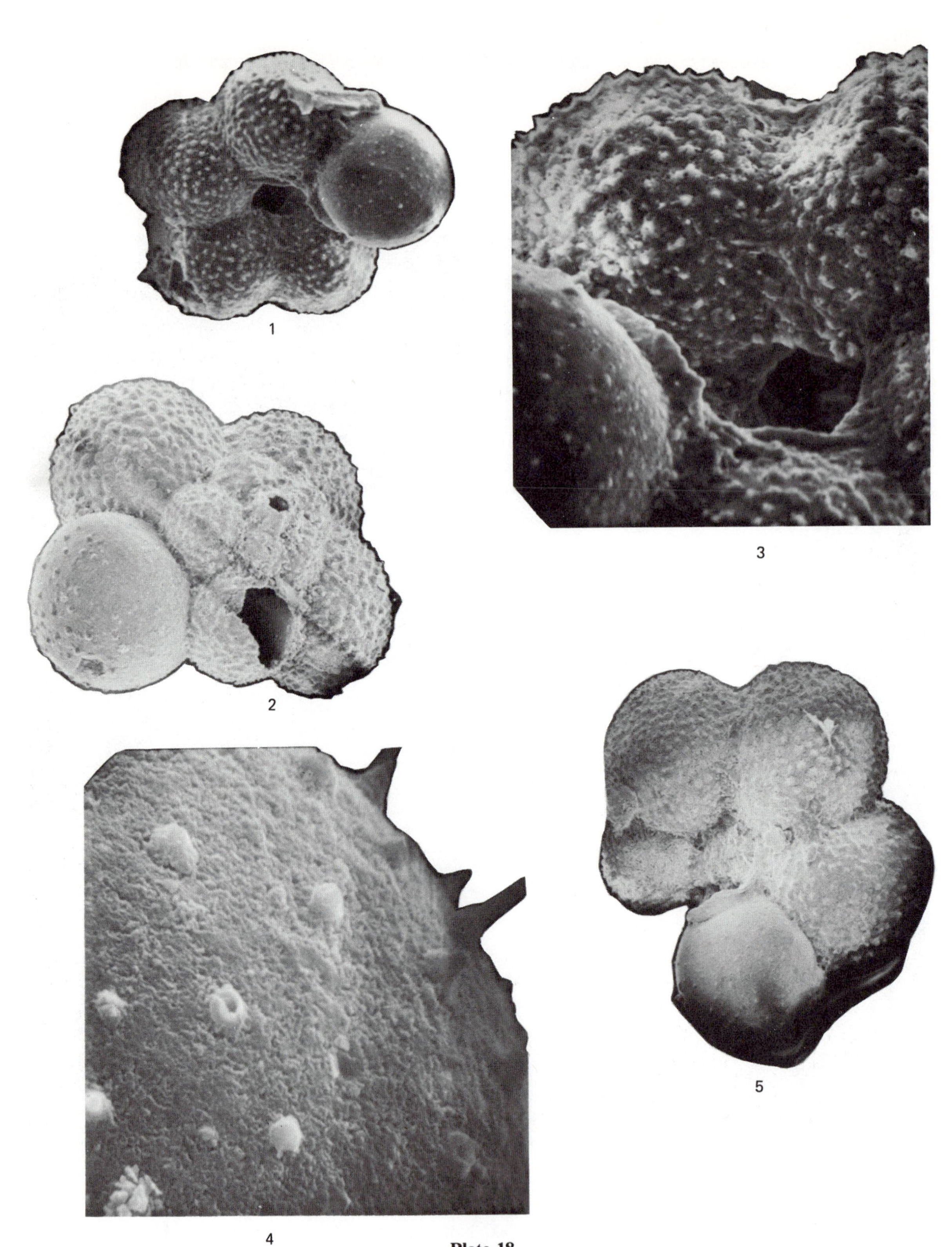

Plate 18
(SEM Illustrations)

Fig. 1. *Globigerina exumbilicata* n.sp., ventral view; about X 383. Note deep umbilicus and distinct apertural lip. T 3-67-12, 192 cm.

Fig. 2. *G. exumbilicata* n.sp., dorsal view; X 272. T 3-67-12, 192 cm.

Fig. 3. *G. exumbilicata* n.sp., same as Fig. 1; about X 1063. Note deep umbilicus and umbilical teeth. T 3-67-12, 192 cm.

Fig. 4. *G. exumbilicata* n.sp., ventral view; about X 383. T 3-67-12, 192 cm.

Fig. 5. *G. exumbilicata* n.sp., last whorl; about X 3400. Illustration shows broken spines and hollow spine bases. T 3-66-S.8, 5 cm.

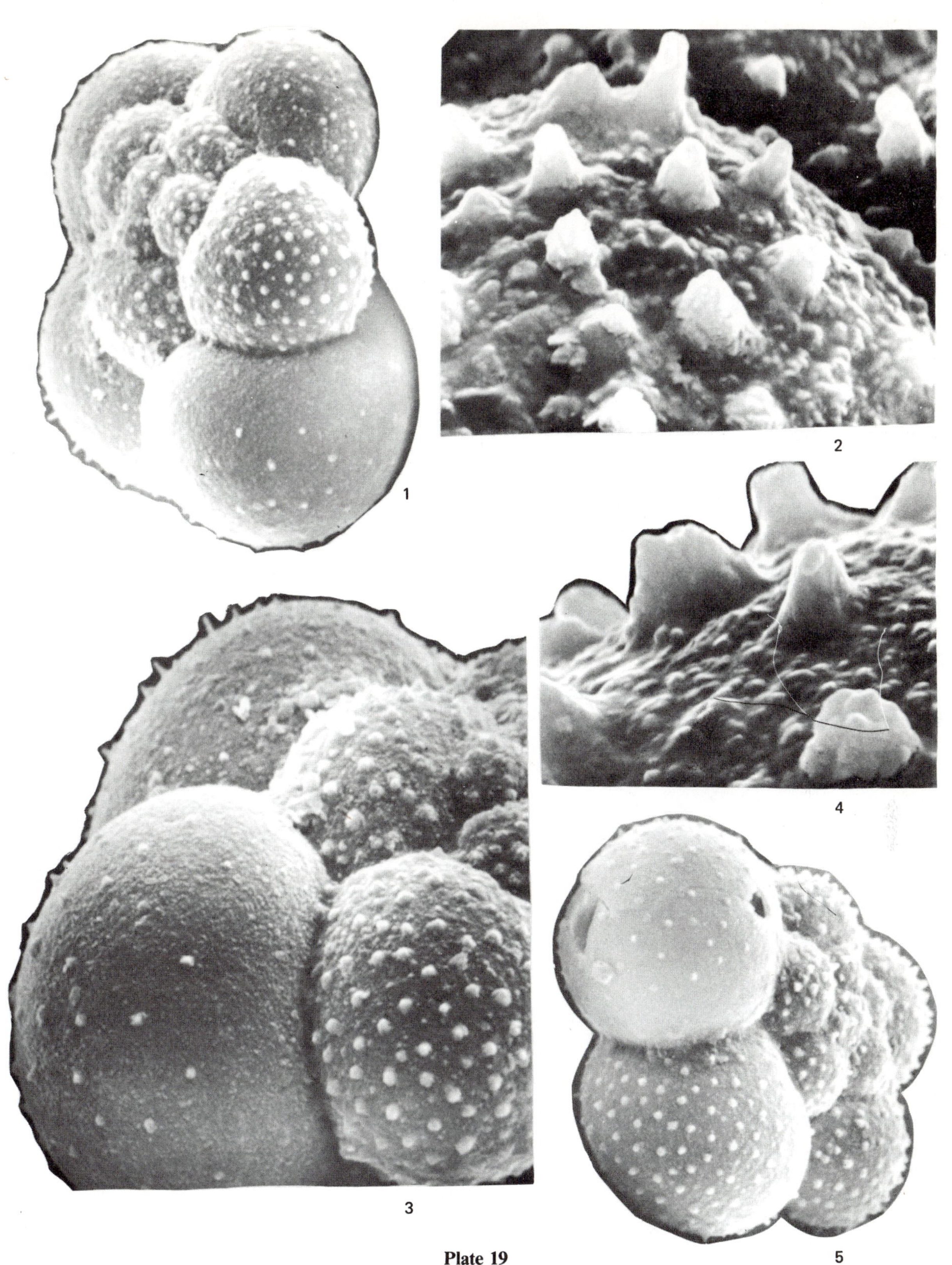

Plate 19
(SEM Illustrations)

Fig. 1. *Globigerina exumbilicata* n.sp., dorsal view; X 510. T 3-67-12, 25 cm.

Fig. 2. *G. exumbilicata* n.sp., same as Fig. 5; about X 4675. Last whorl showing broken spines. T 3-67-12, 25 cm.

Fig. 3. *G. exumbilicata* n.sp., same as Fig. 1; X 850. Dorsal view of last and penultimate whorls showing broken spines.

Fig. 4. *G. exumbilicata* n.sp., same as Fig. 5; about X 8925. T 3-67-12, 25 cm.

Fig. 5. *G. exumbilicata* n.sp., dorsal view; about X 578. T 3-67-12, 25 cm.

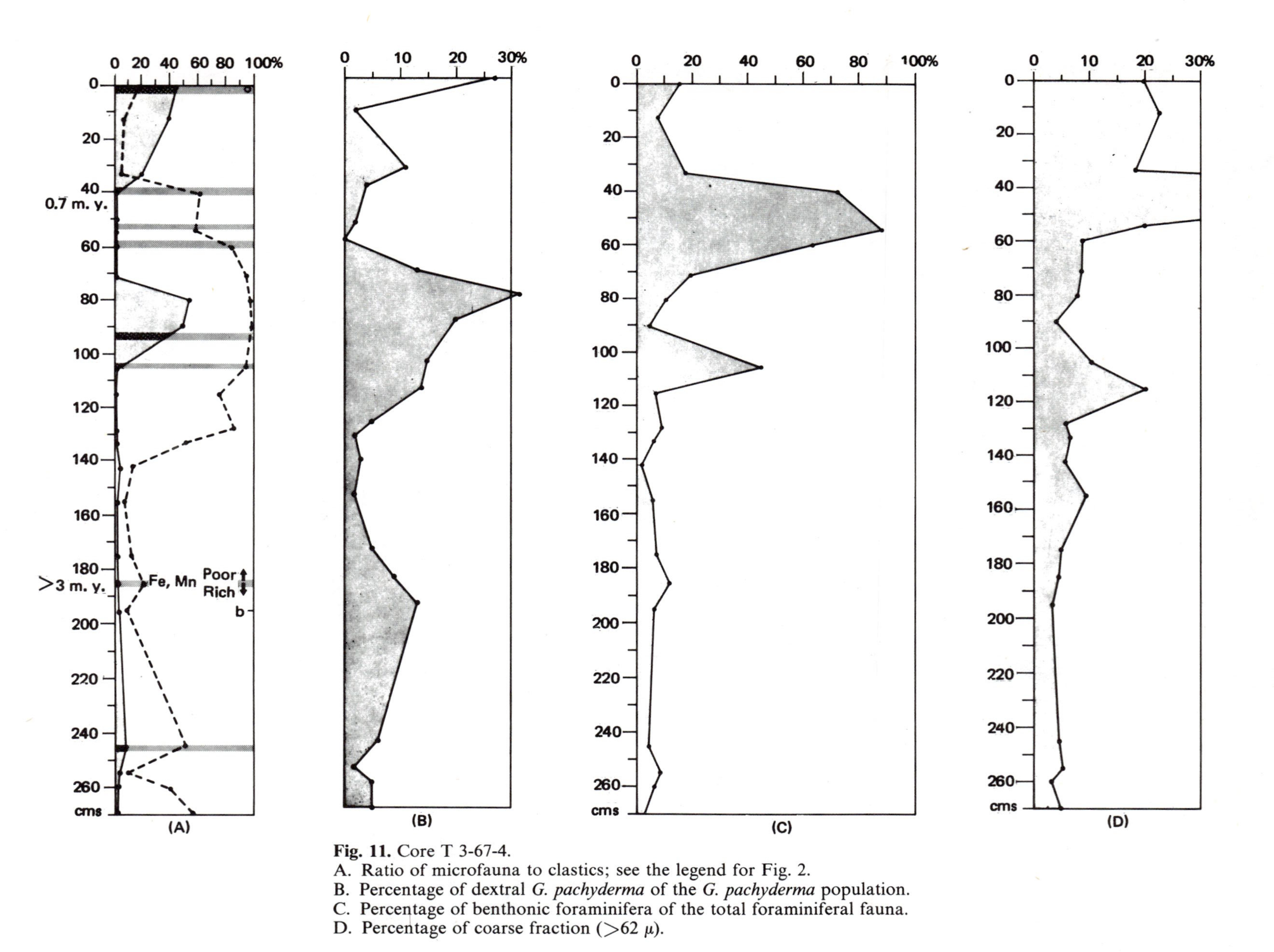

Fig. 11. Core T 3-67-4.
A. Ratio of microfauna to clastics; see the legend for Fig. 2.
B. Percentage of dextral *G. pachyderma* of the *G. pachyderma* population.
C. Percentage of benthonic foraminifera of the total foraminiferal fauna.
D. Percentage of coarse fraction (>62 μ).

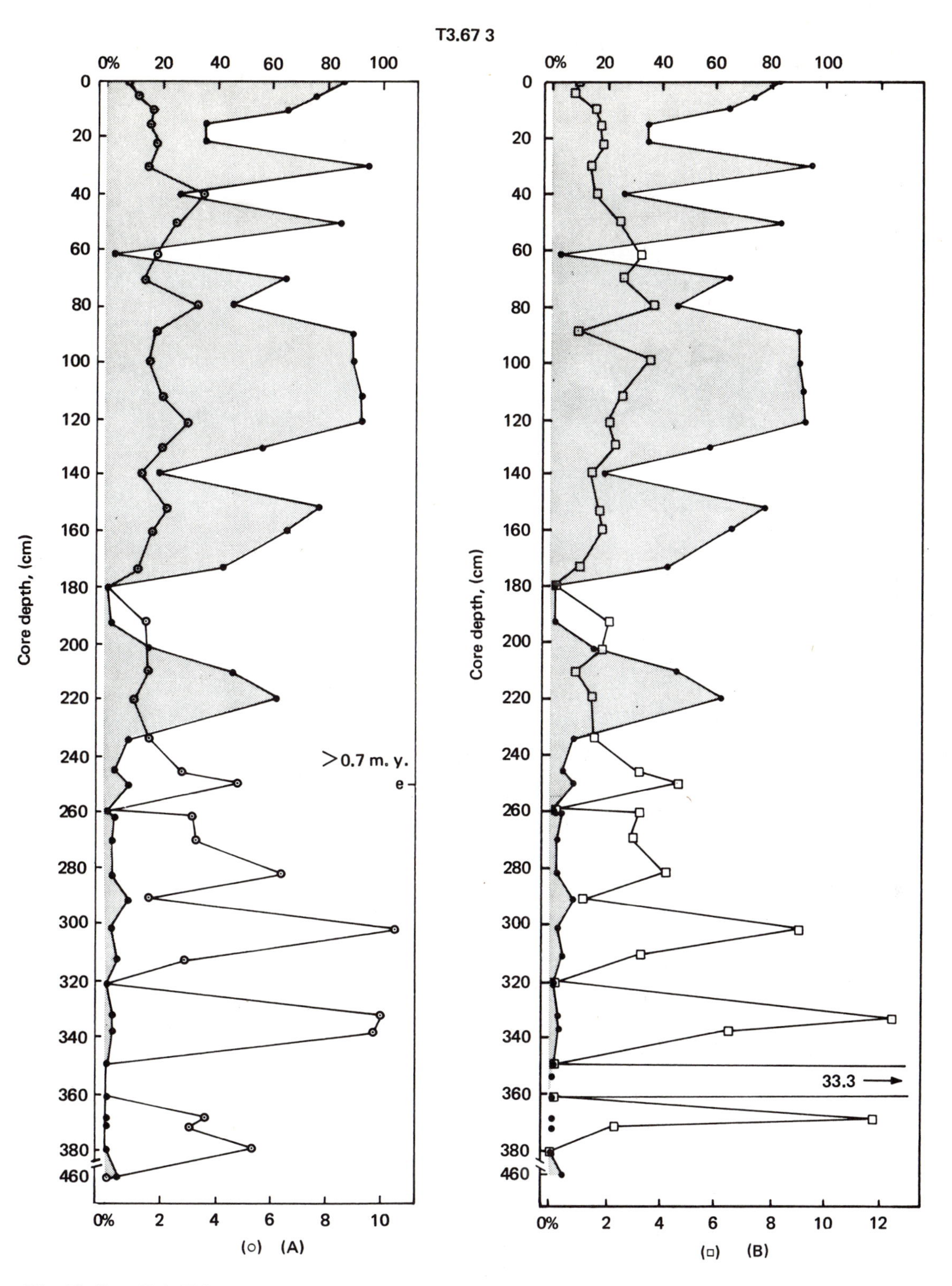

Fig. 12. Core T 3-67-3.

A. Ratio of microfauna to clastics in the coarse fraction (>62 μ). Shaded areas represent the microfauna, white areas the clastics; values are indicated on the abscissa at the top of the diagram.

Occurrence of low-latitude foraminifera:

e = *Globigerinoides ruber*

Circles represent the percentage of dextral *G. pachyderma;* values are indicated on the abscissa at the bottom of the diagram.

B. See the legend for diagram A.

Squares represent the percentage of dextral *G. occlusa.*

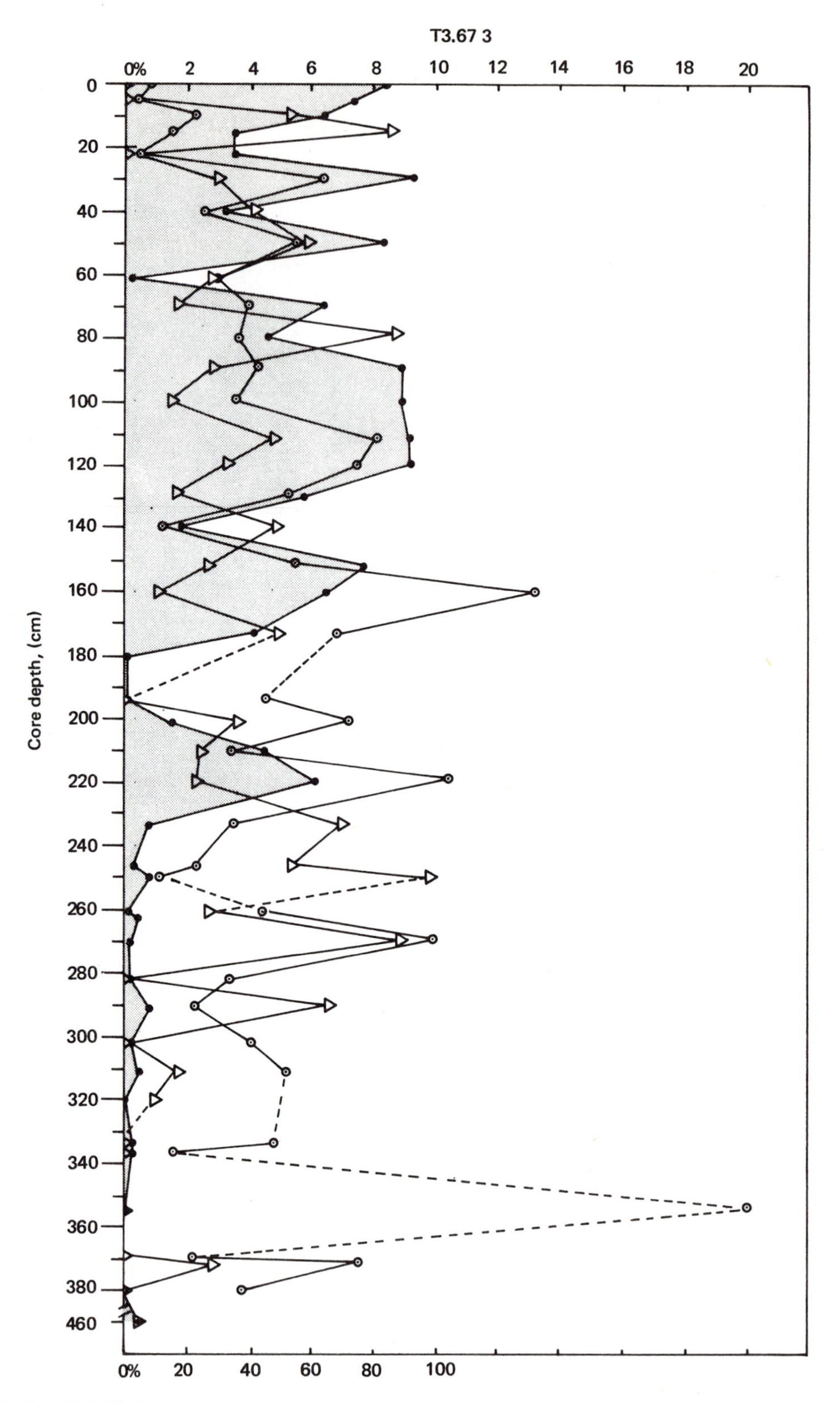

Fig. 13. Core T 3-67-3.
Ratio of microfauna to clastics in the coarse fraction (>62 μ). Shaded areas represent the microfauna, white areas the clastics; values are indicated on the abscissa at the bottom of the diagram.

Open circles are the percentage of *Globigerina* sp., cf. *G. pachyderma* A (possessing 5 to 5½ chambers in the final whorl) of the total adult planktonic foraminiferal population; values are indicated on the abscissa at the top of the diagram.

Triangles are the percentage of dextral *Globigerina* sp., cf. *G. pachyderma* A (possessing 5 to 5½ chambers in the final whorl); values are indicated on the abscissa at the top of the diagram.

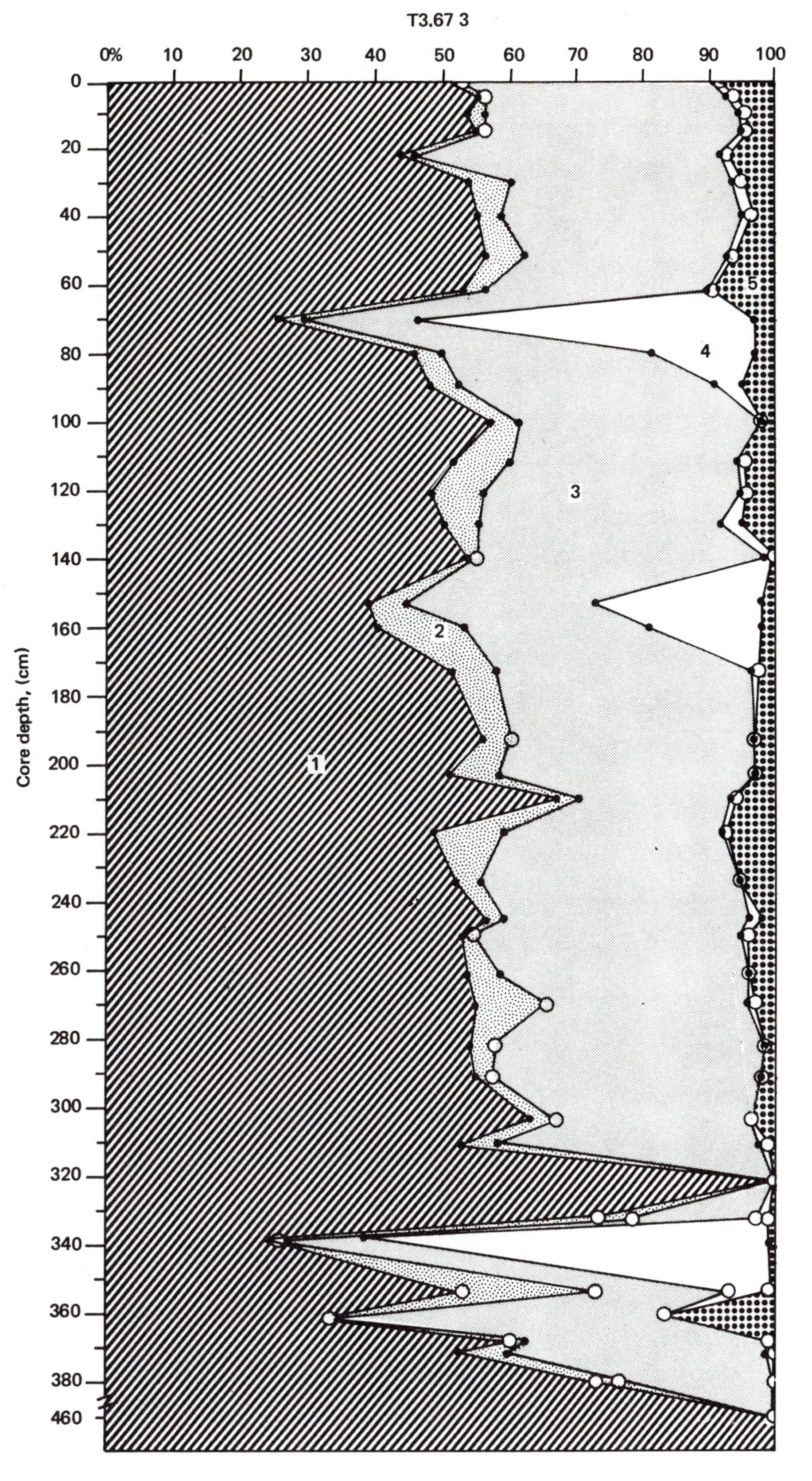

Fig. 14. Core T 3-67-3.
Cumulative percentage of planktonic foraminifera; open circles signify less than 30 specimens; dark circles signify more than 30 specimens per sample.

- *G. pachyderma*
- *G.* sp., cf. *G. pachyderma* A (with 5 to 5½ chambers in final whorl)
- *G. occlusa*
- *G. quinqueloba* and *G. exumbilicata*
- All other planktonic foraminifera.

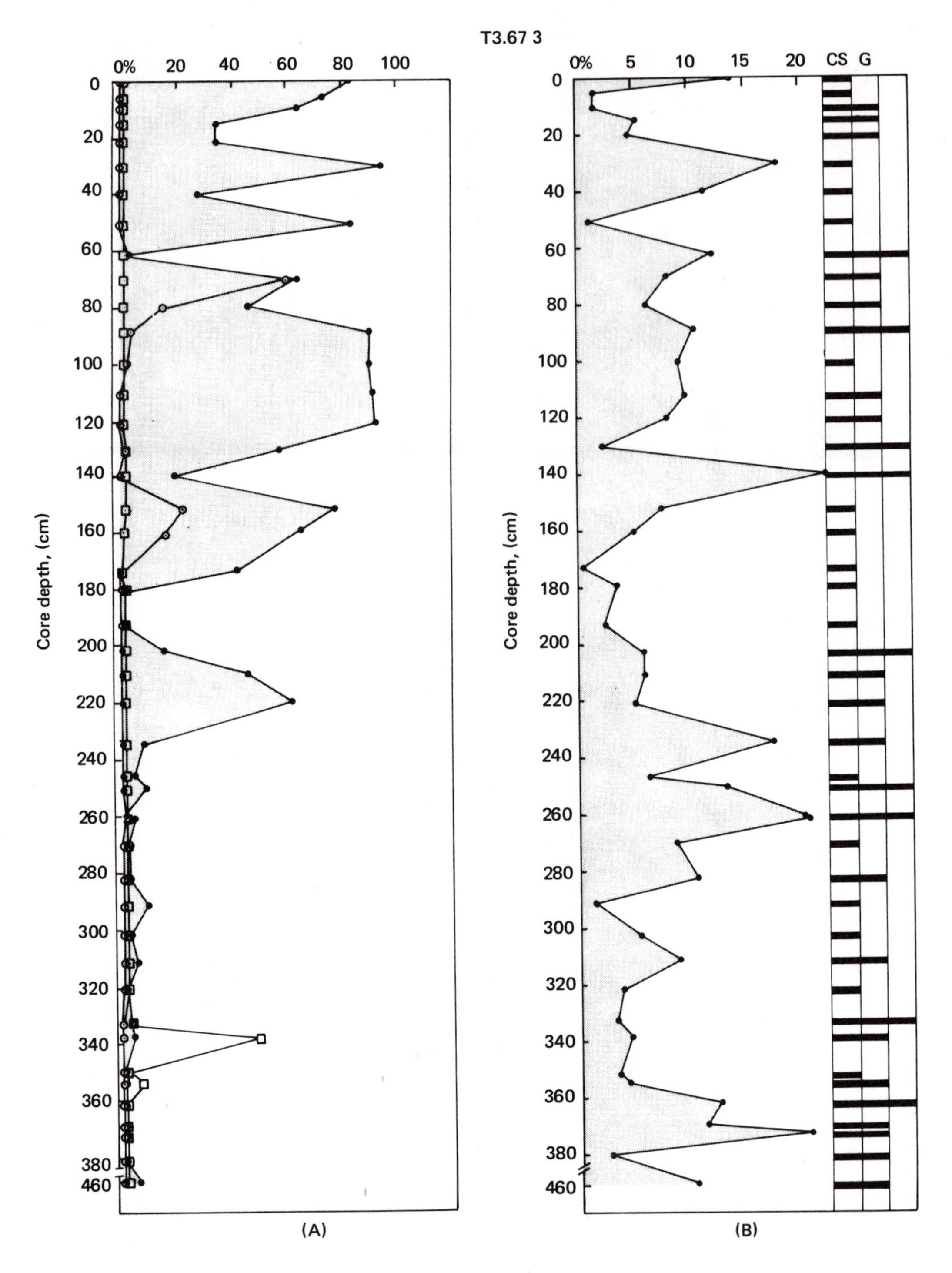

Fig. 15. Core T 3-67-3.

A. Ratio of microfauna to clastics in the coarse fraction (>62 μ). Shaded areas represent the microfauna, white areas the clastics.
Open circles represent the percentage of *G. quinqueloba* of the total adult planktonic foraminiferal population.
Open squares represent the percentage of *G. exumbilicata* of the total adult planktonic foraminiferal population.

B. Percentage of coarse fraction (>62 μ).
CS = coarse sand; G = granules; P = pebbles.

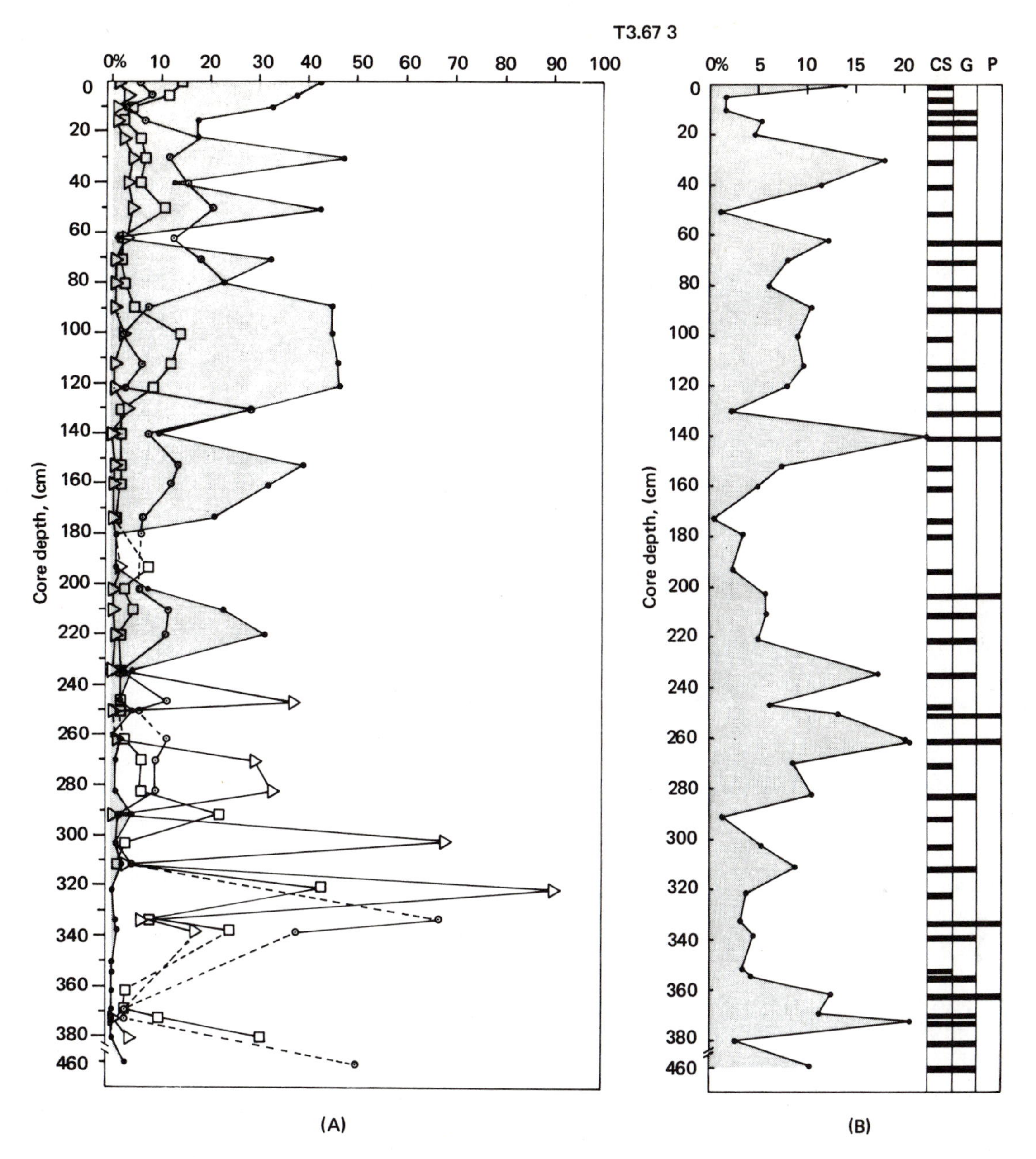

Fig. 16. Core T 3-67-3.

A. Ratio of microfauna to clastics in the coarse fraction (>62 μ). Shaded areas represent microfauna, white areas the clastics.

Triangles represent the percentage of benthonic foraminifera of the total foraminiferal population.

Open circles represent the percentage of juvenile planktonic foraminifera of the total planktonic foraminiferal population.

Squares represent the percentage of planktonic foraminiferal shell fragments of the total planktonic foraminiferal population.

B. Percentage of coarse fraction (>62 μ). CS = coarse sand; G = granules; P = pebbles.

Several sinistrally coiled specimens were observed in foraminifera-poor zones from the Alpha Rise (Herman, 1969; Fig. 4).

Globorotalia inflata (d'Orbigny)

Globigerina inflata d'Orbigny, in Barker-Webb and Berthelot, *Histoire naturelle des Iles Canaries, "Foraminifères,"* 2, pt. 2, Zool.: 134, pl. 2, Figs. 7 through 9.

Rare specimens were observed in foraminifera-poor zones from the Alpha Rise (Herman, 1969;

Fig. 4). The lowest Antarctic surface water temperatures in which this species was found living was −0.40°C (Boltovskoy, 1969a).

Globoquadrina dutertrei (d'Orbigny)
(Plate 4, Fig. 21)
Globigerina dutertrei d'Orbigny, 1839, in *de la Sagra, Histoire physique, politique et naturelle de l'Ile de Cuba "Foraminifères,"* 8:84, pl. 4, Figs. 19 through 21.

Counts of this species may include *G.* cf. *pachyderma* A; it is present in low frequencies throughout the cores, constituting <0.1% of the planktonic foraminiferal population. Its coiling direction is predominantly sinistral. The lowest Antarctic surface water temperatures in which this species was found living was 0.33°C (Boltovskoy, 1966).

Rare specimens of other low-latitude planktonic foraminifera were occasionally recorded (Herman, 1970a; Herman et al., 1971; Figs. 2 and 4).

Pteropoda

Limacina helicina (Phipps)
(Plate 4, Figs. 25 through 29)
Clio helicina Phipps, C. J. A voyage toward the North Pole undertaken by his Majesty's command, London, 1774:4.

Rare broken specimens were observed in Alpha Rise and Northwind Escarpment cores. They are most abundant in the shallow T 3 and D.st.A. core sediments constituting up to 5% of the total planktonic population (Figs. 2 and 4). The small shell (largest diameter 0.65 mm) with distinct umbilicus, has a whitish smooth surface. Two varieties can be distinguished, a low spired (Plate 4, Figs. 25 and 27) and a somewhat higher spired form (Plate 4, Figs. 28 and 29). Both lack the vertical striation described by Meisenheimer (1905) as well as the circumbilical keel. The latter structure, however, is developed in the adult shell only (Tesch, 1948) and the largest Arctic specimens observed measured less than 0.7 mm in diameter.

L. helicina was recorded in plankton tows in the Arctic by Meisenheimer (1905) and mentioned by Zenkevitch (1963). Inasmuch as only the juvenile shell portion was preserved, the species identification is tentative.

Benthonic Foraminifera

The percentages of benthonic foraminifera vary considerably, particularly in the foraminifera-poor zones, ranging from 0.1 to 5% and attaining a maximum of 97% in core D.st.A. 2 at 142 cm (Herman, 1969).

Zones in which this ratio exceeds 5%, two distinct benthonic assemblage prevail:

In the first (assemblage A), solution-resistant robust miliolids, *Cibicides wüellerstorfi*, and arenaceous species are associated with *G. pachyderma, G.* cf. *pachyderma* A, *G. paraobesa*, and *G. occlusa*; furthermore, the relative abundances of solution-susceptible *G. quinqueloba, G. exumbilicata*, and juvenile tests is low and that of shell fragments is high (Figs. 14 through 16). These levels are considered to represent periods of low sedimentation and/or corrosive deep water. Evidence of selective dissolution is present at various levels and is most obvious in sediments of unit III, in foraminifera-rich zones of unit I as well as in cores raised from depths greater than 2500 m (Tables 4 through 6; Herman, 1969).

In the second assemblage (B), shallow-water *Elphidium* spp. accompanied by deep-water forms such as *Stetsonia horvathi, Sphaeroidina bulloides*, and *Bolivina arctica* abound, constituting together with other benthonic species up to 97% of the foraminiferal fauna. Inasmuch as there is no evidence of differential solution at these levels, it is suggested that the temporary reduction in planktonic foraminiferal productivity was caused by drastic alteration in the environment, possibly resulting from decreased surface water salinities. Interestingly most levels that contain high percentages of benthonic foraminifera belonging to assemblage B correspond to times of reversals of the earth's magnetic field. These levels are prevalent in foraminifera-poor zones, interpreted as representing mild periods, times when the Arctic Ocean was free of a permantnt ice cap (Fig. 2).

Annotated List of Species

Following is an alphabetized reference list of common benthonic foraminifera:

Alveolophragmium subglobosum (G. O. Sars) 1868 (1869), *Forh. Vid. Selsk. Christiania*, 250.
This species is present together with other arenaceous forms in unit III sediments in Alpha Rise cores and in scattered core samples in Northwind

Table 3 Distribution and Coiling Direction of *G. pachyderma* and *G. quinqueloba*

Core no.	Depth in core		Herman	Hunkins et al. 1971a
T3/67-12	180 cm	*G. pachyderma*	2% dextral	50% dextral
		G. quinqueloba	45% dextral	not recorded
	190 cm	*G. pachyderma*	3% dextral	about 48% dextral
		G. quinqueloba	54% dextral	not recorded
T3/67-9	169 cm	*G. pachyderma*	0% dextral	50% dextral
		G. quinqueloba	52% dextral	not recorded
	170 cm	*G. pachyderma*	1% dextral	about 40% dextral
		G. quinqueloba	57% dextral	not recorded
	190 cm	*G. pachyderma*	2% dextral	about 30% dextral
		G. quinqueloba	27% dextral	not recorded

Escarpment sediments (Herman, 1970a; Herman et al., 1971).

Bolivina arctica Herman, 1972, *J. Foraminiferal Res.*, in press a.
(Plate 4, Figs. 30 and 31; Plate 20, Figs. 1 through 7.)
The distribution of this species is shown in Tables 4 through 7.

Bolivina arctica was referred to as *Bolivina* sp., and *Bolivina* sp. A in earlier publications (Herman, 1969, 1970a; Herman et al., 1971).

Bulimina aculeata d'Orbigny, 1826, *Ann. Sci. Nat., Paris, Ser. 1*, 7(7):269.
(Plate 21, Figs. 1 through 3.)
Distribution: highest frequencies were recorded in Northwind Escarpment sediments; in Alpha Rise sediments its occurrence is scattered and rare.

Cassidulina teretis Tappan 1951, *Contrib. Cushman Found. Foraminiferal Res.*, 2(1):7, pl. 1, Figs. 30a–c.
Distribution: highest frequencies were recorded in Northwind Escarpment sediments; in other regions rare specimens occur sporadically.

Cibicides wüellerstorfi (Schwager) 1866, *Novara Exped.* 1857–1859. *Geol. Theil*, 2 (2):258, pl. 7, Figs. 105–107.
(Plate 4, Figs. 32 through 34; Plate 21, Figs. 4 through 6; Plate 22, Fig. 5)
This species predominates in deep stations and in core levels where the solution-susceptible tests are rare or absent (Tables 4 through 6).

Cyclammina pusilla Brady, 1881, *Quart. J., Microscop. Sci.*, London, 1881, n. s., 21:53.
Distribution: highest frequencies were recorded in unit III sediments, in the Alpha Rise cores.

Elphidium clavatum Cushman, 1930, *U.S. Nat. Mus. Bull.* 104(7):20, pl. 7, Fig. 10.
Distribution: this species occurs sporadically in unit II sediments, attaining highest frequencies in sediments that contain coarse ice-rafted rock fragments; scattered occurrences were also recorded in unit I sediments (Table 4).

Elphidium incertum (Williamson) 1858, *Recent Foraminifera of Great Britain*: 44, pl. 3, Fig. 82.
Distribution: this taxon occurs sporadically together with other species of *Elphidium* in sediments that contain coarse ice-rafted detritus; highest frequencies were recorded in unit II sediments (Table 4).

Eponides tumidulus (Brady) var. *horvathi* Green, 1960, *Micropaleontology*, 6:74, pl. 1, Figs. 5a, b, c.
(Plate 4, Figs. 38 through 40.)
Together with *C. wüellerstorfi*, *Pyrgo* sp., and few other solution-resistant forms, this species is common in sediments containing partially dissolved assemblages (Tables 4 through 6).

Fissurina cucullata Silvestri, 1902, *Accad. Pont. Romana Nuovi Lincei, Mem., Roma*, 1902, 19:146.
Distribution: scattered occurrences were recorded in Alpha Rise sediments (Tables 4 and 6).

Fissurina kerguelenensis Parr, 1950, *B.A.N.Z. Antarctic Res. Exped. 1929–1931, Repts.*, Adelaide, 1950 (Zool., Bot.), ser. B, 5 (6):305, pl. 8, Fig. 7.
Distribution: scattered occurrences of rare specimens were recorded in Alpha Rise sediments (Table 4).

Glomospira gordialis (Jones and Parker), 1860, *Geol. Soc. London, Quart. J.*, London, England, 16:304, fig. not given.
Distribution: together with other arenaceous species it was recorded frequently in unit III sediments (Herman, 1970a).

Oolina hexagona (Williamson), 1848, *Ann. Mag. Nat. Hist.*, ser. 2, 1:20, pl. 2, Fig. 23.
Distribution: this taxon occurs infrequently in

Alpha Rise core sediments, as shown in Tables 4 and 6.

Oolina longispina (Brady), 1881, *Quart. J. Microscop. Sci.*, London, England, 21:61.
Challenger Expedition 1873–1876 Repts. London, *Zool.* 9(22):444, 454, pl. 56, Figs. 33 through 36.
The distribution of this species is shown in Tables 4 and 6.

Patellina corrugata Williamson, 1858, *Roy. Soc. London*, England, 46, pl. 3, Figs. 86 through 89, 89a.
Distribution: scattered occurrences of rare specimens were recorded in Alpha Rise and Northwind Escarpment sediments.

Pseudoeponides umbonatus (Reuss), 1851, *Z. Deutsch. Geol. Ges.*, 3:75, pl. 5, Figs. 35a–c.
(Plate 4, Figs. 34 and 35; Plate 22, Figs. 1 and 2.)
This species is common in both foraminifera-rich and foraminifera-poor zones of units II and I. It occurs together with other solution-resistant forms such as *C. wüellerstorfi*, *E. tumidulus horvathi*, and robust miliolids in deep stations (Tables 4 through 6).

Pullenia bulloides (d'Orbigny), 1826, *Ann. Sci. Nat.*, 7:293; *1846, Foram. Bass. Tert. Vienne*: 107, The distribution of this species is shown in Tables 4 through 6.

Sphaeroidina bulloides d'Orbigny, 1826, *Ann. Sci. Nat.*, 7(1):267; Modèles, no. 65.
Distribution: highest frequencies were recorded in unit II sediments (Tables 4 through 6).

Stetsonia horvathi Green, 1960, *Micropaleontology*, 6(1):72, pl. 1, Fig. 6.
(Plate 4, Figs. 37 and 38; Plate 22, Fig. 6.)
The distribution of this species is shown in Tables 4 through 6.

Virgulina davisi Chapman and Parr, 1937, *Australian Antarctic Expedition* 1911–1914, *Sci. Repts.*, Sydney, 1937, ser. C. (Zool., Bot.), 1(2):88.
Distribution: this species was recorded in both foraminifera-poor and foraminifera-rich layers (Table 6).

Virgulina loeblichi Feyling-Hanssen, 1954, *Norsk Geol. Tidsskr.*, 33:191, pl. 1, Figs. 14–18.
(Plate 22, Figs. 3 and 4.)
The distribution of this species is shown in Tables 4 through 6.

Rare specimens of the following taxa were sparingly observed: *Buccella* cf. *frigida*, *Cassidulina norcrossi*, *Cassidulina* sp., *Cibicides* sp., *Cornuspira* sp., *Dentalina* spp., *Discorbis* sp., *Elphidium* sp., *Lagena* spp., *Nonion* sp., and *Robertina* sp.

PALEOSALINITIES AND PALEOTEMPERATURES

Oxygen Isotope Ratios of Foraminifera

Among the factors conditioning and modifying the surface water salinity in the Arctic, precipitation, river runoff, evaporation, drift-ice, inflow and subsequent mixing of Atlantic and Pacific water, and convective mixing are considered the most important. The oxygen isotopic composition of Arctic surface water is determined by the same factors; evaporation results in concentration of ^{18}O, while precipitation and river inflow causes its dilution (Craig and Gordon, 1965). The ranges of present-day surface water temperatures and salinities are -1.6 to 0°C and <29 to $\sim 34‰$ respectively (Treshnikov, 1959; Coachman and Aagaard, this volume).

As a result of slow vertical mixing, present-day surface salinities are about 4.8‰ lower than those of the bottom water (Treshnikov, 1959). $^{18/16}O$ analysis of Arctic water indicates that the ^{18}O depletion closely follows the salinity profile with $\delta^{18}O$ changing by 0.8‰ with 1‰ change in salinity in the upper 350 m (van Donk and Mathieu, 1969). This surface sea water is depleted in ^{18}O by about 4‰, whereas at greater depth the $\delta^{18}O$ approaches values close to SMOW.

The $^{18/16}O$ ratios of calcareous tests from foraminifera-rich and foraminifera-poor layers in cores T 3-67-9 and T 3-67-12 were measured by Grazzini (in Herman et al., 1971). The foraminifera were hand-picked and treated with commercial Clorox solution, the CO_2 was extracted from the $CaCO_3$ according to the method described by McCrea (1950).

To calculate the temperatures from the $^{18/16}O$, the equation of Epstein et al. (1953), modified by Craig (1965) was used;

$$t = 16.9 - 4.2\,(\delta s - \delta w) + 0.13\,(\delta s - \delta w)^2$$

[where $\delta s = \delta^{18}O$ of sample and $\delta w = \delta^{18}O$ of sea water.]

The established relation,

$$\delta s\,^{18}O = \left(\frac{^{18}O/^{16}O \text{ sample}}{^{18}O/^{16}O \text{ standard}} - 1\right) \times 1000$$

depends on the water temperature (t,°C) as well as

on the isotopic composition of the water in which the organisms lived.

When sufficient shell material was available, planktonic tests were analyzed separately from the benthonic forms. Moreover, when samples yielded adequate amounts of *G. quinqueloba* and *G. pachyderma,* their tests were analyzed separately. As shown in Figs. 17 and 18, the former registered consistently lower $\delta^{18}O$ values than the latter. This can be readily explained by their differing life habitats. *G. quinqueloba* an epiplanktonic form, incorporates the $CaCO_3$ necessary for test building at the surface, in the isotopically lighter, low-salinity water, whereas the thick-shelled *G. pachyderma,* known to inhabit greater water depths secretes the early part of its test at surface, and the outer thick cortex at greater depth, in denser more saline water, hence its higher $^{18/16}O$ ratio. Determinations of $\delta^{18}O$ made on composite samples of benthonic and planktonic foraminifera indicate that, with one exception, the bottom water was enriched in ^{18}O, attesting to the restricted vertical mixing and subsequent density stratification of the Arctic water (Figs. 17 and 18).

As mentioned previously, changes in the $\delta^{18}O$ of the sea water are reflected in the oxygen isotopic ratios of the calcareous foraminiferal tests. Interestingly, the greatest shift toward positive values occurs at 65 cm depth in core T 3-67-9 and at 40 cm depth in core T 3-67-12 (Figs. 17 and 18); both samples come from foraminifera-rich horizons and were deposited during the Brunhes epoch, when sea water was colder and/or more saline than today. Below 85 cm $\delta^{18}O$ values are more negative than in the core "tops," suggesting slightly warmer and/or less saline water in the Arctic basin between approximately 2 and 0.7 m.y. ago than those of today.

Estimates of past water temperatures and salinities based on variations in the $^{18/16}O$ ratios of planktonic foraminiferal tests are in agreement with those deduced from faunal analyses and suggest that average surface water temperatures have varied between about -2 and $+0.5°C$, with salinities ranging from <29 to $\sim 34‰$ during the time interval represented by the investigated cores (Herman, unpublished; Herman et al., 1971). Temperature minima and salinity maxima correspond to the upper Pleistocene, probably the Riss or Würm Glacial, whereas lowest salinities and/or higher temperatures were recorded by planktonic foraminifera between about 2 and 0.7 m.y.b.p., during the Matuyama epoch.

Paleoclimatic and Paleo-oceanographic Record

Faunal analyses, oxygen isotope measurements, and lithologic characteristics were the basis for reconstructing the paleo-oceanographic and paleoclimatic history of the Arctic Ocean back to mid-Pliocene. As mentioned earlier, three major climatic units were distinguished.

In unit III sediments, deposited earlier than 2.4 m.y.b.p., the fauna constitutes less than 1% of the coarse fraction (Figs. 2, 11, and 19 through 21).

Planktonic foraminifera are dominated by sinistral *G. pachyderma, G.* cf. *pachyderma* A, *G. paraobesa,* and *G. occlusa,* some of which are corroded. Juvenile specimens generally constitute 1 to 2% of the planktonic assemblages in this unit but are more abundant in units II and I.

The coiling direction of *G. pachyderma* is predominantly sinistral (94% sinistral, Table 2). Similar values were obtained for the foraminifera-rich zones of unit I (Herman 1969; Fig. 12). Arenaceous foraminifera (*G. gordialis, C. pusilla,* and *A. subglobosum*) dominate the benthonic foraminiferal assemblages. The composition and character of the preserved fauna suggests differential solution of the less-resistant limy tests. In the shallow core T 3-67-4, *G. exumbilicata* and *G. quinqueloba* together, constitute up to 55% of the planktonic fauna (Figs. 2 and 11), indicating that these taxa inhabited the Arctic earlier than 2.4 m.y.b.p. Their absence or low frequency in deep-water sediments is possibly due to dissolution. No $^{18/16}O$ measurements were attempted on foraminiferal tests of this unit. It is tentatively concluded that unit III sediments were deposited during a period with climates similar to those that prevailed during the deposition of foraminifera-rich beds of unit I. Lower than present sedimentation rates and/or corrosive deep water may account for the selective solution of the less-resistant limy tests and the impoverished character of the fauna. These deductions regarding paleoenvironments should be regarded as tentative and subject to modification as new data become available. The unit II/III boundary is marked by simultaneous lithologic and faunal changes and occurs near a magnetic reversal. The lithologic change from dark brown Mn and Fe oxide-rich silty-lutite below to tan silts with higher percentages of coarse ice-rafted detritus above the boundary is accompanied by faunal alterations.

Sediments of unit II deposited between about

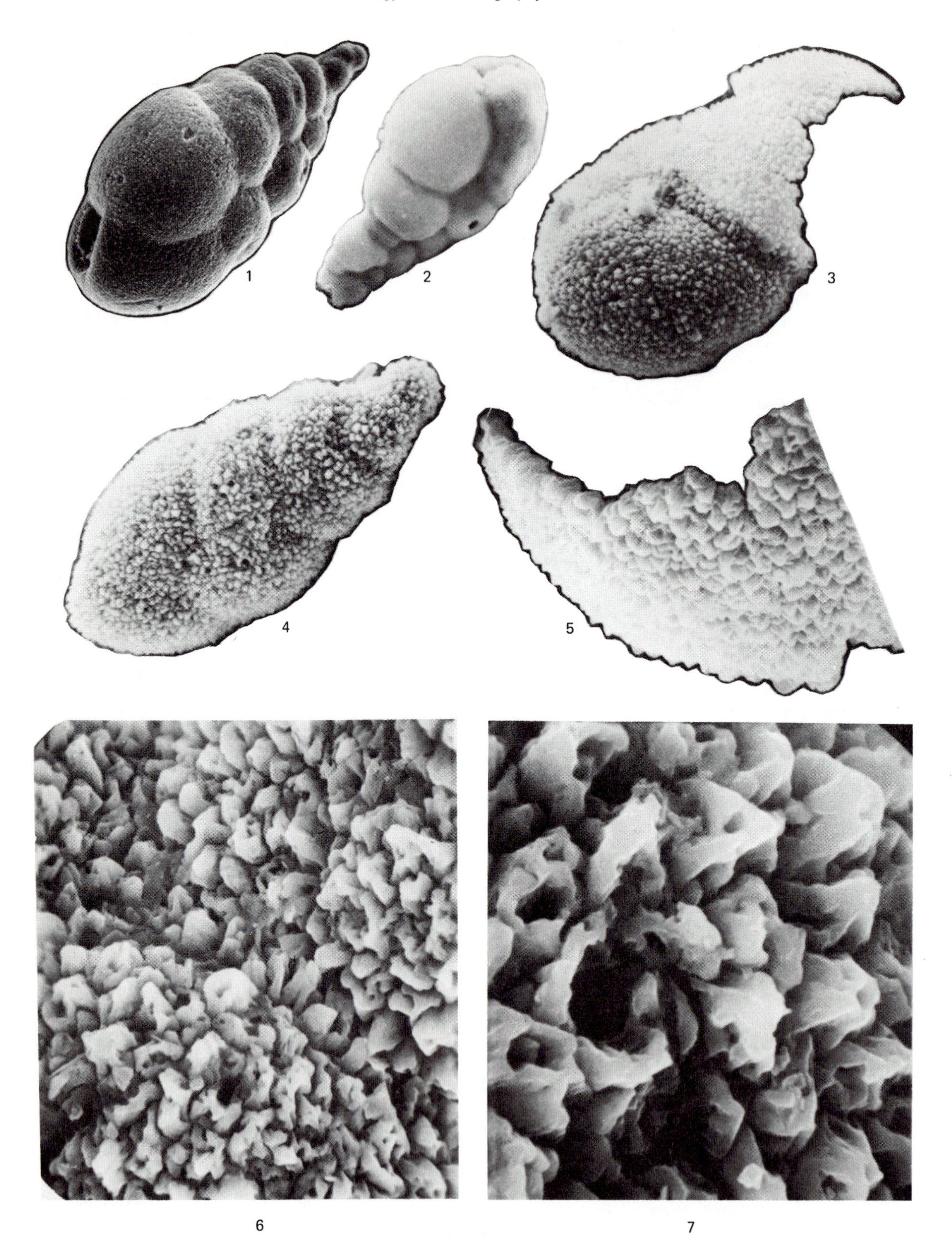
1
2
3
4
5
6
7

2.4 and 0.7 m.y. ago, during the Matuyama epoch, are poor in both foraminifera and in Fe and Mn oxides, but contain one foraminifera-rich layer. Two faunal assemblages are distinguished in this unit, assemblage α and assemblage B.

In assemblage α planktonic foraminifera constitute >90% of the foraminiferal fauna. The relative abundance of *G. exumbilicata* and *G. quinqueloba* is high in shallow cores T 3-67-4, T 3-67-9, D.st.A. 2, and D.st.A. 5 and decreases considerably in deeper cores, suggesting that their low frequency is due to solution. Data on their present-day distributional patterns and $^{18/16}O$ determinations indicate that both *G. quinqueloba* and *G. exumbilicata* are eurytherm and euryhaline species (Herman 1970a; Herman et al., 1971; Herman, 1972b). Low surface water salinities are believed to have determined the dominance of these two species in Matuyama time. *G. pachyderma, G.* cf. *pachyderma* A, *G. occlusa,* and *G. paraobesa* were not able to proliferate in low salinities, whereas *G. quinqueloba* and *G. exumbilicata* adapted to the new environment more readily. Occasionally low-latitude foraminifera were observed. Assemblage α was deposited during mild intervals with moderate to low salinities.

In assemblage B, discussed in a preceding section, benthonic deep-water calcareous species including *B. arctica, S. horvathi,* and *S. bulloides,* frequently accompanied by shallow-water *Elphidium* spp. constitute up to 97% of the foraminiferal fauna. The dominant planktonic taxa in these zones are *G. quinqueloba* and *G. exumbilicata.* Dextral *G. pachyderma, G.* cf. *pachyderma,* and *G. occlusa* attain highest frequencies in layers containing assemblage B fauna (Tables 4 through 6; Figs. 11 through 13, 15, 16, 19 through 21). Furthermore these horizons are characterized by the presence of abundant coarse ice-rafted detrital minerals. Assemblage B is thought to have been deposited during warmer episodes than assemblage α. Temperature increase enhanced river and glacial melt-water supply to the Arctic and caused further decrease in surface water salinities. The drastic alteration in the environment lead to reduction in planktonic productivity and the consequent dominance of benthonic foraminifera.

Paleontologic and oxygen isotope data indicate that Matuyama was a time of equable and higher global temperatures than the preceding Gauss and the following early Brunhes (e.g., Hays and Opdyke, 1967; Donahue, 1967; Herman, 1970a; Bandy et al., 1971; Olsson, 1971; Hopkins, 1972, and others).

One of the important effects of the worldwide climatic amelioration (warming) in the polar regions was increased ice melting. As long as continental glaciers existed, calved, and melted into the Arctic, the water did not warm up appreciably. The main consequence of increased cold fresh-water discharge into the Arctic was surface-water freshening. Consequently, faunal changes probably reflect the effects of salinity oscillations rather than temperature changes. It is assumed that during most of the Matuyama epoch the Arctic Ocean was free of permanent pack-ice, and the debris-laden ice drifted unimpeded across the ocean, melting and releasing to the sea floor the incorporated detritus. The unit II/I climatic boundary, defined by faunal and lithologic changes, approximately corresponds to the last major polarity reversal of the earth's magnetic field, marking the Brunhes/Matuyama boundary (Herman, 1970a; Herman et al., 1971; Figs. 2, 11, 12, 19 through 21).

The sediments of unit I were laid down in the last 0.7 million years during the Brunhes epoch, a time of conspicuous climatic fluctuations, as indicated by temporal variations in the faunal composition and in the fauna/mineral ratio (Herman, 1969, 1970a; Figs. 2, 4, 11 through 21). Foraminifera-rich and foraminifera-poor beds alternate, the former representing conditions similar to those prevailing today (permanent ice-cap) and contain cold-water sinistral *G. pachyderma* and *G. occlusa* almost exclusively. However, *G. quinqueloba* and

Plate 20
(SEM Illustrations)

Fig. 1. *Bolivina arctica,* side view; X 287. Illustration shows aperture and sutures. T 3-67-3, 245 cm.
Fig. 2. *B. arctica,* side view; X 323. T 3-66-S.8, 5 cm.
Fig. 3. *B. arctica,* front view; X 612. T 3-66-S.8, 5 cm.
Fig. 4. *B. arctica,* side view; about X 510. T 3-66-S.8, 5 cm.
Fig. 5. *B. arctica,* same as Fig. 3; X 1530. T 3-66-S.8, 5 cm.
Fig. 6. *B. arctica,* same as Fig. 3; X 2550. T 3-66-S.8, 5 cm.
Fig. 7. *B. arctica,* same as Fig. 3; X 5100. T 3-66-S.8, 5 cm. Note closely packed, hollow, conical structures and intervening pores in Figs. 6 and 7.

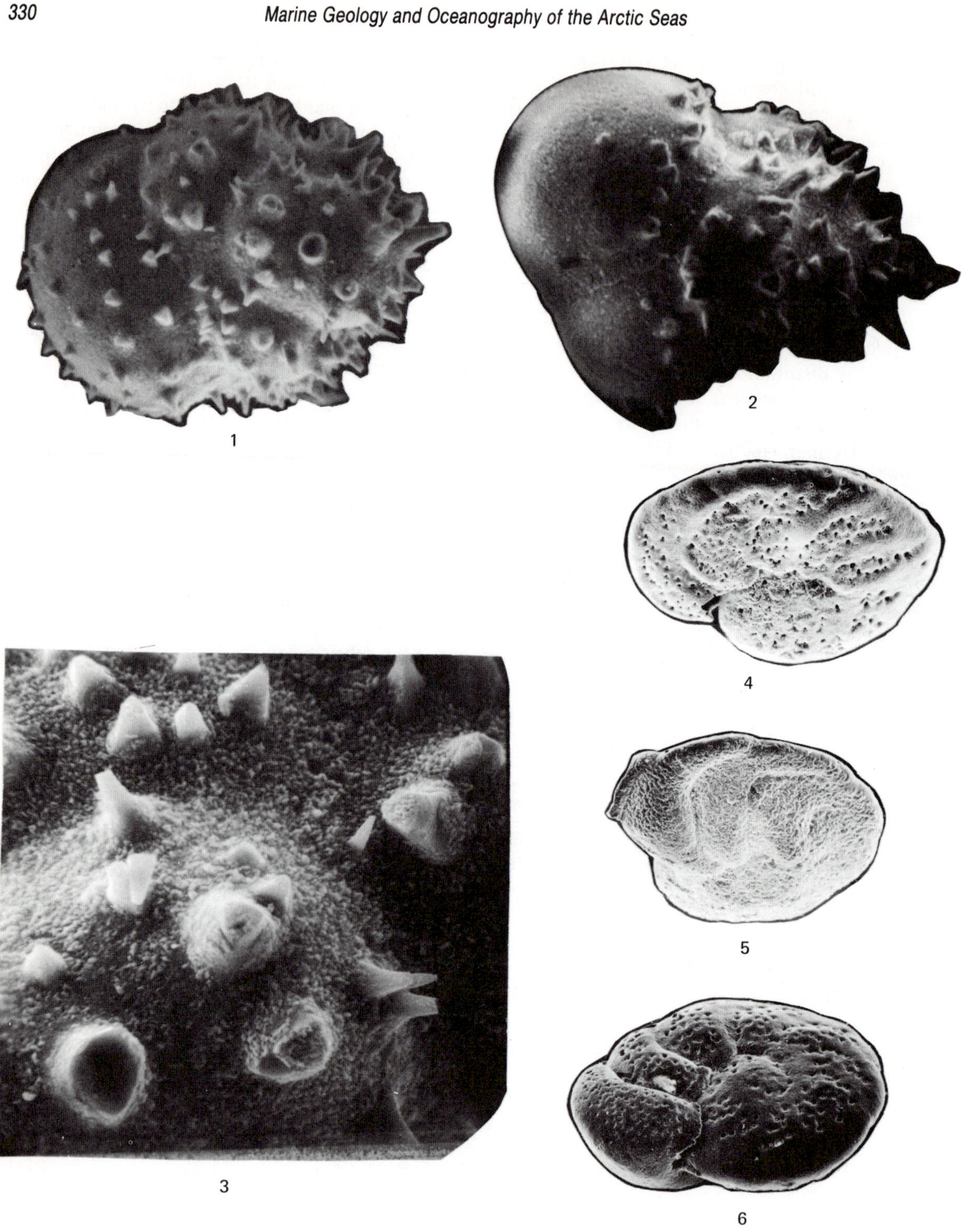

Plate 21
(SEM Illustrations)

Fig. 1. *Bulimina aculeata*, side view; X 360. T 3-66-S.8, 5 cm.
Fig. 2. *B. aculeata*; about X 360. T 3-66-S.8, 5 cm.
Fig. 3. *B. aculeata*, same as Fig. 1; X 900. T 3-66-S.8, 5 cm.
Fig. 4. *Cibicides wüellerstorfi*; about X 72. T 3-67-3, 30 cm.
Fig. 5. *C. wüellerstorfi*; about X 72. T 3-67-3, 30 cm.
Fig. 6. *C. wüellerstorfi*; about X 72. T 3-67-3, 30 cm.

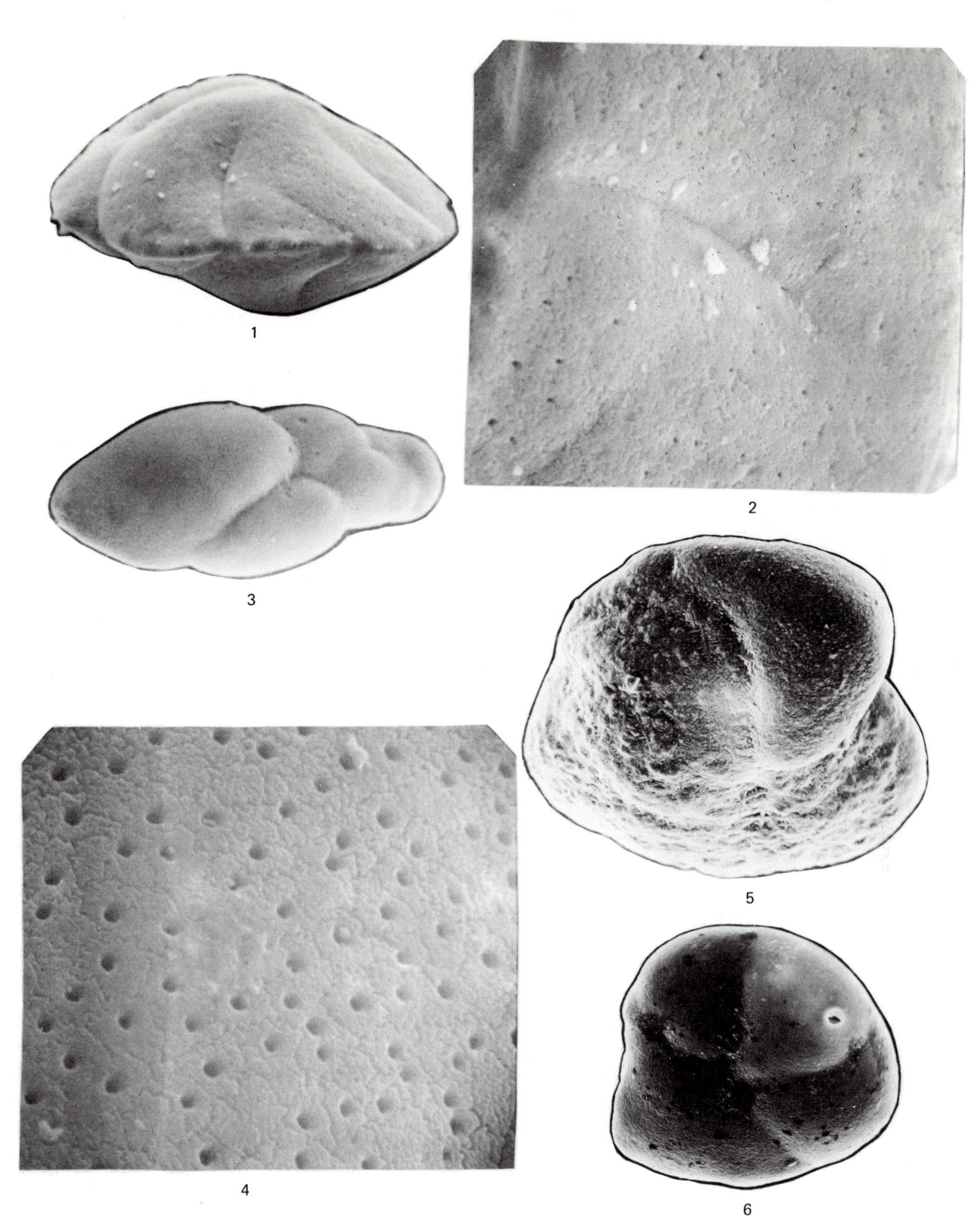

Plate 22
(SEM Illustrations)

Fig. 1. *Pseudoeponides umbonatus*; X 255. T 3-66-S.8, 5 cm.
Fig. 2. *P. umbonatus*, same as Fig. 1; X 850. T 3-66-S.8, 5 cm.
Fig. 3. *Virgulina loeblichi*, side view; X 383. T 3-66-S.8, 5 cm.
Fig. 4. *V. loeblichi*, same as Fig. 3; X 3825. T 3-66-S.8, 5 cm.
Fig. 5. *Cibicides wüellerstorfi*; X 170. T 3-67-3, 30 cm.
Fig. 6. *Stetsonia horvathi*; X 425. T 3-67-3, 300 cm.

Table 4 Abundance of Benthonic Foraminifera

	Top	4–5	10	15	22	30	40	49	70	80	89	98	112	121	130	140	151
Bolivina arctica						...				...			...	ooo	...	...	...
Cassidulina sp.																	
Cibicides wüellerstorfi	—	—	—	...	—	■	xxx	xxx	...	...	...	■	ooo		...	ooo	...
Dentalina sp.		...	...		ooo					ooo	ooo				...		
Elphidium clavatum																	
Elphidium incertum																	
Elphidium sp.									...		...				...		ooo
Eponides tumidulus horvathi	—	xxx	...	...	xxx	xxx	—	—			...						
Fissurina cucullata				...				...			ooo				...		
Fissurina kerguelenensis				...				...			ooo				...		...
Glandulina sp.		...															
Lagena sp.													ooo	ooo			
Nonion sp.																	
Oolina hexagona						...	ooo				ooo		...				
Oolina longispina															ooo		
Pseudoeponides umbonatus	—	—	...	...	—	—	■	■	...	xxx	...	...	xxx	—	—	...	...
Pyrgo sp.	...						ooo										
Quinqueloculina sp.	ooo	—	—	—	xxx	—	...	...	...	...	...						
Sphaeroidina bulloides																	
Stetsonia horvathi						xxx	xxx	xxx	...					—	■		—
Virgulina loeblichi		...			...		ooo	...	...		...				...		ooo
Miliolids											ooo						
Arenaceous spp. and fragments	...																

Key: ooo Present; ... 1 to 5 Rare; — 6 to 10 Frequent; xxx 11 to 25 Common; ■ 26 to 100 Abundant; ▨ >100 Very abundant

Table 5 Abundance of Benthonic Foraminifera

	0	6	8	10	12	18	20	24	25	33	35	40	42	43	49	54	55	60	62	70	75	79	82	85	86	91	99
Bolivina arctica																				...	...		...				...
Cassidulina sp.			...	ooo		...	...	...		...	...	...	...						...	...	...	...	...			...	...
Cibicides sp.	...										ooo				ooo		...	...									
Cibicides wüellerstorfi	xxx	—		...	...	—	xxx	...																...	...		
Dentalina sp.													ooo											...		ooo	
Elphidium sp.	...										...			ooo		...				ooo				...	...	...	
Enthoselenia sp.												ooo											...		...	...	
Pseudoeponides umbonatus	—	—	...	...		xxx	■	xxx	...	...	...	ooo	ooo	...	...	...	—	—	...	xxx	xxx	...	ooo				
Eponides tumidulus horvathi	—	—	...	...	xxx	xxx	xxx	—	...	...	...			...	...		ooo		...	—			...				
Eponides sp.																									...		
Lagena sp.	...	...	...	...					...	...	ooo	ooo							...					...	ooo		
Sphaeroidina bulloides												...	...	...		—		...		—	...		...	...			
Stetsonia horvathi	—	ooo		...		—	xxx	...	—		—	...	...	—		xxx	—	xxx	...	xxx		—	xxx	...	—	—	—
Virgulina loeblichi				...		...			...		ooo																
Virgulina sp.	...	...	...																								
Miliolids	xxx	xxx	—	...	...	...	ooo			xxx	...	...	...	ooo			...										
Arenaceous spp.	...																...			...							
Arenaceous fragments																											

Key: ooo Present; ... 1 to 5 Rare; — 6 to 10 Frequent; xxx 11 to 25 Common; ■ 26 to 100 Abundant; ▨ >100 Very abundant in Core T3-67-3, Core Depth in cm

160 173 193 200 210 219 233 245 250 261 270 282 291 301 311 321 333 337 349 354 361 368 372 380 460

in Core T3-67-11, Core Depth in cm

104 109 115 120 130 135 140 144 145 145½ 150 160 169 172 175 182 185 188 192 194 200 202 210 216 224 230 234 250 260 315 400

Table 6 Abundance of Benthonic Foraminifera

	0	12	19	25	32	42	50	55	69	78	82	85	90	100	110	120	131	140
Bolivina arctica						...		...	...			...	...	—	...	...	—	...
Buccella frigida																	...	...
Cassidulina sp.												...			...			...
Cibicides sp.						...						...	...					...
Cibicides wüellerstorfi	...	xxx			...		...				...	...	—					
Dentalina sp.	...		...				...	...							...			...
Elphidium sp.		ooo										...		...	...	...	xxx	■
Eponides tumidulus horvathi	xxx	...			...	...												
Enthoselenia sp.																		
Fissurina cucullata												...		...	...	...	...	—
Lagena sp.	...	...								...							xxx	...
Oolina hexagona															...		...	
Oolina longispina															...		...	...
Pseudoeponides umbonatus	xxx	■	—	...	—	xxx												
Pyrgo sp.	—	—							...					■	■	...	■	■
Sphaeroidina bulloides			—	—	...	■	...	—	...	...	—	—	—					
Stetsonia horvathi	...		—	...	...	■	—	...	xxx	...	—	—	■	...	—	—	—	▨
Virgulina davisi	...						...					—	...			...	...	...
Virgulina loeblichi											...							
Virgulina sp.			...		...		...							...		...		...
Miliolids	xxx	...	...	...											■		ooo	ooo
Arenaceous spp.	...					...												...
Arenaceous fragments																		

Key: ooo Present; ... 1 to 5 Rare; — 6 to 10 Frequent; xxx 11 to 25 Common; ■ 26 to 100 Abundant; ▨ >100 Very abundant

G. exumbilicata attain high frequencies at the commencement and end of some of these cold periods (Figs. 2, 11, and 15) and the percentage of dextral warm-water tolerant *G. pachyderma* is highest in these zones. This suggests temporary warming with subsequent increased river and glacial melt-water supply, followed by formation of low-salinity surface water to which euryhaline-eurytherm *G. exumbilicata* and *G. quinqueloba* were able to adapt best (Herman, 1970a; Herman et al., 1971; Figs. 2, 11, 15, and 19 through 21). In the foraminifera-poor beds which are believed by the author to represent seasonally pack-ice free periods, the dominant sinistral *G. pachyderma* and *G. occlusa* are accompanied by *G. paraobesa, G.* cf. *pachyderma* A, *G. quinqueloba,* and *G. exumbilicata.* Occasionally low-latitude foraminifera occur (Herman, 1964, 1969, 1970a; Figs. 2 and 4).

The Brunhes cold-"warm" temperature fluctuations represented in most cores by 4 to 6 foraminifera-rich, foraminifera-poor sequences are possibly correlative with the classic Donau, Günz, Mindel Riss and Würm Glacials and intervening interglacials, respectively.

Results and Their Implications

Recent studies of deep-sea cores allow further elaboration on the evolution of Arctic paleo-oceanography and paleoclimatology proposed by Herman (1970a) and adapted by Hunkins et al. (1971a). The well-documented progressive late Cenozoic global refrigeration, which was accompanied by expansion of continental ice sheets and eustatic sea level drop was coupled with continental uplift (Damon, 1968; Hamilton, 1968; Tanner, 1965, 1968). These combined terrestrial effects gradually reduced influx of warm Pacific and Atlantic water into the Arctic, converting it into an inland sea. Lowering of temperatures below a certain critical value permitted astronomical factors (essentially decreased summer insolation at mid and high latitudes) to trigger the major global glaciation and the ensuing glacial-interglacial oscillations which characterize the upper Pleistocene

in Core T3-67-12, Core Depth in cm

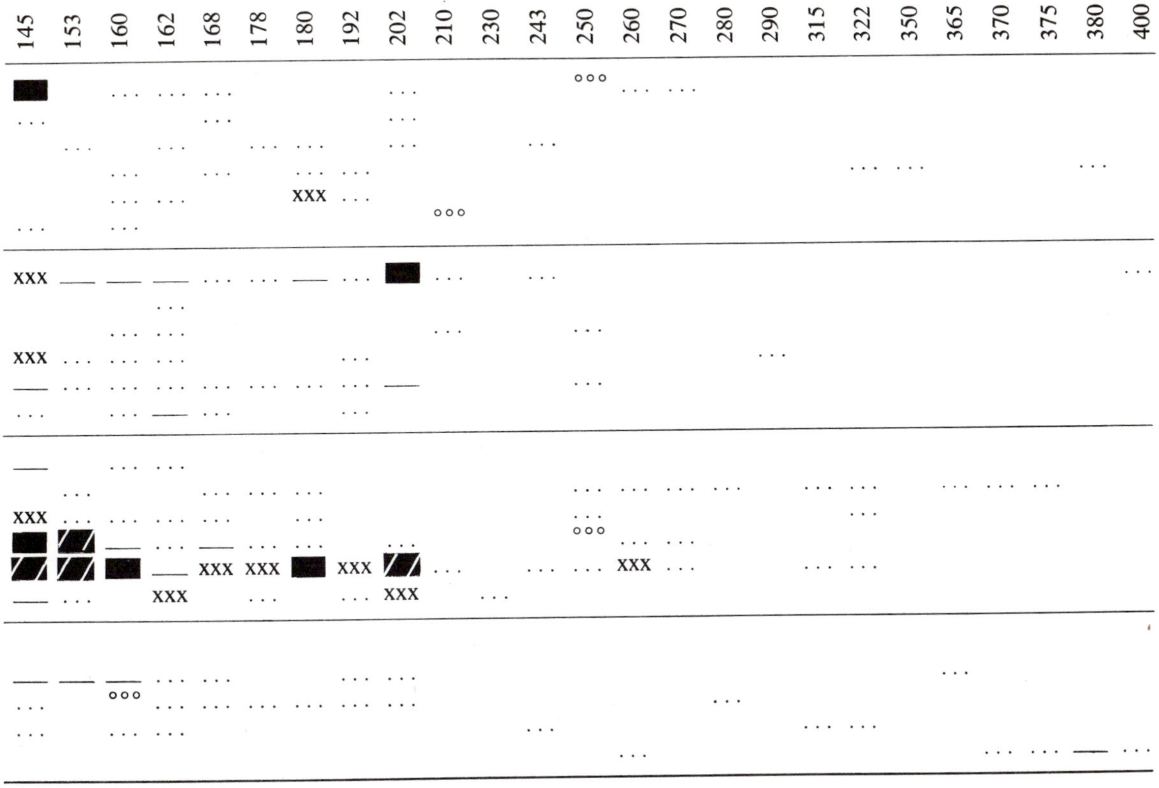

145	153	160	162	168	178	180	192	202	210	230	243	250	260	270	280	290	315	322	350	365	370	375	380	400
■		...	...	...				...				ooo	...	...										
...				...				...																
	...		...		...	...		...			...													
		...		...		...	...											...	...				...	
		...	...			xxx	...		ooo															
...		...																						
xxx	—	—	—	...	...	—	...	■	...		...													...
			...																					
		...	...						...			...												
xxx	...	...	...				...									...								
—	...	...	...	...	...	...	...	—				...												
...		...	—	...			...																	
—		...	...																					
	...			...	...	...						...	...	...	...		...	...		...	...	...		
xxx	...	...	...	...		...						...						...						
■	▨	—	...	—	...	...		...				ooo	...	...										
▨	▨	■	—	xxx	xxx	■	xxx	▨	...		...	...	xxx	...			...	...						
—	...		xxx		...		...	xxx		...														
—	—	—	...	...			...	...												...				
...		ooo	...	...	...	...	...	...							...									
...		...	...								...						...	...						
													...							...	...		—	...

(The Milankovitch hypothesis, Milankovitch, 1930; also discussed by Zeuner, 1959; elaborated and modified by Emiliani and Geiss, 1959; Fairbridge, 1961, 1970; Emiliani, 1966a; Van den Heuvel and Buurman, this volume and bibliography therein).

The low- and mid-latitude periodicity of changes from glacial to interglacial climates observed by Emiliani (1966b) and its coincidence with calculated insolation variations (Milankovitch, 1930; Van den Heuvel and Buurman, this volume) are more difficult to discern in the Arctic deep-sea record. The Brunhes severe cold/"warm" cycles are represented by 4 to 6 foraminifera-rich/foraminifera-poor sequences (Figs. 2, 4, and 12 through 21) possibly correlative with the classic Donau, Günz, Mindel Riss and Würm Glacials and intervening interglacials, respectively. The former were deposited during permanent pack-ice covered times, the latter in seasonally pack-ice free periods (Herman, 1969, 1970a).

Global Matuyama temperatures were higher than those which prevailed throughout most of the Brunhes epoch. The effect of regional increase in air temperature on the Arctic water was mainly freshening of the surficial water layers. Euryhaline and eurytherm species (*G. quinqueloba* and *G. exumbilicata*) predominated in mild periods, and benthonic foraminifera belonging to assemblage B during the warmest episodes. The paucity of the fauna in unit III sediments (late Gauss) precludes definitive paleoclimatic reconstruction of this time interval. It is tentatively suggested that climates were similar to those that prevailed during the deposition of foraminifera-rich layers of unit I.

Discussion

Recently, Hunkins et al. (1971a, 1971b) published results for several Arctic cores that are at variance with the author's data for the same cores. Hunkins et al. (1971a, 1971b) report occurrences of dextral *G. pachyderma* in excess of 60% between 700,000 to 450,000 years B.P. (Fig. 22). Their data suggest that surface water temperatures were 9 to 12°C higher than those prevailing today. A warming of such order of magnitude should have

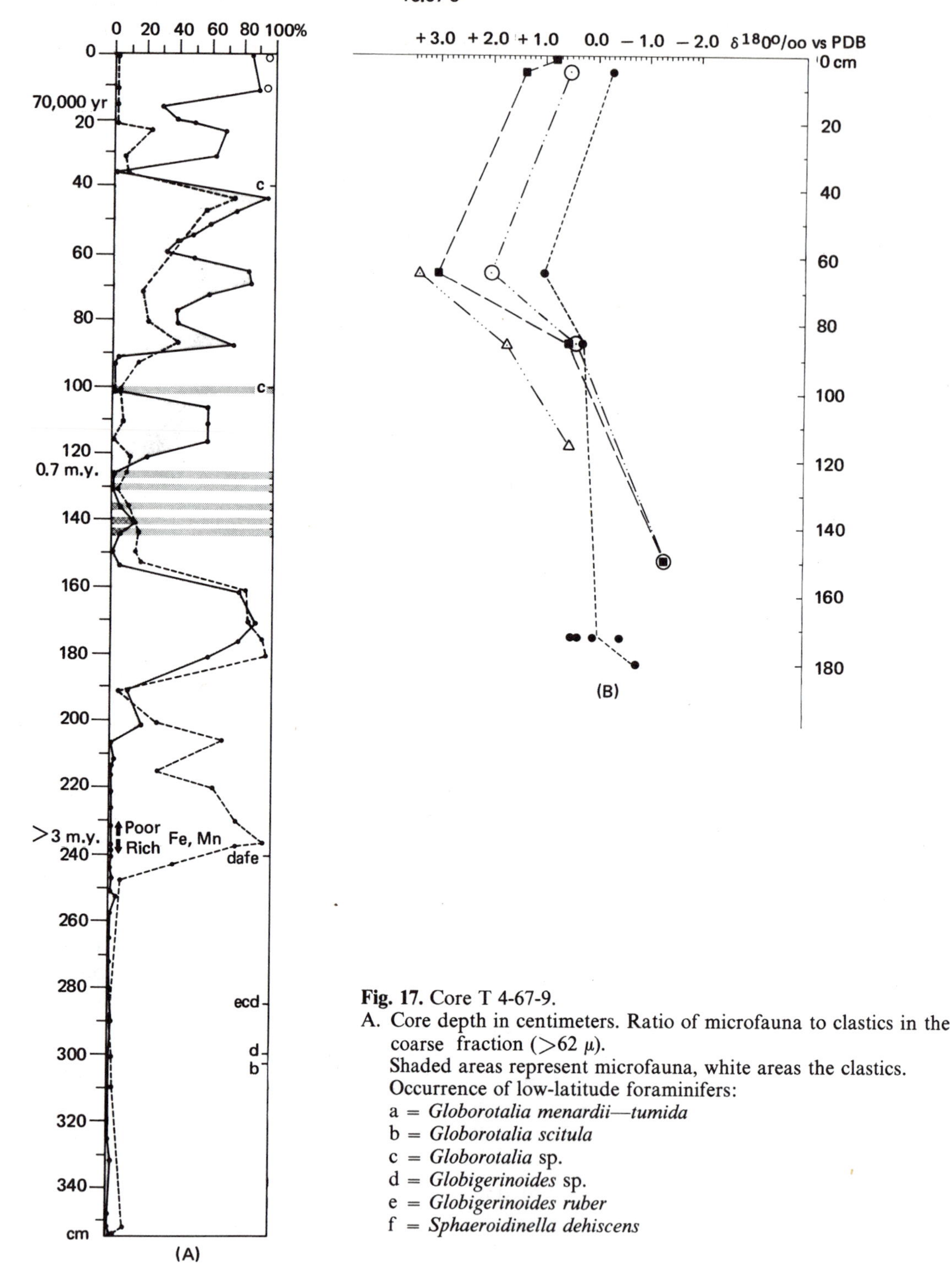

Fig. 17. Core T 4-67-9.

A. Core depth in centimeters. Ratio of microfauna to clastics in the coarse fraction (>62 μ).
Shaded areas represent microfauna, white areas the clastics.
Occurrence of low-latitude foraminifers:
a = *Globorotalia menardii—tumida*
b = *Globorotalia scitula*
c = *Globorotalia* sp.
d = *Globigerinoides* sp.
e = *Globigerinoides ruber*
f = *Sphaeroidinella dehiscens*

Circles: Pteropods.
Dashed line represents the percentage of *G. quinqueloba* complex out of the total planktonic population.
Dark horizontal lines are zones in which benthonic foraminifers are in excess of 10%.

B. Variation in $\delta^{18}O$
1. *G. pachyderma*
2. *G. quinqueloba* complex
3. Average planktonic species
4. Total planktonic and benthonic foraminifera Modified after Herman et al. (1971).

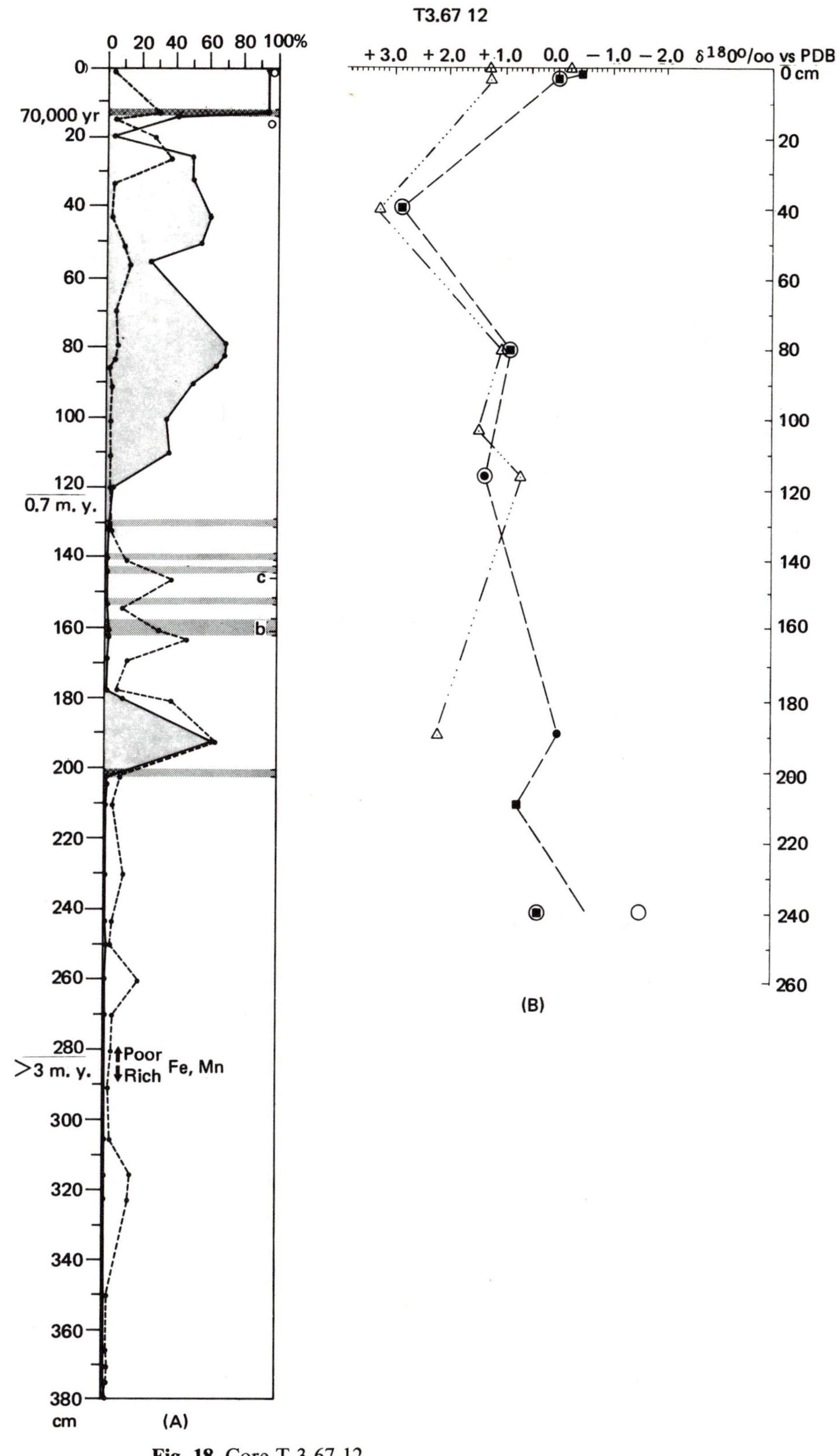

Fig. 18. Core T 3-67-12.
A. Ratio of microfauna to clastics. See the legend for Fig. 17.
B. Variation in $\delta^{18}O$. See the legend for Fig. 17.

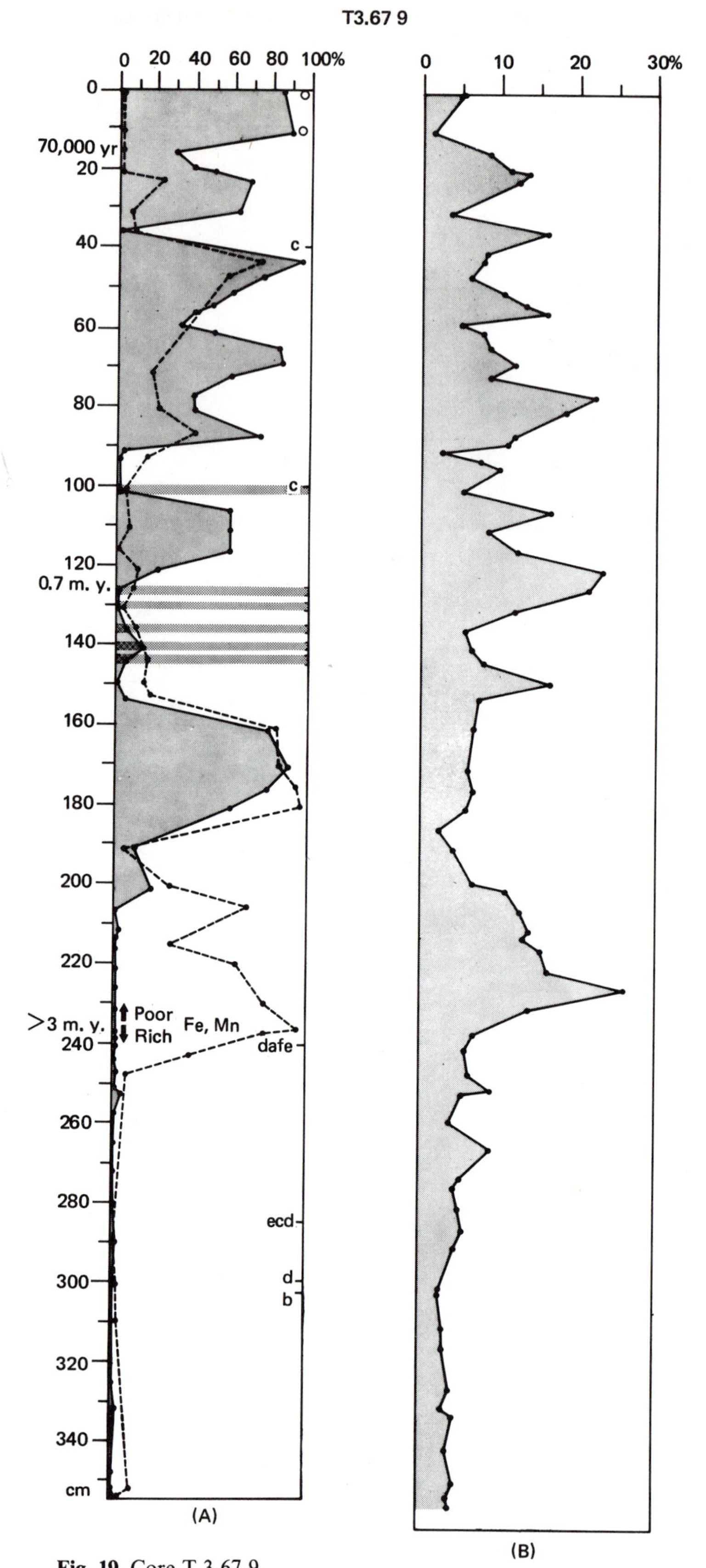

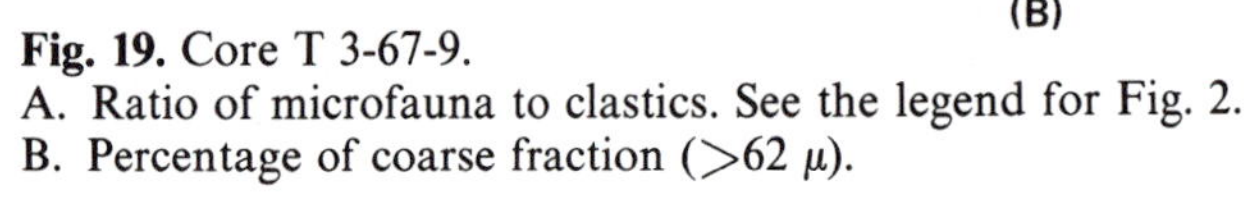

Fig. 19. Core T 3-67-9.
A. Ratio of microfauna to clastics. See the legend for Fig. 2.
B. Percentage of coarse fraction (>62 μ).

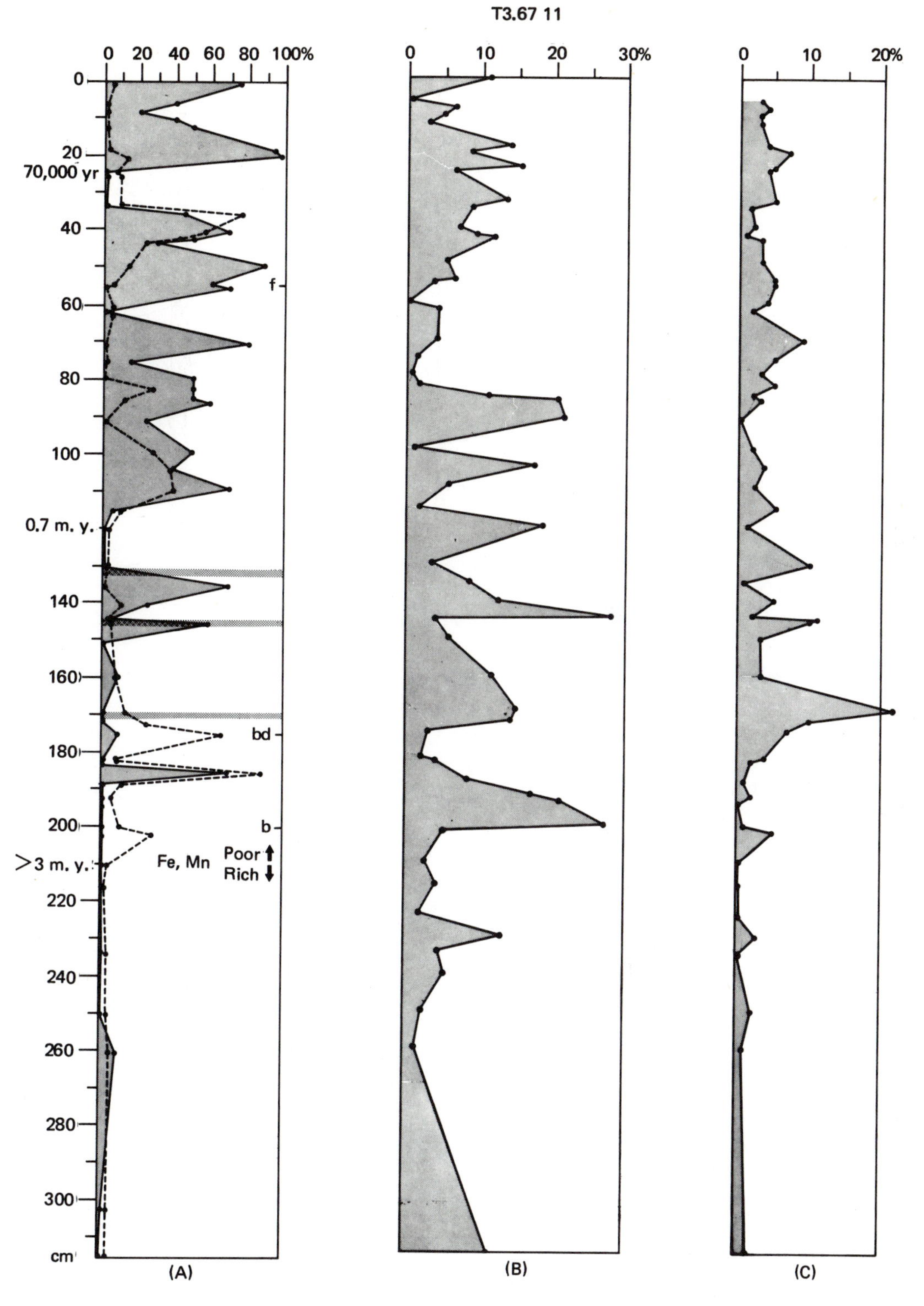

Fig. 20. Core T 3-67-11.
A. Ratio of microfauna to clastics. See the legend for Fig. 2.
B. Percentage of coarse fraction (>62 μ).
C. Percentage of benthonic foraminifera of the total foraminiferal fauna.

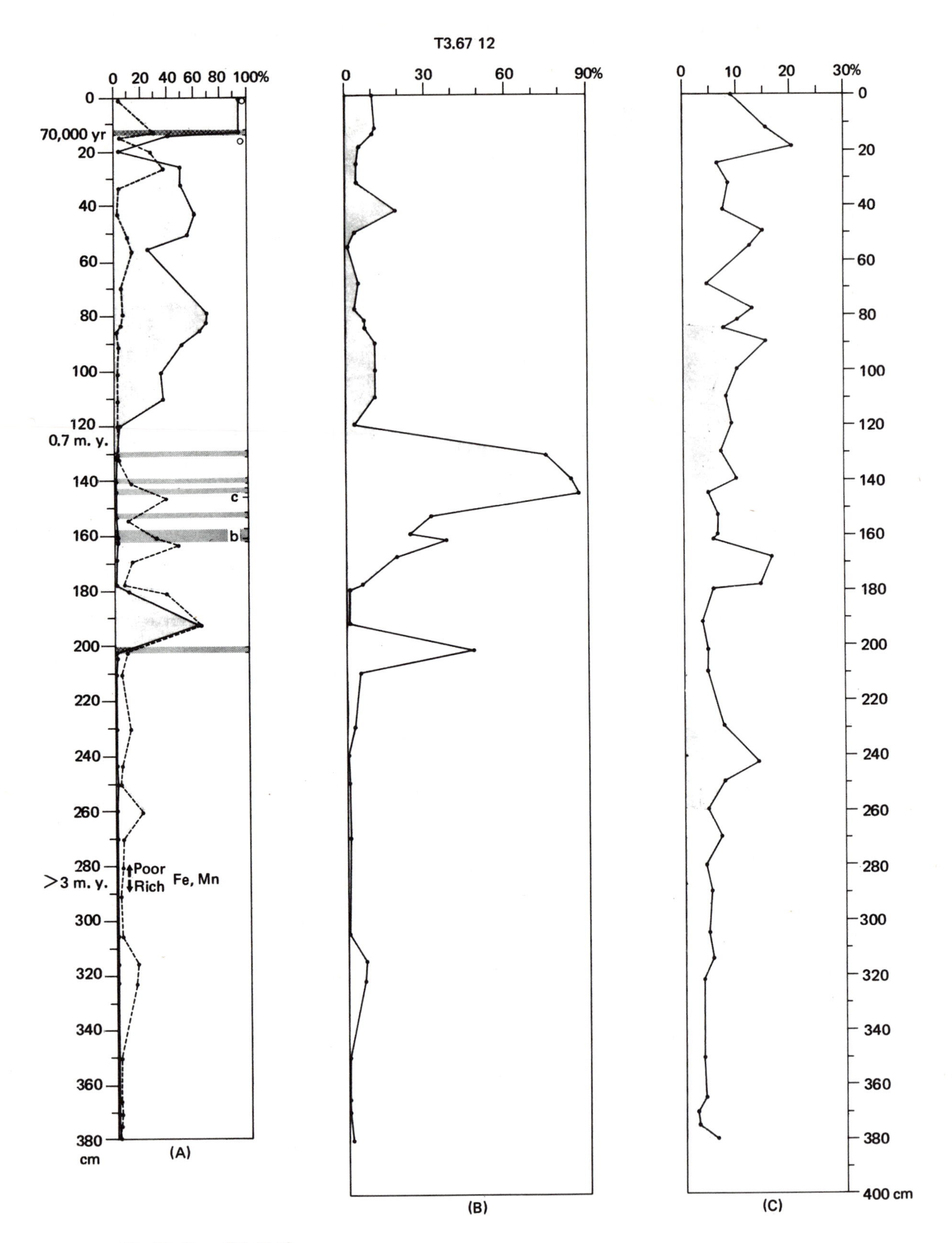

Fig. 21. Core T 3-67-12.
A. Ratio of microfauna to clastics. See the legend for Fig. 2.
B. Variation in the percentage of benthonic foraminifera out of the total foraminiferal population.
C. Variation in the percentage of coarse fraction (>62 μ).

Table 7 Variation in the Percentage of *Bolivina arctica*

Core T3-67-4			Core T3-67-12		
	Depth in core (cm)	Percentage		Depth in core (cm)	Percentage
Unit I	Top*	1.2	Unit I	Top	0
	12	33		12	0
	30	28		13	0
	40	11		19	0
Unit II	54	8.5		25	0
	60	27		32	0
	71	13		42	0.8
	90	31		50	0
	105	12		55	8
	115	12		70	6
	133	0		78	0
	142	0		82	0
	155	0		85	3
	175	0		90	2
	185	0		110	6
				120	5
			Unit II	130	1.6
				160	5
			Absent in unit III		

* The "real" top of this core appears to be missing.

considerably affected the Arctic environment and its fauna as well as global climates. Arctic cold water assemblages dominated by sinistral *G. pachyderma* and related taxa, intermittently accompanied by euryhaline-eurytherm *G. quinqueloba* and *G. exumbilicata,* would have been replaced by lower-latitude species, forms that are known to inhabit today the temperate and subarctic marine zones. Among the planktonic foraminifera known to live today in regions where surficial water temperatures range between about 7 and 10°C, the more common forms are *G. bulloides* d'Orbigny, *G. inflata* (d'Orbigny), *Globorotalia truncatulinoides* (d'Orbigny), *Globorotalia scitula* (Brady), *G. uvula* (Ehrenberg), and *G. glutinata* (Egger), (Boltovskoy, 1969a, 1969b) to name a few. However, there is no record of other low-latitude forms during the period in which Hunkins et al. (1971a) found high frequencies of dextral *G. pachyderma.* Scattered occurrences of a few low-latitude planktonic species generally constituting less than 1% of the total population, have been reported by Herman (1964 through 1971); however, because of their low frequencies and irregular distribution in cores, no satisfactory explanation for their presence in Arctic sediments has been offered to date (Herman, 1964 through 1971). It should be emphasized that while the low-latitude planktonic foraminifera occur throughout the Arctic cores and are generally found in foraminifera-poor zones (Herman, 1964 through 1971), Hunkins' et al. "dextral *G. pachyderma*" is restricted to one horizon (Fig. 22). As mentioned in a previous section, variations in $^{18/16}O$ ratios of planktonic and benthonic foraminiferal tests from Arctic Ocean cores T 3-67-9 and T 3-67-12 suggest that surface water temperatures varied between approximately -2 and $+0.5$°C, and salinities ranged from <29 to $\sim 34‰$ during the last 2 m.y. (Herman et al., 1971; Herman, unpublished). The two cores utilized for $^{18/16}O$ determinations are among those also described by Hunkins et al., (1971a, 1971b). In contrast with the data of Hunkins' et al., geologic and paleontologic evidence indicates that severe world-wide cooling commenced 700,000 to 900,000 years B.P. This global refrigeration resulted in the initiation of widespread continental glaciations at mid latitudes which marked the onset of the "Glacial Pleistocene."

Hunkins et al. (1971a, 1971b) report the presence of "small percentages of *G. quinqueloba*" in the Arctic cores. I have found in the same cores that *G. quinqueloba* is a major faunal component constituting up to 99% of the total planktonic foraminiferal fauna (Herman, 1970a, 1971a, 1971b, 1971c; Herman et al., 1971). A comparison of the data available may help elucidate the discrepancy in our results. I have chosen two cores to illustrate the differences (Table 3). Although morphologic differences between *G. pachyderma* and *G. quinqueloba* are striking, even to the nonspecialist (Plates 5 through 19), it appears that Hunkins et al. (1971a) have included *G. quinqueloba* and *G. exumbilicata* in their counts of low-latitude dextral *G. pachyderma;* this is believed to account for their anomalous results. Finally it should be noted that Hunkins et al. (1971a) conclusions are not supported by their own data, instead, these authors adapted my interpretation (Herman, 1970a) of Arctic climatic evolution based on data they criticized (Hunkins et al., 1971a, 1971b).

Climatic Record of Other Marine Regions

Simultaneous paleomagnetic, faunal, and absolute age determinations are generally being used to establish correlations of climatic events over wide oceanic areas.

According to the majority of researchers, the record of climatic changes based on the aforementioned parameters indicates that the major glacial-interglacial cycles were broadly synchronous throughout the world (North Atlantic: Emiliani, 1955–1966; Ericson et al., 1961; Arctic: Herman, 1969, 1970a; North Pacific: Olsson, 1971; Mediterranean: Herman, 1971a, in press b; Red Sea: Herman-Rosenberg, 1965; Herman, 1968b, 1971a; South Pacific and Antarctic Oceans: Bandy 1968; Bandy et al., 1971; Blackman and Somayajulu, 1966; Donahue, 1967; Hays and Opdyke, 1967).

Oxygen isotope ratios of planktonic foraminiferal tests accompanied by quantitative micropaleontologic analyses indicate that the oceans were less affected by minor temperature fluctuations than were small inland seas such as the Mediterranean and the Red Seas. Furthermore, the amplitudes of these oscillations were most pronounced in temperate areas (e.g., Emiliani and Flint, 1963; Herman, 1965, 1968, 1971a; Vergnaud-Grazzini and Herman-Rosenberg, 1969; Emiliani, 1971).

Meteorologic data indicate that present-day major changes in atmospheric circulation are global in extent. It has also been shown that there is a high positive correlation between atmospheric pressure pulsations in the two polar regions, and consequently a certain parallelism exists between ice conditions in the Arctic and the Antarctic (Defant, 1961; Lamb, 1964).

Magnetic Reversals and Their Relation to Climatic Changes

The effect of magnetic reversals on planetary climates and on the earth's biota has been discussed by several authors (Uffen, 1963; Simpson, 1966; Hays and Opdyke, 1967; Harrison, 1968; Wollin et al., 1971a, 1971b; and many others). Uffen (1963) suggested that increased radiation during polarity reversals directly affected organisms, whereas Harrison (1968) proposed that magnetic reversals modulate climates, which in turn cause faunal alterations.

In the Arctic simultaneous lithologic and faunal changes believed by the author to reflect major alterations in environment occur near magnetic reversals. In the T 3 cores the unit III/II climatic boundary was drawn near a magnetic reversal, interpreted here to correspond to the Gauss/Matuyama magnetic boundary. Likewise, the unit II/I climatic boundary was drawn near the Matuyama/Brunhes magnetic boundary (Herman, 1970a; Herman et al., 1971; Fig. 2).

In addition to the correspondence between major polarity reversals and environmental changes, anomalous faunal assemblages occur at several core levels, particularly in foraminifera-poor beds (Figs. 2, 11, 16, 20, and 21). In these beds the percentage of planktonic foraminifera drops drastically (to a minimum of 3% in core D.st.A. 2 at 142 cm depth) and benthonic foraminifera predominate. Coarse sand, granules, and pebble-size ice-rafted detritus abound and are followed by zones dominated by euryhaline-eurytherm *G. quinqueloba* and *G. exumbilicata* (Figs. 2, 4, 5, 11, 15, 16, and 19 through 21). Inasmuch as there is no evidence of differential solution at these levels, it is suggested that the temporary reduction in planktonic foraminiferal productivity was caused by drastic alteration in the environment.

An inspection of the paleomagnetic polarity changes of several Arctic sediment cores shows a remarkable correlation in time between geomagnetic polarity reversals and climatic changes (Opdyke, in Hunkins et al., 1971a; Fig. 2). These observations, albeit limited, lend support to Harrison's (1968) hypothesis of the effect of magnetic reversals upon planetary climates.

Summary and Conclusions

Distributional patterns of extant planktonic foraminifera and pteropods and their relationship to known environmental factors supplemented by oxygen isotope measurements, and sediment characteristics were the basis for reconstructing the paleo-oceanographic and paleoclimatic history of the Arctic Ocean back to mid-Pliocene. Based on data available the following conclusions may be drawn:

1. Normal and low salinity cycles alternated and were superimposed upon temperature fluctuations.

2. Salinity changes appear to have been more important than temperature fluctuations in determining the composition of the Arctic fauna in Matuyama time, whereas temperature changes characterize the Brunhes epoch.

3. Average surface water temperatures have varied between about -2 and $+0.5$°C, and

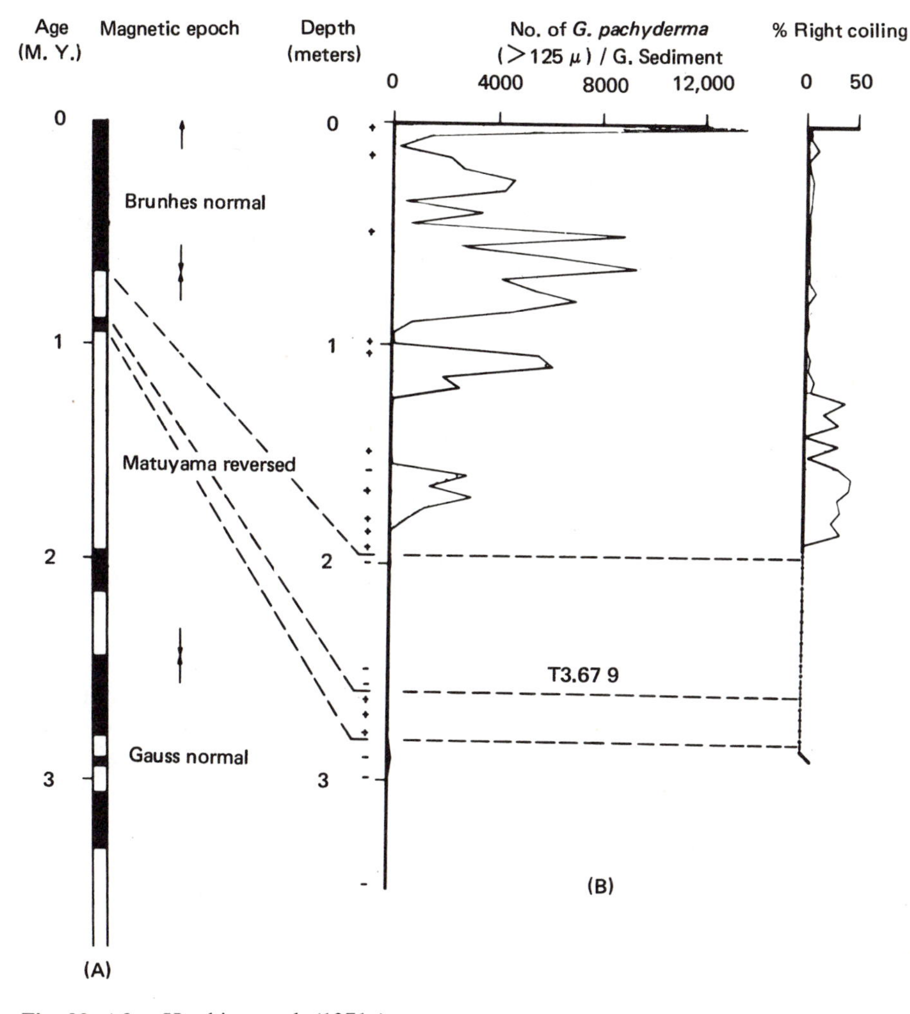

Fig. 22. After Hunkins et al. (1971a).
A. Paleomagnetic time scale.
B. Fluctuation in abundance of *G. pachyderma* and its coiling direction in core T 3-67-9.

salinities ranged from <29 to ~34‰ during the time interval represented by the investigated cores.

4. Lowest temperatures correspond to the upper Pleistocene, probably the Riss or Würm Glacial, whereas higher temperatures and lower salinities were recorded by foraminifera during the Matuyama epoch.

It is assumed that the Arctic Ocean was free of permanent pack-ice during the Matuyama epoch, whereas during the Brunhes epoch, permanent pack-ice covered and seasonally pack-ice free periods alternated.

5. Presence of ice-rafted detritus in core bottoms indicates that high-latitude glaciation was underway prior to 3 million years ago.

6. An apparent correspondence between geomagnetic polarity reversals and climatic changes exists.

7. Three major climatic units have been recognized, and tentatively correlated with three magnetic epochs. The oldest, unit III, is thought to correspond to the Gauss normal epoch; unit II, to the Matuyama reversed polarity epoch; and the youngest, unit I, to the Brunhes normal epoch.

8. The record of global climatic changes indicates that the late Cenozoic major glacial-interglacial cycles were broadly synchronous throughout the world and were most pronounced in mid latitudes. The polar regions remained relatively cold throughout much of the interglacial periods.

Acknowledgments

I am indebted to David B. Ericson, Charles L. Drake, David M. Hopkins, Philip E. Rosenberg, and other colleagues who in the course of the past years have been willing to share their knowledge and who through constructive criticisms have helped me formulate the ideas put forward in this paper, and to Philip E. Rosenberg for critically reviewing the manuscript.

I would like to thank Orville L. Bandy, K. E. Green, Ken Hooper, James Ingle, James P. Kennett, Richard K. Olsson, Frances L. Parker, S. van der Spoel, and Ruth Todd for discussions of the foraminifera and pteropod species and for sending comparative material; Maurice Ewing, former director of Lamont-Doherty Geological Observatory, for organizing the sea-going expedition during which the cores were raised; K. Hunkins, H. Kutschale, G. Mathieu, G. Rasmussen, and their colleagues for collecting the cores; R. R. Capo for sampling and shipping the cores and the core photographs; Peter Barnes (U.S.G.S.) for collecting the plankton tows aboard the U.S.C.G.C. *Glacier*; R. E. Garrison, U.C.S.C., for making the scanning electron microscope available; and G. Wolery for technical assistance. All foraminifera with the exception of those listed below were photographed with a JEOLCO J.M.S.-2 instrument at U.C.S.C. Arthur L. Cohen, W.S.U. Electron Microscope Laboratory, made the SEM available and G. E. Garner provided technical assistance. The following five micrographs were taken with an Etec autoscan at W.S.U.; Plate 20, Fig. 1; Plate 21, Figs. 4, 5, and 6; Plate 22, Fig. 5. Virgil Reynolds assisted with the photographs, and Patricia M. Wilson typed the manuscript.

This research was partially supported by the W.S.U. Research Committee Grant No. 11 C-2454-372, the Graduate School Development Fund 14N-2940-0020, and the Office of Marine Geology of the U.S.G.S.

The coring operations and curatorial services were supported by grants ONR (NOOO 14-67-A-0108-0004), and NSF (GA-29460) held by LDGO.

References

Anderson, G. J. 1963. Distribution patterns of recent foraminifera of the Bering Sea. *Micropaleontology*, 9(3):305–317.

Arrhenius, G. 1963. Pelagic sediments. In: M. N. Hill, ed. *The Sea*, Vol. 3 New York, New York: Interscience Publishers, pp. 655–718.

Bandy, O. L. 1968. Paleoclimatology and Neogene planktonic foraminiferal zonation. In: Committee Mediterranean Neogene Stratigraphy, *Proceedings of the 4th Session, Bologna, 1967, Giornale di Geologia (2) 35, 2.* Bologna, pp. 277–290.

———. 1969. Relationships of Neogene planktonic foraminifera. In: P. Brönnimann and H. H. Renz, eds. *Proceedings of the First International Conference on Planktonic Microfossils*, Geneva, 1967. Leiden: E. J. Brill, 1, pp. 46–57.

———, E. A. Butler, and R. C. Wright. 1969. Alaskan upper Miocene marine glacial deposits and the *Turborotalia pachyderma* datum plane. *Science*, 166:607–609.

———, and R. E. Casey. 1969. Major late Cenozoic planktonic datum planes, Antarctica to the tropics. *Antarctic J.U.S.*, 4(5):170–171.

———, and J. C. Ingle, Jr. 1970. Neogene planktonic events and radiometric scale, California. *Geol. Soc. Am.*, Special Paper 124:131–172.

———, R. E. Casey, and R. C. Wright. 1971. Late Neogene planktonic zonation, magnetic reversals, and radiometric dates, Antarctic to the tropics. *Antarctic Res. Ser.*, 15:1–26.

Barash, M. S. 1971. The vertical and horizontal distribution of planktonic foraminifera in Quaternary sediments of the Atlantic Ocean. In: B. M. Funnell and W. R. Riedel, eds. *The Micropaleontology of Oceans.* London: Cambridge University Press, pp. 433–442.

Bé, W. A. 1960. Some observations on Arctic planktonic foraminifera. *Contrib. Cushman Found. Foram. Res.*, 11:64–68.

Beal, M. A. 1969. Bathymetry and structure of the Arctic Ocean. Doctoral thesis, Oregon State University, Corvallis, 204 pp.

Berry, R. W., and W. D. Johns. 1966. Mineralogy of the clay-sized fractions of some North Atlantic—Arctic Ocean bottom sediments. *Geol. Soc. Am. Bull.*, 77(2):183–195.

Blackman, A., and B. L. K. Somayajulu. 1966. Pacific Pleistocene: Faunal analyses and geochronology. *Science*, 154:886–889.

Boltovskoy, E. 1962. Planktonic foraminifera as indicators of different water masses in the South Atlantic. *Micropaleontology*, 8(3):403–408.

———. 1966. Zonación en las latitudes altas del Pacifico sur segun los foraminiferos planctonicos vivos. *Revista del Museo Argentino de Ciencias Naturales "Bernardino Rivadavia." Hidrobiologia*, 2:1–56.

———. 1969a. Foraminifera as hydrological indicators. In: P. Brönnimann and H. H. Renz, eds. *Proceedings of the First International Conference on Planktonic Microfossils*, Vol. 2. Geneva, 1967. Leiden: E. J. Brill, pp. 1–14.

———. 1969b. Living planktonic foraminifera at the 90°E meridian from the Equator to the Antarctic. *Micropaleontology*, 15(2):237–255.

———. 1971. Ecology of the planktonic foraminifera

living in the surface layer of Drake Passage. *Micropaleontology*, 17(1):53–68.

Bostrom, K. 1970. Deposition of manganese-rich sediments during glacial periods. *Nature*, 266:629–630.

———. 1971. Origin of manganese-rich layers in Arctic sediments. *Second International Symposium on Arctic Geology*, San Francisco, 1971 (abstract), pp. 9–10.

Brady, H. B. 1881. On some Arctic foraminifera from soundings obtained on the Austro-Hungarian North-Polar expedition of 1872–1874. *Ann. Mag. Natural Hist.*, 8(5):393–418.

Brooks, C. E. P. 1949. *Climate Through the Ages*. New York, New York: McGraw-Hill Book Company, 395 pp.

Burwell, R. L., Jr. 1972. Chilean glacial chronology 20,000 to 11,000 carbon-14 years ago: Some global comparisons. *Science*, 176:1118–1119.

Carroll, D. 1970. Clay minerals in Arctic Ocean sea-floor sediments. *J. Sedimentary Petrol.*, 40(3):814–821.

Charlesworth, J. K. 1957. *The Quaternary Era*. Vols. 1 and 2. London: Edward Arnold Ltd., 1700 pp.

Coachman, L. K., and C. A. Barnes. 1962. The movement of Atlantic water in the Arctic Ocean. *Arctic*, 16:11–16.

Cooper, S. C. 1964. Benthonic foraminifera of the Chukchi Sea. *Cushman Found. Foram. Res. Contrib.*, 15:79–104.

Cox, A. 1968. Polar wandering, continental drift, and the onset of Quaternary glaciation. *Meteorol. Monographs*, 8(30):112–125.

———. 1969. Geomagnetic reversals. *Science*, 163:237–245.

———, R. R. Doell, and G. B. Dalrymple. 1964. Reversals of the earth's magnetic field. *Science*, 144:1537–1543.

Craig H. 1965. The measurement of oxygen isotope paleotemperatures. In: *Stable Isotopes in Oceanographic Studies and Paleotemperatures.* Pisa: Spoleto, Consiglio Nazionale delle Ricerche, 23 pp.

———, and L. I. Gordon. 1965. Isotopic oceanography. In: *Symposium in Marine Geochemistry*, Publ. 3. Narragansett Marine Laboratory, University of Rhode Island, pp. 277–374.

Cromie, W. J. 1961. Preliminary results of investigations on Arctic drift station "Charlie." In: G. O. Raasch, ed. *Geology of the Arctic*, Vol. 1 Toronto: University of Toronto Press, pp. 690–708.

Crowell, J. C., and L. A. Frakes. 1970. Phanerozoic glaciation and the causes of ice ages. *Am. J. Sci.*, 268:193–224.

Cushman, J. A. 1948. Arctic foraminifera. *Cushman Lab. Foram. Res. Spec. Pub.*, 23. 79 pp.

Damon, P. E. 1968. The relationship between terrestrial factors and climate. *Meteorol. monographs*, 8(30):106–111.

Dansgaard, W., S. J. Johnsen, J. Moller, and C. C. Langway, Jr. 1969. One thousand centuries of climatic record from Camp Century on the Greenland ice sheet. *Science*, 166:377–381.

Defant, A. 1961. *Physical Oceanography*, Vol. 1. London: Pergamon Press, 729 pp.

Degens, E. T. 1965. *Geochemistry of Sediments.* Englewood Cliffs, N.J.: Prentice-Hall, 342 pp.

Demenitskaya, R. M., and K. L. Hunkins. 1970. Shape and structure of the Arctic Ocean. In: Arthur E. Maxwell, ed. *The Sea*, Vol. 4. New York, New York: Wiley-Interscience, pp. 223–249.

Donahue, J. G. 1967. Diatoms as indicators of Pleistocene climatic fluctuations in the Pacific sector of the Southern ocean. In: M. Sears, ed. *Progress in Oceanography*, Vol. 4. New York, New York: Pergamon Press, pp. 133–140.

Donk, J. van, and G. Mathieu. 1969. Oxygen isotope compositions of foraminifera and water samples from the Arctic Ocean. *J. Geophys. Res.*, 74(13):3396–3407.

Emery, K. O. 1949. Topography and sediments of the Arctic Basin. *J. Geol.*, 57:512–521.

Emiliani, C. 1955. Pleistocene temperatures. *J. Geol.*, 63(6):538–573.

———. 1964. Paleotemperature analysis of the Caribbean cores A254-BR-C and CP-28. *Geol. Soc. Am. Bull.*, 75:129–143.

———. 1966a. Isotopic paleotemperatures. *Science*, 154:851–857.

———. 1966b. Paleotemperature analysis of Caribbean cores P6304-8 and P6304-9 and a generalized temperature curve of the past 425,000 years. *J. Geol.*, 74(2):109–124.

———. 1967. The Pleistocene record of the Atlantic and Pacific oceanic sediments; correlations with the Alaskan stages by absolute dating; and the age of the last reversal of the geomagnetic field. In: Mary Sears, ed. *Progress in Oceanography*, Vol. 4. New York, New York: Pergamon Press, 219–224.

———. 1971. The amplitude of Pleistocene climatic cycles at low latitudes and the isotopic composition of glacial ice. In: Karl K. Turekian, ed. *The Late Cenozoic Glacial Ages.* New Haven, Connecticut: Yale University Press, pp. 183–197.

———, and J. Geiss. 1959. On glaciations and their causes. *Geol. Rundschau*, 46:576–601.

———, and R. F. Flint. 1963. The Pleistocene record. In: M. N. Hill, ed. *The Sea*, Vol. 3. New York, New York: Interscience Publishers, pp. 888–927.

Enbysk, B. J. 1970. Distribution of foraminifera in the northeast Pacific. Doctoral thesis, University of Washington, Seattle, 231 pp.

Epstein, S., R. Buchsbaum, H. Lowenstam, and H. C. Urey. 1953. Revised carbonate-water isotopic temperature scale. *Geol. Soc. Am. Bull.*, 64:1315–1325.

———, R. P. Sharp, and A. J. Gow. 1970. Antarctic ice sheet: stable isotope analyses of Byrd Station cores and interhemispheric climatic implications. *Science*, 168:1570–1572.

Ericson, D. B. 1959. Coiling direction of *Globigerina pachyderma* as a climatic index. *Science*, 130:219–220.

———. 1963. Cross-correlation of deep-sea sediment cores and determination of relative rates of sedimentation by micropaleontological techniques. In: M. N. Hill, ed. *The Sea*, Vol. 3. New York, New York: Interscience Publishers, pp. 832–842.

———, and G. Wollin. 1959. Micropaleontology and lithology of Arctic sediment cores. In: *Geophysical Research Papers 63*, "Scientific Studies at Fletcher's Ice Island,"—T-3, *1952–1955*. 1:50–58.

———, M. Ewing, G. Wollin, and B. C. Heezen. 1961. Atlantic deep-sea sediment cores. *Geol. Soc. Am. Bull.*, 72:193–286.

———, M. Ewing, and G. Wollin. 1963. Pliocene-Pleistocene boundary in deep-sea sediments. *Science*, 139: 727–737.

———, M. Ewing, and G. Wollin. 1964. Sediment cores from the Arctic and subarctic seas. *Science*, 144:1183–1192.

Fairbridge, R. W. 1961. Convergence of evidence on climatic change and Ice Ages. *Ann. N. Y. Acad. Sci.*, 95(1):542–579.

———. 1970. World paleoclimatology of the Quarternary. *Rev. Geog. Physique Geol. Dynamique* (2), 12(2):97–104.

Geological Map of the Arctic. 1960. First International Symposium on Arctic Geology: *Alberta Society Petroleum Geology*, 1.

Green, K. E. 1960. Ecology of some Arctic foraminifera. *Micropaleontology*, 6(1):57–78.

Griffin, J. J., H. Windom, and E. D. Goldberg. 1968. The distribution of clay minerals in the world Ocean. *Deep-Sea Res.*, 15:433–459.

Gurianova, E. F. 1970. Special features in the fauna of the Arctic Ocean and their value in understanding the history of the formation of the fauna. In: *The Arctic Ocean and Its Shores during the Cenozoic* (transl.). Leningrad: Hydrometeorological Publishers, pp. 126–161.

Hamilton, W. 1968. Cenozoic climatic change and its cause. *Meteorol. Monographs*, 8(30):128–133.

Harrison, C. G. A. 1968. Evolutionary processes and reversals of the earth's magnetic field. *Nature*, 217:46–47.

Hattersley-Smith, G. 1960. Some remarks on glaciers and climate in Northern Ellesmere Island. *Geografiska Annaler*, 1:45–48.

———. 1963. The Ward Hunt ice shelf: Recent changes of the ice front. *J. Glaciol.*, 4(34):415–424.

———, A. P. Crary, and R. L. Christie. 1955. Northern Ellesmere Island, 1953 and 1954. *Arctic*, 8(1):3–36.

Hays, J. D., and N. D. Opdyke. 1967. Antarctic radiolaria, magnetic reversals, and climatic change. *Science*, 158:1001–1011.

Herman, Y. 1964. Temperate water planktonic foraminifera in Quaternary sediments of the Arctic Ocean. *Nature*, 201(4917):386–387.

———. 1966. Climatic changes in Quaternary cores from the Mediterranean and Red Sea basins recorded by (1) pteropods, (2) planktonic foraminifera. *Second International Oceanographic Congress* (abstract), 156–157.

———. 1968a. Late Cenozoic climatic changes in the Central Arctic Basin. *Geol. Soc. Am. Bull.* (abstract), 134.

———. 1968b. Evidence of climatic changes in Red Sea cores. Seventh INQUA Congress Proceedings, Vol. 8. In: R. B. Morrison and H. E. Wright, Jr., eds. *Means of Correlation of Quaternary Sequences*, pp. 325–348.

———. 1969. Arctic Ocean Quaternary microfauna and its relation to paleoclimatology. *Paleogeogr. Paleoclim. Paleoecol.*, 6:251–276.

———. 1970a. Arctic paleo-oceanography in late Cenozoic time. *Science*, 169:474–477.

———. 1970b. Late Cenozoic Arctic oceanography. *Trans. Am. Geophys. Union*, 51(4) (abstract), p. 333.

———. 1971a. Vertical and horizontal distribution of pteropods in Quarternary sequences. In: B. M. Funnell and W. R. Riedel, eds. *The Micropaleontology of Oceans.* Cambridge: Cambridge University Press, pp. 463–486.

———. 1971b. Late Cenozoic biostratigraphic and paleoecologic studies of Arctic Ocean deep-sea cores. *Second International Symposium on garctic Geology*, San Francisco, 1971 (abstract), pp. 25–26.

———. 1971c. Arctic paleo-oceanography in late Cenozoic time. *Science*, 174:963.

———. 1972a. South Pacific Quaternary paleo-oceanography. *24th International Geologic Congress*, Montreal, Canada (abstract), p. 260.

———. 1972b. *Globorotalia truncatulinoides:* a paleo-oceanographic indicator. *Nature*, 238:394–395.

———. In press a. *Bolivina arctica,* new Pleistocene benthonic foraminifera from Arctic Ocean sediments, *J. Foram. Res.*

———. In press b. Quarternary Eastern Mediterranean sediments: micropaleontology and climatic record. In: Daniel J. Stanley, ed. *The Mediterranean Sea: A Natural Sedimentation Laboratory.* Stroudsburg, Pennsylvania: Dowden, Hutchinson and Ross.

———, and C. Vergnaud-Grazzini. 1969. Late Cenozoic climatic changes in Arctic Basin deep-sea cores as indicated by microfaunal and paleotemperature ($^{18/16}O$) analyses. *Seventh INQUA Congress*, Paris (abstract), p. 68.

———, C. Vergnaud-Grazzini, and C. Hooper. 1971. Arctic paleotemperatures in late Cenozoic time. *Nature*, 232(5311):466–469.

———, and M. C. Metz. 1972. Staining technique for recent and fossil calcareous invertebrates. *J. Paleontol.*, 46(1):152.

Herman-Rosenberg, Y. 1965. Etudes des sédiments Quaternaires de la Mer Rouge. *Annales Institut Oceanographique.* Paris: Masson & Cie. 42(3), pp. 343–415.

Hopkins, D. M. 1967. The Cenozoic history of Beringia, a synthesis. In: David M. Hopkins ed. *The Bering Land Bridge.* Stanford: Stanford University Press, pp. 451–484.

———. 1972. The paleogeography and climatic history of Beringia during late Cenozoic time. *Inter-Nord,* 12:121–150.

———, R. W. Rowland, and W. W. Patton. 1972. Middle Pleistocene mollusks from St. Lawrence Island and their significance for the paleo-oceanography of the Bering Sea. *Quat. Res.,* 2(2):119–134.

Hunkins, K., A. W. H. Bé, N. D. Opdyke, and G. Mathieu. 1971a. The late Cenozoic history of the Arctic Ocean. In: Karl K. Turekian, ed. *The Late Cenozoic Glacial Ages.* New Haven: Yale University Press, pp. 215–237.

———, A. W. H. Bé, N. D. Opdyke, and T. Saito. 1971b. Arctic paleo-oceanography in late Cenozoic time. *Science,* 174:962.

Kennett, J. P. 1968. Latitudinal variation of *Globigerina pachyderma* (Ehrenberg) in surface sediments of the southwest Pacific Ocean. *Micropaleontology,* 14(3):305–318.

———, and N. D. Watkins. 1970. Geomagnetic polarity change, volcanic maxima and faunal extinction in the South Pacific. *Nature,* 227(5261):930–934.

Khoreva, I. M. 1970. Foraminifera and stratigraphy, marine Quaternary depositions; the western shore of the Bering Sea. In: *The Arctic Ocean and Its Shores During the Cenozoic* (transl.). Leningrad: Hydrometeorological Publishers, pp. 548–551.

Koenig, L. S., K. R. Greenaway, M. Dunbar, and G. Hattersley-Smith. 1952. Arctic ice islands. *Arctic,* 5(2):67–103.

Ku, T. L., and W. S. Broecker. 1967. Rates of sedimentation in the Arctic Ocean. In: M. Sears, ed. *Progress in Oceanography,* Vol. 4. London: Pergamon Press, pp. 95–104.

———, W. S. Broecker, and N. Opdyke. 1968. Comparison of sedimentation rates measured by paleomagnetic and the ionium methods of age determination. *Earth Planet. Sci. Lett.,* 4(1):1–16.

Kutschale, H. 1966. Arctic Ocean geophysical studies: The southern half of the Siberia Basin. *Geophys.,* 31(4):683–710.

Lamb, H. H. 1964. Climatic changes and variations in the atmospheric and ocean circulations. *Geolog. Rundschau,* 54:486–504.

Lin'kova, T. I. 1965. Some results of paleomagnetic study of Arctic Ocean floor sediments. In: *The Present and Past of the Geomagnetic Field* (transl. by E. R. Hope). Moscow: Nauka, pp. 279–291.

Loeblich, A. R., Jr., and H. Tappan. 1953. Studies of Arctic foraminifera. *Smithsonian Misc. Collections,* 121(7), 150 pp.

McCrea, J. M. 1950. On the isotopic chemistry of carbonates and a paleo-temperature scale. *J. Chem. Phys.,* 18:849–857.

Meisenheimer, J. 1905. Die arktischen Pteropoden. In: F. Romer and F. Schaudinn, eds. *Fauna Arctica,* Band IV (1906), Lieferung II. Jena: G. Fisher, pp. 407–430.

Milankovitch, M. 1930. Mathematische Klimalehre und Astronomische Theorie der Klimaschwankungen. In: W. Köppen and R. Geiger, eds. *Handbuch der Klimatologie,* Berlin: 1, Pt. A: 176 pp.

Naidu, A. C., D. C. Burrell, and D. W. Hood. 1971. Clay mineral composition and geologic significance of some Beaufort Sea sediments. *J. Sedimentary Petrol.,* 4(3):691–694.

Nansen, F. 1902. Oceanography of the North Polar Basin. *The Norwegian North Polar Expedition, 1893–1896.* 3, Christiania, Science Results. London: Longmans, Green and Company, 427 pp.

———. 1904. The bathymetrical features of the North Polar Seas. In: F. Nansen. ed. *The Norwegian North Polar Expedition, 1893–1896.* London: Longmans, Green and Company, 4(13), pp. 1–232.

Olsson, R. K. 1971. Pliocene-Pleistocene planktonic foraminiferal biostratigraphy of the Northeastern Pacific. In: A. Farinacci, ed. *Proceedings of the Second International Planktonic Conference,* Vol. 2. Roma: Edizioni Tecnoscienza, pp. 921–928.

Parker, F. L. 1971. Distribution of planktonic foraminifera in recent deep-sea sediments. In: B. M. Funnell and W. R. Riedel, eds. *The Micropaleontology of Oceans.* Cambridge: Cambridge University Press, pp. 289–309.

Phleger, F. B. 1952. Foraminifera distribution in some sediment samples from the Canadian and Greenland Arctic. *Cushman Found. Foram. Res. Contrib.,* 3:80–89.

Raasch, G. O. 1961. *Geology of the Arctic,* Vol. 1. Toronto: University of Toronto Press, 732 pp.

Saks, V. N., N. A. Belov, and N. N. Lapina. 1955. Our present concepts of the geology of the central Arctic. *Priroda* (transl.). In: *Defense Research Board of Canada Translation no. T196R,* 1955, 7:13–22.

Selli, R. 1967. The Pliocene-Pleistocene boundary in Italian marine sections and its relationship to continental stratigraphies. In: M. Sears, ed. *Progress in Oceanography,* Vol. 4. London: Pergamon Press, pp. 67–86.

Simpson, J. F. 1966. Evolutionary pulsations and geomagnetic polarity. *Geol. Soc. Am. Bull.,* 77(2):197–203.

Smith, P. B. 1963. Possible Pleistocene-Recent boundary in the Gulf of Alaska based on benthonic foraminifera. *U.S. Geol. Surv. Prof. Pap. 475-C*:C73–C77.

Spoel, S. van der. 1967. Euthecosomata. A group with remarkable developmental stages (Gastropoda, Pteropoda). *J. Noorduijn en zoon N. V.* Gorinchem, 375 pp.

Steuerwald, B. A., D. L. Clark, and J. A. Andrew. 1968. Magnetic stratigraphy and faunal patterns in Arctic Ocean sediments. *Earth Planet. Sci. Lett.*, 5:79–85.

Stschedrina, Z. G. 1959. The dependence of the distribution of foraminifera in the seas of the USSR on the environmental factors. *International Congress Zoology, 15th*, London 1958, Proceedings, Section 3, Paper 30, pp. 218–221.

———. 1962. Foraminifery Zalivov Belogo Morya (foraminifera of the White Sea bays). In: L. A. Zenkevitch, ed. *Biology of the White Sea; Reports. Moscow: White Sea Biology Station of State University*, 1:51–69.

Sverdrup, H. U. 1950. Physical oceanography of the North Polar Sea. *Arctic*, 3:178–186.

———. 1956. Oceanography of the Arctic. In: *The Dynamic North.* Washington: Technical Assistant to the Chief of Naval Operations for Polar Projects (Op-03A3), 1(5):1–31.

———, M. W. Johnson, and R. H. Fleming. 1942. *The Oceans: Their Physics, Chemistry, and General Biology.* Englewood Cliffs, N.J.: Prentice-Hall, 1087 pp.

Tanner, W. F. 1965. Cause and development of an ice age. *J. Geol.*, 73(3):413–430.

———. 1968. Cause and development of an ice age. *Meteorol. Monographs*, 8(30):126–127.

Tesch, J. J. 1948. The Thecosomatous pteropods; the Indo-Pacific. *Dana Report 30.* Copenhagen: Carlsberg Foundation, 44 pp.

Todd, R., and D. Low. 1966. Foraminifera from the Arctic Ocean off the Eastern Siberian Coast. *U.S. Geol. Surv. Prof. Pap. 550-C*:C79–C85.

Treshnikov, A. F. 1959. Oceanography of the Arctic Basin. *International Oceanographic Congress* (abstract), pp. 522–523.

Uffen, R. J. 1963. Influence of the earth's core on the origin and evolution of life. *Nature*, 198:143.

Vergnaud-Grazzini, C., and Y. Herman-Rosenberg. 1969. Etude paleoclimatique d'une carotte de Méditerranée orientale. *Rev. Geograph. Phys. Geolog. Dynamique*, 9(3):279–292.

Vilks, G. 1969. Recent foraminifera in the Canadian Arctic. *Micropaleontology*, 15(1):35–60.

Wentworth, C. K. 1922. A scale of grade and class terms for clastic sediments. *J. Geol.*, 30:377–392.

Weyl, P. K. 1968. The role of the oceans in climatic change: A theory of the ice ages. *Meteorol. Monographs*, 8(30):37–62.

Wollin, G., D. B. Ericson, W. B. F. Ryan, and J. H. Foster. 1971a. Magnetism of the earth and climatic changes. *Earth Planet. Sci. Lett.*, 12:175–183.

———, D. B. Ericson, and W. B. F. Ryan. 1971b. Variations in magnetic intensity and climatic changes. *Nature*, 232(5312):549–550.

Zenkevitch, L. 1963. *Biology of the Seas of the USSR.* New York, New York: Interscience Publishers, 955 pp.

Zeuner, F. E. 1959. *The Pleistocene Period.* London: Hutchinson Scientific and Technical, 447 pp.

Chapter 14
Atmospheric Circulation during the Onset and Maximum Development of the Wisconsin/Würm Ice Age

HUBERT HORACE LAMB [1]

Abstract

Meteorological arguments and methods of analysis of the field data are used to derive and present the probable characteristics of the general atmospheric circulation over the northern hemisphere at the start and at the maximum development of the Last Ice Age.[2]

Introduction

Recent studies (Lamb et al., 1966; Lamb and Woodroffe, 1970) used values for summer and winter surface temperatures prevailing at various well-marked climatic stages during and since the Last Ice Age, suggested by paleobotanical and oceanographic research, to obtain estimates of upper air temperature and 1000 to 500 millibar (mb) thickness distribution over the northern hemisphere. From these, regions of recurrent cyclonic and anticyclonic development could be calculated and then probable prevailing surface pressure and wind patterns were derived.

The results appear to throw fresh light on the climatic behavior of the times concerned and may even help solve the problem of how much open water there was in the Arctic Ocean during the Ice Age. The dryness of the boreal period in Europe is seen as associated with anticyclonic development, in summer and winter alike, at the right-hand side of a jet stream guided strongly to the northeast and north over Greenland and the East Greenland Sea by the thermal gradient at the limit of the still great American ice sheet. This effect presumably declined progressively through the period to 4000 B.C., as the Laurentide ice sheet dwindled to nothing.

The large-scale geography of the Last Glacial maximum around 20,000 to 17,000 B.C. is sufficiently well established from the abundant morainic and other evidence and the numerous field studies that have been made of this evidence over many years. Disagreement between different workers' diagnoses regarding the extent of glaciation in central Asia, more particularly on the high ground northeast of Lake Baikal (50 to 60°N, 10 to 120°E) and on the Himalayas, Hindu Kush-Pamirs-Tien Shan, and the Tibetan highlands hardly matters as regards establishing the broad-scale thermal characteristics of the surfaces of the northern hemisphere that must have been extensive enough to influence the atmospheric circulation pattern. The atmospheric circulations that prevailed in summer and winter around the time of maximum glaciation some 20,000 years ago can therefore be derived in the manner described.

Blüthgen (1966, p. 561) summarizes the conclusions of field research with the statement that

[1] Climatic Research Unit, School of Environmental Sciences, University of East Anglia, Norwich, England.

[2] In this chapter Last Ice Age refers only to the Würm (Wisconsin) Glacial.

the Last Glaciation was about the longest and coldest of the Pleistocene glaciations. It apparently lasted 50,000 to 60,000 years, though with several distinct maxima (see Fig. 1). There is some evidence (e.g., Penny et al., 1969) that the coldest time in the Würm, and in Britain the greatest extent of the ice in this glaciation, was as late as 15,000 B.C. The maximum spread of the ice sheets during the Würm was not as great as in the earlier Pleistocene glaciations, possibly because the Würm climates became so cold that there was too little moisture in the air.

The gross features of the glacial maximum here mapped lasted for 3000 to 5000 years or more, though no doubt with superimposed shorter-term fluctuations. The time scale of the Allerød warm epoch and the 600 years of severe glacial re-advance climate that followed appears to be the same as that of the build-up from Roman times to the early medieval warm epoch around A.D. 1000 to 1200 and the subsequent cold climate, which culminated around A.D. 1700 (Lamb, 1965). Both these oscillations, around 10,000 to 8000 B.C. and A.D. 0 to 1900, were preceded by a somewhat similar oscillation, of which there is evidence both in the paleobotany of Europe (see, for instance,

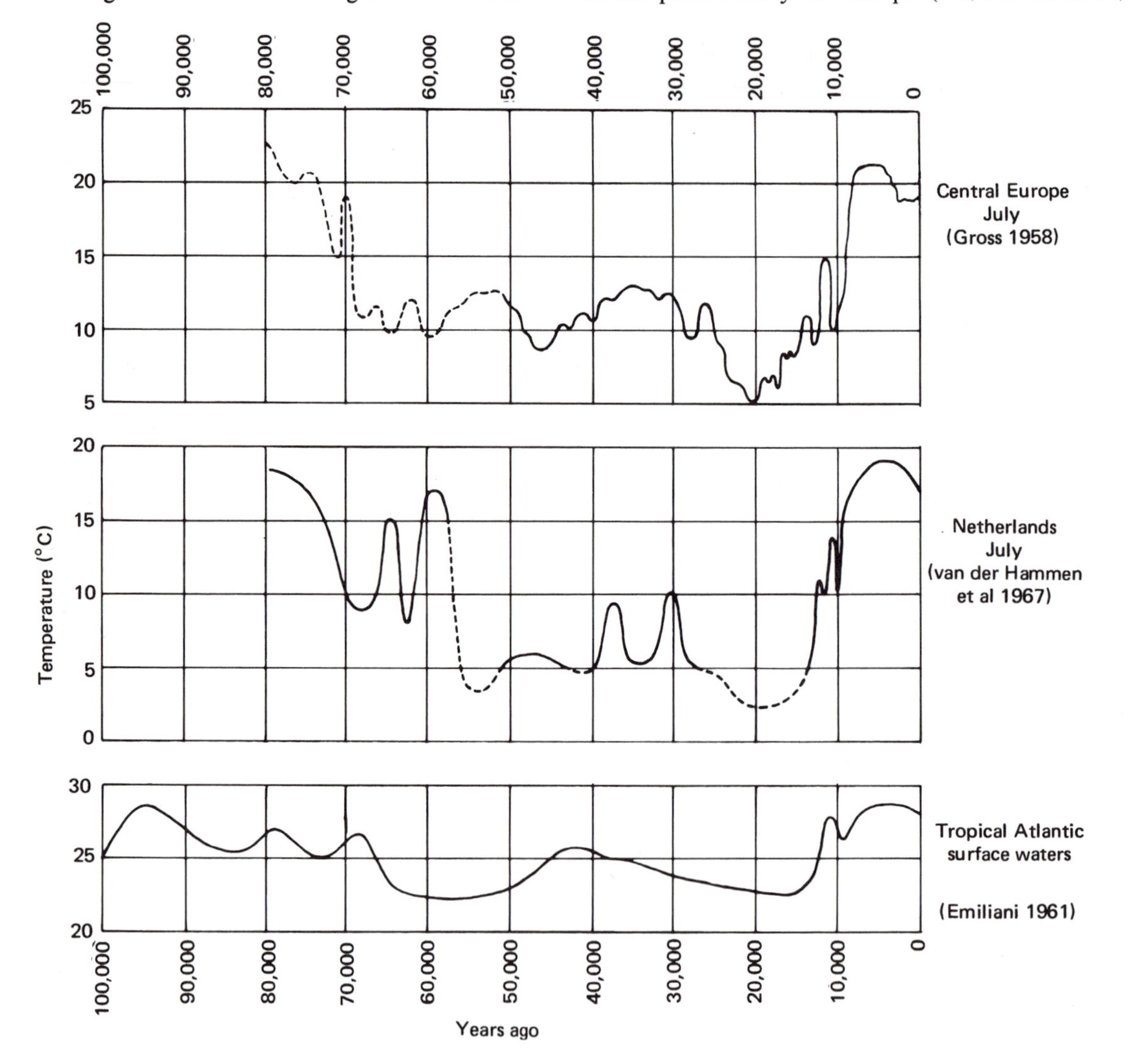

Fig. 1. Course of prevailing temperatures through the Last Ice Age and since. (a) July averages in central Europe after Gross (1958), derived partly from floristic evidence and partly from faunal, including molluscan evidence. (b) July averages in the Netherlands (after van der Hammen et al., 1967), derived partly from evidence of soil formations and partly from the flora. (c) All-year averages of sea-surface temperature characteristic of the tropical Atlantic (after Emiliani, 1961; derived from the $^{18/16}O$ ratios in the tests of planktonic foraminifera found in the ocean bed deposits.

Flint, 1957) and in the Greenland ice (Dansgaard and Johnsen, 1969), occupying the immediately preceding 2000 years.

The repetitions of this time scale, of which further instances both before and since the ones here touched on are suspected, suggests the operation of some regular variation in the sun's radiation output or (less plausibly because of the varying glacial geography) some regular cycle in the circulation of the oceans.

The Sequence of the Last Glaciation

Present understanding of the course of the Last Ice Age is illustrated by the three curves in Fig. 1. These represent average surface water temperatures in the Caribbean and equatorial Atlantic (Emiliani, 1961), average July temperatures in central Europe (Gross, 1958), and average July temperatures in the Netherlands (van der Hammen et al., 1967). There is substantial agreement about the dating of the last and main climax of the Würm Glaciation and on the course of the temperature curves since, though Emiliani's diagnosis of the general history of the temperature of the tropical Atlantic shows only one of the two late glacial warm stages recognized in Europe and in the Greenland ice.

The shape of the temperature curve about the main cooling stage in the early glacial seems broadly agreed upon, especially as to its abruptness, but the dates suggested range from about 57,000 to 70,000 years ago. Such ages are near the limit of the radiocarbon dating technique.

There seems to be agreement that traces of moss assemblages, tree remains, and other flora and fauna found in England from periods about 57,000 to 59,000 years ago (Coope et al., 1961; Dickson, 1967; Simpson and West, 1958), organic layers in sediments in New York State estimated to be 52,000 to 64,000 years old (Muller, 1964) and in Denmark from 55,000 to 59,000 years ago (Andersen, 1961), and similar deposits found in the Netherlands (Zagwijn, 1961) represent one or more early warmer interstadials.

Two warm interstadials have been named, successively, Amersfoort and Brørup, after the places at which evidence was found in the early Würm deposits. They seem, from the establishment of pine forest in England, with some birch and a little spruce, and in New York State (pine and spruce) and in southeastern Quebec (spruce), to have brought summer temperatures nearly up to present levels (July average probably +12°C or even a little higher in Cheshire, England). The winter temperatures evidently remained low: those typical of regions where the same tree species flourish today are low (a mean January temperature in the range −10° to −15°C might be indicated for Cheshire, England). The interstadial climate therefore seems to have been highly continental in these continental margin areas, probably indicating the continued presence of an already formed extensive ice sheet not far away and remarkably little (zonal) wind flow from the Atlantic over Europe.

Probable Atmospheric Circulation of the Glacial Onset

The evidence cited above suggests that at some rather uncertain time between about 60,000 and 70,000 years ago ice sheets were quickly established over northern and northwestern Europe and over some extensive hinterland in North America, presumably enveloping the Hudson's Bay region, within some 1000 to 5000 years. These ice sheets evidently acquired enough thickness to survive a warm interstadial, lasting some centuries or longer, which brought summer temperatures high enough for the re-establishment of forest in southern Britain and southeastern Quebec, though the ice sheets continued to exert a strong control on winter temperatures.

Such a rapid constitution of the ice sheets at a time when the seas were initially almost as warm as now seems to demand a great prevalence of northerly surface winds and a meridionally extended upper cold trough, in most months of the year, over the sector Norwegian Sea—British Isles—Scandinavia and a similar, or even more marked, meridional trough and surface northerly wind regime over Hudson's Bay and the western plains sector of North America (Fig. 2). These southward-extended troughs in the upper westerlies persistently occupying positions west of their counterparts of the present day would presumably require either a much weakened or a much expanded circumpolar vortex, or both, with the main thermal gradients in the continental sectors at lower latitudes than now. The rapidity of establishment of the ice sheets, though they need not have attained more than a fraction of their final thickness to survive a 500- to 1500-year-long warm interstadial, seems to indicate the character of the atmospheric circulation that must have prevailed during the glacial onset, because only one type of circulation pattern seems capable of supplying the moisture required to build the ice sheets fast in the areas where they formed.

Recent research (e.g., Hunkins and Kutschale, 1967; Ku and Broecker, 1967) indicates that, contrary to Ewing and Donn's (1956) hypothesis about the origin of the former great ice sheets, the Arctic Ocean has probably been largely ice-covered throughout the last 70,000 years. Moreover, even if there had been open water in the Arctic, the water evaporated into the atmosphere from a sea surface at a temperature around 0 to +5°C is only one-fifth to one-sixth of that evaporated from seas at 25 to 30°C under similar conditions of windiness. So it seems clear that the main moisture supply did not come from the oceans in high latitudes.

The picture here developed of a highly meridional circulation during the onset of the Würm Ice

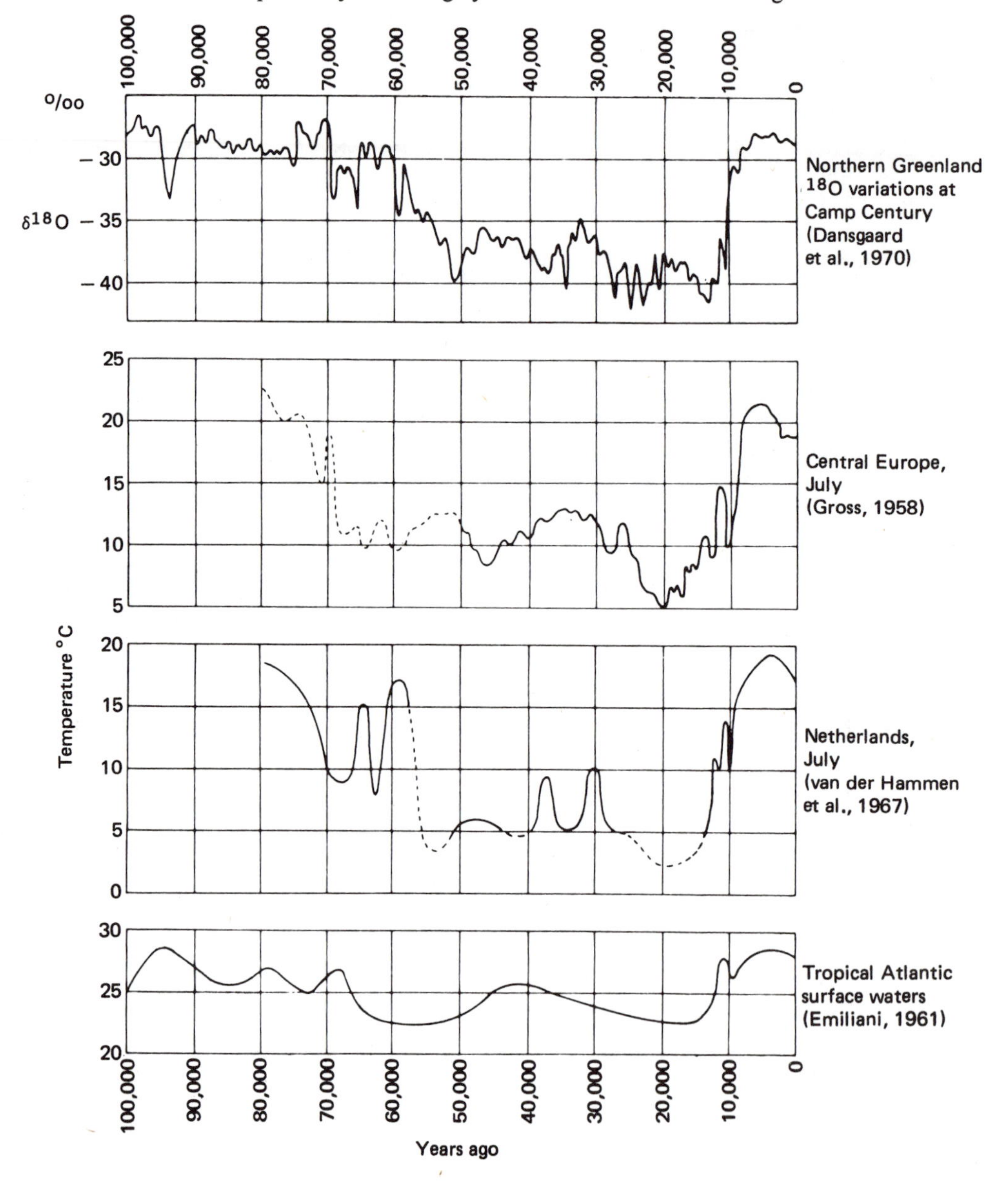

Fig. 2. Suggested pattern of Northern Hemisphere circulation during the onset of the Würm/Wisconsin Glaciation, during formation of the Laurentide (northeastern America) and Scandinavian ice caps. Thin continuous line, suggested average course of 5300 isopleth of 1000 to 500 mbar thickness in January; short arrows, prevailing surface wind directions; broad shaded arrows, most frequent tracks of surface cyclones.

Age, with great frequency of northerly surface winds in sectors near 90 to 100°W and 10°W to 10°E, differs only in degree from the anomaly that apparently produced the recent cold centuries between A.D. 1400 and 1900: It was evidently more extreme as well as more persistent than the latter. The upper cold troughs presumably extended farther south, and were concentrated with higher frequency over the sectors mentioned, than in the recent Little Ice Age.

With this proposed prevailing wind pattern, the beginnings of the Laurentide ice sheet in northern Canada would find a straightforward explanation. The prevailing surface northerly winds would represent one side of a belt of frequent cyclonic activity steered northward and northeastward from the Gulf of Mexico and the western Atlantic toward Labrador and recurving northwestward over the Canadian Arctic. This would represent diversion of the present normal cyclone sequences crossing the Atlantic to the northward tracks described over and near North America. Hence, the main moisture-bearing winds from between south and west would be diverted from their present trans-Atlantic paths and steered north as southerly surface winds toward Arctic Canada: They would doubtless deposit snow from September to May or June over the district about 55 to 70°N with a frequency unknown today. Deposition might be mainly as snow, even in summer, if the reason for the anomalous circulation prevailing was that the strength of the radiation supply was for some reason materially less than now. Thus, the moisture source for the Canadian ice sheet appears to have been the Gulf of Mexico and the tropical western Atlantic.

Examination of the rainfall, and the water-equivalent of the snow, in each month of the year on the lands about the Hudson's Bay region between 55 and 70°N, taking only the months with the most cyclonic and southerly surface wind flow patterns between 1955 and 1968 (i.e., those cases in recent years which came nearest to the suggested patterns of glacial onset), showed that the total downput was typically 150 to 300%, and sometimes 400 to 500%, of normal over substantial areas east, north, and northwest of Hudson's Bay. This regime therefore appears capable of raising the annual total downput from the present 20-to-60-cm average values across the region to perhaps 60 to 200 cm. This would presumably cut Weertman's (1964) estimate of 15,000 to 30,000 years for the growth time required by a major ice sheet to

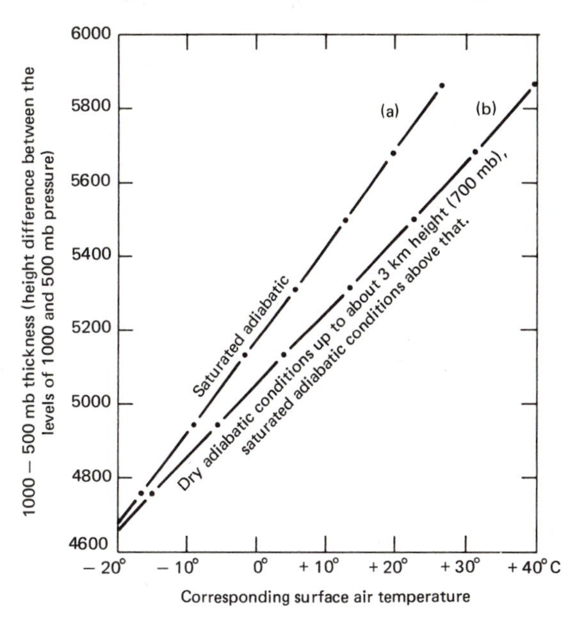

Fig. 3. Thickness values, 1000 to 500 mbar, corresponding to various surface-air temperatures under conditions of convective equilibrium. (a) With saturation of the air, i.e., saturated adiabatic lapse rate of temperature throughout the layer from 1000 to 500 mbar. (b) With saturation above about the 3 km level, i.e., dry adiabatic lapse rate of temperature through the layer from 1000 to 700 mbar; saturated adiabatic lapse rate between 700 and 500 mbar.

5000 to 10,000 years. In the case of the former Laurentide ice sheet it is easy to envisage the still more rapid covering of a large area by ice, actually with an ice sheet of initially rather flat profile, because of the vast, nearly uniform terrain which seems likely to have provided not just one but many starting points, where snow accumulation would first survive a few summers and then grow: The growth rate would not depend on the building up and unstable flow of an ice dome.

Initially, the eastern fringe of North America would retain the warmth of a temperate climate with this circulation arrangement, and might even become warmer for a time, as a result of the increased frequency of southerly winds on the eastern side of the track of frequent cyclonic storms, and these conditions could return in the warmer interstadials. Gradually, however, growth of the continental ice sheet would strengthen the thermal gradients near its border, tending to increase the zonal index and wavelength in the upper westerlies. These tendencies, and the growth of the ice sheet itself, would ultimately displace the northbound cyclone tracks and the belt of southerly surface winds farther east, out over the Atlantic.

The other proposed upper cold trough, also meridionally extended, in the European sector, should be understood as a dynamically induced accompaniment (downstream wave in the upper westerlies) of the American trough. It would suggest a similar prevalence of cyclonic developments steered northward and northeastward over Central and Eastern Europe, depositing snow at most times of the year over the region between northern Scandinavia and Novaya Zemlya. Parts of Russia or western Siberia would initially remain warm, with prevalent southerly winds.

Reconstruction of the Atmospheric Circulation Characterizing the Last Glacial Maximum

The surface air temperatures prevailing in July and January, taken as approximately the warmest and coldest months everywhere, were deduced mainly from botanical and marine biological evidence.

Upper air temperatures have been expressed in terms of the thickness of the layer of the atmosphere between the levels where pressure is 1000 and where it is 500 mb. It is the expansion and contraction of such a deep layer as this, as the

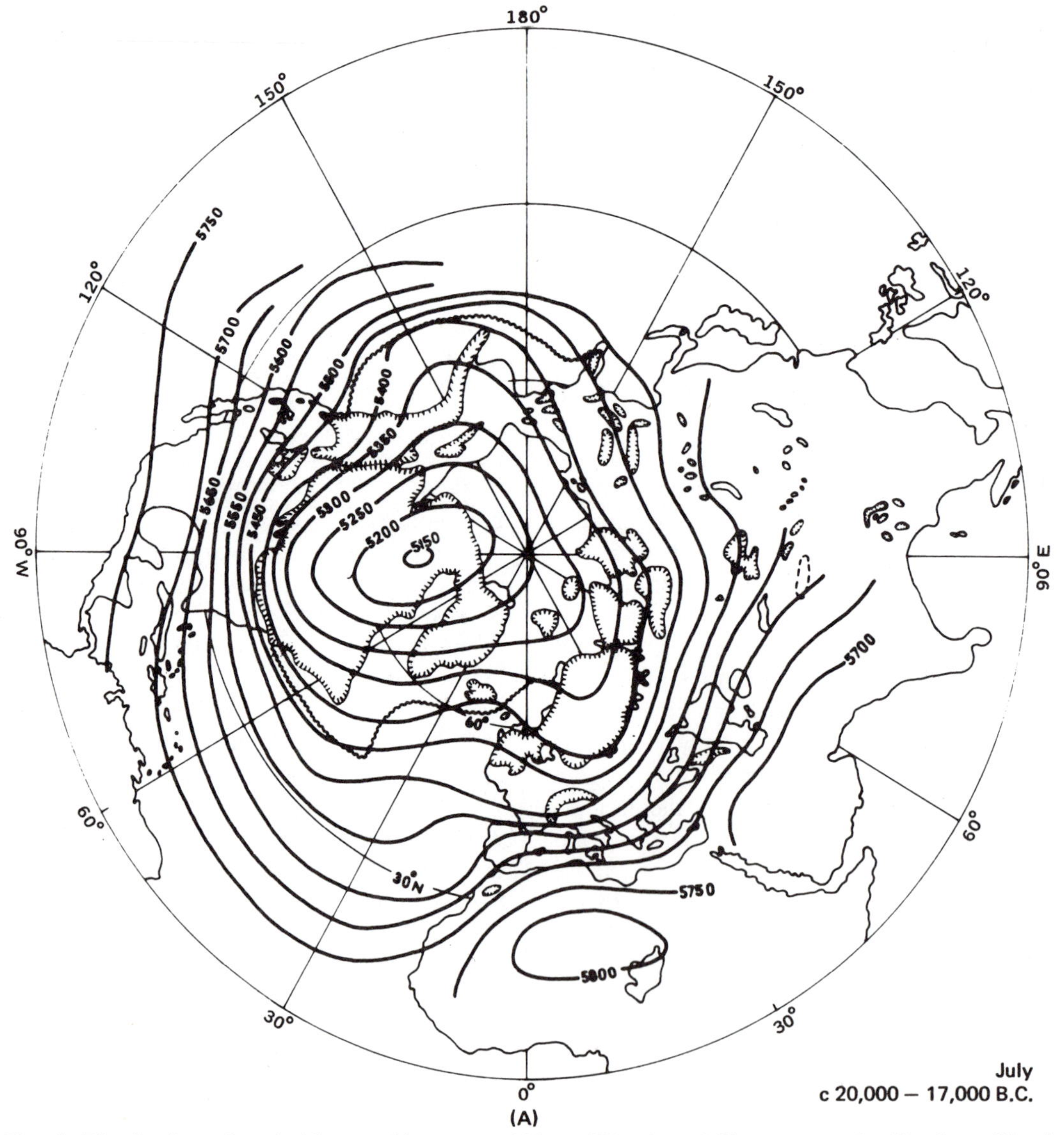

Fig. 4. Distribution of probable monthly mean 1000 to 500 mbar thickness over the Northern Hemisphere derived from surface temperatures indicated by field evidence for (a) July and (b) January during the Würm Ice Age maximum extent of glaceration.

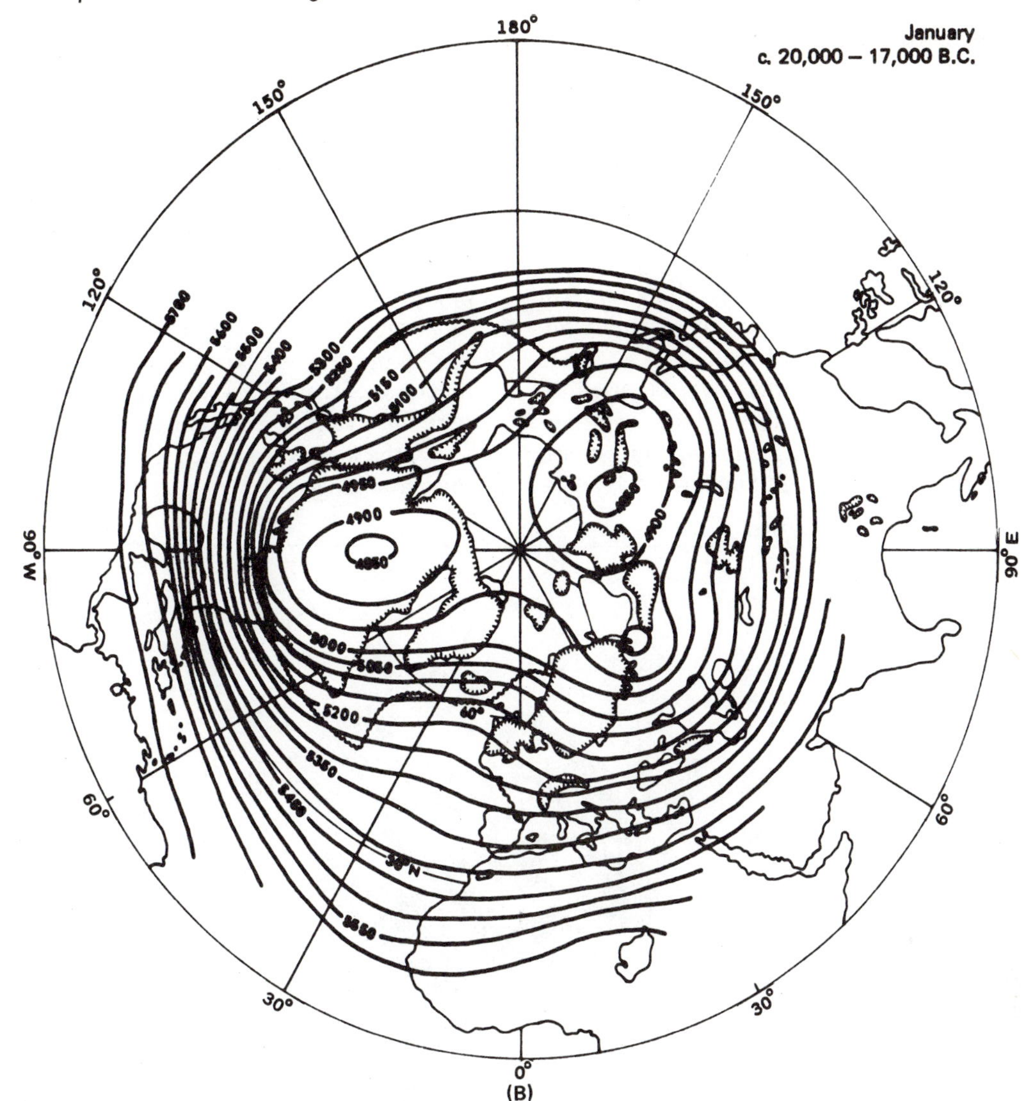

(B)

mean temperature of the air column rises and falls, which calls into being the broad-scale pressure distribution aloft that determines the main flow through a great range of depth of the atmosphere (cf. Lamb, 1964, 1966, 1972). The association between 1000- and 500-mb layer thickness and representative surface air temperatures corresponding to dry and saturated adiabatic lapse rates—the equilibrium conditions produced by convection in unsaturated and saturated air, respectively—is demonstrated in Fig. 3. Lapse rates over warm ocean, and above the level of the cloud base over heated land in summer, are generally near the saturated adiabatic rate. Less fall of temperature with height has been presumed (higher thicknesses for a given surface temperature), and in some cases *much* less than the saturated adiabatic lapse rate (i.e., much higher thicknesses), near the southwestern ocean limits of ice caps where warm air would habitually flow over the cold surface and over the continental lowlands in winter (because of strong radiation cooling of the surface). In all difficult areas appeal has been made to comparable situations or terrains in the world as it is today.

One further final stage was necessary in the representation of the 1000-to-500 mb thickness distribution, which itself approximates to the prevailing pattern of wind flow throughout the middle ranges of the atmosphere (i.e., middle and upper troposphere). The pattern derived from the surface temperatures by the arguments described above was at a number of points too angular to represent a probable pattern of the strong flow of the mainstream of the upper winds. The patterns were therefore smoothed, reference being made to the strongest curvatures at present occurring in the upper westerlies before and behind such barriers as the Rocky Mountains and on encountering regions of strongest thermal contrast at the present day, e.g., prairies–Hudson's Bay and Siberia–Arctic Ocean in summer, Gulf of Alaska–cold surface of Canada, and Norwegian Sea–north Russia in winter. These adjustments amounted to a correction of the lapse rates assumed over the regions affected.

Figs. 4(a) and (b) show the resulting maps of 1000-to-500 mb thickness for the Julys and Januarys around the final ice maximum.

These thickness patterns may be used to deduce the areas favorable for cyclonic and anticyclonic development, using computations based on the development theory of Sutcliffe (1947). The areas favoring cyclonic development are found on modern maps to correspond to:

(1) areas of most frontal activity (on the warmer side of the monthly mean low pressure region),

(2) areas where frontal wave development or the occlusion process repeatedly occur, and

(3) the paths of warm front waves and other cyclonic features "plunging" southeastwards.

Anticyclonic development areas correspond to:

(1) the rear sides of cyclonic circulations where outbreaks of polar air are frequent,

(2) the main subsidence regions within anticyclones and ridges of high pressure at the surface.

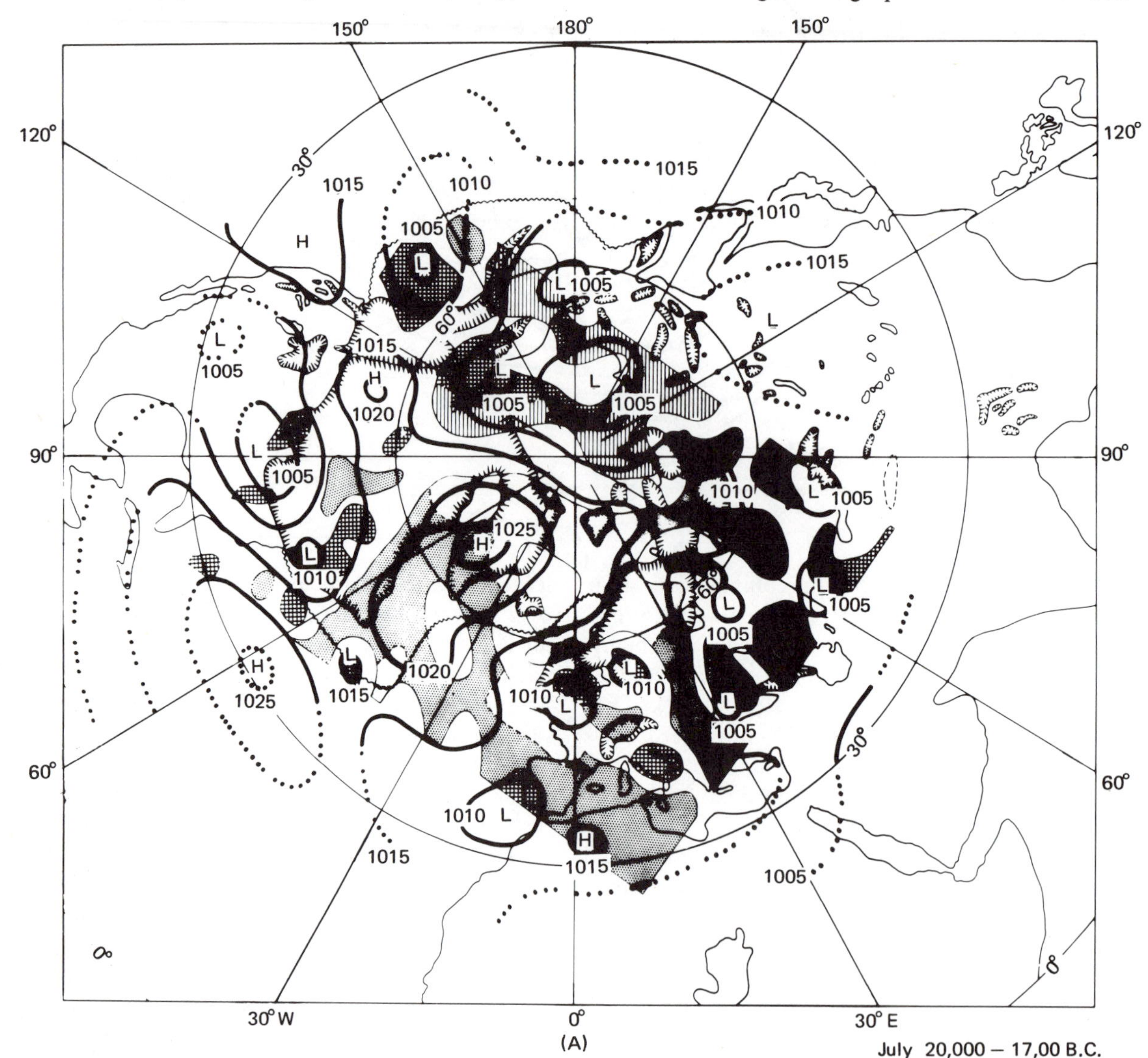

Fig. 5. Derived (computed) distribution of surface-level cyclonic (hatched) and anticyclonic (stippled) development and putative mean isobars for (a) July (b) January Würm/Wisconsin Ice Age maximum.

NB: Cross-hatching and heavy stipple are *strong* indications of the type of development concerned.

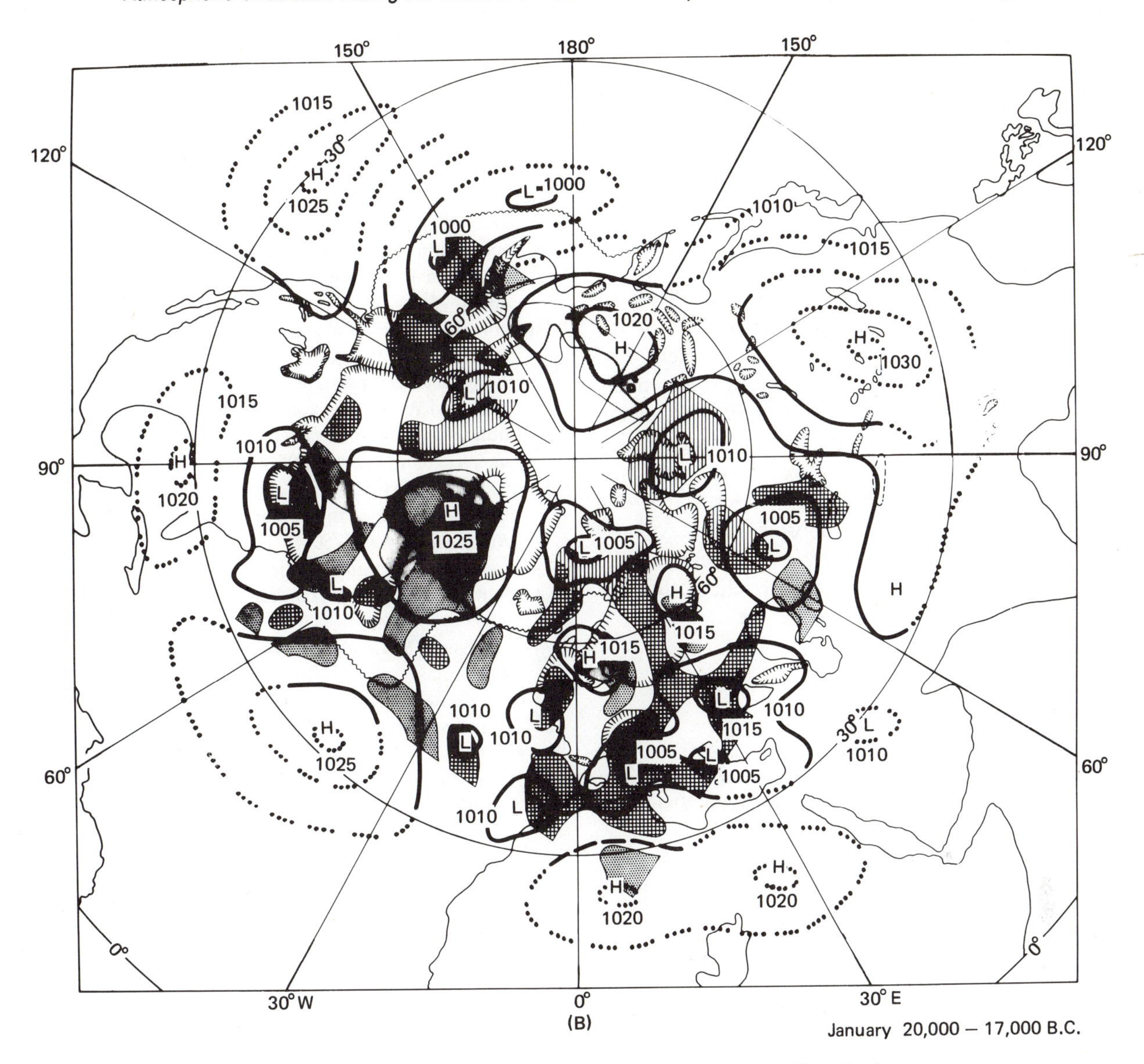

(B)

January 20,000 – 17,000 B.C.

It therefore appears permissible to sketch the prevailing isobar patterns indicated, by combining the above results with steering of the surface depressions and anticyclones by the main currents of the thermal winds in the upper troposphere [i.e., the run of the isopleths in the zones of strong gradient in Figs. 4(a) and (b)]. The results are the isobars on the surface maps in Figs. 5(a) and (b). Putative pressure values have been entered on these isobars only to indicate probable variations of intensity from one epoch to another (see the maps for the late glacial and Postglacial stages put forward in Lamb et al. (1966) and Lamb and Woodruff (1970).

It must be remembered that mean sea level pressure is really fictitious wherever the ice-sheet surface was high above sea level.

Conclusions

The maps give the following indications regarding the atmospheric circulation during the Last Ice Age:

(1) displaced center of the circumpolar vortex and surrounding pressure zones, with the surface polar anticyclone generally occupying the region around Baffin Land and west to northwest Greenland.

(2) great meridionality of the mean surface wind and pressure distributions, particularly in the Atlantic sector, with the polar anticyclone extending far south towards mid Atlantic and blocking most of the west to east, or southwest to northeast, progression of cyclonic systems characteristic of Postglacial climates (including the present day).

(3) little seasonal change of vigor, or of latitude, of the main circulation features (rather as in the southern hemisphere temperate and polar zones today). This applies especially to America and the Western Atlantic; some weakening and a more northern position of the cyclonicity in summer over Europe do appear.

(4) a cyclonic regime over Western Siberia both in summer and winter.

(5) reversal of the circulation over the inner Arctic, probably in the ocean as well as in the atmosphere, by comparison with the present day. Cyclonic (counterclockwise) rotation seems likely to have predominated over the polar region between north Greenland and the coast of Alaska. This cyclonic circulation, if it really occurred, may well have maintained an area of open water, due to upwelling (which would bring saline Atlantic water toward the surface), somewhere in the central Arctic, but it presumably had nothing to do with the causation of the ice age.

References

Andersen, Sv. Th. 1961. Vegetation and its environment in Denmark in the Weichselian glacial. *Danmarks geologiske Undersøgelse, II Raekke*, No. 75.

Blüthgen, J. 1966. *"Allegemeine Klimageographie."* In *"Lehrbuch der allgemeinen Geographie,"* Band II. Berlin: Walter de Gruyter.

Coope, G. R. Shotton, F. W., and Strachan, I. 1962. A late Pleistocene fauna and flora from Upton Warren, Worchestershire. *Philosoph. Trans. Royal Soc. London, Ser. B*, 244:379–418.

Dansgaard, W., and Johnsen, S. J. 1969. A flow model and a time scale for the ice core from Camp Century, Greenland. *J. Glaciol.*, 8:215–223.

———, Johnsen, S. J., Møller, J., and Langway, C. C. 1969. One thousand centuries of climatic record from Camp Century on the Greenland ice sheet. *Science*, 166:377–380.

Dickson, J. 1967. The British moss flora of the Weichselian Glacial. *Rev. Palaeobotany Palynology*, 2:245–253.

Emiliani, C. 1961. Cenozoic climatic changes as indicated by the stratigraphy and chronology of deep-sea cores of *Globigerina*-ooze facies. *Ann. N.Y. Acad. Sci.*, 95(1):521–536.

Ewing, M., and W. L. Donn. 1956a. A theory of Ice Ages, I. *Science*, 123:1061–1066.

———, and W. L. Donn. 1956b. A theory of Ice Ages, II. *Science*, 127:1159–1162.

Flint, R. F. 1957. *Glacial and Pleistocene Geology*. New York: Wiley, 553 pp.

Gross, H. 1958. Die bisherigen Ergebnisse von C-14 Messungen. *Eiszeitalter und Gegenwart*, 9:155–187.

Hammen, van der, T., G. C. Maarleveld, J. C. Vogel, and W. H. Zagwijn. 1967. Stratigraphy, climatic succession and radiocarbon dating of the last glacial in the Netherlands. *Geologie en Mijnbouw*, 46:79–95.

Hunkins, K. and H. Kutschale. 1967. Quaternary sedimentation in the Arctic Ocean. In: M. Sears, ed. *Progress in Oceanography*, pp. 4:89–94.

Ku, T. L. and W. S. Broecker. 1967. Rates of sedimentation in the Arctic Ocean. In: M. Sears, ed. *Progress in Oceanography*, pp. 4:95–104.

Lamb, H. H. 1964. Climatic changes and variations in the atmospheric and ocean circulations. *Geolog. Rundschau*, 54:486–504.

———. 1965. The early medieval warm epoch and its sequel. *Palaeogeogr. Palaeoclim. Palaeoecol.* 1:13–37.

———. 1966. *The Changing Climate*. London: Methuen, 236 pp.

———. 1972. *Climate: Present, Past and Future*. Vol. 1. London: Methuen, 613 pp.

———, R. P. W. Lewis, and A. Woodroffe. 1966. Atmospheric circulation and the main climatic variables. In: *Proceedings of the International Symposium on World Climate 8000-0 B.C.* London: Royal Meteorological Society, pp. 174–217.

———, and A. Woodroffe. 1970. Atmospheric circulation during the Last Ice Age. *Quaternary Res.*, 1(1):29–58.

Muller, E. H. 1964. Quaternary sections at Otto, New York. *Am. J. Sci.*, 262:461–478.

Penny, L. F., G. R. Coope, and J. A. Catt. 1969. Age and insect fauna of the Dimlington silts, East Yorkshire. *Nature*, 224:65–67.

Simpson, I. M., and R. G. West. 1958. On the stratigraphy and palaeobotany of a late Pleistocene deposit at Chelford, Cheshire. *New Phytologist*, 57:239–250.

Sutcliffe, R. C. 1947. A contribution to the problem of development. *Quart. J. Meteorol. Soc.*, 73:379–383.

Weertman, J. 1964. Rate of growth or shrinkage of nonequilibrium ice sheets. *U.S. Army C.R.R.E.L. Res. Rep.*, 145.

West, R. G. 1968. *Pleistocene Geology and Biology*. London: Longmans, 377 pp.

Zagwijn, W. H. 1961. Vegetation, climate and radiocarbon datings in the Late Pleistocene of the Netherlands. *Mededelingen van de Geologische Stichting*, 14:15–45.

Chapter 15
Possible Causes of Glaciations

EDWARD PETER JACOBUS VAN DEN HEUVEL [1]
AND PETER BUURMAN [2]

Abstract

Evidence of glaciation in the earth's history is briefly reviewed and possible cosmic and terrestrial causes of climatic change are discussed. The late Tertiary-Pleistocene glaciations appear to belong to one large Ice Age, with a duration of over 7 million years, which is presently underway. Throughout this Ice Age parts of the polar regions have been continuously glaciated. The duration of this Ice Age resembles that of other large Ice Ages in the earth's history (cf. Table 1). The only plausible cause of large Ice Ages seems to be a combination of continental uplift, mountain building, and thermal isolation of one or both of the poles, as suggested by Ewing and Donn. The large variations in mid-latitude glaciations during the Pleistocene, on a time scale of about 40,000 years, may have been triggered by insolation variations of the type calculated by Milankovitch. This is evidenced by the observed time correlation between insolation variations, oxygen isotope temperatures, and oscillations in sea level during the Pleistocene. The observed correlation between interglacial high sea levels and the precession on one hand, and the large-amplitude variations in isotopic temperatures and the tilt of the ecliptic plane on the other, seem to confirm theories in which the contributions of precession and tilt are given different weights (Broecker, 1966). A possible physical explanation for these different weights is suggested.

[1] Sterrewacht "Sonnenborgh", Rijksuniversiteit, Utrecht, The Netherlands, and Astrophysical Institute, Vrije Universiteit, Brussels, Belgium.

[2] Department of Soil Science and Geology, Agricultural University, Wageningen, The Netherlands.

Introduction

During the past two decades it has become clear that the Antarctic ice sheet as well as high-latitude glaciers in the Northern Hemisphere have begun their expansion probably as early as 7 million years ago and that large glaciers already existed >2 million years ago in Iceland, North America, and Argentina. At the same time, abundant confirmation has been found of the hypothesis of continental drift. Accurate radioactive dating and paleotemperature measurements with the $^{18/16}O$ method have furnished quantitative data on Pleistocene and Tertiary temperatures.

In the light of these findings, this paper attempts to review critically ideas on the origins of glaciations. Inevitably, in an area in which much is still uncertain, this review will be somewhat colored by personal opinions.

Before discussing the possible causes we shall briefly review the observed facts that have to be explained by any theory on the origin of glaciations.

The review is based on literature post dating 1959. For earlier references we refer to Emiliani and Geiss (1959). For an updated review of the most important literature on Pleistocene and late

Table 1 Ice Ages in the History of the Earth
(partly after Holmes, 1965; and Dunn et al., 1971)

Ice Age	Time (millions of years)	Main glaciated areas
Late Cenozoic (present)	7 to 0 (50 to 0 for parts of Antarctica)	Antarctica, Greenland, and periodically, parts of Northern Eurasia, North America, South America, and New Zealand
Upper Paleozoic	250 to 200	Australia, South Africa, South America, India, and Antarctica
Late Precambrian	750 to 700 (Sturtian) 650 to 570 (Marinoan)	Australia, Spitsbergen, North Atlantic area, U.S.S.R., Africa, northeastern Asia, and possibly South America
Huronian (or part of the Late-Precambrian Ice Age?)	800 (?)	Cobalt, Ontario; Witwatersrand, Transvaal
Bothnian	1000	Finland, western Australia (?)
Damara	1200 (?)	Chuos District, southwestern Africa
(No name)	1500 (?)	Medicine Bow Mountains, Wyoming
Gowganda	2200	southern Canada

Tertiary geology we refer to Flint (1971); for general reading about paleoclimatology and possible causes of Ice Ages we refer to the books by Shapley (1953), Brooks (1971), Schwarzbach (1963), Nairn (1964), and Lamb (1971).

The Observational Record on Glaciation

Pre-Tertiary Ice Ages

The "normal" climatic state of the earth seems to be tropical to subtropical from the equator to the poles (Brooks, 1971). This normal warm situation is at times interrupted by periods on the order of several tens of millions years during which parts of the earth's surface are glaciated. Such long periods of variable glaciation we shall call Ice Ages. The shorter periods of maximum extent of continental glaciers within an Ice Age we shall call glaciations.

Table 1 summarizes the Ice Ages in the geological history of the earth, recognized to date (partly after Holmes, 1965; and Dunn et al., 1971). Of the pre-Tertiary Ice Ages the Carboniferous (Upper Paleozoic) is the best documented. Its abundant traces have been found in South America, South Africa, India, and Australia, and have furnished one of the first important arguments for the idea that these continents formed one single land mass (Gondwanaland) in the past (Wegener, 1915, 1924; du Toit, 1937). Paleomagnetic data (McElhinny et al., 1968) show that the center of the ice cap, which is now located in South Africa, was close to the magnetic South Pole during the Carboniferous (Fig. 1). Theoretical considerations concerning the generation of the earth's magnetic field (Elsässer, 1950; Runcorn, 1954; Parker, 1970, 1971) indicate that the average axis of the magnetic dipole is expected to coincide with the rotational axis. Paleomagnetic measurements are therefore expected to give the positions of the *geographic* South Pole. The absence of traces of the Late Paleozoic Ice Age in the Northern Hemisphere is explained by the fact that most of the other land masses were located near the equator at that time, and the North Pole was located in the present mid-Pacific Ocean. The late Paleozoic Ice Age probably lasted about 50 million years and the weight of the ice cap may have caused the break-up of Gondwanaland into its present pieces (Gough, 1970).

Of the older Ice Ages shown in Table 1, only the Late Precambrian has been well documented. Abundant information comes from Australia (Dunn et al., 1971), where the Late Precambrian Ice Age extended between 750 and 570 million years ago. Dunn et al. find that it consisted of at least two major stages of very long duration: the first (Sturtian) stage extended for about 50 million years, while the second (Marinoan) stage extended for approximately 80 million years. Both left immense deposits of glacial sediments—in places 5500 m thick—covering about half of the Australian continent. It was probably the most widespread of all known Ice Ages, as it also left glacial sediments in Britain, Scandinavia, Spitsbergen, Greenland, China, and possibly South Africa (cf.

Roberts, 1971). Due to its long duration, it is possible that the different maxima were recorded as separate Ice Ages in various parts of the world. For instance, in China four Ice Ages were recorded in the late Precambrian and one in the early Cambrian (cf. Dunn et al., 1971).

For ice ages preceding the late Precambrian similar uncertainties exist. From the amounts of sediment deposited, however, it seems that each Ice Age probably also lasted for over 10 million years.

Tertiary Evidence of Glaciation

During the Mesozoic (~230 to ~70 m.y.b.p.) some 85% of the earth's surface was covered by oceans and shallow seas. Only continents of low elevation (mostly less than 2000 m) existed. There were no glaciers, and even in the polar regions temperatures probably exceeded 10°C, as no $^{18/16}O$ ocean bottom temperatures lower than 14°C have been recorded from sediments deposited during these periods (Emiliani, 1961). The absence of coral reefs in polar regions and their widespread distribution in equatorial zones indicate polar temperatures below 20°C and equatorial temperatures around 29°C for this period (cf. Newell, 1972). Similar temperatures existed between the late Precambrian and the Carboniferous (Brooks, 1971). The onset of the Pleistocene glaciations seems to have started more than 30 m.y.b.p., and perhaps as far back as 70 m.y.b.p. Paleontological evidence of changes in fauna and flora indicates a temperature decrease of 8 to 10°C in middle northern latitudes during the Tertiary (Colbert, 1953; Barghoorn, 1953). The $^{18/16}O$ analyses of equatorial Pacific Ocean bottom sediments (Emiliani, 1961) indicate a temperature decrease of Pacific bottom water from 14°C in Cretaceous, to 11°C in middle Oligocene, and 3°C in the late Pliocene time. As this bottom water is presently formed by Antarctic melt-water, the Tertiary temperature decrease suggests that the formation of high-latitude glaciers in the Southern Hemisphere may have already started between 70 and 30 m.y.b.p. During the same period, due to continental uplift, many of the shallow Mesozoic seas were being drained and the continental area increased to about 29% (of the earth's surface) by late Pliocene time.

Figure 2 shows the gradual Tertiary temperature decrease, as inferred from fossil plants in various parts of the world.

Studying 18 Cenozoic subantarctic ocean bottom sediment cores, raised between 90 and 160°W longitude, Margolis and Kennett (1970) found ice-rafted detritus in sediments of Eocene (53 to 36 m.y.b.p.) and Oligocene (36 to 22 m.y.b.p.) age. No ice-rafted detritus was found in lower and middle Miocene deposits in this area, suggesting that warmer climates prevailed during these time intervals. From upper Miocene (7

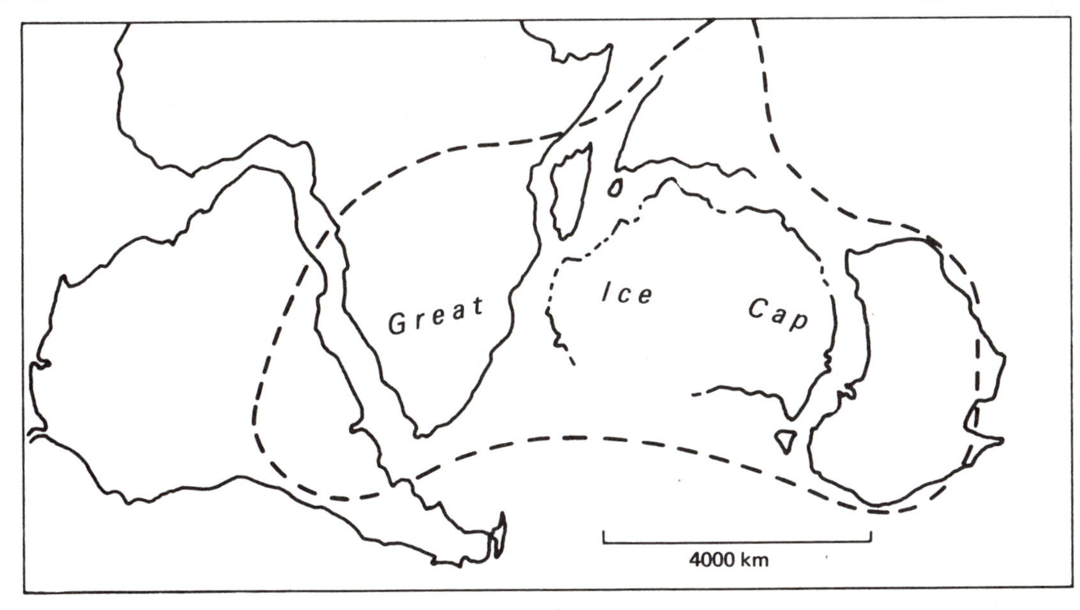

Fig. 1. The upper Paleozoic (Carboniferous) ice cap covered parts of South Africa, South America, India, Australia, and Antarctica; at that time, these continents most probnea (Veeh and Chappel, 1970); (B) after Gough, 1970).

m.y.b.p.) on, ice-rafted detritus was continuously present in the studied cores. Diversity of foraminiferal species in sediments confirmed these climatic trends; paleoclimatic curves for New Zealand also confirm the Eocene-Oligocene low as well as the lower and middle Miocene high temperatures (references in Margolis and Kennett, 1970). Although these findings agree with the general trends for the Northern Hemisphere (Fig. 2), the middle Miocene high temperature was apparently higher in the Southern than in the Northern Hemisphere. Tertiary transgressions and regressions in western Europe suggest a relation with the amount of Antarctic ice during this period.

The presence of ice-rafted detritus in a core raised from latitude 32°59′S (87°57′W longitude) in Eocene sediments is interesting, as even today icebergs in this region do not reach north of 45°S latitude. This may indicate that, due to continental drift, the position of Antarctica and the surrounding ocean bottom in those times was quite different from today. From the distribution of fossil coral reefs (Newell, 1972) it may be concluded that during the late Mesozoic the equator along these longitudes must have been situated north of its present position, and the South Pole was located nearer the above-mentioned site. The observation that Antarctica was continuously glaciated for the last 7 m.y. confirms earlier findings by Armstrong et al. (1968) and Hays and Opdyke (1967). Hays and Opdyke found that warm water radiolaria began to disappear about 5 m.y.b.p. in subantarctic cores and have been absent in the last 2.4 m.y. On continental Antarctica, data from McMurdo Sound indicate the continuous presence of glaciers for over 4 m.y. (cf. Flint, 1971, p. 711). Mercer (1969) found glacial deposits in Argentina covered by a lava bed more than 2 m.y. old.

In the Northern Hemisphere glacial deposits in Iceland about 2 m.y. old were first noticed by Rutten and Wensink (1960). In Sierra Nevada, California, glacial deposits about 3 m.y. old were found (Curry 1966). Herman (1970), studying sediment cores from the Arctic Ocean, found ice-rafted detritus in sediments deposited >3 m.y.b.p. From the fauna and lithology of these cores, it appears that increased iceberg and shelf-ice production and melting occurred between 2.4 and 0.7 m.y.b.p. (Herman, this volume). Herman found that the Arctic Ocean was free of permanent pack-ice between 2.4 and about 0.7 m.y.b.p., after which successive ice-covered and ice-free conditions existed.

The continuous glaciation of both polar caps of the earth for the past 6 to 7 m.y., together with the presence of Antarctic glaciers in early Eocene time, make the present (late Cenozoic) Ice Age similar in duration to the other large Ice Ages listed in Table 1.

The Pleistocene Glaciations

Mid-latitude glaciation seems to have started not much earlier than 0.7 m.y.b.p. and was accompanied by freezing of the Arctic Ocean (Herman, 1970). The increase in glacier ice volume as compared with the present volume, during the maximum of Pleistocene glaciation as esti-

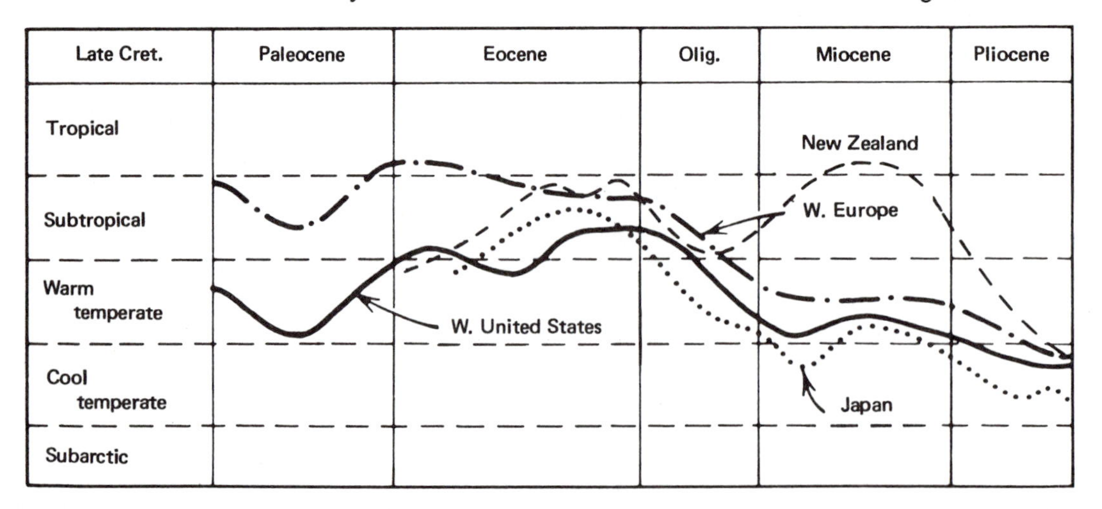

Fig. 2. Climatic cooling through Cenozoic time, inferred from fossil plants in four parts of the world (adapted from Flint, 1971, Fig. 16-3; and Margolis and Kennett, 1970, Fig. 2).

mated by Flint (1971, p. 84), is given in Table 2.

Through sea level lowering by about 100 to 130 m, the Pleistocene glaciations had profound influence on the distribution of flora and fauna, even in equatorial regions such as the Indonesian Archipelago. There seem to have been at least six large continental glaciations during the Pleistocene (Flint, 1971, pp. 624–625), alternating with interglacial periods during which climates and sea level were similar to present. In the Alps the Pleistocene glaciations were named Würm, Riss, Mindel, Günz, Donau and Biber. During interglacials, sea level probably never exceeded present levels by more than 6 to 10 m (cf. Emiliani, 1969), indicating that most of the Antarctic ice cover did not melt. Had the ice cover melted, it would have caused a subsequent rise in sea level of about 60 m. However, parts of the Greenland ice sheet may have melted, for this would not have raised the sea level by more than 10 m. All available evidence (Flint, 1971, p. 715) indicates that during the past 1.4 m.y. temperatures on the entire Antarctic continent did not rise above 0°C. The bottom 10 to 15% of the Greenland ice has an age of >100,000 years (Dansgaard et al., 1969), which means that it already existed during the Last Interglacial. Unfortunately, the ages of the Pleistocene glaciations are still a matter of dispute. The range of reliable dating extends back to about 120,000 years. Figure 3 shows that the $^{18/16}O$ isotopic temperature curve, as derived from Caribbean and Mediterranean bottom sediments (Emiliani, 1969), agrees well with the $^{18/16}O$ determinations from a core of Greenland ice, dated by the ^{14}C method (Dansgaard et al., 1971), as well as with data from continental Europe and North America, dated by various radioactive isotope methods. Simultaneously with the Pleistocene glaciations of the Northern Hemisphere, parts of the Southern Hemisphere were also glaciated, notably southern Chile and Argentina and the southern part of New Zealand (cf. Flint, 1971, pp. 687–710).

Table 2 Estimated Volume of Land Ice Today and During Maximum Pleistocene Glaciation (somewhat speculative, after Flint, 1971, p. 84)

	Volume (10^6 km^3 water)		Sea level equivalent (m)	
Ice sheet	Today	Glacial Age	Today	Glacial Age
Antarctica	21.50	23.84	59	66
Greenland	2.38	4.01	6	11
Laurentide ice sheet (North America)	—	27.01	—	74
Cordilleran ice sheet (North America)	—	3.25	—	9
Scandinavian ice sheet	—	12.21	—	34
All other ice	0.18	1.04	0.5	3
Total	24.06	71.36	65	197

For the past 400,000 years the most reliable continuous record available is Emiliani's (1966, 1969) generalized $^{18/16}O$ isotopic temperature curve obtained from Caribbean deep-sea sediment cores (Fig. 4). Dating was carried out with the ^{14}C and $^{231}Pa/^{230}Th$ methods. In the high-temperature parts of the curve the relative maxima coincide very well with independently dated time intervals of interglacial high sea levels (Broecker et al., 1968; Emiliani, 1969; Emiliani and Rona, 1969), which are indicated by arrows in Fig. 4. The curve reflects the variations in surface water $^{18/16}O$ ratio at low latitudes. According to Dansgaard's and Tauber's (1969) interpretation, some 70% of the $^{18/16}O$ variation in ocean water does not reflect water temperature, but is determined by the volume of glacial ice. At times of maximum continental glaciation, the ^{18}O content of the ocean was highest inasmuch as ^{16}O-enriched water was removed from the ocean and precipitated on land and tied up in glaciers. The temperature minima in Emiliani's curve would therefore probably reflect the maximum extent of continental glaciers rather than minimum ocean water temperatures. As we are mainly concerned here with the dating of the glaciations, we have represented Emiliani's curve in its original unmodified form. The curve shows that during the last 400,000 years the amount of glacial ice varied with a periodicity of roughly 40,000 years (Emiliani and Geiss, 1959; Emiliani, 1966; Broecker, 1966; Van den Heuvel, 1966a). Apart from the Würm (Weichsel, Wisconsin) Glaciation, the correlation of continental glaciations with the minima of Emiliani's curve is not yet firmly established.

The Problem of Causes

From the evidence in the foregoing section, it appears that any theory on the origin of glaciations should explain two facts:

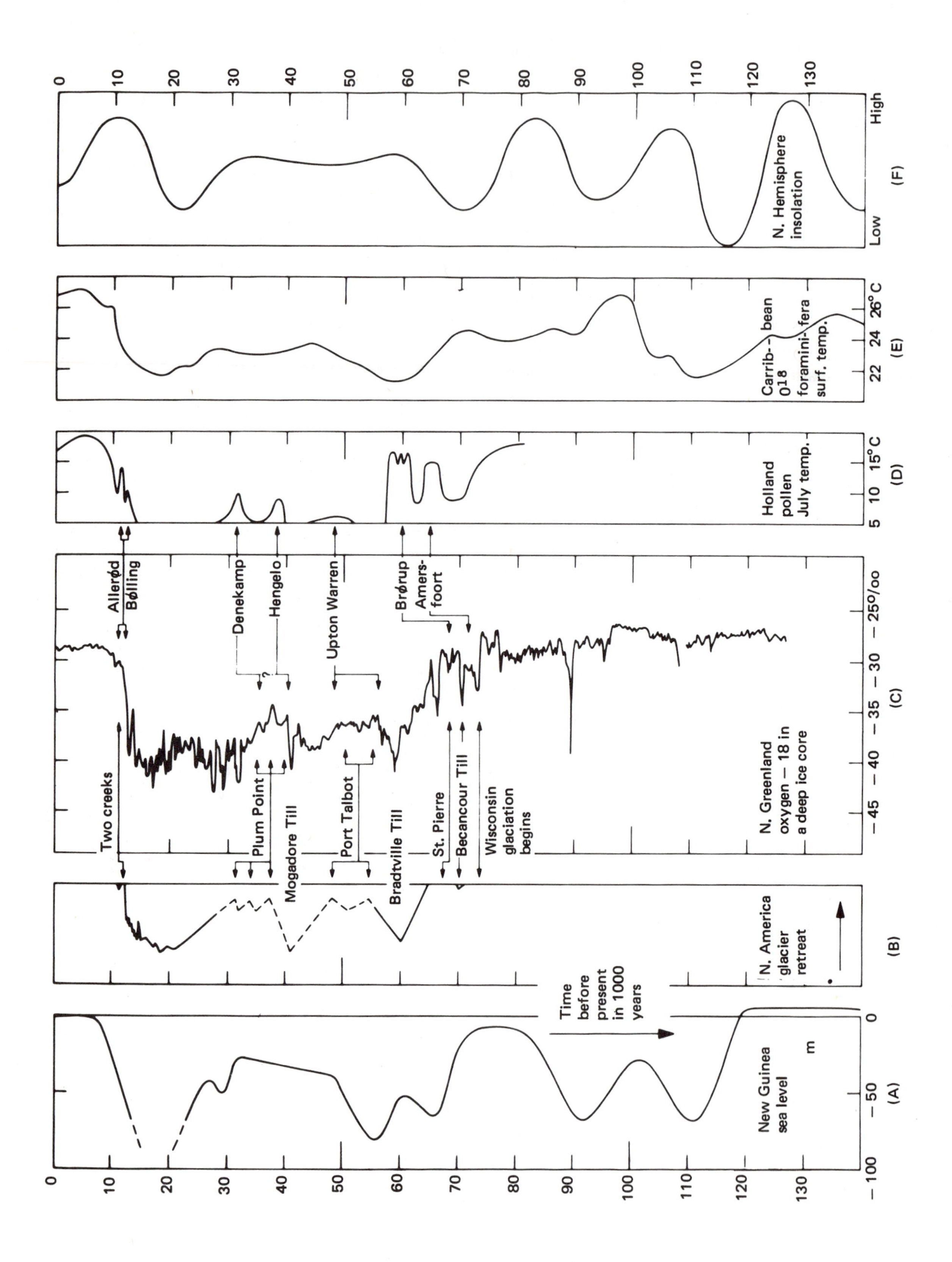
New Guinea sea level
m
(A)
N. America glacier retreat
(B)
Time before present in 1000 years
N. Greenland oxygen – 18 in a deep ice core
(C)
Holland pollen July temp.
(D)
Carrib-bean O18 foramini-fera surf. temp.
(E)
N. Hemisphere insolation
(F)
Low
High
Two creeks
Plum Point
Mogadore Till
Port Talbot
Bradtville Till
St. Pierre
Becancour Till
Wisconsin glaciation begins
Allerød
Bølling
Denekamp
Hengelo
Upton Warren
Brørup
Amers-foort
−45 −40 −35 −30 −25‰
5 10 15°C
22 24 26°C
−100 −50 0

(1) The high-latitude cooling for periods on the order of 10 to 100 million years;
(2) For the present (late Cenozoic) Ice Age: the large variations in the amount of continental glacial ice, on a time scale on the order of 40,000 years.

As equatorial temperatures apparently did not drop by more than a few degrees Centigrade since the Mesozoic, while the polar temperatures decreased by more than 14°C. The Ice Age condition seems to mean a *steepening of the temperature gradient between the equator and the poles, rather than a uniform cooling of all parts of the earth* (cf. Brooks, 1971).

Furthermore, it should be explained that during the Pleistocene the amount of glacial ice apparently varied between two well-defined limits: the *interglacial* stages with ice sheets only on Antarctica and (parts of) Greenland, and the *glacial* stages in which the Scandinavian mountains, Canada, and the northern United States, as well as New Zealand and the southern Andes were centers of large continental ice sheets.

The Heat Balance of the Earth and Possible Cosmic and Terrestrial Factors Affecting Climate

The overall heat balance of the earth is of primary importance to the problem of climatic change. This balance is largely determined by the albedo, which in turn is influenced by the distribution of oceans and continents, by cloudiness, sea roughness, and by the amount of glaciated surface area. Table 3 lists the albedos of various types of surface. The present over-all albedo of the earth, as determined from the reflection of earth light by the moon, is about 35% (Fritz, 1949; Danjon, 1954; Franklin, 1967). (We do not use values obtained from satellite observations, as these might be less reliable, cf. de Vaucouleurs, 1970). Sixty-five percent of the radiation not directly reflected is absorbed by the atmosphere, oceans, and continents, and is finally radiated outward, mostly in the form of long-wave radiation. It is this process of absorption and re-emission which determines the equilibrium temperature of the earth. It is clear that changes in the albedo will affect the heat balance and the temperature equilibrium. The present heat balance of the earth, according to Möller (1950), is depicted in Fig. 5. According to calculations by Wexler (1953), an increase of the albedo from 35 to 43% will change the heat balance in such a way that the mean equilibrium temperature will drop by about 8°C; hence, a decrease by about 1°C for 1% change in albedo. Furthermore, Simpson (1938) calculated that a ±10% change in solar radiation will induce changes of ±6°C in the equilibrium temperature. Although these calculations are only rough, first-order approximations, they show that climatic changes, and possibly glaciations, may be induced by terrestrial as well as by cosmic factors.

Table 3 Albedo for Various Types of Surfaces (after Fritz, 1951; and Dietrich and Kalle, 1957, p. 126)

Type of surface	Albedo (%)
Vegetation, ground	8 to 15
Calm sea	8
Rough sea	40
Clouds	60 to 70
Ice and snow	50 to 80

The following cosmic and terrestrial factors have been suggested as possible causes of glaciations:

(1) *Cosmic*
 (a) Secular changes in the earth's orbit (Milankovitch, 1920, 1930, 1941);

Fig. 3. Climatic variations during the last 130,000 years, estimated by various methods (from Dansgaard et al., 1971).
(A) Sea-level changes in New Guinea (Veeh and Chappel, 1970);
(B) A^{14}C-dated stratigraphic study of Pleistocene deposits in the Ontario and Erie basins, showing the advances and retreats of the edge of the Laurentian ice sheet (Goldwaith et al., 1965);
(C) The δ (^{18}O) variations in the Camp Century ice core (Dansgaard et al., 1971);
(D) A^{14}C-dated pollen study from Holland (Van der Hammen et al., 1967);
(E) Oxygen-isotope study of deep-sea cores showing part of the generalized temperature curve for the surface water of the Central Caribbean (see Fig. 4, after Emiliani, 1966);
(F) Northern Hemisphere summer insolation curve (W. S. Broecker, 1968; cf. the curve for ϕ +45° in Fig. 7).

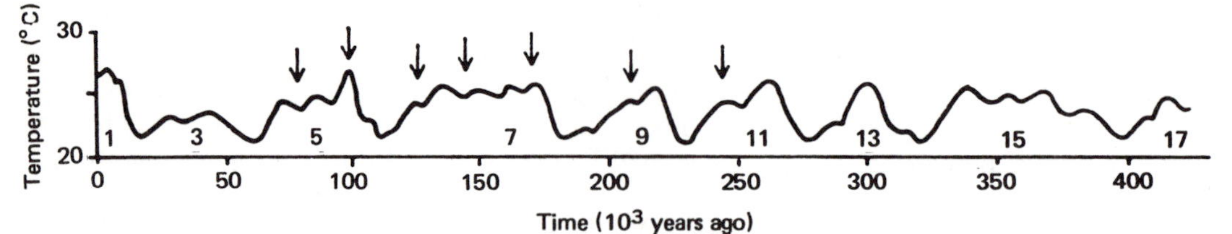

Fig. 4. Generalized temperature curve for the surface water of the Central Caribbean, as derived from oxygen-isotope studies of deep-sea cores (Emiliani, 1966). Dating was carried out with the ^{14}C and $^{231}Pa/^{230}Th$ methods. Arrows indicate independently dated interglacial high sea levels (after Emiliani, 1969).

(b) Variations in the energy output of the sun (Öpik, 1953a, 1953b, 1965);
(c) The passage of the solar system through an interstellar cloud (cf. Krook, 1953);
(d) Variations in solar activity (Wexler, 1953; Willett, 1953; Bray, 1968, 1971);
(e) Variations in the tidal strength (Petterson, 1914);

(2) *Terrestrial*
(a) Mountain building and continental uplift, which may have changed the atmospheric circulation pattern (cf. Emiliani and Geiss, 1959);
(b) Continental drift, which may have caused isolation of the poles from the oceanic circulation system (cf. Ewing and Donn, 1956a, 1956b);
(c) Changes in the atomospheric carbon-dioxide content (Roberts, 1971), or
(d) Changes in the amount of vulcanic dust in the atmosphere (Wexler, 1953).

As to the terrestrial factors, factors 2(a), and (d) are partially due to the same cause, *viz.*, continental drift. These factors have about the right time scale for explaining the large Ice Ages. (We define as "time scale" the time interval required for the development of significant changes, e.g., for a continent to move over a distance comparable to its size). However, a decrease in the carbon dioxide content of the atmosphere—"anti greenhouse effect"—may only have been effective during the late Precambrian Ice Age, as the main phase of atmospheric carbon-dioxide reduction occurred between 2 billion and 500 m.y.b.p. (Cloud, 1968; Cloud and Gibor, 1970). If we were to adopt this effect as a cause of glaciations, it might imply that one would have to look for a different cause for each individual glaciation. Because this does not seem to be a sound scientific principle, we shall not consider variations in atmospheric carbon dioxide content as a general cause of glaciation. As to cosmic factors, the interstellar cloud hypothesis seems unlikely because the hypothetical cloud which would have caused the Würm (Weichsel) Glaciation could not have travelled over a distance larger than several light years and thus it should still be observable. However, this is not the case. The evolutionary variations of the solar energy output, as suggested by Öpik seem very unlikely (Van den Heuvel, 1966b, 1966c); particularly the required initial solar hydrogen content of 30% only is entirely unacceptable.* For these reasons the only reasonable causes remaining are 1(a), (d), and (e), and 2(a), (b), (d), or a combination of these. We shall therefore consider them in more detail.

Cosmic Causes of Climatic Change

Solar Activity

The climatic effects of solar activity are still poorly understood. Some evidence of correlation between climate and solar activity during the past 2500 years was found by Bray (1968, 1971). Dansgaard et al. (1969) found in the $^{18/16}O$ record of Greenland ice, periodicities of 120 and 940 yr, which they ascribed as due possibly to solar activity. Nevertheless, such activity would have had only small climatic effects, as the temperature changes involved most probably did not exceed a few degrees Centigrade (the average reduction in mean temperatures during the "Little Ice Age" of the 17th and 18th centuries). Therefore, the solar activity seems to be an insufficient explanation of a drop of 14°C in polar temperatures since the Mesozoic. However, it may have triggered in-

* Recently, the negative results of efforts to detect neutrinos from the solar interior have led a number of astrophysicists to reconsider the possibility of instabilities in the solar core. These considerations (Dilke and Gough, 1972; Ezer and Cameron, 1972) suggest that possibly, with time intervals on the order of 250 m.y. instability and mixing occurs in the solar core. This would cause a decrease in solar energy output by about 5% over a time interval on the order of 6 m.y. Hence, it might cause an Ice Age of considerable duration. However, before conclusions about this interesting possibility can be drawn, the physics of the suggested instability process should be carefully worked out in detail.

creased glaciation on an already cool earth. An extensive though rather speculative, discussion of solar activity as a possible cause of glaciation was given by Willett (1953). Triggering effects will be considered under a separate heading.

Variations in Insolation Produced by Secular Changes in the Earth's Orbit

These variations do not affect the total amount of *yearly received* solar radiation on the entire globe; they only influence the *distribution* of solar radiation over different latitudes. Milankovitch (1941) systematically explained how, at different latitudes, the insolation varies, as a consequence of secular variations of the elements e, ϵ, and Π of the earth's orbit, due to the variable gravitational attraction of other planets. Here e denotes the orbital eccentricity (see Fig. 6), ϵ is the obliquity of the ecliptic plane (equals the angle between the ecliptic and the equatorial plane), and Π is the longitude of the perihelion (equals the angle, seen from the sun, between the directions to the equinox γ and to the perihelion P, measured in the direction of the earth's orbital motion).

The insolation variations can be computed accurately as far back in history as one wishes. According to Van Woerkom's (1953) revision, e varied during the past million years between 0.0 and 0.053 in a period of roughly 96,600 years; ϵ varied between 21°.8 and 24°.4 in a period of nearly 41,000 years; and Π increased by 360° in a period ranging from 13,500 to 29,000 years, with an average of 21,000 years. The main contribution to the last-mentioned periodicity comes from the precession (P = 25,700 years); the remainder is due to the motion of the line of apsides in the ecliptic plane.

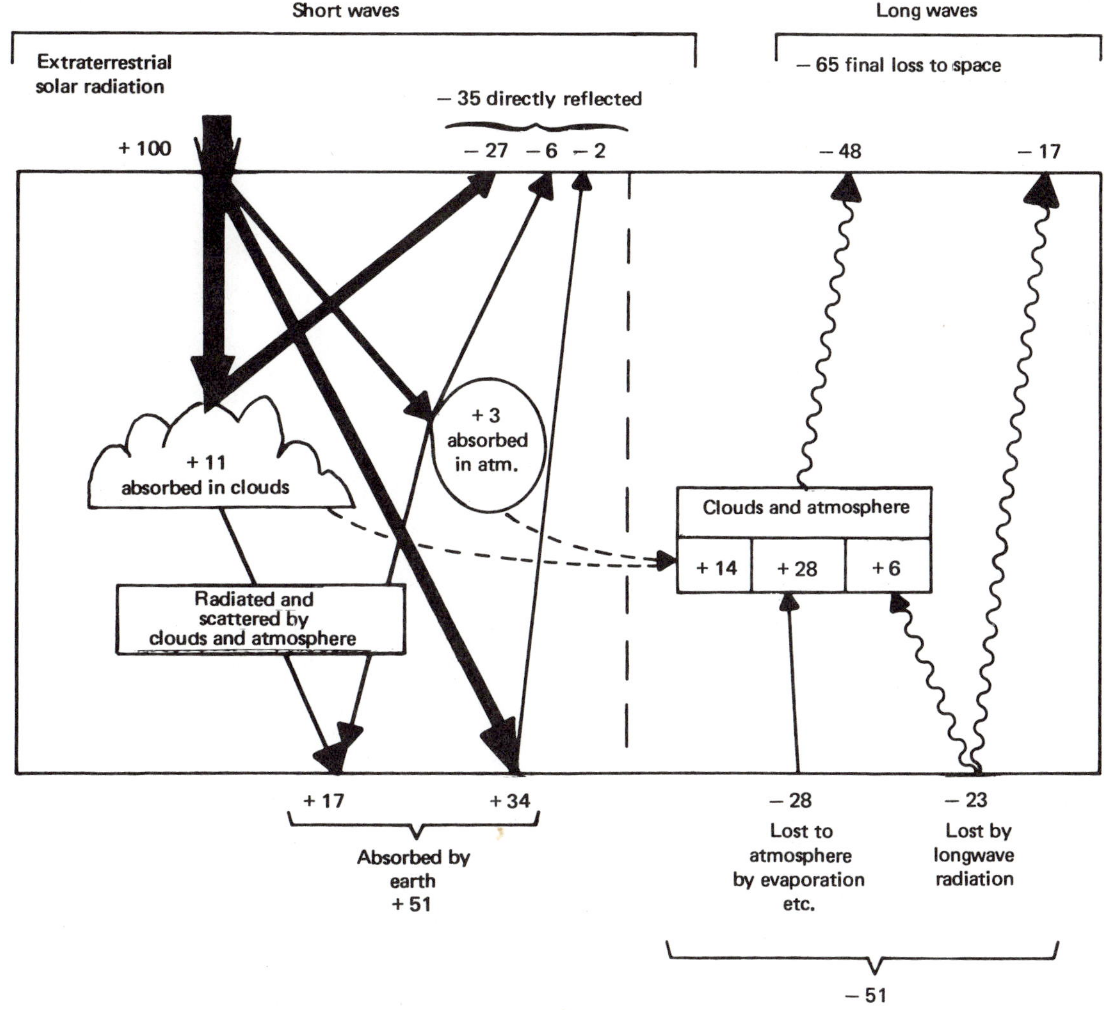

Fig. 5. Average radiation balance of the earth's atmosphere (after Möller, 1950, and Wexler, 1953).

The Astronomical Climate The orbital variations have two effects:

(1) *The ϵ variations* have equal effects in both hemispheres. If ϵ is larger than at present, the higher latitudes (above 43°) in both hemispheres receive more solar radiation during the year than when ϵ is smaller (this can be easily understood by taking the poles as example). At the poles an increase of ϵ by 1° gives an increase in summer isolation of 4.01%, at 70° latitude of 3.18%. The situation with minimal ϵ is therefore favorable for the occurrence of increased glaciation at high latitudes.

(2) *The Π variations*: we assume $e = 0$. If $\Pi = 0°$, summer and winter have equal length on both hemispheres (see Fig. 6; we define the astronomical summer as the time interval during which the sun is above the equator). However, 5250 years later, when $\Pi = 90°$, the northern summer will last longer than the winter. The northern summer will be long and cool, the winter short and mild. The southern summer will be warm and the winter cold. The difference in duration between summer and winter is

$$\Delta T = \frac{4e}{\pi} \sin \Pi . 365{,}24 \text{ days} \qquad (1)$$

For $e = 0.05$, $\Pi = 90$ or $270°$, one has $\Delta T = 25$ days.

If $\Pi = 180°$, there will again be no difference between the two hemispheres. If $\Pi = 270°$, conditions are the reverse of conditions for 90°: the northern summer will be short and hot, the southern long and cool.

The effects of the Π-variations vary with latitude and are strongest at *lower* latitudes (see Fig. 7).

The Occurrence of Ice Ages Milankovitch postulated that the *mean summer insolation* determines the growth or melting of glaciers. Thus during periods with hot summers, snow which fell during the winter would melt. Conversely reduction of summer temperature would favor preservation and growth of glaciers. Therefore a situation with ϵ low, $\Pi = 90°$, and e large is according to Milankovitch most favorable for the occurrence of Northern Hemisphere glaciation. The situation with ϵ low, $\Pi = 270°$, and e large is most favorable for a Southern Hemisphere glaciation. As a consequence of the different periodicities in the variations of ϵ, e, and Π, the situations of extreme insolation are expected to occur at irregular intervals of time.

Computations by Milankovitch (1941) show that at latitudes higher than about 70° the insolation variations are mainly controlled by the ϵ variations. Hence, roughly a 41,000-year periodicity in insolation occurs (see Fig. 7), with

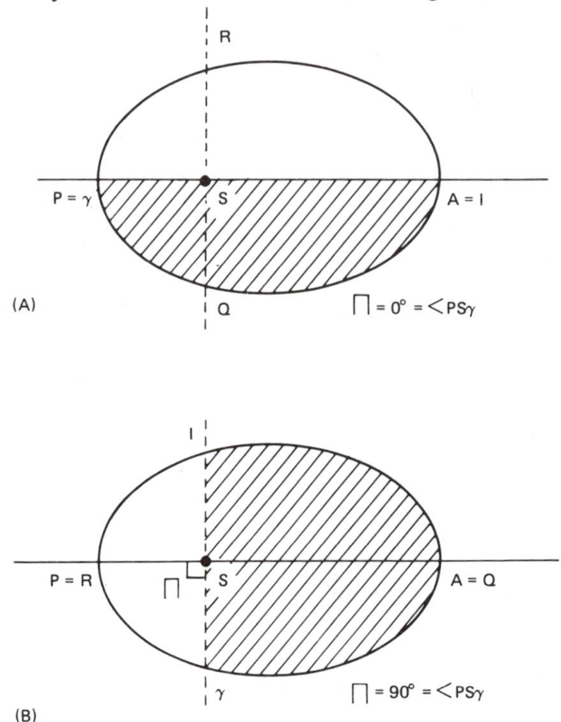

Fig. 6.
(A) If the perihelion length Π = PSγ = 0°, the sun is in the equatorial plane when the earth is at point P of its orbit. (γ is the equinox, S is the sun.) The sun reaches its maximal angular height above the equatorial plane when the earth is at point Q (mid-summer), and is again in the equatorial plane when the earth is at point A (aphelion). Midwinter in the Northern Hemisphere occurs if the earth is at the point R. Thus if $\Pi = 0°$, the length of summer (area PQA) is equal to the length of winter (area ARP). The same occurs if $\Pi = 180°$. In that case $A = \gamma$ and $P = \lambda$, while R and Q are interchanged.
(B) If $\Pi = 90°$, mid-summer in the Northern Hemisphere occurs when the earth is at aphelion A and mid-winter when it is at perihelion P. Therefore, the northern summer (area $\gamma A1$) will be long and cool, the winter short and mild. For the Southern Hemisphere the winter is long and cold, the summer short and hot. The situation for $\Pi = 270°$ is just opposite to that for $\Pi = 90°$. As a consequence of the precession and the proper motion of the perihelion the angle Π increases by 360° in a time-interval of about 21,000 years.

simultaneous extremes for the Northern and Southern Hemispheres. On the other hand, the Π variations dominate below about 40° latitude. Therefore Southern Hemisphere maxima occur on the average 10,500 years later than Northern Hemisphere maxima (see Fig. 7). Between 40 and 70° latitude no clear correlation between Northern and Southern insolation curves is found. Milankovitch postulated that the insolation variations at 65°N (where most of the Pleistocene ice sheets were located) were the main contributors to glaciation. Here the ϵ variations are still somewhat dominant over the Π variations and roughly a 41,000-year periodicity occurs in both hemispheres (cf. Van Woerkom, 1953). The magnitude of insolation variations at 65°N can be judged from the fact that in extreme conditions the mean summer insolation could be reduced to the amount presently received at 82°N, or increased to that presently received at 60°N. This indicates that *the effects are not small.*

Variations in Tidal Strength

Because of changes in the moon's orbit, the tidal strength varies with a periodicity of about 1800 years. Petterson (1914) found evidence for the occurrence of severe climatic conditions during periods of strong tides (e.g., the 15th to the 18th centuries) and milder conditions during periods of weak tidal action (e.g., around A.D. 500). Petterson suggested that during periods of strong tides, the tidal waves carry warm Atlantic Ocean water into the Arctic Ocean, causing accelerated iceberg production and spreading of pack-ice to lower latitudes. The subsequent cooling of surface ocean water will cause extreme climatic conditions on adjacent continents. Karlstrom (1955) found a 3500-year periodicity in the extent of the Laurentide ice sheet. Such a periodicity is almost equal to two 1800-year periods, suggesting a possible relation with tidal strength.

Possible Correlation between Pleistocene Temperatures and Astronomical Periodicities

As noted earlier, Emiliani's generalized temperature curve shows a periodicity of about 40,000 years. This period is almost equal to the period of variation of the obliquity ϵ of the plane of the eliptic (Emiliani, 1966; Broecker and van Donk, 1970). Van den Heuvel (1966a) noted that a period of approximately 12,000 to 13,000 years was also present in Emiliani's generalized curve, and suggested the precession as a possible cause. Subsequently, the same period of 13,000 years was found by Dansgaard et al. (1969) in the $^{18/16}O$ content of the Greenland ice cap. In this ice cap periodicities of 120 and 940 years were also found (cf. Fig. 3), which were ascribed as due possibly to solar activity. It should be noted, however, that the 940-year periodicity almost equals half the periodicity of tidal strength. Broecker et al. (1968) and Emiliani (1969) found that the *interglacial* high sea levels coincide very closely with the times at which the perihelion of the earth's orbit coincides with the Northern Hemisphere summer solstice (arrows, Fig. 4). Emiliani suggests that this second-order oscillation of his temperature curve is due mainly to significant melting of the Greenland ice cap after interglacial conditions are established. (If melting of the Antarctic were the main cause, coincidence of high sea levels would be expected when the perihelion *coincides with the Southern Hemisphere summer solstice.*)

Veeh and Chappel (1970) found a correlation between variations in sea level over the past 230,000 years (as derived from uplifted coral reef terraces in New Guinea, dated with the ^{14}C and ^{230}Th methods), and variations in insolation at 45° northern latitude. The latter variations are mainly regulated by the precession, which seems to confirm Emiliani's findings. These observations indicate that Pleistocene temperatures were in some way influenced by variations in summer insolation (cf. Fig. 3).

The Importance of Insolation Variations as a Cause of Glaciations

The variations in the earth's orbit have continued throughout the earth's history and were apparently unable to induce glaciation during epochs when polar temperatures were high, such as the Mesozoic. As these variations do not affect the total amount of yearly received solar radiation, it is not expected that they would cause large temperature changes on the entire globe. Moreover, with periodicities of less than 100,000 years, insolation variations are ruled out as a cause of high-latitude glaciation during periods of 10 to 100 million years. As a prime cause of the Ice Ages such variations cannot be of importance. On the other hand, on an earth with glaciated polar caps, they may induce an increase or a decrease in the amount of polar ice. Thus, by means of the albedo effect, insolation variations might possibly trigger the earth to settle at a lower mean temperature for some time. We shall consider this possibility in more detail under a separate heading.

Terrestrial Factors

Continental Uplift and Mountain Building

Large Ice Ages have developed only during periods of high land elevation and mountain building (cf. Emiliani and Geiss, 1959). This holds true for the late Cenozoic as well as for the Carboniferous Ice Age. Uplift of continents since the late Cretaceous increased the continental area above sea level from less than 20% to about 29% at the present time (cf. Brooks, 1971; Newell, 1972). Greater atmospheric turbulence due to increased global relief may have caused added ocean water evaporation and cloudiness, thus increasing the earth's albedo (Table 3). Furthermore, the presence of high mountains favors the formation of glaciers, even in tropical areas. The largely latitudinal direction of the Eurasian mountain chains—which were all formed since the late Cretaceous—also prevents atmospheric heat transport from tropical to polar regions in these parts of the world. It is clear, therefore, that during Mesozoic times, with a small continental area and only lowland elevations, the situation was extremely favorable for the development of warm "maritime" climates all over the world, whereas during the Tertiary conditions became gradually favorable for the development of continental climates, with increased seasonal contrasts and reduced yearly mean temperatures, especially on continents. Although uplift and mountain building seem to be necessary conditions for the occurrence of glaciation, they alone do not seem to be sufficient, as periods of high land elevation and mountain building have also existed without inducing glaciation. During the Devonian (~400 to ~350 million years ago), for instance, an epoch of increased uplift existed without glaciation. It did, however, induce cooling, as evidenced by a sharp reduction in coral diversity (Newell, 1972). The Caledonian orogeny (Silurian) was not accompanied by glaciation either. According to Ewing and Donn (1956a, 1956b), the additional prerequisite apart from uplift that has to be fulfilled for a glaciation to occur is isolation of the poles from the oceanic heat exchange system.

Continental Drift and Thermal Isolation of the Poles

As the heat capacities of equal volumes of water and air have a ratio of about 3000/1, it is clear that the oceans have an enormous heat storage capacity in comparison with the atmosphere. Most solar heat is absorbed by the oceans in a belt around the equator, from there it is distributed toward the poles by ocean currents. On their way back these currents carry cold Arctic water toward the equator, thus continuously reducing the temperature gradient between the equator and the poles. Inasmuch as ocean water temperature determines the air temperature in the surroundings, warm currents induce mild winters and little seasonal contrast in coastal areas, even within the polar circle (e.g., northern Norway), whereas cold currents may induce severe winters, even at middle latitudes (e.g., southern Chile).

Ewing and Donn argue that the main cause of the Tertiary temperature decrease at high latitudes was that, due to continental drift, the polar regions became isolated from the oceanic heat exchange system. Further, they contend that if the poles were located in open oceans, as during the Mesozoic, warm ocean currents would be able to reach them, thereby keeping the polar air temperatures above freezing all year long, even though the amount of winter solar radiation in these regions is extremely low. If, on the other hand, a large continent (such as Antarctica) is located at the pole, or if a pole is surrounded by continents (e.g., the North Pole at present) ocean currents would not be able to reach the polar regions. Consequently the polar climates would be regulated by insolation, which is low in summer and absent in winter. Cooling of polar regions to below freezing would result and glaciation would ensue. Thus, thermal isolation of the poles would cause the establishment of a large temperature gradient between the equator and the poles, while free access of ocean water to the poles would reduce temperature gradients. Confirmation of the importance of thermal isolation of a pole comes from the fact that the Carboniferous Ice Age also occurred when a large continent (Gondwanaland) was located at the South Pole. Other epochs of continental uplift probably did not always induce glaciations because the poles were freely accessible to warm ocean currents, which prevented them from cooling below freezing point. It should be emphasized that, although for polar isolation, continental drift no doubt played an important part, uplift must also have played a considerable role. The early Tertiary uplift of the Bering Strait, for example, closed off the access of Pacific water

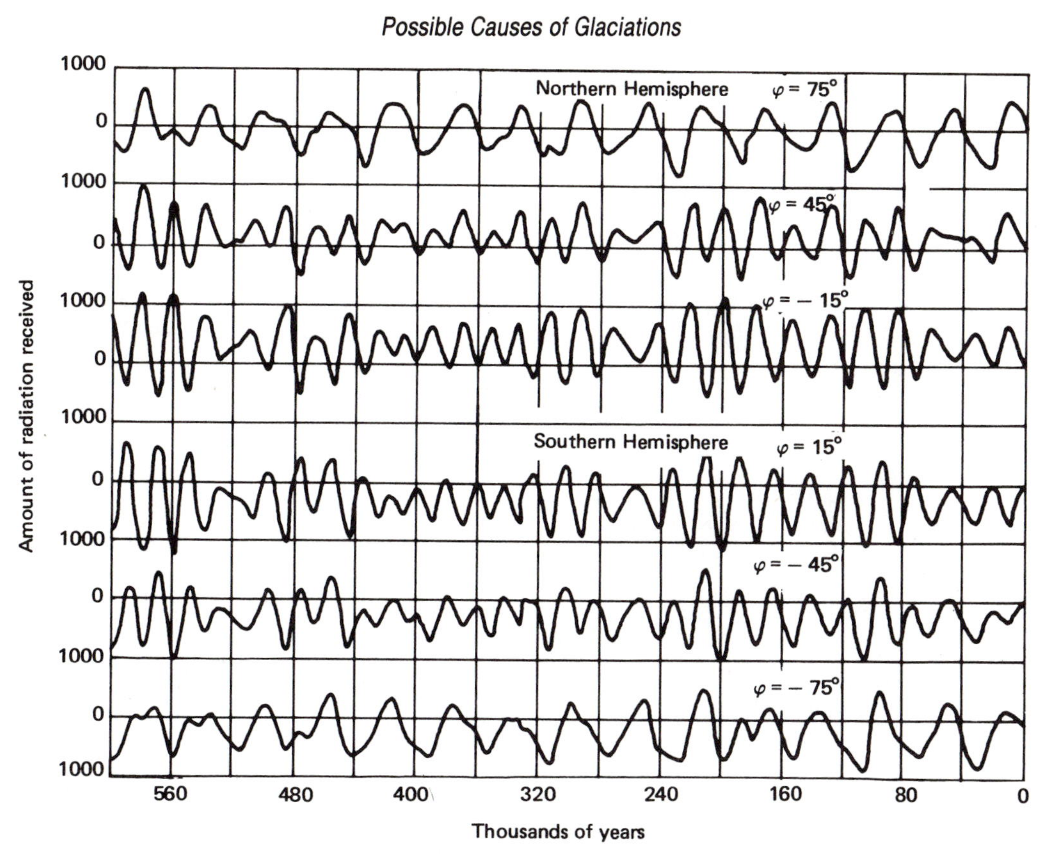

Fig. 7. Milankovitch's (1941) insolation curves for the past 600,000 years. The figure represents the curves for geographical latitudes $\varphi = \pm 15°$, $\varphi = \pm 45°$,° and $\varphi = \pm 75°$. Note that the curves for $\varphi = +15°$ and $\varphi = -15°$ show opposite variations; the amounts of radiation are given in "canonic units" (Milankovitch, 1941). The curves for $\varphi = -75°$ and $\varphi = -75°$ show maxima at about the same time. The curves for $|\varphi| \leq 40°$ are mainly regulated by the precession and show opposite variations in the two hemispheres.

to the Arctic Ocean for long time intervals (cf. Flint, 1971, p. 538). With a depth of only 40 m the Bering Strait is still practically closed for oceanic heat exchange.

Most Probable Causes of the Large Ice Ages

Having investigated the possible cosmic and terrestrial mechanisms, we found that all known cosmic mechanisms are insufficient to cause cooling of polar regions by over 14°C, for periods on the order of 10 to 100 m.y.* From the arguments presented in the foregoing paragraph, and in view of the time scales involved, it seems most likely that continental uplift, together with thermal isolation of one or both poles, was the main cause of large Ice Ages. On the other hand, these terrestrial mechanisms have time scales far too large to explain the alternating glacial and interglacial stages, with a periodicity of roughly 40,000 years.

Since many theories have been proposed to explain these "short-term" climatic variations, we shall consider them in a separate section.

* See footnote on page 366.

Possible Causes of the Pleistocene Glaciations

Before dealing with specific theories, we shall briefly review the relevant observations on the growth and development of continental ice sheets during the Pleistocene.

The Role of Mountains, Precipitation, Melting, and Albedo

The northern European ice cap started to form on the Scandinavian mountains, whence it invaded the lowlands up to Minsk (Russia) to the east, and the Netherlands to the south. Figure 8 illustrates the development of the Scandinavian ice cap according to Flint (1971, p. 598). In North America, Argentina, and New Zealand also, the ice caps first formed on continental mountains and elevated plateaus, and from there invaded the lowlands and fused to form gigantic ice sheets. In the terminal phases the sequence of events took place in reverse direction, the mountains being deglaciated last. The location of all major Pleistocene ice sheets close to open oceans (Siberia did not have much of an ice cap) stresses the importance of precipitation. Glaciers can grow only if there is a net excess of snowfall over melting during the year. Reconstruction of wind directions during the Pleistocene glaciations—e.g., from the direction of wind-blown sand dunes (cf. Flint, 1971, pp. 244–250)—indicates that the prevailing wind directions did not deviate much from the present day. The predominantly westerly winds in northwestern Europe, and the southwesterly winds over the eastern and midwestern United States, carry moist oceanic air masses far across the continents. On their way over mountains they unload their moisture in the form of snow. The growth of glaciers may result from either (a) increased precipitation or (b) reduced melting. At present precipitation in Scandinavia is already quite heavy (300 cm per year in southern Scandinavia); hence a large increase over the present amount does not seem very likely. As melting occurs mainly during the summer, a decrease in mean summer temperatures maintained over a sufficiently long period of time might be adequate for inducing a net growth of glaciers. Some areas of the world are indeed so critical that a small decrease in mean summer temperature would induce glaciation. Barry (1966) estimates from the rates of precipitation and melting that a decrease of 3.3°C in the mean temperature of Labrador-Ungava, maintained over 20,000 years, would be sufficient to build up a 2000-m-thick ice cap. Under present-day conditions with polar ice caps and mountain glaciers, relatively small changes in mean summer temperatures would probably cause extensive continental glaciations. Furthermore if glaciers with a reflectivity of 50 to 80% were to replace land and water surfaces with a reflectivity of 8 to 15%, the heat intake of the earth would be reduced and further cooling would ensue. This in turn would induce further growth of glaciers. By such a self-amplifying *albedo effect*, a relatively small change in cosmic or terrestrial parameters may *trigger* the onset of extensive continental glaciation. This *essentially nonlinear behavior* of ice and glaciers, requiring only a relatively small temperature drop as a trigger for inducing a large glaciation, plays an important part in most theories on the origin of the Pleistocene glaciations.

Theories of Pleistocene Glaciations

In view of the difficulties involved in nonlinear calculations and due to the complexity of the problem involving irregularly shaped continents and relief, coupling between oceanic and atmospheric circulation and heat exchange, evaporation, precipitation, albedo, etc., all existing theories are descriptive and by no means sufficiently quantitative. As the time for realistic nonlinear climatological computations seems to be still far off, objective evaluation of the different theories is difficult. The only manner of assessing their validity is to compare their predictions with the observed climatological record of the Pleistocene. We shall try to do this here by reviewing three important theories.

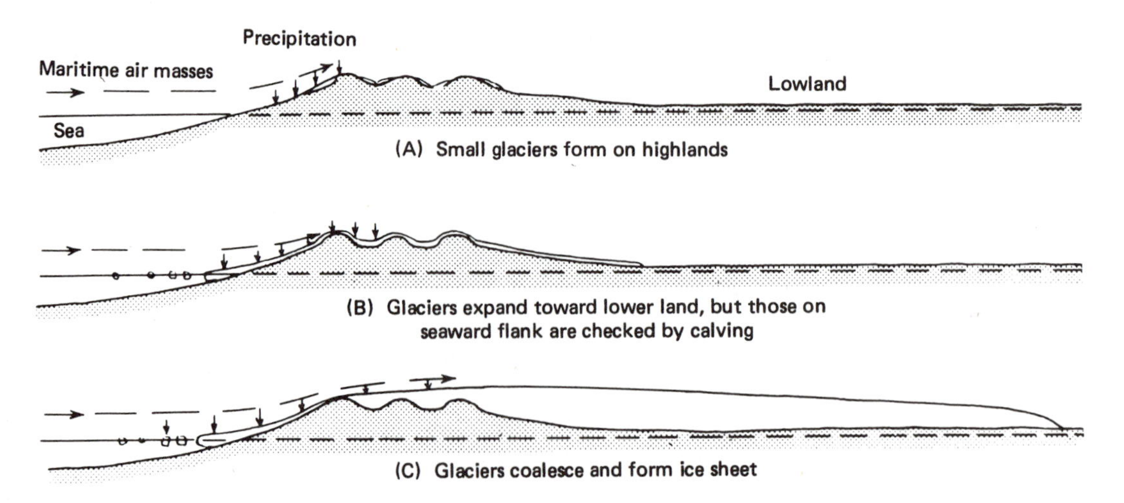

Fig. 8. Idealized development of the Scandinavian ice sheet. Length of the section about 800 km. Not to scale. (After Flint, 1971, p. 598).

The Ewing and Donn theory. In this theory the Arctic Ocean is the main regulator of the Pleistocene glaciations (similar cyclic theories have been proposed before, with the Arctic Ocean playing a less crucial role, e.g., cf. Willett, 1953). According to Ewing and Donn, if the Arctic Ocean were ice-free, evaporation would cause increased precipitation on the surrounding continents. Mountain glaciers would grow and fuse to form large ice sheets. These ice sheets would influence the climate and atmospheric circulation in their surroundings. Evidence from Greenland and Antarctica indicates that very stable anticyclonic conditions exist above a large ice sheet. Above the warmer oceanic regions on the other hand, the atmospheric pressure is generally low. In such a situation off-land winds would prevail which would cause cooling of the surrounding oceans. The cooling of the ocean would also induce glaciation in the Southern Hemisphere.

Due to the drop in the sea level the Gulf Stream would no longer reach the Arctic Ocean and this ocean would finally freeze. This would cause a decrease in precipitation, consequently melting and evaporation of the ice sheets become larger than feeding. From here on, the volume of ice sheets begins to decrease, although due to plastic ice flow, the area may remain constant for some time. During this melting stage the surrounding oceans remain cold, and hence no increase in precipitation would occur. When the ocean temperature finally begins to rise, the area of the ice caps would have become too small to have significant influence on climates, and melting would continue in spite of increased precipitation. After the melting of the ice sheets the Gulf Stream would be deflected northward again and the Arctic Ocean ice would also begin to melt. The "interglacial" stage, with a still-frozen Arctic Ocean, would correspond to the present situation. When the Arctic Ocean finally becomes ice-free, a new glacial cycle would start.

Although this nonlinear picture involving only terrestrial factors seems attractive, there are a number of problems:

(1) The Arctic Ocean should remain open until the maximum volume of continental ice is attained. However, the maximum of the Würm (Wisconsin, Weichsel) Glaciation occurred about 20,000 years ago (cf. Dansgaard et al., 1969, 1971) and from Arctic Ocean bottom sediments it appears that the Arctic Ocean has probably been frozen for the past 70,000 to 100,000 years (Herman, 1970). The continuous presence of at least part of the Greenland ice sheet over the past 120,000 years also suggests that during this period the Arctic Ocean did not warm up much above its present temperature.

(2) It is hard to see why the ice cover of the Arctic Ocean did not melt immediately after the termination of the Last Glaciation, some 10,000 years B.P. The present climatic situation, with a frozen Arctic Ocean and glaciated Greenland, has been quite stable for the past 10,000 years.

The Milankovitch theory (with possible modifications). As mentioned earlier (and shown in Fig. 3), most probably Pleistocene temperatures and high sea levels show periodicities similar to the secular variations in insolation computed by Milankovitch (1920, 1930, 1941) and Van Woerkom (1953). This indicates that insolation variations in some way must have played an important part in the occurrence of the Pleistocene glaciations. Milankovitch attempted with the help of a rough atmospheric model to translate the local insolation variations to local temperature variations. This attempt was criticized by a number of people, notably Simpson (1938), who found that atmospheric temperature variations caused by variations in insolation would be much too small to cause glaciation. In fact, neither Milankovitch's nor Simpson's calculations were correct, as both calculations involved linear approximations, while, as pointed out previously, the response of high-latitude climates to small temperature changes is essentially nonlinear. Furthermore, neither of these two authors took all relevant factors (especially the oceans) sufficiently into account. The same holds true for the computations by Shaw and Donn (1968). The latter authors showed that, starting from a realistic atmospheric model, the insolation variations computed by Milankovitch could have induced temperature variations of about ± 2°C with respect to the present day temperatures at 65°N. From this Shaw and Donn argued that insolation variations are not sufficient as a cause of glaciations, because the Little Ice Age of the 17th and 18th centuries, which involved a drop in mean temperatures by about 2°C over one century, did not trigger the onset of a real glaciation. However, this argument is not valid, as it shows only that the cumulative effect of a 2°C temperature decrease *over one century* is insufficient to cause a complete glaciation. Insolation variations would, however, lower the mean temperatures by 2°C for thousands of years. The cumulative effects in this case, for the oceanic circulation system as well, might be sufficient to induce a large glaciation. In fact, the *observed* correlations mentioned earlier strongly suggest that insolation variations were indeed an important trigger in Pleistocene temperature variations. The large amplitude with a 40,000-year periodicity in Emiliani's curve (cf. also Van den Heuvel, 1966a) together with the synchroneity of Northern and Southern Hemisphere glaciations, suggest that insolation variations at latitudes higher than about 65° had the greatest influence. This seems reasonable for two reasons: (1) at high latitudes many permanent glaciers are present, hence their growth might be rather simply induced, and (2) here secular variations in insolation have the largest amplitudes.

The precession apparently also influenced Pleistocene climates, as interglacial high sea levels varied with the precession period (Broecker et al., 1968; Emiliani, 1969; Veeh and Chappel, 1970) (the term "precession" here includes apsidal motion). High sea levels occurred at times when the perihelion passage coincided with the summer solstice (June) in the Northern Hemisphere, indicating that partial melting of the Greenland ice sheet probably occurred during such times (Emiliani, 1969). Emiliani argues that this ice sheet is more sensitive to insolation changes than the Antarctic ice sheet, as it is not centrally located with respect to the pole and extends to lower latitudes. However, it is not certain that an explanation in terms of local insolation is sufficient. At latitudes higher than 65°C the precession has only a minor influence on insolation, whereas at latitudes lower than 40° it has a large effect (Fig. 7). An alternative explanation might be that when the perihelion passage occurred in June, the increased insolation caused heating of the north equatorial parts of the oceans and induced the Gulf Stream to extend its influence further north. This may have caused increased melting of the Greenland ice. The summer insolation changes due to the precession are as large as 7%, and in times of high orbital eccentricity may become as large as 15%. Therefore their influence on climates is certainly not expected to be negligible (in fact, in pre-Pleistocene times the effects of the precession are found notably in varves of the Tertiary Green River Formation; Bradley, 1929).

Apparently no appreciable melting of the Antarctic ice sheet took place in times when the perihelion passage occurred in December (southern summer solstice). This fact may be due to the entirely different distribution of oceans and continents in the Southern Hemisphere. Antarctica is surrounded by a wide belt of cold ocean water (the Antartic convergence); furthermore, in the vast Southern Hemisphere oceans the currents are much less channelled and restricted by coast lines than in the narrow North Atlantic. For this reason their path and extent will be much less sensitive to changes in equatorial heating. These two factors suggest that changes in insolation in the equatorial belt have far less effect on the Antarctic than on the Greenland ice sheet.

It is interesting to note that, in combining the effect of the precession on the equatorial oceans with the effect of obliquity on high-latitude glaciation, one arrives at Broecker's (1966) semi-empirical modification of Milankovitch's theory, in which precession and obliquity are given different weights. The above-mentioned correlations indicate that such a theory seems most plausible.

The occurrence of *half* the precession period in Emiliani's and Dansgaard's $^{18/16}O$ records (Figs. 3 and 4) seems puzzling. The $^{18/16}O$ cycle in both records is well in phase with the record of extremes in the June sun-earth distance (cf. Dansgaard et al., 1969; Van den Heuvel, 1966a). In terms of Dansgaard and Tauber's (1969) modification of Emiliani's curve, maximum $^{18/16}O$ means maximum volume of continental ice. Maximum June sun-earth distance is expected to coincide with maximum glaciation on the Northern Hemisphere and minimum June sun-earth distance with maximum glaciation on the Southern Hemisphere. The double wave therefore indicates that at maximum Northern as well as at maximum Southern Hemisphere glaciation the total volume of continental ice apparently was larger than in intermediate periods. Its occurrence therefore does not conflict with Milankovitch's theory.

Wilson's 1964 theory assumes cyclic surging of the Antarctic ice sheet. In a growing ice sheet the vertical temperature gradient is mainly determined by the rate of precipitation (Robin, 1962). A drop in precipitation will reduce the temperature gradient and may cause the glacier bottom to reach the pressure melting point. In such a case the glacier may start to advance at high speed. This kind of glacier instability has often been observed, notably in Spitsbergen (Hollin, 1965). Wilson's theory suggests that after a period of growth, a drop in precipitation in Antarctica triggers the ice sheet to become unstable and to flow out over the surrounding ocean. It then covers an area four times larger, and its thickness is reduced from 2000 to 500 m. Due to the sudden albedo increase and the cooling of the oceans, glaciation would also spread to the Northern Hemisphere. The ice shelf would displace enough water to cause a sudden 20-to30-m rise in sea level.

After the ice shelf has melted, the oceans warm up and finally the Northern Hemisphere glaciation would disappear. During interglacials the Antarctic ice sheet would slowly build up again to a thickness of over 2000 m, when a drop in precipitation may again cause it to surge. According to Wilson, the decrease in precipitation might be due to insolation variations. Van den Heuvel (1966d) pointed out that in such a case a coincidence between Northern Hemisphere summer insolation minima and a northern (as well as southern) glaciation is to be expected. The reason for this is that low precipitation in Antarctica is expected in times of minimum winter insolation, which coincides with times of minimum summer insolation in the Northern Hemisphere.

The attractive aspects of this theory are:

(1) It gives a simple explanation for the synchroneity of Northern and Southern Hemisphere glaciation.

(2) It also explains simply how the earth, despite the albedo effect, could have returned to interglacial conditions after glaciation had been established. This is very important, as none of the other theories give a simple explanation for this occurrence.

(3) It does not conflict with observed astronomical periodicities in the Pleistocene as recorded by $^{18/16}$O.

However, one may wonder whether surging on the scale of the entire Antarctic ice sheet is possible. Furthermore, the Pleistocene $^{18/16}$O record, showing the effects of precession as well as obliquity, indicates gradual rather than sudden climatic changes. A crucial test for the theory is to examine whether each glaciation was preceded by a sudden sea level rise. Although the present evidence makes interglacial high sea levels exceeding 10 m above the present level unlikely (Emiliani, 1969), some evidence of sudden sea level rises by the end of interglacials may exist (Hollin, 1965). Clearly, further research for testing this interesting theory is required.

Other theories. For a brief review of other theories concerning Pleistocene glaciations we refer to Flint (1971, pp. 789–809). Most of these theories involve some modifications of Milankovitch's theory (Broecker, 1966), or involve unknown factors such as solar activity during the Pleistocene (cf. Willett, 1953). One of the omissions in all existing theories that involve insolation changes is the neglect of the influence of low-latitude insolation variations on the ocean current system. The merit of Ewing and Donn is to have pointed out the importance of the oceans in connection with glaciation.

Summary and Conclusions

1. The occurrence throughout the history of the earth of periods of high-latitude glaciation on the order of 10 to 100 m.y. (Table 1), alternating with warm periods on the order of 100 to 200 m.y., is most probably due to a combination of continental uplift and isolation of one or both poles from the oceanic heat exchange system in the sense of Ewing and Donn. Such an explanation agrees with all observational evidence of the well-documented upper Paleozoic and late Cenozoic (present) Ice Ages.

2. No cosmic causes of climatic change with time scales on the order of 10 to 100 million years are known.*

3. The present situation with permanent ice sheets in Greenland and Antarctica has lasted throughout the Pleistocene and has two fairly stable climatic states: the "glacial" state in which large parts of the North American, Eurasian, and South American continents are glaciated, and the "interglacial" (present) state. These two states alternate with a roughly 40,000-year periodicity. The switch-over from one state to the other can be triggered by relatively small changes in high-latitude temperatures. The observed correlation between Pleistocene isotopic temperatures and variations of high-latitude summer insolation in both hemispheres suggests that the triggering is provided by secular variations in summer insolation.

4. The roughly 13,000-year modulating period in the $^{18/16}$O ratio of Greenland ice and of Caribbean deep-sea cores, as well as the occurrence of interglacial high sea levels at times when perihelion passage and northern summer sol-

* See footnote on page 366.

stice coincided (Figs. 3 and 4), suggests that the precession played a part. Variable heating of ocean water in the equatorial belt may have induced these lower-amplitude climatic variations at high northern latitudes.

5. The small-amplitude 940- and 120-year periodicities in the $^{18/16}O$ ratio of the Greenland ice indicate that variations in tidal strength and solar activity may cause minor climatic changes.

6. Tertiary transgressions and regressions in Europe are probably related to variations in glaciation of the Antarctic continent.

Acknowledgments

We are grateful to Drs. W. Dansgaard, C. Emiliani, D. I. Gough and R. F. Flint for the permission to reproduce figures from their publications, and to Dr. Yvonne Herman for valuable comments on the manscript.

References

Armstrong, R. L., W. Hamilton, and G. H. Denton. 1968. Glaciation in Taylor Valley, Antarctica, older than 2.7 million years. *Science*, 159:187–189.

Barghoorn, E. S. 1953. Evidence of climatic change in the geological record of plant life. In: H. Shapley, ed. *Climatic Change.* Cambridge: Harvard University Press, pp. 234–248.

Barry, R. G. 1966. Meteorological aspects of the glacial history of Labrador-Ungava with special reference to atmospheric vapour transport. *Geograph. Bull. Ottawa*, 8:319–340.

Bradley, W. H. 1929. The varves and climate of the Green River·Epoch. In: *Shorter Contributions to General Geology.* Washington, D.C.: U.S. Geological Survey, pp. 87–119.

Bray, J. R. 1968. Glaciation and solar activity since the fifth century B.C. and the solar cycle. *Nature*, 220:672–674.

——— 1971. Solar-climate relationships in the post-Pleistocene. *Science*, 171:1242–1243.

Broecker, W. S. 1966. Absolute dating and the astronomical theory of glaciation. *Science*, 151:299–304.

———. 1968. In defense of the astronomical theory of glaciation. *Meteorological Monographs*, 8(30):139–141.

———, D. L. Thurber, J. Godart, T. L. Ku, R. K. Matthews, and K. J. Mesolella. 1968. Milankovitch hypothesis supported by precise dating of coral reefs and deep-sea sediments. *Science*, 159:297–300.

———, and J. van Donk. 1970. Insolation changes, ice volumes and the ^{18}O record of deep sea cores. *Rev. Geophys. Space Phys.*, 8:169–198.

Brooks, C. E. P. 1971. *Climate through the ages.* New York: Dover, 395 pp.

Cloud, P. 1968. Atmospheric and hydrospheric evolution on the primitive Earth. *Science*, 160:729–736.

———, and A. Gibor. 1970. The oxygen cycle. *Sci. Am.*, 223(September):111–123.

Colbert, E. H. 1953. The record of climatic changes as revealed by vertebrate paleoecology. In: H. Shapley, ed. *Climatic Change.* Cambridge: Harvard University Press, pp. 248–271.

Curry, R. R. 1966. Glaciation about 3,000,000 years ago in the Sierra Nevada. *Science*, 154:770–771.

Danjon, A. 1954. Albedo, color and polarization of the earth. In: G. P. Kuiper, ed. *The Earth as a Planet.* Chicago: University of Chicago Press, pp. 726–738.

Dansgaard, W., S. J. Johnsen, J. Møller, and C. Langway. 1969. One thousand centuries of climatic record from Camp Century on the Greenland ice sheet. *Science*, 166:377–381.

———, and H. Tauber. 1969. Glacier oxygen-18 content and Pleistocene ocean temperatures. *Science*, 166:499–502.

———, S. J. Johnsen, H. B. Clausen, and C. C. Langway, Jr. 1971. Climatic record revealed by the Camp Century ice core. In: Karl A. Turekian, ed. *The Late Cenozoic Glacial Age.* New Haven: Yale University Press, pp. 37–56.

Dietrich, G., and K. Kalle. 1957. *Allgemeine Meereskunde.* Berlin: Geb. Borntraeger, 492 pp.

Dilke, F. W. W., and D. O. Gough. 1972. The solar spoon. *Nature*, 240:262–264, 293–294.

Dunn, P. R., B. P. Thomson, and K. Rankama. 1971. Late Pre-Cambrian glaciation in Australia as a stratigraphic boundary. *Nature*, 231:498–502.

Du Toit, A. L. 1937. *Our Wandering Continents.* Edinburgh, Scotland: Oliver and Boyd, 366 pp.

Elsässer, W. M. 1950. The earth's interior and geomagnetism. *Rev. Modern Phys.*, 22:1–35.

Emiliani, C. 1961. The temperature decrease of surface sea water in high latitudes and of abyssal-hadal water in open oceanic basins during the past 75 million years. *Deep-Sea Res.*, 8:144–147.

———. 1966. Paleotemperature analysis of Caribbean cores P 6304-8 and P 6304-9 and a generalized temperature curve for the past 425,000 years. *J. Geol.*, 74:109–127.

———. 1969. Interglacial high sea levels and the control of Greenland ice by the precession of the equinoxes. *Science*, 166:1503–1504.

———, and J. Geiss. 1959. On glaciations and their causes. *Geol. Rundschau*, 46:576–601.

———, and E. Rona. 1969. Caribbean cores P 6304-8

and P 6304-9: new analysis of absolute chronology. A reply. *Science*, 166:1551–1552.

Ewing, M. and W. L. Donn. 1956a. A theory of Ice Ages, I. *Science*, 123:1061–1066.

———, and W. L. Donn. 1956b. A theory of Ice Ages, II. *Science*, 127:1159–1162.

Ezer, D., and A. G. W. Cameron. 1972. A mixed-up sun and solar neutrinos. *Nature*, 240:178–182.

Flint, R. F. 1971. *Glacial and Quaternary Geology*. New York: John Wiley and Sons. 892 pp.

Franklin, F. A. 1967. Two-color photoelectric photometry of the earth's light. *J. Geophys. Res.*, 72:2963–2967.

Fritz, S. 1949. The albedo of the planet earth and of clouds. *J. Meteorol.*, 6:277–282.

———. 1951. Solar radiant energy and its modifications by the earth and its atmosphere. In: T. F. Malone, ed. *Compendium of Meteorology*, American Meteorology Society, pp. 243–251.

Goldthwait, R. P., A. Dreimanis, J. L. Forsyth, P. F. Karrow, and G. W. White. 1965. Pleistocene deposits of the Erie lobe. In: H. E. Wright, Jr. and D. G. Frey, eds. *The Quaternary of the United States*. Princeton: Princeton University Press, pp. 85–97.

Gough, D. I. 1970. Did an ice cap break Gondwanaland? *J. Geophys. Res.*, 75:4475–4477.

Hammen, T. van der, et al. 1967. Stratigraphy climatic succession and radiocarbon dating of the last glacial in the Netherlands. *Geologie en Mijnbouw*, 46:79–94.

Hays, J. D., and N. D. Opdyke. 1967. Antarctic radiolaria, magnetic reversals, and climatic change. *Science*, 158:1001–1011.

Herman, Y. 1970. Arctic paleo-oceanography in late Cenozoic time. *Science*, 169:474–477.

Hollin, J. T. 1965. Wilson's theory of ice ages. *Nature*, 208:12–16.

Holmes, A. 1965. *Principles of Physical Geology*. London: Nelson, 1288 pp.

Karlstrom, T. N. V. 1955. Late Pleistocene and recent glacial chronology of south-central Alaska. *Geol. Soc. Am. Bull.*, 66:1581–1582.

Krook, M. 1953. Interstellar matter and the solar constant. In: H. Shapley, ed. *Climatic Change*. Cambridge: Harvard University Press, pp. 143–146.

Lamb, H. H. 1971. *Climate: Present, past and future, 1: Fundamentals and Climate now*. London: Murray. 624 pp.

Margolis, S. V., and J. P. Kennett. 1970. Antarctic glaciation during the Tertiary recorded in sub-Antarctic deep-sea cores. *Science*, 170:1085–1087.

McElhinny, M. W., J. C. Briden, D. L. Jones, and A. Brock. 1968. Geological and geophysical implications of paleomagnetic results from Africa. *Rev. geophys.*, 6:201–238.

Mercer, J. H. 1969. Glaciation in southern Argentina more than two million years ago. *Science*, 164:823–825.

Milankovitch, M. 1920. *Théorie mathématique des phénomènes thermiques produits par la radiation solaire*. Paris: Gauthier-Villars, 339 pp.

———. 1930. Mathematische Klimalehre und astronomische Theorie der Klimaschwankungen. In: W. Köppen and R. Geiger, eds. *Handbuch der Klimatologie*. 1(A), 176 pp.

———. 1941. Kanon der Erdbestrahlung und seine Anwendung auf das Eiszeitenproblem. *Academie Royale Serbie Edition Spéciale*, 133 (section des sciences mathématiques et naturelles, 33), 633 pp.

Möller, F. 1950. Der Wärmehaushalt der Atmosphäre. *Experientia*, 6:361–367.

Nairn, A. E. M. 1964. ed. *Problems of Paleoclimatology*. Proceedings of the NATO Paleoclimates Conference, 1963. London: Interscience, 705 pp.

Newell, N. D. 1972. The evolution of coral reefs. *Sci. Am.* 226 (June): 54–65.

Öpik, E. J. 1953a. A climatological and astronomical interpretation of the Ice Ages and of the past variations of terrestrial climate. *Armagh Observ. Contrib.*, 9:1–79.

———. 1953b. Disturbances in dwarf stars caused by nuclear reactions and gas diffusion. *Mémoires de Société Royale des Sciences de Liège, 8e*, 14:187–199; *Armagh Observ. Contrib.*, 12, 12 pp.

———. 1965. Climatic change in cosmic perspective. *Icarus*, 4:289–307.

Parker, E. N. 1970. The origin of magnetic fields. *Astrophys. J.*, 160:383–404.

———. 1971. The generation of magnetic fields in astrophysical bodies. IV. The solar and terrestrial dynamos. *Astrophys. J.* 164:491–509.

Petterson, O. 1914. Climatic variations in historic and prehistoric time. *Svenska hydrografisk-biologiska Kommissionens Skrifter 5*.

Roberts, J. D. 1971. Late Precambrian glaciation: an anti-greenhouse effect? *Nature*, 234:216.

Robin, G. de Q. 1962. The ice of the Antarctic. *Sci. Am.*, 207(September):132–142.

Runcorn, S. K. 1954. The Earth's core. *Am. Geophys. Union, Trans.* 35:49–63.

Rutten, M. G. and H. Wensink. 1960. Paleomagnetic dating, glaciations and the chronology of the Plio-Pleistocene in Iceland. *International Geological Congress Norden, 21st*, 4:62–71.

Schwarzbach, M. 1963. *Climates of the Past—An Introduction to Paleoclimatology* (transl. by R. O. Muir) London: van Nostrand, 328 pp.

Shapley, H. 1953. ed. *Climatic Change*. Cambridge: Harvard University Press, 318 pp.

Shaw, D. M., and W. L. Donn. 1968. Milankovitch radiation variations, a quantitative evaluation. *Science*, 162:1270–1272.

Simpson, G. C. 1938. Ice Ages. *Nature*, 141:591–598.

Van den Heuvel, E. P. J. 1966a. On the precession as a

cause of Pleistocene variations of the Atlantic Ocean water temperatures. *Geophys. J. Royal Astron. Soc.* 11:323–336.

———. 1966b. On climatic change in cosmic perspective. *Icarus*, 5:214–215.

———. 1966c. On climatic change in cosmic perspective. 2. *Icarus*, 5:218–219.

———. 1966d. Ice shelf theory of Pleistocene glaciations. *Nature*, 210:363–365.

Van Woerkom, A. J. J. 1953. The astronomical theory of climate change. In: H. Shapley, ed, *Climatic Change.* Cambridge: Harvard University Press, pp. 147–157.

Vaucouleurs, G. de. 1970. Photometrie des Surfaces Planetaires. In: A. Dolfuss, ed. *Surfaces and Interiors of the Planets.* London: Academic Press, pp. 225–316.

Veeh, H. H., and J. Chappel. 1970. Astronomical Theory of climatic change: support from New Guinea. *Science*, 167:862–865.

Wegener, A. 1915. *Die Entstehung der Kontinente und Ozeane.* Braunschweig, Sammlung Vieweg, 23, 94 pp.

———. 1924. *The origin of the continents and the oceans.* London: Methuen, 212 pp.

Wexler, H. 1953. Radiation balance of the Earth as a factor in climatic change. In: H. Shapley, ed. *Climatic Change.* Cambridge: Harvard University Press, pp. 73–105.

Willett, H. C. 1953. Atmospheric and oceanic circulation as factors in glacial-interglacial changes of climate. In: H. Shapley, ed. *Climatic Change.* Cambridge: Harvard University Press, pp. 51–71.

Wilson, A. T. 1964. Origin of Ice Ages: an ice shelf theory for Pleistocene glaciation. *Nature,* 201:147–149.

Index

Index